PHYSICAL CONSTANTS

Avogadro's number	6.02217×10^{23} particles/mole
Electronic charge	4.80325×10^{-10} esu
	1.60219×10^{-19} coulomb
Electron mass	9.10956×10^{-28} gm
Atomic mass unit	1.66053×10^{-24} gm
Gas constant	8.31434×10^{7} erg/mole deg
	1.9872 cal/mole deg
	0.08206 liter atm/mole deg
Faraday constant	96486.7 coulomb/mole
	23061 cal/volt mole
Boltzmann constant	1.38062×10^{-16} erg/deg
Planck constant	6.6262×10^{-27} erg sec

ENERGY CONVERSION FACTORS

	ergs/molecule	kJ/mole	kcal/mole	electron volts/molecule
ergs/molecule	1	6.0222×10^{13}	1.4393×10^{13}	6.2415×10^{11}
kJ/mole	1.6605×10^{-14}	1	0.23901	1.0364×10^{-2}
kcal/mole	6.9478×10^{-14}	4.184	1	4.336×10^{-2}
electron volts/molecule	1.6022×10^{-12}	9.6487×10^{1}	23.061	1

UNIVERSITY CHEMISTRY

Third Edition

UNIVERSITY CHEMISTRY

Third Edition

BRUCE H. MAHAN
University of California, Berkeley

ADDISON-WESLEY PUBLISHING COMPANY
Reading, Massachusetts · Menlo Park, California
London · Amsterdam · Don Mills, Ontario · Sydney

This book is in
THE ADDISON-WESLEY SERIES IN CHEMISTRY

Consulting Editor
Francis T. Bonner

ISBN 0-201-04405-6
FGHIJKLMN-RN-898765432

Preface
to the Third Edition

In preparing this third edition of *University Chemistry* I have continued to be guided by the idea that there is a body of skills and information that chemists and those who use chemistry will almost certainly need to master, and which matches the preparation of first-year college students. Accordingly, the emphasis in a freshman chemistry course should be on leading the student to a thorough understanding of the basic ideas and mathematical operations of stoichiometry and equilibrium, and on providing a useful introduction to the electronic structure of atoms and molecules, chemical kinetics, thermodynamics, states of matter, and the descriptive chemistry of the elements. As substantial progress has been made in systematizing and clarifying the conceptual aspects of chemistry there has been a tendency to introduce increasing amounts of this material at the freshman level, often at the expense of eliminating significant portions of descriptive chemistry from the course. I believe that it would be a mistake to carry this to the point where students and instructors lose sight of the fact that chemistry is basically motivated by the desire to expose and understand the properties of matter in bulk. Conceptual material should illuminate factual material, not eliminate it.

Accordingly, in this third edition I have incorporated new material on descriptive inorganic and organic chemistry. In addition, discussions of the practical aspects of electrochemistry and the application of thermodynamics to heat engines are included. The chapter on the nucleus has been enlarged considerably by the addition of material on radiometric geochronology, the chemical effects of radiation, the synthesis of elements in stars, and the use of nuclear fission and fusion as energy sources.

I have chosen to retain the conventional mixture of CGS and thermochemical units employed in previous editions, since these are employed in most reference books encountered by the beginning student. However, an appendix on SI units has been added, and at appropriate places throughout the book results are given in two or more sets of units. Consequently, students should become familiar with SI units and their relation to those of other systems.

As in the past, I have been encouraged and assisted in preparing this revision by the comments and suggestions of many students and instructors. To all those who have been kind enough to offer this help I would like to extend my sincere thanks.

Berkeley, California   B.H.M.

Preface to the First Edition

This is a textbook of general chemistry intended for students who have had an introductory high-school chemistry course. Its design is based on my experience with such a group, which included students who had had a "standard" high-school course, one of the two newer courses generated by Chem Study and CBA, and a few who had had a second high-school course designed for advanced placement in college. Writing a book which emphasizes fundamental principles and builds on previous experience presents some difficult problems. It has been my experience, one shared by other university teachers, that at this time there is no one block of the traditional elementary material so well understood by the majority of freshman students that it can be entirely omitted from the beginning college course. On the other hand, it is still possible to take advantage of a student's previous training by assuming a general familiarity with the most fundamental chemical concepts. The ability to assume such a background allows the instructor to treat the elementary material in a critical manner—an approach which is generally appreciated by students who might otherwise consider review material to be boring. This is the course I have followed in writing this book. All the important elementary concepts are discussed, but from what I hope is a more penetrating point of view than the one already familiar to the student.

A further advantage afforded by the student's previous training is that time and space become available for new material, usually treated only in the upper division and graduate physical, inorganic, and organic chemistry courses. The concepts of elementary thermodynamics, reaction rates, and chemical-bond theory are most appropriate in the general chemistry course, for they can be immediately applied to systematize the descriptive material of inorganic and organic chemistry. There is a danger here to be avoided: there is more material which might be presented than can be assimilated by the average beginning student. Just "covering" a great deal of advanced material can leave the student with no real command of the subject, but instead with a blasé attitude that he has "had all that," which can inhibit his success in more advanced courses. Consequently, I have made an effort to discuss only those advanced concepts which are either useful in the general chemistry course or which help to give a picture of what presently concerns the professional chemist.

The organization of this book departs somewhat from the recent standard of first presenting a detailed description of the electronic structure of atoms. Instead, the first four chapters are principally concerned with the macroscopic properties of matter, the origins of the atomic theory, and chemical arithmetic. Then follow five chapters which emphasize the characterization of chemical reactions and systems. Of these, Chapters 5, 6, and 7 treat, in increasing depth, the problem of chemical equilibrium. Thus the first seven chapters contain virtually all of the fundamental material of quantitative chemistry, and their position is particularly appropriate if the laboratory work in the course is to be quantitative in nature. Chapter 8 is an introduction to chemical thermodynamics which unifies the earlier phenomenological treatments of colligative properties and chemical equilibria. The thermodynamic concepts introduced here are used repeatedly throughout the chapters on descriptive chemistry. Chapter 9 treats the problem of chemical reaction rates and emphasizes the idea of reaction mechanisms.

The next two chapters, 10 and 11, are concerned with the electronic structure of atoms and the nature of chemical bonding. A part of Chapter 10 is devoted to the historical development of the quantum theory in an attempt to show that our knowledge of atomic structure and the quantum theory was gained by deduction from experiments, and not *ex cathedra*, as some elementary texts seem to imply. Deciding what can be said about chemical bonding that is simple, useful, and essentially correct is very difficult. I have tried to present a simplified point of view, while emphasizing its approximate nature and occasional inadequacies.

The final chapters contain the descriptive material of inorganic and organic chemistry. Their general thesis is that chemistry makes sense; that there are relationships, trends, and similarities in chemical behavior which can be understood in terms of chemical bonding, thermodynamics, and the periodic table. These chapters also contain certain extensions of the conceptual material covered earlier; for instance, the bonding in the boron hydrides, and the magnetic properties of matter.

The organization of this book is governed principally by my feeling that students should be supplied with the background that allows them to do quantitative laboratory work as early in the course as possible. I have, however, tried to write in a manner which will accommodate the preferences of others. Chapters 10 and 11 on atomic structure and bonding may be treated immediately after

Chapter 1, if the instructor prefers this organization. The material on thermodynamics can be delayed and presented just previous to the descriptive material of Chapter 13 *et seq*. Much of the discussion in Chapter 3 on the structures of solids can be combined with descriptive chemistry or with the material on chemical bonding.

This book is primarily intended for serious students of science, including those majoring in biology, engineering subjects, and premedicine, as well as physics and chemistry. Calculus is used in and after Chapter 6; concurrent enrollment in the introductory calculus course will provide an entirely adequate mathematical background.

Sincere thanks go to Professors Jerry A. Bell, Francis T. Bonner, and Paul B. Dorain for reading the manuscript and making many kind and helpful suggestions concerning the material. The entire manuscript was typed by Mrs. Nancy Monroe, whose cheerful help and cooperation are greatly appreciated. A considerable portion of this book was written while I was a visitor at Oxford University, and I would like to acknowledge the hospitality extended to me by the fellows of the Queen's College, and particularly the many kindnesses of Dr. and Mrs. J. W. Linnett. It has been my privilege to associate with two outstanding chemistry faculties. I would like to thank my teachers at Harvard University and my colleagues at the Berkeley campus of the University of California for the stimulation and instruction I have received from them.

Berkeley, California   B.H.M.
January 1965

Contents

STOICHIOMETRY AND THE BASIS OF THE ATOMIC THEORY

Our whole understanding of chemical phenomena is based on the atomic theory of matter. It is a theory remarkable for the detailed accuracy with which it describes a seemingly unknowable part of our physical world, and it stands as the most important collection of ideas in science. Throughout this book we will be continually drawing on the theory to help us organize and understand chemical behavior. Faced with a repeated demonstration of the usefulness, detail, and subtlety of the atomic concept, it is natural to wonder how such ideas were generated. In this chapter we will briefly outline the origins of the atomic theory and show how its development was connected with the growth of chemical science.

1.1 THE ORIGINS OF THE ATOMIC THEORY

The "atomism" of the Greek philosophers of 400 B.C. provides an interesting and enlightening contrast to our present-day atomic theory. The Grecian atom was designed to resolve a logical conflict: on one hand there was the *observation* that natural objects are in a constant state of change; on the other there was the unshakable *faith* that there must be a permanence associated with things which are real. The Greeks felt that this philosophical impasse could be avoided if invisible atoms were accepted as the permanent constituents of the universe, and if observable changes were interpreted in terms of their motions.

Now it is clear that the phenomena of mixing, evaporation, erosion, and precipitation can be readily "explained" in terms of an atomic picture which is not at all detailed. With but little elaboration the atomic idea encompasses many properties of matter. The existence of solids requires only that certain atoms have extensions with which they can interlock themselves to form an unyielding mass. The atoms of liquids need only be smooth to flow over one another, while the taste of some chemicals arises from sharp edges of their atoms slashing at the tongue. While some of these ideas are of remarkable accuracy (the enzyme molecules in raw pineapple do indeed "slash" at the tongue by destroying the protein structure), they are nothing but conjecture. The Greek atomism lacked the essential feature of a scientific theory: it was not supported or tested by critical experiments. Since it was a construct of conjecture, it could be demolished by more conjecture. The objections which arose concerned the simplicity of atoms and the complexity of nature. How could something so small and inanimate be responsible for things which live? How could the variety of nature arise from particles which, the Greeks felt, differed only slightly from one another? How could any body, being a collection of particles moving in chaos, have predictable behavior? These are questions which still concern us and which stimulate the constant refinement of the modern atomic theory. They are also questions which, by 40 B.C., led to the abandonment of atomism as an active philosophy. One conjecture had been toppled or seriously shaken by others, and so the situation remained for nearly 2000 years.

Surely it is correct to say that the logical basis for belief in the existence of atoms was supplied by Dalton, Gay-Lussac, and Avogadro, in work published in the early 1800's. What is it, though, that distinguishes the contributions of these men from the ineffectual speculations of the preceding 2000 years? Why is it that Dalton is called the father of the atomic theory, when for a century and a half previous to his work such distinguished men as Boyle and Newton had used the particulate description of matter? The quality which Dalton held in common with Gay-Lussac and Avogadro was a concern for the testing of an idea by performance of a quantitative experiment, and the success he shared with them was the demonstration that divers experimental data can be summarized by a limited set of generalizations on the behavior of matter. Dalton's contribution was not that he proposed an idea of astounding originality, but that he clearly formulated a set of postulates concerning the nature of atoms; a set of postulates which emphasized *weight* as a most fundamental atomic property.

On the basis of the crude experimental data available to him, Dalton suggested that there are indivisible atoms; the atoms of different elements have different weights; and atoms combine in a variety of simple whole number ratios to form compounds. We recognize today that these postulates are not all exactly correct, but they were the first rationalization of the quantitative laws of chemical combination. Inasmuch as the combining laws comprised the first convincing experimental demonstration that Dalton's ideas were essentially correct, they form the experimental basis for the atomic theory. Let us examine each of these laws, with respect to both its role in the development of the atomic theory, and the extent to which it is held valid today.

Questions. Do you feel that any of the following phenomena constitute qualitative evidence for the existence of atoms: sharp edges on crystals; dissolution of solids in liquids; high compressibility of gases but not of liquids and solids; the suspension of small particles in liquids and gases; the occurrence of chemical change? Can you think of other models that will explain any or all of them?

The Law of Definite Proportions

In a given compound, the constituent elements are always combined in the same proportions by weight, regardless of the origin or mode of preparation of the compound. To those familiar with the atomic theory this law is in obvious agreement with the principle that each molecule of a given compound contains the same *number* of atoms of each constituent element. Since the atoms of each element can be assigned a definite average weight, the composition by weight of a given compound is some definite value fixed by the atomic weights and the molecular formula. The law of definite proportions had been established experimentally before Dalton published his atomic theory in 1807, and the consistency of the theory with existing experiments was clearly in its favor and hastened its acceptance. However, the law of definite proportions is by no means a proof of the validity of the atomic theory. Our argument demonstrating the consistency of the atomic theory and the law of definite proportions might be more critically stated by saying that *if* there are atoms, and *if* compound formation involves interaction of these atoms in some specific way, then we might *expect* that all molecules of a given compound contain the same numbers of atoms. Then, *if* all atoms of a given element have the same weight, a compound must have a definite composition by weight. It was Dalton's position that each of these conditional statements was true; but proof of this requires more than just the fact that the consequence of all of them together is consistent with experiment. We can say, however, that it is difficult to imagine any theory not based on the atomic concept which could explain the law of definite proportions without even more serious conjecture.

Considering the importance of the law of definite proportions to the development of the atomic theory, it is surprising to find that this "law" is in many cases only a rough approximation to observed behavior. In the first place, the composition by weight of any compound depends on the atomic weights of its constituent elements. For elements having more than one isotope, the atomic weight is an *average number* whose value depends on the isotopic composition, and this may vary noticeably, depending on the source of the element. Therefore the atomic weight of an element and the weight composition of its compounds are necessarily subject to variations, and consequently the law of definite proportions is not strictly followed. One of the most serious offenders is boron, whose atomic weight may range from 10.82 to 10.84 as a result of natural variation in the ratio of abundance of the B^{11} and B^{10} isotopes. Fortunately, the variations of natural isotopic composition associated with most elements are smaller than this, and become troublesome only in the most precise work.

There is, however, another source of more serious violations of the law of definite proportions. While it is true that compounds composed of simple discrete molecules display a definite atomic and weight composition, it is also an experimental fact that there is an obvious variation in the relative numbers of atoms in ionic solids such as zinc oxide, cuprous sulfide, and ferrous oxide. For example, the composition of cuprous sulfide can range from $Cu_{1.7}S$ to Cu_2S. Materials in which the atomic composition is variable are called **nonstoichiometric compounds,** and the most extreme examples of this behavior are found among the sulfides and oxides of the transition metals.

Let us see how the atomic theory accommodates the existence both of stoichiometric and nonstoichiometric compounds. First consider a compound which consists of simple, discrete molecules, like nitric oxide, NO. Now it is clear that in order to make the atomic composition of nitric oxide depart from a 1/1 ratio, we must in some way change the atomic composition of *each* nitric oxide molecule. But the smallest possible change which we can make in a nitric oxide molecule is to add to it either one atom of nitrogen or one atom of oxygen. This results in the formation of N_2O or NO_2, both of which we recognize as compounds whose chemical properties are quite distinct from those of nitric oxide. Accordingly, we conclude that no change in the atomic composition of nitric oxide is possible without creating a new chemical species. The atomic and weight compositions of nitric oxide are therefore constant, and this and other molecular compounds obey the law of definite proportions.

Solid compounds that contain no discrete molecules present an entirely different situation. It is possible to prepare crystals of TiO with an atomic ratio of 1/1, yet if the conditions of preparation are varied, crystals of composition ranging from $Ti_{0.75}O$ to $TiO_{0.69}$ can be obtained. All these crystals have the same spatial arrangement of ions, as x-ray studies show. Depending on the preparation of the crystal, varying fractions of the titanium (II) and oxide ions are absent from sites in the crystal lattice that could be occupied, and titanium (II) oxide does not obey the law of definite composition. Such variation in atomic composition can occur without affecting the chemical properties, because titanium (II) oxide contains no discrete molecules, and the change in the ratio of atoms in the crystal as a whole does not cause a change in its crystal structure. In contrast, the electrical and optical properties of the crystal are sensitive to its atomic composition, for the resistivity and color of nonstoichiometric compounds change markedly as the atomic ratio varies.

Figure 1.1 indicates schematically how nonstoichiometry can occur through lattice vacancies (as in TiO) or through extra interstitial atoms (as in ZnO). Note that the ability of an atom to assume more than one oxidation state

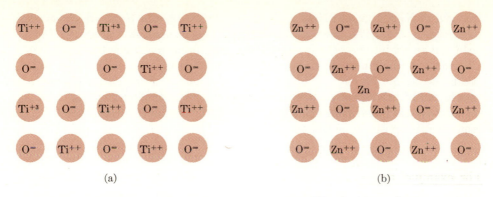

A schematic representation of nonstoichiometry in ionic solids. In (a) one Ti^{++} is missing, and electrical neutrality is maintained by the presence of two Ti^{+3} ions. In (b), a neutral zinc atom occupies an interstitial position in the crystal structure.

FIG. 1.1

provides a mechanism for creating charge neutrality in the crystal, even though some ions of one charge are missing. This is the reason why nonstoichiometry is so common in transition metal compounds. *several ox. states*

We should note that some of the compounds used by the early nineteenth-century chemists to "prove" the law of definite proportions were in fact non-stoichiometric! The variations in composition fell within the rather large experimental uncertainty of the early chemical analyses. Thus this "law" which was so important to the development of the atomic theory, and which is today the basis for virtually every stoichiometric calculation, is only an approximation, originally "proved" by data inadequate to disclose its failures. This is a common situation in physical science: laws are derived from experiments, and have a validity determined by the accuracy of the experiment and the number of cases investigated. As more accurate experiments are done in more varied situations laws may need refinement, or may have to be discarded in favor of a more general concept. It has been useful to retain the idea of definite proportions, with a realization of its limitations.

Questions. A compound is frequently *defined* as a substance which has a definite elemental composition by weight. Apparently this definition cannot include nonstoichiometric "compounds." Can you think of some other properties which can be used to identify and define compounds? Is a precise definition really necessary?

The Law of Multiple Proportions

If two elements form more than one compound, then the different weights of one which combine with the same weight of the other are in the ratio of small whole numbers. The oxides of nitrogen provide a very satisfactory demonstration of this principle: the weights of nitrogen which combine with 16 grams (gm) of oxygen in N_2O, NO, and NO_2 are respectively 28, 14, and 7 gm, which stand in the ratio 4:2:1. The law was demonstrated experimentally only *after* Dalton had advanced his hypothesis that compound formation involved combination of atoms in whole number ratios. This idea, taken together with the postulate that the atoms of each element had a characteristic weight, constitutes a *prediction* of the law of multiple proportions. Little time elapsed between the prediction and its experimental verification, and this timely success strengthened the scientific position of Dalton's ideas.

✳ The Law of Equivalent Proportions

Consider two substances A and B, either elements or compounds, which can react with each other and with a third substance C. Now a constant weight of C will react with different weights of A and B, and the ratio of the reacting weight of A to that of B is some number, generally not an integer; let us call this ratio R. When A reacts directly with B, the law of equivalent proportions says the ratio r of the reacting weight of A to that of B either is equal to R, or is a simple multiple or fraction of R. That is, $r = nR$, where n is an integer or a ratio of integers. An illustration of this law will make its meaning considerably clearer. Nitrogen (A) and oxygen (B) react with hydrogen (C) to form ammonia (NH_3) and water (H_2O) respectively. One gram of hydrogen will react with 4.66 gm of nitrogen to form ammonia, and with 8.0 gm of oxygen to form water. Therefore R is $4.66/8.00 = 0.583$. Now nitrogen and oxygen can form any one of five compounds; let us consider the reaction

$$N_2 + O_2 \rightarrow 2NO,$$

in which r, the ratio of the reacting weight of nitrogen to that of oxygen, is $\frac{28}{32} = 0.875$. According to the law of equivalent proportions,

$$0.875 = 0.583n,$$

where n is an integer or integer ratio. Solving for n we find

$$n = \frac{0.875}{0.583} = \frac{3}{2}.$$

For other compounds of nitrogen and oxygen, n has different values, but each is the ratio of integers.

While it is difficult to state the law of equivalent proportions concisely, an analysis of it supplies significant support for the atomic theory. If we have a fixed weight of substance C, the atomic theory tells us that we have a fixed, though unknown, number of particles. If we allow C to react with substance A, we have in effect counted out a number of particles of A which is related by integers to the number of particles of C, *if* compound formation involves combination of atoms in simple whole number ratios. Again, combination of our fixed weight of C with B "counts out" a number of particles of B which is again some simple multiple of the number of C-particles. Thus, the number of A- and B-particles "selected" in this way must stand in the ratio of integers.

A little reflection shows that the ratio R is merely the ratio of selected A-particles to B-particles, multiplied by the ratio of the weights of these particles. The direct reaction of A with B also "selects" A- and B-particles in an integer ratio, which when multiplied by the ratio of weights of A- and B-particles becomes the ratio r. Since both R and r are related by the same factor to integer ratios, they must be related to each other by a multiplier which is the ratio of whole numbers. This multiplier is equal to the factor n used in the previous paragraph. Once again the assumption of characteristic atomic weights and of compounds involving atoms in small whole number ratios allows us to advance an argument, tedious but not elaborate, which rationalizes a general observation of chemical behavior.

Compounds have a definite composition by weight; the different weights of one element which combine with a fixed weight of another stand in integer ratios; the weights of two chemicals which combine with each other and with a third are related by integers. Does any one of these laws of chemical composition *prove* the atomic theory? Not at all. By itself, each stands as a narrow observation, only hinting of its probable cause. Taken together they are generalizations of three *apparently independent* aspects of natural behavior which find a *unified* explanation in the atomic theory. Rather than any one piece of evidence, it was the ability to relate and explain an accumulation of observations which led to acceptance of the atomic theory.

Question. The equivalent weight of an element is that weight which combines with 8 gm of oxygen. Is the following an accurate restatement of the law of equivalent proportions: Elements combine with each other in proportion to their equivalent weights, or in small integer multiples thereof?

1.2 DETERMINATION OF ATOMIC WEIGHTS AND MOLECULAR FORMULAS

Given the existence of atoms of characteristic weight, and given their tendency to combine in the ratio of small whole numbers, we consider the problem of determining *both* atomic weights and molecular formulas. Now it is clear that if one of these were known, the other could be determined. For example, if it were known that in cupric oxide there is exactly one oxygen atom for every copper atom, then the experimental fact that in CuO there are 63.5 gm of copper for every 16 gm of oxygen would mean that the average relative weights of a copper atom and an oxygen atom were in the ratio of 63.5 to 16. In other words, the atomic weight of copper would be 63.5 on a scale which took the atomic weight of oxygen as 16. So it is simple to see how atomic weights can be determined if formulas are known, and of course the process is easily reversed; if atomic weights are known, formulas can be determined readily. The problem confronting the chemists of the early nineteenth century was how to determine atomic weights and formulas *simultaneously*, since both were unknown.

The approach taken by Dalton is one often used in scientific investigation: if there is no information to the contrary, make the simplest possible assumption and pursue its consequences. Thus, Dalton's "rule of greatest simplicity": if two elements form only one compound with each other, a molecule of this compound contains only one atom of each element. In 1805, water was the only known compound of hydrogen and oxygen, so its formula was taken as HO. This assumption, together with the fact that water contains 8 gm of oxygen for each gram of hydrogen, forced the conclusion that the atomic weight of oxygen was 8 on a scale which took the atomic weight of hydrogen as unity. A risky supposition had led to a false conclusion, but Dalton could offer no better solution to the problem of atomic weights and molecular formulas.

Work of Gay-Lussac, published in 1808, supplied what became the eventual basis for the establishment of molecular formulas. An investigation of gaseous reactions showed that the combining volumes, measured under conditions of constant temperature and pressure, stood approximately in the ratios of small integers. The cases examined by Gay-Lussac included the reaction of ammonia with carbonic acid, where he found two different reactions which we would write:

$$NH_3(g) + H_2CO_3(aq) = NH_4HCO_3(aq),$$
$$2NH_3(g) + H_2CO_3(aq) = (NH_4)_2CO_3(aq).$$

The volumes of ammonia consumed in these two reactions are in a 1/2 ratio; thus the law of multiple proportions finds its analog in the volume relationships of reacting gases. Pursuing the idea that reacting volumes were always related by integers, Gay-Lussac interpreted the measurements of others to show that ammonia is composed of three volumes of hydrogen to one of nitrogen, and that in nitrous oxide, nitric oxide, and nitrogen dioxide (N_2O, NO, and NO_2) there were respectively 2, 1, and $\frac{1}{2}$ volumes of nitrogen for each volume of oxygen. The apparently relentless appearance of integer relationships seemed to Gay-Lussac and others to support the atomic theory, but these results found least favor with the man who, we might have expected, should have welcomed them most.

Dalton saw that Gay-Lussac's observations, if correct, implied that the numbers of particles contained in equal volumes of gases were either equal or integral multiples of one another. Dalton could find two serious objections to this conclusion. First, it was known that the density of the water vapor which resulted from the reaction of hydrogen with oxygen was *less* than the original density of oxygen. With the correct molecular formulas at hand, this result is not difficult to explain. We merely recognize that since the formula of water is H_2O, and that of oxygen is O_2, the weight of an oxygen molecule is greater than the weight of a water molecule. Then, if one liter of each of these gases contains equal numbers of molecules, the density or the weight per liter of oxygen *should* be greater than the density of water vapor. The difference in density is entirely explained by the fact that elemental oxygen contains two oxygen atoms per molecule, and the water molecule but one. However, to someone conditioned to think of a chemical reaction only as an *addition* of one atom to another, to someone who pictured the formation of water as

$$H + O = HO,$$

there appeared only one way for water vapor to be less dense than oxygen: *fewer* molecules of water than of oxygen had to be contained in equal volumes.

Dalton's second objection was based on the following argument. It was observed that from equal volumes of nitrogen and oxygen, two volumes of nitric oxide would be produced. If equal volumes of different gases contained equal numbers of particles, and if, as Dalton felt, each particle of an elemental gas was an indivisible atom, we would be forced to write

$$\text{nitrogen } + \text{ oxygen } \rightarrow \text{ nitric oxide,}$$
$$1 \text{ volume} + 1 \text{ volume} \rightarrow 2 \text{ volumes,}$$
$$n \text{ atoms } + n \text{ atoms } \rightarrow 2n \text{ molecules.}$$

The first two lines represent experimental facts, the last, Dalton pointed out, is an impossibility. The reaction of n indivisible atoms can never produce more than n new particles. This reasoning is based on a most arbitrary assumption: the "particles" of elements are individual atoms. Dalton chose not to question the validity of this assumption, but instead *rejected* the "equal volumes-equal numbers" idea and the data on which it was based.

An argument in favor of the "equal volumes-equal numbers" hypothesis was advanced by Amadeo Avogadro in 1811. By combining this idea with his novel suggestion that gaseous elements could consist of *polyatomic* molecules, Avogadro successfully reconciled the combining volume data with the concept of the indivisible atom. He pointed out that once it is admitted that nitrogen and oxygen can be polyatomic, the volume relations accompanying the formation of nitric oxide can be explained by saying

1. nitrogen + oxygen $\rightarrow$ nitric oxide;
2. 1 volume + 1 volume $\rightarrow$ 2 volumes;
3. n molecules + n molecules $\rightarrow 2n$ molecules;
4. $N_2 + O_2 \rightarrow 2NO$;
5. $N_4 + O_4 \rightarrow 2N_2O_2$.

Lines 1 and 2 are experimental facts. Line 3 combines the "equal volumes-equal numbers" hypothesis with the idea that elemental gases can be polyatomic; it avoids the necessity of splitting atoms by offering the opportunity to divide polyatomic molecules. Lines 4 and 5 contain two suggestions (there are many others) of molecular formulas which are consistent with the combining volume data. Avogadro's work answers Dalton's objections to the "equal volumes-equal numbers" principle, but only at the cost of introducing *another* hypothesis: the existence of polyatomic elements. Moreover, as shown by line 5 above, the formulas of these polyatomic elements are left undetermined; consequently we are still without a scheme for the unequivocal determination of atomic weights and molecular formulas.

In the years between 1811 and 1858 the problem of determining the atomic weight scale became more and more vexing. Solutions were proposed, only to be abandoned when they failed to account for all of a growing body of experimental facts. Eventually there were those who felt that it was impossible ever to determine atomic weights and molecular formulas. The permanent solution required only a slight extension of Avogadro's reasoning, and this Stanislao Cannizzaro supplied in 1858.

Cannizzaro's Analysis

Cannizzaro based his method of atomic weight determination on the idea that a molecule must contain a whole number of atoms of each of its constituent elements. Given this, it is clear that in one gram molecular weight of a compound there must be at least one gram atomic weight of a given element, or otherwise some integral multiple of this weight. Therefore, if a series of the compounds of this element are analyzed, and the weights of the element contained in one gram molecular weight of the various compounds compared, it should eventually become obvious that all these weights *are integral multiples of some number which is very probably the gram atomic weight.* To use this method, there must be a way to find the molecular weights of the compounds. For this, Cannizzaro turned to Avogadro's principle: since under the same conditions equal volumes of gases contain equal numbers of molecules, the weights of these equal volumes must stand in the same ratio as the weights of their molecules, or as their molecular weights. With such a set of relative molecular weights available, Cannizzaro *defined* the molecular weight of hydrogen gas to be 2, and thus fixed the absolute values of all others.

Table 1.1 Atomic Weight of Oxygen

Compound	Molecular weight relative to $H_2 = 2$	Weight of oxygen in one gram molecular weight
Water	18	16
Nitric oxide	30	16
Nitrous oxide	44	16
Nitrogen dioxide	46	32
Sulfur dioxide	64	32
Carbon dioxide	44	32
Oxygen	32	32
Ozone	48	48

Let us illustrate Cannizzaro's procedure by outlining the determination of the atomic weight of oxygen. First, the molecular weights of a number of gaseous compounds of oxygen are determined by measuring their densities relative to hydrogen gas at the same temperature and pressure. For example, the density of nitric oxide gas is almost exactly fifteen times that of hydrogen; therefore the molecular weight of nitric oxide is 30 on a scale in which hydrogen gas is *defined* to be 2. Similar experiments on the other oxygen containing gases provide the molecular weight data in Table 1.1.

The second step is the determination of the weight of oxygen contained in one gram molecular weight of each compound. This follows easily from a chemical analysis, which gives the composition of the compound in percent or fraction by weight. The weight fraction of oxygen multiplied by the molecular weight gives the desired quantity. To pursue our example, when one gram of nitric oxide is decomposed to the elements, 0.533 gm of oxygen is recovered. Consequently in 30 gm of nitric oxide there are $30 \times 0.533/1 = 16$ gm of oxygen. Repetition of this procedure gives the weight of oxygen contained in one gram molecular weight of each of the gases in Table 1.1.

The third and final step is to examine these data for integral relationships. The smallest weight of oxygen found in one gram molecular weight of a compound is 16 gm, and all others are its integral multiples; so we might hastily conclude that 16 is the atomic weight of oxygen. However, all numbers in the table are also multiples of 8, and we might argue that 8 is the atomic weight of oxygen. But to adopt 8 as the atomic weight would force us to the very unlikely conclusion that all compounds of oxygen available to us contain even numbers of oxygen atoms, since the odd multiples of 8 (8, 24, 40) are absent from the table. It seems, then, that the proof that the atomic weight of oxygen is 16 lies in the repeated failure to find a compound which contains 8, 24, or 40 gm of oxygen in one gram molecular weight.

It is apparent that analogous procedures will yield the atomic weights of other elements each of which forms a series of gaseous compounds. Thus by clever combination of the law of multiple proportions and the "equal volumes-equal numbers" principle, Cannizzaro removed the dilemma associated with the atomic weight-molecular formula and supplied a lasting chemical basis for the atomic theory.

Let us review the arguments which led to a quantitative atomic theory. We have first the proposals of Dalton: there are indivisible atoms; those of a given element are alike in weight; the atoms of different elements have different weights; atoms combine in a variety of simple whole number ratios to form compounds. These proposals find strong support in the laws of chemical combination. Then come Gay-Lussac's observations of combining gaseous volumes; these suggest only that a fixed volume of different gases contains numbers of molecules which are in the ratio of integers. Avogadro's statement that these numbers are equal is an assumption, and this assumption is at the heart of Cannizzaro's procedure for finding atomic weights. Thus the original arguments which led to the definition of the chemical atomic weight scale were not free from conjecture, but involved hypotheses which have since been justified by a wealth of experimental data.

Other Guides to the Atomic Weights

Cannizzaro's method was initially limited to determining the atomic weights of elements which formed gaseous compounds. We turn now to two other principles which helped to establish a complete atomic weight scale: the Dulong and Petit law of atomic heat capacity and Mendeleev's law of chemical periodicity.

In 1819, Dulong and Petit measured the specific heats* of a number of metals, and found that the values for the various materials differed considerably. They then sought to calculate the heat required to raise not a fixed weight, but a *fixed number of atoms* one degree. One gram atomic weight of the different elements contains the same number of atoms; thus multiplication of the specific heat of each element by its gram atomic weight gives the heat required to raise a fixed number of atoms of all substances one degree celsius. Naturally this procedure requires an accurate table of atomic weights, and the best available to Dulong and Petit contained several entries which, because of the atomic weight-molecular formula confusion, differed from the true values by simple numerical factors. However, the product of the specific heat and *some* of the atomic weights was a constant which on our present scale has the approximate value 6. Dulong and Petit *assumed* that this was a universal relation, and that failure to obey this law indicated an incorrect atomic weight. Furthermore, they found that it was possible to "correct" the offending atomic weights by multiplying them by integer ratios, and once this was done all elements obeyed the relation

specific heat (cal/gm-deg) $\times$ atomic weight (gm/gm atom)

$$\cong 6 \text{ (cal/deg-gm atom).}$$

The "correction" applied to some of the atomic weights was not so arbitrary as might first appear, for at the time it was generally accepted that molecular formulas were uncertain, and that atomic weights could be in error by factors which involved small integers. Thus Dulong and Petit found a relationship which today we still accept as correct, within certain well-understood limits: the heat capacity of one mole of atoms in the solid state is approximately 6 calories per degree celsius. Table 1.2 shows that a variety of solid elements obey this law, and that the deviations characteristically fall among solids which contain the very light elements.

---

* The *specific heat* is the number of calories (cal) required to raise one gram of material one degree celsius. The term *heat capacity* generally refers to the heat required to raise *one mole* of material one degree celsius.

Table 1.2 Heat Capacities of Some Solid Elements

Element	Atomic weight	Specific heat (cal/deg-gm)	Heat capacity (cal/deg-mole)
Lithium	6.9	0.92	6.3
Beryllium	9.0	0.39	3.5
Magnesium	24.3	0.25	6.1
Carbon (diamond)	12.0	0.12	1.4
Aluminum	27.0	0.21	5.7
Iron	55.8	0.11	6.1
Silver	107.9	0.056	6.0
Lead	207.2	0.031	6.4
Mercury	200.6	0.033	6.6

The following example illustrates how, despite its approximate nature, the law of Dulong and Petit can lead to an accurate atomic weight. A careful chemical analysis shows that in a compound of copper and chlorine there are 0.3286 gm of chlorine and 0.5888 gm of copper. Given that the atomic weight of chlorine is 35.46, what is the atomic weight of copper? Our data allow us to calculate that one gram atomic weight of chlorine will combine with $0.5888 \times 35.46/0.3286 = 63.54$ gm of copper. If the empirical formula of the compound is CuCl, the atomic weight of copper is 63.54. But if the formula is Cu_2Cl or $CuCl_2$ the atomic weight of copper is either 31.77 or 127.08. Clearly, a rough value of the atomic weight will allow us to choose between the more precise alternatives. The measured specific heat of copper is 0.093 cal/deg-gm. Applying the law of Dulong and Petit we find

$$\text{atomic weight} \cong \frac{6}{0.093} = 64,$$

which allows us to choose 63.54 as the correct atomic weight, and CuCl as the formula of the chloride.

The periodic table also served as a guide to the rough values of the atomic weights. The table published by Mendeleev in 1869 ordered the elements according to the atomic weights determined largely by the method of Cannizzaro. The table contained gaps which corresponded to undiscovered elements, or to those whose atomic weights were unknown or miscalculated. These gaps (or their absence) aided subsequent atomic weight assignments. The atomic weight of uranium had first been taken as 120 on the basis of chemical analysis and an *assumed* formula of its oxide. Mendeleev saw that there was no place in this table for an element with atomic weight between that of tin (119) and antimony (122), and so suggested that the correct atomic weight of uranium was nearer $2 \times 120 = 240$, as in fact it is. Similarly, he corrected the value initially assigned to indium from 76, which would have forced it between arsenic and selenium, to $76 \times 3/2 = 114$, which is quite close to the presently accepted value.

Precise Atomic Weights

Our discussion has emphasized the principles involved in the construction of the atomic weight scale; there remains the problem of how to establish the best values of the atomic weights. There are three important methods: accurate determination of reacting or combining weights, accurate determination of gas densities, and mass spectrometry. The first two are refinements of the classical methods already discussed, while the third is a totally different approach.

Table 1.3 Relations between atomic weight scales*

	Old physical scale	Old chemical scale	New unified scale
O^{16}	16 exactly	15.99560	15.99491
O		16 exactly	15.9994
C^{12}	12.00382	12.00052	12 exactly
C		12.011	12.0115

*After E. A. Guggenheim, *J. Chem. Ed.*, **38**, 86 (1961).

Before taking up these methods of atomic weight determination in detail, we should discuss the basis of the numerical values in the atomic weight scale. The atomic weights are a set of relative numbers whose absolute value depends on some defined standard. In the past, physicists have defined the mass of the most abundant isotope of oxygen, O^{16}, as 16 exactly, while chemists defined the average atomic weight of the naturally occurring mixture of oxygen isotopes (O^{16}, O^{17}, O^{18}) as exactly 16. These two definitions led to two different atomic weight scales. According to the physicists' scale, the atomic weight of the mixture of oxygen isotopes was not 16, but 16.0044. Thus to convert a physicist's atomic weight to a chemist's atomic weight it was necessary to multiply by $16.0000/16.0044 = 1/1.000275$. In 1961 both scales were abandoned and replaced by a unified scale based on a new standard. The atomic weight of the most abundant isotope of carbon, C^{12}, is now defined to be 12 exactly. The very small changes caused by this new definition are summarized in Table 1.3.

1. The combining weight method has given us most of the accepted atomic weights. The principle is simple: some reaction involving the element of unknown atomic weight and another element whose atomic weight is known is carried out quantitatively. This establishes a combining or reacting weight ratio between the two elements. Since atoms combine in the ratio of small integers, the combining weight ratio is either exactly equal to, or is an integral multiple of, the ratio of atomic weights. For example, it is found that 1.292 gm of pure silver react with 0.9570 gm of bromine to form silver bromide. The ratio of reacting weights, $1.292/0.9570 = 1.350$, is equal to the ratio of atomic weights, since the formula of silver bromide is AgBr. Thus, taking Ag = 107.87, we find Br = $107.87/1.350 = 79.90$. A variety of chemical reactions are used to establish atomic weights, and in each case greatest care must be

taken to see that the reagents are of highest purity, and that the reactions proceed quantitatively to give compounds of definite, known composition.

To obtain accurate atomic and molecular weights from gas density measurements it is first necessary to realize that Avogadro's principle is not strictly true for gases near atmospheric pressure. Because of forces between molecules, which are different in different gases, equal volumes of various gases do not contain *exactly* equal numbers of molecules. Avogadro's principle becomes exact only in the limit of very low pressure, and gas densities measured under these conditions do stand in the ratio of the molecular weights of the gases.

How can we find the density a gas would have if no forces acted between the molecules? Experiments and the kinetic theory of gases cause us to be confident that in the limit of zero pressure, that is, when the molecules are far apart and the forces between them are negligible, the gas density δ should be *exactly* proportional to the pressure P:

$$\delta = \alpha P \qquad \text{(no intermolecular forces).}$$

Here it is helpful to think of the proportionality constant α as the density that a gas would have at 1 atm pressure ($P = 1$) if there were no forces between the molecules. Thus, it is just the quantity we want. When gases are at finite pressure, and intermolecular forces are important, there must be a correction term added to the equation which shows that the density is no longer strictly proportional to the first power of the pressure. The simplest type of correction term would be of the form βP^2, where β is a positive or negative constant that depends on the existence of intermolecular forces. The equation for the density now is

$$\delta = \alpha P + \beta P^2,$$

which we can rewrite as

$$\delta/P = \alpha + \beta P.$$

Now we see that in order to find α, we need only plot the measured ratio of the density to pressure as a function of pressure, and find the intercept of the resulting straight line. Such plot is shown in Fig. 1.2. From it we can find that the ideal density of CO_2 gas at 1 atm and 273.1°K is 1.9635 gm/liter.

2. Such ideal densities can be used to compute molecular weights. For example, the ideal density of neon gas is found to be 0.9004 gm/liter, while that of oxygen is 1.428 gm/liter, both at 1 atm and 273.1°K. Since equal volumes of gas contain equal numbers of molecules, the ratio of gas densities must be the ratio of molecular weights. Taking the molecular weight of oxygen to be 31.999, we find

$$\frac{0.9004}{1.428} \times 31.999 = 20.18$$

for the atomic weight of neon.

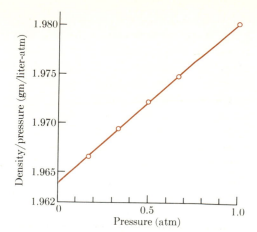

FIG. 1.2 Determination of the ideal density of carbon dioxide gas at 273.1°K.

Questions. The ideal density of CO_2 found from Fig. 1.2 is 1.9635 gm/liter at 1 atm and 273.1°K. What is the molecular weight of CO_2? What is the atomic weight of carbon?

3. The mass spectrographic method is capable of giving the most precise atomic weights. Once again the principle of the method is simple; its execution requires the greatest care to be successful. As shown in Fig. 1.3, the mass spectrograph consists of three major parts: a source of gaseous ions, an evacuated dispersing region in which ions of different charge-to-mass ratio are forced to travel different paths, and a detector which locates the trajectories followed by the different ions. The ion source is most often a small chamber in which gaseous atoms or molecules are bombarded with a beam of energetic electrons obtained from a heated filament. The collision of the electrons with the molecules produces positive ions, some of which are only fragments of the original molecule. For instance, the bombardment of water vapor gives not only H_2O^+, but OH^+, O^+, and H^+. Some of the positive ions are accelerated by an electric field, pass

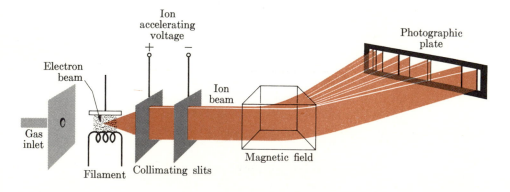

FIG. 1.3   Schematic drawing of a mass spectrograph.

through a slit, and enter a highly evacuated flight tube in which they are subject to a magnetic field. This field causes the ions to follow circular trajectories whose radii are given by $1/r^2 = (B^2/2V)(e/m)$, where B is the value of the magnetic field, V is the voltage difference through which the ions were accelerated, and e/m is the ratio of the ionic charge to the ionic mass. Those ions having the same ratio of charge to mass follow the same path; the paths of greatest radius are taken by ions whose charge-to-mass ratio is smallest.

The detector, which is a photographic plate, intercepts the ion trajectories and records the position and intensity of each ion beam. For each image found on the plate, the radius of curvature of the corresponding ion path can be found, and in turn, the charge-to-mass ratio of the ion computed. The most accurate comparisons can be made between the masses of fragments which fall close to each other on the photographic plate. For instance, the ions $^{16}O^+$, $^{12}CH_4^+$, $^{12}CDH_2^+$, $^{12}CD_2^+$, $^{14}NH_2^+$, and $^{14}ND^+$ all have a mass of 16, approximately. As Fig. 1.4 shows, the mass spectrograph is capable of resolving the small mass differences which exist among these fragments. This mass spectrogram permits accurate calculation of the mass of the deuterium atom in terms of the mass of the hydrogen atom, and if the mass of hydrogen is known, the masses of O^{16} and N^{14} can be calculated in terms of the mass of C^{12}.

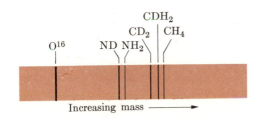

Mass spectrum of some ions of approximate mass 16.   **FIG. 1.4**

We see that the mass spectrograph permits the accurate determination of the masses of the isotopes of a given element. But as we noted in Section 1.1, for elements having more than one isotope, the atomic weight is an average number whose value depends on the relative amounts of each isotope. Thus, accurate determination of the relative abundances of the isotopes of an element is necessary before its chemical atomic weight can be calculated. This isotopic abundance determination is also done with the mass spectrograph, but instead of recording the ions photographically, the intensity of each ion beam is measured electrically. For example, carbon is found to consist of 98.892% C^{12}, and 1.108% C^{13}. On the unified atomic weight scale, C^{12} has a mass of 12 exactly, and C^{13} has a mass of 13.00335. Thus, the atomic weight of the isotopic mixture is

$$12.0000 \times 0.98892 = 11.8670$$
$$13.0033 \times 0.01108 = \underline{0.1441}$$
$$12.0111$$

Considering the great precision which mass spectrometry offers in the determination of both isotopic mass and abundance, one might ask why many of our atomic weights are listed only to four significant figures. Often the limitation is set not by the accuracy with which the determination can be done, but by natural variations in the isotopic content of the elements themselves. There is no point to listing the atomic weight of an element to five significant figures when the natural variations in isotopic composition cause variation in the fourth significant figure.

Question. Can you think of any reasons why the isotopic composition of elements should differ in different sources?

1.3 THE MOLE CONCEPT

The fundamental unit of chemical thought is the atom or molecule; it is therefore no surprise that the ability to measure and express the number of molecules present in any chemical system is of foremost importance. The number of molecules present in a system occupies a central position in all chemical reasoning, and virtually all the equations of theoretical chemistry contain the number of molecules as an important factor. While it is possible today to detect the presence of single atoms, any direct attempt to count the enormous number of atoms in even the smallest chemical system would occupy the total population of the world for many centuries. The practical solution to counting large numbers of atoms is much less imposing; we need only use the most fundamental of laboratory operations, weighing.

Our discussion of the development of the atomic theory led to the conclusion that equal numbers of atoms are contained in one gram atomic weight of each element, and that the same number of molecules is found in one gram molecular weight of any compound. The terms "gram atomic weight" and "gram molecular weight" are awkward and tend to conceal the fact that the reason they are used is to refer to a fixed number (Avogadro's number,* 6.022×10^{23}) of particles. It is more convenient to use the term *mole* to stand for the amount of material which contains this number of particles. As formal definitions we have

The number of carbon atoms in exactly 12 gm of C^{12} is called Avogadro's number, N. One mole is the amount of material which contains Avogadro's number of particles.

These definitions emphasize that the mole refers to a *fixed number of any type of particles*. Thus it is correct to refer to a mole of helium, a mole of electrons, or a mole of Na^+, meaning respectively Avogadro's number of atoms, electrons, or ions. On the other hand, phrases like "one mole of hydrogen" can be ambiguous, and should be restated as "one mole of hydrogen atoms" or "one mole

---

* Avogadro's number has been determined by a variety of techniques some of which are described in Appendix A.

of hydrogen molecules." It is a matter of common practice among chemists, however, to let the name of the element stand for its most common form. Thus one mole of O_2 is frequently referred to as one mole of oxygen, whereas one mole of O is called one mole of oxygen atoms.

From the definitions we see that the weight of one mole of material is always equal to its atomic (or molecular) weight in grams. Thus, to measure out some multiple or fraction of Avogadro's number of particles we have only to weigh out the appropriate multiple or fraction of the atomic or molecular weight. That is,

$$\text{number of moles} = \frac{\text{weight (gm)}}{\text{weight of one mole (gm/mole)}} = \frac{\text{mass}}{\text{molar mass}}$$

$$= \frac{\text{weight}}{\text{atomic (molecular) weight}}.$$

Despite the fact that the number of moles is often measured by weighing, it is more profitable to think of one mole as a fixed number of particles rather than a fixed weight. One mole is *always* Avogadro's number of particles, but the weight which contains one mole differs for various substances.

Question. We make frequent use of "conversion factors" to change from one set of units to another. The atomic mass unit (amu) is defined as $\frac{1}{12}$ the mass of a C^{12} atom. Do you think the statement that Avogadro's number is the conversion factor changing atomic mass units to grams is correct?

Some confusion can arise concerning the "molecular weights" of substances in which no discrete molecules exist. For example, in solid sodium chloride there are no identifiable NaCl molecules, only ions of sodium and chlorine. Nevertheless, it is common to use the term "molecular weight of sodium chloride" as if this substance were composed of NaCl molecules. In this context the words "molecular weight of NaCl" mean only the weight of material that contains 6.02×10^{23} ions of each type, and carry no implication about the existence of molecules in the crystal. Strictly we should refer to the *formula weight* of a substance like NaCl, and write

$$\text{number of moles} = \frac{\text{weight}}{\text{formula weight}}.$$

However, the words "molecular weight" are frequently substituted for "formula weight."

1.4 THE CHEMICAL EQUATION

A chemical reaction is a mechanical process in the same sense as is raising a weight, for in a chemical reaction, a mechanical system of atoms is converted from one state to another. For example, the reactants in the transformation

$$2Ag^+(aq) + H_2S(aq) = Ag_2S(s) + 2H^+(aq)$$

are atoms of silver, hydrogen, and sulfur in a particular mechanical arrangement or grouping which we describe as an aqueous solution of silver ions and an aqueous solution of hydrogen sulfide molecules. The products are just the same atoms in different arrangement—the silver and sulfur atoms are grouped together as ions in a solid substance which is precipitated from the solution, and the hydrogen atoms are present as ions in the aqueous solution. The role of the chemical equation is to describe the chemical process both qualitatively and quantitatively in a way which is at once accurate and brief.

To describe a reaction qualitatively, we want the symbols to describe the state or condition of the reactants and products as they occur under the conditions of the reaction. In the example above, the solution of hydrogen sulfide contains H^+, HS^-, and $S^=$ ions, as well as H_2S molecules. However, for most purposes it would needlessly complicate our description if we listed each of these species as a reactant. Investigations of aqueous solutions of hydrogen sulfide show that most of the sulfur-containing material is present as undissociated hydrogen sulfide molecules, and therefore this solution is most conveniently represented by the symbol $H_2S(aq)$. In contrast, in a solution prepared by mixing sodium hydroxide and hydrogen sulfide, the sulfur species in greatest concentration might be HS^-; if so, this symbol would be the best description of the solution. In general, when one of the reacting atoms is present in more than one form, the species in highest concentration is used in writing the chemical reaction. The point is simply to describe the system as accurately as possible without sacrificing brevity.

Another qualitative principle used in writing chemical equations is that chemical species which are not used or produced by the reaction are not included in the equation, even though they may be present in the reacting system. In our example, the silver ion may have been obtained as an aqueous solution of silver nitrate, but the nitrate ion is left unchanged by the reaction; so including it in the equation would serve no purpose.

The stoichiometric coefficients which appear in a chemical equation express the quantitative aspect of a chemical reaction. In many cases these coefficients represent only the *relative numbers* of molecules which participate in a chemical reaction. Thus in this context it is quite correct to write

$$\tfrac{1}{2}N_2 + \tfrac{3}{2}H_2 = NH_3.$$

The equation says only that in the synthesis of ammonia, the number of ammonia molecules produced is twice the number of nitrogen molecules, and two-thirds the number of hydrogen molecules used.

In other instances, chemical equations are used to describe the behavior of individual molecules. For example, the recombination of oxygen atoms to form molecules occurs in the presence of O_2 by two successive chemical transformations. In the first step, one oxygen atom reacts with one oxygen molecule to give ozone:

$$O + O_2 \rightarrow O_3.$$

The second step is the reaction of another oxygen atom with O_3 to give two oxygen molecules:

$$O + O_3 \rightarrow 2O_2.$$

Since in writing these equations we are using chemical symbols to stand for individual atoms and molecules, it is *not* proper to divide the last equation by 2 to give

$$\tfrac{1}{2}O + \tfrac{1}{2}O_3 \rightarrow O_2.$$

In the context used, this would imply the existence of "half-molecules," which is nonsense. It is always clear from context whether the chemical symbols stand for individual molecules or a collection of molecules.

The chemical equation is an example of a conservation equation, since it expresses the fact that in a chemical reaction the number of atoms of each element is constant, or is conserved. This is, of course, the principle used to "balance" the equation. Furthermore, the equation represents the fact that net electrical charge is neither created nor destroyed by chemical reactions. In the equation describing the reaction between silver ion and hydrogen sulfide we see that the net charge on the reactants *and* the products is the same: $+2$. We shall find in Chapter 7 that the principle of charge conservation is very helpful in balancing oxidation-reduction equations.

1.5 STOICHIOMETRIC CALCULATIONS

Now let us draw on the definition of the mole and the principle of atom conservation to attack the problem of weight relations in chemical reactions. To illustrate the procedure which we shall find can cope with the most difficult problems, let us choose a simple example. The chemical equation

$$2CO + O_2 = 2CO_2$$

states that $2n$ molecules of carbon monoxide and n molecules of oxygen are used to form $2n$ molecules of carbon dioxide, where n is any number. If n is taken as 6.022×10^{23}, then the equation says that two moles of carbon monoxide plus one mole of oxygen can be converted to two moles of carbon dioxide. This argument illustrates the supremely important fact that since a mole contains a definite number of particles, *anything that can be said about relative numbers of molecules or atoms may be said about relative numbers of moles.* This statement is the basis for all stoichiometric calculations.

Now let us solve the elementary problem of finding the weight of carbon dioxide which can be obtained from the combustion of 12.0 gm of carbon monoxide with excess oxygen. The general method of attack on this and similar problems involves use of the most valuable piece of information about the reaction, the chemical equation. In this example the chemical equation tells us directly that:

2 moles of carbon monoxide yield 2 moles of carbon dioxide;

so we can say that, in general,

moles of carbon monoxide used = moles of carbon dioxide produced.

Since there is one mole of carbon atoms in both one mole of carbon monoxide and one mole of carbon dioxide, these word equations are simply statements that carbon atoms are conserved in the chemical reaction. To effect the desired calculation, we need only apply the expression for the number of moles in an arbitrary weight:

$$\text{moles of CO} = \frac{\text{weight of CO}}{\text{molecular weight}} = \frac{12.0}{28.0} = 0.429 = \text{moles of CO}_2,$$

$$\text{weight of CO}_2 = \text{number of moles of CO}_2 \times \text{weight of one mole of CO}_2$$

$$= \frac{12.0}{28.0} \times 44.0 = 18.9 \text{ gm of CO}_2.$$

Note the equivalence of this procedure (the "mole method") to the so-called "proportion method." Following the latter we would write

$$2CO + O_2 = 2CO_2,$$

$$\frac{12.0}{2 \times 28.0} = \frac{\text{weight of CO}_2}{2 \times 44.0},$$

$$\text{weight of CO}_2 = \frac{12.0}{28.0} \times 44.0 = 18.9 \text{ gm}$$

for no apparent reason other than that it produces the correct answer. The "mole method" is preferable because in its algebraic statements it uses information derived directly from the principle of atom conservation (i.e., the moles of CO equal the moles of CO_2 because both contain the same number of carbon atoms). This advantage becomes more apparent as the problems become more difficult.

Another advantage of solving stoichiometric problems in terms of moles is found if we ask for the weight of oxygen consumed in the above reaction. Reference to the chemical equation shows that the number of moles of oxygen consumed is just one-half the number of moles of carbon monoxide used. Therefore

$$\text{moles of O}_2 = \tfrac{1}{2} \times \text{moles of CO}$$

$$= \tfrac{1}{2}(0.429) = 0.214 \text{ moles},$$

$$\text{weight of oxygen} = 32.0 \text{ gm/mole} \times 0.214 \text{ moles} = 6.85 \text{ gm}.$$

This procedure is more efficient than setting up a new proportion involving, this time, oxygen and carbon monoxide. Since the stoichiometric coefficients in a chemical equation are generally small integers (in this case 1 and 2), the number of moles of one substance can be found from the number of moles of another by a simple multiplication involving the ratio of small integers. The number of moles of any substance can, at any time, be converted to the corresponding weight by multiplication by the weight of one mole.

The utility of the mole concept is further illustrated by the problem of determining the *empirical* formulas of compounds. The empirical formula expresses only the *relative numbers of atoms* of each element, and implies nothing about how many atoms are actually in one mole of compound. Empirical formulas are found experimentally by measuring the weights of the elements necessary to *synthesize* a certain amount of the compound, or alternatively by *analyzing* a sample of the material for the weights of its constituents. The results of such determinations are often expressed as composition in percent by weight. For example, a certain sulfide of iron contains 46.5% iron and 53.5% sulfur by weight. The problem of converting these data to an empirical formula is easily solved if we remember two things: empirical formulas represent relative numbers of atoms of the elements, and anything we can say about relative numbers of atoms may be said for relative numbers of moles of atoms. A calculation of the relative number of moles of iron and sulfur will therefore lead us to the empirical formula.

The relative numbers of moles of atoms can be found if we imagine we have one gram of the iron sulfide. Then we have

$$1.00 \times 0.465 = 0.465 \text{ gm of iron,}$$
$$1.00 \times 0.535 = 0.535 \text{ gm of sulfur,}$$

which is just

$$\frac{0.465 \text{ gm}}{55.8 \text{ gm/mole}} = \text{moles of iron} = 0.00833 \text{ moles}$$

and

$$\frac{0.535 \text{ gm}}{32.1 \text{ gm/mole}} = \text{moles of sulfur} = 0.0166 \text{ moles.}$$

The relative number of moles, and of atoms, is just

$$\frac{\text{moles Fe}}{\text{moles S}} = \frac{0.00833}{0.0166} = \frac{\text{atoms Fe}}{\text{atoms S}} = \frac{1}{2}.$$

The empirical formula of the sulfide is FeS_2.

In the foregoing we have emphasized the relationship between the weight of a substance and the corresponding number of moles. Weighing is not the only method by which we can count molecules. Avogadro's law states that under conditions of constant temperature and pressure, equal volumes of gases contain equal numbers of particles. Experimental investigations show that at one atmosphere pressure and a temperature of 273.1°K (conditions known as standard temperature and pressure, STP), one mole of any gas occupies a

101.3 kPa

volume which is very close to 22.4 liters. Therefore, the number of moles in any gas sample can be found by comparing its volume under standard temperature and pressure conditions with 22.4 liters. We have then

$$\text{moles of gas} = \frac{\text{volume under STP conditions}}{\text{volume of one mole under STP conditions}}$$
$$= \frac{V(\text{STP}) \text{ liters}}{22.4 \text{ liters/mole}}.$$

To illustrate the use of the standard molar volume in stoichiometric calculations let us consider the thermal decomposition of potassium chlorate, $KClO_3$, according to the equation

$$KClO_3(s) = KCl(s) + \tfrac{3}{2}O_2(g).$$

A certain sample of $KClO_3$ when decomposed yielded 637 cubic centimeters (cc) of oxygen gas, measured at 273°K and one atmosphere pressure. Our object is to find the original weight of the $KClO_3$ and the weight of the KCl produced. Remembering that there are 1000 cc in 1 liter, we can immediately calculate that the number of moles of oxygen produced is

$$\text{moles of } O_2 = \frac{0.637 \text{ liter}}{22.4 \text{ liters/mole}} = 0.0284 \text{ mole.}$$

From the number of moles of oxygen we can calculate the number of moles of potassium chlorate by means of an equation of atom conservation. By recognizing that the numbers of oxygen atoms in the $KClO_3$ consumed and the O_2 produced are the same we can write

$$\text{oxygen atoms in reactant} = \text{oxygen atoms in product,}$$
$$3 \times \text{moles of } KClO_3 = 2 \times \text{moles of } O_2,$$

since $KClO_3$ and O_2 contain, respectively, 3 and 2 atoms per chemical unit. We can reach the same expression by a different path. The chemical equation tells us that for each mole of potassium chlorate decomposed, $\tfrac{3}{2}$ moles of oxygen are produced. Therefore the number of moles of potassium chlorate multiplied by $\tfrac{3}{2}$ must equal the number of moles of oxygen. Thus

$$\tfrac{3}{2} \times \text{moles of } KClO_3 = \text{moles } O_2,$$

which is, apart from the transposition of the 2, the same as the equation derived from the atom conservation principle. Proceeding, we write

$$\text{moles of } KClO_3 = \tfrac{2}{3} \times \text{moles of } O_2$$
$$= \tfrac{2}{3} \times 0.0284 = 0.0189.$$

Since the formula weight of $KClO_3$ is 122.5, we find

$$\text{weight of } KClO_3 = \text{moles of } KClO_3 \times \text{formula weight}$$
$$= 0.0189 \times 122.5$$
$$= 2.32 \text{ gm.}$$

Furthermore, since the numbers of moles of $KClO_3$ and KCl are equal,

$$\text{weight of } KCl = 0.0189 \times 74.55$$
$$= 1.41 \text{ gm,}$$

where 74.55 is the formula weight of KCl.

The procedure used for solving nearly every stoichiometric problem is to set up an equation based on atom conservation (or mole balance) and apply to it the expression which relates the number of moles to the weight and molecular weight of the substances involved. However, an appreciation of this procedure and the ability to use it come only from experience. Therefore we will end our discussion with a series of examples typical of problems encountered in chemical practice.

Example 1.1 A sample of pure calcium metal weighing 1.35 gm was quantitatively converted to 1.88 gm of pure CaO. If the atomic weight of oxygen is taken to be 16.0, what is the atomic weight of calcium?

The formula of calcium oxide tells us immediately that

$$\text{moles O} = \text{moles Ca,}$$
$$\frac{1.88 - 1.35}{16.0} = \text{moles O} = 0.033,$$
$$0.033 = \text{moles Ca} = \frac{1.35}{\text{atomic weight of Ca}},$$
$$\text{atomic weight of Ca} = \frac{1.35}{0.033} = 40.9.$$

Example 1.2 In the gravimetric determination of phosphorus, an aqueous solution of dihydrogenphosphate ion, $H_2PO_4^-$, is treated with a mixture of ammonium and magnesium ions to precipitate magnesium ammonium phosphate, $MgNH_4PO_4 \cdot 6H_2O$. This is heated and decomposed to magnesium pyrophosphate, $Mg_2P_2O_7$, which is weighed. The reactions are

$$H_2PO_4^- + Mg^{++} + NH_4^+ + 6H_2O = MgNH_4PO_4 \cdot 6H_2O + 2H^+,$$
$$2MgNH_4PO_4 \cdot 6H_2O = Mg_2P_2O_7 + 2NH_3 + 13H_2O.$$

A solution of $H_2PO_4^-$ yielded 1.054 gm of $Mg_2P_2O_7$. What weight of NaH_2PO_4 was present originally?

The answer comes immediately if we apply the principle that the number of phosphorus atoms (or the number of moles of phosphorus) is conserved. Thus

$$\text{moles of P} = \text{moles of NaH}_2\text{PO}_4 = 2 \times \text{moles of Mg}_2\text{P}_2\text{O}_7,$$

$$\text{moles of Mg}_2\text{P}_2\text{O}_7 = \frac{1.054}{222.5} = 0.004737.$$

Therefore,

$$\text{moles of NaH}_2\text{PO}_4 = 2 \times 0.004737 = 0.009474,$$

$$\text{weight of NaH}_2\text{PO}_4 = 0.009474 \times 119.9 = 1.136 \text{ gm.}$$

Regardless of the number of reactions which lead from reactant to product, the reacting weights of the reagents are related by the principle of atom conservation. It is not even necessary to know the sequence of reactions, as we demonstrate in the next example.

Example 1.3 A sample of K_2CO_3 weighing 27.6 gm was treated by a series of reagents so as to convert all of its carbon to $K_2Zn_3[Fe(CN)_6]_2$. How many grams of this product were obtained?

Since all the carbon in the reactant is found in the product, the number of moles of carbon is conserved. Moreover, each mole of $K_2Zn_3[Fe(CN)_6]_2$ contains $2 \times 6 = 12$ moles of carbon atoms, so we can write

$$\text{moles of carbon} = 12 \times \text{moles of K}_2\text{Zn}_3[\text{Fe(CN)}_6]_2$$

$$= \text{moles of K}_2\text{CO}_3 = \frac{27.6}{138}.$$

Thus

$$\text{moles of K}_2\text{CO}_3 = 12 \times \text{moles of K}_2\text{Zn}_3[\text{Fe(CN)}_6]_2,$$

$$\frac{27.6}{138} = 12 \times \frac{\text{weight of K}_2\text{Zn}_3[\text{Fe(CN)}_6]_2}{698},$$

$$\text{weight of K}_2\text{Zn}_3[\text{Fe(CN)}_6]_2 = 11.6 \text{ gm.}$$

The empirical formulas of compounds of hydrogen and carbon are found by combustion analysis, as we illustrate next.

Example 1.4 One gram of a gaseous compound of carbon and hydrogen gives upon combustion 3.30 gm of carbon dioxide and 0.899 gm of water. What is the empirical formula of the compound?

The empirical formula is the simplest set of whole numbers which expresses the *relative* numbers of atoms in the compound. The relative numbers of atoms are the same as the relative numbers of moles of each element, and hence we can proceed as follows: The formula of carbon dioxide is CO_2, so the number of moles of carbon dioxide is equal to the number of moles of carbon in the sample of the compound. The formula of water is H_2O; thus two times the number of moles of water is equal to the number of moles of hydrogen atoms in the sample. The algebraic expressions of these facts are

$$\text{moles of C} = \text{moles of CO}_2 = \frac{3.30}{44.0} = 0.0750,$$

$$\text{moles of H} = 2 \times \text{moles of H}_2\text{O} = 2 \times \frac{0.899}{18.0} = 0.099,$$

$$\frac{\text{atoms of C}}{\text{atoms of H}} = \frac{\text{moles of C}}{\text{moles of H}} = \frac{0.075}{0.099} = \frac{3}{4}.$$

The empirical formula of the compound is C_3H_4. One molecule of the compound might have the formula C_3H_4 or C_6H_8 or some higher multiple. The molecular weight of the compound must be found before its molecular formula can be determined.

In a separate experiment, the density of a gaseous sample of the compound is found to be 1.78 gm/liter under conditions of standard temperature and pressure. The volume of one mole of any gas is, under these conditions, 22.4 liters, so we find that the molecular weight of the unknown is

$$1.78 \text{ gm/liter} \times 22.4 \text{ liters/mole} = 39.9 \text{ gm/mole}.$$

Since the molecular weight is 40, the molecular formula must be C_3H_4. This gas is methyl acetylene, usually written CH_3CCH.

Example 1.5 One volume of a gaseous compound of hydrogen, carbon, and nitrogen gave upon combustion 2 volumes of CO_2, 3.5 volumes of H_2O, and 0.5 volume of N_2, all measured at the same temperature and pressure. What is the empirical formula of the compound? Can the molecular formula be found from these data?

This problem requires a direct application of Avogadro's principle: under the same conditions of temperature and pressure, equal volumes of gases contain the same number of molecules. Noting that each nitrogen molecule contains two nitrogen atoms, and each water molecule two hydrogen atoms, we find that the relative numbers of carbon, hydrogen, and nitrogen atoms in the unknown are

$$C : H : N = \text{volume of CO}_2 : 2 \times \text{volume of H}_2\text{O} : 2 \times \text{volume of N}_2 = 2 : 7 : 1$$

The empirical formula is C_2H_7N. This is also the molecular formula, since the data tell us that in one volume (or x moles) of compound, there are enough carbon atoms to make 2 volumes (or $2x$ moles) of CO_2. Thus each mole of unknown contains two moles of carbon, and no more. The compound referred to is ethyl amine, more commonly written $CH_3CH_2NH_2$.

Example 1.6 A carefully purified sample of potassium chlorate, $KClO_3$, weighing 4.008 gm, was quantitatively decomposed to 2.438 gm of potassium chloride, KCl, and oxygen. The potassium chloride was dissolved in water and treated with a silver nitrate solution. The result was a precipitate of silver chloride, $AgCl$, weighing 4.687 gm. Under further treatment the silver chloride was found to contain 3.531 gm of silver. What are the atomic weights of silver, chlorine, and potassium relative to $O = 15.999$?

This problem will require several steps, so it is advisable to plan our procedure. Examination of the equations

$$KClO_3 = KCl + \tfrac{3}{2}O_2, \qquad Cl^- + Ag^+ = AgCl$$

shows that if we knew how many moles of KCl were produced, we would in turn know the number of moles of silver chloride and silver involved. Knowing the number of moles of silver and its weight, we can calculate its atomic weight. This is our first objective. Then, from the atomic weight of silver, the weight, *and* the number of moles of silver chloride, we can find the atomic weight of chlorine. Finally, the number of moles of potassium chloride, its weight, and the atomic weight of chlorine suffice to calculate the atomic weight of potassium. This whole plan is based on the expression moles = weight/molecular weight, and the realization that two of these quantities are required to calculate the third.

Let us put the plan into effect. Since oxygen is the only element whose atomic weight is known, we must use a relation involving it to find the number of moles of KCl. From the first chemical equation we see that

$$\text{moles of } O_2 = \tfrac{3}{2} \times \text{moles of KCl},$$

$$\frac{\text{weight } O_2}{\text{molecular weight of } O_2} = \frac{4.008 - 2.438}{31.998} = 0.04907 = \frac{3}{2} \times \text{moles of KCl},$$

$$\text{moles of KCl} = 0.03271.$$

By the second equation,

$$\text{moles of KCl} = \text{moles Ag},$$

$$0.03271 = \frac{3.531}{\text{atomic weight of Ag}},$$

$$\text{atomic weight of Ag} = 107.9.$$

Also

$$\text{moles of KCl} = \text{moles AgCl},$$

$$0.03271 = \frac{4.687}{\text{formula weight of AgCl}},$$

$$\text{formula weight of AgCl} = 143.3.$$

The atomic weight of chlorine can now be found from the formula weight of silver chloride and the atomic weight of silver:

$$\text{atomic weight of Cl} = 143.3 - 107.9 = 35.4.$$

Finally

$$\text{moles KCl} = \frac{2.438}{\text{formula weight of KCl}} = 0.03271,$$

$$\text{formula weight of KCl} = 74.53,$$

$$\text{atomic weight of K} = 74.53 - 35.4 = 39.1.$$

This problem, though lengthy, required no new principles. The ability to proceed in an orderly fashion through an involved problem follows from a thorough understanding of the relation between number of moles, weight, and molecular weight.

A rather tricky type of problem, easily solved by the atom conservation principle, involves the reaction of a mixture of substances to give a single product.

Example 1.7 A 1.000-gm mixture of cuprous oxide, Cu_2O, and cupric oxide, CuO, was quantitatively reduced to 0.839 gm of metallic copper. What was the weight of CuO in the original sample?

The key to this problem is to make use of the fact that copper atoms are conserved. Letting the weight of CuO equal w, and the weight of Cu_2O equal $1.000 - w$, we have

$$\text{moles of Cu in oxides} = \text{moles of metallic Cu},$$
$$\text{moles of CuO} + 2 \times \text{moles of Cu}_2\text{O} = \text{moles of Cu},$$
$$\frac{w}{\text{formula weight of CuO}} + 2 \times \frac{1.000 - w}{\text{formula weight of Cu}_2\text{O}}$$
$$= \frac{0.839}{\text{atomic weight of Cu}}, \qquad w = 0.55 \text{ gm of CuO.}$$

A very similar problem involves the reaction of a mixture of metals with hydrogen ion, as in the next example.

Example 1.8 A mixture of aluminum and zinc weighing 1.67 gm was completely dissolved in acid and evolved 1.69 liters of hydrogen, measured at 273°K and 1 atm pressure. What was the weight of aluminum in the original mixture?

From the equations

$$Zn + 2H^+ = Zn^{++} + H_2,$$
$$Al + 3H^+ = Al^{+3} + \tfrac{3}{2}H_2,$$

we see that from one mole of zinc we get one mole of hydrogen gas, while one mole of aluminum gives $\frac{3}{2}$ moles of hydrogen. Thus, for the moles of hydrogen formed we can write

$$\text{moles of Zn} + \tfrac{3}{2} \text{ moles Al} = \text{moles of H}_2.$$

Then, if w equals the weight of Al,

$$\frac{1.67 - w}{65.4} + \frac{3}{2}\frac{w}{27.0} = \frac{1.69}{22.4}, \qquad w = 1.24 \text{ gm of Al.}$$

1.6 CONCLUSION

In this chapter we have exposed the chemical basis for our belief in the atomic theory of matter, and shown how to use the relation between weight and number of atoms to do stoichiometric calculations. We have stressed the concept of the mole, for this is the fundamental unit used in dealing with all properties of collections of molecules. A thorough understanding of the mole concept is absolutely essential to the study of chemistry, and the stoichiometric calculations we have discussed, together with the problems at the end of this chapter, provide an immediate opportunity to test one's understanding of this important idea.

Apart from the factual and mathematical aspects of atomic weights and the atomic theory, there are a number of illustrations of important scientific principles in this chapter. The story of how the atomic weight scale was developed shows how scientific ideas are generated. Many experiments are done, and comparison of the results allows formulation of empirical laws which have a validity limited by the accuracy of the experiments and the variety of situations investigated. An attempt to find a unified explanation of a number of experimental laws leads to a theory. In the development of a theory there may be ideas advanced that are incorrect and which must be modified or removed as more experiments are done. It is at this stage that a dispassionate evaluation of existing experimental data and the execution of decisive experiments is most important. It is a rare event when the results of only one experiment are enough to bring the acceptance or rejection of a theory as a whole. Therefore, our firm acceptance of scientific ideas of great scope, such as the atomic theory, is generally based on the existence of an overwhelming collection of experimental results that individually do not prove, but are consistent with, the theory. We shall find further illustrations of these points in subsequent chapters.

SUGGESTIONS FOR FURTHER READING

Beveridge, W. I. B., *The Art of Scientific Investigation*. New York: Random House, 1961.

Cragg, L. H., and R. P. Graham, *An Introduction to the Principles of Chemistry*. New York: Rinehart and Co., 1954.

Kieffer, W. F., *The Mole Concept in Chemistry*. New York: Reinhold, 1962.

Nash, L. K., *The Atomic-Molecular Theory*. Cambridge, Mass.: Harvard University Press, 1950.

Nash, L. K., *Stoichiometry*. Reading, Mass.: Addison-Wesley, 1966.

Partington, J. R., *A Short History of Chemistry*. New York: Harper and Brothers, 1960.

PROBLEMS

1.1 An oxide of antimony is found to contain 24.73% oxygen. What is its empirical formula?

1.2 When 0.210 gm of a compound containing only hydrogen and carbon was burned, 0.660 gm of CO_2 was recovered. What is the empirical formula of the compound? A determination of the density of this hydrocarbon gave a value of 1.87 gm/liter at 273.1°K and 1 atm. What is the molecular formula of the compound?

1.3 A sample of europium dichloride, $EuCl_2$, weighing 1.00 gm is treated with excess aqueous silver nitrate, and all the chloride is recovered as 1.29 gm of AgCl. What is the atomic weight of europium?

1.4 A sample of an oxide of iron weighing 1.60 gm was heated in a stream of hydrogen gas until it was completely converted to 1.12 gm of metallic iron. What is the empirical formula of the iron oxide?

1.5 When barium bromide, $BaBr_2$, is heated in a stream of chlorine gas, it is completely converted to barium chloride, $BaCl_2$. From 1.50 gm of $BaBr_2$ just 1.05 gm of $BaCl_2$ is obtained. Calculate the atomic weight of barium from these data.

1.6 A 0.578-gm sample of pure tin is treated with gaseous fluorine until the weight of the resulting compound is constant at a value of 0.944 gm. What is the empirical formula of the tin fluoride formed? Write an equation for its synthesis.

1.7 A certain metal forms two chlorides which contain 85.2% and 65.8% of the metal. (a) Show that these compounds are consistent with the law of multiple proportions. (b) What are the simplest formulas of the compounds, and what is the corresponding atomic weight of the metal? (c) Considering the other possible formulas, what other atomic weights are possible? (d) Refer to the periodic table and determine the atomic weight of the metal.

1.8 From the following isotopic masses and abundances calculate the atomic weight of magnesium.

Isotope	Abundance	Mass
24	78.60%	23.993
25	10.11%	24.994
26	11.29%	25.991

1.9 A sample of a metal oxide weighing 7.380 gm is decomposed quantitatively to give 6.840 gm of the pure metal. The specific heat of the metal is found to be 0.0332 cal/gm. Calculate the accurate atomic weight of the metal, and the empirical formula of the oxide.

1.10 By measuring the density of a certain elemental gas at several pressures, the value of 1.787 gm/liter was found for the ideal density at 1-atm pressure and 273.1°K. The ideal density of oxygen gas under the same conditions is 1.428 gm/liter. What is the molecular weight of the unknown gas? Consult the periodic table for "gaps" at appropriate fractions of this molecular weight, and construct an argument that this element does not exist as a diatomic or polyatomic molecule. Is this molecular weight an atomic weight?

1.11 Equal weights of zinc metal and iodine are mixed together and the iodine is completely converted to ZnI_2. What fraction by weight of the original zinc remains unreacted?

1.12 A 4.22-gm sample of a mixture of $CaCl_2$ and NaCl was treated to precipitate all the calcium as $CaCO_3$, which was then heated and converted to pure CaO. The final weight of the CaO was 0.959 gm. What was the percentage by weight of $CaCl_2$ in the original mixture?

1.13 An alloy of aluminum and copper was treated with aqueous HCl. The aluminum dissolved according to the reaction $Al + 3H^+ \rightarrow Al^{+3} + \frac{3}{2}H_2$, but the copper remained as the pure metal. A 0.350-gm sample of the alloy gave 415 cc of H_2 measured at 273.1°K and 1-atm pressure. What is the weight percentage of Al in the alloy?

1.14 A sample of pure lead weighing 2.07 gm is dissolved in nitric acid to give a solution of lead nitrate. This solution is treated with hydrochloric acid, chlorine gas, and ammonium chloride. The result is a precipitate of ammonium hexachloroplumbate, $(NH_4)_2PbCl_6$. What is the maximum weight of this product that could be obtained from the lead sample?

1.15 A 0.596-gm sample of a gaseous compound containing only boron and hydrogen occupies 484 cc at 273.1°K and 1-atm pressure. When the compound was ignited in excess oxygen all its hydrogen was recovered as 1.17 gm of H_2O, and all the boron was present as B_2O_3. What is the empirical formula, the molecular formula, and the molecular weight of the boron-hydrogen compound? What weight of B_2O_3 was produced by the combustion?

1.16 A sample of an unknown oxide of barium gave upon exhaustive heating 5.00 gm of pure BaO and 366 cc of oxygen gas measured at 273.1°K and 1-atm pressure. What is the empirical formula of the unknown oxide? What weight of oxide was present initially?

1.17 A mixture of KBr and NaBr weighing 0.560 gm was treated with aqueous Ag^+ and all the bromide ion was recovered as 0.970 gm of pure AgBr. What was the fraction by weight of KBr in the original sample?

CHAPTER 2

THE PROPERTIES OF GASES

Our discussion in Chapter 1 showed how important the study of gases was to the development of the atomic theory. An understanding of gaseous behavior is a fundamental part of modern chemistry as well. According to Avogadro's principle, measuring the volume of a gas is equivalent to counting the number of molecules in that volume, and the importance of this type of measurement cannot be overemphasized. Moreover, many industrially important elements and compounds are gases under the conditions of use. But quite aside from the historical or practical importance of gases, there is another reason for studying them. The business of the chemist is to relate the properties of matter in bulk to the properties of individual molecules. The kinetic theory of gases is a most satisfactory example of the successful explanation of macroscopic phenomena in terms of molecular behavior. By pursuing the mathematical consequence of the fact that a gas consists of a large number of particles that collide with the walls of a containing vessel, it is possible to derive Boyle's law, and to gain a more thorough understanding of the concept of temperature. By trying to account for the failure of gases to obey Boyle's law exactly, we can learn about the sizes of molecules and the forces which they exert on each other. Thus the study of this simplest state of matter can introduce us to some of the most universally useful concepts of physical science.

In general, the volume of any material, solid, liquid, or gas, is determined by the temperature and the pressure to which it is subjected. A mathematical relationship exists between the volume of a given amount of material and the values of the pressure and temperature; this mathematical relation is called an **equation of state** and can be written symbolically as

$$V = V(T, P, n).$$

This equation is read: *V is some function of temperature, pressure, and the number of moles of material.* In the case of liquids or solids, equations of state may be algebraically very complicated, and may differ considerably in algebraic form from one substance to another. Gases are unique, however, in that the equations of state of all gases are very nearly the same. We shall see later that this simplification is due to the fact that in the gaseous state molecules are essentially independent of one another, and that consequently the detailed nature of individual molecules does not strongly affect the behavior of the gas as a whole. For the present, however, we will address ourselves to the problem of the determination and expression of the gaseous equation of state.

Inevitably the determination of an equation of state for gases involves a measurement of pressure, or the force per unit area, which a gas exerts on the walls of the containing vessel. Commonly the pressure of gases is expressed in units of atmospheres or millimeters of mercury, rather than in units which are more obviously related to force and area. In order to establish the relation between the atmosphere or millimeter as pressure units and the more fundamental idea of force per unit area, we need only examine how pressure is measured experimentally.

FIG. 2.1   A mercury barometer. The atmospheric pressure is proportional to the height *h*.

The force per unit area exerted by the earth's atmosphere is commonly measured by the device called a barometer, shown in Fig. 2.1. The vertical tube containing mercury is completely evacuated of all gases, except for a very small amount of mercury vapor. The height of the column of mercury above the lower mercury surface is determined by the requirement that the force per unit area due to mercury in the column be equal to the force per unit area exerted by the surrounding atmosphere on the mercury surface. Under ordinary atmospheric conditions at sea level this height is in the neighborhood of 760 millimeters (mm). Therefore, the arbitrary definition is made that one standard atmosphere corresponds to 760 mm of mercury, when the mercury is at 0°C.

Let us now calculate what one atmosphere is when expressed in terms of force per unit area. Consider a barometer tube whose cross-sectional area is 1 cm^2. Then the force exerted by the mercury column on this area is equal to the mass of the mercury in the tube times the acceleration due to gravity. In turn, the mass of mercury in the tube is the volume of mercury times the density of mercury at 0°C. We have then

$$
\begin{aligned}
\text{force} &= \text{mass} \times \text{acceleration} \\
&= \text{density of Hg} \times \text{height} \times \text{area} \times \text{acceleration} \\
&= 13.59 \text{ gm/cm}^3 \times 76.00 \text{ cm} \times 1.000 \text{ cm}^2 \times 980.7 \text{ cm/sec}^2 \\
&= 1.013 \times 10^6 \text{ gm·cm/sec}^2 = 10.13 \text{ kgm m/sec}^2 \\
&= 1.013 \times 10^6 \text{ dynes} = 10.13 \text{ newtons.}
\end{aligned}
$$

This is the force exerted by a column of mercury 760 mm high and of 1-cm^2 cross sectional area. Therefore, it is also the force per unit area (one square centimeter) that corresponds to one atmosphere pressure. We have, then, that

$$
\begin{aligned}
1 \text{ atm} &= 760.0 \text{ mm of mercury} \\
&= 1.013 \times 10^6 \text{ dynes/cm}^2 = 1.013 \times 10^5 \text{ newtons/m}^2
\end{aligned}
$$

Increasingly, the awkward unit "mm Hg" is being replaced by its equivalent, the torr (after Torricelli, inventor of the barometer). Thus in practical laboratory work, pressure is most often measured and expressed in torr or millimeters of mercury, but for calculational purposes it is frequently convenient to use units of atmospheres. Recently, SI (System Internationale) units have come into frequent use, particularly in Europe. In this system, the unit of pressure is the newton per square meter, and one newton per square meter is called a Pascal (abr. Pa). Thus we see that

$$
\begin{aligned}
1 \text{ atm} &= 1.013 \times 10^5 \text{ N/m}^2 \\
&= 1.013 \times 10^5 \text{ Pa.}
\end{aligned}
$$

Although we shall not make extensive use of SI units in this book, it is important to become familiar with them, since they will be found with increasing frequency in the scientific literature as time passes.

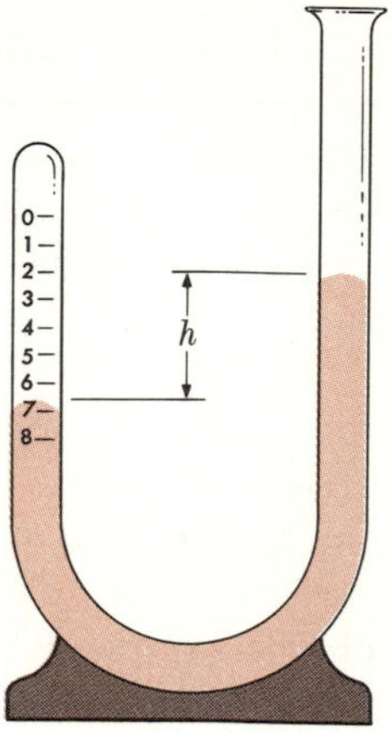

FIG. 2.2   A U-tube employed in demonstrating Boyle's law.

Boyle's Law

The mathematical relation that exists between the pressure and volume of a fixed amount of gas at a fixed temperature was discovered by Robert Boyle in 1662. Boyle trapped a quantity of air in the closed end of a U-tube, using mercury as the containing fluid, as shown in Fig. 2.2. In this type of experiment, the pressure that exists in the closed tube is equal to the pressure of the atmosphere plus the pressure exerted by the column of mercury of height h. By pouring mercury into the longer tube the pressure on the gas can be increased, and the corresponding decrease in the gas volume noted. Boyle discovered that the product of the pressure and the volume of a fixed amount of gas was of approximately constant value. He also noted that warming a gas increased its volume when the pressure was kept constant. However, he did not investigate this phenomenon further, possibly because the idea of temperature was not well defined at the time. Nevertheless, Boyle's observation of the qualitative effect of warming a gas was important, because it showed that in order to make meaningful determinations of the relation between pressure and volume, the temperature of the surroundings had to be kept constant during the experiment.

Very often in experimental investigations data are obtained as sets of numbers (such as simultaneous values of P and V) which depend on each other in some unknown way. A very useful and convenient technique for discovering the relationship between a series of simultaneous values of pressure and volume is to plot the data on a rectangular coordinate system having pressure and volume as coordinate axes. A smooth curve that passes through the experimentally determined points may then indicate the mathematical connection between the two variables. Figure 2.3 shows some experimental data plotted in this manner. The curve generated by the data appears to be a rectangular hyperbola with the coordinate axes as asymptotes. Since the algebraic equation which corresponds to a hyperbola is known to have the form $xy = $ constant, we can deduce that for a fixed amount of gas at a constant temperature, $\underline{PV = \text{constant}}$, which is in fact Boyle's law. Repeating the experiment at a series of different temperatures generates a family of hyperbolas, each characteristic of a particular value of the temperature. Since the temperature is a constant along each line, these curves are called **isotherms.**

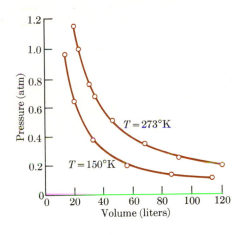

Pressure-volume isotherms for an ideal gas. **FIG. 2.3**

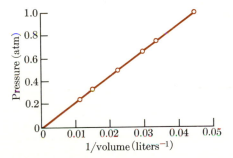

Pressure of an ideal gas as a function of reciprocal volume. **FIG. 2.4**

Plotting the pressure as a function of the volume is often a useful way of representing the behavior of a gas, but it has the disadvantage that it is difficult to tell by using the eye how close to a perfect hyperbola each experimental curve is. Consequently, it is difficult to tell whether a gas obeys Boyle's law exactly or just approximately. This problem can be solved by plotting pressure as a function of the reciprocal of the volume, as shown in Fig. 2.4. Since Boyle's law can be written as

$$P = \frac{k_{m,t}}{V},$$

where $k_{m,t}$ is a constant whose value depends on temperature and the amount of gas, a gas that obeys Boyle's law should give a straight line when pressure is plotted as a function of the reciprocal of the volume. Since deviations from a straight line are readily detected by eye, it is easy to tell how closely a gas follows Boyle's law, by plotting the data in this manner.

FIG. 2.5 Pressure-volume product as a function of pressure for an ideal gas.

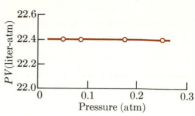

Another even more useful way of treating such experimental data is to plot the *product* of pressure and volume as a function either of pressure or the reciprocal of volume. Figure 2.5 shows that the result of this plot should be a straight line of zero slope for a gas that follows Boyle's law exactly. The experimental data show that gases do in fact obey Boyle's law quite closely over the range of pressures investigated. Any deviations are due to the forces which molecules exert on each other, and tend to vanish as the density of the gas becomes small. In the limit of very low pressure, all gases obey Boyle's law exactly.

The Law of Charles and Gay-Lussac

When the dependence on temperature of the volume of a gas at fixed pressure is investigated experimentally it is found that the volume increases linearly with increasing temperature. This relation is known as the law of Charles and Gay-Lussac, and can be expressed algebraically as

$$V = V_0(1 + \alpha t).$$

Here V is the volume of a fixed amount of gas at constant pressure, V_0 is the volume it occupies at the temperature of zero degrees on the celsius scale, α is a constant that has the value of approximately $\frac{1}{273}$ for all gases, and t is the temperature on the celsius scale. This equation states that the volume of a gas increases linearly with its temperature. That this statement can be made as an experimental fact implies that we have some previous knowledge of how to measure temperature.

Common experience provides us with a qualitative concept of temperature. To create a quantitative temperature scale we must select some easily measurable property of matter that depends on what we recognize as "hotness," and define temperature in terms of the value of this property. The most familiar thermometric property is the length of the column of mercury that extends into a capillary tube from an otherwise closed bulb. The position of the mercury meniscus can be marked when the bulb of this thermometer is immersed in an ice-water mixture, and when it is surrounded by the vapor of boiling water at one atmosphere pressure. These two positions can be *arbitrarily* defined as 0- and 100-degree points respectively. The distance between these two marks can then be divided by 99 equally spaced lines and a working temperature scale created.

The division of the scale into *equal* units is very significant, for by so doing we say that temperature is something that *increases linearly with the length of the mercury column*. The same procedure could be followed using some other working liquid, such as alcohol, to make a second thermometer. If these two different thermometers were both placed in the same water-ice bath, they would both read zero degrees. If both were placed in the vapor of boiling water, they would read 100 degrees. However, if they were both placed in the same room where the mercury thermometer read exactly 25 degrees, the alcohol thermometer would indicate a temperature slightly different from 25 degrees. This behavior would, in general, be repeated at any other temperature on the scale except for the calibrating points of 0 and 100 degrees, since in order for the two thermometers to read the same at all temperatures, the equations of state of mercury and alcohol would have to be exactly the same. Due to the intrinsic differences in the molecular structure of the two liquids, these or any other pair of liquids do not expand by exactly the same amount for a given change in temperature. Consequently, if we wish to use a liquid to define our temperature scale, we must be careful to specify which liquid is being used.

For gases, the temperature dependence of volume is considerably simpler than for liquids. Even without a temperature scale it is possible to determine that the volume of any gas at the temperature of boiling water is 1.366 times its volume at the temperature of an ice-water mixture. The important fact here is that the proportionality constant *is the same for all gases*. A similar measurement can be carried out in which the ratio of the volume of a gas at the boiling point of water to its volume at the boiling point of ether is measured. In this case the volume ratio is 1.295 for all gases. The fact that all gases behave the same when subjected to a given change in temperature suggests that the properties of gases should be used to *define* a temperature scale. That is just what is done. The equation previously given as expressing the law of Charles and Gay-Lussac, $V = V_0(1 + \alpha t)$, can be rewritten in the following way:

$$ t = \frac{V - V_0}{V_0 \alpha} = \frac{1}{\alpha} \left(\frac{V}{V_0} - 1 \right). $$

The second equation can be interpreted as saying that there is such a thing as the temperature, t, which is a quantity that *increases linearly with the volume of a gas, by definition*. That is, the "law" of Charles and Gay-Lussac really is not a law, but is actually a *definition* of temperature.

Actually, not all gases behave in *exactly* the same way when their temperature is changed, but the differences diminish as the pressure is lowered, and are generally so slight as to be negligible in most instances. While gas thermometers

Table 2.1 Comparison of thermometers*

Constant-volume hydrogen thermometer, $t(P)$	Constant-volume air thermometer, $t(P)$	Platinum resistance thermometer, $t(R)$	Thermocouple, $t(emf)$	Mercury thermometer, $t(l)$
0	0	0	0	0
20	20.008	20.240	20.150	20.091
40	40.001	40.360	40.297	40.111
60	59.990	60.360	60.293	60.086
80	79.987	80.240	80.147	80.041
100	100	100	100	100

*After M. W. Zemansky, *Heat and Thermodynamics*. New York: McGraw-Hill, 1951.

can be used to define a temperature scale, other devices which are more convenient to use are employed in practical temperature measurements. The resistance of a platinum wire under a constant tension, and the voltage produced by a platinum-rhodium thermocouple are useful thermometric properties. In Table 2.1 these thermometers are compared with two gas thermometers and a mercury thermometer. The hydrogen gas thermometer is considered to define the scale, and the readings each of the thermometers would show if immersed in various temperature baths are given. These are the readings that would result in each case from using a scale obtained by dividing the change of the property being measured into 100 equal units between the freezing point and boiling point of water. The fact that all the thermometers do not read the same as the hydrogen gas thermometer merely indicates that resistance, voltage, and liquid density do not change in a way which is strictly linearly related to temperature as defined by the hydrogen gas scale.

The Absolute Temperature Scale

The relation between temperature and gaseous volume can be simplified by defining a new temperature scale. Starting from Charles' law we can write

$$V = V_0(1 + \alpha t) = V_0 \frac{1/\alpha + t}{1/\alpha}.$$

For the ratio V_1/V_2 of the volume of gas at two different temperatures t_1 and t_2 we get

$$\frac{V_1}{V_2} = \frac{1/\alpha + t_1}{1/\alpha + t_2}.$$

Since by direct experiment it is found that $1/\alpha = 273.15$ when t is expressed in celsius degrees,

$$\frac{V_1}{V_2} = \frac{273.15 + t_1}{273.15 + t_2}.$$

The form of this equation suggests that it would be very convenient to define a new temperature scale by the equation

$$T = 273.15 + t. \tag{2.1}$$

The temperature T is called **absolute temperature** or the temperature on the kelvin scale, and is denoted by °K. Using the kelvin scale, the relation between the temperature and volume for a fixed amount of gas at constant pressure assumes the very simple form

$$\frac{V_1}{V_2} = \frac{T_1}{T_2}$$

or

$$\frac{V}{T} = \text{constant.} \tag{2.2}$$

The implication of this last statement is that the volume of a gas decreases as T, the absolute temperature, decreases and would become zero when $T = 0$. This suggests that $T = 0°K$ or, by Eq. (2.1), $t = -273.15°C$ is the lowest possible temperature, since any lower temperature would correspond to a negative volume of gas. Actually at very low temperatures Eq. (2.2) cannot be tested experimentally, since all gases condense to liquids as the temperature approaches zero on the kelvin scale. Nevertheless, much more detailed arguments show that $-273.15°C$ or $0°K$ is the lowest conceivable temperature, and that in practical experiments this lowest temperature cannot be reached, but only approached very closely. The lowest temperature which has been reached is generally acknowledged to be $0.0014°K$, although $0.0001°K$ may have been attained in other less well-controlled experiments.

The Ideal Gas Equation

Experimental measurements have shown that at constant temperature, PV is a constant, and at constant pressure, V is proportional to T. We now wish to combine these relationships into one equation which expresses the behavior of gases. According to Boyle's law

$$PV = C'(T, n),$$

where $C'(T, n)$ is a constant that depends on temperature and the number of moles of gas. From Charles' law we know that at constant pressure, the volume of a fixed amount of gas is directly proportional to the absolute temperature. Therefore, the dependence of $C'(T, n)$ on temperature must be

$$C'(T, n) = C(n)T,$$

where $C(n)$ is a parameter that depends only on the number of moles of gas n. This must be true, since we can now write

$$PV = C(n)T, \tag{2.3}$$

which is consistent with both the laws of Boyle and of Charles. By a slight rearrangement of Eq. (2.3), we can write

$$\frac{PV}{T} = C(n).$$

A gas that obeys this equation of state, which incorporates the laws of Boyle and of Charles and Gay-Lussac, is called an **ideal gas.** This result may also be written

$$\frac{P_1V_1}{T_1} = \frac{P_2V_2}{T_2}. \tag{2.4}$$

Equation (2.4) is a symmetric form of the gas laws that is easy to remember. It can be used to calculate the volume V_2 of a gas under the arbitrary conditions P_2 and T_2 from a knowledge of its volume V_1 at pressure P_1 and temperature T_1.

Example 2.1 A certain sample of gas has a volume of 0.452 liter measured at 87°C and 0.620 atm. What is its volume at 1 atm and 0°C?

Letting $V_1 = 0.452$ liter, $P_1 = 0.62$ atm, $T_1 = 273 + 87 = 360$°K, $P_2 = 1$ atm and $T_2 = 273$°K, we find from Eq. (2.4)

$$V_2 = V_1 \times \frac{T_2}{T_1} \times \frac{P_1}{P_2} = 0.452 \times \frac{273}{360} \times \frac{0.620}{1.00}$$
$$= 0.213 \text{ liter.}$$

Rather than remember or refer to Eq. (2.4) it is often safer and simpler to proceed by an intuitive method. We know that since the final temperature is less than the initial temperature, we must multiply V_1 by a ratio of temperatures less than unity to obtain V_2:

$$V_2 \propto V_1 \times \frac{273}{360}.$$

Also, the final pressure is higher than the initial pressure. This must act to decrease the final volume, and we must multiply V_1 by a ratio of pressures less than unity to find V_2. Thus

$$V_2 = V_1 \times \frac{273}{360} \times \frac{0.620}{1.00},$$

which is exactly the expression obtained by a mechanical use of Eq. (2.4).

Earlier we remarked that measuring the volume of a gas at a known pressure and temperature is equivalent to counting molecules. Under conditions of constant temperature and pressure, the volume is proportional to the number of moles of gas. Consequently, comparison of the volume of any sample of gas with the volume occupied by one mole under the same conditions tells us the number of moles or molecules in the sample. As we noted in Chapter 1, it is found experimentally that <u>one mole of any</u> *ideal* <u>gas occupies 22.414 liters,</u> measured at 1-atm pressure and 273.15°K. This volume is called the standard molar volume, and 273.15°K and 1-atm pressure are called standard temperature and pressure conditions (abbreviated STP). Therefore, to calculate the number of moles of gas in an arbitrary sample we write

$$\text{number of moles of gas} = \frac{\text{volume of gas}}{\text{volume of one mole}}$$

$$= \frac{V(\text{STP})}{22.4},$$

where $V(\text{STP})$ is the volume in liters that the gas sample would occupy under standard pressure and temperature conditions. This volume, $V(\text{STP})$ can be calculated from the volume measured under any conditions of temperature and pressure by using Eq. (2.4).

There is a somewhat more convenient method for calculating the number of moles of a gas in a sample having the volume V at the arbitrary pressure P and temperature T. By Eq. (2.3)

$$\frac{PV}{T} = C(n),$$

where $C(n)$ is a constant that depends only on the amount of gas in the sample. We have already remarked that at constant pressure and temperature, volume is proportional to the number of moles of gas. Therefore, we can rewrite $C(n)$ in terms of a new constant R and the number of moles of gas n:

$$C(n) = nR.$$

Consequently

$$\frac{PV}{T} = nR,$$

$$PV = nRT. \tag{2.5}$$

The constant R is known as the **universal gas constant,** <u>and is independent</u> <u>of pressure, temperature, or the number of moles in the sample.</u> If the numerical value of R were known, measurements of P, V, and T could be used to calculate n, the number of moles of gas in any sample.

We can evaluate R from information already available to us. Since one mole of gas occupies a volume of 22.41 liters at 1-atm pressure and 273.15°K, we can write

$$R = \frac{(1 \text{ atm})(22.414 \text{ liters})}{(1 \text{ mole})(273.15 \text{ deg})} = \boxed{0.08206 \, \frac{\text{liter-atm}}{\text{mole-deg}}}.$$

Note that the numerical value of R depends on the units used to measure pressure, volume, and temperature. The expression $PV = nRT$ is obeyed by all gases in the limit of low densities and high temperatures—"ideal" conditions under which the forces between molecules are of minimum importance. Consequently, Eq. (2.5) is known as the perfect gas law, or the ideal gas equation of state.

Example 2.2 Calculate the number of moles in a sample of an ideal gas whose volume is 0.452 liter at 87°C and 0.620 atm.

In Eq. (2.5) we set $P = 0.620$ atm, $V = 0.452$ liter, and $T = 360°K$:

$$n = \frac{PV}{RT} = \frac{(0.620)(0.452)}{(0.0821)(360)}$$

$$= 0.00948 \text{ mole.}$$

In Example 2.1 we found that the volume of this same sample of gas under conditions of standard temperature and pressure was 0.213 liter. Therefore we can also compute the number of moles by

$$n = \frac{V(\text{STP})}{22.4} = \frac{0.213}{22.4} = 0.0095 \text{ mole.}$$

The use of the ideal gas equation of state is an alternative to the procedure of finding the gaseous volume under standard conditions and dividing by the standard volume of one mole.

Dalton's Law

Suppose a mixture of two ideal gases, A and B, is contained in a volume V at a temperature T. Then, since each gas is ideal, we can write

$$P_A = n_A \frac{RT}{V}, \qquad P_B = n_B \frac{RT}{V}.$$

That is, in the mixture each gas exerts a pressure that is the same as it would exert if it were present alone, and this pressure is proportional to the number of moles of the gas present. The quantities P_A and P_B are called the *partial pressures* of A and B respectively. According to Dalton's law of partial pressures, the total pressure, P_t, exerted on the walls of the vessel is the sum of the partial pressures of the two gases:

$$P_t = P_A + P_B = (n_A + n_B)\left(\frac{RT}{V}\right).$$

The expression can be generalized so as to apply to a mixture of any number of gases. The result is

$$P_t = \sum_i P_i = \frac{RT}{V} \sum_i n_i, \qquad (2.6)$$

where i is an index that identifies each component in the mixture and the symbol $\sum_i$ stands for the operation of adding all the indexed quantities together. Another useful expression of the law of partial pressures is obtained by writing

$$P_A = n_A \frac{RT}{V},$$

$$P_t = \frac{RT}{V} \sum_i n_i,$$

$$\frac{P_A}{P_t} = \frac{n_A}{\sum_i n_i},$$

$$P_A = P_t \left(\frac{n_A}{\sum_i n_i} \right). \tag{2.7}$$

The quantity $n_A/\sum_i n_i$ is called the mole fraction of component A, and Eq. (2.7) says that the partial pressure of any component, such as component A, is the total pressure of the mixture multiplied by $n_A/\sum_i n_i$, the fraction of the total moles which are component A.

Use of the Gas Laws

It is essential for every chemist to have a thorough understanding of the gas laws and to be able to apply them to a variety of problems. The following examples are chosen to illustrate the ways in which the gas laws are used in chemical practice.

Example 2.3 An ideal gas at 1-atm pressure was contained in a bulb of unknown volume V. A stopcock was opened which allowed the gas to expand into a previously evacuated bulb whose volume was known to be exactly 0.500 liter. When equilibrium between the bulbs had been established, it was noted that the temperature had not changed, and that the gas pressure was 530 mm. What is the unknown volume, V, of the first bulb?

Since the gas is ideal and the temperature constant, we can use Boyle's law:

$$P_1 V_1 = P_2 V_2,$$
$$760 V_1 = 530(0.5 + V_1),$$
$$V_1 = 1.15 \text{ liter}.$$

The ideal gas equation can be used to help calculate molecular weights from gas density measurements, which we illustrate next.

Example 2.4 It is found that 0.896 gm of a gaseous compound containing only nitrogen and oxygen occupies 524 cc at a pressure of 730 mm and a temperature of 28.0°C. What is the molecular weight and molecular formula of the gas?

The molecular weight can always be calculated from a knowledge of the number of moles which correspond to a given weight of material. In this problem, the number of moles of the gas can be found by using the ideal gas equation of state:

$$n = \frac{PV}{RT} = \frac{(\frac{730}{760})(\frac{524}{1000})}{(0.0821)(301)} = 0.0204 \text{ mole.}$$

Since the units of the gas constant are liter-atmospheres per mole-degree, care has been taken to express the measured pressure, volume, and temperature in units of atmospheres, liters, and degrees kelvin, respectively. The molecular weight of the gas can now be found to be

$$\frac{0.896}{0.0204} = 43.9 \text{ gm/mole.}$$

The only combination of the atomic weights of nitrogen and oxygen which adds to 44 is $2 \times 14 + 16$, which means the molecular formula of the gas is N_2O.

The following is a simple illustration of the use of Dalton's law of partial pressures.

Example 2.5 The valve between a 5-liter tank in which the gas pressure is 9 atm and a 10-liter tank containing gas at 6 atm is opened, and pressure equilibration ensues at a constant temperature. What is the final pressure in the two tanks?

Let us imagine the gases in the two tanks to be distinguishable, and call them components a and b. Then, when the connecting valve is opened, each expands to fill a total volume of 15 liters. The partial pressures of the two components after the expansion are

$$P_a = \frac{5 \times 9}{15} = 3 \text{ atm,} \qquad P_b = \frac{10 \times 6}{15} = 4 \text{ atm.}$$

According to the law of partial pressures, the total pressure is

$$P = P_a + P_b = 3 + 4 = 7 \text{ atm.}$$

Our final example combines the use of Dalton's law and the ideal gas equation of state.

Example 2.6 A sample of PCl_5 weighing 2.69 gm was placed in a 1.00-liter flask and completely vaporized at a temperature of 250°C. The pressure observed at this temperature was 1.00 atm. The possibility exists that some of the PCl_5 may have dissociated according to the equation

$$PCl_5(g) = PCl_3(g) + Cl_2(g).$$

What are the partial pressures of PCl_5, PCl_3, and Cl_2 under these experimental conditions?

The solution to this problem involves several steps. In order to decide whether the PCl_5 has dissociated at all, let us first calculate the pressure which would have been observed *if* no PCl_5 had dissociated. This can be calculated from the number of moles of PCl_5 used, together with the volume and temperature of the flask. Since the molecular weight of PCl_5 is 208, the number of moles of PCl_5 initially in the flask is

$$n = \frac{2.69}{208} = 0.0129.$$

The pressure corresponding to this number of moles would be

$$P = \frac{nRT}{V} = \frac{(0.0129)(0.082)(523)}{1} = 0.553 \text{ atm.}$$

Since the observed pressure is higher than this, some dissociation of PCl_5 must have occurred.

Using the law of partial pressures we can write

$$P_{PCl_5} + P_{PCl_3} + P_{Cl_2} = P_t = 1.00 \text{ atm.}$$

We now notice that

$$P_{Cl_2} = P_{PCl_3}, \qquad P_{PCl_5} = 0.553 - P_{Cl_2},$$

since one mole of PCl_3 and one mole of Cl_2 are produced every time one mole of PCl_5 dissociates. Therefore we can write Dalton's law as

$$0.553 - P_{Cl_2} + P_{Cl_2} + P_{Cl_2} = 1.00,$$

$$P_{Cl_2} = 0.447 \text{ atm,}$$

and

$$P_{PCl_3} = 0.447 \text{ atm,} \qquad P_{PCl_5} = 0.106 \text{ atm.}$$

2.2 THE KINETIC THEORY OF GASES

In the introduction to this chapter we stated that one of the occupations of a chemist is to relate the properties of bulk matter to the properties of individual atoms. In this section we shall see that simple assumptions about the structure and behavior of atoms in the gas phase lead to a molecular theory of gases that is consistent with several observed macroscopic properties.

In order to develop a molecular theory of gases we must first assume that we can represent a gas by a simple "model." A model is some imaginary construct or picture which incorporates only those features that are thought to be important in determining the behavior of the real physical system. These features are often selected intuitively, or sometimes on the basis of mathematical convenience. The validity of any model can be determined only by comparing predictions based on it with actual experimental facts.

A most important feature of our model is that the gaseous particles, whether atoms or molecules, behave like point centers of mass, which most of the time exert no force on one another. This assumption is suggested by measurements of the densities of solids, which show that the actual volume displaced by a single molecule is only about 10^{-23} cc, while for a gas at 1-atm pressure, the volume per molecule is $(22.4 \times 10^3)/(6 \times 10^{23}) = 4 \times 10^{-20}$ cc. Since the actual volume of a molecule is so much smaller than the volume per molecule in the gaseous state, we can justifiably assume that molecules are point particles that behave independently except for brief moments when they collide with each other. Furthermore, since gas molecules exert force on each other only during the brief instants when they collide, all the obvious macroscopic properties of a gas must be consequences primarily of the independent *motion* of the molecule. This is the reason that the idea we are about to develop is called the kinetic theory of gases.

Derivation of Boyle's Law

In the following pages we shall present three derivations of Boyle's law. The first derivation is very straightforward, and gives the correct result, but is perhaps unconvincing because of some obvious oversimplifications. The second derivation is similar to the first, but avoids a major simplification, and is therefore more elaborate. The third derivation employs a totally different point of view, avoids all objectionable oversimplifications, and is, therefore, much more convincing. It has the drawback that it is less pictorial and mechanical than the other derivations, but in addition to its rigor, it has the advantage that it can be extended to deal with nonideal gases. The purpose of presenting three derivations of Boyle's law is to demonstrate that it is the *methods* and *thought processes* which are employed, as well as the final result of a derivation, that are useful and revealing.

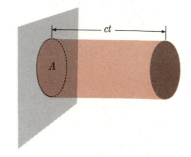

FIG. 2.6 The imaginary cylinder of base area A and altitude ct which contains the molecules which will collide with A in time t.

Consider N molecules, all of the same mass m, contained in a cubical vessel of volume V. We want to compute the pressure, or force per unit area, on the walls due to molecular impacts. To do this, we first make a major assumption: all molecules in the vessel move along the three cartesian coordinates perpendicular to the walls of the box and have the same speed c. Now we center our attention on an imaginary cylinder which extends perpendicularly from one of the walls, as shown in Fig. 2.6. The base of this cylinder has the arbitrary area A. We choose the altitude to be ct, where c is the molecular speed, and t

is an arbitrary but short length of time. The cylinder has the following important property: it contains all the molecules which will strike the wall in a time t, since the molecules located at the top of the cylinder and moving toward the wall will just travel the distance ct in time t. Those closer to the wall will reach it in less time.

The force experienced by a molecule in a wall collision is given by Newton's second law,

$$f = ma,$$

where a is the acceleration experienced by the molecule. Since acceleration is defined as the change in velocity per unit time, we can use Newton's second law in the form

$$f = ma,$$

$$f = m\frac{\Delta c}{\Delta t} = \frac{\Delta(mc)}{\Delta t},$$

force = change in momentum per unit time.

Rather than calculate $\Delta(mc)/\Delta t$, the change in momentum per unit time, we shall compute the change in momentum of the molecule *per collision* and multiply this times the number of wall collisions per unit time. That is,

force = change in momentum per impact × impacts per unit time

The change in momentum that occurs in one impact can be obtained by subtracting the momentum a molecule has after a wall collision from its momentum before the collision. Initially a molecule traveling toward a wall has momentum mc; after the collision its velocity is assumed to be exactly reversed in direction but unchanged in magnitude. The final momentum is therefore $-mc$, and the change in momentum, the final value minus the initial value, is

$$\Delta(mc) = -mc - mc = -2mc.$$

This is the change in momentum of the molecule, and the change in momentum imparted to the wall is the negative of this, or $2mc$, since momentum is conserved in any collision.

The number of collisions with the area A in time t can now be calculated very simply. The volume of the cylinder is Act, and since the number of molecules per unit volume is N/V, the total number of molecules in the collision cylinder is $NAct/V$. Of these, however, only one-sixth are moving toward the wall since only one-third move along any one of the three coordinate axes, and only one-half of these move in the correct direction. Accordingly, the number of molecules hitting A *per unit time* is

$$\frac{1}{6}\frac{N}{V}\frac{Act}{t} = \frac{1}{6}\frac{NAc}{V}.$$

Thus for the force on A we get

$$f = 2mc \times \frac{1}{6}\frac{NAc}{V} = \frac{1}{3}\frac{NAmc^2}{V}.$$

The pressure is the force per unit area f/A, so

$$P = \frac{f}{A} = \frac{1}{3}\frac{Nmc^2}{V} \qquad \text{or} \qquad PV = \tfrac{1}{3}Nmc^2 = \tfrac{2}{3}N(mc^2/2).$$

We can patch up our incorrect assumption that all molecules have the same speed c by replacing c^2 in the above expression by the average value $\overline{c^2}$. Thus we get

$$PV = \tfrac{2}{3}N(\overline{mc^2}/2). \tag{2.8}$$

This looks very much like Boyle's law. In fact, *if* it is true that $\tfrac{1}{2}\overline{mc^2}$, the average kinetic energy of gas molecules, is constant at constant temperature, then Eq. (2.8) does express Boyle's law exactly: the product of the pressure and the volume for an ideal gas is a constant which depends on the number of molecules in the sample.

The foregoing derivation actually does give the correct result, but the assumption that all molecules move only parallel to the coordinate axes or perpendicular to the walls is not correct and tends to shake our confidence in the result. Fortunately this assumption can be eliminated, and doing so gives us the opportunity to use calculus in the derivation.

FIG. 2.7   An oblique collision cylinder of slant height ct and base area A. All molecules in it moving toward the wall with directions specified by θ and ϕ will collide with the wall during the time t.

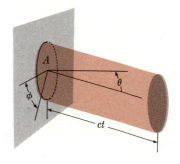

Consider the cylinder shown in Fig. 2.7. The area of its base is A, and its slant height is ct, where c is the molecular speed, and t is a short arbitrary time. The axis of the cylinder is located by the angle θ from the direction perpendicular to the wall, and the angle ϕ. Molecules in it that are moving parallel to its axis with speed c have a component of velocity perpendicular to the wall of $c\cos\theta$, and upon striking the wall acquire a new perpendicular component $-c\cos\theta$. The momentum imparted to the wall in one such collision is therefore $2mc\cos\theta$.

Now we must find the number of molecules in the cylinder which move parallel to its axis. This is simply the volume of the cylinder $Act \cos \theta$, times the number of molecules per unit volume N/V, times the fraction of molecules moving in the direction specified by the small range of angles θ, $\theta + d\theta$, and ϕ, $\phi + d\phi$. This fraction is

$$\frac{d\phi \sin \theta \, d\theta}{4\pi},$$

which is found by dividing $r^2 \sin \theta \, d\theta \, d\phi$, the area on the surface of a sphere of radius r corresponding to the differential angles $d\phi$ and $d\theta$, by the total area of the sphere $4\pi r^2$. Consequently, the momentum change per unit area and time (that is, the pressure) due to the molecules in this cylinder is

$$\left(\frac{2mc \cos \theta}{At}\right)(Act \cos \theta)\left(\frac{N}{V}\right)\left(\frac{d\phi \sin \theta \, d\theta}{4\pi}\right)$$

or

$$\frac{Nmc^2}{2\pi V} \cos^2 \theta \sin \theta \, d\theta \, d\phi.$$

To get the total pressure due to all possible orientations of the cylinder, we must add (by integration) the values of the trigonometric terms for all allowed values of θ and ϕ. The angle θ can range from 0 to $\pi/2$ before the imaginary cylinder hits the wall, whereas ϕ can run from 0 to 2π. We must evaluate

$$\frac{Nmc^2}{2\pi V} \int_0^{2\pi} d\phi \int_0^{\pi/2} \cos^2 \theta \sin \theta \, d\theta.$$

The integral over ϕ simply gives 2π. The integral over θ can be evaluated by noting that $d(\cos \theta) = -\sin \theta \, d\theta$, so if we let $x = \cos \theta$, we get

$$\int_0^{\pi/2} \cos^2 \theta \sin \theta \, d\theta = -\int_1^0 x^2 \, dx = -\frac{x^3}{3}\Big|_1^0 = \frac{1}{3}.$$

Our expression for the total pressure is then

$$P = \frac{Nmc^2}{2\pi V}(2\pi)(\tfrac{1}{3})$$

or, rearranging and replacing c^2 by $\overline{c^2}$, we get

$$PV = \frac{2}{3}\frac{Nm\overline{c^2}}{2},$$

which is the result obtained by the more elementary method.

We can use the technique just employed to calculate the rate at which molecules strike a unit area of a wall from all directions. The contribution from a cylinder with orientation ϕ, θ is the volume of the cylinder $Act \cos \theta$, multiplied by the number of molecules per unit volume N/V and the fraction moving along θ, ϕ toward the wall, $\sin \theta\, d\theta\, d\phi / 4\pi$, divided by the time t and area A:

$$\left(\frac{Act}{At} \cos \theta \right) \left(\frac{N}{V} \right) \left(\frac{\sin \theta\, d\theta\, d\phi}{4\pi} \right).$$

To get the total rate at which molecules hit a unit area of the wall, we integrate over allowed angles:

$$\text{wall collision rate} = \frac{Nc}{4\pi V} \int_0^{2\pi} d\phi \int_0^{\pi/2} \cos \theta \sin \theta\, d\theta$$

$$= \frac{Nc}{4V}. \tag{2.9}$$

We can replace c by the average speed $\bar{c}$, and by so doing obtain an exact expression for the wall collision rate. Note that the elementary method first employed to derive Boyle's law would give $N\bar{c}/6V$ for the wall collision rate, which is too small. Its success in the Boyle's law derivation came from compensating errors.

We will now derive the equation of state for a gas by a third, much more rigorous, but somewhat simpler method. This last technique, which involves use of the *virial theorem*, exposes most clearly the minimum necessary assumptions involved in deriving Boyle's law, and has the very important feature that it provides a basis for rigorously treating nonideal gases. Before we derive the equation of state, we must prove the virial theorem.

Consider S_x, the product of the x-component of momentum p_x times the coordinate x:

$$S_x \equiv x p_x.$$

The derivative with respect to time of S_x is

$$\frac{dS_x}{dt} = \frac{dx}{dt} p_x + x \left(\frac{dp_x}{dt} \right). \tag{2.10}$$

Now we ask for the average value of dS_x/dt over a long time period τ. This can be obtained by adding up all the values of dS_x/dt during this time, and dividing by τ. Since the values of dS_x/dt can be considered continuous, we can perform the addition (and the averaging) by integration. The average we want is

$$\left\langle \frac{dS_x}{dt} \right\rangle = \frac{1}{\tau} \int_0^\tau \frac{dS_x}{dt}\, dt = \frac{1}{\tau} \int_0^\tau dS_x = \frac{S}{\tau} \Big|_0^\tau$$

$$= \frac{S_x(\tau) - S_x(0)}{\tau}.$$

For mechanical systems in which the motion of the particles is confined, the quantity S_x has some finite upper limit. For a gaseous molecule confined to a finite volume, the coordinates have obvious upper limits set by the vessel walls. Also, the momentum has a finite upper limit set by the finite total kinetic energy. Consequently the quantity $S_x(\tau) - S_x(0)$ must also be finite and, in fact, may be quite small. Since the period of time over which the averaging is performed can be made arbitrarily long, the quantity $[S_x(\tau) - S_x(0)]/\tau$ can be made to vanish. Thus the long time average of dS_x/dt vanishes, and from Eq. (2.10) we get

$$\left\langle \frac{dS_x}{dt} \right\rangle = 0 = \left\langle \frac{dx}{dt} p_x \right\rangle + \left\langle x \frac{dp_x}{dt} \right\rangle$$

or

$$\left\langle p_x \frac{dx}{dt} \right\rangle = -\left\langle x \frac{dp_x}{dt} \right\rangle.$$

Now since, by definition,

$$v_x = dx/dt, \qquad p_x = mv_x,$$

and the force F_x is dp_x/dt, we can substitute to get

$$\langle mv_x^2 \rangle = -\langle xF_x \rangle.$$

This argument has been carried out for the x-component of one particle, and is the same for all three components and all other particles in a gas. Thus we can add the terms for the other components and N particles to get

$$\left\langle \sum_N (mv_x^2 + mv_y^2 + mv_z^2) \right\rangle = -\left\langle \sum_N (xF_x + yF_y + zF_z) \right\rangle,$$

where $\sum_N$ stands for the operation of adding up the quantities for the N particles. The left-hand side is just twice the average kinetic energy of the collection of particles, so we have

$$\langle \mathrm{KE} \rangle = -\frac{1}{2} \left\langle \sum_N (xF_x + yF_y + zF_z) \right\rangle. \tag{2.11}$$

The left-hand side is the average kinetic energy of all the particles, and the right-hand side is called the virial. Equation (2.11) is a statement of the **virial theorem.**

Now that we have the virial theorem, it is a simple matter to derive the equation of state for an ideal gas. In such a system the only forces on the molecules are those due to the walls of the container. For the cubical box shown in Fig. 2.8 we must evaluate the product of the force and coordinate for each of the six faces. Taking the quantity xF_x as an example, we see that it vanishes for all faces except those perpendicular to the x-axis, since a wall parallel to the x-axis cannot exert a net force in the x-direction if the gas is at rest. The contribution to the virial of the wall at $x = 0$ is zero, while the wall at $x = L$

FIG. 2.8

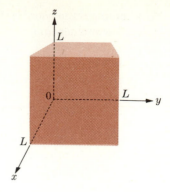

A cubical gas container with faces at $x = 0, L$, $y = 0, L$, and $z = 0, L$.

contributes $LF_x = -L(L^2P)$, where P is the gas pressure. A similar argument holds for the other walls, with the result that

$$\langle\text{KE}\rangle = -\tfrac{1}{2}L(F_x + F_y + F_z) = \frac{L}{2}(L^2P + L^2P + L^2P) = \tfrac{3}{2}PL^3 = \tfrac{3}{2}PV.$$

Therefore we get

$$PV = \tfrac{2}{3}\langle\text{KE}\rangle = \tfrac{2}{3}N\left\langle\frac{mc^2}{2}\right\rangle,$$

which is exactly the result obtained earlier. Note that the important assumption made in this derivation is that the only forces that contribute to the virial are those that the molecules exert on the walls. This appears then to be the *necessary* condition for a gas to behave *exactly* according to Boyle's law: the forces between the molecules must be negligible. Since molecules always exert forces on each other when they are close, Boyle's law is obeyed exactly only in the limit of zero pressure.

Temperature, Energy, and the Gas Constant

Our theory cannot provide a way to evaluate the constant $\tfrac{1}{2}m\overline{c^2}$; this must be done by comparing the theory to the results of some experiment. To see how this is accomplished, let us return to Eq. (2.8) and write the number of molecules N as the product of Avogadro's number N_0 and n, the number of moles of gas in the sample:

$$N = nN_0,$$

$$PV = n\tfrac{2}{3}N_0\frac{m\overline{c^2}}{2}.$$

But by experiment we know that $PV = nRT$, and therefore

$$RT = \tfrac{2}{3}N_0\frac{m\overline{c^2}}{2}.$$

Now the quantity $\overline{mc^2}/2$ is the average translational kinetic energy of a single molecule, so $N_0(\overline{mc^2}/2)$ is the *total* kinetic energy of one mole of gas. We have the remarkably simple result that

$$N_0 \frac{\overline{mc^2}}{2} = \frac{3}{2} RT,$$

translational kinetic energy of a mole of gas $= \frac{3}{2}RT$. $\qquad$ (2.12)

From this we see that temperature is a parameter related to the total kinetic energy of translation of the gas particle. If we divide both sides of Eq. (2.12) by Avogadro's number, we get

$$\frac{\overline{mc^2}}{2} = \frac{3}{2} \frac{R}{N_0} T = \frac{3}{2} kT, \qquad (2.13)$$

where $k = R/N_0$, the gas constant *per molecule*, is called Boltzmann's constant. This equation tells us that temperature is a measure of the *average* kinetic energy of a single molecule.

According to Eq. (2.12) the quantity nRT and consequently PV must have the units of energy. We have been expressing both of these factors in units of liter-atmospheres, which is an uncommon and perhaps unrecognizable energy unit. To assure ourselves that pressure times volume indeed has the units of energy, we need only write

$$\text{pressure} \times \text{volume} = (\text{force/area})(\text{area} \times \text{length})$$
$$= \text{force} \times \text{length}.$$

Since work or energy is defined as the product of force and distance, we see that PV actually does have the units of energy. Let us calculate the value of 1 liter-atm of energy in more common units. Since 1 atm is 1.013×10^6 dynes/cm^2, and 1 liter is 10^3 cc,

$$\begin{aligned} 1 \text{ liter-atm} &= (1.013 \times 10^6 \text{ dynes/cm}^2) \times 10^3 \text{ cm}^3 \\ &= 1.013 \times 10^9 \text{ dyne-cm} \\ &= 1.013 \times 10^9 \text{ ergs.} \end{aligned}$$

Noting that 1 joule $= 10^7$ ergs, we find that 1 liter-atm is equal to 1.013×10^2 joules, or to 24.4 cal. We can use these conversion factors to calculate the value of R in more familiar energy units. The value of R in ergs per mole-degree is

$$0.08206 \frac{\text{liter-atm}}{\text{mole-deg}} \times 1.013 \times 10^9 \frac{\text{ergs}}{\text{liter-atm}} = 8.313 \times 10^7 \frac{\text{ergs}}{\text{mole-deg}}.$$

Other values for R in different units are given in Table 2.2.

Now we are in a position to evaluate Boltzmann's constant k, and to calculate the average speed of gaseous molecules. We know that the value of the

Table 2.2 The gas constant R

0.08206	liter-atm/mole-deg
1.987	cal/mole-deg
8.313	joules/mole-deg
8.313×10^7	ergs/mole-deg

gas constant R is 8.313×10^7 ergs/deg-mole, so for k we find

$$k = \frac{8.313 \times 10^7}{6.022 \times 10^{23}} = 1.380 \times 10^{-16} \text{ ergs/molecule-deg.}$$

We have chosen to express Boltzmann's constant in cgs units so as to be able to compute molecular speeds in cm/sec. Let us calculate $\sqrt{\overline{c^2}}$ for a nitrogen molecule at room temperature. We write

$$\tfrac{1}{2}m\overline{c^2} = \tfrac{3}{2}kT,$$

$$\sqrt{\overline{c^2}} = \sqrt{3kT/m} \tag{2.14}$$

$$= \left(\frac{3 \times 1.38 \times 10^{-16} \times 298}{4.65 \times 10^{-23}}\right)^{1/2}$$

$$= 5.1 \times 10^4 \text{ cm/sec.}$$

The quantity $\sqrt{\overline{c^2}}$ is called the root-mean-square speed of a molecule, c_{rms}, and is not the same as the average speed $\overline{c}$. The difference between the two, however, is so small that for most purposes they can be equated to each other. Equation (2.14) shows that the root-mean-square speed depends on the mass of the molecule, and repetition of our calculation gives the root-mean-square speed of a hydrogen molecule at 25°C as 19.3×10^4 cm/sec. Table 2.3 gives c_{rms} for some other molecules.

Equation (2.13) shows that if two gases are at the same temperature, their molecules have the same average kinetic energy, so we can write

$$\tfrac{1}{2}m_1\overline{c_1^2} = \tfrac{1}{2}m_2\overline{c_2^2}, \qquad \frac{\overline{c_1^2}}{\overline{c_2^2}} = \frac{m_2}{m_1},$$

$$\frac{\overline{c_1}}{\overline{c_2}} \cong \left(\frac{m_2}{m_1}\right)^{1/2}. \tag{2.15}$$

Table 2.3 Root-mean-square speeds of molecules at 298°K

Argon	4.31×10^4 cm/sec	Hydrogen	1.93×10^5 cm/sec
Carbon dioxide	4.11×10^4	Oxygen	4.82×10^4
Chlorine	3.23×10^4	Water	6.42×10^4
Helium	1.36×10^5	Xenon	2.38×10^4

When taken at the same temperature, lighter molecules move faster, on the average, than heavier molecules, and the ratio of the average molecular speeds is very nearly equal to the square root of the inverse ratio of molecular masses.

In the derivation of Boyle's law we showed that the frequency of wall collisions is proportional to molecular speed, and hence inversely proportional to the square root of molecular mass. Consequently, lighter molecules collide with the walls of their container more frequently than do heavier molecules at the same temperature. On the other hand, the change in momentum per wall collision is proportional to $m\bar{c}$, and by taking account of Eq. (2.15) we see that $m\bar{c}$ increases proportionally to the square root of the molecular mass. Thus, while lighter molecules collide more frequently with the vessel walls, heavier molecules experience a greater change in momentum per collision. These two factors exactly cancel each other, and gas pressure is independent of the nature of the molecules.

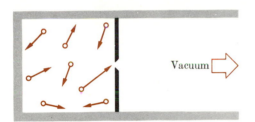

Vacuum

A schematic representation of a molecular effusion apparatus. The diameter of the hole is smaller than the distance that molecules travel between collisions. Consequently molecules pass independently, not collectively, through the hole.

FIG. 2.9

Effusion and Diffusion

There are two simple experiments which make the mass dependence of the average molecular speed directly observable. Consider first the apparatus shown in Fig. 2.9. A gas is separated from a vacuum chamber by a wall which has a very small hole in it. If the hole is small enough, there will be no "pouring" or collective mass flow of gas into the vacuum. Instead, individual molecules will pass through the hole independently only if their trajectories cause them to approach the wall area where the hole is. The rate of passage of molecules through the hole, which is the *effusion* rate, is just the rate at which molecules strike a unit area of the wall times the area A of the hole. From Eq. (2.9) we get

$$\text{effusion rate} = \text{rate of wall collisions per cm}^2 \times \text{hole area}$$
$$= \frac{1}{4}\frac{N}{V}\bar{c} \times A.$$

Since $\bar{c}$, the average molecular speed, is inversely proportional to the square root of molecular mass, we should have

$$\text{effusion rate} \propto m^{-1/2}.$$

This is observed experimentally. In particular, if an equimolar mixture of H_2 and N_2 is allowed to effuse through a hole, we can expect

$$\frac{\text{rate of effusion of } H_2}{\text{rate of effusion of } N_2} = \frac{\bar{c}_{H_2}}{\bar{c}_{N_2}} = \sqrt{\frac{m_{N_2}}{m_{H_2}}}$$

$$= \sqrt{\frac{28}{2}} = 3.7.$$

Thus the gas which passes through the hole should be richer in H_2, and the gas remaining in the vessel should be richer in N_2. This result is indeed found experimentally.

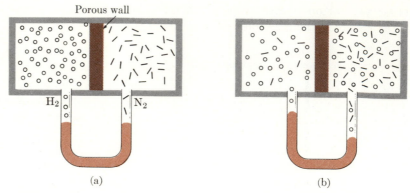

FIG. 2.10

Diffusional mixing of H_2 and N_2. (a) Initial state, and (b) some time later.

The second type of experiment which demonstrates the difference in molecular speeds is gaseous *diffusion*. Figure 2.10 shows an apparatus in which hydrogen and nitrogen gas, initially at the same pressure and temperature, are separated by a porous wall. The porous wall prevents a mass flow of gas, but does allow molecules to pass from one chamber to the other. It is observed that the initial diffusive flow of hydrogen from left to right is more rapid than the flow of nitrogen from right to left.

The explanation of the diffusive flow rate is more complicated than the explanation of molecular effusion, since diffusion involves the effects of collisions between molecules, whereas effusion does not. However, the dependence of diffusion rate on molecular mass can be deduced as follows. As each gas starts to diffuse through the porous plug, it transfers momentum to the plug. Initially, the pressures of the gases on each side of the plug are equal, and this means that the momentum imparted by each gas to the plug must be the same. Since the gases are flowing, the momentum imparted per unit time by each to the plug is the product of the flux of molecules J through the plug and the average momentum carried by the molecules, $m\bar{c}$. Thus

$$\text{momentum transferred/sec by } H_2 = J_{H_2} m_{H_2} \bar{c}_{H_2},$$
$$\text{momentum transferred/sec by } N_2 = J_{N_2} m_{N_2} \bar{c}_{N_2}.$$

These momenta are equal, since the pressures are equal, so

$$J_{H_2} m_{H_2} \bar{c}_{H_2} = J_{N_2} m_{N_2} \bar{c}_{N_2},$$

$$\frac{J_{H_2}}{J_{N_2}} = \frac{(m\bar{c})_{N_2}}{(m\bar{c})_{H_2}}.$$

Now we make use of the fact that $\bar{c} \propto m^{-1/2}$, and get

$$\frac{\text{rate of diffusion of } H_2}{\text{rate of diffusion of } N_2} = \frac{J_{H_2}}{J_{N_2}} = \sqrt{\frac{m_{N_2}}{m_{H_2}}}. \qquad (2.16)$$

The result is that the ratio of rates of diffusion is inversely proportional to the square root of the mass ratio, just as was true for effusion. Note, however, that the inverse mass dependence of effusion and diffusion arises in two different ways. In effusion, the molecular flux is *directly* proportional to molecular speed, and hence inversely proportional to the square root of molecular mass. In diffusion, the molecular flux is *inversely* proportional to the molecular momentum $m\bar{c}$, and hence inversely proportional to the square root of molecular mass.

The fact that the rate of diffusion is greater for lighter gases can be made the basis of a purification procedure. If we had an equimolar mixture of hydrogen and nitrogen, and allowed it to diffuse through a porous wall into a vacuum, the gas which initially diffused through the barrier would be enriched in hydrogen. The enrichment factor would be the ratio of the rates of diffusion of the two gases, or as Eq. (2.16) shows, a factor of 3.7. If the enriched sample were collected and allowed to diffuse through another porous barrier, further enrichment could be achieved. It is by an elaboration of this process that the U^{235} isotope is separated from U^{238}. Gaseous UF_6 diffuses through thousands of porous barriers until an acceptable enrichment of $U^{235}F_6$ occurs. The enrichment factor at each barrier is only

$$\left(\frac{238 + 114}{235 + 114}\right)^{1/2} = 1.004,$$

and consequently many barriers and careful collection and recycling of the gas are required to achieve a useful isotopic separation.

2.3 THE DISTRIBUTION OF MOLECULAR SPEEDS

As we implied in our kinetic derivation of Boyle's law, not all gaseous molecules travel at the same speed. To obtain a more detailed picture of gaseous behavior, it might seem desirable to know the speed of each molecule. However, this is definitely impossible. Just to write down the 6×10^{23} values of the molecular speeds that occur in a mole of gas at one particular instant would require a stack of paper reaching past the moon. It is even more discouraging to realize that these data would be valid for less than 10^{-9} seconds. Because of collisions, the speed of each gaseous molecule changes billions of times every second. Consequently, we must abandon the idea of ever knowing the speed of each molecule in even a modest-sized sample of gas.

There is still a useful approach to the problem, however. We can take advantage of the large numbers of molecules present in any gas sample and make a statistical prediction of *how many of them* have a particular speed. This approach is similar to that used in actuarial problems. Without detailed information, it is impossible to say *who* will die in a given week, but *the number* of people who will die in the same period is statistically predictable. In a gas, despite the constant collisional "exchange" of speeds, the number of molecules with any particular speed, for instance in the small range between c and $c + \Delta c$, is a constant. Consequently, it is possible to specify the *distribution of molecular speeds:* the fraction $\Delta N/N$ of the molecules which have speeds between each value of c and $c + \Delta c$.

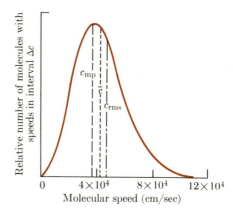

FIG. 2.11   Molecular speed distribution for oxygen at 273°K.

The molecular speed distribution, derived by statistical considerations and confirmed by experiment, is represented graphically in Fig. 2.11. The value of the ordinate is proportional to the fraction of molecules which have speeds in a narrow range Δc centered about each value of c. Note that there are relatively few molecules with very high or very low speeds. The value of c for which $\Delta N/N$ is a maximum is called the most probable speed, c_{mp}. The distribution curve is not symmetrical about its maximum, and as a result the average speed $\bar{c}$ is slightly larger than c_{mp}, and the root-mean-square speed $c_{rms} = \sqrt{\bar{c^2}}$ is larger still. However, exact calculation shows that these speeds are related by $c_{mp} : \bar{c} : c_{rms} = 1 : 1.13 : 1.22$, and for many purposes can be considered identical.

Very often it is useful to know the fraction of molecules that have speeds between two different values c_1 and c_2; this number is equal to the area under the distribution curve between c_1 and c_2. With this in mind, we can see from Fig. 2.11 that most gaseous molecules have speeds that are near the average speed $\bar{c}$.

Figure 2.12 shows how the distribution of speeds changes as the gas temperature is raised. The values of c_{mp}, $\bar{c}$, and c_{rms} all increase, and the distribution curve becomes broader. In other words, at higher temperatures there are more molecules with greater speeds than at low temperatures. The temperature dependence of the distribution curve is helpful in explaining the effect of temperature on chemical reaction rates. Consider the possibility that in order to react, a molecule must have a speed greater than c_a, shown in Fig. 2.12. The area under the distribution curve for speeds greater than c_a is very small at low temperatures, and thus very few molecules meet the requirement for reaction. As the temperature is increased the distribution curve broadens, and the area under the curve corresponding to speeds greater than c_a increases. Thus at higher temperatures more molecules satisfy the criterion for reaction, and the reaction rate increases.

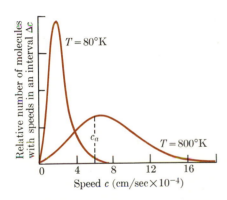

Molecular speed distribution for oxygen at two temperatures.

FIG. 2.12

The Maxwell-Boltzmann Distribution Function

The mathematical form of the speed distribution function was first derived by Clerk Maxwell and Ludwig Boltzmann in 1860. Their expression for $\Delta N/N$ is

$$\frac{\Delta N}{N} = 4\pi \left(\frac{m}{2\pi kT}\right)^{3/2} e^{-mc^2/2kT} c^2 \, \Delta c, \qquad (2.17)$$

where m is the molecular mass, k is Boltzmann's constant, T is the absolute temperature, and $e = 2.71\ldots$ is the base of natural logarithms. We will not derive this equation, for to do so requires moderately elaborate mathematics.

It is useful, however, to analyze the expression, and see that the dependence of $\Delta N/N$ on c is the product of two factors. One is

$$e^{-(1/2)(mc^2/kT)}$$

and the other, apart from the constants, is c^2.

The exponential factor is a special example of Boltzmann's factor $e^{-\epsilon/kT}$, with $\epsilon = \frac{1}{2}mc^2$. It is a general and very important feature of all systems that the fraction N_ϵ/N of molecules with energy ϵ is proportional to $e^{-\epsilon/kT}$. Thus, at any particular temperature there tend to be fewer molecules with high energies than with low energies.

Even without using detailed mathematics, it is possible to justify the exponential form of the Boltzmann factor in the speed distribution law. We know that the kinetic energy can be expressed in terms of the squares of its velocity components:

$$\tfrac{1}{2}mc^2 = \tfrac{1}{2}m\dot{x}^2 + \tfrac{1}{2}m\dot{y}^2 + \tfrac{1}{2}m\dot{z}^2.$$

If the motion of molecules is truly random, the value of one velocity component is *independent* of the values of the others. Therefore, if we say that the speed distribution function depends on the energy $\frac{1}{2}mc^2$, the function must be consistent with two facts:

1. The kinetic energy of a molecule is the sum of its component kinetic energies.

2. The probability of observing a particular magnitude of one velocity component is independent of the values of other velocity components.

Now the probability of observing two independent events is equal to the product of the probabilities of observing each of them separately. Therefore our requirements on the speed distribution function amount to saying that the kinetic energies associated with the velocity components must be additive, but the probabilities of observing their individual values must be multiplicative. We can see that the Boltzmann factor satisfies these requirements, for

$$e^{-mc^2/2kT} = e^{-(m/2kT)(\dot{x}^2+\dot{y}^2+\dot{z}^2)} = (e^{-m\dot{x}^2/2kT})(e^{-m\dot{y}^2/2kT})(e^{-m\dot{z}^2/2kT}),$$

where the three factors on the right-hand side are the probabilities of observing the individual values of the velocity components. The argument we have just outlined can be used as a basis for a proof that the exponential function alone satisfies the requirements for the distribution law.

The origin of the c^2-factor in the distribution law lies in the fact that there are more "ways" in which a molecule can have a high speed than a low speed. For instance, there is only one way in which a molecule can have zero speed: it must not be moving along the x-, y-, or z-axis. But if a molecule has a finite speed, say 100 m/sec, it may move in either direction along the x-axis at 100 m/sec, and not along y or z, or it may move along y at 100 m/sec and not along x or z, or it may move with a velocity component of 57.7 m/sec along each of the axes. Any combination of velocity components which satisfies the relation $\dot{x}^2 + \dot{y}^2 + \dot{z}^2 = c^2 = (100)^2$ is possible. As the speed of a molecule increases, the number of possible combinations of its velocity components which are consistent with its speed increases proportionally to c^2.

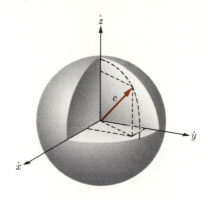

Graphical representation of $\dot{x}^2 + \dot{y}^2 + \dot{z}^2 = c^2$.   **FIG. 2.13**

To see this argument more clearly, we need only plot the equation $\dot{x}^2 + \dot{y}^2 + \dot{z}^2 = c^2$ on a coordinate system in which $\dot{x}$, $\dot{y}$, $\dot{z}$ are the coordinate axes. Figure 2.13 shows that this equation generates a spherical surface of radius c. This surface contains all the values of $\dot{x}$, $\dot{y}$, and $\dot{z}$ which are consistent with a speed c. Therefore, the number of possible ways a molecule can have speed c should be proportional to the number of points on the surface, or to the surface area. Since the surface is a sphere, its area and the number of ways in which the speed c can occur are proportional to c^2. Thus the Maxwell-Boltzmann distribution has in it two opposing factors. The c^2-factor favors the presence of molecules with high speeds, and is responsible for the fact that there are few molecules with speeds near zero. The Boltzmann factor, $e^{-mc^2/2kT}$, favors low speeds and limits the number of molecules which can have high speeds.

Experimental Verification of the Speed Distribution

Figure 2.14 is a simplified diagram of an apparatus that has been used to determine the form of the speed distribution. The vapor of cesium is contained in an oven O which has a very narrow horizontal slit cut in its side. Cesium atoms pass through the slit into the surrounding chamber, in which a vacuum is maintained. In the absence of gravity the only cesium atoms passing through the slits S_1 and S_2 and to the detector would have horizontal trajectories. Because of the action of gravity, however, the atoms are deflected downward as they travel toward the detector D. The deflection is greatest for the atoms which travel with the slowest speeds. To see that this is true, recognize that the time that an atom with speed c takes to travel the distance l from the oven to the

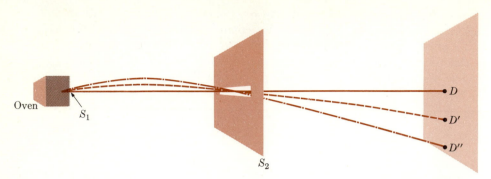

FIG. 2.14   Schematic diagram of apparatus for determining the molecular speed distribution.

detector is $t = l/c$, and the deflection s produced by the acceleration of gravity g during this time is $s = \frac{1}{2}gt^2 = \frac{1}{2}g(l/c)^2$. Thus, by measuring the signal reaching the detector when it is placed at the various positions D, D', D'', etc., the relative number of atoms with each speed can be determined. The results found in this way are in excellent agreement with the mathematical form of the speed distribution found by statistical calculations.

2.4 HEAT CAPACITIES

In Chapter 1 we defined the heat capacity of a substance to be equal to the amount of heat in calories required to raise the temperature of one mole of the substance one degree centigrade. Now we shall see that the molecular kinetic theory leads to a very satisfactory prediction and interpretation of the experimental heat capacities of many gases.

First we must remark that our definition of heat capacity is incomplete. It is found experimentally that the measured value of the heat capacity depends on how a gas is heated. In particular, if a gas is heated with its volume held constant, the measured heat capacity is smaller than when the gas is heated with the pressure fixed. Let us call these two heat capacities C_V and C_P, the heat capacity at constant volume and at constant pressure, respectively.

When a gas is heated, energy is added to it. This added energy must appear as kinetic energy of the molecules, or as work done by the gas in expanding against an external pressure, or as both. In order for a gas to perform work, it *must* expand, for work is always the product of a force and a displacement. If the volume of a gas is held constant, there is no displacement and no work done. Therefore, any energy we add to a gas at constant volume must appear as kinetic energy of the molecules. The kinetic theory tells us that for an ideal gas of monatomic particles,

$$E = \tfrac{3}{2}RT.$$

Table 2.4 Heat capacity ratios, C_P/C_V, for some gases

Gas	C_P/C_V	Gas	C_P/C_V
He	1.66	H_2	1.41
Ne	1.64	O_2	1.40
Ar	1.67	N_2	1.40
Kr	1.68	CO	1.40
Xe	1.66	NO	1.40
Hg	1.67	Cl_2	1.36

Therefore if we increase the energy by an amount ΔE, we must make a change in temperature ΔT, where

$$\Delta E = \tfrac{3}{2}R\,\Delta T.$$

But $\Delta E/\Delta T$ is the increase in energy per degree per mole, or the heat capacity at constant volume:

$$C_V = \frac{\Delta E}{\Delta T}, \qquad C_V = \tfrac{3}{2}R.$$

Thus C_V for an ideal monatomic gas is $\tfrac{3}{2}R$, or about 3 cal/mole-deg.

When the temperature is raised at constant pressure, the kinetic energies of the molecules increase and the gas does work by virtue of its increase in volume. If we remember that the product PV has the units of work, the amount of work done by the gas expansion can be calculated easily. It is simply equal to $\Delta(PV)$, the change in the pressure-volume product. But at constant pressure we can write

$$\Delta(PV) = P\,\Delta V = P(V_2 - V_1) = PV_2 - PV_1.$$

For one mole of gas, $PV = RT$ and

$$PV_2 - PV_1 = RT_2 - RT_1 = R\,\Delta T.$$

Thus, the "extra" heat capacity due to the expansion of the gas is $\Delta(PV)/\Delta T = R$, and so

$$C_P = C_V + R$$
$$= \tfrac{3}{2}R + R = \tfrac{5}{2}R,$$

$$C_P/C_V = \tfrac{5}{3} = 1.67.$$

The heat-capacity ratio C_P/C_V can be measured experimentally, and Table 2.4 shows that the values found for the monatomic gases agree well with the predictions of the kinetic theory. It is also clear, however, that the heat-capacity ratios for the diatomic gases are consistently less than 1.67, and we must now explore the reasons for these deviations.

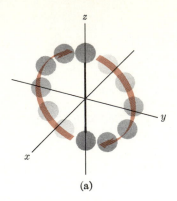
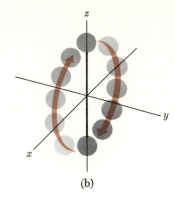

(a)   (b)

FIG. 2.15   Rotational motion of a diatomic molecule. (a) Rotation about the x-axis. (b) Rotation about the y-axis.

First we note that C_V, the heat capacity that is due to the translational motion of the molecules, is equal to $\frac{3}{2}R$, and that there are three independent velocity components associated with the translational motion. Therefore we can infer that each of the three independent translational motions contributes $\frac{1}{2}R$ to the heat capacity. On this basis we might expect that if any other type of motion is available to gas molecules, there will be additional contributions to the heat capacity, coming in units of $\frac{1}{2}R$.

Figure 2.15 shows that in addition to the three translational motions, a diatomic molecule can rotate about its center of mass in two mutually perpendicular and independent ways. Assigning $\frac{1}{2}R$ as the heat-capacity contribution of each of these motions we get

$$C_V = \tfrac{3}{2}R + \tfrac{1}{2}R + \tfrac{1}{2}R = \tfrac{5}{2}R,$$
$$C_P = C_V + R = \tfrac{7}{2}R,$$
$$C_P/C_V = \tfrac{7}{5} = 1.40.$$

Thus this intuitive argument accounts in large measure for the observed heat-capacity ratios of the diatomic gases.

Were we to stop the analysis here, we would be guilty of overlooking the fact that the atoms of a diatomic molecule are not rigidly held at a fixed distance from each other, but vibrate about a well-defined average separation distance. This vibrational motion is independent of the rotations and translations, and apparently should contribute to the total heat capacity of the molecule. That it does not make an appreciable contribution for most diatomic molecules is a fact that can be explained only by analyzing the vibrational motion using quantum mechanics instead of the Newtonian laws of motion. This analysis is beyond our scope, but its result is the prediction that vibrational motion can contribute any amount to the heat capacity between 0 and R, and the latter value is approached only at high temperatures for most molecules.

The ideal gas equation of state, $PV = nRT$, while of pleasing simplicity, is restricted in its application. It is an accurate representation of the behavior of gases only when they are at pressures not much greater than one atmosphere, and at temperatures well above their condensation points. In other words, the ideal gas equation is an approximation to more accurate equations of state which must be used when gases are at high pressures and low temperatures. These more accurate equations are naturally more complicated mathematically, and thus more difficult to use. Nevertheless, their study is actively pursued, for the form of these more accurate equations of state can tell us much about the forces that molecules exert on each other.

The quantity $z = PV/nRT$ is called the compressibility factor of a gas. If a gas were ideal, z would be equal to unity under all conditions. Experimental data, some of which appear in Fig. 2.16, show clearly that z may deviate considerably from its ideal value, which is approached only in the range of low pressures. Moreover, deviations from ideal behavior may cause z to be greater or less than unity, depending on the temperature and the pressure.

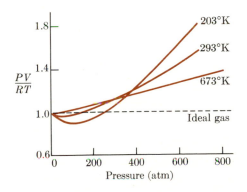

Compressibility factor for nitrogen as a function of pressure.

FIG. 2.16

An empirical equation of state generated intuitively by van der Waals in 1873 reproduces the observed behavior with moderate accuracy. For n moles of gas, the van der Waals equation is

$$\left(P + \frac{n^2 a}{V^2}\right)(V - nb) = nRT,$$

where a and b are positive constants, characteristic of a particular gas. While it is only one of several expressions that are used to represent the behavior of gases over wide ranges of pressure and temperature, it is perhaps the simplest to use and interpret. At low gas densities V tends to become much larger than nb, and n^2a/V^2 tends toward zero. Under such conditions it is a good approximation to write

$$P + \frac{n^2 a}{V^2} \cong P,$$

$$V - nb \cong V,$$

and so the van der Waals equation for one mole of gas reduces to $PV = RT$ at low pressures.

To analyze the van der Waals equation further, let us for simplicity specify $n = 1$ mole, and then rearrange it to the form

$$z = \frac{PV}{RT} = \frac{V}{V - b} - \frac{a}{RT} \frac{1}{V}.$$

Now we can see that as the volume diminishes, the terms on the right-hand side of the equation become large. If the temperature is high, however, the second term will tend to be small, and we will have

$$z = \frac{PV}{RT} \cong \frac{V}{V - b} > 1.$$

This reproduces the "positive" deviations from ideality observed at high temperature and pressure. On the other hand, at room temperatures and moderate densities the approximation

$$\frac{V}{V - b} \cong 1$$

holds, and the term proportional to a becomes important; so we have

$$z = \frac{PV}{RT} \cong 1 - \frac{a}{RT} \frac{1}{V}.$$

Thus the compressibility factor is less than unity, as is observed for many gases at moderate densities and low temperatures.

Intermolecular Forces

Now we must find an explanation of the origin and significance of the van der Waals constants a and b. The constant b has the units of volume per mole, and according to Table 2.5 has a value of about 30 cc/mole for many gases. To a rough approximation, 30 cc is the volume that one mole of gas occupies when it is condensed to a liquid. This in turn suggests that b is somehow related to the volume of the molecules themselves. Comparison of the simplified van der Waals equation $P(V - b) = RT$ with $PV = RT$ further supports this view. In deriving the ideal gas equation of state we assumed that the molecules were mass points which had available to them the whole geomet-

Table 2.5 Van der Waals constants

Gas	a(liter^2-atm/mole^2)	b(liter/mole)
H_2	0.2444	0.02661
He	0.03412	0.02370
N_2	1.390	0.03913
O_2	1.360	0.03183
CO	1.485	0.03985
NO	1.340	0.02789
CO_2	3.592	0.04267
H_2O	5.464	0.03049

rical volume of the container. If the molecules are not points, but are of finite size, each must exclude a certain volume of the container from all the others. If we call this "excluded"" volume b, then we might say that the "true" volume available for molecular motion is $V - b$, and that consequently the equation $PV = RT$ should be written as $P(V - b) = RT$. Thus the effect of finite molecular size is to make the observed pressure greater for a given volume than is predicted by the ideal gas law.

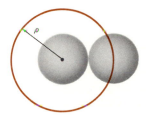

Excluded volume due to finite molecular size. **FIG. 2.17**

Let us assume that molecules are impenetrable spheres of diameter ρ, and ask how this diameter is related to the van der Waals b-factor. Figure 2.17 shows that the presence of one molecule excludes a volume of $\frac{4}{3}\pi\rho^3$ from the center of any other molecule. For a collection of molecules, we can regard half of them as excluding a certain volume from the other half, so the total excluded volume is

$$\frac{N}{2}\left(\tfrac{4}{3}\pi\rho^3\right) = \tfrac{2}{3}\pi\rho^3 N = nb,$$

where for a 1-mole sample, N equals Avogadro's number. Thus by determining the van der Waals b-factor experimentally we can obtain an estimate of the size of a molecule.

To interpret the factor a/V^2 in the van der Waals equation of state we note once again that the pressure of a gas arises from a transport of momentum to the walls of the container. If there are attractive forces between molecules, this momentum transport is somewhat impeded by the interaction of molecules nearing the walls with the molecules "behind" them in the bulk of the gas. In effect, attractive forces cause molecules nearing the walls to transfer some of their momentum to other gas molecules rather than to the walls. We can expect the magnitude of this "negative pressure" effect to be jointly proportional to the densities of each of interacting *pairs* of molecules, or to $(N/V)^2$. For one mole of gas this can be written as a/V^2, where a is a proportionality constant greater than zero that measures the strength of the attractive intermolecular forces.

Because of attractive intermolecular forces alone, the *actual* pressure of an imperfect gas is lower than that predicted by the ideal gas law. Therefore we should add the term a/V^2 to the actual pressure P to obtain $[P + (a/V^2)]$, a quantity which when multiplied by volume gives the ideal pressure-volume product $(PV)_{\text{ideal}} = RT$. This argument rationalizes the way in which the term a/V^2 appears in the van der Waals equation.

We should also note that if there are attractive forces between molecules, two molecules can become bound to each other to form an associated molecular pair, or *dimer*. The "bond" between such molecules is very weak, so that under ordinary conditions only a small fraction of the gaseous molecules are present as dimers. For each dimer formed, the net number of free particles decreases by one. According to the kinetic theory, the gas pressure is proportional to the number of free particles, regardless of their mass. Thus, if an appreciable number of molecules are dimerized, the actual number of free particles will be smaller than the stoichiometric number of molecules, and the observed pressure will be less than the "ideal" value of RT/V. This is the same conclusion that we reached previously by using a different argument.

A slight rearrangement of the van der Waals equation will make clear how the types of deviation from ideal behavior depend on temperature. For one mole we write

$$\frac{PV}{RT} = \frac{V}{V-b} - \frac{a}{VRT}.$$

Letting $V/(V-b) \cong 1 + b/V$, we get

$$\frac{PV}{RT} = 1 + \left(b - \frac{a}{RT}\right)\frac{1}{V}. \tag{2.18}$$

This shows clearly that to a first approximation deviations from ideal behavior are proportional to $1/V$, and that the magnitude and sign of the deviations depend on the size of the molecules, the strength of the attractive forces between them, and the temperature. At high temperatures the quantity PV/RT will tend to be greater than unity, while the opposite will be true at low temperature.

Question. At the so-called Boyle temperature, the effects of the repulsive and attractive intermolecular forces just offset each other, and a nonideal gas behaves ideally. From Eq. (2.18), express the Boyle temperature in terms of the van der Waals constants a and b. What are the Boyle temperatures of He and N_2?

Equation (2.18) gives an adequate representation of gas behavior only in a rather limited range of densities. A simple extension of this equation that can be fitted to experimental data over a much wider range of densities is the virial equation of state:

$$\frac{PV}{RT} = 1 + \frac{B(T)}{V} + \frac{C(T)}{V^2} + \frac{D(T)}{V^3} + \cdots$$

The quantities $B(T)$, $C(T)$, etc., are called the second, third, etc., virial coefficients, and depend only on temperature and the properties of the gas molecules. The second virial coefficient $B(T)$ represents the contributions that interactions between pairs of molecules make to the equation of state, while the third virial coefficient $C(T)$ measures the effects due to the simultaneous interaction of three molecules. In the simple van der Waals model where molecules are pictured as rigid spheres that attract each other weakly, the second virial coefficient is $(b - a/RT)$.

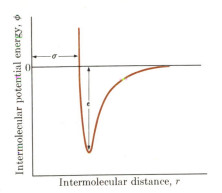

Graphical representation of the Lennard-Jones intermolecular potential energy function.   **FIG. 2.18**

The van der Waals model for molecular interactions is admittedly very crude, for we cannot seriously expect that molecules are impenetrable spheres of well-defined diameter. Fortunately, the experimental determinations of the virial coefficients have led to a more detailed and satisfying picture of intermolecular forces. All molecules attract each other when they are separated by distances of the order of a few angstrom units, and the strength of these attractive forces decreases as the intermolecular distance increases. When molecules are brought very close together they repel each other, and the magnitude of this repulsive force increases very rapidly as the intermolecular separation decreases. These phenomena are often represented by plotting the mutual potential energy of a pair of molecules as a function of the distance between their centers of mass.

Figure 2.18 shows the general form of the potential energy used to describe the interaction between two uncharged spherical molecules. The force between the two molecules at any separation is equal to the negative slope of the potential energy curve at that point. We see that if the zero of potential energy is taken as that of two infinitely separated molecules, the potential energy becomes negative as the molecules are brought together. After reaching a minimum value the potential energy rises abruptly as the molecules are brought still closer, and the force between them becomes repulsive.

An algebraic representation of the intermolecular potential energy curve is

$$\phi = 4\epsilon\left[\left(\frac{\sigma}{r}\right)^{12} - \left(\frac{\sigma}{r}\right)^{6}\right],$$

and is called the Lennard-Jones potential function. In this expression, r is the separation of molecular centers, and the parameter ϵ is equal to the minimum value of the potential energy, or to the "depth" of the potential energy "well," as Fig. 2.18 shows. The distance parameter σ is equal to the minimum distance of approach of two molecules colliding with zero initial kinetic energy. In a sense it is a measure of the diameter of the molecules. Actually the true diameter of a molecule is an ill-defined quantity, because two molecules can approach each other to the distance at which their initial kinetic energy of relative motion is converted entirely to potential energy. If their initial kinetic energy is large, then their distance of closest approach can be somewhat smaller than σ.

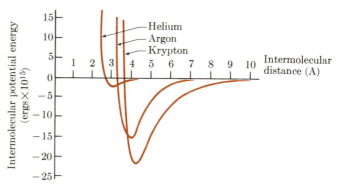

FIG. 2.19   Lennard-Jones intermolecular potential energy function for He, Ar, Kr.

The values of the parameters ϵ and σ depend on the nature of the interacting molecules. In general, both parameters increase as the atomic number of the interacting atoms increases. Figure 2.19 shows the potential energy curves for three of the inert gases. Note that ϵ is of the same order of magnitude or somewhat smaller than kT at room temperature. This means that the average kinetic energy of gas molecules is larger than the largest possible value of the attractive potential energy of a molecular pair. Because molecules are generally far apart at ordinary pressures the *average* potential energy of interaction is much less than the average kinetic energy, and consequently it is the latter that is largely responsible for the observed behavior of gases.

Determination of the equation of state of an imperfect gas can lead to a description of the potential energy of interaction between molecules. In this section we shall find that the viscosity of gases, their ability to conduct heat, and the rate at which they diffuse into each other all can be related to the frequency of molecular collisions, and thus to the forces between molecules. Therefore study of these phenomena can provide additional information about the nature of intermolecular potential energy.

The thermal conductivity of a gas is the rate at which it transports energy from a surface at a higher temperature to one at a lower temperature. The total rate of energy transport depends on the area of the surfaces, the temperature difference ΔT and the distance Δd between them, so the *specific thermal conductivity κ* is defined by

$$\text{rate of energy transport per unit area} = -\kappa \frac{\Delta T}{\Delta d} \text{ (ergs/cm}^2\text{-sec).} \qquad (2.19)$$

Defined this way, κ depends on the properties of the gas alone.

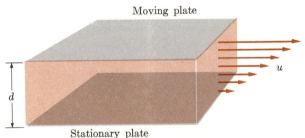

Moving plate

u

d

Stationary plate

Gas between a moving and a stationary plate.   **FIG. 2.20**

The viscosity of a fluid represents an internal friction that causes the effects of a motion through the fluid to be transmitted in a direction *perpendicular* to that of the motion. Consider Fig. 2.20, which shows a gas confined between a stationary and a moving plate. The motion of the upper plate causes the adjacent layer of gas to move *as a whole* with a velocity u. Layers of gas successively farther from the moving plate also move, but with steadily diminishing velocity. Because this motion is transmitted through the gas, the stationary plate feels a force in the direction of the motion of the upper plate. Experiments show that this force per unit plate area is given by

$$\text{force per unit area} = -\eta \frac{\Delta u}{\Delta d} \text{ (dynes/cm}^2\text{),} \qquad (2.20)$$

where $\Delta u / \Delta d$ is the amount by which the mass velocity u changes with distance

Δd from the moving plate, and η is a proportionality constant called the *coefficient of viscosity*. Its value depends only on the nature of the gas.

There is another way of looking at the viscosity phenomenon which makes the physical situation clearer. The reason why the stationary plate feels a force is that the gas molecules at the upper moving plate acquire an extra momentum mu in the direction of the plate motion. If these molecules could proceed unhindered to the stationary plate, they could transfer their extra momentum to it, and thereby exert a force on it. They are, to a certain extent, prevented from doing this by collisions which tend to randomize the direction and amount of their extra momentum. The viscosity coefficient is a measure of how efficient momentum transport is.

We have noted earlier that when a partition separating two different gases at the same pressure is removed, the gases move into each other and mix by diffusion. The net rate at which one gas moves across an imaginary surface into the other gas depends on the area of the surface, the concentration change per unit length, and the nature of the gases. Thus

$$\text{rate of diffusion per unit area} = -D \frac{\Delta n}{\Delta d} \text{ (molecules/sec-cm}^2\text{)}, \qquad (2.21)$$

where $\Delta n/\Delta d$ is the concentration change per unit length, and D is a proportionality constant, called the *diffusion coefficient*, which depends on the nature of the gases.

Just as thermal conductivity is the transport of energy, and viscosity is caused by the transport of momentum, diffusion is the transport of matter. We should be able to use the kinetic theory of gases to express the transport coefficients κ, η, and D in terms of more fundamental quantities of molecular motion. It is clear that collisions between molecules must be important in the determination of the values of the transport coefficients. We know, for example, that even though the average speed of a molecule is roughly the speed of sound, it takes a long time for a malodorous substance to diffuse across a room. That rapid transport of material does not occur must be a consequence of intermolecular collisions.

The Mean Free Path

Let us see how collisions affect the motion of molecules. In Fig. 2.21 a particular gas molecule has been singled out and its trajectory plotted. Each segment of its trajectory between collisions is called a free path. It is clear that since these free paths are of finite length, the progress of the molecule in any one direction is inhibited. We are interested in computing the average value of the length of these free paths, or the *mean free path*.

To accomplish this we consult Fig. 2.22, where the motion of one "hard-sphere" molecule relative to its fellows is represented. The molecule of interest

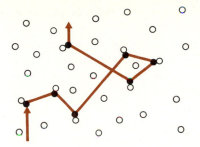

moves through the gas and collides with any other molecule whose center-to-center distance is less than ρ, the molecular diameter. Thus the interesting molecule sweeps out per unit time a collision cylinder whose cross-sectional area is $\pi\rho^2$, and whose average altitude is $\bar{c}$, the average speed of the molecule. A collision will occur with any molecule whose center lies in this cylinder.

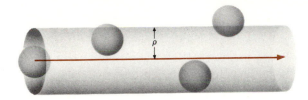

Collision cylinder swept out by　　FIG. 2.22
a molecule. Molecules whose
centers are within the cylinder
would undergo collision.

Thus, if n is the average number of molecules per cc, the average number of collisions per second experienced by the molecule of interest is

collisions/sec = volume of collision cylinder × molecules/unit volume = $\pi\rho^2\bar{c}n$.

The mean free path λ is the average distance traveled by the molecule per unit time divided by the average number of collisions per unit time:

$$\lambda = \frac{\bar{c}}{\pi\rho^2\bar{c}n} = \frac{1}{\pi\rho^2 n}. \tag{2.22}$$

For a numerical estimate of ρ, the molecular diameter, we can use the Lennard-Jones potential parameter σ, which experiments show is roughly 3×10^{-8} cm for many molecules. We find that at 1-atm pressure ($n = 3 \times 10^{19}$ molecules/cc) the mean free path is

$$\lambda = \frac{1}{\pi(3 \times 10^{-8})^2 \times 3 \times 10^{19}} \cong 10^{-5} \text{ cm}.$$

Here we find the reason for the low rate of diffusion and the poor thermal conductivity of gases. Any one molecule in a gas at atmospheric pressure can travel only 10^{-5} cm before its trajectory is interrupted, perhaps reversed.

Now let us see how the diffusion coefficient is related to the mean free path. As defined by Eq. (2.21) the diffusion coefficient has the units cm^2/sec, since the units of $\Delta n/\Delta d$ are molecules/cm^4. Now if we imagine that D must increase with the mean free path and the molecular speed, then the units of D tell us that

$$D\left(\frac{cm^2}{sec}\right) \propto \lambda\bar{c}\left(\frac{cm^2}{sec}\right).$$

That is, the diffusion constant is proportional to the average molecular speed $\bar{c}$ and the mean free path λ.

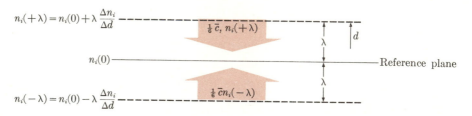

$$n_i(+\lambda) = n_i(0) + \lambda\frac{\Delta n_i}{\Delta d}$$

$$\tfrac{1}{6}\bar{c},\ n_i(+\lambda)$$

$$n_i(0)$$

Reference plane

$$\tfrac{1}{6}\bar{c}n_i(-\lambda)$$

$$n_i(-\lambda) = n_i(0) - \lambda\frac{\Delta n_i}{\Delta d}$$

FIG. 2.23 The transport of molecules across a reference plane. The quantities $n_i(+\lambda)$ and $n_i(-\lambda)$ are the concentrations of molecules at a distance of λ above and below the reference plane, respectively.

It is not difficult to determine the proportionality constant between D and $\lambda\bar{c}$. We imagine a physical situation in which a gas has uniform pressure, but contains a radioactive isotope whose concentration n_i is a function of the distance d, as shown in Fig. 2.23. The rate at which this isotope diffuses is the difference between the rates at which its molecules pass through the median reference plane, from above and from below. If there are n_i isotopic molecules per unit volume, one-third have velocities along the d-coordinate, and half of these have velocities in the $+d$-direction, the other half in the $-d$-direction. The rate at which molecules cross a unit area in one direction is just $n/6$ times their speed $\bar{c}$, or $n\bar{c}/6$. However, molecules which cross the reference plane from above have, on the average, experienced their last collision at a distance λ, or one mean free path, above the plane. At this point the concentration of isotopic molecules is

$$n_i(0) + \lambda\frac{\Delta n_i}{\Delta d},$$

where $n_i(0)$ is the concentration of molecules in the reference plane. The number of molecules/cm^2-sec crossing the reference plane from above is then

$$\frac{1}{6}\bar{c}\left[n_i(0) + \lambda\frac{\Delta n_i}{\Delta d}\right].$$

A similar argument applies to molecules crossing the reference plane from below, except that at a distance λ below the reference plane, the concentration of molecules is

$$n_i(0) - \lambda \frac{\Delta n_i}{\Delta d}.$$

Therefore the upward flux is

$$-\frac{1}{6} \bar{c} \left[n_i(0) + \lambda \frac{\Delta n_i}{\Delta d} \right]$$

We now take the difference between upward and downward fluxes:

$$\text{net flux} = \frac{1}{6} \bar{c} \left[n_i(0) - \lambda \frac{\Delta n_i}{\Delta d} \right] \times \frac{1}{6} \bar{c} \left[n_i(0) + \lambda \frac{\Delta n_i}{\Delta d} \right]$$

$$= -\frac{1}{3} \lambda \bar{c} \frac{\Delta n_i}{\Delta d}.$$

By definition,

$$\text{net flux} \equiv -D \frac{\Delta n}{\Delta d},$$

so we get

$$D = \tfrac{1}{3} \lambda \bar{c}.$$

Now, in addition to the functional relation between D, λ, and $\bar{c}$, we have the proportionality constant. Note one very important point: the mean free path λ is determined by the *total* concentration of all molecules ($\lambda = 1/n_t \pi \rho^2$), not just by the concentration of the diffusing isotope.

Question. Would you expect gases to mix by diffusion more rapidly at lower than at higher pressure? Would the mixing rate increase or decrease with increasing temperature?

The viscosity effect depends on the ability of the gas to transport momentum. Thus it is to be expected that the expression for the viscosity coefficient η should involve the mass of the molecules. Dimensional analysis of Eq. (2.20),

$$\text{force/area} = -\eta \frac{\Delta u}{\Delta d}, \tag{2.20}$$

$$\text{gm-cm/sec}^2\text{-cm}^2 = \eta \text{ cm/sec-cm},$$

shows that η has the dimension of gm/cm-sec. This can be achieved by writing

$$\eta \propto nm\bar{c}\lambda, \tag{2.23}$$

$$\text{gm/sec-cm} \propto (\text{molecules/cm}^3)(\text{gm/molecule})(\text{cm/sec})(\text{cm}).$$

This result, obtained by dimensional analysis alone, shows how the viscosity coefficient of a gas depends on its mass, concentration, mean free path, and

Table 2.6 Lennard-Jones potential constants*

Gas	Determined from viscosity		Determined from virial coefficients	
	$\epsilon/k(°K)$	$\sigma(A)$	$\epsilon/k(°K)$	$\sigma(A)$
He	10.2	2.58	6.03	2.63
Ar	124	3.42	120	3.41
Xe	229	4.06	221	4.10
H_2	33.3	2.97	29.2	2.87
N_2	91.5	3.68	95.1	3.70
O_2	113	3.43	117	3.58

*The parameter ϵ is divided by Boltzmann's constant k to facilitate the comparison of the magnitude ϵ, the attractive interaction between molecules, with their kinetic energy of motion kT.

average speed. Note that the product of concentration and mean free path reduces to $1/\pi\rho^2$, where ρ is the molecular diameter.

We can obtain the numerical value of the proportionality constant in Eq. (2.23) by the same technique that was used to find the diffusion coefficient. The quantity being transported is the momentum mu. The average values of this momentum at distances of one mean free path above and below an arbitrary reference plane are respectively

$$m\left(u + \lambda \frac{\Delta u}{\Delta d}\right) \quad \text{and} \quad m\left(u - \lambda \frac{\Delta u}{\Delta d}\right),$$

where u is the average velocity in the reference plane. In each of these planes, $n/6$ of the molecules (per cc) are moving in one direction along d, with average speed $\bar{c}$. Therefore, to get the net momentum transport through the reference plane, we multiply each of the above quantities by $n\bar{c}/6$, and take the difference between them, thereby obtaining

$$\text{momentum transport rate per unit area} = -\frac{1}{3}nm\bar{c}\lambda\frac{\Delta u}{\Delta d}.$$

By definition,

$$\text{momentum transport rate per unit area} = -\eta\frac{\Delta u}{\Delta d},$$

so we have

$$\eta = \tfrac{1}{3}nm\bar{c}\lambda,$$

which gives us a direct numerical relation between the gas viscosity coefficient and the molecular parameters.

Question. If η for H_2 gas is 0.93×10^{-4} g/cm-sec at 298°K, what is the approximate diameter of the hydrogen molecule?

By recognizing that thermal conductivity is energy transport by molecular motion, we can determine the way in which the thermal conductivity coefficient κ depends on molecular properties. We select an arbitrary reference plane in a gas which has been subjected to a temperature difference. The temperature in this reference plane is T, and the corresponding average energy of the molecules is $c_v T$, where c_v is the heat capacity *per molecule*. At distances of one mean free path above and below the reference plane, the average energies of a molecule are

$$c_v \left(T + \lambda \frac{\Delta T}{\Delta d} \right) \quad \text{and} \quad c_v \left(T - \lambda \frac{\Delta T}{\Delta d} \right).$$

Above the reference plane $n/6$ molecules/cc are proceeding downward with an average speed $\bar{c}$, and below the reference plane, the same number of molecules is moving upward with the same speed. The net rate of energy transport is then the difference between the average energy at these two planes, multiplied by $n\bar{c}/6$. We get

$$\text{rate of energy transport per unit area} = -\frac{1}{3} n\bar{c}c_v\lambda \frac{\Delta T}{\Delta d}$$

$$\equiv -\kappa \frac{\Delta T}{\Delta d}.$$

Therefore

$$\kappa = \tfrac{1}{3}n\bar{c}c_v\lambda,$$

which is the relation between the thermal conductivity coefficient and the molecular properties.

Question. Which should have the higher thermal conductivity, He^4 or He^3? Given that D_2 and He have nearly the same collision diameter, which should have the higher thermal conductivity?

Derivation of the Virial Equation of State

We can use the virial theorem to indicate how the virial equation of state for nonideal gases is derived. All that is necessary is to include in the virial the contributions from the forces between the molecules. From the virial theorem we get

$$\langle KE \rangle = \text{virial from walls} + \text{virial from intermolecular forces}$$
$$= \tfrac{3}{2}PV - \tfrac{1}{2}\left\langle \sum rF(r) \right\rangle_{\text{molecules}}. \tag{2.24}$$

Here we have assumed that the force $F(r)$ that molecules exert on each other depends only on their separation r, and has its only component along r. These intermolecular forces can come from pairs of molecules interacting, and triplets of molecules, and quartets, and so on. For simplicity we will consider only the most important interaction which is the one between pairs. Then the quantity $rF(r)$ must be summed over all molecular pairs.

We proceed in the following manner. If N is the total number of molecules, the total number of pairs of molecules is $\frac{1}{2}N(N-1) \cong \frac{1}{2}N^2$. The factor $\frac{1}{2}$ compensates for the fact that $N(N-1)$ counts each molecular pair twice. To calculate the virial we need the number of pairs which are separated by a distance r, or more properly, lie in a spherical shell of radius r and thickness Δr. This number is proportional to the volume of this shell $4\pi r^2 \, \Delta r$, divided by the total volume V of the vessel. We also expect that the number of pairs separated by r will depend on the potential energy $\phi(r)$ of molecular interaction at r through a Boltzmann factor $\exp[-\phi/kT]$. Thus we write

$$\frac{1}{2}\left\langle \sum_{\text{pairs}} rF(r) \right\rangle = \frac{1}{2} \sum_r \frac{N^2}{2} \frac{4\pi r^2 \, \Delta r}{V} rF e^{-\phi/kT}, \tag{2.25}$$

where now we can calculate the intermolecular virial by summing over all values of r. This can be done by either calculus or numerical calculation, but is, in general, a difficult process which we will not attempt. Instead, we will merely introduce the symbol $\frac{3}{2}\beta$ to stand for this calculation, and get

$$\frac{1}{2}\left\langle \sum rF(r) \right\rangle_{\text{pairs}} = \frac{3}{2} \frac{N^2}{V} \beta \equiv \frac{3}{2} \frac{n^2}{V} B.$$

In the last step we have defined n^2B as $N^2\beta$, where n is the number of moles. We now substitute our result in Eq. (2.24):

$$\langle \text{KE} \rangle = \frac{3}{2} PV^* - \frac{3}{2} \frac{n^2}{V} B.$$

Rearranging and evaluating $\langle \text{KE} \rangle$, we get

$$PV = \frac{2}{3} \langle \text{KE} \rangle + \frac{n^2}{V} B,$$

$$PV = nRT + \frac{n^2}{V} B,$$

$$\frac{PV}{nRT} = 1 + \frac{n}{V} \frac{B}{RT} = 1 + \frac{n}{V} B'. \tag{2.26}$$

The right-hand side of Eq. (2.26) contains the first two terms of the virial equation of state. We see clearly now that the first correction term nB/V to the ideal gas equation does indeed arise from interaction between *pairs* of molecules. In addition, we have an explicit expression by which it is possible to calculate the second virial coefficient B from the intermolecular potential. We can also see that if we were to include the contributions of triplets of molecules, a term proportional to $N(N-1)(N-2) \cong N^3$ and to $1/V^2$ would appear in the virial, and correspondingly, a term n^2C/V^2 would be added to Eq. (2.26).

Questions. Evaluate the second virial coefficient for a gas of hard spheres from Eq. (2.25). In this particular case, $\phi = 0$ for all accessible r, and $F = 0$ for all r axcept $r = \rho$, the molecular diameter, so only one value of r contributes to the sum in Eq. (2.25). The quantity $\Delta r F$ may be set equal to kT, the average energy with which molecules collide. Compare your result with the expression for the van der Waals constant b.

2.7 CONCLUSION

In this chapter we have examined the relations that describe the behavior of gases, and again have come upon an illustration of the origin and significance of scientific laws. The laws of Boyle and of Charles and Gay-Lussac, like the laws of chemical combination, are derived from experimental measurements, and are accurate only in certain restricted experimental situations. To deal accurately with a greater variety of situations, the ideal gas laws are amended or modified, again on the basis of experimental measurements, and the van der Waals, or virial equations of state replace $PV = nRT$. As a practical matter, however, it is often acceptable to use the ideal gas equation of state for approximate calculations, providing the deviations from ideal gas behavior are not severe.

The kinetic theory of gases is an example of the successful explanation of the behavior of a macroscopic system in terms of the properties of its microscopic constituents. That is, we can claim to "understand" why gases obey Boyle's law, because we can show that this law is a mathematical consequence of picturing gaseous molecules as small bodies in incessant random motion. We can say, as a result of the insight provided by the kinetic theory of gases, that gas pressure increases as volume decreases, because molecules collide with the walls of a smaller container more frequently. Similarly, the reason gas pressure increases with increasing temperature is that raising the temperature increases the average speed of the molecules, and thereby increases the rate of wall collisions and the average momentum change upon collision. Moreover, we have seen how a theory relates a variety of observed phenomena to each other. The equation of state and the processes of diffusion, viscosity and thermal conductivity all are shown to be consequences of the properties of molecules by the kinetic theory of gases.

Within the structure of the kinetic theory of gases, we have encountered several representations, or "models" of what molecules are like. As noted in Section 2.2, a model is a construct of the human mind, and incorporates only the most important features of the real entity it represents. To derive Boyle's law, we used a model that pictures molecules as point masses. While clearly unrealistic, this model is useful because it possesses *only* the properties required to derive Boyle's law. Indeed, it is the success of this simple model that *shows* us that mass and motion are the molecular properties responsible for Boyle's law.

To account for other macroscopic behavior, it is necessary to use a more refined model for molecules. Picturing molecules as spheres of finite volume that attract each other leads to explanations of deviations from the ideal gas law, and to an understanding of viscosity, thermal conductivity, and diffusion. A further refinement is to introduce an algebraic form such as the Lennard-Jones expression to represent the potential energy of interaction between molecules. This most detailed model allows us to account for the subtle features of equations of state and transport properties.

Which model is best? This question does not have a single answer, for the best model is often the one which allows us to discuss most simply the properties in which we are interested. To find the equation of state for a low density gas we need only the point mass model: the Lennard-Jones model is needlessly complicated for this problem. Only to discuss the finer details of equations of state and transport processes do we need the most refined picture of molecules. Throughout our study of chemistry we shall use a variety of models for molecules, some of them of extreme simplicity. We must be careful to recognize the simplifying features, the justification for introducing them, and the limitations they impose.

SUGGESTIONS FOR FURTHER READING

Cowling, T. G., *Molecules in Motion*. London: Hutchinson's University Press, 1950.

Hildebrand, J. H., *An Introduction to Molecular Kinetic Theory*. New York: Reinhold, 1963.

Hill, T. L., *Lectures on Matter and Equilibrium*. Menlo Park, Calif.: W. A. Benjamin, 1966.

Kauzmann, W., *Kinetic Theory of Gases*. Menlo Park, Calif.: W. A. Benjamin, 1966.

Present, R. D., *Kinetic Theory of Gases*. New York: McGraw-Hill, 1958.

PROBLEMS

2.1 An ideal gas at 650 mm pressure occupied a bulb of unknown volume. A certain amount of the gas was withdrawn and found to occupy 1.52 cc at 1-atm pressure. The pressure of the gas remaining in the bulb was 600 mm. Assuming that all measurements were made at the same temperature, calculate the volume of the bulb.

2.2 A sample of nitrogen gas is bubbled through liquid water at 25°C and then collected in a volume of 750 cc. The total pressure of the gas, which is saturated with water vapor, is found to be 740 mm at 25°C. The vapor pressure of water at this temperature is 24 mm. How many moles of nitrogen are in the sample?

2.3 When 2.96 gm of mercuric chloride is vaporized in a 1.00-liter bulb at 680°K, the pressure is 458 mm. What is the molecular weight and molecular formula of mercuric chloride vapor?

2.4 Gaseous ethylene, C_2H_4, reacts with hydrogen gas in the presence of a platinum catalyst to form ethane, C_2H_6, according to

$$C_2H_4(g) + H_2(g) = C_2H_6(g).$$

A mixture of C_2H_4 and H_2 known only to contain more H_2 than C_2H_4 had a pressure of 52 mm in an unknown volume. After the gas had been passed over a platinum catalyst, its pressure was 34 mm in the same volume and at the same temperature. What fraction of the molecules in the original mixture was ethylene?

2.5 A gaseous compound which contains only carbon, hydrogen, and sulfur is burned with oxygen under such conditions that the individual volumes of the reactants and products can be measured at the same temperature and pressure. It is found that 3 volumes of the compound react with oxygen to yield 3 volumes of CO_2, 3 volumes of SO_2, and 6 volumes of water vapor. What volume of oxygen is required for the combustion? What is the formula of the compound? Is this an empirical formula or a molecular formula?

2.6 A gaseous compound known to contain only carbon, hydrogen, and nitrogen is mixed with exactly the volume of oxygen required for the complete combustion to CO_2, H_2O, and N_2. Burning 9 volumes of the gaseous *mixture* produces 4 volumes of CO_2, 6 volumes of water vapor, and 2 volumes of N_2, all at the same temperature and pressure. How many volumes of oxygen are required for the combustion? What is the molecular formula of the compound?

2.7 A mixture of methane, CH_4, and acetylene, C_2H_2, occupied a certain volume at a total pressure of 63 mm. The sample was burned to CO_2 and H_2O, and the CO_2 alone was collected and its pressure found to be 96 mm in the same volume and at the same temperature as the original mixture. What fraction of the gas was methane?

2.8 Scandium (Sc) metal reacts with excess aqueous hydrochloric acid to produce hydrogen gas. It is found that 2.41 liters of hydrogen, measured at 100°C and 722-mm pressure are liberated by 2.25 gm of scandium. Calculate the number of moles of H_2 liberated, the number of moles of scandium consumed, and write a balanced net equation for the reaction that occurred.

2.9 A mixture of hydrogen and helium is prepared such that the number of wall collisions per unit time by molecules of each gas is the same. Which gas has the higher concentration?

2.10 A good vacuum produced in common laboratory apparatus corresponds to 10^{-6}-mm pressure at 25°C. Calculate the number of molecules per cc at this pressure and temperature.

2.11 By assuming that the molecular diameter is given by the σ of the Lennard-Jones potential, and that $\bar{c}$ is equal to the root-mean-square speed, calculate the number of collisions experienced per second by a nitrogen molecule in a gas at 25°C and at pressures of 1 atm, 0.76 mm, and 7.6×10^{-6} mm. Repeat the calculation for He at 1 atm.

2.12 A balloon made of rubber permeable to hydrogen in all its isotopic forms is filled with pure deuterium gas (D_2 or H_2^2) and then placed in a box that contains pure H_2. Will the balloon expand or contract?

2.13 Calculate the root-mean-square speed, in cm/sec and at 25°C, of a free electron and of a molecule of UF_6.

2.14 By using the σ of the Lennard-Jones potential as an estimate of molecular diameter, calculate the mean free path of a nitrogen molecule at 25°C and at the following pressures: 1 atm, 1 mm, 10^{-6} mm.

2.15 In the derivation of Boyle's law by the kinetic theory, we assumed that the molecules collide only with the walls of the vessel, and not with each other. How must the mean free path and the distance between the walls compare in order for this assumption to be valid? At what gas pressure is this relation satisfied for molecules of 3-A diameter and at 25°C in a 10-cm cubical vessel?

2.16 It is a matter of common experience that liquids become more viscous as their temperature is lowered. Assuming the concentration of molecules to be fixed, would you expect this to be true in general for gases? Why?

2.17 Explain why gaseous helium is a better heat conductor than is xenon. If H_2 and D_2 (H_2^2) both have the same molecular diameter, which should be a better conductor of heat?

2.18 One of the first triumphs of the kinetic theory of gases was the correct prediction that the coefficient of viscosity η should be independent of gas pressure at a fixed temperature. On the other hand, the expression for η derived in this chapter contains n, the concentration of molecules, as an explicit factor. Why is it then that η is independent of pressure or concentration?

2.19 The first evidence that the noble gases are monatomic elements involved interpretation of their measured heat capacities. Explain how such data can lead to such a conclusion.

2.20 Equation (2.18) is an approximate equation of state for real gases. Examine this equation and then give a condition which, if satisfied by the temperature, leads to the prediction that $PV/RT = 1$, despite gas imperfection. The temperature which satisfies this condition is called the Boyle point. By using data given in Table 2.5, calculate the Boyle temperature for nitrogen.

2.21 The Maxwell-Boltzmann distribution function can be written

$$\frac{\Delta N}{N} = \frac{2}{\sqrt{\pi}} \left(\frac{mc^2}{2kT}\right) e^{-mc^2/2kT} \left[\frac{m(\Delta c)^2}{2kT}\right]^{1/2}.$$

Plot $mc^2/2kT$ as a function of $mc^2/2kT$, and by consulting a table of the exponential function plot $e^{-mc^2/2kT}$ as a function of $mc^2/2kT$. By examining these two curves, sketch the product of the two functions.

2.22 Real gases follow the equation of state $PV = RT$ only when their pressure is very low. Using the data given for CO_2 and O_2 below, show graphically that for a constant temperature of 0°C, PV is not a constant as predicted by the ideal gas law. This is best done by plotting PV as a function of P on a scale sufficiently expanded to show variations in PV. From your graph determine the value that RT should have for all ideal gases at 0°C. Determine from your graph the constants in the empirical equation of state $PV = A + BP$ for CO_2.

O_2, P(atm)	PV(liter-atm)	CO_2, P(atm)	PV(liter-atm)
1.0000	22.3939	1.00000	22.2643
0.7500	22.3987	0.66667	22.3148
0.5000	22.4045	0.50000	22.3397
0.2500	22.4096	0.33333	22.3654
		0.25000	22.3775
		0.16667	22.3897

What percent error in the volume of one mole of CO_2 at 1-atm pressure would be made by using the ideal value of PV and ignoring gas imperfections?

2.23 By a procedure similar to the one used in Problem 2.22, the value of PV for an ideal gas at 100°C has been found to be 30.6194. If the empirical relation $PV = j + kt$, where t is the temperature in degrees celsius, is assumed to hold, determine the values of j and k for an ideal gas from the information you have. From these values of j and k determine R, and the value of T, the absolute temperature at 0°C.

CHAPTER 3

THE PROPERTIES OF SOLIDS

The outstanding macroscopic properties of gases are compressibility and fluidity. We have seen that the kinetic molecular theory accounts for this macroscopic behavior in terms of a microscopic picture whose central feature is the chaotic motion of independent molecules—molecules which rarely exert appreciable forces on each other. In contrast, the most noticeable macroscopic features of crystalline solids are rigidity, incompressibility, and characteristic geometry. We shall find that the explanation of these macroscopic properties in terms of the atomic theory involves the idea of a lattice: a permanent ordered arrangement of atoms held together by forces of considerable magnitude. Thus the extremes of molecular behavior occur in gases and solids. In the former we have molecular chaos and vanishing intermolecular forces, and in the latter we have an ordered arrangement in which the interatomic forces are large.

3.1 MACROSCOPIC PROPERTIES OF SOLIDS

There are substances such as sodium chloride, sugar, and elementary sulfur that not only possess the properties of rigidity and incompressibility, but also occur naturally in characteristic geometric forms. These are the crystalline solids, and they are to be distinguished from the amorphous materials such as glass, rubber, plexiglass, or any of the other plastics. These amorphous ma-

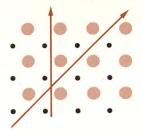

A two-dimensional lattice of spherical atoms. Resistance to shear is different in the indicated directions.

FIG. 3.1

terials possess *some* of the mechanical properties commonly associated with the word "solid," but do not occur in regular characteristic shapes. Even more important, amorphous materials are *isotropic;* their properties, such as mechanical strength, refractive index, and electrical conductivity, are the *same in all directions.* This is a feature which they hold in common with liquids and gases. Crystalline solids are quite different, for they are *anisotropic;* their mechanical and electrical properties do depend, in general, on the direction along which they are measured.

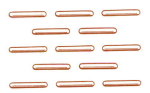

A possible packing scheme for long molecules.

FIG. 3.2

The anisotropy of crystals is important, for it, perhaps more than any other macroscopic property, provides a strong indication of the existence of ordered atomic lattices. Consider Fig. 3.1, which shows a simple two-dimensional lattice consisting of only two kinds of atoms. Mechanical properties such as the resistance to a shearing stress might be quite different in the two directions indicated. Deformation of the lattice along one of the directions involves displacing rows that are made up of alternate types of atoms, while in the other direction each of the displaced rows consists of one type of atom. Thus even though the constituents of the lattice are spherical atoms, the crystal itself may be anisotropic. Contrast this with the situation found in liquids and gases, where the "arrangements" of the particles are random and disordered. Due to the molecular chaos, all directions are equivalent, and all properties of liquids and gases are the same in all directions.

The properties of some crystals reflect the asymmetry of their constituent molecules. Figure 3.2 represents a situation in which long thin molecules are packed parallel to each other in the crystal lattice. It is inevitable that the lattice arrangement is one which will magnify this molecular anisotropy into a macroscopically observable property. A well-known substance in which this effect appears is the mineral asbestos, which has properties reflecting the long fiberlike structure of its individual molecules.

There is another macroscopic property which permits a clear distinction between crystalline and amorphous substances. Crystalline solids have sharp melting points; the mechanical properties of a crystal change only slightly up to a characteristic temperature at which it melts abruptly and becomes a liquid. Amorphous materials do not have sharp melting points. For example, as its temperature is raised, glass gradually and continuously softens and starts to flow. The absence of a sharp melting point suggests that glass and most other amorphous materials are best thought of as *liquids* at all temperatures. Indeed, we shall find that the atomic arrangement in amorphous substances has none of the order found in the crystalline lattices, but rather shows a disorder which is characteristic of liquids. Consequently, the word "solid" is used in the most restrictive scientific sense to refer to crystalline materials only.

Crystal Sizes and Shapes

The crystalline state is easy to recognize in many instances, particularly in the naturally occurring minerals. The well-defined characteristic angles and faces of the natural quartz crystals shown in Fig. 3.3 suggest that they are a consequence of an ordered internal lattice. At other times, solid materials occur as powder, lumps, or agglomerates that in many ways resemble amorphous substances, but when an individual particle is examined under a microscope, the characteristic angles of a crystal may become obvious. Therefore, we must be careful to distinguish between *amorphous* and *polycrystalline* solids. In the latter, individual crystals with their ordered atomic lattices exist, but are so small as to be unrecognizable except under a microscope. Metals often occur in the polycrystalline condition. Figure 3.4 shows the etched surface of a sample of copper. The boundaries of the individual crystal grains are obvious, even if rather irregular. Since the individual crystals are randomly oriented, a metallic sample may appear to be isotropic, even though a single crystal is anisotropic.

FIG. 3.3

Natural quartz crystals.

Etched surface of copper showing microcrystalline structure.

FIG. 3.4

The size of the crystals of a given substance can vary enormously, and is profoundly influenced by the conditions under which the crystal is formed. In general, slow growth from a solution which is only slightly supersaturated favors the formation of large crystals. For this reason the natural crystals of minerals formed by geological processes are often very large. On the other hand, the crystals formed in laboratory precipitation reactions are very small, for they are often formed very rapidly from solutions which are greatly supersaturated. For example, when the slightly soluble salt barium sulfate is precipitated by mixing aqueous solutions of barium chloride and sulfuric acid, the solid barium sulfate particles are so small and ill-formed as to be virtually unrecognizable as crystals, even under microscopic examination. The quality of these crystals can be improved, however, by letting them "age" in the presence of their saturated solution. During the aging process, the smaller imperfect crystals, being relatively unstable, tend to dissolve, and the larger more stable crystals tend to grow. Thus, as a result of a continuous redissolving and reprecipitation, a precipitate of virtually amorphous material can be converted to a polycrystalline substance.

It appears that the *sizes* of crystals reflect the conditions of growth, rather than the internal constitution of the crystal. The shape of a crystal is more characteristic of the material itself, but is still subject to some influence by the growth conditions. For example, sodium chloride crystals that are carefully grown by suspending a seed crystal in a slightly supersaturated solution are invariably cubic in shape, as shown in Fig. 3.5(a). On the other hand, sodium chloride crystals grown at the bottom of a beaker are square plates whose thickness is never greater than half their width, as shown in Fig. 3.5(b). This shape is a consequence of the fact that crystals resting on a surface can use all four sides to grow in the horizontal directions, but can use only the top to grow vertically. Thus the growth rate in either of the two horizontal orientations is twice that in the vertical direction.

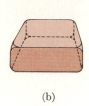

(a)   (b)   (c)

FIG. 3.5   Shapes of sodium chloride crystals grown under differing conditions.

The influence of the environment can be even more subtle. When sodium chloride crystals are grown suspended in a solution which contains urea, they take the form of the regular octahedra shown in Fig. 3.5(c). While the appearance of sodium chloride crystals at one time as cubes and at another as octahedra may seem to suggest that crystal shape is unrelated to the internal arrangement of the atoms, this is not true. In the first place, the cube and the octahedron are related geometrically, for an octahedron can be formed from a cube by cutting off the corners as illustrated in Fig. 3.6. Second, the angle between the octahedral faces of all sodium chloride crystals is always the same, and is never altered by changing external conditions. The invariance of the angles between a given set of crystal faces is a universal property of solids. It appears, then, that subject to external conditions, crystals may assume a variety of shapes, but that the angles between two characteristic faces are always the same, and thus must be determined by fixed geometry of the lattice itself.

FIG. 3.6   Relation between the cubic and octahedral faces of a crystal. The octahedron can be obtained by shearing off the corners of the cube.

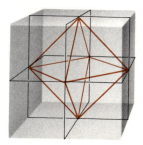

3.2 TYPES OF SOLID

We have emphasized that the existence of an ordered lattice is responsible for two of the characteristic macroscopic properties of crystalline solids: their anisotropy and their characteristic geometry. It seems inevitable that the more detailed features of solids should be closely related to the nature of the forces which hold the crystal lattice together. Accordingly, we find it useful to distinguish between ionic, metallic, molecular, and covalent network solids, and to associate a set of characteristic properties with each of these bond types.

The object of this classification scheme is to provide a framework for recognizing and systematizing similarities and differences in the properties of the various solids. If definite macroscopic properties can be associated with each bond type, we might eventually be able to use macroscopic behavior as a diagnostic tool to determine the type of bond which exists in new substances. However, we must remember that this classification scheme is somewhat arbitrary, and thus there may be substances that do not clearly fit into one of the four classes.

Ionic Crystals

In ionic crystals the repeating units of the lattice are positively and negatively charged fragments arranged so that the potential energy of the ions in the lattice positions is lower than when the ions are infinitely separated. There are many types of stable ionic lattice arrangements. One of the most common, which occurs in sodium chloride and many other alkali halides, is shown in Fig. 3.7. It is important to note that each ion of a given sign occupies an equivalent lattice position, and that there are no discrete groups of atoms, or molecules, in the crystal. In effect, each ion of a given sign is bonded by the Coulomb force to *all* ions of the opposite sign in the crystal. The amount of energy necessary to evaporate some typical ionic crystals to their separated ions is, as Table 3.1 shows, of the order of 200 kcal/mole. This is a relatively large binding energy, and it is responsible for the fact that ionic crystals have vanishingly small vapor pressure at room temperature, and melt and boil only at relatively high temperature.

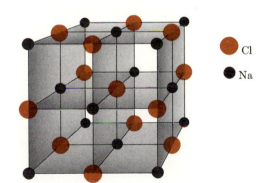

Cl

Na

The sodium chloride structure.   FIG. 3.7

Ionic crystals generally tend to be hard and brittle, and we can find an explanation for this in the nature of the Coulomb forces between the ions. In order to distort a perfect ionic crystal, two planes of ions have to be displaced relative to each other. Depending on the nature of the moving planes and their direction of motion, a displacement may bring ions of the same charge close to each other. When this occurs, the cohesive forces between the two planes are replaced with a strong Coulomb repulsion, and as a result the crystal fractures.

Table 3.1 Cohesive energy of crystals

Ionic crystals	Energy required to separate ions (kcal/mole)
LiF	246.7
NaCl	186.2
AgCl	216
ZnO	964

Molecular crystals	Energy required to separate molecules
Ar	1.56
Xe	3.02
Cl_2	4.88
CO_2	6.03
CH_4	1.96

Covalent network crystals	Energy required to separate atoms
C(diamond)	170
Si	105
SiO_2	433

Metallic crystals	Energy required to separate atoms
Li	38
Ca	42
Al	77
Fe	99
W	200

However, there are planes which can be moved relative to one another without bringing ions of the same sign into opposition. Examination of Fig. 3.8 shows that there are planes in sodium chloride that are made up of "sheets" of sodium ions alternating with "sheets" of chloride ions. Motion of these planes in a direction parallel to these rows of ions does not bring ions of the same charge directly opposite each other. Consequently, slippage of these planes relative to each other is the easiest way of distorting the crystal. However, even along this most favored direction, distortion of the crystals requires a large force, for any motion of ions relative to each other is resisted by the large Coulomb forces which tend to hold the ions in their lattice sites.

Another identifying characteristic of ionic crystals is that they are electrical insulators at low temperatures, but good electrical conductors when they are melted. Once again, the ionic bonding model provides a simple explanation.

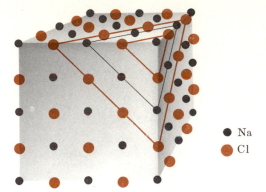

The sodium chloride structure. The planes marked consist of alternating sheets of sodium ions and chloride ions.

FIG. 3.8

● Na

● Cl

In the solid state, there seems to be no obvious mechanism by which an ion can move under the influence of an electric field without a considerable expenditure of energy. In the perfect crystal, all ions occupy well-defined positions, and in moving from its lattice site through the crystal, an ion would follow a tortuous path which brings it close to ions of the same charge. In the liquid state, however, the arrangement of ions is more disordered and less dense, and their motion under the influence of an electric field is greatly facilitated.

Molecular Crystals

In molecular crystals, the repeating unit is a chemically identifiable atom or molecule which does not carry a net charge. The cohesion of these crystals is a consequence of the van der Waals forces that we discussed in Chapter 2. Van der Waals forces are considerably weaker than the attractive Coulomb force between two ions, and consequently the binding energy of molecular crystals is relatively small, as the examples in Table 3.1 show. Since so little energy is required to separate the individual molecules, these crystals tend to be rather volatile, and have low melting and boiling points. As we discussed in Chapter 2, the magnitude of van der Waals forces can vary considerably, depending on the number of electrons and the polarity of the molecules. As a result, even though a solid which is volatile is very likely to be a molecular crystal, not *all* molecular crystals are volatile.

Molecular crystals generally tend to be soft, compressible, and easily distorted. These properties too are consequences of the relatively weak intermolecular forces, and their nondirectional nature. That is, all nonpolar molecules have an attraction for each other whose magnitude is not very sensitive to molecular orientation. Thus two planes of a molecular crystal can be moved past each other without significant diminution of the attractive forces between them. Since the energy of the intermediate positions is not much greater than that of the stable positions, the distortion requires little energy expenditure.

Molecular crystals are, in general, good electrical insulators. The molecules themselves have no net electrical charge and thus cannot transport electricity. Moreover, the very existence of discrete molecules implies that the electrons tend to be localized about a specific set of nuclei. Consequently, there are no charged particles, either ions or electrons, that are free to move in an electric field and conduct electricity.

Covalent Network Solids

Crystals in which all the atoms are linked by a continuous system of well-defined electron-pair bonds are called covalent network solids. The most familiar example is the diamond crystal, in which each carbon atom is covalently bonded to four other atoms, as Fig. 3.9 demonstrates. The result is a rigid three-dimensional network which links each atom to all the others. In effect, the entire crystal is a single molecule.

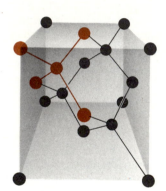

FIG. 3.9   The diamond structure. Each atom is linked directly to four others.

In some crystals there are two-dimensional infinite covalent networks. The best-known example is the graphite structure shown in Fig. 3.10. Each carbon atom is covalently bonded to three others in such a way that all atoms in a single plane are linked in a sheetlike structure. In the graphite crystal, these infinite sheets of atoms are packed in a layer structure in which the attractive forces between different layers are of the van der Waals type.

Table 3.1 shows that the energy needed to separate typical network solids into their constituent atoms can be as great as 200 kcal/mole. Consequently, these materials, like ionic substances, are extremely involatile and have very high melting points. Moreover, covalent bonds have very noticeable directional properties. That is, a central atom forms strong covalent bonds with its neighbors only if they occupy certain fairly specific locations. Therefore any significant distortion of a covalent network solid involves the breaking of covalent bonds, which requires considerable amounts of energy. As a result, network solids are the hardest and most incompressible of all materials.

With respect to volatility and mechanical properties, infinite covalent network crystals are very similar to ionic solids. Therefore, the fact that a substance is very hard and has a high melting and boiling point does not tell us what type of bonding exists in the crystal. However, we can use electrical properties to distinguish between ionic and covalent network solids. Both types are electrical insulators at low temperatures. Ionic substances become good electrical conductors only at temperatures above their melting point. The conductivity of a covalent network solid, if noticeable at all, is, in general, quite small, and while it may increase as temperature increases, the conductivity does not rise abruptly when the substance is melted.

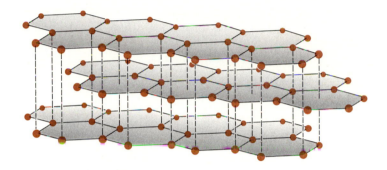

The structure of graphite. The atoms in alternate planes are directly under one another.   FIG. 3.10

Metallic Crystals

Metallic crystals are characterized by their silvery luster and reflectivity, high electrical and heat conductivity, and by the ease with which they can be drawn, hammered, and bent without fracture. Silver, gold, and platinum are substances in which all these properties appear most clearly. On the other hand, most metals are somewhat lacking in one or more of these characteristics. For example, tungsten has a silvery luster, but is brittle and not easily worked, and lead, which is soft and workable, is not a good conductor of electricity.

The electronic structure of metals differs from other substances in that the valence electrons of the metallic atoms are not localized at each atom, but are the general property of the crystal as a whole. Thus in a simplified picture, a metal is thought of as a collection of positive ions immersed in a "sea" of mobile electrons. The qualitative features of this "free-electron" picture are consistent with the general metallic properties. The high electrical conductivity of metals is readily explained if the valence electrons are free to move in an applied electric field. The high thermal conductivity of metals is also a consequence of the free electrons, which can acquire large thermal kinetic energy, move rapidly through the crystal, and thereby transport heat.

The idea that valence electrons are not localized but shared by all atoms in the crystal is consistent with the mechanical properties of metals. Since there are no highly directional localized bonds, one plane of atoms may be moved over another with relatively little expenditure of energy. Because the valence electrons are not localized and the metallic bonding is not strongly directional, bonding forces need not be completely disrupted when the crystal is distorted.

While the free electron picture is consistent with the general properties of most metals, there can be no doubt that it is an oversimplification. Within the group of metallic elements there is a considerable variation of properties: mercury melts at −39°C, and tungsten at 3300°C; the alkali metals can be cut with a table knife, but osmium is hard enough to scratch glass; as an electrical conductor, copper is 65 times better than bismuth. Understanding these variations in metallic properties requires more elaborate bonding theories; certain aspects of these will be discussed in Chapter 11.

3.3 X-RAYS AND CRYSTAL STRUCTURE

The diffraction of x-rays by crystals is an important phenomenon because it can be used to tell us the relative locations of atoms in a solid. The results of x-ray diffraction studies thereby contribute to our general understanding of molecular structure, and how it is related to chemical and physical properties. Before we treat the details of the x-ray diffraction experiment, we shall find it helpful to analyze the basic aspects of electromagnetic waves.

Electromagnetic Waves

When an electrically charged sphere is suspended in free space, and a second "test" charge brought near, the test charge experiences a force from the charged sphere. This "action at a distance," which is characteristic of electrical systems, is rationalized by imagining that each electric charge is surrounded by an electrical disturbance, or force field, and that this electric field is responsible for the ability of one charge to act on another even though they are physically separated, and in a vacuum. Similarly, the magnetic effects which are produced by an electric current are pictured as the result of a magnetic force field which surrounds each moving electric charge. In describing the properties of electrical systems, it is often more profitable to emphasize the behavior of the electric and magnetic fields, rather than that of the electric charges themselves.

By the end of the nineteenth century, physicists had recognized that a number of optical experiments could be understood if light was pictured as electromagnetic wave motion. According to this picture, light is to be thought of as being produced by the oscillating motion of an electric charge. This oscillation causes the electric field surrounding the charge to change periodically and also

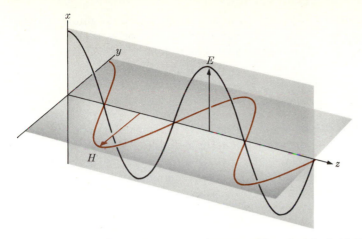

Representation of an electromagnetic wave. The electric field *E* and magnetic field *H* oscillate in perpendicular planes.

FIG. 3.11

produces an oscillating magnetic field. These oscillating electric and magnetic disturbances are radiated or propagated through space—hence the name "electromagnetic radiation." A test charge placed in the path of electromagnetic radiation experiences an oscillating force, first in one direction, then in the opposite direction, then back in the first direction, and so on. This suggests that the electric field of the light is propagated as a wave, and other experiments suggest that this is true also of the magnetic field. Thus at any instant, a snapshot of the electromagnetic wave would look like Fig. 3.11.

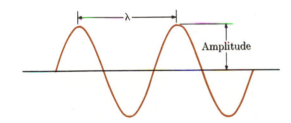

Instantaneous profile of a wave, demonstrating the definitions of amplitude and wavelength.

FIG. 3.12

Consider Fig. 3.12, which shows only the electric component of an electromagnetic wave. The maximum magnitude of the disturbance is called the **wave amplitude,** and the distance between two successive maxima is known as the **wavelength,** denoted by λ. A test charge placed at a wave maximum would experience a maximum electrical force in one direction, and a test charge placed at the wave minimum would also feel the maximum force, but in the opposite direction. Figure 3.12 is only an instantaneous picture of a wave, and a moment later the position of all the wave maxima will have changed uniformly. In other words, the wave maxima are propagated with a **velocity** which we shall call c. Then the number of maxima which reach a stationary point in one second is

$$\frac{c \text{ cm/sec}}{\lambda \text{ cm/cycle}} = \nu \text{ cycles/sec}.$$

The quantity ν is called the **frequency** of the wave. Electromagnetic radiation includes visible, infrared, and ultraviolet light, as well as radio waves and x-rays. These various types of electromagnetic radiation, which have such different effects on matter, are all propagated through a vacuum with a velocity $c = 2.98 \times 10^{10}$ cm/sec. They differ only in wavelength, or equivalently, frequency. While radio waves have wavelengths which range from one centimeter to several meters, and visible radiation lies in the range of 4 to 7×10^{-5} cm, x-rays have wavelengths of approximately 10^{-8} cm.

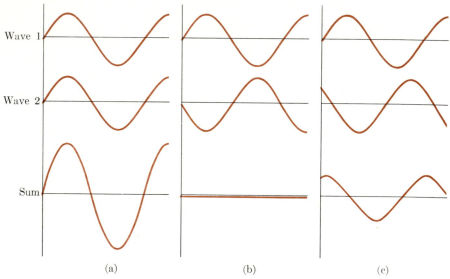

FIG. 3.13   Superposition of waves; (a) in phase; (b) out of phase; (c) small phase difference.

Wave Interference

The intensity of electromagnetic radiation is proportional to the square of its wave amplitude. With this in mind, let us examine what happens when two electromagnetic waves of the same frequency are superimposed. If the two waves are brought together as in Fig. 3.13(a), so that they both reach their maximum amplitude at the same point and at the same time, they are said to be *in phase* with each other. In this situation, the electromagnetic fields created by the waves add, producing an even stronger electromagnetic field, which we might sense as an increase in the intensity of the radiation. This phenomenon is called *constructive interference*.

On the other hand, if the waves are superimposed as in Fig. 3.13(b), so that one wave reaches its positive maximum amplitude just as the other reaches its negative maximum amplitude, the two waves are said to be *out of phase*. In this case, the electric and magnetic fields of the two waves cancel each other, and the intensity of the radiation vanishes. This is known as *destructive interference*. When two waves are not exactly out of phase as in Fig. 3.13(c), some destructive interference still occurs. There is a partial cancellation of the electric and magnetic fields, and the radiation intensity diminishes.

Destructive and constructive interference can be observed in the double-slit diffraction experiment shown in Fig. 3.14. The two slits act as separate radiation sources, each emitting a circular wave pattern, and the instantaneous position of the successive wave maxima are indicated by the two sets of con-

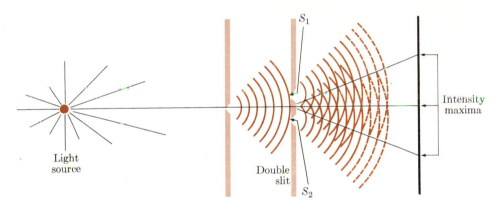

Double-slit diffraction experiment. Constructive interference occurs along the rays indicated and produces intensity maxima.   **FIG. 3.14**

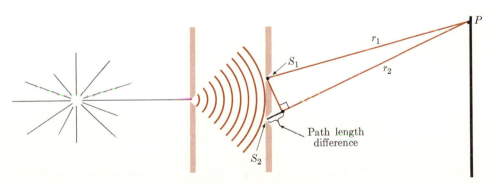

Double-slit diffraction experiment. If path-length difference between r_1 and r_2 is an integral number of wavelengths, constructive interference occurs and the intensity is a maximum at *P*.   **FIG. 3.15**

centric circles. The two patterns become superimposed, and the drawing shows that there are "rays" along which the wave maxima from the two slits are always in phase. When the radiation falls on a screen, maximum intensity is observed at the points where these rays intersect the screen. In the other regions, the radiation intensity is small or zero, since these regions lie along paths where the two waves are out of phase, and interfere destructively.

A slightly different point of view may show more clearly why the double-slit interference phenomenon occurs. From Fig. 3.15 we can see that the waves which leave the two slits travel different distances to reach point P. The two waves are exactly in phase when they leave the slits, and in order for them to be exactly in phase when they reach the point P, the difference in the distance traveled must be exactly equal to an integral number of wavelengths. Examination of Fig. 3.15 suggests that this condition can be met only at certain points on the screen, and these are the points at which maximum intensity in the diffraction pattern is observed.

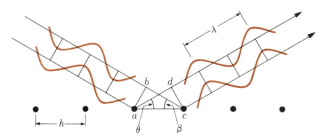

.FIG. 3.16 Diffraction by a row of equally spaced atoms.

X-ray Diffraction

Diffraction patterns are produced whenever light passes through or is reflected by a *periodic* structure that has a regularly repeating feature. The two-slit apparatus of Fig. 3.14 is the simplest of periodic structures. In order for the diffraction pattern to be prominent, the repeat distance of the periodic structure should be about equal to the wavelength of the light used. A crystal lattice is a three-dimensional periodic structure, in which the repeat distance is roughly 10^{-8} cm, the distance between atoms. Thus we should expect, and do indeed find, diffraction patterns produced when x-rays of approximately 10^{-8} cm wavelength pass through crystals.

Let us analyze what happens when x-rays of wavelength λ strike a single plane of atoms, as in Fig. 3.16, and are diffracted at an angle β. Just as in the double-slit experiment, the diffracted waves will produce a maximum intensity at the detector if the difference in the path of adjacent rays is an integral number of wavelengths. If this condition is satisfied, the waves will arrive at the detector in phase. From Fig. 3.16 we see that the difference in the paths followed by adjacent rays is $ad - bc$, and this must equal $m\lambda$, where m is an integer. Thus we have

$$ad - bc = m\lambda, \qquad m = 0, 1, 2, \ldots,$$

$$h(\cos\theta - \cos\beta) = m\lambda.$$

For $m = 0$, this gives $\beta = \theta$. Therefore when the angle of the incident beam is equal to the angle of the diffracted beam, there is a maximum in the intensity at the detector. This tells us that because of the regular periodic repetition of lattice points, a plane of atoms will "reflect," at least partially, an x-ray beam in much the same manner as a mirror reflects ordinary light.

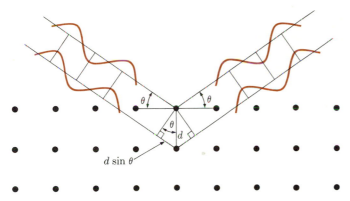

Diffraction from successive planes of atoms. Diffracted waves are in phase if $n\lambda = 2d\sin\theta$.   **FIG. 3.17**

However, because a single plane of atoms reflects only a fraction of the incident x-ray intensity, there is still another condition to be met if a diffraction pattern of appreciable intensity is to be observed. The waves reflected from successive parallel planes of atoms must reach the detector in phase in order to produce an intensity maximum. Figure 3.17 illustrates how the condition for maximum diffracted intensity can be derived. In order for the waves to reach the detector in phase, the difference in the distance they travel must be equal to a whole number of wavelengths, $n\lambda$, where n is an integer. From Fig. 3.17 we find that the path difference for the two waves is $2d\sin\theta$, where d is the spacing between the planes. Thus we have

$$n\lambda = 2d\sin\theta, \qquad n = 1, 2, 3, \ldots, \tag{3.1}$$

for the condition which must be satisfied in order for a maximum in the diffracted intensity to occur.

Equation (3.1) is called the Bragg diffraction equation, after W. L. Bragg, who first derived it and used it to analyze the structure of crystals. The Bragg equation has two important applications. If the spacing d of the planes of the crystal lattice is known, then the wavelength of the x-rays can be calculated from the measured diffraction angle θ. This is the procedure that Moseley used to determine the characteristic x-ray wavelengths emitted by each of the elements in his investigations which led to the determination of atomic numbers. Alternatively, if the x-ray wavelength λ is known, the characteristic interplanar spacings of a crystal can be computed from measurements of the diffraction angles θ. In this way a complete picture of the lattice structure of a crystal can be obtained.

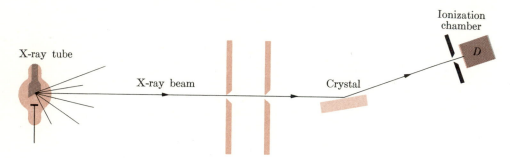

FIG. 3.18   Schematic representation of the Bragg x-ray diffraction apparatus.

Note carefully that the most important factor which enters the derivation of the Bragg equation is the *regular* spacing of the lattice planes. We saw earlier that the reason that a single plane of atoms reflects x-rays most efficiently when the angle of incidence is equal to the angle of reflection is a consequence of the regular spacing of atoms in the plane. Our derivation of the Bragg equation shows that the fact that reflections from parallel planes of the lattice reinforce each other is a consequence of the uniform interplanar spacing. If the arrangement of atoms in the planes or the spacing between parallel planes becomes irregular, as is the case in liquids and amorphous solids, sharp x-ray diffraction patterns are not observed.

The simplest type of apparatus for observing x-ray diffraction is shown in Fig. 3.18. X-rays of a single wavelength impinge on a crystal which is mounted on a rotating platform. The diffracted radiation is detected by the ionization it produces in the chamber D. When the crystal is set at an arbitrary angle with respect to the incident x-ray beam, very little diffracted radiation reaches the detector, since it is likely that at this angle there is no plane of the crystal lattice which satisfies the Bragg equation for maximum diffracted intensity. However, as the crystal is rotated, eventually some set of planes becomes aligned at an angle θ that satisfies Eq. (3.1), and a strong signal appears at the detector. As the crystal is rotated still further this signal disappears, but at some other angle θ' another diffracted signal may appear when a new set of lattice planes satisfies the Bragg equation. As the two-dimensional lattice in Fig. 3.17 shows, there are many sets of parallel planes in a lattice, and so diffracted radiation is observed at many angles. However, only the lattice planes which contain large numbers of atoms will reflect the x-rays appreciably, so in practice, only diffraction from the most important lattice planes is observed.

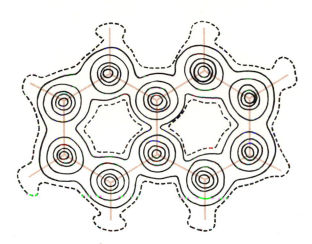

Electron density map of naphthalene. Dashed lines correspond to 0.5 electron charge per cubic angstrom.

FIG. 3.19

X-rays and Electron Density

The measurement of the diffraction angles and use of the Bragg equation leads to a determination of the spacing of the planes of a crystal lattice. So far we have assumed that the lattice planes are made up of identical structureless points whose only feature is the ability to scatter x-rays. In reality, the occupants of the lattice sites may be individual atoms, or what is more likely, may be molecules or groups of molecules of rather complex structure. It is the electrons in these molecules that are responsible for the scattering of the x-rays, and the efficiency of the scattering, and hence the intensity of the diffraction pattern, depends on the number and distribution of the electrons at the lattice sites. The electron distribution is, of course, determined by the structure of the molecules which occupy the lattice sites. Thus by studying not only the angles at which x-rays are diffracted, but also the intensities of the diffracted radiation, it is possible to determine the structure of the molecules which are at the lattice sites.

In the most elegant applications of this technique, it is possible to obtain contour maps of the electron density in very complicated molecules. Figure 3.19 shows the structure of the naphthalene molecule, $C_{10}H_8$, as determined by x-ray diffraction intensity studies. The contours represent lines of constant average electron density, and show conclusively that the geometric structure of naphthalene is

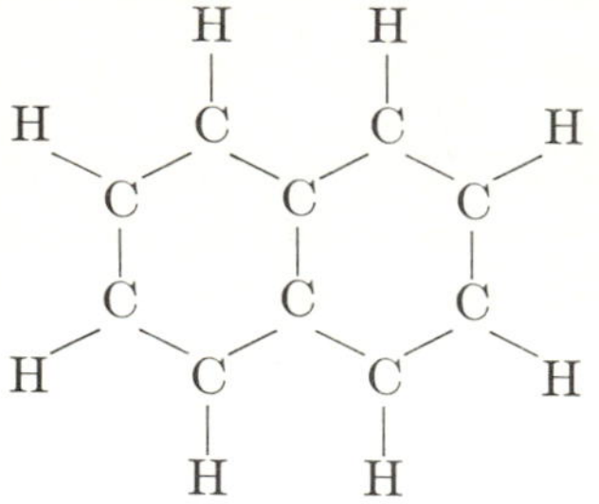

where all the atoms lie in the same plane. The contour maps provide very precise values for the distances between nuclei and the angles between bonds. Consequently, much of our knowledge of molecular structure is obtained from x-ray studies.

Chemical Analysis by X-rays

We have seen that the interplanar spacing of a crystal lattice determines the angles at which strong x-ray diffraction occurs. These interplanar spacings are a most intimate characteristic of the crystal, for they are determined by the size and arrangement of its atoms. Each crystalline compound has its own set of interplanar spacings and thus its own characteristic set of x-ray diffraction angles which, like a fingerprint, can be used to identify the substance.

One of the simplest applications of x-ray identification is to prove or disprove the existence of new compounds. For example, it is well known that cadmium forms an oxide whose formula is CdO. In this compound, cadmium is in the $+2$ oxidation state, which is consistent with all the other chemistry of cadmium. It is possible, however, to prepare an apparently homogeneous substance that, according to chemical analysis, has the empirical formula Cd_2O. Is this substance a true compound in which cadmium is in the $+1$ oxidation state? X-ray examination provides the answer. The x-ray reflections from "Cd_2O" occur only at angles which are characteristic either of metallic cadmium or CdO. It would certainly be expected that "Cd_2O," if it were a true compound, would have different lattice spacing and hence different x-ray reflection angles from either cadmium metal or CdO. Therefore it must be concluded that "Cd_2O" is really not a true compound, but a physical mixture of tiny crystals of cadmium metal and CdO.

Determination of Avogadro's Number

One of the most fundamentally important results of x-ray diffraction studies is the precise determination of the value of Avogadro's number. In principle, the measurement is very simple. X-rays of known wavelength are used to determine the interplanar spacings of a crystal. From this, the volume occupied by one molecule (or atom) can be calculated. Then the measured volume of one mole of the crystal is divided by the volume of one molecule, and the result is Avogadro's number.

One of the keys to this procedure is the knowledge of the wavelength of the x-rays. How is this determined in the first place? The answer is surprisingly simple. An artificial "lattice" is created by carefully drawing closely spaced lines on a surface. Even though these lines are as much as 5×10^{-4} cm apart, much greater than the 10^{-8}-cm wavelength of x-rays, they still can produce an x-ray diffraction pattern if the x-rays are directed almost parallel to the ruled surface. A similar diffraction effect can be seen when visible light grazes a phonograph record. Even though the wavelength of the light is much smaller than the spacing between the record grooves, diffraction is observed if the grazing angle is small enough. In the x-ray experiment, the wavelength can be calculated from the measured diffraction angles and the known spacing of the ruled lattice.

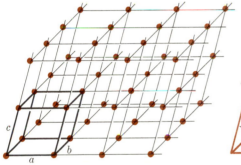

A lattice of points, showing the unit cell in heavy outline, and three fundamental translational vectors a, b, and c.

FIG. 3.20

3.4 THE CRYSTAL LATTICES

In our discussion of x-ray diffraction, we emphasized that the microscopic characteristic that produces sharp diffraction patterns is an *ordered, regularly repeating structure*. Crystalline solids which show these sharp diffraction phenomena can therefore be described in terms of *lattices*, or three-dimensional arrays of points that display a regular repetition pattern. An example of such a lattice is shown in Fig. 3.20. The lattice is characterized by the distance between successive points along each of the three axes indicated, and the angles between these axes.

The Unit Cell

The lattice of points shown in Fig. 3.20 may also be discussed in terms of a basic, simple array of points called the *unit cell*. This unit cell is the smallest unit which, when repeated in three dimensions, will generate the entire crystal. A crystal can thus be thought of as composed of a combination of unit cells, with neighboring cells sharing faces, edges, or corners. The unit cell of the lattice pictured in Fig. 3.20 is shown with its outlines emphasized.

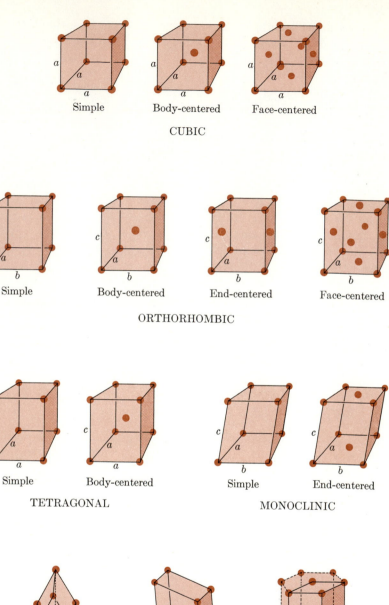

Simple Body-centered Face-centered

CUBIC

Simple Body-centered End-centered Face-centered

ORTHORHOMBIC

Simple Body-centered

TETRAGONAL

Simple End-centered

MONOCLINIC

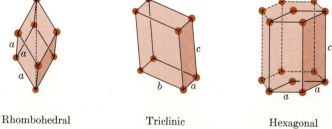

Rhombohedral Triclinic Hexagonal

FIG. 3.21   The unit cells of the fourteen Bravais lattices grouped into the seven crystal systems.

By considering the possible combinations of the lattice point spacings (a, b, c) along each of the lattice axes and the angles (α, β, γ) between these axes, it is possible to generate seven *crystal systems*. The geometric characteristics of these crystal systems are listed in Table 3.2. In 1848, Bravais showed that associated with these seven crystal systems, there were only fourteen crystal lattices. The unit cells for these fourteen Bravais lattices are shown in Fig. 3.21. Every naturally occurring crystal has one of these lattice structures. However, in a natural crystal the lattice sites may be occupied by atoms or more frequently by complex groups of atoms, and hence the unit cell may have a complicated internal molecular structure. Since even in the most complex molecular structures the unit cell is the basic repeating unit of the crystal, the unit cell dimensions can be obtained from x-ray diffraction patterns by use of the Bragg relation, Eq. (3.1). The actual structure within the unit cell is a more difficult problem, however, and is determined from measurements of the *intensities* of the spots in the diffraction pattern.

Table 3.2 The seven crystal systems and fourteen Bravais lattices

Crystal system	Unit cell dimensions and angles	Bravais lattice
Cubic	$a = b = c;\ \alpha = \beta = \gamma = 90°$	Simple Body-centered Face-centered
Orthorhombic	$a \neq b \neq c;\ \alpha = \beta = \gamma = 90°$	Simple Body-centered End-centered Face-centered
Tetragonal	$a = b \neq c;\ \alpha = \beta = \gamma = 90°$	Simple Body-centered
Monoclinic	$a \neq b \neq c;\ \alpha = \gamma = 90° \neq \beta$	Simple End-centered
Rhombohedral	$a = b = c;\ \alpha = \beta = \gamma \neq 90°$	Simple
Triclinic	$a \neq b \neq c;\ \alpha \neq \beta \neq \gamma \neq 90°$	Simple
Hexagonal	$a = b \neq c;\ \alpha = \beta = 90°;\ \gamma = 120°$	Simple

3.5 COMMON CRYSTAL STRUCTURES

When the sites of the various Bravais lattices are occupied or surrounded by polyatomic molecules, quite complex crystal structures can result. However, there are a few types of structures that not only have simple geometric characteristics but also occur quite frequently in natural crystals. In this section we shall examine a few of these common crystal structures.

Closest-Packed Structures

There are many substances whose atomic arrangement can be pictured as a result of packing together identical spheres so as to achieve maximum density. In particular, almost all the metallic elements and many molecular crystals display these "closest-packed" structures. Let us see, then, how the closest packing of spheres can be achieved.

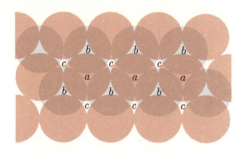

FIG. 3.22   A single layer of closest-packed spheres.

Figure 3.22 shows the closest-packed arrangement of spheres whose centers lie in a single plane. Each sphere is surrounded by six others, which are called its nearest neighbors. For future reference, we shall label each of the sites occupied by the spheres with the letter a. Consideration of Fig. 3.22 suggests that a second layer of closest-packed spheres can be added by placing spheres in the depressions, or "holes," of the first layer. All of these depressions are identical, but they can be divided into two groups. If we place a sphere at a site marked b, we cannot place one in the adjacent sites marked c, and vice versa. Thus all spheres which form the second closest-packed layer must be placed either at the b-sites or the c-sites. The arrangement produced by choosing the b-sites is shown in Fig. 3.23.

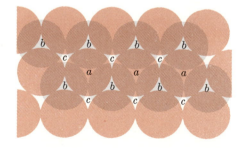

FIG. 3.23   Two layers of closest-packed spheres.

When we attempt to add a third layer, two possibilities confront us. There are once again two types of depression available to accept the third layer, but they are not exactly equivalent. One type of depression is labeled a in Fig. 3.23, for it lies directly above the center of a sphere of the first layer. The other type of depression, denoted by c, lies directly over a hole in the first layer—in fact, directly over a *c*-type hole. Either of these two types of site may be used to accommodate atoms of the third layer, and either choice leads to a closest-packed crystal lattice.

When the spheres of the third layer are placed at the *a*-sites and the sequence of the layers is continued indefinitely as *abababab*, etc., we obtain a *hexagonal closest-packed* structure. This designation is chosen because the atoms in the two *a*-layers occupy the sites of the unit cell of the hexagonal Bravais lattice. The three atoms within the cell (the *b*-layer) do not occupy lattice sites. As Fig. 3.24 shows, when the structure is rotated about an axis perpendicular to the layers and passing through one sphere, the same structure is encountered three times in the course of a complete revolution.

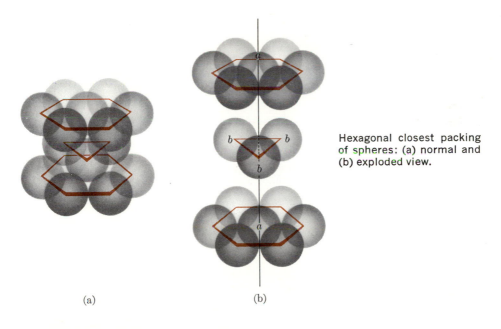

Hexagonal closest packing of spheres: (a) normal and (b) exploded view.

FIG. 3.24

(a)   (b)

The second possibility is that the spheres of the third layer be located at the *c*-sites, and when the sequence *abcabcabc*, etc., is continued indefinitely, the resulting arrangement is called *cubic closest packing*. To see why this lattice is described as cubic, we must rotate the structure as in Fig. 3.25. Then it becomes clear that the basic unit of the cubic closest-packed structure is a cube which contains 14 spheres. The layers which we used to build up the structure run through the cube parallel to diagonals that link opposite corners. Close study of the cubic closest-packed lattice shows that there is a sphere at the center of each face of the cube, and consequently this structure is also known as the *face-centered cubic lattice*. Generally it is difficult to see in a picture both the closest-packed layers and the face-centered cube surfaces, but a three-dimensional model reveals both features clearly.

There are a few properties that are common to both of the closest-packed structures. In both instances 74.0% of the available space is occupied by spheres. In either of the closest-packed lattices each sphere is in direct contact with 12 nearest neighbors—6 which lie in one layer, and 3 from each of the layers above and below. The number of nearest neighbors is called the *coordination number* of the sphere, and our arguments show that 12 is the maximum possible coordination number, since it is found when spheres are packed with maximum density.

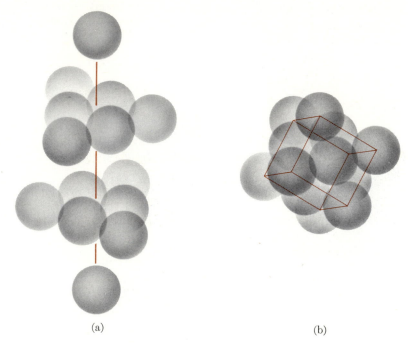

(a)　　　　　　　　　　　　　　　　　(b)

FIG. 3.25　　Cubic closest packing of spheres: (a) generation of unit from closest-packed layers, and (b) rotation to show cubic symmetry.

Solids whose molecules or atoms are essentially spherical in shape and are linked by nondirectional bonds are often found to have one of the closest-packed structures. In particular, all of the noble gases crystallize in either cubic or hexagonal closest-packed structures. The electron cloud in the hydrogen molecule H_2 is nearly spherical in shape and crystals of solid hydrogen have the cubic closest-packed structure. The hydrogen halides HCl, HBr, and HI are also nearly spherical molecules, since they consist of a large spherical halogen atom only slightly distorted by the very small hydrogen atom. Accordingly, all three of these compounds crystallize in a hexagonal closest-packed structure. The tendency of spherical molecules to crystallize in one of the closest-packed structures is apparently a consequence of the fact that, by so doing, they achieve the maximum coordination number, 12, thereby maximizing the strength of the intermolecular van der Waals forces.

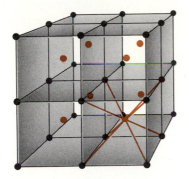

The body-centered cubic lattice. Spheres at cube centers are shaded to emphasize their position.

FIG. 3.26

Table 3.3 gives the crystal structures of many of the metallic elements. A number of these elements are polymorphic: they crystallize in more than one structure. Most of the metals have either the hexagonal or cubic closest-packed structures in which the atoms have the maximum possible coordination number of 12. Several metals are found in the *body-centered cubic* structure illustrated in Fig. 3.26. In this lattice there is a sphere at each corner and in the center of a cube, which is the repeating unit of the lattice. Because each sphere has only 8 nearest neighbors in the body-centered cubic lattice, it is not a closest-packed structure. However, in the body-centered lattice, the spheres occupy 68% of the available space, only slightly less than the 74% characteristic of the closest-packed structures. We shall see in Chapter 11 that the high coordination numbers found in metallic lattices are related to the nature of metallic bonding.

Table 3.3 Crystal structures of the metallic elements

Li I	Be II	B —	I Body-center cubic								
Na I	Mg II	Al III	II Hexagonal closest packed III Cubic closest packed								
K I	Ca III	Sc II	Ti II	V I	Cr I	Mn I	Fe I	Co II	Ni III	Cu III	Zn II
Rb I	Sr III	Y II	Zr II	Nb I	Mo I	Tc II	Ru II	Rh III	Pd III	Ag III	Cd II
Cs I	Ba I	La II	Hf II	Ta I	W I	Re II	Os II	Ir III	Pt III	Au III	Hg —

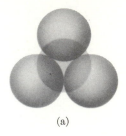 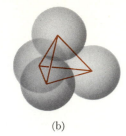

(a)   (b)

FIG. 3.27   Construction of tetrahedral interstitial site in a closest-packed lattice.

Structures Related to Closest-Packed Lattices

The crystal structures of many of the binary compounds of the type AB, AB_2, and A_2B are related in a simple manner to the closest-packed arrangements. Very often atoms of one type, let us say B, can be pictured as spheres that form a closest-packed structure, while the A-atoms occupy the interstices or holes between the close-packed spheres. To pursue this idea, we must examine the geometry of the holes in the closest-packed structures.

One of the basic units in either of the closest-packed structures is one sphere resting upon three others, as is illustrated in Fig. 3.27. The centers of the four spheres in this arrangement lie at the apices of a regular tetrahedron. Consequently, the *space* at the center of this tetrahedron is called a *tetrahedral site*. In any close-packed structure, each sphere is in contact with three others in the layer above it, and with three more in the layer below. As a result, there are two tetrahedral sites associated with each sphere. From these observations, we can see how the crystal structure of a binary compound AB might be related to a closest-packed arrangement. First, imagine that the B-atoms form a close-packed lattice. Then a compound AB could have a structure in which half of the tetrahedral sites were occupied by A-atoms. If the formula of the compound were A_2B, all of the tetrahedral sites could be occupied by A-atoms. As we shall see, both these arrangements are found in nature.

There is a second type of interstitial site in the closest-packed structures. This site is surrounded by six spheres whose centers lie at the apices of a regular octahedron, as illustrated by Fig. 3.28. The existence of these *octahedral sites* in the closest-packed structures is easier to recognize if we realize that each face of a regular octahedron is an equilateral triangle. Figure 3.28 shows, therefore, that an octahedral site can be generated by two sets of three spheres, each set forming, in parallel planes, equilateral triangles with apices pointing in opposite directions. These sets of equilateral triangles appear naturally in the parallel planes of the closest-packed structures.

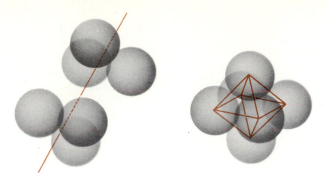

Construction of octahedral interstitial site in a closest-packed lattice.

FIG. 3.28

To see the relation between the octahedral sites and the tetrahedral sites, recall that in building the second layer of a closest-packed structure, spheres were set into only half the depressions in the top of the first layer. The depressions in which spheres of the second layer rest are tetrahedral sites, while the depressions in which no sphere rests form the octahedral sites.

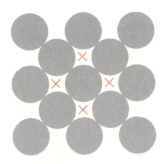

Locations of the octahedral sites in a face-centered cubic lattice.

FIG. 3.29

There is another way of locating octahedral sites. Figure 3.29 shows the face of a face-centered cubic lattice. The center of any regular octahedron falls on an equatorial plane formed by the centers of four spheres; these locations are marked in the drawing. By using Fig. 3.29, it is easy to see that there is one octahedral site for every sphere in the structure, because if we follow one column of spheres vertically, octahedral sites and spheres alternate. This conclusion holds for the hexagonal closest-packed structure as well. Thus there are half as many octahedral sites in a closest-packed structure as there are tetrahedral sites. Consequently a compound AB might be formed with the B-atoms arranged in a closest-packed structure, and A-atoms at all the octahedral sites.

The dimensions of the interstitial sites in closest-packed lattices are related to the size of the spheres used to form the structures. Figure 3.30 shows a cross section through an octahedral site. The radius of the small sphere that occupies the site can be found by simple geometry, for if we let r_1 and r_2 be the radii of the small and large spheres, respectively, the theorem of Pythagoras tells us that

$$2(r_1 + r_2)^2 = (2r_2)^2,$$
$$r_1 + r_2 = \sqrt{2}\, r_2,$$
$$\frac{r_1}{r_2} = 0.414.$$

Thus in order to occupy the octahedral site without disturbing the closest-packed lattice, the radius of the small sphere should be no greater than 0.414 times that of the large spheres. A similar calculation for the tetrahedral sites gives $r_1/r_2 = 0.225$ for the maximum radius ratio, so a tetrahedral site is noticeably smaller than an octahedral site.

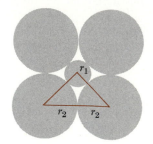

FIG. 3.30   The geometry of an octahedral site in a closest-packed lattice.

With the geometric properties of the closest-packed structures in mind, we are in a position to appreciate the relationships among the structures of many simple compounds. Our approach applies most clearly to compounds made up of monatomic ions, for these ions can be thought of as spheres with characteristic radii. It is true that an actual ion is not a sphere with a *well-defined* radius, but is a spherical charge "cloud" of rapidly decreasing density that, in principle, extends to infinity. Our procedure is to picture ions which have many electrons as large spheres and those with fewer electrons as smaller spheres, and while this is a useful model, its limitations must be remembered.

Consider first Fig. 3.31, the sodium chloride, or rock-salt structure. The chloride ions considered alone form a face-centered cubic lattice, and the sodium ions alone also have a face-centered cubic arrangement. Neither of these interpenetrating lattices is really closest-packed in the model, since the ions along the diagonal of a cube face do not touch each other. However, because the chloride ions are represented by the larger spheres, it is profitable to think of them as forming a cubic close-packed lattice that has been slightly expanded by the presence of the sodium ions. Study of the rock-salt structure shows us

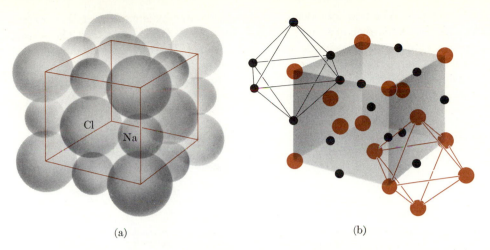

(a)   (b)

The sodium chloride structure: (a) space-filling model showing relative sizes of ions; **FIG. 3.31**
(b) lattice model showing the octahedral coordination about each ion.

that the sodium ions occupy the *octahedral sites* in the chloride close-packed lattice. Thus each sodium ion is surrounded by six chloride ions, and Fig. 3.31 shows that the converse is also true. As we remarked earlier, there is one octahedral site for each sphere in the closest-packed structure, and since there are equal numbers of sodium and chloride ions in sodium chloride, all the octahedral sites in the chloride lattice must be occupied by sodium ions. Many other compounds of the AB-type have the rock-salt crystal structure; this is true even of some in which the ions are polyatomic. A partial list of these compounds is contained in Table 3.4.

Another structure common to several binary 1:1 compounds is the zinc-blende (ZnS) structure illustrated in Fig. 3.32. The simplest point of view in this case is to think of the large sulfide ions as forming a face-centered cubic lattice that is almost closest-packed, with the zinc ions occupying *alternate* tetrahedral sites. Only half of the tetrahedral sites are occupied because the compound has 1:1 stoichiometry and there are two tetrahedral sites associated with each sulfide

Table 3.4 Structures derived from cubic closest packing

Holes used	Fraction filled	Name	Examples
Octahedral	1	Rock salt	Halides of Li, Na, K, Rb; NH_4Cl, NH_4Br, NH_4I, AgF, AgCl, AgBr; oxides and sulfides of Mg, Ca, Sr, Ba.
Tetrahedral	$\frac{1}{2}$	Zinc blende	ZnS, CuCl, CuBr, CuI, AgI, BeS
Tetrahedral	1	Fluorite	CaF_2, SrF_2, BaF_2, PbF_2, HfO_2, UO_2
	1	Antifluorite	Oxides and sulfides of Li, Na, K, and Rb

ion. Because the zinc ions occupy tetrahedral sites, it is clear that they have a coordination number of four, and reference to Fig. 3.32 shows that the coordination number of the sulfide ions is also four. Because the tetrahedral sites are relatively small, the zinc-blende structure is found in 1:1 compounds in which the cation is much smaller than the anion. Table 3.4 lists several of the compounds that have the zinc-blende structure.

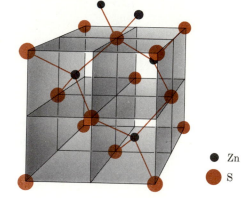

FIG. 3.32 The zinc-blende crystal structure. Both the zinc and the sulfur atoms have a coordination number of four.

● Zn

● S

Some compounds whose stoichiometry is of the type 1:2 are found to crystallize in the fluorite (CaF_2) structure pictured in Fig. 3.33. One way of describing this arrangement is to say that the calcium ions form a face-centered cubic lattice and the fluoride ions occupy *all* the tetrahedral sites. Closely related to this lattice is the *antifluorite* structure characteristic of 2:1 compounds like Na_2O. In the antifluorite structure, it is the *anions* that form the face-centered cubic lattice, and the cations that occupy all the tetrahedral sites. The antifluorite and the fluorite structures are both related to the zinc-blende structure, in which only half the tetrahedral sites in a face-centered cubic lattice are occupied.

Our discussion has shown that there are four structures derived from cubic closest-packing that are found in a large number of common binary inorganic compounds. It is easy to see that a similar group of lattices can be generated from the hexagonal closest-packed structure, and examples of these are common among known compounds. Even the structures of some compounds with more complex stoichiometry can be related to the closest-packed lattices. Thus the packing of spheres provides a simple picture which seems to unify much of structural chemistry, and we shall make some use of this approach in our subsequent discussions of descriptive chemistry.

Local Packing Arrangements

In discussing structures derived from the closest-packed lattices, we noted that the sizes of octahedral and tetrahedral sites were different. In a crystal like ZnS, the very small Zn^{++} ions occupy the small tetrahedral sites in the nearly

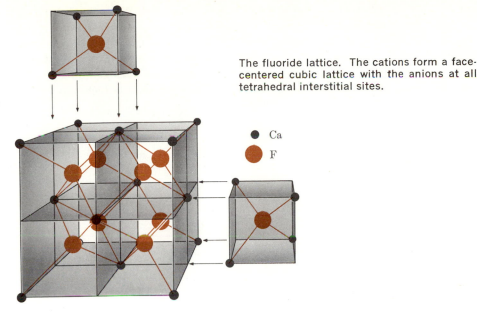

The fluoride lattice. The cations form a face-centered cubic lattice with the anions at all tetrahedral interstitial sites.

FIG. 3.33

● Ca

● F

closest-packed sulfide lattice. In NaCl, however, the ion sizes are more nearly comparable, and thus the positive ions occupy the larger octahedral interstitial sites in the anion lattice. These and similar observations suggest that in other situations, not necessarily derived from the close-packed structures, the coordination number of an atom or ion may be influenced or determined by the ratio of its radius to the radius of the atoms surrounding it.

To explore this idea further, we consider the coordination of a cation by anions, and make the following postulates:

1. Cations and anions try to be as close as their radii permit in order to maximize Coulomb attraction.

2. Anions never approach each other more closely than their ionic radii permit.

3. Each cation tends to be surrounded by the largest possible number of anions that is consistent with the first two postulates.

According to these postulates, a particular coordination number will occur between the values of the cation-to-anion radius ratio at which the anions simultaneously touch each other and the central cation, and the radius ratio at which this condition becomes possible for the next higher coordination number. Thus tetrahedral coordination is stable for cation-to-anion radius ratios of 0.225 to 0.414, at which point it becomes possible to increase the coordination number to six. A simple geometric calculation shows that it is possible to bring three anions and one cation into contact when $r_+/r_- = 0.155$. The coordination number three is stable between this radius ratio and 0.225, at which point tetrahedral coordination becomes possible. Table 3.5 summarizes the range of radius ratios over which coordination numbers of 2, 3, 4, 6, 8, and 12 are expected to be stable.

By using values of the ionic radius assigned to various cations and anions (see Chapter 11) it is possible to predict the cation coordination number in various compounds. In most cases the predicted coordination number is observed. Thus the consideration of radius ratios is a valuable technique which can be used to guess the structures of species in which the bonding is suspected to be electrostatic or nondirectional.

Table 3.5 Local packing arrangements

Cation coordination number	Radius ratio	Geometry	Examples
2	0–0.155	Linear	F^-—H^+—F^-
3	0.155–0.225	Triangular	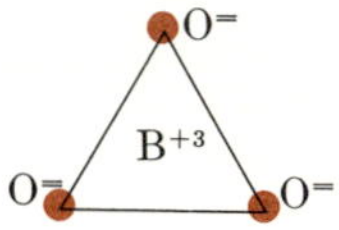
4	0.225–0.414	Tetrahedral	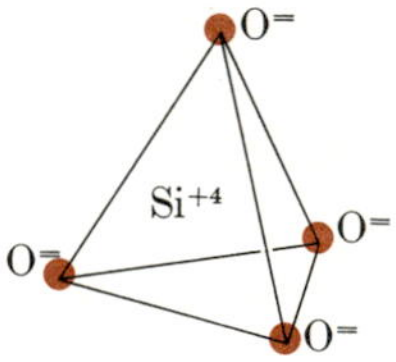
6	0.414–0.732	Octahedral	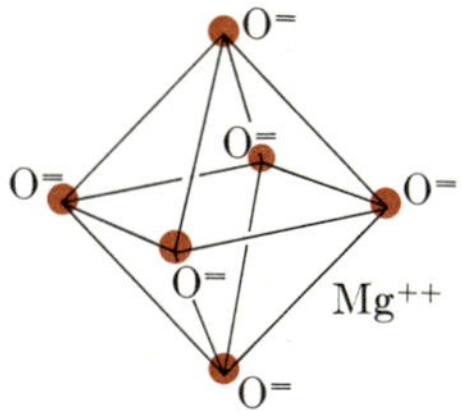
8	0.732–1.0	Cubic	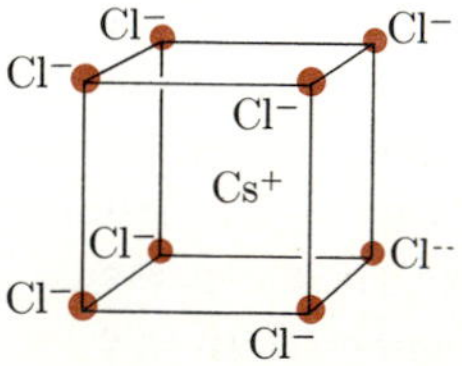

In our discussion up to this point we have tacitly assumed perfect lattice structures for crystalline solids. In fact, naturally occurring crystals have substantial numbers of defects. These defects have an important and sometimes dominating influence on the mechanical, electrical, and optical properties of solids. The study of the ramifications of defects is by itself a major research area encompassing chemistry, engineering, and solid state physics. Here we will indicate the classification, description, and major consequences of the most common crystal imperfections.

Point Defects

Lattice imperfections are classified according to their geometric characteristics. *Point defects* involve only one or two lattice sites directly. *Line defects* have to do with alterations or displacements of a row of lattice sites. *Plane defects* arise when a plane of sites is imperfect.

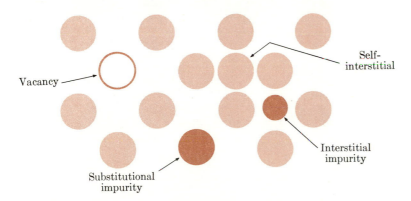

Types of point defect in a simple crystal structure.   **FIG. 3.34**

The four major kinds of point defects that occur in a crystal of one type of atom or molecule are indicated in Fig. 3.34. If an atom is missing from a lattice site there is a *vacancy;* an atom out of place is called *self-interstitial.* A foreign atom occupying a lattice site is called a *substitutional impurity,* whereas one placed off a site is an *interstitial impurity.*

In ionic solids, certain special cases of these point defects occur (see Fig. 3.35). A vacancy at a cation site is frequently accompanied by a vacancy at a nearby anion site. Such paired cation-anion vacancies, called *Schottky defects,* preserve the electrical neutrality of the crystal, and their formation requires relatively little energy. In a *Frenkel defect,* an ion leaves its lattice site and enters an interstitial position. This process also preserves overall electrical neutrality.

Formation of Frenkel defects requires relatively little energy if the anions are large and the cation is small, as in the silver halides, or if the crystal structure is of an open type with large interstices, like the fluorite (CaF_2) structure.

A simple anion vacancy in an ionic crystal creates a local excess of positive charge. An electron can migrate to this site and be bound, in effect replacing the absent negative ion. The presence of a number of such defects imparts a color to an otherwise colorless crystal, and consequently this type of imperfection is called an F-center, after *Farbe*, the German word for color.

In ionic compounds, substitutional impurities can occur relatively easily. For example, the ions Ba^{++} and Sr^{++} have the same charge and are fairly similar in size. Consequently, if $BaSO_4$ is precipitated from a solution containing strontium ions, some of the latter are inevitably incorporated in the newly formed barium sulfate crystals. Such substitutional impurities are very common in ionic compounds of the transition metals, since many of these elements form ions of the same charge and very nearly the same size.

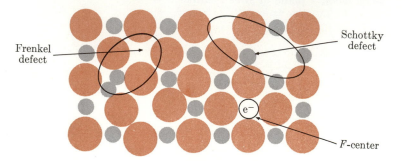

Frenkel defect

Schottky defect

e^-

F-center

FIG. 3.35

Types of point defect in an ionic solid.

As mentioned in Chapter 1, vacancies and interstitial atoms are responsible for the occurrence of nonstoichiometric compounds. Nickel oxide, NiO, is a good example of a compound which has a somewhat variable stoichiometry. When nickel oxide is prepared at relatively low temperature (1100°K) by partial oxidation of excess nickel, its composition is $Ni_{1.0}O_{1.0}$, its color is pale green, and its electrical properties are those of an insulator, as one would expect for a simple ionic compound. If the same substance is treated with excess oxygen at 1500°K, cation vacancies occur, the composition approaches $Ni_{0.97}O_{1.0}$, the color turns black, and nickel oxide becomes an electrical semiconductor.

The deficiency of positive charge which would otherwise accompany cation vacancies is made up by the presence of the appropriate amount of Ni^{+3}, and it is just this factor which is responsible for the electrical conductivity of nonstoichiometric NiO. If a Ni^{+3} ion exists at some point, an electron from elsewhere in the lattice may jump to it, converting it to Ni^{++}, and simultaneously creating a Ni^{+3} ion at a new lattice point. By a series of such electron jumps, charge can migrate through the crystal, and thus nonstoichiometric NiO is not an insulator, but a semiconductor.

It has been found that NiO will dissolve substantial amounts of Li_2O. The lithium ion Li^+ enters the NiO lattice as a substitutional impurity at some of the cation sites. Since Li^+ has only a single positive charge, an equal number of Ni^{+3} ions must be present in the lattice in order to preserve electrical neutrality. Consequently, NiO which has been "doped" with Li_2O is an even better electrical conductor than nonstoichiometric NiO, since the doping procedure permits the introduction of even greater numbers of Ni^{+3} ions.

Figure 3.36 shows the two types of line defects. In an *edge* dislocation an extra half-plane of atoms is present. The rest of the lattice along the edge of this plane is alternatively compressed or expanded in order to accommodate it.

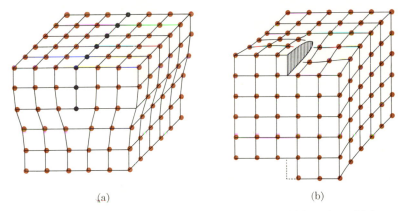

(a)   (b)

Two types of line defect in a crystal lattice. (a) An edge dislocation. (b) A screw dislocation.   **FIG. 3.36**

In a *screw* dislocation, part of a set of lattice planes has been moved one or more lattice units relative to their neighboring lattice planes. In effect, the screw dislocation represents a partially successful attempt to shear the crystal lattice. In the region around the dislocation, the lattice is under a shear stress.

The frequency of occurrence of edge dislocations is expressed as a number of dislocations per unit area. In a normally annealed metal, there may be as many as 10^6 edge dislocations per cm^2. This figure should be compared with the number of atoms per cm^2, which is approximately 10^{15}. In a metal which has been cold-worked, however, the dislocation density can rise to 10^{11} to 10^{12} per cm^2.

Edge dislocations have profound influence on the mechanical properties of matter. If two perfect close-packed planes of atoms are to be displaced relative to each other, the applied shear stress must overcome the attraction of each atom in one plane to its nearest neighbors in the other plane. The stress necessary to do this can be calculated from the known interatomic forces to be on the order of 10^6 psi. The actual required force as measured experimentally is only 10^3 psi or less. The discrepancy between these two numbers arises because of the presence of edge and screw dislocations in normal metal samples. Figure 3.37 indicates how an edge dislocation facilitates the motion of one plane of atoms over another. Because only one row of atoms must move at a time, and because the row which moves is already in a distorted, energetically unstable position, less force is needed to carry out the shear.

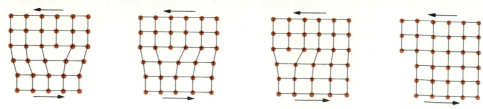

FIG. 3.37   The motion of an edge dislocation under a shearing stress.

Metals can be hardened and strengthened by the inclusion of foreign atoms. Foreign atoms whose size is different from that of the host tend to occupy lattice positions in or near dislocations, where they can be more easily accommodated in the distorted lattice. This makes it more difficult to move the dislocation under an externally applied stress, since it is now energetically more favorable for the defect to remain where the lattice is distorted by impurity atoms. The efficacy of different impurities in strengthening a metal in this manner is directly related to their size. Thus, while addition of 10% zinc increases the strength of copper by 30%, the same amount of beryllium, a much smaller atom, nearly triples the strength.

As pointed out in Section 3.1, solids often have a microcrystalline structure, and the interface between two differently oriented microcrystals is an example of a *plane* or surface defect. Since the lattice planes in two such neighboring microcrystals will tend to be randomly oriented, the surface atoms of the two crystallites will be out of register. Consequently, there will be a transition region between the crystallites in which the atomic spacing will be irregular. In these so-called grain boundary regions, the atoms are less stable and consequently tend to be more reactive. This is why the microcrystalline structure of a metal can be revealed by etching: the atoms in the grain boundaries are preferentially removed, and the microcrystals are left as partially raised structures.

3.7 THERMAL PROPERTIES OF SOLIDS

According to the law of Dulong and Petit, the heat capacity of one gram atom of a solid element is approximately 6.3 cal/deg. We have seen that the application of this rule is not limited to elements, for experiments show that the heat capacities of many solids, elemental or compound, is 6 cal/deg *per mole of atoms.* Of course, there are exceptions to this law. There are many substances whose heat capacities, measured at room temperature, are much smaller than predicted. These exceptional substances are the hard high-melting crystals that are made up of light atoms, such as boron, carbon, and beryllium. Moreover, the Dulong and Petit law fails for all substances if the heat capacity is measured at low temperatures. That is, the heat capacity of a solid is not constant, but depends on temperature in the manner illustrated in Fig. 3.38. At the absolute zero of temperature, the heat capacity of all substances is zero. As the temperature is raised, the heat capacity increases, but at different rates for different substances. Finally, in the limit of very high temperatures, the heat capacity of all solids is 6.3 cal/deg *per mole of atoms.*

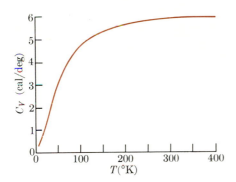

Heat capacity for silver at constant volume, C_V. **FIG. 3.38**

The key to understanding the temperature dependence of the heat capacities of solids was supplied by Einstein in 1905. It is an important argument, for it was one of the first demonstrations of the validity of Planck's hypothesis that atomic systems can exist with only certain discrete energies. Before reconstructing the details of Einstein's treatment, let us examine the general outline of the calculation. The heat capacity measured at constant volume is given by

$$C_V = \frac{\Delta E}{\Delta T},$$

where ΔE is the change in the total energy of a mole of substance produced by the temperature change ΔT. We shall calculate ΔE from the expression

$$\Delta E = N \, \Delta \bar{\epsilon},$$

where N is the number of atoms in the crystal, and $\Delta \bar{\epsilon}$ is the change in the *average* energy of an atom in the crystal produced by the temperature increase ΔT. Our first objective then is to find an expression that tells us how the average energy $\bar{\epsilon}$ depends on temperature.

Imagine, as did Einstein, that a mole of monatomic solid consists of N mass points that can oscillate in the three mutually perpendicular directions with a vibration frequency ν. According to Planck's proposal of quantized energy, an atom oscillating in one direction could have only one of the energies given by*

$$\epsilon_n = nh\nu, \qquad n = 0, 1, 2, 3, \ldots,$$

where h is a proportionality factor between frequency and energy, known as Planck's constant, and n is an integer. Planck's constant has the value 6.6×10^{-27} erg-sec, and the frequency ν is equal to 10^{12} sec^{-1} for many solids. An actual atom in a solid is a three-dimensional oscillator, but for simplicity we shall continue to treat only a one-dimensional oscillator, and correct for this discrepancy later.

At thermal equilibrium, the oscillators in a crystal will be distributed among the various allowed energies according to the Boltzmann distribution law. This is analogous to the situation which exists in gases, where the molecules are distributed over the available translational energies according to the Maxwell-Boltzmann law. The Boltzmann distribution law states that the number N_n of atoms with energy ϵ_n is related to the number N_0 with energy $\epsilon_0 = 0$, by the expression

$$N_n = N_0 e^{-\epsilon_n/kT} = N_0 e^{-nh\nu/kT}. \tag{3.2}$$

The total number of particles, N, is equal to the sum of the numbers in each energy state:

$$N = N_0 + N_1 + N_2 + N_3 + \cdots$$

Substitution of the Boltzmann factor, Eq. (3.2), gives

$$N = N_0 + N_0 e^{-h\nu/kT} + N_0 e^{-2h\nu/kT} + \cdots$$
$$= N_0 \sum_{n=0}^{\infty} e^{-nh\nu/kT}.$$

Now let us compute the total energy due to the oscillation of the atoms in this one direction. We do this by multiplying each energy ϵ_n by the number of particles which have that energy, and adding all these quantities:

$$E = \epsilon_0 N_0 + \epsilon_1 N_1 + \epsilon_2 N_2 + \cdots$$
$$= 0 N_0 + h\nu N_1 + 2h\nu N_2 + \cdots$$
$$= 0 + h\nu N_0 e^{-h\nu/kT} + 2h\nu N_0 e^{-2h\nu/kT} + \cdots$$
$$= N_0 \sum_{1}^{\infty} nh\nu e^{-nh\nu/kT}.$$

---

*Planck's proposal was slightly in error. The energies allowed to an oscillator are given by $\epsilon_n = (n + \frac{1}{2})h\nu$, where n is an integer. The omission of the $\frac{1}{2}$ makes no difference in the problem we are treating.

We are now in a position to calculate $\bar{\epsilon}$, the average energy of a one-dimensional oscillator. It is simply

$$\bar{\epsilon} = \frac{E}{N}$$

or

$$\bar{\epsilon} = \frac{h\nu \sum n e^{-nh\nu/kT}}{\sum e^{-nh\nu/kT}}. \tag{3.3}$$

In order to make any more progress, we must evaluate the sums of these infinite series. The procedure is eased considerably if we make the substitution

$$y = e^{-h\nu/kT}, \qquad \text{which gives} \qquad e^{-nh\nu/kT} = y^n.$$

Then, for the series in the denominator of Eq. (3.3), we get

$$\sum_{n=0}^{\infty} e^{-nh\nu/kT} = \sum_{n=0}^{\infty} y^n = 1 + y + y^2 + y^3 + \cdots$$

$$= \frac{1}{1-y}.$$

This last step can be verified by algebraic long division of 1 by $1 - y$.

We treat the series in the numerator of Eq. (3.3) in the same manner:

$$\sum_{1}^{\infty} n e^{-nh\nu/kT} = \sum_{1}^{\infty} n y^n = y(1 + 2y + 3y^2 + \cdots)$$

$$= \frac{y}{(1-y)^2}.$$

Once again, this last step can be checked by algebraic division.

Equation (3.3) now becomes

$$\bar{\epsilon} = \frac{h\nu y}{1-y} = \frac{h\nu e^{-h\nu/kT}}{1 - e^{-h\nu/kT}}.$$

Multiplying numerator and denominator by $e^{h\nu/kT}$ yields

$$\bar{\epsilon} = \frac{h\nu}{e^{h\nu/kT} - 1}.$$

This is the average energy of a one-dimensional oscillator. The average energy of an atom in a crystal is three times this amount, since the atom vibrates in three directions. Thus

$$\bar{\epsilon}_{\text{atom}} = \frac{3h\nu}{e^{h\nu/kT} - 1} \tag{3.4}$$

is the average energy of the three-dimensional atomic oscillator.

Let us now compute the heat capacity of the solid at high temperatures—high enough so that $h\nu/kT \ll 1$. Under these conditions, the denominator of Eq. (3.4) simplifies considerably. By the fundamental properties of the exponential function,

$$e^{h\nu/kT} = 1 + \frac{h\nu}{kT} + \frac{1}{2}\left(\frac{h\nu}{kT}\right)^2 + \frac{1}{6}\left(\frac{h\nu}{kT}\right)^3 + \cdots$$

But if $h\nu/kT \ll 1$, all terms after the second are very small and may be neglected in comparison with $1 + h\nu/kT$. We have then

$$e^{h\nu/kT} \cong 1 + \frac{h\nu}{kT},$$

which when substituted into Eq. (3.4) gives

$$\bar{\epsilon} = \frac{3h\nu}{e^{h\nu/kT} - 1} = \frac{3h\nu}{h\nu/kT}$$

$$= 3kT.$$

Therefore the total energy of one mole of atoms is

$$E = N\bar{\epsilon} = 3NkT = 3RT,$$

and the heat capacity is

$$C_V = \frac{\Delta E}{\Delta T} = 3R = 6 \text{ cal/mole-deg.}$$

This result is close to the value used in the Dulong and Petit law. The Dulong and Petit constant is usually taken to be approximately 6.3 cal/mole-deg, since it refers to C_P, the heat capacity at constant pressure, which is slightly larger than C_V, the heat capacity at constant volume.

It is more difficult to analyze the behavior of the heat capacity at low temperatures, since it is not possible to simplify Eq. (3.4) appreciably when T is very small. Moreover, a simplification of Eq. (3.4) for low-temperature situations is not particularly useful, for the Einstein theory is only approximate, and its limitations are most serious at low temperatures. Nevertheless, the theory is *qualitatively* correct, and a certain amount can be learned from the graph of the expression

$$E = \frac{3Nh\nu}{e^{h\nu/kT} - 1}$$

as a function of temperature which is shown in Fig. 3.39. The slope of the curve in Fig. 3.39 is $\Delta E/\Delta T$, or exactly equal to the heat capacity. At the higher temperatures, the curve is a straight line of slope $3Nk = 3R$, which corresponds to the limiting Dulong and Petit law. At lower temperatures, the curve becomes progressively flatter, $\Delta E / \Delta T$ becomes smaller, and the heat capacity eventually vanishes at absolute zero temperature.

The important feature of this analysis is the demonstration that when the condition $h\nu / kT \ll 1$ is not satisfied, the heat capacity falls below the limiting value of 6 cal/mole-deg. The quantity $h\nu$ is the difference between the various allowed energies of an oscillator, and if this difference were vanishingly small, we would always have $h\nu / kT \ll 1$ for all finite temperatures, and the heat capacity of solids would always be 6 cal/mole-deg. The failure of the Dulong-Petit law is, therefore, a demonstration of the existence and importance of the separated allowed energies of atomic systems.

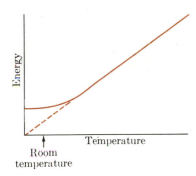

The dependence of the vibrational energy of a crystal on temperature. The slope of the line is the heat capacity, C_V.

FIG. 3.39

3.8 CONCLUSION

The mechanical, electrical, optical, and thermal properties of a solid are determined by the geometry of its crystal lattice and by the nature and strength of the forces that hold it together. Because of the large number of possible crystal lattices and the variety of cohesive forces, the properties of one solid can differ enormously from those of another. Nevertheless, it is possible to detect regularities and relationships in the behavior of solids and to classify them accordingly. Molecular crystals of spherical atoms and molecules tend to assume closest-packed structures and are volatile and mechanically weak. Metals, in general, have a densely packed structure—either of the closest-packed lattices or the body-centered cubic arrangement. The structures of a large number of binary ionic compounds are related to the closest-packed structures; ions of one charge form an expanded lattice of the hexagonal or cubic closest-packed type, with ions of the opposite charge occupying some or all of either the octahedral or tetrahedral holes. We shall find in our study of descriptive chemistry that the ability to recognize the similarities and differences in the crystal structures of solid compounds will help us to understand their chemical and physical properties.

SUGGESTIONS FOR FURTHER READING

Addison, W. E., *Structural Principles in Inorganic Compounds*. London: Longmans, 1961.

Azaroff, L. V., *Introduction to Solids*. New York: McGraw-Hill, 1960.

Hannay, N. B., *Solid State Chemistry*. Englewood Cliffs, New Jersey: Prentice-Hall, 1967.

Harvey, K. B., and G. B. Porter, *Introduction to Physical Inorganic Chemistry*. Reading, Mass.: Addison-Wesley, 1963. Chapter 7.

Moffatt, W. G., G. W. Pearsall, and J. Wulf, *The Structure and Properties of Materials*, Vol. 1. New York: Wiley, 1964.

Moore, W. J., *Seven Solid States*. New York: Benjamin, 1967.

Wells, A. F., *The Third Dimension in Chemistry*. New York: Oxford, 1956.

PROBLEMS

3.1 Many of the features of crystal lattices can be appreciated fully only upon inspection of a three-dimensional model of the structure. Convenient and inexpensive models of crystal lattices can be constructed from gumdrops and toothpicks. Build models of the crystal structures shown in this chapter.

3.2 The density of solid sodium chloride is 2.165 gm/cc, and one mole weighs 58.448 gm. What are the dimensions of a cube that contains one mole of solid NaCl? If the distance between centers of adjacent Na^+ and Cl^- ions is 2.819×10^{-8} cm, how many ions of each charge lie along each edge of the cube? Calculate Avogadro's number from these data.

3.3 Consider a face-centered cubic lattice made up of identical hard spheres of radius R. What are the dimensions of a cubical box that will just enclose the centers of the 14 spheres shown in Fig. 3.25? What is the volume of this box? Calculate the fraction of this volume that the spheres actually occupy.

3.4 The simple cubic lattice consists of eight identical spheres of radius R in contact, placed at the corners of a cube. What is the volume of the cubical box that will just enclose these eight spheres, and what fraction of this volume is actually occupied by the spheres?

3.5 The faces of the cubic crystal of sodium chloride shown in Fig. 3.5(a) are parallel to the faces of the face-centered cubic lattice of Fig. 3.7. Consequently, each of the lattice planes parallel to the cubic faces contains equal numbers of sodium and chloride ions. Is the same true of a plane parallel to the octahedral faces of Fig. 3.5(c)? To answer, compare Figs. 3.6 and 3.7.

3.6 A tetrahedral site in a closest-packed lattice can be generated by placing four spheres of radius R at alternate corners of a cube. Since the spheres are in contact, the length of a diagonal of a face of this cube is equal to $2R$. What is the length of the body diagonal of this cube? The radius of the tetrahedral hole is equal to the difference between half the body diagonal and R. What is the radius of the tetrahedral hole?

3.7 An atom vibrating about a point in a crystal lattice is in some ways similar to a mass vibrating at the end of a spring. For such systems, the vibration frequency increases as mass decreases. From this fact, explain why crystals of light atoms like C, B, and Be have heat capacities that are smaller than 6 cal/mole at room temperature.

3.8 When the NaCl crystal is investigated with x-rays of 0.586-A wavelength, the first Bragg diffraction occurs at $\theta = 5°58'$, and comes from planes of ions which are parallel to the face of the face-centered cubic lattice. Calculate the separation of these planes. What is the smallest distance between sodium and chlorine nuclei in the crystal? What is the smallest distance between chlorine nuclei? Consult Fig. 3.7 for help with this problem.

3.9 To a first approximation, Schottky and Frenkel defects occur without changing the volume of a crystal. Suppose a sodium chloride crystal had 10^{-3} atom fraction of (a) Frenkel defects and (b) Schottky defects. Calculate the change in density for these two cases from the ideal density of 2.165 gm/cc.

3.10 A certain sample of cuprous sulfide is found to have the composition $Cu_{1.92}S$, because of incorporation of Cu^{++} ions in the lattice. What is the ratio of Cu^{++} to Cu^+ in this crystal?

3.11 Copper has a face-centered cubic structure with a unit-cell edge length of 3.61 A. What is the size of the largest atom which could fit into the interstices of the copper lattice without distorting it?

3.12 The heat capacity C_v is defined as the derivative of the energy with respect to temperature, or $C_v = dE/dT$. Using the expression $\bar{\epsilon} = E/N$, and Eq. (3.4), evaluate C_v by differentiation. Verify the fact that C_v goes to zero as T goes to zero.

CHAPTER 4

LIQUIDS AND SOLUTIONS

In previous chapters we have remarked that solids and gases represent the extreme states of behavior of collections of molecules. The liquid state can be thought of as an intermediate condition in which some of the properties found in either solids or gases are displayed. Liquids, like gases, are isotropic and flow readily under applied stress, but like solids, they are dense, relatively incompressible, and have properties that are largely determined by the nature and strength of intermolecular forces. We shall also find that with respect to molecular order, liquids are intermediate between solids and gases. The fact that liquids are isotropic tells us immediately that they do not have the extended lattice structure and long-range order of solids. Yet the density of a liquid is generally only 10% less than that of its solid phase; this must mean that the molecules in a liquid are packed together with some regularity, and do not exhibit the complete chaos associated with molecules in the gas phase.

The remarkable ability of liquids to act as solvents is one of their most important properties. In the first place, liquid solutions provide an extremely convenient means of bringing together carefully measured amounts of reagents and of allowing them to react in a controlled manner. Second, the nature of the reactions which proceed and the speed at which they occur can be greatly influenced by the properties of the liquid solvent medium. Finally, the physical properties of solutions are interesting and important, because they can be used to determine molecular weights of dissolved substances and to study the nature and strength of forces between solvent and solute molecules.

In this chapter, our emphasis will be on the directly observable macroscopic properties of liquids and solutions, rather than on the behavior of individual molecules. Nevertheless, one of the most engaging and absorbing features of the study of chemistry is the attempt to explain the behavior of bulk matter in terms of molecular properties. Therefore in this section we will outline a molecular picture which will help us to understand and relate phenomena associated with the liquid state.

We have remarked that in a liquid, molecules are close to each other, and that consequently the forces exerted on one molecule by its neighbors are substantial. Thus the problem of analyzing the motion of a single molecule is exceedingly difficult, for each is constantly in "collision," subject to the forces of as many as twelve nearest neighbors. What then can we say about molecular motion in liquids? One of the most revealing observations in this respect was made by the botanist Robert Brown in 1827. Brown discovered that very tiny particles (10^{-4}-cm diameter) suspended in a liquid undergo incessant randomly directed motion. These motions occur without any apparent external cause such as stirring or convection, and are evidently associated with an intrinsic property of all liquids. A wealth of experimental observation has confirmed the idea that this Brownian motion is a direct manifestation of the thermal motion of molecules. When suspended in a liquid, a very small particle constantly experiences collisions with all the molecules surrounding it. If the particle is small enough, so few molecules will be able to collide with it that at any particular instant the number striking it from one side may be different from the number striking it from the other sides; consequently the particle will be displaced. Subsequently, another unbalance of collisional forces may occur, this time displacing the particle in a different direction. The great majority of these displacements are so small that they cannot be detected individually, but the motion which is observed is a result of many of the smaller random displacements. In essence, a Brownian particle is a "molecule" large enough to be observable, but small enough to execute observable random thermal motion.

Analysis of the motion of Brownian particles shows that their average kinetic energy is $\frac{3}{2}kT$. Since each particle is to be considered as one of the molecules of the liquid, we can conclude that the average kinetic energy of a molecule in a liquid is also $\frac{3}{2}kT$—exactly the same as the kinetic energy of a gaseous molecule at the same temperature. Even more detailed considerations have led to the conclusion that the kinetic energies of molecules in the liquid phase are distributed over a very wide range of values according to the Maxwell-Boltzmann distribution law, Eq. (2.17). In other words, in liquids, as in gases, molecules are in incessant random motion; the average *kinetic* energy and the fraction of molecules with any particular value of the kinetic energy are the same for either phase at the same temperature. However, a molecule in a liquid is always subject to the forces of its neighbors, and consequently its *potential* energy is lower, and its undeflected trajectories shorter, than if it were in the gas phase.

Further insight into the microscopic nature of liquids comes from the temperature dependence of their densities, or molar volumes. When a substance changes from a solid to a liquid, then, in most cases, the molar volume *increases* abruptly by about 10%, and as the liquid is warmed to even higher temperatures this expansion continues. This increase in volume upon melting must be a consequence of a general "separation" of the molecules and a slight lessening of intermolecular forces.

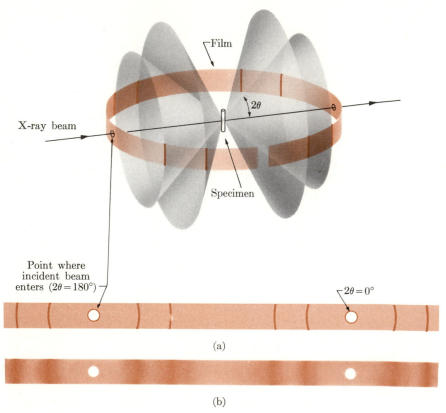

FIG. 4.1 Comparison of the x-ray diffraction patterns of (a) a powdered solid, and (b) its liquid. (After L. H. Van Vlack, *Elements of Materials Science*, 2nd ed. Reading, Mass.: Addison-Wesley, 1964.)

What is the detailed nature of the melting process? X-ray studies provide considerable information. We noted in Chapter 3 that crystalline solids produce sharp x-ray diffraction patterns, and that this sharpness is a consequence of the extended long-range order of the crystal lattice. What, then, is the nature of the x-ray diffraction pattern of a liquid? If the arrangement of atoms in a liquid were completely random, we would expect no diffraction pattern at all, only an almost uniform scattering of radiation. If the arrangement of atoms has some degree of order in the liquid, we might expect a diffraction pattern that somewhat resembles that of a solid. To see the most significant difference between a solid and its liquid, the x-ray pattern of a powdered, or microcrystalline, solid must be compared with the diffraction pattern of its liquid, as in Fig. 4.1. In the powdered sample the crystals are small and randomly oriented, and consequently some crystals are oriented so as to satisfy the Bragg diffraction condition. The result is a sharp line-diffraction pattern shown schematically in Fig. 4.1. The diffraction pattern of the liquid sample also shows a definite structure, and the intensity maxima correspond approximately to those found in the polycrystalline solid. This must mean that in a liquid, a certain amount of regularity in the packing of the atoms exists. The most important qualitative feature of the diffraction pattern of the liquid, however, is that the lines are *diffuse*, not sharp like those of the solid. A diffuse diffraction pattern means that the diffraction angle θ is not well defined. If we refer now to the Bragg diffraction equation

$$n\lambda = 2d \sin \theta,$$

we see that if the diffraction angles θ are not well defined, it must mean that the distance d between the repeating units is not always constant. In other words, the diffuse x-ray diffraction patterns of liquids show that at any one time, there are atoms separated by *various* values of d, and it is in this sense that the structure of a liquid is disordered. The uncertainty or variability of the separation of atoms means that a regular lattice unit which would repeat itself indefinitely simply does not exist.

While the separation of atoms in the liquid state has no single well-defined value, their diffuse x-ray patterns can be interpreted to yield the probability that an atom will be found at any distance r from another atom. Let ρ be equal to the number of atoms per cubic centimeter. Then the quantity $4\pi r^2 \rho \, \Delta r$ is the number of atoms in the spherical shell of radius r and thickness Δr. The probability of finding an atom at a particular distance r from the center of another atom is proportional to $4\pi r^2 \rho$; this quantity is plotted in Fig. 4.2. We see that $4\pi r^2 \rho$ is zero when r is less than the van der Waals diameter of an atom, but rises very rapidly to a sharp maximum for r in the vicinity of the van der Waals diameter. This indicates that a large fraction of the atoms are in what might be described as an almost close-packed situation. At somewhat larger values of r, the probability of finding an atom decreases, since the presence of the nearest neighbors to an atom *tends* to prevent there being any atoms located at, let us say, one and one-half the van der Waals diameter. However, note that the probability of finding an atom does not fall to zero at any value of r. This is a consequence of the imperfect packing, which allows atoms to be separated by any distance. The several maxima in the radial distribution curve show that there are most probable separation distances which are approximately equal to one or two times the van der Waals diameter of the atoms.

It is helpful to compare the broad maxima in the probable locations of atoms in the liquid phase with the very sharp exact locations of atoms in the corresponding crystalline solid. The latter are shown in Fig. 4.2 as vertical lines whose location is set by the crystal lattice and whose height is proportional to the number of atoms at the particular separation distance. Thus the set of vertical lines represents the radial distribution function of the solid. Comparison of the radial distributions for the two phases supports the idea that in the liquid phase atoms are packed together in a disordered manner, and that liquids do not have the extended lattice structure of solids.

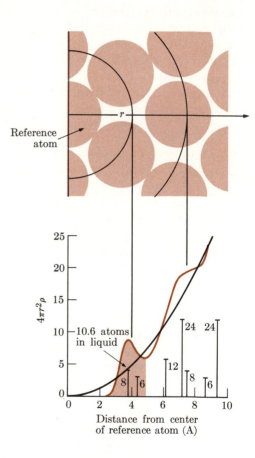

FIG. 4.2   Atomic distribution curve for liquid sodium calculated from x-ray diffraction data. The upper drawing explains the two maxima in the curve in terms of "shells" of atoms around the reference atom. The dashed curve represents a uniform distribution of atoms, and the vertical lines show the positions and numbers of neighboring atoms in solid sodium.   (After G. W. Castellan, *Physical Chemistry.* Reading, Mass.: Addison-Wesley, 1964.)

Figure 4.3 is a schematic comparison of the structures of a solid and a liquid and illustrates the ideas we have just discussed. In the liquid, there are regions in which the arrangement of the atoms is nearly perfect closest packing. However, there are atoms in other regions that have only five or four nearest neighbors, instead of six. This irregularity in packing introduces gaps or "holes" into what might otherwise have been a perfect closest-packed structure. Due to the incessant random motion of the molecules, these "holes" are not of definite size or shape, and they can spontaneously appear, be distorted, and move from one place to another. Since the introduction of these holes increases the average distance between molecules, compared to what it would be in a solid, the average intermolecular potential energy of a liquid must be higher than that of a solid. It is precisely for this reason that the latent heat of fusion must be supplied to melt a solid.

This picture of the destruction of the solid lattice upon melting is consistent with the existence of sharp melting temperatures. It is not possible to introduce the disordered liquid structure gradually into the solid lattice over a range of temperatures. Order is a property associated with the arrangement of many atoms, and one cannot have a structure which is at the same time ordered and disordered. Thus melting and freezing are *cooperative phenomena* which involve a concerted rearrangement of large numbers of atoms. Melting occurs abruptly when atoms acquire enough thermal energy to destroy the energetically more stable crystal lattice in favor of the more disordered liquid structure.

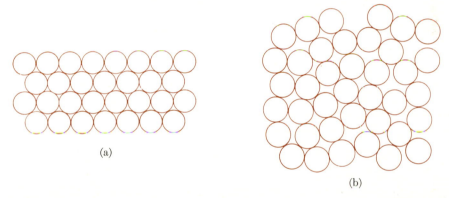

(a)

(b)

Schematic view of structure in (a) a crystal and (b) a liquid. (After G. W. Castellan, *Physical Chemistry*. Reading, Mass.: Addison-Wesley, 1964.)   **FIG. 4.3**

There are other properties of liquids which can be readily explained in terms of their disordered structure. Consider, for example, the fluidity of liquids. At the freezing temperature, a solid and liquid both contain the same type of molecule at the same temperature. Yet the solid structure is rigid, and the liquid yields to an applied stress. To explain this we need only recall that in order to deform a *perfect* crystal lattice, *large numbers* of atoms must be displaced relative to one another *at the same time*. Because so many atoms must move at once, deformation of a solid is opposed by strong intermolecular forces. However, if defects exist in the crystal lattice, the difficulty of producing a deformation is reduced considerably. The defects provide low-energy paths by which atoms can be displaced. In a liquid, of course, these defects are present in profusion. That is, the intrinsically disordered structure of a liquid provides many paths by which groups of atoms can be moved past one another without a serious increase in the existing average interatomic separations. In effect, the irregularities or holes in the structure provide a flow mechanism in which only a few molecules need move simultaneously, and the intermolecular forces which resist such motion are consequently relatively small. A molecule near a hole may move into it, and the vacated site in turn may be occupied by another molecule, and so on. Thus molecular displacement occurs without a serious disturbance of the liquid structure. The spontaneous diffusive mixing which occurs when two liquids are brought into contact proceeds by a similar mechanism. As holes appear, disappear, and change shape, the molecules of the two liquids can intermingle simply as a result of their ever-present thermal kinetic energy.

Question. Can you explain why atomic size is generally not a critical factor in determining the solubility of substances in liquids, whereas it is very important in determining their solubility in solids?

4.2 PHASE EQUILIBRIA

A large part of this chapter is concerned with situations in which two phases exist together in a closed container. If no *net* conversion of one phase to the other is occurring, the two phases are said to be *in equilibrium* with each other. A thorough understanding of both the qualitative and quantitative aspects of physical and chemical equilibria is absolutely essential to the study of chemistry. Fortunately, a study of phase equilibria provides a number of simple illustrations of the important general features of all equilibria which we will use repeatedly throughout this book. Before discussing phase equilibria in detail, we shall examine the energetic relations between the phases.

Energetics of Phase Changes

In order to accomplish or to describe a controlled experiment, a scientist starts by isolating or defining the part of the physical universe in which he is interested. This part of the universe under investigation is called the *system;* all other external entities which may influence the behavior of the system are known as the *surroundings.* In this section our systems will be pure materials that can be interconverted between the liquid, gaseous, and solid states by appropriate modification of their surroundings.

Anyone who has stepped from a swimming pool into a brisk breeze knows that when water (the system) evaporates, it absorbs heat from its surroundings (the skin, in this case). The same effect may be experienced with any other liquid which has a low boiling temperature. Some liquids, such as ethyl chloride, can freeze the skin upon evaporation, and are used as local anesthetics. It is also well known that when a gas condenses to a liquid, it releases heat to its surroundings. The absorption of heat upon evaporation and its evolution upon condensation are direct demonstrations that the energy of a liquid is lower than that of a gas at the same temperature. In order for a liquid to evaporate, work must be done against the attractive forces between the molecules, and this requires that energy be supplied as heat from the surroundings. Conversely, when a vapor condenses, the system goes to a state of lower energy, and thus energy is transferred as heat from the system to its surroundings.

The amount of heat absorbed when one mole of liquid evaporates is of interest, since it is a measure of the intermolecular potential energy. For instance, when one mole of water is completely vaporized at 25°C, it absorbs 10,519 cal from its surroundings. One way of representing this is

$$H_2O(l) + 10,519 \text{ cal} \rightarrow H_2O(g).$$

However, there is another way of describing the energetics of this process which has the advantage of emphasizing the change which takes place in the system. The amount of heat *absorbed by the system* in any change which takes place *at constant pressure* is called the **enthalpy change** of the system, and is symbolized by ΔH. If a system absorbs heat, its enthalpy increases, and ΔH is a positive number; if the system evolves heat, its enthalpy decreases, and ΔH is a negative number. Thus enthalpy may be thought of as a sort of heat content of the system; indeed, the words "heat content" and "enthalpy" are sometimes used synonymously. With these definitions in mind, we can write

$$H_2O(l) \rightarrow H_2O(g), \qquad \Delta H = 10,519 \text{ cal/mole},$$

to indicate that when one mole of water evaporates at a constant pressure, 10,519 cal are absorbed by the system. For the condensation process we have

$$H_2O(g) \rightarrow H_2O(l), \qquad \Delta H = -10,519 \text{ cal/mole},$$

where ΔH is negative, since the system loses heat to its surroundings.

Now let us pursue the suggestion that the enthalpy of vaporization, which we symbolize by ΔH_{vap}, is a measure of the potential energy of attraction between molecules. Liquids in which the attractive forces between molecules are very strong should have large values of ΔH_{vap}. In fact, there should be a general parallelism between ΔH_{vap} and ϵ, the minimum value of the potential energy of two molecules determined by gas imperfections. Table 4.1 compares ϵ and ΔH_{vap} for a number of gases, and a perusal of the values shows that ϵ and ΔH_{vap} are quite definitely related. Both factors increase as the number of electrons in the molecules increases.

The direct conversion of a solid to a vapor is called *sublimation*. In the sublimation process, heat must be supplied to the system in order to overcome the attractive forces between molecules in the solid state, and the amount of heat absorbed by the system when one mole of solid sublimes is known as the *enthalpy*

Table 4.1 Enthalpy of fusion and vaporization

Substance	ΔH_{fus} (kcal)	ΔH_{vap} (kcal)	ϵ^* (kcal)	T_b (°K)
O_2	0.106	1.63	0.225	90
N_2	0.172	1.33	0.181	77
H_2	0.028	0.216	0.073	20
He	0.005	0.020	0.020	4
Ar	0.265	1.56	0.236	87
Xe	0.490	3.02	0.440	166
CH_4	0.225	1.95	0.272	112

* The energy that corresponds to the minimum of the Lennard-Jones potential for the interaction of a pair of molecules, expressed in kcal/mole.

of sublimation, ΔH_{sub}. Thus for the ice-water vapor conversion we have

$$H_2O(s) \rightarrow H_2O(g), \qquad \Delta H = \Delta H_{sub} = 11{,}955 \text{ cal.}$$

Closely related to the enthalpies of sublimation and of vaporization is the *enthalpy of fusion,* the heat absorbed when one mole of solid is converted to liquid at a constant pressure. For the ice-water transition we have

$$H_2O(s) \rightarrow H_2O(l), \qquad \Delta H = \Delta H_{fus} = 1436 \text{ cal.}$$

Note that the direct conversion of solid to vapor is equivalent to melting the solid first, and then allowing the liquid to evaporate. Since the initial and final conditions of the system are the same, the values of ΔH for the two processes must be equal. Thus

$$\Delta H_{sub} = \Delta H_{fus} + \Delta H_{vap}.$$

We can see that this relation holds for water, since in this case we have $1436 + 10{,}519 = 11{,}955$, and experimental data for other substances confirm its general validity.

Examination of the data in Table 4.1 shows that ΔH_{fus} is always considerably less than ΔH_{vap} for a particular substance. Our picture of the solid, liquid, and gaseous states is entirely in accord with this fact. We have remarked that in a liquid the molecules are somewhat more loosely packed than in a solid. This small diminution in density slightly decreases the effect of the attractive forces between molecules, and consequently a relatively small amount of energy is required to convert a solid to a liquid. The evaporation of a liquid separates the molecules entirely, and thus reduces essentially to zero the attractive forces between molecules. Evaporation produces a far more profound change in molecular environment than does melting, and consequently ΔH_{vap} is larger than ΔH_{fus}.

Question. Plot ΔH_{vap} as a function of ϵ for the substances in Table 4.1. Can you interpret the significance of any approximate numerical relation you might find between ϵ and ΔH_{vap}?

A liquid of relatively low boiling temperature placed in a container open to the atmosphere will eventually evaporate entirely. Remembering that molecules in the liquid are "bound" by attractive forces to their neighbors, we might ask why some are able to overcome these forces and leave the liquid spontaneously. The answer lies in a consideration of the possible magnitudes of molecular kinetic energies, for these, as we have already mentioned, range from very low to very high values, and are distributed according to the Maxwell-Boltzmann law. Therefore, even if the average potential energy which binds the molecules to the liquid is substantial, there are always some molecules which have enough kinetic energy to overcome the binding forces and enter the vapor. According to the Maxwell-Boltzmann law, the fraction of the molecules which have kinetic energies greater than some minimum value ϵ, the value required for the molecules to leave the liquid, is proportional to the Boltzmann factor, $e^{-\epsilon/kT}$. Therefore, as long as the temperature remains constant, the fraction of liquid molecules with enough kinetic energy to evaporate remains the same, and evaporation continues. If the vessel is open to the atmosphere, vapor molecules are swept away, and evaporation continues until no liquid is left.

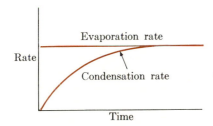

The time dependence of evaporation and condensation rates for a liquid evaporating into a closed container.

FIG. 4.4

Now let us analyze what happens when a liquid is placed in a closed evacuated container. Immediately the liquid starts to evaporate at a rate which is primarily determined by the fraction of molecules which have enough kinetic energy to overcome attractive forces and leave the surface. Initially the rate of condensation is zero, for there are no molecules in the vapor. As long as the temperature stays constant, evaporation continues at a constant rate, and the number of molecules in the vapor phase increases. Concurrently, the rate of condensation starts to increase, for as the pressure of the vapor grows, the number of gas molecules which collide with and reenter the liquid surface also increases.

The time dependence of the evaporation and condensation rates is shown in Fig. 4.4. As the condensation rate grows, it eventually becomes equal to the rate of evaporation. At this time the number of molecules which enter and which leave the vapor per unit time is the same, and consequently the pressure of the vapor stops increasing and remains constant. If the system is left undisturbed at a fixed temperature, evaporation and condensation continue at equal rates, and the pressure of the vapor remains unchanged. This, then, is a situation of equilibrium between the two phases. Note particularly that at equilibrium, evaporation and condensation do not stop, but that the constancy of the equilibrium vapor pressure is a consequence of these opposing processes proceeding at *equal* rates. Thus we say that phase equilibrium is *dynamic* in nature.

Suppose now that the liquid-vapor system is at equilibrium in a cylinder closed by a movable piston. What will happen if we suddenly raise the piston and increase the volume of the cylinder by a small amount, and at the same time, keep the temperature of the system constant? The immediate effect is to lower the pressure of the vapor, thereby removing the system from the equilibrium state. Because there are fewer molecules of vapor per unit volume, the number of collisions per unit time with the liquid surface is lowered, and the rate of condensation decreases. Nevertheless, the rate of evaporation does not change, for the expansion in no way alters the state of the liquid. Thus the consequence of the disturbance is that the rate of evaporation is once again greater than the rate of condensation—a condition which will inevitably cause an increase in the number of vapor molecules, a subsequent equality of the evaporation and condensation rates, and the *restoration of phase equilibrium*.

Now let the system, initially at equilibrium, be subjected to a sudden decrease in its volume. The initial consequence is to increase the concentration of molecules in the vapor and thus increase the rate of condensation. Once again, the rate of evaporation remains unchanged. With the condensation rate greater than the evaporation rate, the number of molecules in the vapor starts to decrease, and continues to do so until the rates of the opposing processes become equal and equilibrium is restored. Thus, regardless of the direction in which the system is displaced, it inevitably returns *unaided* to the equilibrium state.

Note particularly that the equilibrium vapor pressure is set by the equality of the condensation and evaporation rates, and that the evaporation rate, determined only by the fraction of molecules which have enough energy to leave the liquid, is always constant at a fixed temperature. Therefore, at a fixed temperature, the equilibrium vapor pressure is always the same, regardless of the direction from which the system approached equilibrium.

The Equilibrium State

Now it is possible to draw from our discussion of liquid-vapor equilibrium some very useful generalizations that apply to all situations of physical or chemical equilibrium. We have emphasized that the potential energy of molecules in the liquid state is lower than that of molecules in the gas phase. We have also observed that a liquid left undisturbed in a closed container inevitably moves toward a state of equilibrium in which there is a definite concentration of molecules in the vapor phase. Thus it follows that at equilibrium the system is not in a condition of minimum energy, for the energy of the system can always be lowered by condensing the vapor entirely. This conclusion may seem strange, for all our experience with simple mechanical systems suggests that they seek an equilibrium condition in which their energy is as low as possible; that is, objects fall, clocks run down, a stirred liquid stops moving. All these phenomena can be summarized by saying that mechanical systems seek a resting place of minimum energy.

Almost the same thing can be said about molecular systems. It is clear that *one* of the driving forces that determines the behavior of molecular systems is the tendency to seek a state of the lowest possible energy. After all, this is the reason that a gas condenses or that a liquid freezes. But it is also certain that the tendency toward minimum energy cannot be the *only* factor governing the behavior of molecular systems; if it were, no gases would exist at any temperature. There is another driving force, just as important as the energy factor. To put it briefly, it is the tendency of systems to assume a state of maximum molecular chaos, or disorder.

Perhaps the simplest demonstration of this tendency is the observation that an ideal gas will expand spontaneously into an evacuated space. Surely the gas does not do so in order to achieve a state of lower energy, for we have seen that the energy of an ideal gas depends only on its temperature, and this need not change during the expansion. However, when a gas occupies the larger volume, all molecules have more space available to them, and it is more difficult to predict the exact position of any one of them. Whenever the detailed arrangement of the molecules is unknown or unknowable, we say that the system is disordered. Thus we can justifiably describe the expansion of an ideal gas by saying that it increases the variety of positions available to molecules, and thereby increases the disorder of the system.

The evaporation of a liquid provides still another illustration of the tendency toward maximum disorder. In the liquid state, the motion of any one molecule is somewhat limited by the presence of its neighbors. The molecules are arranged in a manner such that if we know where one is, we can predict the location of others with some certainty by using the radial distribution function. This possibility is diminished considerably for molecules in the gas phase, where at any instant the distribution of the molecules is completely random. Thus we can classify the gas phase as a condition of greater molecular chaos than the liquid or solid phases.

Now if the tendency of systems to move toward a state of maximum molecular chaos were all-important, all materials would eventually evaporate or dissociate entirely, and there would be no solids or liquids at any temperature. Thus, on one hand, we have the drive toward lowest energy, that can be followed by allowing molecules to associate in one of the condensed phases, and on the other hand, we have the drive toward molecular chaos, which could be achieved by the evaporation or separation of molecules into independent units. The condition of equilibrium must be one which is the best compromise between these two conflicting drives toward maximum chaos and minimum energy. Consequently we have two ways of interpreting the condition of liquid-vapor equilibrium. From one point of view, equilibrium represents the situation in which the rates of evaporation and condensation are equal. From the other viewpoint, equilibrium is the condition of most favorable compromise between the natural tendencies of the system to reach minimum energy and maximum chaos.

By now it should be clear that the concept of molecular chaos is important not only in the description of the nature of atomic arrangements in pure phases, but also in understanding the factors that are responsible for phase changes and phase equilibria. In fact, molecular chaos is a concept useful in the analysis of any phenomenon, chemical or physical, which involves collections of molecules. We shall find in Chapter 8 that it is possible to define and determine experimentally a property of a system that measures molecular chaos. This property is called **entropy.** A full appreciation of what entropy is and how it depends on the properties of systems requires the thermodynamic arguments presented in Chapter 8, but for the present we shall find the qualitative association between entropy and molecular chaos to be sufficient. What we have said about the nature of solids, liquids, and gases suggests that the entropy of a liquid is greater than that of a solid, and that the entropy of a gas is greater than that of either a liquid or solid. Our remarks about the natural tendency of systems to reach a state of molecular chaos can be rephrased to say that systems have a tendency to reach a state of maximum entropy.

To summarize our discussion, we can list four important features of all equilibria which have been illustrated by the liquid-vapor equilibrium:

1. Equilibrium in molecular systems is dynamic and is a consequence of the equality of the rates of opposing reactions.

2. A system moves spontaneously toward a state of equilibrium. If a system initially at equilibrium is perturbed by some change in its surroundings, it reacts in a manner which restores it to equilibrium.

3. The nature and properties of an equilibrium state are the same, regardless of how it is reached.

4. The condition of a system at equilibrium represents a compromise between two opposing tendencies: the drive for molecules to assume the state of lowest energy and the urge toward molecular chaos or maximum entropy.

Temperature Dependence of Vapor Pressure

Experimental measurements show that the equilibrium vapor pressure of a liquid increases as the temperature increases. Data that illustrate this point are shown in Fig. 4.5. In the temperature range in which the vapor pressure is small, it is relatively insensitive to the temperature, but the vapor pressure grows at an increasing rate as the temperature is raised. The temperature at which the equilibrium vapor pressure becomes equal to 1 atm is called the normal boiling temperature, or the boiling point. In the boiling process, bubbles of vapor form throughout the bulk of the liquid. In other words, evaporation occurs *anywhere* in the liquid, not just at the upper surface. The reason that this occurs only when the vapor pressure equals the atmospheric pressure is easy to understand. In order for a bubble to form and grow, the pressure of the vapor inside the bubble must be at least equal to the pressure exerted on it by the liquid. This in turn is equal to the pressure of the atmosphere plus the very small pressure due to the weight of the liquid above the bubble. Therefore, bubble formation and boiling occur only when the vapor pressure of the liquid is equal to the pressure of the atmosphere.

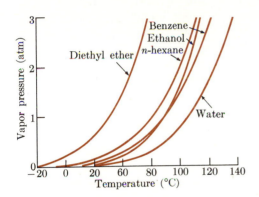

The vapor pressures of several liquids plotted as a function of temperature.

FIG. 4.5

The initiation of a bubble in the bulk of a pure liquid is a very difficult process, since it requires that many molecules with kinetic energies greater than that required for vaporization must be close to one another. Hence the fact that the liquid reaches the boiling temperature is no guarantee that boiling will occur. If it does not, continued addition of heat will cause the liquid to become superheated; that is, to reach a temperature greater than its boiling point. When bubble formation in a superheated liquid finally occurs, it does so with almost explosive violence, because the vapor pressure in any bubble formed greatly exceeds atmospheric pressure, and the bubbles tend to expand rapidly. Such violent boiling, called *bumping*, can be avoided by introducing agents which initiate bubbles in the liquid as soon as the boiling temperature is reached. Porous pieces of ceramic material which evolve small bubbles of air into which evaporation can occur serve very well in this application.

From Fig. 4.5 we can see that comparison of boiling points is a convenient, if approximate, way of evaluating the relative volatilities of liquids. That is, liquids that boil at low temperatures usually have vapor pressures that are greater at all temperatures than the vapor pressures of liquids that boil at much higher temperatures. However, some of the curves in Fig. 4.5 cross each other, showing that there are exceptions to the correspondence between vapor pressure and normal boiling temperature. Another point of some importance is the correlation between ΔH_{vap} and the normal boiling temperature. Reference to Table 4.1 shows that liquids with high boiling temperatures have relatively large enthalpies of vaporization. Thus a liquid in which the intermolecular forces are large will, in general, have a high boiling temperature.

Phase Diagrams

A solid, like a liquid, can exist in equilibrium with its vapor in a closed container. Thus at any fixed temperature, each solid has a characteristic fixed vapor pressure. The vapor pressure of a solid increases with increasing temperature, and it is informative to plot the vapor pressure of a solid and that of its liquid on the same diagram, as is done in Fig. 4.6. The vapor pressure of the solid increases more rapidly as the temperature is raised than does the vapor pressure of the liquid. Therefore, there is an intersection of the two vapor-pressure curves. At the temperature corresponding to the intersection, the liquid and solid phases are in equilibrium and have the same vapor pressure. It is not difficult to construct an argument which shows that in this condition, liquid and solid must be in equilibrium with each other.

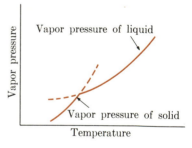

FIG. 4.6   Vapor pressure of a solid and its liquid, as a function of temperature.

Consider the apparatus shown in Fig. 4.7. One bulb contains a solid, the other its liquid, and the two bulbs are connected so that vapor can pass freely from one to the other. Now let both bulbs be immersed in a bath at temperature T_1. If the vapor pressure of the solid at T_1 is less than that of the liquid, then gas will flow from the bulb containing the liquid to that containing the solid. As this flow persists, the liquid evaporates, and the vapor condenses as solid. This continues until all the liquid has evaporated. Alternatively, if the apparatus is held at a temperature T_2 at which the vapor pressure of the solid is greater than that of the liquid, the vapor flows from the solid to the liquid. This is accompanied by evaporation of the solid and condensation of the vapor to the liquid, and the process continues until all the solid is consumed.

Clearly, the solid and liquid are not in equilibrium with each other at temperature T_1 or T_2, for if they were in equilibrium, the system would not change, and both phases would remain indefinitely. In contrast, if the temperature of the apparatus is set at T_0, at which the vapor pressures of the liquid and solid are the same, the pressure is uniform, and there is no tendency for vapor to flow from one chamber to the other. Thus both the liquid and solid phases remain indefinitely. This persistence of the state of the system indicates that the solid and liquid phases are in equilibrium at a temperature at which their vapor pressures are the same. This situation is a specific example of an important general principle: if each of two phases is in simultaneous equilibrium with a third, then the two phases are in equilibrium with each other.

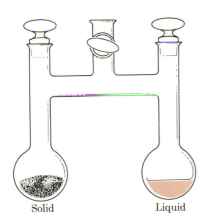

Solid   Liquid

Apparatus for the equilibration of two phases not in contact.

FIG. 4.7

The temperature at which liquid, vapor, and solid are in simultaneous equilibrium with one another is called the *triple-point temperature*. The triple point is usually very close to what is known as the *freezing point*, which is the temperature at which liquid, vapor, and solid are in simultaneous equilibrium *in the presence of 1-atm pressure of air*. For example, liquid water and ice are simultaneously in equilibrium with water vapor only at a temperature of 0°C, in the presence of 1-atm pressure of air. When air is completely eliminated from the container, water, ice, and water vapor are simultaneously in equilibrium only at a temperature of 0.0098°C. This temperature is called the triple point of water, and we see that it differs only slightly from the normal freezing point.

Let us consider the equilibrium among water vapor, ice, and liquid water in more detail. The vapor pressure of water at the triple point is 4.579 mm. What would happen if the external pressure applied to the system by a piston were to be increased above 4.579 mm? First of all, the water vapor initially present would be completely converted to liquid and solid. It is found experimentally that as the pressure on the system is further increased, the temperature of the system must be lowered, *in order for both ice and water to remain at equilibrium*. For each pressure *greater* than 4.579 mm, there is only one particular value of the temperature at which ice and water can both be present at equilibrium, and as the pressure increases the temperature necessary to maintain equilibrium decreases. Another way of stating this is to say that as the pressure applied to ice increases, its melting temperature decreases.

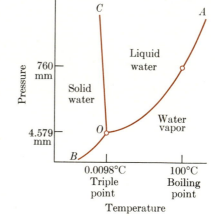

FIG. 4.8   The phase diagram for water (not drawn to scale).

Figure 4.8 is called the phase diagram for water. The lines represent simultaneous values of pressure and temperature at which two phases may be present at equilibrium. At the temperatures and pressures which lie on the line OA, liquid water and its vapor can be at equilibrium. Along the line OB, ice and its vapor are at equilibrium, while the sets of temperature and pressure at which both ice and liquid water are in equilibrium lie along OC. Only at the pressure and temperature corresponding to the triple point (0.0098°C, 4.579 mm) are ice, water, and water vapor simultaneously present at equilibrium. The areas between the curves represent temperatures and pressures at which only one phase can exist.

Figure 4.8 shows that when ice melts, its vapor pressure is considerably less than 760 mm. This is the behavior which is observed for most substances. However, carbon dioxide and iodine are solids whose vapor pressures reach 760 mm at temperatures which are *lower* than their triple-point temperature, and consequently these substances evaporate at 1-atm pressure without ever melting. The phase diagram for carbon dioxide is shown in Fig. 4.9. The temperature at which the vapor-pressure curve intersects the 760-mm line is known as the *normal sublimation temperature*. Solid carbon dioxide evaporating at 1-atm pressure maintains a constant temperature of −78.1°C. Carbon dioxide can be liquefied only by raising the applied pressure to at least 3880 mm (5.1 atm); at this pressure carbon dioxide melts at a temperature of −56.6°C.

It is difficult to give a definition which tells clearly and briefly how solutions differ from mixtures and compounds, in spite of the fact that solutions are among the most familiar substances in nature. However, it is often true that the most common concepts are the most difficult to define precisely. A solution is a homogeneous substance that has, over certain limits, a continuously variable composition. The word "homogeneous" sets a true solution apart from a mechanical mixture, for mixtures have macroscopic regions which have distinct and different composition and properties. The properties and composition of a solution are uniform, as long as the solution is not examined at the molecular level. There are substances, however, which cannot be clearly classified as solutions or mixtures. A solution of soap in water has a cloudy appearance due to particles which consist of many soap molecules collected together. Such a substance has properties and composition which might be described as either inhomogeneous or homogeneous depending on the experiment which is to be done. Therefore there is no sharp dividing line between mixtures and solutions.

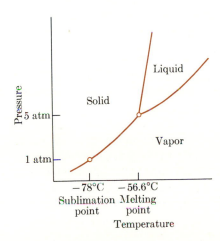

The phase diagram for carbon dioxide (not drawn to scale).

FIG. 4.9

The requirement that solutions have continuously variable composition distinguishes them from most compounds. However, as we saw in Chapter 1, many solid materials which we commonly think of as compounds actually show variable composition. Cuprous sulfide and ferrous oxide are examples of compounds which might also be thought of as solutions. No matter how carefully we make our definitions of solution, mixture, and compound, we must expect to find certain substances which cannot be uniquely classified as one of these. There is no reason to expect nature to be cooperative and produce only substances which are easily classified, and since this is the case, the most useful definitions are often the shortest, rather than the most exhaustive.

Types of Solutions

Just as the variables, pressure, volume, and temperature, were used to describe the state or condition of pure gases, liquids, and solids, these and certain other variables must be used to describe solutions. First, some statement must be made about what chemically important constituents are present in the solution. A solution of ethyl alcohol (C_2H_5OH) and water really contains three elements, hydrogen, oxygen, and carbon. However, because there is a quantitative relationship (the law of definite composition) between the amounts of carbon, hydrogen, and oxygen in ethyl alcohol, and a similar relationship between the amounts of hydrogen and oxygen in water, the composition of the solution can be completely described by specifying only the quantities of alcohol and water used to prepare the solution. The substances used to specify the composition of a solution are known as *components*. One of the components, usually the one present in greatest quantity, is called the *solvent;* any other component is called a *solute*.

There are many possible types of solute-solvent pairs. A mixture of two gases satisfies our definition of a solution, but the properties of gaseous mixtures have been treated by Dalton's law in Chapter 2, and we shall not consider them again here. Other types of solutions that are important are:

1. liquid in liquid,
2. solid in liquid,
3. gas in liquid,
4. liquid in solid,
5. gas in solid,
6. solid in solid.

Of these, the first three are common, and the last three, called *solid solutions*, occur less frequently. Mercury dissolved in zinc, hydrogen gas dissolved in palladium metal, and zinc dissolved in copper are examples of solutions in which a liquid, gas, and solid, respectively, are dissolved in a solid. Apart from their mechanical properties, solid solutions do not differ greatly from the solutions of liquids which we shall discuss in detail.

Concentration Units

In addition to the qualitative statement of what components are present in a solution, some specification of the amount of each component must be made. Usually only the relative amounts of the components are specified, since the properties of solutions do not depend on the absolute amounts of material present. The relative amount of a substance is known as its *concentration* and is expressed in five common sets of units:

1. Mole-fraction units. The mole fraction of component 1 is the number of moles of component 1 divided by the total number of moles of *all* the components of the solution. For a solution made up of two components, 1 and 2, we have:

$$\text{mole fraction of component 1} = \frac{n_1}{n_1 + n_2},$$

$$\text{mole fraction of component 2} = \frac{n_2}{n_1 + n_2},$$

where n_1 and n_2 are the number of moles of components 1 and 2 in the solution. Commonly, the symbol x stands for the composition expressed in mole-fraction units. That is, in our example

$$x_1 = \frac{n_1}{n_1 + n_2}, \qquad x_2 = \frac{n_2}{n_1 + n_2}.$$

For an arbitrary number (i) of components we would have

$$x_1 = \frac{n_1}{n_1 + n_2 + \cdots + n_i}, \qquad x_2 = \frac{n_2}{n_1 + n_2 + \cdots + n_i}.$$

It is always true that the sum of all the mole fractions is unity:

$$x_1 + x_2 + \cdots = 1.$$

Mole-fraction units are useful when it is desirable to emphasize the relation between some concentration-dependent property of a solution and the relative numbers of molecules of solute and solvent.

2. Molality. The molality of a solution is defined as the number of moles of solute in 1000 gm of *solvent*. Molality is commonly symbolized by the lower-case letter m. Thus a 1-m aqueous solution of sodium chloride contains one mole of sodium ions and one mole of chloride ions in 1000 gm of water. Molality is a useful unit in calculations dealing with the freezing and boiling points of solutions, but the fact that it is difficult to weigh out liquid solvents makes molality an inconvenient unit for common laboratory work.

3. Molarity. This is the most common concentration unit. The molarity of a solution is the number of moles of solute in 1 *liter of solution*. The symbol for molarity is M. A 0.2-M solution of $BaCl_2$ contains 0.2 mole of the salt barium chloride in a liter of solution. The concentration of barium ion is 0.2 M as well, but the concentration of chloride ion is 0.4 M, since there are two moles of chloride ion in every mole of barium chloride. Molarity is a very convenient unit for laboratory work, since aqueous solutions of known molarity can be prepared easily by weighing out small amounts of solute and measuring the volume of solution in calibrated containers. However, since the volume of a solution depends on temperature, the concentration expressed in units of molarity also depends on temperature. This is a disadvantage which the mole-fraction and molality units do not have.

4. Formality. The formality of a solution is the number of gram-formula weights of solute per liter; the symbol for this unit is F. Formality is very similar to molarity, and its use avoids the difficulty of assigning a molecular weight to something (such as NaCl) that contains no discernible molecules. If the formula of sodium chloride is written as NaCl, a 1-F NaCl (one-formal) solution contains 58.5 gm of sodium chloride in one liter of solution. When an actual molecule (and hence molecular weight) exists, molarity and formality become identical. For solutions of ionic substances, or materials for which only empirical formulas are known, formality would seem to be the preferred unit of concentration. However, most chemists avoid the use of formality as a concentration unit in the interest of uniform notation. This book, for instance, will always refer to a solution which contains 58.5 gm of sodium chloride per liter as a 1-M NaCl solution, even though no sodium chloride molecules exist in solution.

5. Normality. The equivalent weight of any material is the weight which would react with, or be produced by, the reaction of 7.999 gm of oxygen or 1.008 gm of hydrogen. The normality of a solution is the number of gram-equivalent weights of solute in one liter of solution. The equivalent weight of zinc ion, for example, is $(65.38/2.016) \times 1.008$, since 65.38 gm of zinc metal will produce 2.016 gm of hydrogen gas in a reaction with any acid. A one-normal zinc ion solution (1-N Zn^{++}) therefore contains 32.5 gm of Zn^{++} in one liter of solution. Normality is a convenient unit for certain calculations of quantitative analysis.

Example 4.1 A solution is prepared by dissolving 2.50 gm of NaCl in 550 gm H_2O. The density of the resulting solution is 0.997 gm/ml. What is the molality, molarity, and mole fraction of NaCl?

$$\text{moles NaCl} = \frac{2.50 \text{ gm}}{58.44 \text{ gm/mole}} = 0.0428 \text{ moles},$$

$$\frac{\text{moles NaCl}}{\text{gm solvent}} \times 1000 = \text{molality} = 0.0778 \text{ m}.$$

The volume of the solution is

$$\frac{550 \text{ gm}}{0.997 \text{ gm/ml}} = 552 \text{ ml} = 0.552 \text{ l}.$$

Thus the molarity of NaCl is

$$\frac{0.0428 \text{ moles}}{0.552 \text{ l}} = 0.0776 \text{ M}.$$

The number of moles of water is

$$\frac{550 \text{ gm}}{18.0 \text{ gm/mole}} = 30.55,$$

so the mole fraction of NaCl is

$$X_{\text{NaCl}} = \frac{0.0428}{30.55 + 0.0428} = 1.4 \times 10^{-3}.$$

4.4 THE IDEAL SOLUTION

Let us consider a solution made up of a volatile solvent and a nonvolatile solute. It is found experimentally that the vapor pressure of the solvent depends on the concentration of the solvent in the solution, often in a very complicated manner. However, there are some solutions, those that can be formed from their components with no evolution or absorption of heat, for which the relation between vapor pressure and concentration is very simple. For these solutions the vapor pressure of the solvent is proportional to its mole fraction, and the proportionality constant is simply the vapor pressure of pure solvent. That is,

$$P_{\text{solution}} = P_{\text{solvent}} = P_1 = P_1^0 x_1 = P_1^0 \left(\frac{n_1}{n_1 + n_2} \right). \qquad (4.1)$$

In this equation, P_1 is the actual vapor pressure of the solvent (component 1), P_1^0 is the vapor pressure of component 1 when it is pure, x_1 is the mole fraction of component 1 (the solvent) in the solution, and n_1, n_2 are the number of moles of components 1 and 2, respectively, in the solution. Any solution whose vapor pressure depends on concentration according to Eq. (4.1) is called an **ideal solution.** The relation $P_1 = x_1 P_1^0$ is called **Raoult's law,** and thus a definition of an ideal solution is that it is one that obeys Raoult's law.

Let us examine the mathematical consequences of Eq. (4.1). First we will define the vapor-pressure lowering of the solvent as ΔP, where

$$\Delta P = P_1^0 - P_1.$$

Substitution of Eq. (4.1) for P_1 gives $\Delta P = P_1^0 - x_1 P_1^0 = (1 - x_1)P_1^0$. If the solution has only two components, 1 and 2, we have

$$x_1 + x_2 = 1,$$

so

$$x_2 = 1 - x_1,$$

and therefore

$$\Delta P = x_2 P_1^0. \tag{4.2}$$

This simple result will be very useful when we examine the effect of the solute on the boiling point and freezing point of the solution. In itself, however, it suggests a method for the determination of molecular weights of dissolved substances. We can write

$$\Delta P = P_1^0 \left(\frac{n_2}{n_1 + n_2} \right) = P_1^0 \left(\frac{W_2/MW_2}{W_1/MW_1 + W_2/MW_2} \right). \tag{4.3}$$

If a solution is made up by adding a certain known weight W_2 of a substance 2, whose molecular weight MW_2 is unknown, to a known weight W_1 of a solvent whose molecular weight MW_1 and vapor pressure P_1^0 are known, then measurement of ΔP, the vapor-pressure lowering, permits calculation of MW_2, the molecular weight of the unknown.

Example 4.2 The vapor pressure of water at 20°C is 17.54 mm. When 114 gm of sucrose are dissolved in 1000 gm of water, the vapor pressure is lowered by 0.11 mm.

$$\Delta P = P_1^0 \frac{n_2}{n_1 + n_2},$$

$$0.11 = 17.54 \left[\frac{114/MW}{114/MW + (1000/18.0)} \right],$$

$$MW = 325.$$

The formula of sucrose is actually $C_{12}H_{22}O_{11}$, which corresponds to a molecular weight of 342.

Boiling and Freezing Points of Solutions

It is informative to compare the vapor pressures of the ideal solution of a non-volatile solute and the pure solvent itself as a function of temperature. Figure 4.10 shows that at each temperature the vapor pressure of the solution is lower than that of the pure solvent, and as predicted by Eq. (4.2), the difference increases as temperature and vapor pressure increase. The intersection of the solution vapor-pressure curve with the line corresponding to 760 mm defines the normal boiling point of the solution. It is clear from Fig. 4.10 that the boiling point of the solution is higher than that of the pure solvent.

The intersection of the vapor-pressure curve of the solution with the vapor-pressure curve of the pure solid solvent is the freezing point of the solution, and should be the temperature at which the first crystals of the solvent appear when a solution is cooled. As the crystallization continues, the composition of the solution changes because solvent molecules are being removed from the solution. Therefore, the temperature at which crystals *first* appear is defined as the freezing point of a solution of a given concentration. From Fig. 4.10 it is clear that the freezing point of the solution is lower than that of the solvent, and that the freezing point of the solution decreases as the concentration of solute increases.

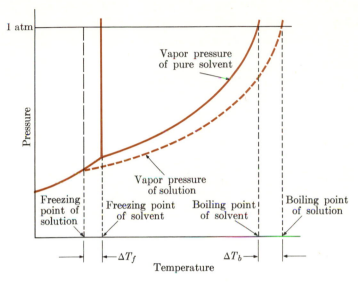

Diagram showing the lowering of the vapor pressure, increase of the boiling temperature, and depression of the freezing point that occur when a nonvolatile solute is dissolved in a volatile solvent.

FIG. 4.10

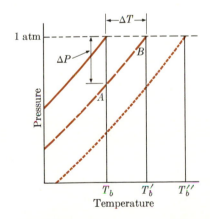

Relation between vapor-pressure lowering and boiling-point elevation. Vapor pressures of the pure solvent and of solutions of two concentrations are shown.

FIG. 4.11

By referring to Fig. 4.11, we can derive the quantitative relationship between the elevation of the boiling point ΔT, and the concentration of the solute. For small values of ΔT and ΔP, the segment AB of the vapor-pressure curve can be considered to be straight line. Then, by the properties of similar triangles,

ΔT is proportional to ΔP. For ideal solutions, ΔP is, by Eq. (4.2), proportional to x_2, the mole fraction of the solute. If K_b' is the proportionality constant which relates x_2 to ΔT, we have

$$\Delta T = K_b' x_2 = K_b' \left(\frac{n_2}{n_1 + n_2} \right). \qquad (4.4)$$

We can simplify this equation if we restrict our treatment to very dilute solutions. For a dilute solution, $n_2 \ll n_1$; therefore in Eq. (4.4) we can make the approximation that

$$\frac{n_2}{n_1 + n_2} \cong \frac{n_2}{n_1};$$

that is, we can neglect n_2 in the denominator if it is much smaller than n_1. We can write further that

$$\frac{n_2}{n_1} = \frac{w_2/MW_2}{w_1/MW_1},$$

where w_1 and w_2 refer to the weights of components 1 and 2 which are present in solution, and MW_1 and MW_2 refer to their molecular weights. Therefore, Eq. (4.4) becomes

$$\Delta T = K_b' \left(\frac{w_2/MW_2}{w_1/MW_1} \right). \qquad (4.5)$$

Experiments show that for ideal solutions the value of the constant K_b' depends only on the identity of the *solvent*, and not at all on the solute. The quantity MW_1 is also a property of the solvent only, and thus it is reasonable to combine MW_1 with K_b' to obtain a new constant. We define the constant K_b by the equation

$$1000\, K_b = K_b' MW_1.$$

Then substitution into Eq. (4.5) gives us

$$\Delta T = K_b \left(\frac{w_2/MW_2}{w_1} \right) 1000.$$

The factor in parentheses is the number of moles of component 2 per gram of component 1. If this quantity is multiplied by 1000, it becomes equal to the number of moles of solute per 1000 gm of solvent, or the molality, m. Thus we obtain the final form,

$$\Delta T = K_b m, \qquad (4.6)$$

for the dependence of the boiling point on concentration.

The constant K_b is called the molal boiling-point elevation constant; it is equal to the increase in the temperature of the boiling point of a 1-m solution.

Table 4.2 Molal boiling-point and freezing-point constants

Solvent	Boiling point	K_b	Freezing point	K_f
Acetic acid	118.1	2.93	17	3.9
Benzene	80.2	2.53	5.4	5.12
Chloroform	61.2	3.63	—	—
Naphthalene	—	—	80	6.8
Water	100.0	0.51	0	1.86

As we mentioned, K_b is a quantity characteristic only of the solvent, and values of K_b for different liquids are given in Table 4.2. These constants are obtained by measuring the boiling points of solutions of known concentration.

Example 4.3 Exactly 1.00 gm of urea dissolved in 75.0 gm of water gives a solution that boils at 100.114°. The molecular weight of urea is 60.1. What is K_b for water?

For the molality of the solution we obtain

$$\frac{1.00}{60.1} \frac{1000}{75.0} = m = 0.222.$$

Since $\Delta T = 0.114°C$, we find that

$$K_b = \frac{\Delta T}{m} = \frac{0.114}{0.222} = 0.513.$$

The phenomenon of boiling-point elevation provides a simple method for determining the molecular weights of soluble materials. A weighed amount of material of unknown molecular weight is dissolved in a weighed amount of a solvent whose boiling-point elevation constant is known. The measured increase in the boiling point permits calculation of the molality of the solution, and from this the molecular weight of the solute can be obtained.

Example 4.4 A solution prepared by dissolving 12.00 gm of glucose in 100 gm of water is found to boil at 100.34°C. What is the molecular weight of glucose?

The boiling-point elevation constant for water is 0.51. Therefore the molality of the solution is

$$m = \frac{\Delta T}{K_b} = \frac{0.34}{0.51} = \frac{0.67 \text{ mole}}{1000 \text{ gm } H_2O}.$$

The solution as prepared would contain 120 gm of glucose in 1000 gm of water, which, as we have just learned, corresponds to 0.67 mole. Therefore

$$\text{molecular weight} = \frac{120}{0.67} = 179.$$

This answer agrees very well with the exact value of 180, which can be derived from the formula of glucose, $C_6H_{12}O_6$.

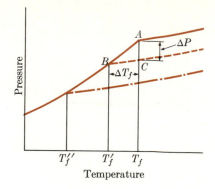

FIG. 4.12   Relation between vapor-pressure lowering and freezing-point depression for a solution of a nonvolatile solute in a volatile solvent. Vapor pressures of the pure solvent and of solutions of two concentrations are shown.

Reference to Fig. 4.12 shows that the change in the freezing temperature of the solution is related to the change in vapor pressure produced by addition of the solute. If the figure ABC is taken to be a triangle, then the depression of the freezing point ΔT is directly proportional to ΔP, the lowering of the vapor pressure. Use of Eq. (4.2) then gives us

$$\Delta T = K_f' x_2, \tag{4.7}$$

where x_2 is the mole fraction of the solute and K_f' is a proportionality constant. Once again we limit ourselves to dilute solutions, and hence we can make the approximation that

$$x_2 = \frac{n_2}{n_1 + n_2} \cong \frac{n_2}{n_1} = \frac{w_2/MW_2}{w_1/MW_1}. \tag{4.8}$$

If we define the freezing point depression constant K_f by

$$1000\, K_f = K_f' MW_1, \tag{4.9}$$

then Eqs. (4.7), (4.8), and (4.9) yield

$$\Delta T = K_f \left(\frac{w_2/MW_2}{w_1} \right) 1000$$
$$= K_f m, \tag{4.10}$$

when m is the molality of the solution. Experiments show that K_f is characteristic of the *solvent only*, and does not depend on the nature of the solute. The values of K_f for several liquids are given in Table 4.2.

Freezing-point depression measurements can be used to determine the molecular weights of dissolved substances, by a procedure quite analogous to that employed in the boiling-point elevation experiments. A weighed amount of a solute of unknown molecular weight is dissolved in a weighed amount of a liquid whose freezing-point depression constant is known. The freezing point of the solution is measured, the freezing-point depression and the molality of the solution are calculated, and the molecular weight of the solute is found from the weights of solute and solvent, and the molality.

Example 4.5 The freezing temperature of pure benzene is 5.40°C. When 1.15 gm of naphthalene are dissolved in 100 gm of benzene, the resulting solution has a freezing point of 4.95°C. The molal freezing-point depression constant for benzene is 5.12. What is the molecular weight of naphthalene?

The molality of the solution is

$$m = \frac{\Delta T}{K_f} = \frac{5.40 - 4.95}{5.12} = 0.088.$$

The weight of naphthalene in 1000 gm of solvent is 11.5 gm. Consequently the molecular weight of naphthalene is

$$\frac{11.5}{0.088} = 130 = \text{molecular weight of naphthalene.}$$

The molecular formula of naphthalene is $C_{10}H_8$, which corresponds very well to the result obtained by experimental determination of molecular weight.

In our treatment of both freezing-point depression and boiling-point elevation we have used Eq. (4.2), which refers to a system with only one type of solute species. How should we treat solutions in which two or more distinct solute species are present? Such solutions occur when a salt such as NaCl or $BaCl_2$ is dissolved in water. A 1-m solution of sodium chloride contains one mole of sodium ions and one mole of chloride ions in 1000 gm of solvent. The total concentration of solute particles is 2 m, and such a solution shows a freezing-point depression of $2K_f$. Thus, in general, the molality to be used in Eqs. (4.6) and (4.10) is the *total* molality of all solute species.

Example 4.6 When 3.24 gm of mercuric nitrate, $Hg(NO_3)_2$, are dissolved in 1000 gm of water, the freezing point of the solution is found to be −0.0558°C. When 10.84 gm of mercuric chloride, $HgCl_2$, are dissolved in 1000 gm of water, the freezing point of the solution is −0.0744°C. The molal freezing-point depression constant for water is 1.86. Are either of these salts dissociated into ions in aqueous solutions?

From the freezing-point data we find that the molality of the mercuric nitrate solution is

$$m = \frac{\Delta T}{K_f} = \frac{0.0558}{1.86} = 0.03.$$

But the number of moles of $Hg(NO_3)_2$ in 1000 gm of water was

$$\text{moles } Hg(NO_3)_2 = \frac{3.24}{324} = 0.01.$$

This must mean that mercuric nitrate is dissociated into Hg^{++} and NO_3^- in aqueous solution.

The freezing-point data show that the molality of the mercuric chloride solution is

$$m = \frac{\Delta T}{K_f} = \frac{0.0744}{1.86} = 0.040.$$

The number of moles of mercuric chloride dissolved in 1000 gm of solvent is

$$\text{moles of HgCl}_2 = \frac{10.84}{271} = 0.040.$$

Therefore, mercuric chloride must be present in solution largely as undissociated $HgCl_2$ molecules.

Osmotic Pressure

We have seen that the lowering of the solvent vapor pressure by a nonvolatile solute, and its consequences of boiling-point elevation and freezing-point depression can be used to determine the molecular weights of dissolved substances. The phenomenon of *osmotic pressure* is also associated with vapor-pressure lowering, and can also be used to determine molecular weights of solute molecules. In addition, it is profoundly important in the operation of living systems.

The osmotic pressure phenomenon involves a semipermeable membrane, that is, some film which has pores large enough to allow the passage of small solvent molecules but small enough to prevent large solute molecules of high molecular weight from passing through. When a solution is separated from its pure solvent by a semipermeable membrane, as in the apparatus shown in Fig. 4.13, one observes that some of the pure solvent passes through the membrane into the solution. The flow stops and the system reaches equilibrium after the meniscus has risen to a height which depends on the concentration of the solution. Under these equilibrium conditions, the solution is under a greater hydrostatic pressure than the pure solvent. The height of the meniscus, multiplied by the density of the solution and the acceleration due to gravity, gives the extra pressure on the solution, and this is the osmotic pressure π.

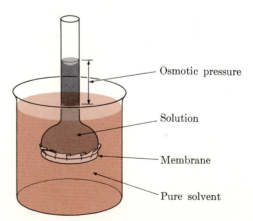

Osmotic pressure

Solution

Membrane

Pure solvent

FIG. 4.13 A simple apparatus for exhibiting the osmotic pressure phenomenon.

From experimental measurements on dilute solutions of known concentration, it has been found that the relation between osmotic pressure and concentration is simply

$$\pi = cRT, \tag{4.11}$$

where c is the concentration of solute, R is the gas constant, and T is the temperature in degrees kelvin. If c is expressed in moles/liter, and R is taken as 0.082 liter-atm/mole-deg, the osmotic pressure π is expressed in atmospheres. Equation (4.11) will be derived simply and rigorously by using the methods of thermodynamics in Chapter 8. For the present, we will only call attention to the relation between osmotic pressure and solvent vapor-pressure lowering by the following argument.

Consider a pure solvent and a corresponding solution of a nonvolatile solute placed in the two compartments of the apparatus shown in Fig. 4.7. Because the vapor pressure of the solution is lower than that of the pure solvent, the pure solvent will tend to evaporate, flow to the solution chamber, and condense in the solution. This is entirely analogous to the flow of pure solvent through the semipermeable membrane into the solution chamber in the osmotic pressure experiment, and occurs for exactly the same reason. We can anticipate that we could stop the flow of pure solvent into the solution by doing something that *raised the vapor pressure of the solution* back to a value equal to the vapor pressure of the pure solvent. This can be done and, in fact, is done in the osmotic-pressure experiment, by exerting an external hydrostatic pressure on the solution.

According to Eq. (4.11), the osmotic pressure corresponding to a solute concentration of 1 mole/liter would be

$$\pi = 0.082 \, \frac{\text{liter-atm}}{\text{mole-deg}} \times 273 \, \text{deg} \times 1 \, \text{mole/liter}$$

$$= 22.4 \, \text{atm}.$$

It is possible to measure precisely pressures of less than 10^{-3} atm. Consequently, the osmotic pressure due to $10^{-4} \, M$ solutions is readily detectable. This very great sensitivity of osmotic pressure is used to advantage in the determination of the molecular weight of biologically important molecules. These substances tend to be rather insoluble; however, it is often possible to measure the osmotic pressure of their very dilute solutions and, by knowing the weight of material dissolved, calculate their molecular weight.

Example 4.7 An aqueous solution containing 5.0 gm of horse hemoglobin in 1 liter of water shows an osmotic pressure of 1.80×10^{-3} atm at 298°K. What is the molecular weight of horse hemoglobin?

By Eq. (4.11)

$$c = \frac{\pi}{RT} = \frac{1.80 \times 10^{-3}}{0.082 \times 298}$$

$$= 0.74 \times 10^{-4} \, \text{mole/liter}.$$

Since 5.0 gm/liter corresponds to 0.74×10^{-4} mole/liter,

$$\text{molecular weight} = \frac{5.0}{0.74 \times 10^{-4}} = 68{,}000.$$

Solutions of Two Volatile Components

If two volatile liquids are mixed to form a solution, and there is no evolution or absorption of heat, the solution is ideal, and both components follow Raoult's law over the entire range of concentrations. That is,

$$P_1 = x_1 P_1^0, \qquad P_2 = x_2 P_2^0.$$

The vapor pressure of the solution is simply the sum of the partial pressures of each component:

$$P_T = P_1 + P_2 = x_1 P_1^0 + x_2 P_2^0.$$

Figure 4.14 shows how the partial and total vapor pressures depend on concentration, for an ideal solution of two volatile components.

It is important to note that the composition of a vapor in mole-fraction units is not the same as the composition of the liquid solution with which it is in equilibrium. For example, consider a mixture of benzene (component 1) and toluene (component 2). Let us choose a liquid mixture in which the mole fraction of benzene (x_1) is 0.33, and that of toluene (x_2) is 0.67. Then at 20°C we have

$$P_1^0 = 75 \text{ mm} \qquad \text{and} \qquad P_2^0 = 22 \text{ mm},$$

so that

$$P_1 = 0.33 \times 75 = 25 \text{ mm},$$
$$P_2 = 0.67 \times 22 = 15 \text{ mm},$$

and

$$P_T = P_1 + P_2 = 40 \text{ mm}.$$

The composition of the vapor in mole-fraction units can be obtained by using Dalton's law. In the vapor, we have

$$x_1 = \frac{P_1}{P_T} = \frac{25}{40} = 0.63 \qquad \text{and} \qquad x_2 = \frac{P_2}{P_T} = \frac{15}{40} = 0.37.$$

The vapor is nearly twice as rich in benzene as the liquid is. This is a specific illustration of the general fact that when an ideal solution is in equilibrium with its vapor, the vapor is always richer than the liquid in the more volatile component of the solution. This fact can be made the basis for a procedure which will separate the components of the solution. Suppose that the vapor (63% benzene, 37% toluene) from the previous example was collected, condensed to a liquid, and then allowed to evaporate so as to come into equilibrium with its vapor. What is the composition of this new vapor? We have as before

$$P_1^0 = 75 \text{ mm} \quad \text{and} \quad P_2^0 = 22 \text{ mm}.$$

But now the mole fractions of components 1 and 2 in the liquid are 0.63 and 0.37, respectively. Therefore their vapor pressures are given by

$$P_1 = (0.63)(75) = 47 \text{ mm},$$
$$P_2 = (0.37)(22) = 8.1 \text{ mm},$$
$$P_T = 47 + 8.1 = 55 \text{ mm}.$$

From these data we can calculate the mole fractions in the vapor phase to be

$$x_1 = \frac{P_1}{P_T} = \frac{47}{55} = 0.85$$

and

$$x_2 = \frac{P_2}{P_T} = \frac{8.1}{55} = 0.15.$$

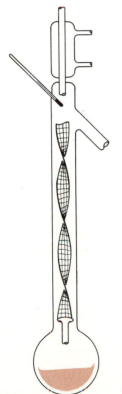

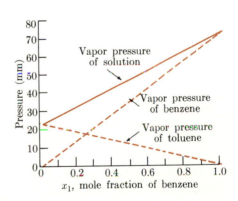

Vapor pressure as a function of composition for an ideal solution of benzene and toluene.

FIG. 4.14

A laboratory distillation apparatus which employs a twisted metal gauze to increase liquid-vapor contact in the column.

FIG. 4.15

Thus the second evaporation has produced a vapor still richer in benzene, the more volatile component. It is clear that if we were to collect the vapor once again, condense it, and then allow a small amount of it to evaporate, we would obtain a vapor even richer in benzene. Repetition of the evaporation-condensation process therefore tends to produce a vapor that is very nearly pure benzene, and a liquid that is very nearly pure toluene. This is the process that goes on in the distillation apparatus shown in Fig. 4.15. As the material rises through the column it undergoes many evaporation-condensation cycles, and thus the material that reaches the top of the column is richest in the most volatile component.

4.5 NONIDEAL SOLUTIONS

A nonideal solution is formed when the process of mixing its components is accompanied by the evolution or absorption of heat. Nonideal solutions do not obey Raoult's law, and indeed, the dependence of their vapor pressures on concentration can be quite complicated. For the purpose of discussion, it is convenient to divide nonideal solutions into two groups: those whose formation is accompanied by the evolution of heat, and those that are formed with the absorption of heat.

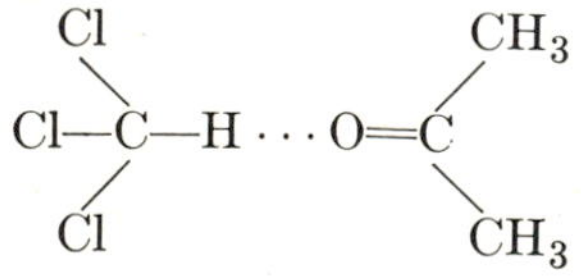

FIG. 4.16   Hydrogen bond interaction between chloroform and acetone.

The evolution of heat upon mixing indicates that in the solution the components have found a situation of lower energy than in their pure states. This behavior occurs when the molecular structure of the components is such that there are stronger attractive forces between unlike molecules than between molecules of the same kind. A specific example of such a pair of molecules is chloroform, $CHCl_3$, and acetone, $(CH_3)_2CO$. When molecules of chloroform and acetone are brought together, the single hydrogen on the chloroform molecule is strongly attracted to the oxygen atom in the acetone molecule, as shown in Fig. 4.16. This type of interaction is known as hydrogen bonding, and will be discussed more completely in Chapter 12. Hydrogen bonding does not occur in pure chloroform or pure acetone, because there are no oxygen atoms in chloroform and the hydrogen atoms in acetone do not have the proper electrical characteristics needed to form strong hydrogen bonds.

Since the heat evolution indicates that the molecules in solution are in a situation of low energy, it is not surprising that the vapor pressure of each component is lower than would be predicted by Raoult's law. Figure 4.17 shows the vapor pressures as a function of concentration for chloroform-acetone mixtures. Such a solution is said to display negative deviations from Raoult's law, since at each concentration, the vapor pressure of each component is less than that predicted by Raoult's law. Note, however, that at each end of the concentration scale, the component which is in excess deviates from Raoult's law only very slightly. Since the component in excess is always taken to be the solvent, we can assert that to an acceptable approximation, the *solvent* in a dilute solution obeys Raoult's law.

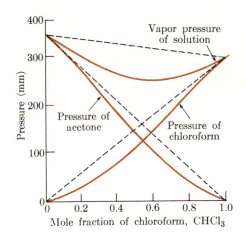

Vapor pressure as a function of composition for acetone-chloroform solutions. Behavior expected if the solutions were ideal is shown by dashed lines.

FIG. 4.17

Absorption of heat during mixing indicates that the component molecules in solution have higher energy than they do in their pure states. In other words, the attractive forces between unlike molecules are weaker than those between molecules of the same kind. Since the molecules in such a solution are in a condition of high energy, it is not surprising they should have an increased tendency to escape from solution, and that the vapor pressure of each of the components is greater than that predicted by Raoult's law. Solutions which show these positive deviations from Raoult's law are often the result of mixing a liquid consisting of polar molecules with nonpolar molecules. In the solution, the strong attraction between two polar molecules is replaced by the weaker attraction between polar and nonpolar molecules, and this is an energetically unfavorable situation. The acetone-carbon disulfide system provides an example of this type of behavior. Carbon disulfide is a linear molecule that is not polar, since its atoms are arranged symmetrically (S=C=S). Acetone has a dipole moment, and when it is mixed with carbon disulfide the vapor pressures

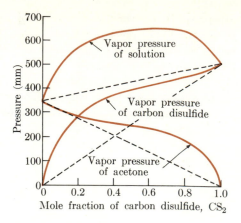

FIG. 4.18 Vapor pressure as a function of composition for acetone-carbon disulfide solutions. Behavior expected if the solutions were ideal is shown by dashed lines.

of both components exceed the predictions of Raoult's law, as is shown in Fig. 4.18.

In our discussions of ideal solutions, we stated that the vapor in equilibrium with such solutions is always enriched in the more volatile of the two components, i.e., the one with the lower boiling point. This simple rule does not always apply to nonideal solutions, however. In fact, consideration of Fig. 4.18 shows that at mole fractions of 0.65 or greater for carbon disulfide, the vapor phase is actually less concentrated in carbon disulfide than the liquid phase, even though *pure* carbon disulfide is more volatile than pure acetone. At mole fractions of carbon disulfide of less than 0.65 the solution behaves normally. Right at the 0.65 CS_2 concentration, the composition of the vapor is exactly the same as that of the liquid, and consequently no amount of distillation can purify such a mixture. Solutions which distill with no change in composition are called *azeotropes*. There are many azeotropic mixtures, and one of the most extensively studied is the solution of hydrochloric acid in water. If any solution of hydrochloric acid is boiled long enough at atmospheric pressure its composition will change until it becomes 20.22% hydrochloric acid by weight. This technique can be used to prepare solutions of known acid concentration for use in quantitative analysis. However, it is important to note that the composition of the azeotrope depends on the atmospheric pressure, and the value of 20.22% applies only to a solution boiling at 760 mm.

4.6 SOLUBILITY

While there are many pairs of substances which, like water and ethyl alcohol, can be mixed in any proportions to form homogeneous solutions, it is a matter of common experience that the capacity of a solvent to dissolve a given solute is often limited. When a solvent placed in contact with an excess of solute attains and maintains a constant concentration of solute, the solute and solution are at equilibrium, and the solution is said to be saturated. The solubility of a substance in a particular solvent at a given temperature is the concentration of the solute in the saturated solution. In other words, the solubility of a solute is the dissolved concentration characteristic of the state of equilibrium between the solute and the solution. It is difficult to overemphasize the importance of the concept of solubility to chemistry; it is the basis of innumerable laboratory and industrial processes that prepare, separate, and purify chemicals, and is the controlling factor in a variety of geological and other natural phenomena. The solubility of a substance in a particular solvent is controlled principally by the nature of the solvent and solute themselves, but also by the conditions of temperature and pressure. Let us analyze these factors, first limiting ourselves to the case of ideal solutions.

Two liquids that form an ideal solution are always miscible in any proportions and thus have infinite solubility in each other. The reason for this is easy to see if we recall two facts. First, limited solubility and a saturated solution result only when a solute and its solution reach equilibrium. Second, the equilibrium state is a compromise between a natural tendency toward minimum energy and maximum molecular chaos. Now the mixing of two ideal liquids is always accompanied by an increase in entropy or molecular chaos, because in the solution, the solute molecules are spread randomly throughout the solvent, rather than being nearly closest packed as they are in the pure solute. That is, even if we could locate one solute molecule in solution, we could not predict what the identity of its nearest neighbors was, as we could if the molecule were in the pure solute phase. Consequently the solution has a higher entropy than the pure solvent and solute, and the tendency toward maximum molecular chaos favors the mixing of the two liquids. Moreover, the fact that there is no *energy* change in the mixing process means that the tendency toward minimum energy does *not* restrict the solution process. Consequently, the two liquid components of an ideal solution can mix in any proportion.

Consider now a solid substance dissolving in a liquid solvent. The solid is such that when melted, it is converted to a liquid that in turn can form an ideal solution with the solvent. The dissolution of the solid can be pictured as occurring in two hypothetical stages:

$$\text{solid solute} \rightarrow \text{liquid solute} \rightarrow \text{solute in solution.}$$

We have just specified that the second of these steps does not involve any energy change, for the solution formed is ideal. In contrast, the first step does involve the absorption of energy in the amount ΔH_{fus} per mole of solute. Consequently, while the tendency toward maximum entropy favors the dissolution of the solid, the tendency toward minimum energy favors the solid remaining undissolved. Therefore the solubility of the solid is limited, and a saturated solution which represents the best compromise between maximizing entropy and minimizing energy is formed. Since ΔH_{fus} is related to the strength of the attractive forces between solute molecules, we can deduce that the magnitudes of these same forces determine the solubility of the solid in ideal solutions.

By using some care, we can extend our arguments to nonideal solutions. Two liquids which mix with the evolution of heat will be infinitely soluble in each other, for both energy and entropy effects favor their mixing. Two liquids which mix with the absorption of heat *may* have limited solubility in each other, for if the mixing process is energetically unfavorable, the tendency toward maximum molecular chaos may or may not be sufficient to allow the liquids to mix in all proportions. Likewise, the solubility of a solid is likely to be small if it enters the solution only with considerable absorption of heat. On the other hand, if the dissolution of the solid is accompanied by evolution of heat, the solubility of the solid may be quite high. Even with these generalizations it is difficult to predict or even rationalize qualitatively the solubilities of substances that form markedly nonideal solutions, for the energy and entropy changes that accompany the mixing of strongly interacting molecules are subtle and difficult to anticipate.

Temperature Effects

If the enthalpy change that accompanies the dissolution of a solute is known, it is possible to predict the effect of a change in temperature on the solubility. To do this, we need only recall one of the general principles of equilibrium discussed in Section 4.2. There we noted that if a system initially at equilibrium is perturbed by some change in its surroundings, it reacts in a manner which restores it to an equilibrium state. A very useful restatement and extension of this observation is known as Le Chatelier's principle: *If anything is done to a system initially in equilibrium that would result in a change of any of the factors that determine the state of equilibrium, the system will adjust itself in such a way as to minimize that change.* To see how to use LeChatelier's principle, let us apply it to liquid-vapor equilibrium.

Let us consider a liquid and its vapor at equilibrium at some particular temperature. If heat were delivered to this system, and no other change allowed to occur, the temperature would rise. This would cause a departure from equilibrium, since the *existing* vapor pressure is not the equilibrium vapor pressure at the higher temperature. The induced temperature change could be *minimized*, however, if the system reacted, that is, if liquid and vapor were interconverted in a way that absorbs heat. Le Chatelier's principle tells us that this is the way the system will react, and we also know that the system absorbs heat if the liquid evaporates. Therefore, the prediction based on Le Chatelier's principle and a positive value for ΔH_{vap} is that as heat is added, the liquid will evaporate, and consequently the vapor pressure will increase. After the addition of heat the new position or state of equilibrium corresponds to increased vapor pressure and temperature. This prediction is consistent with the observed fact that the vapor pressure of a liquid increases as the temperature increases.

The form of the argument we have just used is in no way limited to liquid-vapor equilibrium. We might generalize and say that for any reaction which has a *positive* ΔH, an increase in temperature will result in an increase in product concentration at the expense of reactant concentration. The converse must also be true, for it corresponds to reversing the direction in which a reaction is written: if ΔH is *negative*, increasing the temperature decreases the concentration of products in favor of the reactants. We can use this general conclusion to predict the temperature dependence of solubility. If the enthalpy change that accompanies the solution reaction ΔH_{sol} is positive, that is, if heat is absorbed when the reaction

$$\text{solvent} + \text{solute} \rightarrow \text{solution}$$

takes place, the solubility of the solute increases as temperature increases. If, on the other hand, ΔH_{sol} is negative, then the solubility of the solute decreases as temperature increases. Figure 4.19 illustrates the temperature dependence of the solubility of KNO_3, $NaCl$, and Na_2SO_4. The prediction based on Le Chatelier's principle is substantiated by these data, for ΔH_{sol} is $+8.5$, $+1.3$, and -5.5 kcal/mole, respectively, for KNO_3, $NaCl$, and Na_2SO_4.

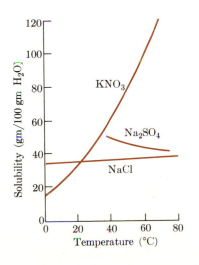

The temperature dependence of the solubility of three salts in water.

FIG. 4.19

Two-Component Phase Diagrams

In Section 4.2 we discussed phase diagrams for systems of one chemical component. With such diagrams, we could describe the conditions of temperature and pressure under which we could expect to have one phase, or two or three phases present in equilibrium. Such phase diagrams are also very useful in analyzing the behavior of systems which have two or more chemical components. When we have two components, there are three state functions that must be specified in order to describe the system: temperature, pressure, and the concentration (mole fraction) of one of the components. Since there are three variables, the phase diagram should be three-dimensional. It is common practice, however, to consider the total pressure as fixed; thus we can have a two-dimensional phase diagram with concentration and temperature as variables.

Figure 4.20 shows the phase diagram of the bismuth-cadmium system at a fixed pressure. The point A is the freezing point of pure bismuth, 273°C, while point B is the freezing temperature of pure cadmium, 323°C. At temperatures above 323°C we have, regardless of composition, only one phase present: a liquid solution of bismuth and cadmium. At temperatures below 140°C, two pure solid phases are present: crystalline bismuth and crystalline cadmium. At intermediate temperatures, one, two, or three phases may be present, depending on the temperature and composition.

Phase diagrams are constructed from experiments in which mixtures are liquefied and allowed to cool, and the temperature as a function of time determined. Examples of such cooling curves are shown in Fig. 4.20(b), and their relation to the phase diagram (Fig. 4.20(a)) is indicated. If pure bismuth is melted and allowed to cool, the temperature at first falls, and then when the freezing point of bismuth is reached, the temperature remains constant until all the bismuth solidifies, and then it starts to fall again. This abrupt leveling off of the cooling curve is characteristic of the freezing of a pure substance.

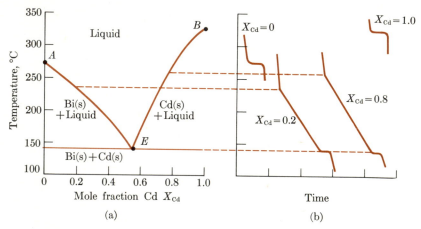

FIG. 4.20   (a) Phase diagram and (b) cooling curves for the cadmium-bismuth system.

If a liquid having a mole fraction X_{Cd} of cadmium of 0.2 were allowed to cool, nothing would happen until the temperature reached 240°C, the intersection of $X_{Cd} = 0.2$ and the curve AE of Fig. 4.20(a). At this point, pure solid bismuth would just begin to crystallize from the liquid, and solid bismuth would be in equilibrium with a solution containing 0.2 mole fraction Cd. The cooling curve shows a change in slope at this temperature, as the latent heat of crystallization of bismuth is released, and reduces the rate of temperature decrease. In contrast to the behavior of a pure substance, the solution does not freeze completely at one temperature. As bismuth crystallizes, the liquid becomes richer in cadmium, and therefore has a lower freezing temperature. Consequently, in order to continue to crystallize bismuth, the temperature must continue to be lowered. At 140°C, the cooling curve does level off, and at this one temperature, the solution that remains solidifies abruptly. The temperature at which this occurs is the *eutectic temperature*, and the abrupt leveling off of the cooling curve is called the *eutectic halt*.

If liquids of other compositions which are near to 0.2 mole fraction cadmium are cooled, the first crystallization of bismuth occurs at a temperature given by the intersection of the composition with the curve AE. In fact, the curve AE represents all conditions of temperature and concentration at which pure solid bismuth and a liquid solution can exist in equilibrium. We can indeed say that the curve AE describes the freezing point depression of bismuth by cadmium over a range of compositions indicated. For small mole fractions of Cd, the curve AE and hence the freezing point depression ΔT_f varies linearly with X_{Cd}, as has been discussed earlier. For larger mole fractions of Cd, this is not the case.

We can now analyze what happens when a solution with 0.8 mole fraction Cd is cooled. At the temperature corresponding to the intersection of $X_{Cd} = 0.8$ and the curve BE, the first solid appears, and is found to be pure crystalline cadmium. The cooling curve shows a change of slope characteristic of the gradual crystallization of a solid from a solution. At 140°C there is a *eutectic* halt, and complete solidification.

The curve BE represents all pairs of temperatures and concentrations at which pure solid cadmium is in equilibrium with a cadmium-bismuth solution. It is, therefore, the curve of the freezing-point depression of cadmium by bismuth. As a solution from which cadmium is crystallizing is cooled, the solution becomes increasingly rich in bismuth. The concentration of the solution left behind at any temperature can be found from the curve BE. As the temperature is further lowered, the point that describes the system moves along the curve BE toward the point E.

The curve BE represents temperatures and concentrations at which pure solid cadmium is in equilibrium with a Bi-Cd solution, and the curve AE gives the temperatures and concentrations at which pure solid bismuth is in equilibrium with the solution. Their intersection at point E must therefore be a situation at which cadmium and bismuth as pure solids are in equilibrium with a liquid solution. Note that this occurs only at one temperature and for one liquid phase composition. This point is called the *eutectic point*. Any further cooling of this eutectic mixture below the eutectic temperature produces abrupt crystallization of the liquid to mixed crystals of cadmium and bismuth. It is this crystallization of *both* solids that produces the eutectic halt in the cooling curves. If a solution of the eutectic composition is cooled, no crystallization occurs until the eutectic temperature is reached, at which point both solids crystallize abruptly.

Question. The curves *AE* and *BE* are said to describe the freezing point depression of bismuth and cadmium, respectively. Could they just as well be said to represent the solubility of bismuth and cadmium in the liquid?

There is another important set of conclusions which we can draw from the cadmium-bismuth phase diagram. In the region of the diagram where there is only one phase, the liquid solution, we can vary both the temperature and the composition *independently* of each other without producing or destroying any phases. It is said, therefore, that in this one phase region, there are two *degrees of freedom*. Along the lines *AE* and *BE*, where two phases are present in equilibrium, there is only one degree of freedom. If the temperature is changed, the liquid composition must change in the way prescribed by the curves if the two phases are to be retained. Thus temperature, but not composition, or composition, but not temperature, may be varied independently. Finally, at the eutectic point where three phases are present, there are no degrees of freedom, since a change in liquid composition or temperature will cause at least one phase to disappear.

It is clear, then, that the number of independent variables available to us is related to the number of phases present. It is also found to depend on the number of chemical components in the system. The relation between the number of components C, the number of phases P, and the number of degrees of freedom F is given by the Gibbs phase rule

$$F = C - P + 2 \qquad \text{(pressure variable)},$$
$$F = C - P + 1 \qquad \text{(pressure fixed)}.$$

Note that in the one-phase region of a two-component system with the pressure fixed, $F = 2$. In the two-phase region $F = 1$, and at the three-phase eutectic point $F = 0$. The phase rule is of considerable use in the analysis of complicated phase diagrams involving multicomponent alloys.

The existence of chemical compounds can be inferred from phase diagrams. Figure 4.21 shows that in the zinc-magnesium system, there are two eutectic points, and an intermediate maximum freezing point at 0.33 mole fraction magnesium. A cooling curve at this composition shows one point at which the cooling stops and the temperature remains constant while the liquid solidifies completely to produce only one solid phase. These are the cooling characteristics of a pure compound, which in this case has the formula $MgZn_2$.

Once the existence of the compound $MgZn_2$ is recognized, the appearance of the rest of the phase diagram is readily understood. The diagram is in effect two simple phase diagrams back to back: one corresponds to the Zn-$MgZn_2$ system, the other to the $MgZn_2$-Mg system. The phase rule can be applied to each of these two-component systems just as we have done for the cadmium-bismuth system. Note, however, that at the exact composition which corresponds to $MgZn_2$, there is only one component present, and consequently at

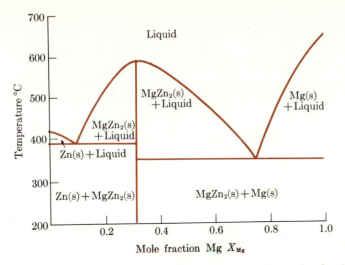

The phase diagram for the magnesium-zinc system, showing the formation of the compound $MgZn_2$.

FIG. 4.21

the freezing point of $MgZn_2$, there are two phases, and the number of degrees of freedom is

$$F = C - P + 1 = 1 - 2 + 1 = 0.$$

This is as expected for the freezing of a pure substance at a fixed pressure.

4.7 CONCLUSION

In this chapter we have explored some of the phenomena and concepts associated with liquids and their solutions. All of these ideas find repeated use in the study of chemistry. We shall see that an appreciation of the kinetic theory of liquids will help us understand some of the factors that determine the rates of chemical reactions which take place in solution. The colligative properties of solutions—the vapor-pressure lowering, boiling-point elevation, and freezing-point depression—offer ways of determining the molecular weights of dissolved substances. These methods are used actively in chemical research, both in organic chemistry and in the study of polymeric molecules of high molecular weight. Of most general importance, however, are the ideas associated with phase equilibria. In subsequent chapters we shall be continually concerned with chemical equilibria, sometimes in systems of considerable complexity. Regardless of the chemical complexity, the general nature of the equilibrium state and the general factors that determine the concentrations that exist at equilibrium are always the same. Consequently, the ideas generated in our discussion of phase equilibria will be of use to us in analyzing all situations of equilibrium.

SUGGESTIONS FOR FURTHER READING

Barrow, G. M., *Physical Chemistry*, 3rd ed. New York: McGraw-Hill, 1973. Chapters 19, 20.

Campbell, J. A., *Chemical Systems*. San Francisco: W. H. Freeman and Co., 1970.

Hildebrand, J. H., and R. E. Powell, *Principles of Chemistry*, 7th ed. New York: Macmillan, 1964. Chapter 16.

Hill, T. L., *Matter and Equilibrium*. Menlo Park, Calif.: W. A. Benjamin, 1966.

Kieffer, W. F., *The Mole Concept in Chemistry*. New York: Reinhold, 1962. Chapter 5.

Sienko, M. J., and R. A. Plane, *Chemistry: Principles and Properties*. New York: McGraw-Hill, 1966. Chapters 7 and 8.

PROBLEMS

4.1 What are the molality and the molarity of a solution of ethanol, C_2H_5OH, in water if the mole fraction of ethanol is 0.05? Assume that the density of the solution is 0.997 gm/ml.

4.2 Concentrated nitric acid is 69% by weight HNO_3 and has a density of 1.41 gm/ml at 20°C. What volume and what weight of concentrated nitric acid are needed to prepare 100 ml of 6-M acid?

4.3 Calculate how many milliliters of 0.10-M $KMnO_4$ are required to react completely with 0.01 mole of oxalate ion, $C_2O_4^=$, according to the reaction

$$2MnO_4^- + 5C_2O_4^= + 16H^+ = 2Mn^{++} + 10CO_2 + 8H_2O.$$

4.4 Exactly 100 gm of a certain solution contain 10 gm of NaCl. The density of the solution is 1.071 gm/ml. What is the molality and the molarity of NaCl?

4.5 The boiling point of a solution of 0.402 gm of naphthalene, $C_{10}H_8$, in 26.6 gm of chloroform is 0.455°C higher than that of pure chloroform. What is the molal boiling-point elevation constant for chloroform?

4.6 The vapor pressure of a dilute aqueous solution is 23.45 mm at 25°C, whereas the vapor pressure of pure water at this temperature is 23.76 mm. Calculate the molal concentration of solute, and use the tabulated value of K_b for water to predict the boiling point of the solution.

4.7 What weight of ethylene glycol, $C_2H_6O_2$, must be included in each 1000 gm of aqueous solvent to lower the freezing point to -10°C?

4.8 When 1.00 gm of sulfur is dissolved in 20.0 gm of naphthalene, the resulting solution freezes at a temperature 1.28C° lower than pure naphthalene does. What is the molecular weight of sulfur?

4.9 The freezing-point depression constant for mercuric chloride, $HgCl_2$, is 34.3. For a solution of 0.849 gm of mercurous chloride (empirical formula $HgCl$) in 50 gm of $HgCl_2$, the freezing-point depression is 1.24°C. What is the molecular weight of mercurous chloride in this solution? What is its molecular formula?

4.10 Ten liters of dry air were bubbled slowly through liquid water at 20°C, and the observed weight loss of the liquid was 0.172 gm. By assuming that 10 liters of saturated water vapor were formed in the experiment, calculate the vapor pressure of water at 20°C.

4.11 Ethanol and methanol form a solution that is very nearly ideal. The vapor pressure of ethanol is 44.5 mm, and that of methanol is 88.7 at 20°C. (a) Calculate the mole fractions of methanol and ethanol in a solution obtained by mixing 60 gm of ethanol with 40 gm of methanol. (b) Calculate the partial pressures and the total vapor pressure of this solution, and the mole fraction of ethanol in the vapor.

4.12 At 20°C, the vapor pressure of pure benzene is 22 mm, and that of pure toluene is 75 mm. What is the composition of the solution of these two components that has a vapor pressure of 50 mm at this temperature? What is the composition of the vapor in equilibrium with this solution?

4.13 The solubility of borax ($Na_2B_4O_7 \cdot 10H_2O$) in water increases as the temperature increases. Is heat evolved or absorbed as this salt dissolves? Is the ΔH of the dissolution process positive or negative?

4.14 At 55°C, ethanol has a vapor pressure of 168 mm, and the vapor pressure of methyl cyclohexane is 280 mm. A solution of the two, in which the mole fraction of ethanol is 0.68, has a total vapor pressure of 376 mm. Is this solution formed from its components with the evolution or absorption of heat?

CHAPTER 5

CHEMICAL EQUILIBRIUM

In this chapter we shall explore the consequences of the fact that chemical reactions are reversible, and that in closed chemical systems there eventually occurs a state of equilibrium between reactants and products. In so doing, we will be starting to develop concepts which will lead us eventually to a *quantitative* expression of "chemical reactivity." The concentrations which exist when a chemical system reaches equilibrium reflect the intrinsic tendency of the atoms to exist either as reactant or product molecules. Thus by learning to describe the equilibrium state quantitatively, we will be able to replace qualitative statements about "the tendency of a reaction to go" with definite numerical expressions of the extent of conversion of reactants to products.

5.1 THE NATURE OF CHEMICAL EQUILIBRIUM

In Chapter 4 we found that the existence of a characteristic equilibrium vapor pressure for a condensed phase is a consequence of the fact that the evaporation process is *reversible*. A liquid or solid that has been completely vaporized can, by an appropriate change in conditions, be completely recondensed. Both evaporation and condensation can occur, and for each substance there is a set of conditions—particular values of temperature and vapor pressure—at which evaporation and condensation occur at equal rates. Under these conditions both phases remain indefinitely, and we say that the system is at equilibrium.

Chemical reactions, like phase changes, are reversible. As a consequence, there are conditions of concentration and temperature under which reactants and products exist together at equilibrium. To illustrate our point, and to emphasize the close connection between phase equilibria and chemical equilibria, we consider the thermal decomposition of calcium carbonate,

$$CaCO_3(s) \rightarrow CaO(s) + CO_2(g). \tag{5.1}$$

By carrying out this reaction in an open vessel which allows carbon dioxide to be swept away, complete conversion of calcium carbonate to calcium oxide can be effected. On the other hand, it is well known that calcium oxide reacts with carbon dioxide, and if the pressure of CO_2 is high enough, the oxide can be completely converted to the carbonate by

$$CaO(s) + CO_2(g) \rightarrow CaCO_3(s). \tag{5.2}$$

This is, of course, just the reverse of reaction (5.1). Thus we must look upon reactions (5.1) and (5.2) as *reversible* chemical processes, a fact which we denote by either of the following notations:

$$CaCO_3(s) = CaO(s) + CO_2(g),$$
$$CaCO_3(s) \rightleftharpoons CaO(s) + CO_2(g).$$

This chemical system is closely analogous to the "physical" system consisting of a condensed phase and its vapor. Just as a liquid and its vapor come to equilibrium in a closed container, there are certain values of the temperature and pressure of CO_2 at which $CaCO_3$, CaO, and CO_2 remain indefinitely. When pure $CaCO_3$ is heated in a closed vessel, it begins to decompose according to reaction (5.1). As the CO_2 accumulates, its pressure increases, and eventually reaction (5.2) begins to occur at a noticeable rate, a rate which increases as the pressure of CO_2 increases. Finally, the rates of the decomposition reaction and its reverse become equal, the pressure of carbon dioxide remains constant, and the system has reached equilibrium.

In the discussion of phase equilibrium in Chapter 4 we used liquid-vapor equilibrium to illustrate four characteristics of all equilibrium situations. Let us review each of these characteristics and see how they are exemplified by chemical equilibria.

The first feature of the equilibrium state is that it is dynamic; it is a permanent situation maintained by the equality of the rates of two opposing chemical reactions. That is, when the system $CaCO_3$, CaO, CO_2 comes to equilibrium with respect to the reaction

$$CaCO_3(s) = CaO(s) + CO_2(g),$$

we assert that $CaCO_3$ continues to be converted to CaO and CO_2, and that CO_2 and CaO continue to form $CaCO_3$. It is not difficult to prove this statement by an experiment. Some pure $CaCO_3$ is decomposed in a closed system, and allowed to come to equilibrium with CaO and a certain pressure of CO_2. Then the reactor is connected to another flask which contains, at the same temperature and pressure, some CO_2 which has been "labeled" with radioactive carbon, C^{14}. This operation in itself does not disturb the equilibrium between CO_2 and the solids, since the pressure and temperature are always constant. After a while, some of the solid is withdrawn and examined for radioactivity. The characteristic radiation of C^{14} is found in the $CaCO_3$, which indicates that some $C^{14}O_2$ reacted with CaO to form $CaC^{14}O_3$, even though the system was always at equilibrium. While this occurred, some $CaCO_3$ must have dissociated to CaO and CO_2 in order to maintain a constant pressure of CO_2. Thus, even though no net change in composition occurred, the opposing reactions went on, and the equilibrium conditions were maintained by a dynamic balance. This type of experiment can be performed with many systems, and the results always indicate that dynamic balance of opposing reaction rates is a characteristic of systems at equilibrium.

Our second generalization is that systems move toward an equilibrium state spontaneously; that a system can be removed from equilibrium only by some outside influence, and once it is left to itself, the disturbed system returns to an equilibrium state. We must be careful to understand the meaning of the word "spontaneously." In this context it means that the reaction proceeds at some finite rate *without* the action of outside influences such as changes in temperature or pressure. This assertion that systems proceed naturally toward equilibrium cannot be proved by a *single* simple example, for it is a generalization based on the observation of many different systems under many different conditions. We can, however, rationalize this behavior with a simple argument. A system moves toward the equilibrium state because the rate of reaction in the forward direction exceeds the rate of the reverse reaction. In general, it is found that the rate of a reaction decreases as the concentrations of reactants decrease, just as the rate of condensation of a vapor decreases as its pressure decreases. Therefore, as reactants are converted to products, the rate of the forward reaction decreases and that of the reverse reaction increases. When the two rates become equal, net reaction ceases, and a constant concentration of all reagents is maintained. In order for the system to move away from equilibrium, the rate of the forward or reverse reaction would have to change, and this does not happen if external conditions such as pressure and temperature are kept constant. Thus systems move toward equilibrium because of an imbalance of reaction rates; at equilibrium these rates are equal, and there is no way for the undisturbed system to move away from equilibrium.

The third generalization about equilibrium is that the nature and properties of the equilibrium state are the same, regardless of the direction from which it is reached. It is easy to see that this applies to our example of the $CaCO_3$, CaO, CO_2 system, for at each temperature there is a fixed value of the equilibrium CO_2 pressure at which the rate of evolution of CO_2 equals the rate of its conversion to $CaCO_3$. It does not matter if this pressure is attained by allowing $CaCO_3$ to decompose, or by allowing CO_2 to react with pure CaO. The rates of the forward and reverse reactions become equal, and net reaction ceases when the equilibrium pressure of CO_2 is reached, whether this pressure is approached from above or below the equilibrium value.

A certain amount of care is necessary in applying the third generalization to chemical systems. Consider the reaction

$$PCl_5(g) = PCl_3(g) + Cl_2(g).$$

It is found experimentally that the equilibrium concentrations are the same when one mole of pure PCl_5 decomposes in a fixed volume or when one mole of PCl_3 and one mole of Cl_2 are mixed and react in the same volume. If in another experiment one mole of PCl_3 and *two* moles of Cl_2 are mixed, a new equilibrium state is reached. To approach this new equilibrium state from the opposite direction we would have to mix one mole of PCl_5 with one mole of Cl_2. That is, our assertion that the equilibrium state is the same regardless of how it is approached presupposes that a fixed number of atoms of each element per unit volume are involved.

The fact that the nature of the equilibrium state is independent of the direction from which it is approached is often used as a criterion for chemical equilibrium. There are chemical reactions that are exceedingly slow. How then can we distinguish between the truly time-invariant concentrations of reagents that exist at equilibrium, and a situation far from equilibrium which is changing so slowly that we cannot detect any net reaction? If bringing together pure "reactants" and then pure "products" leads to the same concentrations of all reagents when all apparent net reaction ceases, then we can be sure the time-invariant situation is one of true equilibrium. If the situations reached from the product side and reactant side are different, then equilibrium has not been reached, and the reaction is merely very slow.

The fourth generalization is that the equilibrium state represents a compromise between two opposing tendencies: the drive for molecules to assume the state of minimum energy and the urge toward a state of maximum molecular chaos or entropy. It is not difficult to analyze the equilibrium situation for the reaction

$$CaCO_3(s) = CaO(s) + CO_2(g)$$

in these terms. In solid $CaCO_3$, the carbon and oxygen atoms are in a highly ordered condition: they are grouped as carbonate ions, $CO_3^=$, which occupy well-defined sites in the crystal lattice. The chemical reaction corresponds to "freeing" a fragment of the $CO_3^=$ group as a CO_2 gaseous molecule. These gaseous molecules can move anywhere within the volume of the container, and their positions at any instant can be regarded as randomly distributed. Consequently, the CO_2 molecules have more entropy in the gas phase than they do when they constitute part of the $CO_3^=$ grouping in the ionic lattice. If the urge toward maximum molecular chaos were dominant, $CaCO_3$ would decompose completely to CaO and CO_2. Experiments show, however, that energy is absorbed by the system when CO_2 is evolved from $CaCO_3$. Consequently, the change in the system that satisfies the drive to maximize entropy violates the tendency to minimize energy, and vice versa. Therefore, the equilibrium pressure of CO_2 over a mixture of $CaCO_3$ and CaO represents the best compromise between the two opposing tendencies.

There are many other chemical reactions in which the influence of entropy and energy is easy to discern. The simplest example is a dissociation reaction of a gaseous molecule:

$$H_2(g) = 2H(g).$$

The drive toward maximum entropy favors the dissociation reaction, for this process converts ordered pairs of atoms into free atoms that can move independently and which are randomly distributed in space at any instant. On the other hand, the dissociation requires energy to break the chemical bond between atoms, and consequently the drive toward minimum energy favors the molecules' remaining undissociated. In general, reactions in which molecules are fragmented and bonds are broken are favored by the drive to maximize entropy, but opposed by the tendency to minimize energy.

There are reactions for which the energy and entropy changes are much less obvious. For example, the entropy change for the reaction

$$N_2 + O_2 = 2NO$$

is nearly zero, and it is not possible to discern by inspection whether the entropy factor favors products or reactants. Similarly, it is not possible to see by inspection that the energy of the reactants is lower than that of the products. To analyze the energy and entropy effects for such a reaction, we must use the quantitative methods of thermodynamics described in Chapter 8. We shall find there that it is possible to evaluate energy and entropy changes quantitatively, and to use this information to predict the extent to which a reaction will proceed from reactants to products.

While the general nature of phase equilibria and chemical equilibria is the same, the manner in which the equilibrium state is specified in the two instances is superficially different. Situations of phase equilibria can often be described simply by saying that a certain compound melts at a particular temperature and pressure, or that the vapor pressure has a certain value at a given temperature. In contrast, to specify situations of chemical equilibrium it is often necessary to give the concentrations of *several* reagents at each temperature. Fortunately, for each chemical reaction there is a single function which compactly expresses all possible situations of equilibrium at a particular temperature. This quantity, the equilibrium constant, is of foremost importance in chemistry, and is discussed in detail in the next section.

It is an experimental fact that the pressure of CO_2 in equilibrium with solid CaO and $CaCO_3$ is a function only of the temperature of the reaction mixture. Once the system has reached equilibrium, either solid CaO or $CaCO_3$ may be added or removed, and so long as *some* of each solid is present in the system, the pressure of CO_2 will remain constant. Thus in order to characterize the equilibrium state of this system, it is only necessary to state the equilibrium pressure of CO_2.

The situation is somewhat different when there are several dissolved or gaseous reagents involved in a chemical reaction. Let us consider the reaction between hydrogen gas and iodine vapor to form gaseous hydrogen iodide:

$$H_2(g) + I_2(g) = 2HI(g).$$

It is easy to demonstrate that this reaction is reversible, and that the same state of equilibrium can be reached either by starting from pure H_2 and I_2 as reactants, or by decomposing pure HI. Now, while the equilibrium state of the $CaCO_3$, CaO, CO_2 system can be characterized by one number, the pressure of CO_2, in the H_2, I_2, HI system, it is found that there are infinitely many sets of pressures of the reagents that can exist at equilibrium. This variety of equilibrium compositions arises because it is possible to reach equilibrium by starting the reaction with equal pressures of H_2 and I_2, or with H_2 in excess of I_2, or vice versa. For any choice of initial concentrations an equilibrium state is reached, but for each choice the concentration of the individual reagents at equilibrium is different. There is, however, a simple relationship between the concentrations of reagents at equilibrium. It is found experimentally that despite the infinite variety of pressures of individual reagents, the quantity

$$\frac{P_{HI}^2}{P_{H_2}P_{I_2}} \tag{5.3}$$

is a constant, dependent only on the temperature. Table 5.1 contains some of the experimental data that demonstrate this fact. Thus, while there is a large number of sets of pressures or concentrations that can exist at equilibrium, there is a *single* universal relationship that is satisfied when the system is at equilibrium.

Equation (5.3) is a simple example of an **equilibrium constant.** In general, experiments show that for a reaction of the form

$$aA + bB = cC + dD,$$

the concentrations of the reactants and products at equilibrium are related by the requirement that the function

$$\frac{[C]^c[D]^d}{[A]^a[B]^b} = K \tag{5.4}$$

Table 5.1 Equilibrium in the H_2, I_2, HI reaction

Partial pressure (atm)			
H_2	I_2	HI	$K = P_{HI}^2/P_{H_2}P_{I_2}$
0.1645	0.09783	0.9447	55.41
0.2583	0.04229	0.7763	55.19
0.1274	0.1339	0.9658	54.67
0.1034	0.1794	1.0129	55.31
0.02703	0.02745	0.2024	55.19
0.06443	0.06540	0.4821	55.16

Data of Taylor and Crist, *J. Am. Chem. Soc.,* **63,** 1377 (1941).
In the first four experiments HI was formed from its elements.
In the last two, equilibrium was approached by decomposing HI.
The temperature was 698.6°K.

be a fixed constant whose value depends on the temperature and the identity of the reactants and products. Should any of the reagents be a gas, its pressure can be used in place of its concentration in Eq. (5.4).

In the equilibrium-constant expression, the concentrations of the reaction *products*, each raised to a power equal to its stoichiometric coefficient in the chemical reaction, appear in the *numerator*, and the concentrations of the *reactants*, each raised to the appropriate power, appear in the *denominator*. Why is there such a thing as the equilibrium constant, and why does it have this form? It is possible to answer this question by using the methods of thermodynamics or reaction kinetics which are presented in Chapters 8 and 9. For the present, we will take the existence and form of the equilibrium constant as experimental facts. We must remark, however, that the expression of Eq. (5.4) holds only if we are dealing with reagents that are ideal gases or are present as solutes in an ideal solution. Thus Eq. (5.4) might be called the ideal law of chemical equilibrium.

There are several matters concerning the use of equilibrium constants that should be carefully noted. First, how are we to write the equilibrium-constant expression for reactions such as

$$CaCO_3(s) = CaO(s) + CO_2(g)$$

that involve *pure* solids like CaO and $CaCO_3$? Direct application of Eq. (5.4) would suggest that the appropriate expression is

$$K' = \frac{[CO_2][CaO]}{[CaCO_3]}.$$

In this instance, the solid phase is a *mixture* of individual microscopic crystals of pure CaO and $CaCO_3$, and it is conventional not to include the concentrations of pure solids in the equilibrium-constant expression. In the first place, the *concentration* of a pure solid in itself is a *constant*, and is not changed by the chemical reaction or by addition or removal of the solid. Moreover, it is an experimental fact that neither the *amount* of $CaCO_3$ nor that of CaO affects the equilibrium pressure of CO_2, so long as *some* of each solid is present. Consequently, we can include the constant concentrations of the pure solids in the equilibrium constant itself and write

$$[CO_2] = \frac{[CaCO_3]}{[CaO]} K' \equiv K, \qquad [CO_2] = K.$$

Thus the equilibrium constant for the decomposition of $CaCO_3$ is equal to the concentration (or pressure) of the carbon dioxide at equilibrium. The same principle applies to the reaction

$$Cu^{++}(aq) + Zn(s) = Cu(s) + Zn^{++}(aq).$$

The equilibrium constant is

$$\frac{[Zn^{++}]}{[Cu^{++}]} = K.$$

The metals do not appear in the equilibrium expression, for they are pure solids of constant composition.

Another important matter is illustrated by the relation between the equilibrium constant for the reaction

$$2H_2(g) + O_2(g) = 2H_2O(g)$$

and that for the simpler reaction

$$H_2(g) + \tfrac{1}{2}O_2(g) = H_2O(g).$$

For the first reaction we have

$$K_1 = \frac{[H_2O]^2}{[H_2]^2[O_2]},$$

while for the second reaction we would write

$$K_2 = \frac{[H_2O]}{[H_2][O_2]^{1/2}}.$$

Comparison of these expressions shows that

$$K_2 = K_1^{1/2}.$$

In general, if a reaction is *multiplied* by a certain factor, its equilibrium constant must be *raised* to a power equal to that factor in order to obtain the equilibrium constant for the new reaction.

A closely related problem is the relationship between the equilibrium constants for a reaction such as

$$2NO + O_2 = 2NO_2$$

and its reverse,

$$2NO_2 = 2NO + O_2.$$

For the reaction as first written we get

$$K_1 = \frac{[NO_2]^2}{[NO]^2[O_2]},$$

while the equilibrium constant for its reverse is

$$K_2 = \frac{[NO]^2[O_2]}{[NO_2]^2}.$$

Comparison of the two expressions shows that

$$K_2 = \frac{1}{K_1}.$$

That is, the equilibrium constants for a reaction and its reverse are reciprocals of each other. We could have obtained this result by following the previous rule, that is, multiplying the forward reaction by -1, and raising K_1 to the -1 power.

Often it is necessary to add two reactions together to obtain a third reaction. The equilibrium constant of the third reaction is related to the equilibrium constants of the two component reactions, as the following example illustrates:

$$2NO(g) + O_2(g) = 2NO_2(g), \qquad K_1 = \frac{[NO_2]^2}{[NO]^2[O_2]};$$

$$2NO_2(g) = N_2O_4(g), \qquad K_2 = \frac{[N_2O_4]}{[NO_2]^2};$$

$$2NO(g) + O_2(g) = N_2O_4(g), \qquad K_3 = \frac{[N_2O_4]}{[NO]^2[O_2]}.$$

Comparing the three equilibrium constants, we see that $K_3 = K_1K_2$. Thus, when two or more reactions are *added*, their equilibrium constants must be *multiplied* to give the equilibrium constant of the overall reaction.

The magnitude of the equilibrium constant depends on the units used to express the concentrations of the reagents. For example, consider the ammonia synthesis reaction at $673°K$:

$$N_2 + 3H_2 = 2NH_3.$$

When the equilibrium constant is written in terms of the pressures of the reagents expressed in units of atmospheres, we have

$$K_P(\text{atm}) = \frac{(P_{NH_3})^2}{(P_{N_2})(P_{H_2})^3} = 1.64 \times 10^{-4}.$$

However, if the pressure units to be used are torr (mm Hg), we would employ a value of K_P given by

$$K_P(\text{torr}) = K_P(\text{atm})\ \frac{\left(\dfrac{760\ \text{torr}}{\text{atm}}\right)^2}{\left(\dfrac{760\ \text{torr}}{\text{atm}}\right)\left(\dfrac{760\ \text{torr}}{\text{atm}}\right)^3},$$

$$K_P(\text{torr}) = \frac{K_P(\text{atm})}{(760)^2} = 2.84 \times 10^{-10}.$$

Similarly, we might wish to use concentration units of moles per liter instead of pressure. From the ideal gas law we see that

$$C(\text{moles/liter}) = \frac{n}{V} = \frac{P}{RT},$$

where P is in atmospheres, and R has units of liter-atmospheres per mole-degree. Thus the conversion factor from pressure to concentration is $1/RT$.

Therefore for the ammonia synthesis we have

$$\begin{aligned}
K_C(\text{moles/liter}) &= K_P(\text{atm})\ \frac{(1/RT)^2}{(1/RT)(1/RT)^3}\\
&= K_P(\text{atm})\ (RT)^2\\
&= 1.64 \times 10^{-4} \times 3.05 \times 10^3\\
&= 0.500.
\end{aligned}$$

Note that to convert an equilibrium constant to a new set of units, we multiply the old value by a factor which is found by raising the conversion factor from old to new units to the power Δn, where Δn is the change in the number of moles (final minus initial) of dissolved or volatile reagents.

Interpretation of Equilibrium Constants

The numerical value of the equilibrium constant for a reaction is a concise expression of the tendency for reactants to be converted to products. Because the algebraic form of the equilibrium constant is sometimes moderately complex, some care and experience are required to interpret its numerical value.

In this section we shall examine some simple types of reaction with the purpose of finding what qualitative information can be learned from the value of the equilibrium constant.

As our first example, we choose reactions in which there is only one reagent of variable concentration. These include

$$CaCO_3(s) = CaO(s) + CO_2(g), \qquad K = [CO_2];$$
$$I_2(s) = I_2 \text{ (in CCl}_4 \text{ solution)}, \qquad K = [I_2].$$

For such reactions, the equilibrium constant is simply equal to the equilibrium concentration of a single reagent. For the second of these reactions, the equilibrium constant is just the solubility of iodine in carbon tetrachloride. One way of representing the iodine solubility is to plot the allowed concentrations of dissolved iodine as a function of the amount of solid iodine present, as in Fig. 5.1. The significance of the horizontal line is that if any excess solid iodine is present, the solution is saturated, and the concentration of dissolved iodine must equal K. The meaning of the vertical line is that so long as no solid I_2 is present, the concentration of dissolved I_2 may be any value less than K.

FIG. 5.1 Solubility of I_2 in CCl$_4$. The solid horizontal line represents the equilibrium states of the system, while the dashed line represents a possible path to an equilibrium state.

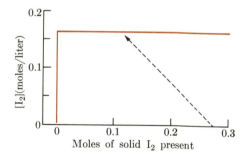

It is informative to use this diagram to represent what happens when a certain amount of I_2 is added to *one liter* of solvent. The initial state in which I_2 is present only as the undissolved solid is represented by a point on the abscissa. As I_2 dissolves, the states through which the system passes fall on a straight line of slope -1, since for each mole of I_2 that enters the solution, the concentration increases by $1\ M$. The final equilibrium situation is at the intersection of this line with the heavy horizontal line. If the initial amount of I_2 is insufficient to form a saturated solution, the sloping line will intersect the ordinate. A little reflection on this example leads us to conclude that any initial situation that is not represented by a point on the heavy line in Fig. 5.1 is not an equilibrium state, and such a system will approach equilibrium by precipitating or dissolving solid iodine. The progress toward equilibrium is represented by a line of negative slope, and the final state of the system by the intersection of this line with the ordinate or the equilibrium line $[I_2] = K$.

Now let us turn to reactions of the type exemplified by

$$Zn(s) + Cu^{++}(aq) = Cu(s) + Zn^{++}(aq), \qquad K = \frac{[Zn^{++}]}{[Cu^{++}]} = 2 \times 10^{37};$$

$$HCl(g) + LiH(s) = H_2(g) + LiCl(s), \qquad K = \frac{[H_2]}{[HCl]} = 8 \times 10^{30};$$

$$\underset{\textit{n}\text{-butane}}{CH_3CH_2CH_2CH_3} = \underset{\text{isobutane}}{CH_3\overset{\overset{\displaystyle CH_3}{\displaystyle |}}{C}HCH_3}, \qquad K = \frac{[\text{isobutane}]}{[\textit{n}\text{-butane}]} = 2.5.$$

In each instance the concentration ratio of products to reactants is a constant at equilibrium. Thus the value of K gives the concentration ratio at equilibrium directly: if the value of K is less than one, the reactant is in dominant concentration, and if K is large, the product is greatly favored. The equilibrium constants depend on temperature, and the values quoted refer to 25°C.

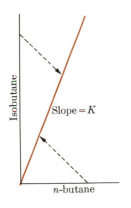

Equilibrium states of the n-butane-isobutane system. The dashed lines represent possible paths to equilibrium.

FIG. 5.2

The possible equilibrium states for such systems can be represented by plotting the concentration of product at equilibrium as a function of the concentration of reactant. The result for the n-butane-isobutane system, shown in Fig. 5.2, is a straight line of slope K. Any point on the line corresponds to a state of equilibrium, while any system represented by a point not on the line is not at equilibrium.

Two paths that a system might follow as it proceeds to an equilibrium state are also shown in Fig. 5.2. If only pure n-butane is present initially, the reaction path starts at a point on the abscissa and is a straight line slope -1, because 1 mole/liter of product appears for each mole/liter of reactant that disappears. If only isobutane is present initially, the reaction path starts on the ordinate and is a line of slope -1 that terminates on the equilibrium line. From these two examples it is clear that any arbitrary mixture of isobutane and n-butane that does not stand in a concentration ratio K is represented by a point some-where off the equilibrium line. The system will proceed to equilibrium by following a path whose slope is -1, and whose direction is determined by whether the initial concentration ratio is greater or less than K.

A third type of reaction that has a simple equilibrium-constant expression is illustrated by

$$BaSO_4(s) = Ba^{++}(aq) + SO_4^=(aq), \qquad K = [Ba^{++}][SO_4^=] = 1 \times 10^{-10};$$

$$NH_4HS(s) = NH_3(g) + H_2S(g), \qquad K = [NH_3][H_2S] = 9 \times 10^{-5}.$$

The numerical magnitude of an equilibrium constant of this type depends on the units chosen for concentration. In this instance, and those that follow, the concentration units are moles/liter. With this in mind, the value of the equilibrium constant can be interpreted without difficulty. A small value of K means that at equilibrium the concentrations of both products must be small, or if one is in large concentration, the other must be in *very* small concentration. For the special case in which the concentrations of the two products are the same, they must both be equal to $K^{1/2}$. In general, the concentrations of the two products are not equal, and the value of K limits only the product of the two concentrations.

These ideas can be understood with the aid of Fig. 5.3, where the equation

$$[Ba^{++}][SO_4^=] = K = 1 \times 10^{-10}$$

is plotted. The equilibrium states lie on a rectangular hyperbola which has the coordinate axes as asymptotes. We see that if the concentration of $SO_4^=$ is made large by addition of Na_2SO_4, the Ba^{++} concentration at equilibrium must become very small.

Equilibrium states of the $BaSO_4$-H_2O system. Dashed line a is the path to equilibrium followed when pure $BaSO_4$ dissolves in water, while b represents the addition of $BaCl_2$ to a solution of H_2SO_4.

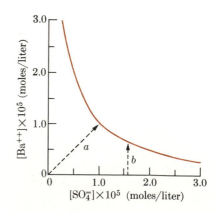

Examples of the paths that systems might follow to reach an equilibrium state are shown in Fig. 5.3. If pure $BaSO_4$ is dissolved in water, the system follows a line of unit slope from the origin, since the concentrations of the two ions increase equally as $BaSO_4$ dissolves. On the other hand, if small amounts of solid $BaCl_2$ are added to a solution of H_2SO_4, the concentration of Ba^{++} rises along a vertical line which eventually intersects the equilibrium hyperbola. The first $BaSO_4$ precipitates at a Ba^{++} concentration corresponding to this point of intersection. Any further addition of Ba^{++} causes more $BaSO_4$ to precipitate, and consequently the concentration of $SO_4^=$ decreases. As this process continues, the point representing the system follows the hyperbola in the direction of increasing $[Ba^{++}]$ and decreasing $(SO_4^=)$.

A fourth type of reaction is of the general form

$$A = 2B, \qquad K = \frac{[B]^2}{[A]}.$$

As specific examples we have

$$H_2(g) = 2H(g), \qquad K_{1000} = \frac{[H]^2}{[H_2]} = 2.1 \times 10^{-21};$$

$$N_2O_4(g) = 2NO_2(g), \qquad K_{298} = \frac{[NO_2]^2}{[N_2O_4]} = 5.7 \times 10^{-3},$$

where the concentrations are expressed in moles per liter. Although there are only two chemical species involved, the interpretation of this type of equilibrium constant is less straightforward than were the previous cases, for K is no longer a simple product or ratio of concentrations. It is still true, however, that if K is a very small number, the reactant will be in predominant concentration at equilibrium, while if K is very large, the product will be favored. The examples above indicate, therefore, that at room temperature, the dissociation of hydrogen to atoms is negligible, but the dissociation of N_2O_4 is noticeable.

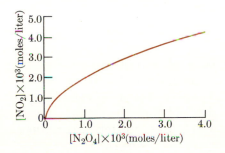

Equilibrium states of the NO_2-N_2O_4 system. The curve is a parabola which has the equation $[NO_2]^2 = K[N_2O_4]$.

FIG. 5.4

Figure 5.4 is a graphical representation of the NO_2-N_2O_4 equilibrium. The equilibrium states lie on a parabola which has the equation

$$[NO_2]^2 = K[N_2O_4].$$

The paths representing the passage of a system to equilibrium are lines of slope -2, for two moles of NO_2 are formed for each mole of N_2O_4 consumed.

The graphical treatment of equilibria which we have just discussed is helpful because it provides a clear way of seeing whether a given set of concentrations corresponds to an equilibrium state, and how a system not at equilibrium will change. Unfortunately, if there are more than two different reagents of variable concentration involved in the reaction, graphical representation of the system is difficult and not particularly informative. Nevertheless, our arguments based on the graphical treatment of simple systems suggest some generalizations that we can express algebraically and apply to all systems. If the concentrations of the reagents that appear in the general reaction

$$aA + bB = cC + dD$$

are such that

$$\frac{[C]^c[D]^d}{[A]^a[B]^b} = K,$$

the system is at equilibrium. If the concentrations are such that

$$\frac{[C]^c[D]^d}{[A]^a[B]^b} < K,$$

the reactants are in excess of the equilibrium values, and the reaction will proceed to equilibrium from left to right as written. On the other hand, if

$$\frac{[C]^c[D]^d}{[A]^a[B]^b} > K,$$

the products are in excess of their equilibrium value, and the reaction proceeds from right to left.

5.3 EXTERNAL EFFECTS ON EQUILIBRIA

In Section 4.6 we made use of LeChatelier's principle, which states that if a system at equilibrium is subjected to a disturbance or stress that changes any of the factors that determine the state of equilibrium, the system will react in such a way as to minimize the effect of the disturbance. LeChatelier's principle is of great help in dealing with chemical equilibria, for it allows us to predict the *qualitative* response of a system to changes in external conditions.

Such qualitative predictions are valuable guides and checks on the quantitative mathematical analysis of equilibria. In this section we shall illustrate the application of LeChatelier's principle and compare its predictions with the results of arguments based more directly on the equilibrium-constant expression.

Concentration Effects

Let us consider first a saturated solution of iodine in carbon tetrachloride which is in contact with excess solid iodine. What is the effect of adding a small amount of pure carbon tetrachloride to the system? The immediate consequence is to remove the system from equilibrium, for immediately after the addition of the pure solvent, the concentration of iodine in the solution is less than the equilibrium value. Thus the addition of solvent does affect the value of a factor that determines the state of equilibrium, and is, therefore, a stress in the sense implied in LeChatelier's principle. The prediction from LeChatelier's principle in this instance is that more solid iodine will dissolve, and thereby minimize the effect of the addition of solvent. Common experience assures us that this is indeed what happens.

The equilibrium-constant expression can be used to make a qualitative prediction of the behavior or response of a system to a disturbance. For the example we are treating, the relation

$$[I_2] = K$$

holds at equilibrium, but immediately after addition of the solvent we have

$$[I_2] < K.$$

This situation can be remedied if the reaction

$$I_2(s) + \text{solvent} = I_2 \text{ (in solution)}$$

proceeds as written from left to right. Thus the predictions based on LeChatelier's principle and on the equilibrium-constant expression are in accord with each other and with experimental fact.

For a somewhat more involved situation, let us turn to the equilibrium between solid barium sulfate and an aqueous solution of its ions:

$$BaSO_4(s) = Ba^{++}(aq) + SO_4^{=}(aq).$$

What is the effect of the addition of a small amount of a concentrated solution of Na_2SO_4? Such an addition causes an immediate and marked increase in the concentration of sulfate ion, and is, therefore, a stress that removes the system from equilibrium. From LeChatelier's principle we predict a reaction in the direction that minimizes the effects of this stress; that is, a reaction that removes sulfate ion from solution. Consequently, addition of the sodium sulfate solution should cause barium sulfate to precipitate from solution.

Let us see how the same prediction can be made by using the equilibrium-constant expression. At equilibrium the relation

$$[Ba^{++}][SO_4^=] = K \qquad (5.5)$$

holds, but immediately after the addition of excess sulfate ion, before any reaction occurs, we must have

$$[Ba^{++}][SO_4^=] > K.$$

It is clear that in order for the system to reach equilibrium, the concentrations of Ba^{++} and $SO_4^=$ must diminish, and this is accomplished by the precipitation of solid barium sulfate. Once again predictions based on the equilibrium constant and LeChatelier's principle are in agreement.

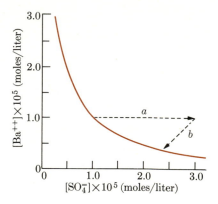

FIG. 5.5   Perturbation of the barium sulfate solubility equilibrium by addition of sulfate ion. Line *a* represents the departure from equilibrium caused by sulfate addition, and line *b* is the path followed by the system to a new equilibrium state.

We can make the conclusions about the behavior of the barium sulfate system more vivid by using the graphical representation of the equilibrium states of the system. Figure 5.5 shows Eq. (5.5) plotted as a rectangular hyperbola. Initially, the system can be represented by a point on this equilibrium curve, but sudden addition of sulfate ion causes the system to follow a horizontal path away from the equilibrium line. When the system is left to itself, it returns to equilibrium by following a path of slope $+1$ on the graph, and the final state is at the intersection of this line with the equilibrium curve. Figure 5.5 shows clearly that the final concentration of barium ion is *less* than the initial concentration, and that the final concentration of sulfate ion is *greater* than the initial concentration, but less than what would have existed after the addition if no precipitation had occurred.

The methods we have been using apply equally well to gaseous equilibria. For instance, if the reaction

$$SO_2(g) + \tfrac{1}{2}O_2(g) = SO_3(g)$$

is initially at equilibrium, what is the effect of suddenly adding oxygen gas to the mixture? Such an addition is a disturbing stress, and from LeChatelier's principle we predict a net reaction that consumes the oxygen—the reaction of SO_2 and O_2 to produce SO_3. By using the equilibrium-constant expression we conclude that immediately after the addition of oxygen, the relation

$$\frac{[SO_3]}{[SO_2][O_2]^{1/2}} < K$$

holds, and can be converted to an equality only if sulfur trioxide is formed from sulfur dioxide and oxygen.

Now we can explore the consequences of changing the volume of the vessel that contains a gaseous system at equilibrium. A change in total volume changes the concentrations of all gases, and consequently *may* cause a departure from equilibrium. This behavior is evident if we imagine the reaction

$$2NO_2(g) = N_2O_4(g), \qquad K = \frac{[N_2O_4]}{[NO_2]^2}$$

initially at equilibrium and ask for the effect of decreasing the volume of the containing vessel by a factor of two. If no reaction occurred, the immediate result would be to double all concentrations, and in such a state we would have

$$\frac{[N_2O_4]}{[NO_2]^2} < K.$$

This must be true, for if all concentrations are doubled, the numerator of the concentration quotient increases by a factor of two, while the denominator increases by a factor of two squared, or four. The concentration quotient then is not equal to the equilibrium constant, and a reaction that increases the concentration of N_2O_4 while decreasing the concentration of NO_2 must occur. Use of LeChatelier's principle confirms this prediction. The stress in this case is an increase in the total concentration of molecules—a stress that can be relieved if the system reacts so as to decrease the net number of molecules. Since two molecules of NO_2 are consumed for each molecule of N_2O_4 formed, the response of the system is to convert NO_2 to N_2O_4.

A volume change does not necessarily remove a gaseous system from equilibrium. If the reaction

$$\underset{\text{Butane}}{CH_3CH_2CH_2CH_3} = \underset{\text{Isobutane}}{CH_3\overset{\overset{\displaystyle CH_3}{\displaystyle |}}{C}HCH_3}$$

is at equilibrium and the volume of the container is halved, each concentration is doubled, but the relation

$$\frac{[\text{isobutane}]}{[\text{butane}]} = K$$

shows that even in this more concentrated situation the system remains at equilibrium. Any change of the concentration of butane is accompanied by an identical change of the isobutane concentration, and the ratio of these two concentrations remains constant. A little reflection on this and the previous example leads to the conclusion that only in the cases of reactions in which the number of moles of gaseous reactants is different from the number of moles of gaseous products does a volume change remove the system from equilibrium.

Temperature Effects

What is the effect of a temperature change on a system initially at equilibrium? In general, the numerical values of equilibrium constants depend on temperature. Therefore, if the temperature of a system initially at equilibrium is changed, some net reaction must occur in order for the system to reach equilibrium at the new temperature. Experiments show that if a reaction is exothermic, that is, if ΔH is negative, its equilibrium constant *decreases* as temperature increases. From the point of view of LeChatelier's principle, an increase of temperature is a stress that is partially relieved by the occurrence of a net reaction that proceeds with absorption of heat by the system. Consider then the reaction

$$2 \text{ NO}_2(\text{g}) = \text{N}_2\text{O}_4(\text{g}), \qquad \Delta H = -13.9 \text{ kcal},$$
$$K_{273} = 76, \qquad K_{298} = 8.8,$$

where the K's are the equilibrium constants at 273°K and 298°K, respectively. The reaction is exothermic, K does in fact decrease, and reactants become more favored as temperature increases. According to LeChatelier's principle also, increasing temperature should favor the formation of reactants from products for this is the direction of reaction that absorbs heat. Thus the prediction based on LeChatelier's principle and the actual observed values of the equilibrium constant are consistent.

As an example of an endothermic reaction, let us inspect

$$\text{N}_2(\text{g}) + \text{O}_2(\text{g}) = 2\text{NO}(\text{g}), \qquad \Delta H = 43.5 \text{ kcal}.$$

By LeChatelier's principle we can predict that increasing temperature should favor formation of nitric oxide, for this process is accompanied by absorption of heat. Actual determination of equilibrium constants confirms this, as we can see:

$$K_{2000} = 4.1 \times 10^{-4}, \qquad K_{2500} = 36 \times 10^{-4}.$$

In the special case of a reaction for which ΔH is zero, a temperature increase can favor neither products nor reactants, for the reaction does not proceed in either direction with absorption of heat. In harmony with this conclusion, the equilibrium constants of reactions for which ΔH is zero are found to be independent of temperature.

5.4 EQUILIBRIA IN NONIDEAL SITUATIONS

We remarked earlier in this chapter that the expression

$$K = \frac{[C]^c[D]^d}{[A]^a[B]^b},$$

holds only in situations where all reactants and products behave like ideal gases or are components of an ideal solution. For reactions between gases at high pressures, or for reactions in markedly nonideal solutions, the equilibrium "constant" expressed in terms of pressures or concentrations is not strictly constant. To analyze these nonideal situations, it is customary to make use of the concept of *activity*. The activity of a substance may be thought of as an *effective* concentration. It has the property that in all situations, the equilibrium-constant expression can be written as

$$\frac{\{C\}^c\{D\}^d}{\{A\}^a\{B\}^b} = K,$$

where the notation $\{A\}$ stands for the activity of A. Thus in terms of the activity, the equilibrium-constant expression has its simple form and is *always* constant at a particular temperature.

How is the activity of a substance determined? For an ideal gas the activity is numerically equal to the pressure in atmospheres, but for a nonideal gas the activity must be calculated from the experimentally determined equation of state. This is a moderately involved process which we will not describe. For our purposes it is enough to remark that for gases at pressures of 1 atm or less and at ordinary temperatures, an error usually of less than 1% is made by considering the activity and the pressure of a gas to be equal.

For electrolytes in very dilute aqueous solution, activity is very nearly numerically equal to the molal or molar concentration. For solutions of approximately 0.1-M concentration, however, activity and concentration differ by from 10 to 50%, depending on the electrolyte concentration and the charge on the ions. One way of determining the activity of such solutions is by freezing-point depression measurements. If solutions were ideal, the freezing-point depression would be directly proportional to the total molal concentration of solute:

$$\Delta T = K_f m.$$

This law is obeyed by very dilute solutions, but the more concentrated solutions deviate from it noticeably. By comparing the observed freezing point of a solution with that expected for an ideal solution of the same concentration, the activity of the solute can be calculated. This procedure establishes a relation between activity and concentration that can be used in dealing with other equilibrium situations.

Any very accurate mathematical analysis of an equilibrium situation must be performed in terms of activities rather than concentrations. However, relating activities to concentrations is often a tedious task, and the accuracy gained by using activities is in many cases not worth the extra effort. In particular, design and analysis of the laboratory work of undergraduate chemistry require performing equilibrium calculations to an accuracy of about 5%, which can be accomplished easily if concentrations are used in place of activities. This is the procedure we shall follow without comment in subsequent sections of this book. It is important to realize that the results of such calculations, while very useful, are usually accurate only to 5%.

5.5 CALCULATIONS WITH THE EQUILIBRIUM CONSTANT

In many situations it is necessary to know the concentrations of the reactants and the products of a reaction when the system is at equilibrium. Rather than measure these numbers directly for each new situation, it is easier to calculate them if the equilibrium constant for the reaction is known. In this section we shall perform some calculations with the equilibrium constant to illustrate the technique and to demonstrate how the value of the equilibrium constant can be interpreted quantitatively. We shall concentrate on some simple gas reactions and reserve for special consideration in Chapter 6 the situations of equilibrium between ions in aqueous solution.

The equilibrium constants of reactions between gases are often expressed in terms of the partial pressures of reagents, and unless otherwise stated, the units are atmospheres. Thus for the reaction

$$N_2O_4(g) = 2NO_2(g),$$

we have

$$K_P = \frac{P_{NO_2}^2}{P_{N_2O_4}} = 0.14,$$

when pressures are expressed in atmospheres, at 25°C.

In dealing with a gaseous system, the experimental information most readily available is the *total pressure* of all the reactants and products, rather than the partial pressures of each. Therefore, it is sometimes useful to express the equilibrium constant in terms of the total pressure. For the N_2O_4-NO_2 reaction we can accomplish this as follows. Imagine starting with a certain number of moles of pure N_2O_4; when equilibrium is reached, a fraction f of the original N_2O_4 has dissociated. Then, $1 - f$ is the fraction of the original N_2O_4 remaining, and is proportional to the number of moles of N_2O_4 present at equilibrium. Similarly, since 2 moles of NO_2 are formed from each mole of N_2O_4 that dissociates, $2f$ is proportional to the number of moles of NO_2 present at equilibrium. Therefore, the total number of moles of all particles present at equilibrium is proportional to $1 - f + 2f$, or to $1 + f$. Now we can write the following expressions for the mole fractions of N_2O_4 and NO_2:

$$X_{N_2O_4} = \frac{1 - f}{1 + f}, \qquad X_{NO_2} = \frac{2f}{1 + f}.$$

By Dalton's law, we can express the partial pressures of NO_2 and N_2O_4 in terms of the *total pressure* P·

$$P_{NO_2} = X_{NO_2}P = \frac{2f}{1 + f} P,$$

$$P_{N_2O_4} = X_{N_2O_4}P = \frac{1 - f}{1 + f} P.$$

We can now return to the equilibrium constant, and use these expressions in place of the partial pressures to get

$$K_P = \frac{\left(\dfrac{2fP}{1 + f}\right)^2}{\left(\dfrac{1 - f}{1 + f}\right) P} = \frac{4f^2 P}{1 - f^2}. \tag{5.6}$$

This is the expression we were seeking: the equilibrium condition expressed in terms of P, the total pressure of reactants and products, and f, the fraction of the original N_2O_4 that dissociated. Bear in mind that K_P is a constant that depends only on temperature. Therefore, as the total pressure P is varied, the fraction of N_2O_4 dissociated must change so as to keep the right-hand side of Eq. (5.6) constant. Suppose, for example, that the volume of the system is increased, so that P decreases. Does f increase or decrease? By LeChatelier's principle we expect more N_2O_4 to dissociate and thus f to increase as P decreases. This is just what Eq. (5.6) says will happen. To see this clearly, we rearrange Eq. (5.6) to

$$P = \frac{K_P}{4}\left(\frac{1}{f^2} - 1\right),$$

from which it is obvious that if P decreases, f must increase if K_P is a constant. Thus, as predicted from LeChatelier's principle, a decrease in total pressure favors the products of this reaction.

Example 5.1 A mixture of N_2O_4 and NO_2 has at equilibrium a total pressure of 1.5 atm. What fraction of N_2O_4 has dissociated to NO_2 at 25°C? The equilibrium constant $K_P = 0.14$. By Eq. (5.6)

$$K_P = \frac{4f^2 P}{1 - f^2}, \qquad 0.14 = \frac{4f^2}{1 - f^2} 1.5;$$

$$f^2 = 0.023, \qquad f = 0.15.$$

If the volume of the system is increased so that the pressure falls to 1 atm, what fraction of the original N_2O_4 is dissociated?

$$0.14 = \frac{4f^2}{1 - f^2} 1, \qquad f^2 = 0.034, \qquad f = 0.18.$$

Thus, f does increase as total pressure decreases.

Occasionally it is possible to measure the partial pressure of one of the substances in an equilibrium mixture directly. Then it may be possible to calculate the partial pressures of the other reagents by using the equilibrium constant. Suppose, for example, the partial pressure of NO_2 in equilibrium with N_2O_4 is found to be 0.23 atm. What pressure of N_2O_4 is present? To solve this problem we use the equilibrium constant in the form

$$K_P = \frac{P^2_{NO_2}}{P_{N_2O_4}} = 0.14.$$

Setting $P_{NO_2} = 0.23$ atm, and carrying out the arithmetic gives

$$P_{N_2O_4} = \frac{(0.23)^2}{0.14} = 0.38 \text{ atm.}$$

Calculations of this type are easy when only two reagents are involved, but we shall find that they can be accomplished even for more complicated situations.

To gain experience with an equilibrium constant of a different form, let us consider the reaction

$$NH_4HS(s) = NH_3(g) + H_2S(g),$$
$$K_P = P_{NH_3} P_{H_2S} = 0.11,$$

where the equilibrium constant is evaluated at 25°C. Since ammonium hydrosulfide is a solid of invariant composition, it is not included in the equilibrium-constant expression. Let us use the value of the equilibrium constant to calculate the pressure of ammonia and hydrogen sulfide in equilibrium with the pure salt at 25°C. Since the gases come from the evaporation of the pure salt in equimolar quantities we have

$$P_{NH_3} = P_{H_2S},$$

$$K_P = P_{NH_3}P_{H_2S} = P_{NH_3}^2 = 0.11,$$

$$P_{NH_3} = 0.33 \text{ atm} = P_{H_2S}.$$

Thus, when the pure salt evaporates into an evacuated container, the pressure of each gas is 0.33 atm at equilibrium.

What would be the effect of injecting some pure ammonia gas into the ammonium hydrosulfide system, after the latter had reached equilibrium? By LeChatelier's principle, this would be a stress that could be relieved by the consumption of ammonia by reaction with hydrogen sulfide. The result of the addition of ammonia, then, would be a decrease in the pressure of hydrogen sulfide at equilibrium. This conclusion can be reached by using the equilibrium-constant expression, for if P_{HN_3} is increased, P_{H_2S} must decrease if their product is to remain constant.

Example 5.2 Some solid NH_4HS is placed in a flask containing 0.50 atm of ammonia. What are the pressures of ammonia and hydrogen sulfide when equilibrium is reached?

Because some ammonia is added, the pressures of ammonia and hydrogen sulfide are not equal when equilibrium is reached. However, we can write

$$P_{NH_3} = 0.50 + P_{H_2S}.$$

Therefore,

$$K_P = P_{NH_3}P_{H_2S} = (0.5 + P_{H_2S})P_{H_2S},$$

$$0.11 = (0.50 + P_{H_2S})P_{H_2S},$$

$$P_{H_2S} = 0.17 \text{ atm},$$

$$P_{NH_3} = 0.50 + 0.17 = 0.67 \text{ atm}.$$

Comparison of the pressure of hydrogen sulfide with that found after the evaporation of the pure solid into vacuum shows that the added ammonia has repressed the evaporation of the solid, as expected according to LeChatelier's principle. Note also that since equilibrium can be approached from either direction, the results of this calculation apply to the case in which a quantity of ammonia equivalent to a pressure of 0.5 atm in this flask is added to the solid already at equilibrium with its vapor.

The two reactions investigated are particularly easy to treat, since each of them involves only two reagents of variable concentration. Nevertheless, our experience with them will help us deal with the reaction

$$PCl_5(g) = PCl_3(g) + Cl_2(g).$$

When pressures are expressed in atmospheres,

$$K_P = \frac{P_{PCl_3}P_{Cl_2}}{P_{PCl_5}} = 11.5,$$

at a temperature of 300°C.

First let us imagine that enough pure solid PCl_5 is placed in a flask so that when the temperature is raised to 300°C, it would all vaporize to give a pressure of P_0 atm, *if no* PCl_5 *dissociated.* If some PCl_5 does dissociate to PCl_3 and Cl_2, we have

$$P_{Cl_2} = P_{PCl_3}, \qquad P_{PCl_5} = P_0 - P_{Cl_2},$$

since for each mole of PCl_5 that dissociates, one mole of Cl_2 and one mole of PCl_3 are formed. With these relations we can write the equilibrium-constant expression entirely in terms of the partial pressure of chlorine and P_0, the pressure of PCl_5 that would exist if no dissociation occurred:

$$K_P = \frac{P_{PCl_3}P_{Cl_2}}{P_{PCl_5}} = \frac{P_{Cl_2}^2}{P_0 - P_{Cl_2}} = 11.5.$$

As a specific case, suppose that P_0 is 1.5 atm. Then at equilibrium,

$$K_P = 11.5 = \frac{P_{Cl_2}^2}{1.5 - P_{Cl_2}},$$

$P_{Cl_2} = 1.34$ atm, $\qquad P_{PCl_3} = 1.34$ atm, $\qquad P_{PCl_5} = 1.50 - 1.34 = 0.16$ atm.

We can compute the fraction f of PCl_5 that dissociated by

$$f = \frac{P_{Cl_2}}{P_0} = \frac{1.34}{1.50} = 0.89.$$

For comparison, let us now assume that enough PCl_5 is used so that P_0 is 3.0 atm. Then

$$K_P = 11.5 = \frac{P_{Cl_2}^2}{3.0 - P_{Cl_2}},$$

$P_{Cl_2} = 2.47$ atm, $\qquad P_{PCl_3} = 2.47$ atm, $\qquad P_{PCl_5} = 0.53$ atm.

For the fraction dissociated we find

$$f = \frac{2.47}{3} = 0.82.$$

Increasing the total amount of material at fixed volume, or increasing the total pressure, decreases the fraction of PCl_5 that dissociates.

What is the effect of adding chlorine gas to the reaction mixture? Qualitatively it is clear that the additional chlorine must repress the dissociation of PCl_5. To demonstrate this quantitatively, let us assume that 1.5 atm of chlorine is added to an amount of PCl_5 sufficient to exert 3.0-atm pressure if none dissociated. Then, if some PCl_5 does dissociate, the stoichiometry of the reaction tells us that

$$P_{PCl_5} = 3.0 - P_{PCl_3}, \qquad P_{Cl_2} = 1.5 + P_{PCl_3}.$$

Substitution of these relations into the equilibrium-constant expression gives

$$K_P = 11.5 = \frac{P_{PCl_3} P_{Cl_2}}{P_{PCl_5}}, \qquad 11.5 = \frac{P_{PCl_3}(1.5 + P_{PCl_3})}{3.0 - P_{PCl_3}},$$

$$P_{PCl_3} = 2.26 \text{ atm}, \qquad P_{Cl_2} = 3.76 \text{ atm}, \qquad P_{PCl_5} = 0.74 \text{ atm}.$$

The expression for the fraction of PCl_5 that dissociated is

$$f = \frac{P_{PCl_3}}{P_0} = \frac{2.26}{3.0} = 0.75.$$

Comparison of these answers with corresponding ones from the previous calculation shows that addition of chlorine does diminish the amount of PCl_5 that dissociates. The fraction dissociated would also be diminished by addition of PCl_3, as a trial calculation will show.

As a final application, we can use the equilibrium-constant concept to provide a heuristic derivation of the equation of state for nonideal gases. Let us assume that nonideality arises from the dimerization of gas molecules,

$$2A = A_2,$$

which is a reaction with a small, but significant equilibrium constant K. If $[A]_0$ is the concentration that A-molecules would have if none dimerized, we can write

$$K = \frac{[A_2]}{[A]^2} = \frac{[A_2]}{([A]_0 - 2[A_2])^2},$$
$$[A_2] = K([A]_0 - 2[A_2])^2 \cong K[A_0]^2,$$

where the approximation is valid if the concentration of dimers is very small. We now regard the gas as a mixture of dimers and monomers, and the total pressure can be found from Dalton's law

$$P = RT([A] + [A_2])$$
$$= RT([A_0] - 2[A_2] + [A_2])$$
$$= RT([A_0] - K[A_0]^2).$$

Letting n stand for the total number of moles of A, we can write

$$P = n\frac{RT}{V}\left(1 - \frac{Kn}{V}\right),$$

$$\frac{PV}{nRT} = 1 - \frac{nK}{V}. \tag{5.7}$$

This is the form of the first two terms of the virial equation of state. Since K is positive, Eq. (5.7) accounts for negative deviations from ideal gas behavior. The positive deviations, which do not arise from dimer formation, are not accounted for in Eq. (5.7). If, however, we had expressed the total concentration of A-molecules not as n/V but as $n/(V - nb)$, both positive and negative deviations from ideality would have been accommodated.

5.6 CONCLUSION

A chemical reaction is a complicated mechanical process in which atoms initially in one arrangement are converted to some other arrangement. This process is reversible, and consequently, in a closed system, reactants and products can exist permanently together in fixed relative amounts that reflect the intrinsic stabilities of the various compounds. The most succinct and meaningful description of these equilibrium situations is contained in the equilibrium constant. Therefore, to deal with the quantitative aspects of chemistry, to understand what is meant by chemical reactivity, an appreciation of the use and significance of the equilibrium constant is absolutely essential. In this chapter, we have investigated some simple chemical equilibria qualitatively and quantitatively, and we shall make continual use of the ideas we have generated. As more complicated equilibria are encountered, it is essential to be guided by previous experience and to attempt to answer the questions we have discussed here: What is the qualitative meaning of the form and magnitude of the equilibrium constant? What do they tell us about the tendency of a reaction to proceed to products? What do the ΔH- and equilibrium-constant expression tell us about how temperature and concentration changes influence the amounts of products and reactants at equilibrium? How can the equilibrium constant be used to calculate the concentrations that exist at equilibrium?

Cragg, L. H., and R. P. Graham, *An Introduction to the Principles of Chemistry.* New York: Rinehart and Co., 1954. Chapter 13.

Daniels, F., and R. A. Alberty, *Physical Chemistry*, 2nd ed. New York: Wiley, 1961. Chapter 9.

Sienko, M. J., and R. A. Plane, *Chemistry: Principles and Properties.* New York: McGraw-Hill, 1966.

PROBLEMS

5.1 Write the equilibrium-constant expression for each of the following reactions.

$$2NOCl(g) = 2NO(g) + Cl_2(g)$$
$$Zn(s) + CO_2(g) = ZnO(s) + CO(g)$$
$$MgSO_4(s) = MgO(s) + SO_3(g)$$
$$Zn(s) + 2H^+(aq) = Zn^{++}(aq) + H_2(g)$$
$$NH_4Cl(s) = NH_3(g) + HCl(g)$$

5.2 For which of the following reactions does the equilibrium constant depend on the units of concentration?

(a) $CO(g) + H_2O(g) = CO_2(g) + H_2(g)$

(b) $COCl_2(g) = CO(g) + Cl_2(g)$

(c) $NO(g) = \frac{1}{2}N_2(g) + \frac{1}{2}O_2(g)$

5.3 Of the metals Zn, Mg, and Fe, which removes cupric ion from solution most completely? The following equilibrium constants hold at room temperature.

$$Zn(s) + Cu^{++}(aq) = Cu(s) + Zn^{++}(aq) \qquad K = 2 \times 10^{37}$$
$$Mg(s) + Cu^{++}(aq) = Cu(s) + Mg^{++}(aq) \qquad K = 6 \times 10^{90}$$
$$Fe(s) + Cu^{++}(aq) = Cu(s) + Fe^{++}(aq) \qquad K = 3 \times 10^{26}$$

5.4 Nitrogen and hydrogen react to form ammonia by the reaction

$$\frac{1}{2}N_2 + \frac{3}{2}H_2 = NH_3, \qquad \Delta H = -11.0 \text{ kcal.}$$

If a mixture of the three gases were in equilibrium, what would be the effect on the amount of NH_3 if (a) the mixture were compressed; (b) the temperature were raised; (c) additional H_2 were introduced?

5.5 Would you expect the equilibrium constant for the reaction

$$I_2(g) = 2I(g)$$

to increase or decrease as temperature increases? Why?

5.6 The equilibrium constants for the following reactions have been measured at $823°K$:

$$CoO(s) + H_2(g) = Co(s) + H_2O(g), \qquad K = 67;$$
$$CoO(s) + CO(g) = Co(s) + CO_2(g), \qquad K = 490.$$

From these data, calculate the equilibrium constant of the reaction

$$CO_2(g) + H_2(g) = CO(g) + H_2O(g)$$

at $823°K$.

5.7 Suggest four ways in which the equilibrium concentration of SO_3 can be increased in a closed vessel if the only reaction is

$$SO_2(g) + \tfrac{1}{2}O_2(g) = SO_3(g), \qquad \Delta H = -23.5 \text{ kcal.}$$

5.8 Solid ammonium carbamate, $NH_4CO_2NH_2$, dissociates completely into ammonia and carbon dioxide when it evaporates, as shown by

$$NH_4CO_2NH_2(s) = 2NH_3(g) + CO_2(g).$$

At 25°C, the total pressure of the gases in equilibrium with the solid is 0.116 atm. What is the equilibrium constant of the reaction? If 0.1 atm of CO_2 is introduced after equilibrium is reached, will the final pressure of CO_2 be greater or less than 0.1 atm? Will the pressure of NH_3 increase or decrease?

5.9 For the reaction

$$H_2(g) + I_2(g) = 2HI(g),$$

$K = 55.3$ at 699°K. In a mixture that consists of 0.70 atm of HI and 0.02 atm each of H_2 and I_2 at 699°K, will there be any net reaction? If so, will HI be consumed or formed?

5.10 Hydrogen and iodine react at 699°K according to

$$H_2(g) + I_2(g) = 2HI.$$

If 1.00 mole of H_2 and 1.00 mole of I_2 are placed in a 1.00-liter vessel and allowed to react, what weight of hydrogen iodide will be present at equilibrium? At 699°K, $K = 55.3$.

5.11 At 375°K, the equilibrium constant K_P of the reaction

$$SO_2Cl_2(g) = SO_2(g) + Cl_2(g)$$

is 2.4 when pressures are expressed in atmospheres. Assume that 6.7 gm of SO_2Cl_2 are placed into a 1-liter bulb and the temperature is raised to 375°K. What would the pressure of SO_2Cl_2 be if none of it dissociated? What are the pressures of SO_2, Cl_2, and SO_2Cl_2 at equilibrium?

5.12 Compute the pressures of SO_2Cl_2, SO_2, and Cl_2 in a 1-liter bulb (at 375°K) to which 6.7 gm of SO_2Cl_2 and 1.0 atm of Cl_2 (at 375°K) have been added. Use the data supplied in Problem 5.11. Compare your answer with that obtained for Problem 5.11 and decide whether they are consistent with LeChatelier's principle.

5.13 The gaseous compound NOBr decomposes according to the reaction

$$NOBr(g) = NO(g) + \tfrac{1}{2}Br_2(g).$$

At 350°K, the equilibrium constant K_P is equal to 0.15. If 0.50 atm of NOBr, 0.40 atm of NO, and 0.20 atm of Br_2 are mixed at this temperature, will any net reaction occur? If so, will Br_2 be consumed or formed?

5.14 The equilibrium constant for the reaction

$$CO_2(g) + H_2(g) = CO(g) + H_2O(g)$$

is 0.10 at 690°K. What is the equilibrium pressure of each substance in a mixture prepared by mixing 0.50 mole of CO_2 and 0.50 mole of H_2 in a 5-liter flask at 690°K?

5.15 At 1000°K, the pressure of CO_2 in equilibrium with $CaCO_3$ and CaO is equal to 3.9×10^{-2} atm. The equilibrium constant for the reaction

$$C(s) + CO_2(g) = 2CO(g)$$

is 1.9 at the same temperature when pressures are in atmospheres. Solid carbon, CaO, and $CaCO_3$ are mixed and allowed to come to equilibrium at 1000°K in a closed vessel. What is the pressure of CO at equilibrium?

CHAPTER 6

IONIC EQUILIBRIA IN AQUEOUS SOLUTIONS

Equilibria between ionic species in aqueous solutions deserve special attention because of their importance in industrial, analytical, and physiological chemistry. The principles used in dealing with problems of ionic equilibria are of course the same as those that apply to other situations of chemical equilibrium. Thus the study of ionic equilibria offers us a chance to learn general principles while studying an important practical application.

The ability to solve equilibrium problems is a result of a thorough under-standing of physical principles and of an intuition that can be acquired only by experience. This chapter should be read with a pencil and paper at hand, and the steps in each derivation and example should be worked out independently of the text. One is not ready to attempt new material until, with the book closed, he can work out the section he has just studied.

6.1 SPARINGLY SOLUBLE SALTS

The problem of finding the equilibrium concentration of a slightly soluble salt involves one of the simplest applications of the principles of chemical equilibrium. Consider the dissolution of solid silver chloride in water, which proceeds by

$$AgCl(s) = Ag^+(aq) + Cl^-(aq). \tag{6.1}$$

When equilibrium between the pure solid and the solution is reached at 25°C, it is found that only 1.67×10^{-5} mole of silver chloride is dissolved in one liter of water. As small as this concentration may seem, it is large enough to be important in many laboratory situations. As a result, we are interested in finding quantitative expressions telling us what the solubilities of salts such as silver chloride are under any conditions.

According to the general procedure for writing equilibrium constants, for Eq. (6.1) we might write

$$K = \frac{[Ag^+][Cl^-]}{[AgCl(s)]}.$$

But the concentration of a pure solid in a pure solid is a constant, and we can immediately simplify our expression by defining a new constant K_{sp} by

$$K_{sp} \equiv K[AgCl(s)]$$
$$= [Ag^+][Cl^-].$$

That is, we include the fixed concentration of the pure solid in the equilibrium constant itself. Consequently, the equilibrium constant for reaction (6.1) is a product of ion concentrations. For this reason, K_{sp} for these reactions is often called the ion product constant, or simply the **solubility product**. Three other such reactions and their equilibrium constants are

$$CaF_2(s) = Ca^{++}(aq) + 2F^-(aq), \qquad K_{sp} = [Ca^{++}][F^-]^2,$$
$$Ag_2CrO_4(s) = 2Ag^+(aq) + CrO_4^=(aq), \qquad K_{sp} = [Ag^+]^2[CrO_4^=],$$
$$La(OH)_3(s) = La^{+3}(aq) + 3OH^-(aq), \qquad K_{sp} = [La^{+3}][OH^-]^3.$$

Table 6.1 Solubility products*

BaSO$_4$	1.1×10^{-10}	Ag$_2$CrO$_4$	1.9×10^{-12}
BaF$_2$	1.7×10^{-6}	Ag$_2$S	1×10^{-51}
BaCO$_3$	1.6×10^{-9}	Fe(OH)$_2$	1.6×10^{-15}
BaCrO$_4$	8.5×10^{-11}	FeS	1×10^{-19}
CaSO$_4$	2.4×10^{-5}	Zn(OH)$_2$	4.5×10^{-17}
Ca$_3$(PO$_4$)$_2$	1.3×10^{-32}	ZnS	4.5×10^{-24}
CaF$_2$	1.7×10^{-10}	ZnCO$_3$	2×10^{-10}
CaCrO$_4$	7.1×10^{-4}	Sn(OH)$_2$	5×10^{-26}
Mg(OH)$_2$	1.8×10^{-11}	SnS	8×10^{-29}
PbS	7×10^{-29}	CuS	4×10^{-38}
PbSO$_4$	1.3×10^{-8}	Cu(OH)$_2$	1.6×10^{-19}
PbCO$_3$	1.5×10^{-13}	Cu(IO$_3$)$_2$	1.3×10^{-7}
PbCrO$_4$	2×10^{-16}	Mn(OH)$_2$	2×10^{-13}
Pb(OH)$_2$	2.8×10^{-16}	MnS	7×10^{-16}
AgBrO$_3$	5.2×10^{-5}	HgS	3×10^{-53}
AgCl	2.8×10^{-10}	CdS	1.4×10^{-28}
AgBr	5.2×10^{-13}	Ni(OH)$_2$	1.6×10^{-16}
AgI	8.5×10^{-17}		

*From W. M. Latimer, *Oxidation Potentials*, 2nd ed., New York, Prentice Hall, Inc., 1952.

These examples demonstrate the practice of not writing any invariable concentrations in the equilibrium-constant expression. Solubility products of a few other salts are given in Table 6.1.

Questions. Suppose the solid phase were impure, as would be the case for a *solid solution* of AgCl and NaCl. Do you expect that the solubility in water of this material would be a function of the concentration of AgCl in the solid phase? Would the simple ion product constant for AgCl describe the equilibrium adequately?

Application of the equilibrium constant to the solubility of electrolytes is usually limited to cases in which the substances are only slightly soluble. There are two principal reasons for this. First, concentrated solutions of electrolytes such as saturated potassium chloride are not ideal solutions, and the simple equilibrium-constant expressions do not apply rigorously to nonideal solutions. Second, in the practical problems of chemical analysis one often takes advantage of the difference in the solubilities of two sparingly soluble salts, and therefore it is for such substances that information derived from the equilibrium constant is of most value. Thus we are in the happy situation of having a theoretical framework that applies best to cases that are of most interest.

There is still another reason for the application of the equilibrium constant to the dissolution of slightly soluble salts. Because of the small quantities of dissolved materials involved, it is very difficult or impossible to measure directly the solubilities of these substances. However, it is possible to detect minute quantities of dissolved ions by measurements of the voltages of electrochemical cells, as we shall see in Chapter 7. These voltage measurements yield the *solubility product* K_{sp} of the salt directly, and from this quantity we must be able to calculate the salt solubility. Let us assume then that we have been provided with the value of the solubility product for silver chloride in water:

$$[Ag^+][Cl^-] = K_{sp} = 2.8 \times 10^{-10}, \tag{6.2}$$

where the concentration unit is mole per liter. How do we calculate from this the solubility of AgCl in pure water?

Reaction (6.1) shows us that for each mole of silver ion dissolving, we have also one mole of chloride ion in solution. There are no other sources of these ions, so in Eq. (6.2) we can make the substitution

$$[Ag^+] = [Cl^-] \quad \text{in otherwise pure water,}$$

to get

$$[Ag^+][Cl^-] = [Ag^+]^2 = 2.8 \times 10^{-10},$$
$$[Ag^+] = 1.7 \times 10^{-5} \ M.$$

This is the concentration of silver ion present when equilibrium between the solution and the excess undissolved solid is reached. From the stoichiometry of reaction (6.1) we see that it is also the maximum number of moles of silver chloride that dissolve in one liter of water, so the solubility of the salt in pure water is $1.7 \times 10^{-5} \ M$.

The calculation of the solubility in pure water of a salt like CaF_2 is nearly as straightforward. The reaction accompanying dissolution and the corresponding solubility product are

$$CaF_2(s) = Ca^{++}(aq) + 2F^-(aq),$$
$$[Ca^{++}][F^-]^2 = K_{sp} = 1.7 \times 10^{-10}.$$

In pure water, the only source of calcium or fluoride ions is the salt itself, and the stoichiometry tells us that twice as many fluoride ions dissolve as calcium ions. Consequently, we can say that

$$[F^-] = 2[Ca^{++}] \quad \text{in otherwise pure water,}$$

and so

$$[Ca^{++}][F^-]^2 = [Ca^{++}](2[Ca^{++}])^2 = 1.7 \times 10^{-10},$$
$$4[Ca^{++}]^3 = 1.7 \times 10^{-10},$$
$$[Ca^{++}] = 3.5 \times 10^{-4} \ M.$$

Since one mole of calcium dissolves for each mole of salt, the solubility of calcium fluoride in pure water is $3.5 \times 10^{-4} \ M$.

In response to this and similar examples, it is often asked why one of the concentrations is doubled *and* squared. The form of the equilibrium constant requires that the concentration of the *fluoride ion* be squared. The stoichiometry requires that the concentration of the calcium ion be doubled *if it is substituted for the fluoride ion concentration.* These two requirements lead to the factor $(2[Ca^{++}])^2$ in the solution of the problem.

Question. When silver chromate, Ag_2CrO_4, dissolves in otherwise pure water, what is the relation between $[Ag^+]$ and $[CrO_4^-]$? The solubility product of Ag_2CrO_4 is 1.9×10^{-12}. Show that its solubility in pure water is $0.78 \times 10^{-4} \ M$.

To see the advantage of knowing the solubility product of a salt, let us calculate the solubility of silver chloride in a solution that contains $0.1 \ M$ $AgNO_3$. So that we can get a qualitative idea of the answer to expect, let us imagine ourselves starting with a saturated solution of $AgCl$ in pure water, and then adding enough solid $AgNO_3$ to reach the eventual concentration of $0.1 \ M$. The addition of Ag^+ in the form of $AgNO_3$ is a "stress" applied to the silver chloride solubility equilibrium, and according to Le Chatelier's principle, the position of this equilibrium must shift so as to relieve the stress. This means that as $AgNO_3$ is added, $AgCl$ must tend to precipitate. Therefore we can conclude that the solubility of $AgCl$ (as measured by the chloride ion concentration) in a solution of $0.1 \ M$ Ag^+ is less than in pure water.

To verify this quantitatively we must calculate the concentration of dissolved chloride ion in the solution containing $0.1 \ M$ $AgNO_3$, and saturated with respect to silver chloride. We could calculate the chloride ion concentration from the solubility-product expression

$$[Cl^-] = \frac{K_{sp}}{[Ag^+]}$$

if we knew the silver ion concentration at equilibrium. For the latter we can surely write

$$[Ag^+] = [Ag^+] \text{ (from } AgNO_3) + [Ag^+] \text{ (from } AgCl).$$

The second term on the right-hand side is bound to be less than $1.7 \times 10^{-5}\ M$, the concentration of silver ion in a saturated solution of AgCl in pure water, according to our argument based on Le Chatelier's principle. The first term on the right-hand side is 0.1 M, which is much larger than the second term. Therefore, to a very good approximation we can write

$$[Ag^+] \cong 0.1\ M$$

and consequently

$$[Cl^-] = \frac{K_{sp}}{[Ag^+]} \cong \frac{2.8 \times 10^{-10}}{0.1},$$
$$[Cl^-] \cong 2.8 \times 10^{-9}.$$

This quantity must be equal to the number of moles of AgCl dissolved in one liter of solution, and thus is the solubility of AgCl in 0.1 M Ag^+. We see that in accordance with Le Chatelier's principle, the solubility is reduced compared with that in pure water. Furthermore, since so little silver chloride is dissolved, the silver ion concentration contributed by the silver chloride is very much less than 0.1 M, and we were justified in neglecting this source of silver ion in comparison with the dissolved $AgNO_3$.

As a further illustration of the use of the solubility product and of the meaning of the word solubility, let us calculate the solubility of CaF_2 in: (1) a solution of 0.1 M $Ca(NO_3)_2$; (2) a solution of 0.1 M NaF.

In considering the dissolution of CaF_2 in a solution of $Ca(NO_3)_2$, we cannot say that the solubility of CaF_2 is equal to the equilibrium concentration of Ca^{++}, for most of the Ca^{++} in the solution is contributed not by CaF_2 but by $Ca(NO_3)_2$. However, we can recognize that CaF_2 is the only source of fluoride ion in the solution, and that for each mole of CaF_2 that dissolves, 2 moles of F^- enter the solution. Therefore the solubility of the salt is

$$\text{solubility} = \tfrac{1}{2}[F^-].$$

Our problem then is to find the fluoride ion concentration that exists at equilibrium.

The equilibrium-constant expression shows us that $[F^-]$ may be found if $[Ca^{++}]$ is known:

$$[Ca^{++}][F^-]^2 = K_{sp}, \qquad [F^-] = \left(\frac{K_{sp}}{[Ca^{++}]}\right)^{1/2}.$$

For the concentration of calcium ion we have

$$[Ca^{++}] = [Ca^{++}] \text{ (from } Ca(NO_3)_2) + [Ca^{++}] \text{ (from } CaF_2).$$

Earlier in this section we found that the calcium ion concentration resulting from CaF_2 dissolving in pure water is $3.5 \times 10^{-4} M$. By Le Chatelier's principle we know that even less CaF_2 will dissolve in $0.1 M Ca(NO_3)_2$, and so the calcium ion *contributed by* CaF_2 in this solution will be even less than $3.5 \times 10^{-4} M$. Therefore, we can neglect the concentration of calcium contributed by CaF_2 *compared with* $0.1 M$, and write

$$[Ca^{++}] \cong 0.1 \; M,$$

which is a good approximation. Substitution into the solubility-product expression

$$[F^-] = \left(\frac{K_{sp}}{[Ca^{++}]} \right)^{1/2}$$

gives

$$[F^-] \cong \left(\frac{1.7 \times 10^{-10}}{0.1} \right)^{1/2} \cong 4.1 \times 10^{-5} \; M,$$

and

$$\text{solubility} = \tfrac{1}{2}[F^-] \cong 2 \times 10^{-5} \; M.$$

As we expected from Le Chatelier's principle, the solubility of CaF_2 is lowered by the presence of an excess of one of its ions. In addition, the fact that calcium ion coming from CaF_2 is only $2 \times 10^{-5} M$ justifies our neglect of this quantity compared with the contribution from $Ca(NO_3)_2$.

Now we consider the solubility of CaF_2 in $0.1 M$ NaF. In this situation we can take the solubility of calcium fluoride to be equal to the concentration of calcium ion at equilibrium:

$$\text{solubility} = [Ca^{++}].$$

As in our previous examples, we can estimate the concentration of the ion in excess, and calculate the concentration of the other ion with the solubility product. For the fluoride ion concentration we have

$$[F^-] = [F^-] \text{ (from NaF)} + [F^-] \text{ (from } CaF_2\text{)}.$$

The first term on the right-hand side is $0.1 M$, and the second must be less than $7 \times 10^{-4} M$, the concentration of F^- in a pure saturated solution of CaF_2. Consequently, we can say

$$[F^-] \cong 0.1 \; M,$$

and by the equilibrium-constant expression,

$$[Ca^{++}] = \frac{K_{sp}}{[F^-]^2} \cong \frac{1.7 \times 10^{-10}}{(0.1)^2} \cong 1.7 \times 10^{-8} \; M.$$

Therefore, the solubility of CaF_2 in $0.1 M$ NaF is $1.7 \times 10^{-8} M$. Once again we can conclude that our decision to neglect the $[F^-]$ contributed by CaF_2 in comparison with $0.1\ M$ was justified, for the actual fluoride ion concentration contributed by CaF_2 ($\approx 2 \times 1.7 \times 10^{-8}\ M$) is much smaller than $0.1\ M$.

Questions. The solubility of Ag_2CrO_4 in pure water is $0.78 \times 10^{-4}\ M$. Will its solubility in $0.05\ M$ $AgNO_3$ be greater or less than $0.78 \times 10^{-4}\ M$? Why? Will $[Ag^+]$ in the resulting solution be greater, less, or approximately equal to $0.05\ M$? Show that the solubility of Ag_2CrO_4 in $0.05\ M$ $AgNO_3$ is $7.6 \times 10^{-10}\ M$.

These examples have shown how the solubility product can be used to calculate the solubility of a salt in any solution of one of its ions. They also show that in effecting these calculations it is a great help if one is guided by Le Chatelier's principle and by comparisons to other simpler situations. These comparisons can lead to very satisfactory approximations that considerably simplify the calculations. Finally, it is most important to use the answer obtained to check the validity of any approximations that are made in the calculation.

Selective Precipitation

In both qualitative and quantitative chemical analysis, it is often necessary to take advantage of differing solubilities to remove only one of several salts from solution. Calculations using solubility products can tell us when a separation of this type is possible. As a practical example, consider the following.

A solution contains $0.1\ M$ Cl^- and $0.01\ M$ $CrO_4^=$. By adding a solution of $AgNO_3$, we wish to precipitate the chloride ion as $AgCl$, and leave the chromate ion in solution. As we can see from the solubility products

$$[Ag^+][Cl^-] = 2.8 \times 10^{-10},$$

$$[Ag^+]^2[CrO_4^=] = 1.9 \times 10^{-12},$$

both $AgCl$ and Ag_2CrO_4 are slightly soluble salts. What will happen when Ag^+ is added slowly to the original solution?

To answer we need only recognize the significance of the solubility product: it is a number which the product of the ion concentrations *can never exceed* at equilibrium. That is, if the product $[Ag^+][Cl^-]$ is greater than 2.8×10^{-10}, the system is not at equilibrium and precipitation of $AgCl$ must occur. On the other hand, the concentration product $[Ag^+][Cl^-]$ can be less than K_{sp}, *but* only if there is no excess solid present. Therefore, upon adding Ag^+ to a solution of Cl^-, no precipitation of $AgCl$ will occur until the Ag^+ concentration becomes high enough so that $[Ag^+][Cl^-] = K_{sp}$. In the case we are considering, no precipitation of $AgCl$ will occur until the concentration of Ag^+ is

$$[Ag^+] = \frac{K_{sp}}{[Cl^-]} = \frac{2.8 \times 10^{-10}}{0.1} = 2.8 \times 10^{-9}\ M.$$

By a similar argument, no precipitation of Ag_2CrO_4 will occur until

$$[Ag^+]^2 = \frac{K_{sp}}{[CrO_4^=]} = \frac{1.9 \times 10^{-12}}{0.01},$$

$$[Ag^+] = 1.4 \times 10^{-5} \ M,$$

if the concentration of $CrO_4^=$ is 0.01 M.

Now we can see what will happen when we add silver ion to a mixture of 0.1 M Cl^- and 0.01 M $CrO_4^=$. No solid will form until the concentration of silver ion reaches 2.8×10^{-9} M, at which point the first precipitation of AgCl will occur. Further addition of Ag^+ will cause more precipitation of AgCl, but no precipitation of Ag_2CrO_4 will occur until the silver ion concentration rises to 1.4×10^{-5} M.*

It is interesting to see how complete the precipitation of the chloride ion is at the time when Ag_2CrO_4 first begins to precipitate. Since $[Ag^+] = 1.4 \times 10^{-5}$ M at this point, the chloride ion concentration must be

$$[Cl^-] = \frac{K_{sp}}{[Ag^+]} = \frac{2.8 \times 10^{-10}}{1.4 \times 10^{-5}} = 2.0 \times 10^{-5}.$$

From this answer we can conclude that the precipitation of chloride ion is essentially complete by the time Ag_2CrO_4 begins to precipitate, for of the original chloride material, less than one part in one thousand is left in solution.

The foregoing calculation demonstrates the basis for using $CrO_4^=$ as an endpoint indicator in the titration of chloride ion by silver ion. The chromate ion in aqueous solution is bright yellow, but a precipitate of silver chromate is dark red. As we have seen, the precipitate of Ag_2CrO_4 appears only after the chloride ion has been essentially completely precipitated by the addition of Ag^+; thus the formation of Ag_2CrO_4 indicates the endpoint of the precipitation titration of the chloride ion.

As another example of the use of selective precipitation in chemical analysis, consider the separation of Zn^{++} and Fe^{++} by controlled precipitation of one of their sulfides. The relevant solubility products are

$$[Zn^{++}][S^=] = 4.5 \times 10^{-24},$$

$$[Fe^{++}][S^=] = 1 \times 10^{-19}.$$

We can see that ZnS is less soluble than FeS. Therefore it may be possible to start with a solution containing both Zn^{++} and Fe^{++} at 0.1-M concentration, and quantitatively precipitate only ZnS, leaving all the Fe^{++} in solution.

---

* We are assuming that essentially no volume change occurs upon adding solution of Ag^+, so that the concentration of $CrO_4^=$ is constant. This is realizable if the Ag^+ solution is sufficiently concentrated.

From the solubility product of FeS we can calculate that in order to *avoid* precipitating FeS, the sulfide ion concentration must be less than

$$[S^=] = \frac{K_{sp}}{[Fe^{++}]} = \frac{1 \times 10^{-19}}{0.1}$$
$$= 10^{-18} \ M.$$

To be safe, a maximum sulfide ion concentration of $10^{-19} \ M$ might be employed. Does this result in essentially complete precipitation of ZnS? At a sulfide concentration of $10^{-19} \ M$, the concentration of Zn^{++} remaining in solution is

$$[Zn^{++}] = \frac{K_{sp}}{[S^=]} = \frac{4.5 \times 10^{-24}}{1 \times 10^{-19}}$$
$$= 4.5 \times 10^{-5} \ M.$$

We see that the concentration of zinc ion left in solution is only a small fraction of the original concentration, and thus it is possible to carry out a quantitative separation in this manner. The problem of how to maintain the sulfide ion concentration at the proper value for the separation can also be solved by arguments involving equilibrium constants, and we shall treat these later in the chapter.

Questions. Silver ion is added to a solution that contains Cl⁻ and I⁻ both at 0.01 *M* concentrations. Which salt precipitates first, AgCl or AgI? What is the value of [Ag⁺] when the first salt starts to precipitate? What is the concentration of the anion of the first precipitate when the second salt just starts to precipitate?

6.2 ACIDS AND BASES

There is perhaps no other class of equilibria as important as that involving acids and bases. As we continue the study of chemistry, we shall find that the classification "acid-base reaction" includes a vast number of chemical changes, so that the principles and practical points that we treat in the following sections are of very general use. Before attacking the mathematical problems of acid-base equilibria, we must devote some time to a discussion of nomenclature and classification of acids and bases.

Arrhenius Theory of Acids and Bases

The classification of substances as acids was at first suggested by their sour taste (Latin *acidus*, sour; *acetum*, vinegar) and alkalis (Arabic *al kali*, ashes of a plant) were taken as those substances that could reverse or neutralize the action of acids. It was thought also that an acid must have, as a necessary constituent, the element oxygen (Greek *oxus*, sour; *gennae*, I produce), but in 1810 Davy demonstrated that hydrochloric acid contained only hydrogen and chlorine.

Shortly thereafter the view was taken that all acids had hydrogen as an essential constituent.

An explanation of why acids had differing strengths was one of the important results of the Arrhenius ionic dissociation theory, developed between 1880 and 1890. The chemical activity and electrical conductivity of solutions of acids were taken to be consequences of their reversible dissociation into ions, one of which was H^+:

$$HCl = H^+ + Cl^-,$$
$$CH_3COOH = CH_3COO^- + H^+.$$

The fact that different acids had different strengths was explained as a result of a variation in the degree of dissociation.

A similar scheme applied to the behavior of bases, which were all thought to produce the hydroxyl ion in solution:

$$NaOH(s) = Na^+(aq) + OH^-(aq),$$
$$Mg(OH)_2(s) = Mg^{++}(aq) + 2OH^-(aq).$$

Thus the proton was responsible for acidic properties, and the hydroxyl ion was responsible for basic properties.

While this point of view was a considerable advance in chemical theory, it led to certain difficulties. The first of these concerned the nature of the proton in aqueous solution, and the second had to do with the fact that substances which did not contain hydroxyl ion were capable of acting as bases. Let us examine these difficulties in turn.

It is generally accepted that one of the reasons water is so excellent a solvent for ionic compounds is that ions in aqueous solution are stabilized by their strong attraction to the water molecule. This attraction is particularly strong because of the asymmetry of the charge distribution in the water molecule. Each ion in aqueous solution is hydrated, or strongly attached to a number of water molecules, generally estimated to be between 4 and 6. The proton is unique among ions in that it has no electrons. Consequently the radius of H^+ is just the nuclear radius, 10^{-13} cm, which is considerably smaller than 10^{-8} cm, the approximate radius of other ions. Therefore the proton should be able to approach and incorporate itself in the electronic system of a solvent molecule to a degree exceeding any other ion. In other words, if ordinary ions are hydrated, the proton should be even more intimately bound to the solvent, and it is not legitimate to think of acid dissociation as producing "free" protons.

There is considerable experimental evidence that the hydrated proton H_3O^+, or the hydronium ion, is particularly stable. The hydronium ion is known to exist in electrical discharges through water vapor, and H_3O^+ has been definitely identified as a distinct species in several crystals. In particular, the crystal of hydrated perchloric acid, sometimes denoted as $HClO_4 \cdot H_2O$, really consists of H_3O^+ and ClO_4^-. Such data suggests that we might take the "true" form of

H^+ in aqueous solution to be H_3O^+. But even this may be an oversimplification. The H_3O^+ species is very probably further hydrated, and there is considerable evidence that indicates three additional water molecules are rather firmly attached to it to form $H_9O_4^+$. While we are not sure of the exact state of H^+ in aqueous solution, we feel sure it is not a free proton. To emphasize the hydration of the proton, we shall represent it in this chapter by $H_3O^+(aq)$, which indicates H_3O^+ with an undetermined number of water molecules attached to it. This notation has a disadvantage, however, because it tends to clutter chemical equations with "extra" molecules of water. Therefore in subsequent chapters we shall find it helpful to abandon the H_3O^+ notation and use simply $H^+(aq)$ to represent the aqueous proton.

The impact this argument has on the acid-base problem is the following. If the proton exists as hydrated H_3O^+, it is not entirely accurate to think of a dissociation as represented by

$$HCl = H^+ + Cl^-.$$

A more realistic view is to think of acid "dissociation" as a *transfer* of a proton from the acid to the solvent:

$$HCl + H_2O = H_3O^+(aq) + Cl^-(aq).$$

This in turn suggests that an acid is not necessarily a substance that *dissociates* to a proton, but rather is a molecule that is capable of *transferring* or donating a proton to another molecule. This is a fruitful concept, as we shall see.

The second difficulty with the Arrhenius picture of acids and bases is that it suggests all basic properties are due to the hydroxide ion. However, there are substances that do not contain the hydroxide ion but can still neutralize acids. For example, in pure liquid ammonia the reaction

$$HCl(g) + NH_3(l) = NH_4^+ + Cl^-$$

proceeds readily, and thus we can look upon ammonia as a base, since it reacts with a known acid, HCl. When sodium carbonate, Na_2CO_3, is dissolved in water a solution that will neutralize acids results. Sodium carbonate cannot by itself dissociate directly to product a hydroxide ion, but its reactions suggest that nevertheless it must be a base. It appears then that a broader view of acids and bases than is afforded by the Arrhenius theory is necessary.

The Lowry-Brønsted Concept

The considerations we have just outlined led in 1923 to a more powerful and general concept of acids and bases, called the Lowry-Brønsted definition: An acid is a species having a tendency to lose or to donate a proton, and a base is a species having the tendency to accept or add a proton. Accordingly the ionization of HCl is pictured as HCl donating a proton (acting as an acid) to water (acting as a base).

$$HCl(aq) + H_2O = H_3O^+(aq) + Cl^-(aq)$$
$$\quad\text{acid} \qquad\qquad \text{base}$$

This reaction is reversible, for Cl^- may accept a proton from H_3O^+ and become HCl(aq). Therefore, the chloride ion must also be a base, and H_3O^+ must also be an acid. Since HCl and Cl^- differ only by a proton, they are called a *conjugate acid-base pair*. Likewise, H_3O^+ and H_2O are a conjugate acid-base pair. To demonstrate this idea we can rewrite our reaction as

$$HCl \ + \ H_2O \ = \ H_3O^+ + \ Cl^-,$$
$$\text{acid 1} + \text{base 2} = \text{acid 2} + \text{base 1},$$

where the numbers indicate the conjugate pairs. The behavior of carbonate ion as a base can be represented by

$$CO_3^= \ + \ H_2O \ = \ HCO_3^- \ + \ OH^-,$$
$$\text{base 1} + \text{acid 2} = \ \text{acid 1} + \text{base 2}.$$

The Lowry-Brønsted definition extends the terms acid and base to include substances besides H^+ and OH^- with the resulting advantage that a large number of reactions can be discussed in the same language and treated mathematically by the same methods.

Strength of Acids and Bases

The Lowry-Brønsted definition suggests that a strong acid has a large tendency to transfer a proton to another molecule, and that a strong base is one with a large affinity for protons. We might then measure acid strength quantitatively by the degree to which reactants are converted to products in a reaction such as

$$HSO_4^- \ + \ H_2O \ = \ H_3O^+ + \ SO_4^=,$$
$$\text{acid 1} + \text{base 2} = \text{acid 2} + \text{base 1}.$$

However, a little reflection shows that the extent to which this reaction proceeds to products is governed not only by the tendency of acid 1 to lose a proton, but by the tendency of base 2 to accept that proton. If the degree of proton transfer depends on the properties of both acid 1 and base 2, it is clear that the only valid way we can compare the strengths of *individual acids* is by measuring their tendencies to transfer a proton to the *same* base. By testing the ability of various acids to transfer a proton to water, we can rank them in order of their acid strength. The quantitative measure of acid strength is the acid dissociation

Table 6.2 Dissociation constants of acids

Acid	Reaction	K (25°C)
Acetic acid	$CH_3COOH + H_2O = H_3O^+ + CH_3COO^-$	1.8×10^{-5}
Formic acid	$HCOOH + H_2O = H_3O^+ + HCOO^-$	1.8×10^{-4}
Hydrogen cyanide	$HCN + H_2O = H_3O^+ + CN^-$	4.8×10^{-10}
Hydrogen sulfate ion	$HSO_4^- + H_2O = H_3O^+ + SO_4^=$	1.2×10^{-2}
Hydrofluoric acid	$HF + H_2O = H_3O^+ + F^-$	6.8×10^{-4}
Nitrous acid	$HNO_2 + H_2O = H_3O^+ + NO_2^-$	4.5×10^{-4}
Chloracetic acid	$CH_2ClCOOH + H_2O = H_3O^+ + CH_2ClCOO^-$	1.4×10^{-3}
Carbonic acid	$CO_2 + 2H_2O = H_3O^+ + HCO_3^-$	4.2×10^{-7}
Bicarbonate ion	$HCO_3^- + H_2O = H_3O^+ + CO_3^=$	$4.8 + 10^{-11}$
Hydrogen sulfide	$H_2S + H_2O = H_3O^+ + HS^-$	1.1×10^{-7}
Hydrogen sulfide ion	$HS^- + H_2O = H_3O^+ + S^=$	1×10^{-14}
Phosphoric acid	$H_3PO_4 + H_2O = H_3O^+ + H_2PO_4^-$	7.5×10^{-3}
Dihydrogen phosphate ion	$H_2PO_4^- + H_2O = H_3O^+ + HPO_4^=$	6.2×10^{-8}
Hydrogen phosphate ion	$HPO_4^= + H_2O = H_3O^+ + PO_4^{-3}$	1×10^{-12}

constant: that is, the equilibrium constant for the reaction

$$HA + H_2O = H_3O^+ + A^-, \qquad K = \frac{[H_3O^+][A^-]}{[HA]}.$$

Table 6.2 contains the dissociation constants of several common acids.

Finally, we note that there is a relation between the strength of an acid and its conjugate base. For example, we say that because HCl has a large tendency to lose a proton, it is a strong acid. But it must also be true that its conjugate base, chloride ion, has only a very small tendency to acquire a proton, and is therefore a weak base. Further reflection shows that in general if an acid is a strong acid, its conjugate base is a weak base, and vice versa.

The Lewis Concept

Another general definition of acids and bases was proposed in 1923 by G. N. Lewis. According to Lewis, an acid is any substance that can accept electrons, and a base is a substance that can donate electrons. Two examples of such acid-base reactions are

$$BF_3 + F^- = BF_4^-,$$
$$\text{acid} \quad \text{base}$$

$$Ag^+ + 2CN^- = Ag(CN)_2^-.$$
$$\text{acid} \qquad \text{base}$$

We see that the Lewis definition extends the acid-base concept to reactions in which protons are not involved.

It is clear that since Lowry-Brønsted bases react by donating electrons to a proton, a Brønsted base is also a Lewis base. However, a Brønsted acid must have a proton available for transfer to another molecule, whereas this is not required by the Lewis definition. In this chapter we are primarily concerned with aqueous solutions in which acid-base phenomena involve proton transfer, and so the Lowry-Brønsted definitions are of most use to us. In other systems where protonic materials are not involved, the Lewis definition is more valuable, and we shall explore this concept in our subsequent discussions of descriptive chemistry.

6.3 NUMERICAL PROBLEMS

We have noted that the equilibrium constant for acid dissociation can be used as a quantitative indication of acid strength. In this section we shall be concerned with the details of the calculation of hydrogen ion concentration using acid and base ionization constants. First we must examine two conventions: one having to do with how ionization-constant expressions are written, the other with how small concentrations are expressed.

We have emphasized that the acid properties of, for example, HSO_4^- are due to the proton transfer reaction

$$HSO_4^- + H_2O = H_3O^+ + SO_4^=.$$

Strictly, the equilibrium constant for this reaction should be written

$$K' = \frac{[H_3O^+][SO_4^=]}{[HSO_4^-][H_2O]}.$$

However, virtually all the solutions that concern us are dilute and therefore the concentration of water ($\sim 55.5\ M$) is always larger than the concentration of the solute. Hence the concentration of water in such solutions can be considered to be constant, and the usual convention of including constant concentrations in the equilibrium constant itself is followed. Accordingly, acid ionization constants are commonly written as

$$K'[H_2O] \equiv K = \frac{[H_3O^+][SO_4^=]}{[HSO_4^-]},$$

and for the reaction

$$NH_3 + H_2O = NH_4^+ + OH^-,$$
$$K = \frac{[NH_4^+][OH^-]}{[NH_3]}.$$

The pH Scale

Because the solutions with which we deal most commonly are dilute, the concentration of hydrogen ion in them is often quite small. As a result, hydrogen ion concentrations in moles per liter are often expressed as negative powers of 10. For example, the hydrogen ion concentration in a saturated solution of CO_2 is $1.3 \times 10^{-4}\ M$, and in a $0.5\ M$ solution of acetic acid it is $3 \times 10^{-3}\ M$. To achieve compactness of notation and brevity of expression it is convenient to give these concentrations in terms of their negative logarithms. Thus we define pH as

$$pH = -\log [H_3O^+].$$

For example, in a saturated solution of CO_2, $[H_3O^+] = 1.3 \times 10^{-4}\ M$, and the pH is given by

$$\begin{aligned}
pH = -\log [H_3O^+] &= -\log [1.3 \times 10^{-4}] \\
&= -(\log 1.3 + \log 10^{-4}) \\
&= -(0.11 - 4) \\
&= 3.89.
\end{aligned}$$

Conversely, a solution that has a pH of 4.5 must have $[H_3O^+] = 3.2 \times 10^{-5}$, as we can see by

$$\begin{aligned}
4.5 &= -\log [H_3O^+], \\
-4.5 &= (0.5 - 5) = \log [H_3O^+], \\
10^{0.5} \times 10^{-5} &= [H_3O^+], \\
3.2 \times 10^{-5} &= [H_3O^+].
\end{aligned}$$

The use of negative logarithms for the expression of concentration is not restricted only to hydrogen ion. For example, pOH is the negative logarithm of the hydroxide ion concentration, and pAg is similarly related to the concentration of silver ion. It is also common to express equilibrium constants as their negative logarithms, and denote such quantities by pK.

Self-ionization of Water

We have seen that water can act as either an acid or a base, so it is not surprising that the reaction

$$\begin{array}{ccccc}
H_2O & + & H_2O & = & H_3O^+ + & OH^- \\
\text{acid 1} & & \text{base 2} & & \text{acid 2} & \text{base 1}
\end{array} \tag{6.3}$$

proceeds to a small but easily measurable extent in pure water. Following the convention of not writing constant concentrations explicitly, we obtain

$$[H_3O^+][OH^-] = K_w \tag{6.4}$$

for the equilibrium-constant expression. The quantity K_w is called the ion product constant for water, and has the value 10^{-14} at 25°C, when concentrations are expressed in moles per liter.

A solution that is neither acidic nor basic has, by definition, equal concentrations of H_3O^+ and OH^-. Thus if

$$[H_3O^+] = [OH^-] \quad \text{and} \quad [H_3O^+][OH^-] = 10^{-14},$$

we obtain

$$[H_3O^+] = [OH^-] = 10^{-7} \, M$$

for the concentrations of H_3O^+ and OH^- in a "neutral" solution at 25°C. Alternatively, we can say that for a neutral solution,

$$pH = pOH = 7.$$

From the ion product constant expression we can also write, in general,

$$pH + pOH = pK_w = 14$$

at 25°C. Thus the pOH can be obtained by subtracting the pH from 14.

The self-ionization of water always contributes to the hydrogen ion and hydroxide ion concentration in a solution, but this is only rarely a complicating factor in calculating the hydrogen ion concentration in solutions of acids and bases. Consider, for example, the problem of calculating the H_3O^+ concentration in a solution prepared by dissolving 0.1 mole of HCl in enough water to make one liter of solution. For simplicity, we will assume that HCl is totally dissociated to its ions. Is the concentration of H_3O^+ equal to 0.1 M? Perhaps not, because the self-ionization of water will also contribute to the H_3O^+ concentration.

It is not difficult to become convinced that the H_3O^+ contributed by the ionization of water is of no importance in this case. In pure water, the concentration of H_3O^+ is $10^{-7} \, M$, as we have noted previously. If H_3O^+ in the form of HCl is added to pure water, the self-ionization of water is subjected to a "stress," and the system must react in a manner that minimizes this stress, according to Le Chatelier's principle. This means that as H_3O^+ is added, the self-ionization of water must diminish, and that *the H_3O^+ contributed by the self-ionization of water must become less than* $10^{-7} \, M$. Therefore, in a 0.1 M solution of HCl, the concentration of H_3O^+ is 0.1 M, because the "extra" amount contributed by the self-ionization of water is bound to be *less* than $10^{-7} \, M$.

Now that we have established that the concentration of H_3O^+ in a 0.1 M HCl solution is 0.1 M, we can calculate the equilibrium concentration of OH^- in the same solution. At equilibrium, the H_3O^+ and OH^- concentrations *always* obey Eq. (6.4):

$$[H_3O^+][OH^-] = K_w = 10^{-14}. \tag{6.4}$$

Since $[H_3O^+] = 0.1\ M$,

$$[OH^-] = \frac{10^{-14}}{[H_3O^+]} = \frac{10^{-14}}{10^{-1}} = 10^{-13}\ M.$$

We see that in an *acidic* solution the concentration H_3O^+ is greater, and the concentration of OH^- less, than in pure water.

The calculation of the OH^- concentration can be used to further justify our assumption that the contribution of the self-ionization of water to the H_3O^+ concentration is negligible. The only source of OH^- in this solution is the self-ionization of water, and by the stoichiometry of reaction (6.3), one H_3O^+ is contributed for each OH^-. Therefore, the $[H_3O^+]$ contributed by the self-ionization of water is equal to $[OH^-]$, which we have seen to be $10^{-13}\ M$. There is no question then that the contribution of the self-ionization of water to the H_3O^+ concentration is completely negligible.

While we have used a solution of a strong acid as our example, it is clear that similar arguments can be applied to a solution of a strong base. In a $0.01\ M$ solution of NaOH, the contribution of the self-ionization of water *to the concentration of* OH^- is negligible, because it must be less than $10^{-7}\ M$. Thus in this solution

$$[OH^-] = 0.01\ M,$$
$$[H_3O^+] = \frac{K_w}{[OH^-]}$$
$$= \frac{10^{-14}}{10^{-2}} = 10^{-12}\ M.$$

These arguments concerning the importance of either the H_3O^+ or OH^- contributed by water relative to the H_3O^+ or OH^- contributed by a dissolved acid or base are very important, for they are used repeatedly to simplify calculations dealing with weak acids and bases. Briefly, we can say that if a dissolved acid itself contributes a concentration of H_3O^+ equal to or greater than $10^{-6}\ M$, the contribution of water to the total H_3O^+ concentration is negligible. A similar statement can be made about bases and the OH^- concentration.

Weak Acids and Bases

Now let us attack the problem of calculating the concentration of H_3O^+ in a pure aqueous solution of a weak acid. Acetic acid is a common laboratory reagent, and is often taken as a "typical" weak acid. The equilibrium we deal with then is

$$CH_3COOH + H_2O = H_3O^+ + CH_3COO^-.$$

Let us abbreviate the formulas for acetic acid and the acetate ion to HOAc and OAc⁻, respectively. Then the equilibrium constant for the ionization is

$$\frac{[\text{H}_3\text{O}^+][\text{OAc}^-]}{[\text{HOAc}]} = K = 1.85 \times 10^{-5}. \tag{6.5}$$

The magnitude of this equilibrium constant lies roughly in the middle of the range of values that apply to weak acids, and this is the reason acetic acid is considered a typical weak acid.

Suppose we have a solution made up by adding C_0 moles of pure acetic acid to enough water to make one liter of solution. Some of this acid will dissociate to H_3O^+ and OAc^-, leaving behind an unknown concentration of undissociated acid. Our problem is to calculate the equilibrium concentrations of H_3O^+, OAc^-, and HOAc by using the ionization-constant expression, Eq. (6.5). To do this we must take an equation in three unknowns, and convert it to an equation in one unknown.

The method we shall use is an intuitive approximate procedure. First, recognize that there are two sources of H_3O^+: the ionization of the acid, and the self-ionization of water. But our experience is that the latter source is often unimportant compared with the former, at least when we deal with solutions of strong acids. As a trial procedure then, let us assume that all the H_3O^+ in solution comes from the ionization of the acid. As an immediate consequence, the stoichiometry of the ionization reaction tells us that

$$[\text{H}_3\text{O}^+] = [\text{OAc}^-], \tag{6.6}$$

since the only source of either of these ions is a reaction which produces them in equal amounts.

Equation (6.6) allows us to convert Eq. (6.5) into an equation in two unknowns. To complete the solution, one more relation is necessary. To obtain it, we recognize that the ionization equilibrium constant is small, and that consequently very little H_3O^+ and OAc^- can exist in equilibrium with undissociated HOAc. This suggests that to a good *approximation*, the equilibrium concentration of HOAc is the same as C_0, the concentration that would be present *if no* HOAc *dissociated*. Therefore as our second assumption we have

$$[\text{HOAc}] = C_0. \tag{6.7}$$

Now let us calculate $[\text{H}_3\text{O}^+]$ for three different values of C_0, and see whether our assumptions are justified in typical situations. First we choose $C_0 = 1\ M$. Then Eqs. (6.5), (6.6), and (6.7) give us

$$K = \frac{[\text{H}_3\text{O}^+][\text{OAc}^-]}{[\text{HOAc}]} \simeq \frac{[\text{H}_3\text{O}^+]^2}{C_0},$$

$$[\text{H}_3\text{O}^+] \simeq (C_0 K)^{1/2} = (1 \times 1.85 \times 10^{-5})^{1/2}$$

$$\simeq 4.3 \times 10^{-3}\ M.$$

This answer is not exact, because we made two approximations in deriving it. However, we can use this approximate answer to check to see if our assumptions were justified.

Our first assumption was that the H_3O^+ contributed by the ionization of water was negligible compared with that derived from the acid. The concentration of H_3O^+ from the ionization of water cannot be greater than 10^{-7} M, and this is much smaller than 4.2×10^{-3} M, so our first assumption and the approximation that $[H_3O^+] = [OAc^-]$ are justified.

Our second assumption was that very little of the HOAc dissociated, so that $C_0 \cong [HOAc]$. Strictly, however, the concentration of HOAc at equilibrium is given by

$$[HOAc] = C_0 - [OAc^-] \cong C_0 - [H_3O^+].$$

But $C_0 = 1$ M, and $[H_3O^+] \simeq 4.3 \times 10^{-3}$. Therefore, our second assumption is justified and the approximation that $C_0 \simeq [HOAc]$ is accurate to better than one percent.

Now let us try out our approximations in the situation where $C_0 = 0.01$ M, a rather dilute solution of acetic acid. Once again, assuming that

$$[H_3O^+] = [OAc^-],$$
$$C_0 = [HOAc] = 0.01\ M,$$

we get

$$\frac{[H_3O^+][OAc^-]}{[HOAc]} \cong \frac{[H_3O^+]^2}{C_0} = K,$$
$$[H_3O^+] \cong (C_0 K)^{1/2} = (10^{-2} \times 1.85 \times 10^{-5})^{1/2}$$
$$\cong 4.3 \times 10^{-4}\ M.$$

Now we must check the validity of our assumptions. As we have argued before, the concentration of H_3O^+ contributed by the self-ionization of water is less than 10^{-7} M, and this is considerably smaller than 4.3×10^{-4} M. Therefore, our assumption that $[H_3O^+] = [OAc^-]$ is justified.

The validity of the second assumption depends on the concentration of H_3O^+ being much less than C_0, for only then can we say that $C_0 \cong [HOAc]$. But we see that $[H_3O^+]$ is as large as 4% of C_0, and therefore there is some question as to whether we were justified in assuming that $[HOAc] = 0.01$ M. However, while we have made an error of about 4% in the concentration of HOAc, the resulting error in the concentration of H_3O^+ is less than this because the concentration of H_3O^+ depends only on the *square root* of the HOAc concentration. Furthermore, in practical laboratory situations we are rarely interested in knowing the concentration of H_3O^+ to better than a few percent, so our second assumption is satisfactory in this case, if only marginally.

Our experience now suggests that if we try to calculate $[H_3O^+]$ for a solution of acetic acid in which $C_0 = 0.001$ M, at least one of our simplifying assumptions will break down. Let us see how serious the failure is, and what can be done about it. We assume that $[H_3O^+] = [OAc^-]$ and $[HOAc] = C_0 = 10^{-3} M$, and find

$$[H_3O^+] \cong (C_0K)^{1/2} = (1.85 \times 10^{-5} \times 10^{-3})^{1/2}$$
$$\cong 1.36 \times 10^{-4} M.$$

Now there is no question that the first assumption is justified, but our answer shows that $[H_3O^+]$ is *not small* compared with C_0, so the assumption that

$$[HOAc] = C_0 - [H_3O^+] \cong C_0$$

is not justified. The error in the HOAc concentration is greater than 10%, and the resulting error in the H_3O^+ concentration is several percent.

There are two ways by which we can avoid this difficulty. One is to use the accurate relation

$$[HOAc] = C_0 - [H_3O^+],$$
$$C_0 = 10^{-3} M$$

in the equilibrium constant, along with $[H_3O^+] = [OAc^-]$. This gives us

$$\frac{[H_3O^+][OAc^-]}{[HOAc]} = \frac{[H_3O^+]^2}{C_0 - [H_3O^+]} = K,$$

which is a quadratic equation in $[H_3O^+]$. Let us rearrange it to

$$[H_3O^+]^2 + K[H_3O^+] - C_0K = 0,$$

and solve it by the quadratic formula

$$[H_3O^+] = \frac{-K \pm \sqrt{K^2 + 4KC_0}}{2}. \tag{6.8}$$

Inserting the values for K and C_0 gives

$$[H_3O^+] = 1.27 \times 10^{-4} M.$$

We see that this answer is nearly 7% less than our approximate answer.

From Eq. (6.8) we can deduce the condition for the validity of our approximate expression for $[H_3O^+]$. If, in Eq. (6.8), $K \ll C_0$, then $K^2 \ll 4KC_0$, in which case we have

$$[H_3O^+] = \frac{-K \pm \sqrt{K^2 + 4KC_0}}{2} \cong -\frac{K}{2} \pm \sqrt{KC_0},$$

and so if $K \ll C_0$, we can further simplify to

$$[H_3O^+] \cong \sqrt{KC_0},$$

which is our approximate expression for $[H_3O^+]$. We see then that in order for this approximation to be valid, the acid must be weak (K small) and fairly concentrated (C_0 large).

The second way to handle problems in which the simplifying assumptions prove inaccurate is to proceed by successive approximations to the correct answer. In our present example, our guess that $[HOAc] = C_0$ proved inaccurate. Let us use the approximate value of H_3O^+ that we found at first to improve our guess of the HOAc concentration. We have

$$[HOAc] = C_0 - [H_3O^+]$$

and $[H_3O^+] \cong 1.36 \times 10^{-4}$. A second, or refined, approximation to [HOAc] would be

$$[HOAc] \cong C_0 - 1.36 \times 10^{-4} = 8.64 \times 10^{-4} \; M.$$

Inserting this into the equilibrium constant gives

$$\frac{[H_3O^+]^2}{8.64 \times 10^{-4}} = 1.85 \times 10^{-5},$$

$$[H_3O^+] \cong 1.26 \times 10^{-4} \; M.$$

This second approximation to the H_3O^+ concentration is nearly the same as the "exact" value obtained from the solution of the quadratic equation. If we did not know the accurate answer to the problem, we could check the validity of this second approximation by using it to further refine our guess of [HOAc], and repeating the calculation. If two successive answers differ negligibly, the final approximation is satisfactory.

Question. Use this second approximation of $[H_3O^+] = 1.26 \times 10^{-4}$ to obtain a third approximation for $[H_3O^+]$. What do you conclude from a comparison of the results of the second and third approximations?

It might seem that the successive approximation method is inferior to the exact solution via the quadratic equation. Actually, the successive approximation procedure is the more useful approach, since it is algebraically and arithmetically simpler, and can be applied to situations in which the exact approach would require the solution of a cubic or quartic equation.

To find the hydroxide ion concentration in a pure solution of a weak base, we make use of approximations very similar to those we have just discussed. The substance methyl amine is a weak base, capable of accepting a proton from water by the reaction

$$CH_3NH_2 + H_2O = CH_3NH_3^+ + OH^-.$$

The corresponding equilibrium constant is

$$\frac{[CH_3NH_3^+][OH^-]}{[CH_3NH_2]} = K = 5.0 \times 10^{-4}.$$

What is the equilibrium concentration of hydroxide ion in a solution prepared by adding 0.1 mole of CH_3NH_2 to enough water to make one liter of solution?

To reduce the problem to one unknown, we make two assumptions. The first is that the concentration of OH^- contributed by the ionization of water is negligible compared with that contributed by the base. Therefore, by the stoichiometry of the ionization reaction

$$[CH_3NH_3^+] = [OH^-].$$

The second assumption is that most of the CH_3NH_2 remains as such, since the equilibrium constant for its conversion to $CH_3NH_3^+$ is small. Therefore, to a good approximation

$$[CH_3NH_2] = 0.1 - [CH_3NH_3^+] \cong 0.1 \ M.$$

Note carefully that these assumptions and approximations are exactly analogous to those made in the treatment of weak acid ionization. Using our two approximations and the equilibrium constant, we obtain an equation for $[OH^-]$:

$$K = \frac{[CH_3NH_3^+][OH^-]}{[CH_3NH_2]} = \frac{[OH^-]^2}{0.1} = 5.0 \times 10^{-4},$$
$$[OH^-] = 7.1 \times 10^{-3} \ M.$$

To justify our assumptions we note that $7 \times 10^{-3} \ M$ is much greater than $10^{-7} \ M$, which is the maximum OH^- concentration ever contributed by the self-ionization of water. The second assumption, that

$$[CH_3NH_2] \cong 0.1 \ M,$$

requires that

$$[OH^-] = [CH_3NH_3^+] \ll 0.1 \ M,$$

which is satisfied, but rather marginally. The validity of the second assumption depends on K being small and the concentration of the base being fairly large. A second approximation to $[OH^-]$ can be obtained by saying

$$[CH_3NH_2] \cong 0.1 - 7.1 \times 10^{-3} = 9.29 \times 10^{-2} \ M,$$

and

$$\frac{[OH^-]^2}{9.29 \times 10^{-2}} = 5.0 \times 10^{-4},$$
$$[OH^-] = 6.8 \times 10^{-3} \ M.$$

This answer differs only slightly from our first result, and therefore is a sufficient approximation for most purposes.

In this section we have discussed specific examples of acid-base equilibria to illustrate the general method by which similar problems can be solved. We have avoided and will continue to avoid supplying general mathematical formulas which give an answer in one step. The reason for this is that any *simple* formula we might give would necessarily be approximate and might fail to apply in a given situation. The only way to be sure that an expression is appropriate for a given problem is to derive it following the procedures we have illustrated, taking full account of the chemical and physical peculiarities of the problem.

6.4 HYDROLYSIS

Hydrolysis is an aspect of acid-base equilibrium that traditionally is treated as a separate, distinct phenomenon, but in fact it requires no concepts beyond those we have already discussed. We have remarked that a weak acid and its anion are a conjugate acid-base pair, and that if an acid is weak, its conjugate base tends to be strong. For example, acetic acid is a moderately weak acid, so acetate ion is a moderately strong base, and will acquire protons in aqueous solution by the reaction

$$OAc^- + H_2O = HOAc + OH^-. \tag{6.9}$$

This reaction represents the hydrolysis of the acetate ion, but we can see that it is nothing more than the "ionization" of a weak base. Therefore it should be possible to calculate the hydroxide ion concentration in a pure solution of NaOAc in the same way that we calculated $[OH^-]$ in a pure solution of CH_3NH_2.

There is one additional matter which must be dealt with. The equilibrium constant for reaction (6.9) is

$$\frac{[HOAc][OH^-]}{[OAc^-]} = K_h, \tag{6.10}$$

where K_h is called the hydrolysis constant. Hydrolysis constants are not often tabulated, because they can be evaluated easily from the ionization constants of the corresponding acid. To do this, we multiply Eq. (6.10) by $[H_3O^+]/[H_3O^+]$ to get

$$\frac{[HOAc][OH^-][H_3O^+]}{[OAc^-][H_3O^+]} = K_h.$$

In the numerator we recognize that the product $[OH^-][H_3O^+]$ is equal to K_w, the ion product constant for water. This gives us

$$\frac{[HOAc]K_w}{[H_3O^+][OAc^-]} = K_h,$$

and it is now easy to see that the remaining factors are equal to the reciprocal of the ionization constant for acetic acid. That is,

$$\frac{[\text{HOAc}]}{[\text{H}_3\text{O}^+][\text{OAc}^-]} = \frac{1}{K_a},$$

so

$$K_h = \frac{K_w}{K_a} \tag{6.11}$$

$$= \frac{10^{-14}}{1.85 \times 10^{-5}}$$

$$= 5.4 \times 10^{-10}.$$

Equation (6.11) is of general validity. It shows that the weaker the acid, the smaller K_a and the more extensively hydrolyzed is the anion. In other words, it is a quantitative demonstration that the weaker the acid, the stronger is its conjugate base, and vice versa.

Now we can proceed with calculation of the hydroxide ion concentration in a solution prepared by dissolving one mole of sodium acetate in enough water to make one liter of solution. Our method, which by now should be familiar, is to neglect the [OH$^-$] contributed by the self-ionization of water and thereby obtain

$$[\text{OH}^-] \cong [\text{HOAc}]$$

from the stoichiometry of the hydrolysis reaction. Furthermore, we recognize that acetate ion is a weak base, so that most of it will remain as the ion, which means

$$[\text{OAc}^-] \cong 1.0 \; M.$$

Substituting these relations into the hydrolysis-constant expression results in

$$\frac{[\text{OH}^-]^2}{1.0} \cong 5.4 \times 10^{-10},$$

$$[\text{OH}^-] \cong 2.3 \times 10^{-5} \; M.$$

This hydroxide ion concentration is sufficiently large for our first assumption to be valid, but sufficiently small for our second assumption to hold, as the reader should demonstrate for himself.

The salts of weak bases are themselves weak acids. For instance, ammonia is a weak base, as we can see from

$$\text{NH}_3 + \text{H}_2\text{O} = \text{NH}_4^+ + \text{OH}^-,$$

$$\frac{[\text{NH}_4^+][\text{OH}^-]}{[\text{NH}_3]} = K_b = 1.8 \times 10^{-5}.$$

As a result, its salts such as NH$_4$Cl will act as weak acids, or be hydrolyzed according to the formula

$$NH_4^+ + H_2O = H_3O^+ + NH_3.$$

The corresponding hydrolysis constant is

$$\frac{[H_3O^+][NH_3]}{[NH_4^+]} = K_h.$$

It can be evaluated numerically by multiplying by $[OH^-]/[OH^-]$ to give

$$\frac{[OH^-][H_3O^+][NH_3]}{[OH^-][NH_4^+]} = K_h,$$

in which we recognize factors equal to K_w and $1/K_b$. Thus

$$K_h = \frac{K_w}{K_b}. \tag{6.12}$$

Equation (6.12) is the general form that relates the ionization constant of a weak base to the hydrolysis constant of its salt. If the base is weak, K_h tends to be large, and the salt tends to be a moderately strong acid, and vice versa.

Now we shall find the concentration of H_3O^+ in equilibrium with a pure solution of NH_4Cl whose chloride ion concentration is $0.1\ M$. We have

$$\frac{[NH_3][H_3O^+]}{[NH_4^+]} = K_h = \frac{K_w}{K_b} = 5.6 \times 10^{-10}.$$

Most of the ammonium ion will remain unchanged, so that $[NH_4^+] \cong 0.1\ M$. Also, most of the H_3O^+ comes from the hydrolysis reaction, so $[NH_3] \cong [H_3O^+]$. As a result

$$\frac{[H_3O^+]^2}{0.1} \cong 5.6 \times 10^{-10},$$

$$[H_3O^+] \cong 7.5 \times 10^{-6}\ M.$$

The reader should verify that the approximations are justified.

6.5 BUFFER SOLUTIONS

So far we have treated only solutions containing a pure weak acid or a pure weak base. In this section, we shall find how to calculate the equilibrium concentrations for solutions which contain a mixture of a weak acid and its salt, or a weak base and its salt. The arguments we use are only slight extensions of what we have learned in the two previous sections.

Let us calculate the concentration of H_3O^+ in a solution prepared by mixing 0.70 mole of acetic acid, and 0.60 mole of sodium acetate with enough water to make one liter of solution. The ionization constant of acetic acid,

$$\frac{[H_3O^+][OAc^-]}{[HOAc]} = 1.85 \times 10^{-5},$$

can be arranged to give

$$[H_3O^+] = \frac{[HOAc]}{[OAc^-]} \times 1.85 \times 10^{-5}, \qquad (6.13)$$

and it is clear that in order to calculate $[H_3O^+]$, we must obtain values for $[OAc^-]$ and $[HOAc]$.

Of the original 0.70 mole of acetic acid added to the solution, some might be lost by dissociation into ions. However, we have argued earlier that acetic acid is a weak acid, and when it is dissolved in pure water, an overwhelming fraction of it remains undissociated. Will this be true when we dissolve it in a solution that already has a large amount of acetate ion in it? Le Chatelier's principle assures us that the added acetate ion will repress the dissociation of the acid by the common ion effect. Therefore, if it is justified to neglect the loss of acetic acid by dissociation in its pure solution, it is even more justified to assume that it is very slightly dissociated in a solution containing excess acetate ion.

Our argument suggests that we can assume $[HOAc] = 0.70\ M$. However, there is one more point to be examined. Acetic acid can be *produced* by the hydrolysis of the excess acetate ion:

$$OAc^- + H_2O = HOAc + OH^-,$$

and this suggests that the concentration of HOAc might be greater than $0.70\ M$. However, we have investigated the hydrolysis of *pure* 1 M sodium acetate, and found that very little HOAc ($\sim 10^{-5}\ M$) is produced. In a solution that already contains acetic acid, the hydrolysis will be repressed, and its contribution to the HOAc concentration will be negligible. Therefore we can conclude that *since the amount of acetic acid added to the solution is fairly large*, the amounts lost by dissociation or gained by hydrolysis must be comparatively small, and we can set the concentration of HOAc equal to $0.70\ M$ to a very good approximation.

To obtain a satisfactory approximation for the acetate ion concentration we first note that its concentration will be fairly large, since sodium acetate is totally dissociated into ions. The loss of acetate ion by hydrolysis is small even in a pure solution of sodium acetate, and is bound to be smaller in a solution where excess acetic acid represses the hydrolysis. The gain of acetate ion from the dissociation of acetic acid is also very small, by a similar argument. Consequently we can set the acetate ion concentration equal to $0.60\ M$.

With satisfactory, if approximate, values for [HOAc] and [OAc⁻] at hand, we can write

$$[H_3O^+] = \frac{[HOAc]}{[OAc^-]} \times 1.85 \times 10^{-5} = \frac{0.70}{0.60} \times 1.85 \times 10^{-5}$$
$$= 2.2 \times 10^{-5} \ M.$$

Thus the concentration of H_3O^+ is lower than in a pure solution of acetic acid at a comparable concentration. This result is consistent with Le Chatelier's principle, which predicts that the addition of excess acetate ion to a solution of acetic acid will repress the dissociation of the acid, and lower the concentration of H_3O^+.

Re-examination of the arguments that led to the solution of this problem will show that the approximations we used require that *both* the acid and its anion be present in substantial concentrations. Therefore, it is safe to use the procedure we have outlined only if the ratio of the acid to salt concentration lies between 0.1 and 10, and the absolute concentration of the acid is numerically much greater than its dissociation constant.

A solution that contains appreciable amounts of both a weak acid and its salt is called a *buffer solution*, and has remarkable and useful properties. Buffer solutions can be diluted without changing the concentration of H_3O^+. The general expression for $[H_3O^+]$, of which Eq. (6.13) is a special case, is

$$[H_3O^+] = \frac{[acid]}{[anion]} \ K_a.$$

The concentration of H_3O^+ depends only on K_a and the *ratio* of the concentrations of acid and anion. When the buffer solution is diluted, the concentration of the acid and anion change, but their ratio remains constant, and $[H_3O^+]$ is unchanged.

Buffer solutions also tend to keep the concentration of H_3O^+ constant even when small amounts of strong acid or strong base are added to them. To illustrate this phenomenon let us calculate the change in the concentration of H_3O^+ that occurs when 1 ml of 1 M HCl is added to 1 liter of (a) pure water and (b) the solution of acetic acid and acetate ion we have just discussed.

In case (a) we add

$$0.001 \text{ liter} \times 1 \text{ mole/liter} = 0.001 \text{ mole}$$

of H_3O^+ to one liter of water. The resulting solution therefore has an H_3O^+ concentration of $10^{-3} \ M$. The addition produces a 10^4-fold change in the concentration of H_3O^+, compared to that in pure water.

When, in case (b), we add 0.001 mole of H_3O^+ to the solution containing acetate ion and acetic acid, the following chemical reaction occurs:

$$OAc^- + H_3O^+ \rightarrow HOAc + H_2O.$$

We are certain of this because the equilibrium constant for this reaction is

$$\frac{[\text{HOAc}]}{[\text{H}_3\text{O}^+][\text{OAc}^-]} = \frac{1}{K_a} = \frac{1}{1.85 \times 10^{-5}} = 5.4 \times 10^4.$$

Since this equilibrium constant is large, virtually all of the added acid reacts with the acetate ion to produce acetic acid. Therefore the new acetic acid concentration is

$$[\text{HOAc}] = 0.70 + 0.001 = 0.701 \ M.$$

Since 0.001 mole of OAc^- reacted with the added H_3O^+, the new acetate ion concentration is

$$[\text{OAc}^-] = 0.60 - 0.001 = 0.599 \ M.$$

As a result, the new concentration of H_3O^+ is given by

$$[\text{H}_3\text{O}^+] = \frac{[\text{HOAc}]}{[\text{OAc}^-]} K_a = \frac{0.701}{0.599} \times 1.85 \times 10^{-5}$$
$$= 2.2 \times 10^{-5}.$$

Within the allowable significant figures, the concentration of H_3O^+ is unchanged by this addition. A similar result is obtained when the effect of adding 0.001 mole of a strong base is calculated. Mixtures of weak acids and their salts resist attempts to change the concentration of H_3O^+ by addition of strong acids or strong bases. By storing excess protons as the weak acid, and excess base as the anion, they are able to modify the effect of any added acid or base. This is the origin of the name *buffer solution*.

Buffer solutions can also be prepared by mixing appreciable amounts of a weak base and one of its salts. Ammonia is a weak base and in aqueous solution produces OH^- by the reaction

$$\text{NH}_3 + \text{H}_2\text{O} = \text{NH}_4^+ + \text{OH}^-.$$

The equilibrium-constant expression for this reaction can be written as

$$[\text{OH}^-] = \frac{[\text{NH}_3]}{[\text{NH}_4^+]} K_b,$$

and we see that the concentration of OH^- depends only on the *ratio* of the concentration of ammonia to that of ammonium ion. Consequently $[\text{OH}^-]$ (and $[\text{H}_3\text{O}^+]$) will be unchanged by any dilution of the solution. Furthermore, excess base is stored in the solution as NH_3 and excess acid as NH_4^+. Therefore, any strong acids or bases added to the solution will be neutralized and the OH^- and H_3O^+ concentrations will be essentially unchanged.

Question. In order for a buffer solution to work satisfactorily, the amount of weak acid and anion or weak base and cation present in solution must be considerably larger than the acid or base additions that must be offset. Explain why this is true.

As a practical illustration of the use of buffer solutions we can reconsider the separation of $0.1\ M$ Zn^{++} and $0.1\ M$ Fe^{++}, which we treated in Section 6.1. We concluded there that in order to precipitate ZnS but not FeS, a sulfide ion concentration of $10^{-19}\ M$ was necessary. This concentration can be maintained by using a saturated solution of H_2S, and an appropriate buffer.

In aqueous solution, H_2S dissociates very slightly into sulfide ions by the reaction

$$H_2S(aq) + 2H_2O = 2H_3O^+ + S^=.$$

The equilibrium constant for this reaction is

$$\frac{[H_3O^+]^2[S^=]}{[H_2S]} = 1.1 \times 10^{-21}.$$

In a *saturated* solution of hydrogen sulfide, $[H_2S] = 0.1\ M$, so we can see that in such a solution, the concentration of sulfide ion can be controlled by setting the concentration of H_3O^+. What concentration of H_3O^+ do we need if $[S^=]$ is to be $10^{-19}\ M$? The equilibrium-constant expression tells us that

$$[H_3O^+]^2 = \frac{[H_2S]}{[S^=]} \times 1.1 \times 10^{-21}.$$

Therefore, if $[H_2S] = 0.1\ M$, and $[S^=]$ is to be $10^{-19}\ M$,

$$[H_3O^+]^2 = \frac{0.1}{10^{-19}} \times 1.1 \times 10^{-21} = 1.1 \times 10^{-3},$$

$$[H_3O^+] = 3.3 \times 10^{-2}\ M.$$

If we maintain the concentration of H_3O^+ at about $0.03\ M$, the separation can be accomplished.

To choose an appropriate buffer solution, we examine the expression

$$[H_3O^+] = \frac{[acid]}{[salt]} K_a.$$

We see that in order to keep the concentration of the acid and its salt comparable, as we should in order to have a good buffer solution, we must have

$$[H_3O^+] \cong K_a.$$

The bisulfate ion, HSO_4^-, is a weak acid with an appropriate ionization constant

$$\frac{[H_3O^+][SO_4^=]}{[HSO_4^-]} = K_a = 1.2 \times 10^{-2},$$

$$[H_3O^+] = \frac{[HSO_4^-]}{[SO_4^=]} \times 1.2 \times 10^{-2}.$$

We see then that in order to maintain the concentration of H_3O^+ at about 0.03 M, we need only dissolve amounts of $NaHSO_4$ and Na_2SO_4 in the ratio of 2.5:1. This buffer will fix the H_3O^+ concentration at the proper value, even though as the ZnS precipitates, H_3O^+ is generated:

$$H_2S + Zn^{++} + 2H_2O = ZnS + 2H_3O^+.$$

Question. The solubility products of $Fe(OH)_3$ and $Zn(OH)_2$ are 4×10^{-38} and 4.5×10^{-17}, respectively. At what pH would the precipitation of $Fe(OH)_3$ be essentially complete while 0.5 M Zn^{++} remained in solution? From Table 6.2 choose an appropriate acid and give the concentrations you would use to make a buffer solution at this pH.

Indicators

Dye molecules whose color depends on the concentration of H_3O^+ provide the simplest way of estimating the pH of a solution. These indicators are themselves weak acids or weak bases whose conjugate acid-base forms are of different color.

For example, the indicator phenol red ionizes according to the equation

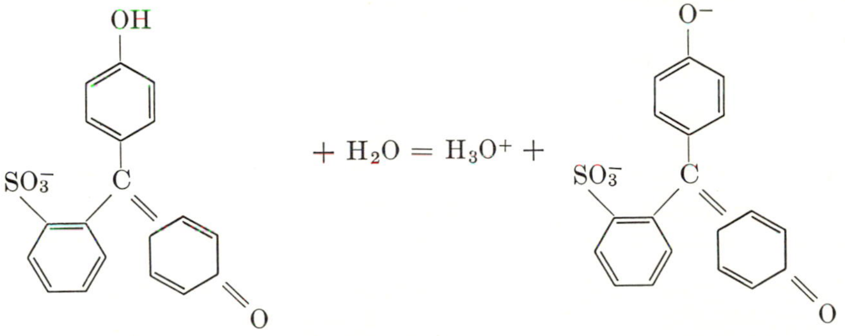

$$+ H_2O = H_3O^+ +$$

Red   Yellow

which we will abbreviate to

$$HIn + H_2O = H_3O^+ + In^-,$$

$$K_I = \frac{[H_3O^+][In^-]}{[HIn]}.$$

If only a very tiny amount of this indicator is added to a solution, the dissociation of the indicator will not affect the concentration of H_3O^+ at all. Quite the reverse happens. The concentration of H_3O^+ in the solution determines the ratio of In^- to HIn by the equation

$$\frac{[In^-]}{[HIn]} = \frac{K_I}{[H_3O^+]}.$$

Table 6.3 pH-ranges of some acid-base indicators

Indicator	pH-interval	Color change: acid to base
Thymol blue	1.2–2.8	red–yellow
Methyl orange	2.1–4.4	orange–yellow
Methyl red	4.2–6.3	red–yellow
Bromthymol blue	6.0–7.6	yellow–blue
Cresol red	7.2–8.8	yellow–red
Phenolphthalein	8.3–10.0	colorless–red
Alizarin yellow	10.1–12.0	yellow–red

The color of the solution will depend on the concentration of H_3O^+, for if $[H_3O^+]$ is large, $[HIn] \gg [In^-]$, and the solution will be red, but if $[H_3O^+]$ is small, $[In^-] \gg [HIn]$, and the solution will be yellow.

There is a natural limitation on the range of pH-values in which a given indicator is useful. The eye can detect changes in color only when the ratio of the concentrations of the two colored forms falls in the range 0.1 to 10. In the case of phenol red we would have

$$\frac{[In^-]}{[HIn]} = 0.1, \qquad \text{solution distinctly red,}$$

$$\frac{[In^-]}{[HIn]} = 1, \qquad \text{solution orange,}$$

$$\frac{[In^-]}{[HIn]} = 10, \qquad \text{solution distinctly yellow.}$$

By referring to the equilibrium-constant expression we see that these three ratios correspond to $[H_3O^+]$ equal to $10K_I$, K_I, and $0.1K_I$ respectively. Therefore the indicator is sensitive to change of pH only in a 100-fold range of H_3O^+ concentration which is centered on the value $[H_3O^+] = K_I$. In order to measure pH in the range of 7 ± 1, we must use an indicator whose acid ionization constant is about 10^{-7}, and likewise for other pH ranges. Table 6.3 gives a list of common indicators and the ranges in which they are effective.

6.6 EXACT TREATMENT OF IONIZATION EQUILIBRIA

There are really only two different types of problem that arise in simple situations of acid-base equilibria. In one case we are concerned with "pure" solutions of a single weak acid or weak base; or, what is the same thing, a solution of a salt of a weak acid, or a salt of a weak base. The second case is that of buffer solutions, which contain appreciable amounts of both a weak acid and its salt or of a weak base and its salt. We have demonstrated intuitive, approximate treatments of these two problems, and have emphasized that there are situations in which our approximations fail. In order to make the conditions for the validity of the approximations more clear, and to provide a means of handling problems for which the approximations fail, we now present an exact method of treating acid-base equilibria.

Consider the ionization of any weak acid HA:

$$HA + H_2O = H_3O^+ + A^-.$$

Accompanying this reaction in aqueous solution we always have

$$2H_2O = H_3O^+ + OH^-.$$

Our general problem is to calculate the four concentrations, $[HA]$, $[A^-]$, $[H_3O^+]$, and $[OH^-]$, given that a certain amount of acid was dissolved to form a specified volume of solution.

Since we have four unknowns, we must find four equations which can be solved simultaneously to give the desired concentrations. Two of these equations are the equilibrium constants

$$\frac{[H_3O^+][A^-]}{[HA]} = K_a, \qquad (6.14)$$

$$[H_3O^+][OH^-] = K_w. \qquad (6.15)$$

We will consider the values of K_a and K_w to be given. To obtain a third equation, we need only recognize that of the original amount of HA added to the solution, all must be present either as undissociated acid HA or as the anion A^-. If we call $[HA]_0$ the total amount of acid added to the solution divided by the volume, we can write

$$[HA]_0 = [HA] + [A^-]. \qquad (6.16)$$

This equation is called the *material balance relation*. It simply says that since no "A-material" is created or destroyed, the sum of the equilibrium concentrations of HA and A^- must equal the concentration HA would have had *if none had dissociated*.

The fourth equation follows from the requirement that the solution be electrically neutral. That is, the total concentration of *positive charge* must equal the total concentration of *negative charge*:

$$[H_3O^+] = [A^-] + [OH^-]. \qquad (6.17)$$

Equation (6.17) is called the *charge balance equation*.

Now let us combine our four equations to find an expression for the concentration of H_3O^+ at equilibrium. We will do this by first finding expressions for $[A^-]$ and $[HA]$ in terms of $[H_3O^+]$, and then substituting these in Eq. (6.14).

From the charge balance equation we obtain

$$[A^-] = [H_3O^+] - [OH^-],$$

which, by use of Eq. (6.15), becomes

$$[A^-] = [H_3O^+] - \frac{K_w}{[H_3O^+]}. \qquad (6.18)$$

This is the first of the necessary relations. Rearrangement of the material balance equation gives us

$$[HA] = [HA]_0 - [A^-].$$

Combining Eq. (6.18) with this results in

$$[HA] = [HA]_0 - [H_3O^+] + \frac{K_w}{[H_3O^+]}. \qquad (6.19)$$

Now we have both $[A^-]$ and $[HA]$ in terms of known constants and $[H_3O^+]$. Substitution of Eqs. (6.18) and (6.19) into Eq. (6.14) produces

$$\frac{[H_3O^+]\left([H_3O^+] - \dfrac{K_w}{[H_3O^+]}\right)}{\left([HA]_0 - [H_3O^+] + \dfrac{K_w}{[H_3O^+]}\right)} = K_a. \qquad (6.20)$$

Equation (6.20) is a cubic equation that can be solved for the exact concentration of H_3O^+. Subsequently, all other concentrations can be found from Eqs. (6.15), (6.18), and (6.19).

The direct solution of a cubic equation is never a pleasant procedure, so it is of interest to find the conditions under which we can simplify Eq. (6.20). First, note that even in solutions of very weak acid, the concentration of H_3O^+ is generally greater than $10^{-6}\ M$. This means that in the numerator of Eq. (6.20), the term

$$[H_3O^+] - \frac{K_w}{[H_3O^+]}$$

is to a good approximation just equal to $[H_3O^+]$, for if $[H_3O^+] > 10^{-6}\ M$, then

$$\frac{K_w}{[H_3O^+]} < 10^{-8}\ M.$$

The same simplification can be made in the denominator of Eq. (6.20), so we get

$$\frac{[H_3O^+][H_3O^+]}{[HA]_0 - [H_3O^+]} = K_a, \quad \text{if} \quad [H_3O^+] \gg \frac{K_w}{[H_3O^+]}. \qquad (6.21)$$

We recognize Eq. (6.21) as a quadratic equation that we encountered in Section 6.3. The conditions for its validity are now clear.

Equation (6.21) can be simplified further, for if $[H_3O^+] \ll [HA]_0$, it becomes

$$\frac{[H_3O^+]^2}{[HA]_0} = K_a, \tag{6.22}$$

which is the familiar simplest approximation to the acid ionization problem. In any problem of weak acid ionization we might first use Eq. (6.22) to obtain an approximate value for $[H_3O^+]$. This approximate answer can be used to decide whether the approximations are valid, and if not, Eq. (6.21) or Eq. (6.20) may be solved by successive approximations.

Example 6.1 Develop the exact expression for the concentration of OH^- in equilibrium with a weak base BOH. We have

$$\frac{[B^+][OH^-]}{[BOH]} = K_b, \qquad [H_3O^+][OH^-] = K_w,$$

and the material balance equation is

$$[BOH]_0 = [B^+] + [BOH],$$

while charge balance requires that

$$[B^+] + [H_3O^+] = [OH^-].$$

Rearrangement of the charge balance equation gives

$$[B^+] = [OH^-] - [H_3O^+] = [OH^-] - \frac{K_w}{[OH^-]}.$$

From this expression and the material balance equation we get

$$[BOH] = [BOH]_0 - [OH^-] + \frac{K_w}{[OH^-]}.$$

These equations for [BOH] and $[B^+]$ convert the ionization-constant expression to

$$\frac{\left([OH^-] - \dfrac{K_w}{[OH^-]}\right)[OH^-]}{\left([BOH]_0 - [OH^-] + \dfrac{K_w}{[OH^-]}\right)} = K_b,$$

which is an exact cubic equation for $[OH^-]$.

Now let us develop an exact expression for the concentration of H_3O^+ in an arbitrary mixture of an acid HA and its salt NaA. The equations

$$\frac{[H_3O^+][A^-]}{[HA]} = K_a, \tag{6.14}$$

$$[H_3O^+][OH^-] = K_w, \tag{6.15}$$

still apply, but now the material balance expression is

$$[HA]_0 + [NaA]_0 = [HA] + [A^-], \tag{6.23}$$

where $[NaA]_0$ is the number of moles of salt added to the solution, divided by the volume of the solution. Equation (6.23) says that all the "A-material" added must be present either as HA or A^-. The charge balance equation for this system is

$$[Na^+] + [H_3O^+] = [A^-] + [OH^-]. \tag{6.24}$$

Upon substitution of $K_w/[H_3O^+]$ for $[OH^-]$, and slight rearrangement, the charge balance equation becomes

$$[A^-] = [Na^+] + [H_3O^+] - \frac{K_w}{[H_3O^+]}. \tag{6.25}$$

Equation (6.25) says that the concentration of A^- is equal to that of Na^+, except for the two correction terms $[H_3O^+]$ and $K_w/[H_3O^+]$. These latter terms are often, but not always, relatively small. The material balance equation

$$[HA] = [HA]_0 + [NaA]_0 - [A^-]$$

can be altered by substitution of Eq. (6.25) for $[A^-]$ to give

$$[HA] = [HA]_0 + [NaA]_0 - [Na^+] - [H_3O^+] + \frac{K_w}{[H_3O^+]}.$$

But since the salt NaA is totally dissociated, $[NaA]_0 = [Na^+]$, and

$$[HA] = [HA]_0 - [H_3O^+] + \frac{K_w}{[H_3O^+]}. \tag{6.26}$$

Substitution of Eqs. (6.25) and (6.26) into Eq. (6.14) results in

$$\frac{[H_3O^+]\left([Na^+] + [H_3O^+] - \frac{K_w}{[H_3O^+]}\right)}{\left([HA]_0 - [H_3O^+] + \frac{K_w}{[H_3O^+]}\right)} = K_a \tag{6.27}$$

which is an exact cubic equation for $[H_3O^+]$.

Equation (6.27) reduces to a simple expression if *both* $[Na^+]$ and $[HA]_0$ are large. That is, when

$$[HA]_0, [Na^+] \gg [H_3O^+]$$

and

$$[HA]_0, [Na^+] \gg \frac{K_w}{[H_3O^+]} = [OH^-],$$

then Eq. (6.27) becomes

$$\frac{[H_3O^+][Na^+]}{[HA]_0} = K_a,$$

$$[H_3O^+] = \frac{[HA]_0}{[Na^+]} K_a = \frac{[acid]}{[salt]} K_a,$$

which is the familiar simple expression for the concentration of H_3O^+ in a buffer solution. We see that in order for our simple expression to be valid, *both* $[Na^+]$ and $[HA]_0$ must be larger than *both* $[H_3O^+]$ and $[OH^-] = K_w/[H_3O^+]$. When only one or two of these conditions are satisfied, Eq. (6.27) may be simplified accordingly.

It is interesting to note that when the concentration of Na^+ is set equal to zero in Eq. (6.27) we obtain Eq. (6.20), which is the exact expression for the concentration of H_3O^+ in a solution of a pure acid. Likewise, setting $[HA]_0 = 0$ gives us an exact expression of the H_3O^+ concentration in a pure solution of the salt NaA. Neither of these conclusions should be surprising, but they reaffirm our confidence that Eq. (6.27) is correct regardless of the concentrations of acid and salt.

6.7 ACID-BASE TITRATIONS

The acid-base titration is one of the most important techniques of analytical chemistry. The general procedure is to determine the amount of, let us say, an acid by adding an equivalent measured amount of a base, or vice versa. In order to see how to design a good acid-base titration experiment, it is useful to calculate the concentration of H_3O^+ at various stages in the titration of 50.00 ml of 1.000 M HCl with 50.00 ml of 1.000 M NaOH.

To make our results applicable to all such titrations, we shall express the progress of the titration by giving the value of f, the fraction of the original acid that has been neutralized. If the original number of moles of HCl is denoted by n_0, f is given by

$$f = \frac{\text{number of moles of base added}}{n_0}.$$

At the start of the titration $f = 0$, and $f = 1$ corresponds to a completely titrated acid, or the equivalence point of the titration.

Before the titration starts, $[H_3O^+] = 1\,M$, and pH $= 0$. To calculate $[H_3O^+]$ for $0 < f < 1$, we let

$$V = \text{original volume of the acid,}$$
$$v = \text{volume of the base added.}$$

Then, since both acid and base have the same concentration, $f = v/V$, and the *amount* of acid at any stage in the titration is

$$H_3O^+ = n_0(1 - f) = n_0\left(1 - \frac{v}{V}\right).$$

To obtain the *concentration* of H_3O^+, we divide by the total volume of the solution

$$[H_3O^+] = n_0\,\frac{(1 - v/V)}{(V + v)} = n_0\left(\frac{1 - f}{V + v}\right).$$

This expression is derived assuming that all of the H_3O^+ comes from the ionization of HCl, and thus it gives $[H_3O^+] = 0$ at the equivalence point. Actually, very near the equivalence point an appreciable amount of the H_3O^+ comes from the ionization of water, and at the equivalence point we have $[H_3O^+] = [OH^-] = 10^{-7}\,M$. Therefore a more accurate expression for $[H_3O^+]$ is

$$[H_3O^+] = [H_3O^+]\,(\text{from acid}) + [H_3O^+]\,(\text{from water})$$
$$= n_0\left(\frac{1 - f}{V + v}\right) + [OH^-]$$
$$= n_0\left(\frac{1 - f}{V + v}\right) + \frac{K_w}{[H_3O^+]}.$$

After the equivalence point is passed, we are essentially adding OH^- to a certain volume of pure water, and the concentration of OH^- can be calculated accordingly.

Question. Why is it legitimate to neglect the term $K_w/[H_3O^+]$ when f is small?

The clearest representation of what happens during the titration is given by the titration curve, a plot of pH as a function of f, which is shown in Fig. 6.1. These curves are drawn from calculations with the equations we have just derived. The most striking feature of the titration curve is the rapid change of pH in the vicinity of the equivalence point. In going from $f = 0.999$ to $f = 1.001$, the pH changes by nearly 8 units, or $[H_3O^+]$ changes by a factor of ten million. Therefore, in an actual experiment any acid-base indicator that changes color anywhere between pH 4 and pH 10 will allow us to locate the equivalence point to within one part in one thousand, or one-tenth of a percent. This situation becomes slightly less favorable as the reagents become more dilute, as Fig. 6.1 shows.

IONIC EQUILIBRIA IN AQUEOUS SOLUTIONS | 6.7

In practical situations, many titrations involve the reaction of a weak acid with a strong base, or of a weak base with a strong acid. Therefore it is important to examine the titration curves for these cases.

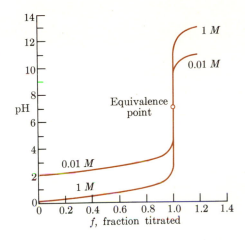

Titration of 1-M and 0.01-M HCl with 1-M and 0.01-M NaOH respectively.

FIG. 6.1

We assume that we are titrating 50.00 ml of 0.1000-M acetic acid with 0.1000 M NaOH. Once again we express the progress of the titration in terms of f, the fraction of the acid titrated, where

$$f = \frac{\text{moles of base added}}{n_0},$$

and n_0 is the original number of moles of acid in the solution. At the start of the titration $f = 0$, and we have a 0.1000 M solution of a pure acid. The value of $[H_3O^+]$ is given by

$$\frac{[H_3O^+]^2}{0.100} = K_a = 1.85 \times 10^{-5},$$

$$[H_3O^+] = 1.36 \times 10^{-3},$$

$$pH = 2.87.$$

As the titration progresses, $0 < f < 1$, and we have appreciable quantities of the acid and its anion present in solution. Under these circumstances,

$$[H_3O^+] = \frac{[HOAc]}{[OAc^-]} K_a, \tag{6.28}$$

as we learned from our discussion of buffer solutions. But the *amount* of OAc$^-$ at any point in the titration is

$$OAc^- = n_0 f,$$

and its concentration is

$$[OAc^-] = \frac{n_0 f}{V + v},$$

where V is the original volume of the acid, and v is the volume of base added. Similarly, the *amount* of HOAc at any time is

$$HOAc = n_0(1 - f)$$

and

$$[HOAc] = n_0 \frac{(1 - f)}{V + v}.$$

Substitution into Eq. (6.28) gives

$$[H_3O^+] = \frac{1 - f}{f} K_a, \qquad 0 < f < 1. \tag{6.29}$$

Thus the concentration of H_3O^+ depends only on K_a and the fraction titrated, and not on the original concentration of the acid.

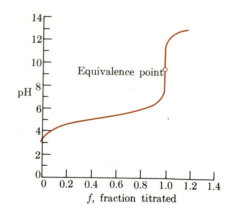

FIG. 6.2 Titration of 0.1-*M* acetic acid with 0.1-*M* sodium hydroxide.

Equation (6.29) is not valid when $f = 1$, since it was derived neglecting the hydrolysis of the anion OAc^-. When $f = 1$, the acid has been completely converted to a solution of $0.0500\ M$ NaOAc, which hydrolyzes according to

$$OAc^- + H_2O = HOAc + OH^-.$$

We know that

$$\frac{[OH^-][HOAc]}{[OAc^-]} = K_h = \frac{K_w}{K_a} = 5.54 \times 10^{-10},$$

and that

$$[HOAc] = [OH^-],$$
$$[OAc^-] = 0.0500\ M.$$

Therefore
$$[OH^-]^2 = 5 \times 10^{-2} \times 5.54 \times 10^{-10},$$
$$[OH^-] = 5.26 \times 10^{-6},$$
$$[H_3O^+] = 1.90 \times 10^{-9},$$
$$pH = 8.72.$$

We see that because of hydrolysis the solution is not neutral at the equivalence point. Beyond the equivalence point the concentration of H_3O^+ can be calculated assuming that we are adding base to a solution of pure water.

The calculations we have just outlined lead to the titration curve shown in Fig. 6.2. It is clear that in this weak acid-strong base titration, the pH changes more slowly in the vicinity of the equivalence point than was the case in the strong acid-strong base titration. Thus it is always more difficult to locate the equivalence point in titrations of a weak acid than in those of a strong acid.

6.8 MULTISTAGE EQUILIBRIA

In the problems of acid-base equilibrium that we have treated so far, we have dealt with solutions in which there was only one weak acid or weak base present, apart from water itself. However, situations in which there is more than one weak acid or base in solution are important, and arise naturally when, for example, an acid can ionize in two or more ways:

$$H_2CO_3 + H_2O = HCO_3^- + H_3O^+, \quad K_1 = 4.2 \times 10^{-7},$$
$$HCO_3^- + H_2O = CO_3^= + H_3O^+, \quad K_2 = 4.8 \times 10^{-11},$$

$$\frac{[H_3O^+][HCO_3^-]}{[H_2CO_3]} = K_1, \quad \frac{[H_3O^+][CO_3^=]}{[HCO_3^-]} = K_2, \quad \frac{[H_3O^+]^2[CO_3^=]}{[H_2CO_3]} = K_1K_2.$$

Thus a solution of carbonic acid is really a mixture of two acids: H_2CO_3 and HCO_3^-. The constant K_1 is called the first ionization constant of carbonic acid, and K_2 is the second ionization constant.

As an example of how a mixture of weak acids is treated mathematically, let us calculate the H_3O^+ concentration in a solution obtained by dissolving 0.02 mole of CO_2 (H_2CO_3) in one liter of water. Because of the possible complications of this system, we had best base our treatment on the exact material and charge balance equations:

$$[H_2CO_3]_0 = [H_2CO_3] + [HCO_3^-] + [CO_3^=], \tag{6.30}$$

$$[H_3O^+] = [HCO_3^-] + 2[CO_3^=] + [OH^-]. \tag{6.31}$$

The coefficient of $[CO_3^=]$ in the charge balance equation is 2, because we are equating positive and negative *charge* concentrations, and the concentration of *charge* contributed by the carbonate ion is twice the concentration of the ion itself.

To obtain an expression for $[H_3O^+]$, we select the charge balance equation and by substitution try to express $[H_3O^+]$ as a function of the known equilibrium constants and of other concentrations that can be guessed. Using the fact that H_2CO_3 is a weak acid, we expect it to be very slightly dissociated, and therefore to have a concentration near to $0.02\ M$. Consequently we will try to express $[H_3O^+]$ as a function of $[H_2CO_3]$. Use of the equilibrium constants gives us

$$[HCO_3^-] = \frac{[H_2CO_3]}{[H_3O^+]} K_1, \qquad [CO_3^=] = \frac{[H_2CO_3]}{[H_3O^+]^2} K_1K_2,$$

$$[OH^-] = \frac{K_w}{[H_3O^+]}.$$

Substitution of these relations in Eq. (6.31) leads to

$$[H_3O^+] = \frac{[H_2CO_3]}{[H_3O^+]} K_1 + 2\frac{[H_2CO_3]}{[H_3O^+]^2} K_1K_2 + \frac{K_w}{[H_3O^+]} \tag{6.32}$$

$$= \frac{[H_2CO_3]}{[H_3O^+]} K_1 \left(1 + \frac{2K_2}{[H_3O^+]}\right) + \frac{K_w}{[H_3O^+]}. \tag{6.33}$$

Equation (6.31) and its direct descendant, Eq. (6.32), have a simple physical interpretation. They say that the observed $[H_3O^+]$ is equal to the sum of three separate contributions:

(1) those molecules of H_2CO_3 that have ionized once, represented by

$$[HCO_3^-] = \frac{[H_2CO_3]}{[H_3O^+]} K_1;$$

(2) those molecules of H_2CO_3 that have ionized twice, which give a contribution to $[H_3O^+]$ equal to

$$2[CO_3^=] = 2\frac{[H_2CO_3]}{[H_3O^+]^2} K_1K_2;$$

(3) the contribution to $[H_3O^+]$ that comes from the ionization of water, which is equal to

$$[OH^-] = \frac{K_w}{[H_3O^+]}.$$

It is important to keep the origin of each of the terms in Eqs. (6.32) and (6.33) in mind in order to find a way of simplifying Eq. (6.33).

Equation (6.33) is an exact cubic equation for $[H_3O^+]$ including that contributed by the ionization of water, and we have argued that in pure solutions of weak acids the ionization of water is a relatively unimportant source of H_3O^+. Therefore, we shall neglect $K_w/[H_3O^+]$ as a first approximation.

The term $2K_1K_2[H_2CO_3]/[H_3O^+]^2$ represents H_3O^+ contributed by acid molecules that have ionized twice to form $CO_3^=$. We can guess that this double ionization is less important than the single ionization to HCO_3^-, just by comparing the values of K_1 and K_2. In the first place, $K_2 \ll K_1$, so the removal of the second proton is more difficult than removal of the first. In the second place, K_1 itself is small, so the concentration of HCO_3^- will be small, and there will be relatively few HCO_3^- ions that can dissociate to form $CO_3^=$. Therefore, we might expect to be able to neglect the second term on the right-hand side of Eqs. (6.32) and (6.33).

We can justify this approximation by another argument. Consider the factor

$$\left(1 + \frac{2K_2}{[H_3O^+]}\right)$$

in Eq. (6.33). The second term in this factor represents the contribution of the second ionization of H_2CO_3 to the concentration of H_3O^+. Is $2K_2/[H_3O^+]$ small compared to unity? We know that the solution is acidic, so that $[H_3O^+] > 10^{-7} \, M$. Therefore

$$\frac{2K_2}{[H_3O^+]} < \frac{2 \times 4.8 \times 10^{-11}}{10^{-7}} \simeq 10^{-3} \ll 1.$$

Thus we see that if the solution is at all acidic, the contribution of the second ionization to the concentration of H_3O^+ is quite small, and Eq. (6.33) reduces to

$$[H_3O^+] = \frac{[H_2CO_3]}{[H_3O^+]} K_1. \qquad (6.34)$$

But this is exactly what we would have obtained if we had started with the expression

$$\frac{[H_3O^+][HCO_3^-]}{[H_2CO_3]} = K_1,$$

and said that $[H_3O^+] \cong [HCO_3^-]$. In other words, *if the second ionization constant is small enough*, we can calculate the H_3O^+ concentration by pretending that we are dealing with a monobasic acid.

Now let us check to see if this approximation is justified. Solving Eq. (6.34) for $[H_2CO_3] = 0.02 \, M$ gives us

$$[H_3O^+] = ([H_2CO_3]K_1)^{1/2} = (0.02 \times 4.2 \times 10^{-7})^{1/2}$$
$$= 9.2 \times 10^{-5} \, M.$$

Therefore,

$$\frac{2K_2}{[H_3O^+]} = \frac{2 \times 4.8 \times 10^{-11}}{9.2 \times 10^{-5}} = 1.0 \times 10^{-6},$$

which is much less than unity, so our simplification of Eq. (6.33) is justified.

Our neglect of the second ionization of H_2CO_3 in calculating $[H_3O^+]$ does not mean that the concentration of $CO_3^=$ is truly zero. It can be calculated most easily from

$$\frac{[H_3O^+]^2[CO_3^=]}{[H_2CO_3]} = K_1K_2 = 2.0 \times 10^{-17}.$$

We know that

$$[H_3O^+] = 9.2 \times 10^{-5},$$

$$[H_2CO_3] = 2 \times 10^{-2},$$

and so

$$[CO_3^=] = \frac{2 \times 10^{-2}}{85 \times 10^{-10}} \times 2.0 \times 10^{-17}$$

$$= 4.7 \times 10^{-11} \ M.$$

Therefore very little carbonate ion exists at equilibrium.

We should reemphasize that the validity of neglecting the contribution of the second ionization to the concentration of H_3O^+ depends on the fact that $K_2 \ll K_1$. Should this inequality not hold, we would have to solve Eq. (6.33) by a method of successive approximations.

Question. Succinic acid $(CH_2)_2(COOH)_2$ is a dibasic acid with $K_1 = 6.5 \times 10^{-5}$ and $K_2 = 3.3 \times 10^{-6}$. Determine whether it is justifiable to neglect the second ionization of the acid in calculating the H_3O^+ concentration in an aqueous solution of 0.1 M succinic acid.

Let us set up the expression for $[H_3O^+]$ in a solution prepared by mixing comparable amounts of H_2CO_3 and $NaHCO_3$. The charge balance equation for such a solution is

$$[H_3O^+] + [Na^+] = [HCO_3^-] + 2[CO_3^=] + [OH^-]. \tag{6.35}$$

At this point, we must realize that both $[H_2CO_3]$ and $[HCO_3^-]$ may be large, and either might be guessed from the specifications of how the solution was made up. Therefore, we may choose to express $[H_3O^+]$ either in terms of $[H_2CO_3]$ or of $[HCO_3^-]$. Choosing the former, we obtain from Eq. (6.35)

$$[H_3O^+] + [Na^+] = \frac{[H_2CO_3]}{[H_3O^+]} K_1 + 2 \frac{[H_2CO_3]}{[H_3O^+]^2} K_1K_2 + \frac{K_w}{[H_3O^+]}.$$

$$\tag{6.36}$$

We can now proceed to simplify this expression. On the left-hand side, $[Na^+] \gg [H_3O^+]$ if appreciable $NaHCO_3$ was used to make up the solution. On the right-hand side, the last two terms *may* be small if $K_2 \ll K_1$, and if $[H_3O^+]$ is greater than $10^{-7} \ M$. Neglecting the first term on the left-hand side and the last two terms on the right-hand side gives

$$[Na^+] = \frac{[H_2CO_3]}{[H_3O^+]} K_1,$$

$$[H_3O^+] = \frac{[H_2CO_3]}{[Na^+]} K_1. \tag{6.37}$$

Equation (6.37) is what we would have obtained if we had treated the mixture of H_2CO_3 and $NaHCO_3$ as a simple buffer solution, neglecting the fact that HCO_3^- can ionize to H_3O^+ and $CO_3^=$. The approximations used to deduce Eq. (6.37) may fail, but even so, Eq. (6.37) can be used to give a first approximation for $[H_3O^+]$, which must then be refined by using Eq. (6.36).

Question. What is the pH of a buffer solution prepared with one mole of H_3PO_4 and 0.5 mole of NaH_2PO_4 in one liter of water? For H_3PO_4, $K_1 = 7.5 \times 10^{-3}$, $K_2 = 6.2 \times 10^{-8}$, and $K_3 = 1 \times 10^{-12}$.

The calculation of the concentration of H_3O^+ in a pure solution of the salt $NaHCO_3$ is a problem somewhat different from those we have been treating. In such a solution the relevant equilibria are

$$HCO_3^- + H_2O = H_3O^+ + CO_3^=,$$
$$HCO_3^- + H_2O = H_2CO_3 + OH^-.$$

In other words, HCO_3^- can act as either an acid or a base. Therefore, we can say that HCO_3^- can react with itself according to

$$2HCO_3^- = H_2CO_3 + CO_3^=.$$

This reaction has an equilibrium constant equal to

$$\frac{[H_2CO_3][CO_3^=]}{[HCO_3^-]^2} = K.$$

By multiplying both numerator and denominator by $[H_3O^+]$, and recognizing two familiar factors we get

$$\frac{[H_2CO_3][CO_3^=][H_3O^+]}{[H_3O^+][HCO_3^-][HCO_3^-]} = K = \frac{K_2}{K_1}.$$

Therefore, if $K_2 \ll K_1$, the conversion of HCO_3^- into H_2CO_3 and $CO_3^=$ will be relatively small. Since $K_2/K_1 \simeq 10^{-4}$, we can say that most of the HCO_3^- remains as such, but that some (10%) does go to form H_2CO_3 and $CO_3^=$.

Now let us develop an exact expression for the concentration of H_3O^+. The charge balance equation for this solution is

$$[Na^+] + [H_3O^+] = [HCO_3^-] + 2[CO_3^=] + [OH^-], \tag{6.38}$$

and the material balance equation is

$$[\text{NaHCO}_3]_0 = [\text{Na}^+] = [\text{H}_2\text{CO}_3] + [\text{HCO}_3^-] + [\text{CO}_3^=]. \qquad (6.39)$$

Note carefully that this particular material balance equation is only valid for a pure solution of NaHCO_3. If we subtract Eq. (6.39) from Eq. (6.38), we get

$$[\text{H}_3\text{O}^+] + [\text{H}_2\text{CO}_3] = [\text{CO}_3^=] + [\text{OH}^-]. \qquad (6.40)$$

We must now express $[\text{H}_2\text{CO}_3]$ and $[\text{CO}_3^=]$ in terms of $[\text{HCO}_3^-]$, the largest of the concentrations, and therefore the easiest to guess. Using the first and second ionization-constant expressions in Eq. (6.40) we obtain

$$[\text{H}_3\text{O}^+] + \frac{[\text{H}_3\text{O}^+][\text{HCO}_3^-]}{K_1} = \frac{[\text{HCO}_3^-]}{[\text{H}_3\text{O}^+]} K_2 + \frac{K_w}{[\text{H}_3\text{O}^+]},$$

$$[\text{H}_3\text{O}^+]^2 \left(1 + \frac{[\text{HCO}_3^-]}{K_1}\right) = [\text{HCO}_3^-]K_2 + K_w,$$

$$[\text{H}_3\text{O}^+]^2 = \frac{[\text{HCO}_3^-]K_2 + K_w}{\left(1 + \dfrac{[\text{HCO}_3^-]}{K_1}\right)}. \qquad (6.41)$$

Now, $[\text{HCO}_3^-]/K_1 \gg 1$ for the usual concentrations of HCO_3^-. Therefore, the denominator in Eq. (6.41) is equal to $[\text{HCO}_3^-]/K_1$ to a good approximation. Also, $K_w = 10^{-14}$, but $K_2[\text{HCO}_3^-] > 10^{-11}$, for $[\text{HCO}_3^-] \geq 0.5\,M$, and under this condition the numerator in Eq. (6.41) is approximately equal to $[\text{HCO}_3^-]K_2$. Therefore, Eq. (6.41) becomes

$$[\text{H}_3\text{O}^+]^2 \simeq \frac{[\text{HCO}_3^-]K_2}{[\text{HCO}_3^-]/K_1} = K_1 K_2,$$

$$[\text{H}_3\text{O}^+] = \sqrt{K_1 K_2}. \qquad (6.42)$$

Thus, if the concentration of HCO_3^- is larger than about $0.1\,M$, the concentration of H_3O^+ in a pure solution of NaHCO_3 is given by the simple expression (6.42). However, even though the concentration of H_3O^+ is independent of the concentration of HCO_3^-, a pure solution of NaHCO_3 is *not* a buffer solution. Should any strong acid or strong base be added to the solution, the material balance expression Eq. (6.39) becomes invalid, and therefore Eq. (6.42) no longer applies.

A pure solution of NaHCO_3 or any other acid salt can be regarded as an equimolar mixture of a weak acid (HCO_3^-) and a weak base (HCO_3^-). The fact that the acid and the base are the same material has no particular importance. If this is true, we should be able to treat an equimolar mixture of a weak acid such as NH_4^+ and a weak base like OAc^- by the method we just applied to a pure solution of HCO_3^-. An equimolar solution of NH_4^+ and OAc^- would be obtained by dissolving the salt ammonium acetate in water. The ions then hydrolyze according to

$$NH_4^+ + H_2O = NH_3 + H_3O^+,$$
$$OAc^- + H_2O = HOAc + OH^-.$$

The charge balance equation for this system would be

$$[NH_4^+] + [H_3O^+] = [OAc^-] + [OH^-]. \qquad (6.43)$$

The material balance equation states that the sum of the concentrations of "ammonia material" must equal the sum of the concentrations of "acetate material":

$$[NH_4^+] + [NH_3] = [OAc^-] + [HOAc]. \qquad (6.44)$$

This expression is a consequence of the fact that in ammonium acetate there is one ammonium ion for each acetate ion.

Subtracting Eq. (6.44) from Eq. (6.43) gives

$$[H_3O^+] + [HOAc] = [NH_3] + [OH^-]. \qquad (6.45)$$

Now we can use the equilibrium constants for the ionization of acetic acid and ammonia to obtain

$$[HOAc] = \frac{[H_3O^+][OAc^-]}{K_a}, \qquad [NH_3] = \frac{[OH^-][NH_4^+]}{K_b} = \frac{K_w}{K_b}\frac{[NH_4^+]}{[H_3O^+]}.$$

Substitution in Eq. (6.45) then gives

$$[H_3O^+] + \frac{[H_3O^+][OAc^-]}{K_a} = \frac{K_w}{K_b}\frac{[NH_4^+]}{[H_3O^+]} + \frac{K_w}{[H_3O^+]},$$

$$[H_3O^+]^2\left(1 + \frac{[OAc^-]}{K_a}\right) = \frac{K_w}{K_b}[NH_4^+] + K_w.$$

If the salt is present in appreciable concentration, $[OAc^-]/K_a \gg 1$ and $[NH_4^+]/K_b \gg 1$, we have

$$[H_3O^+]^2\frac{[OAc^-]}{K_a} = \frac{K_w}{K_b}[NH_4^+],$$

$$[H_3O^+]^2 = \frac{K_w K_a}{K_b}\frac{[NH_4^+]}{[OAc^-]}.$$

If $K_a \cong K_b$, then $[NH_4^+] \cong [OAc^-]$, and

$$[H_3O^+] = \left(\frac{K_w K_a}{K_b}\right)^{1/2}. \qquad (6.46)$$

Equation (6.46) is analogous to Eq. (6.42) and shows that when we have an equimolar mixture of a weak acid and a weak base, the concentration of H_3O^+ is fixed, whether or not the acid and base are the same chemical species.

Questions. What is the principal net reaction by which *both* NH_4^+ and OAc^- are consumed? By using this reaction can you further justify the assertion that $[NH_4^+] \cong [OAc^-]$? Is near equality of K_a and K_b really a necessary condition for this assumption to hold?

Complex Ion Equilibria

In our analysis of multistage equilibria we have been exclusively concerned with the ionization of dibasic acids. However, the methods we have developed are applicable to other equilibria, as the following example will show.

Mercuric chloride, $HgCl_2$, dissolves in water and *remains largely undissociated*. Nevertheless, small amounts of $HgCl^+$, Hg^{++}, and Cl^- are formed by

$$HgCl_2(aq) = HgCl^+ + Cl^-, \qquad K_1 = 3.3 \times 10^{-7},$$
$$HgCl^+ = Hg^{++} + Cl^-, \qquad K_2 = 1.8 \times 10^{-7},$$

where

$$K_1 = \frac{[HgCl^+][Cl^-]}{[HgCl_2]}, \qquad K_2 = \frac{[Hg^{++}][Cl^-]}{[HgCl^+]}.$$

Thus the ionization of $HgCl_2$ is analogous to the dissociation of a dibasic acid.

We can find an expression for $[Cl^-]$ in a pure $HgCl_2$ solution by starting with the charge balance equation:

$$[Cl^-] = [HgCl^+] + 2[Hg^{++}].$$

Here we are neglecting $[H^+]$ and $[OH^-]$, which is essentially the same as assuming the solution is neutral or nearly so. The values of K_1 and K_2 suggest that $HgCl_2$ will be only very slightly dissociated, and therefore in a pure solution the concentration of $HgCl_2$ should be relatively large and easy to estimate. Thus we find expressions for $[HgCl^+]$ and $[Hg^{++}]$ in terms of $[HgCl_2]$:

$$[HgCl^+] = \frac{K_1[HgCl_2]}{[Cl^-]},$$
$$[Hg^{++}] = \frac{K_2[HgCl^+]}{[Cl^-]} = \frac{K_1K_2[HgCl_2]}{[Cl^-]^2}.$$

Next, we substitute these into the charge balance equation to get

$$[Cl^-] = \frac{K_1[HgCl_2]}{[Cl^-]} + 2\frac{K_1K_2[HgCl_2]}{[Cl^-]^2}$$
$$= \frac{K_1[HgCl_2]}{[Cl^-]}\left(1 + \frac{2K_2}{[Cl^-]}\right). \qquad (6.47)$$

Equation (6.47) is a cubic expression for $[Cl^-]$ which may be solved if $[HgCl_2]$ is known.

Let us calculate $[Cl^-]$ for a solution prepared by dissolving 0.1 mole of $HgCl_2$ in enough water to make one liter of solution. Then, since K_1 is small, $[HgCl_2] \cong$ 0.1 M. As a first approximation we neglect the term $2K_2/[Cl^-]$ in Eq. (6.47) and get

$$[Cl^-]^2 = K_1[HgCl_2] = 3.3 \times 10^{-7} \times 10^{-1}, \qquad [Cl^-] = 1.8 \times 10^{-4}\ M.$$

From this answer it appears that our approximation of letting $[HgCl_2]$ equal 0.1 M is justified. Moreover, since $2K_2/[Cl^-] = 1.8 \times 10^{-3}$, neglect of this term compared to unity is also a good approximation. Thus, even though K_1 and K_2 are nearly the same, the second ionization makes a negligible contribution to the chloride ion concentration in this pure solution of $HgCl_2$.

To complete the problem, we note that

$$[Hg^{++}] = K_1 K_2 \frac{[HgCl_2]}{[Cl^-]^2}$$
$$= \frac{3.3 \times 1.8 \times 10^{-7} \times 10^{-7} \times 10^{-1}}{3.3 \times 10^{-8}} = 1.8 \times 10^{-7}\ M,$$

and since $[Hg^{++}]$ is so small,

$$[HgCl^+] = [Cl^-] = 1.8 \times 10^{-4}\ M.$$

It is clear that this problem is very much analogous to the dissociation of a weak dibasic acid. Likewise, an equimolar mixture of $HgCl_2$ and Hg^{++} [from $Hg(NO_3)_2$] is essentially a pure solution of $HgCl^+$ and is analogous to, and can be mathematically treated like, a pure solution of $NaHCO_3$. In other words, fundamentally it does not matter whether we are faced with the stepwise dissociation of a complex species such as $HgCl_2$ or of a weak acid like H_2CO_3. The procedures used to find the equilibrium concentrations are virtually identical.

6.9 CONCLUSION

In this chapter we have developed systematic procedures for calculating concentrations of dissolved substances from equilibrium constants. This type of analysis is of practical value in a number of fields, in particular, analytical, inorganic, and biological chemistry. There is more to be learned from this subject than mechanical facility in making equilibrium calculations, however. The study of ionic equilibria demonstrates how approximations, made with due regard to the physical situation, can ease the business of solving problems. This is an extremely important general idea, for the quantitative problems of chemistry are difficult, and most often can be solved *only* by use of intelligent approximations. Therefore, the ability to sense the useful and valid approximations in any problem is a valuable skill that will be used repeatedly, and study of ionic equilibria can help develop this ability.

SUGGESTIONS FOR FURTHER READING

Bell, R. P., *Acids and Bases* (Methuen's Monograph). New York: Wiley, 1956.

Butler, J. N., *Solubility and pH Calculations*. Reading, Mass.: Addison-Wesley, 1964.

Butler, J. N., *Ionic Equilibrium*. Reading, Mass.: Addison-Wesley, 1964.

Drago, R. S., and N. A. Matwiyoff, *Acids and Bases*. Lexington, Mass.: D. C. Heath and Co., 1968.

King, E. J., *Qualitative Analysis and Electrolytic Solutions*. New York: Harcourt-Brace, 1959.

Nyman, C. J., and R. E. Hamm, *Chemical Equilibrium*. Lexington, Mass.: Raytheon Education Co., 1968.

PROBLEMS

6.1 Electrical conductivity measurements give the solubility of barium sulfate, $BaSO_4$, in pure water as 1.05×10^{-5} mole/liter at 25°C. Calculate the solubility product of barium sulfate.

6.2 Experiments show that in a saturated solution of barium fluoride, BaF_2, in pure water at 25°C, the barium ion concentration is 7.6×10^{-3} mole/liter. What is the concentration of fluoride ion in this solution? What is the solubility product of barium fluoride?

6.3 A saturated solution of lanthanum iodate, $La(IO_3)_3$, in pure water has a concentration of iodate ion equal to 2.07×10^{-3} mole/liter at 25°C. What is the concentration of La^{+3}? What is the solubility product of $La(IO_3)_3$?

6.4 The solubility product of magnesium hydroxide, $Mg(OH)_2$, is 1.8×10^{-11}. What is the solubility of $Mg(OH)_2$ in pure water? What is the concentration of OH^- in the saturated solution? What is the pH of this solution?

6.5 The solubility product of lead sulfate, $PbSO_4$, is 1.8×10^{-8}. Calculate the solubility of lead sulfate in (a) pure water; (b) 0.10 M $Pb(NO_3)_2$ solution; (c) 1.0×10^{-3} M Na_2SO_4 solution.

6.6 The solubility product of calcium fluoride is 1.7×10^{-10}. Calculate the solubility of CaF_2 in a solution of 4×10^{-4} M $Ca(NO_3)_2$ to within 5% accuracy. The method of successive approximations is useful here.

6.7 To a solution that contains 0.10 M Ca^{++} and 0.10 M Ba^{++}, sodium sulfate is added slowly. The solubility products of $CaSO_4$ and $BaSO_4$ are 2.4×10^{-5} and 1.1×10^{-10} respectively. What is the sulfate ion concentration at the instant the first solid precipitates? What is that solid? Neglect dilution and calculate the barium ion concentration present when the first precipitation of $CaSO_4$ occurs. Do you think it should be possible to separate Ca^{++} and Ba^{++} by selective precipitation of the sulfates?

6.8 The solubility product of silver bromate, $AgBrO_3$, is 5.2×10^{-5}. When 40.0 ml of a solution containing 0.100 M $AgNO_3$ is added to 60.0 ml of a 0.200 M $NaBrO_3$ solution, a precipitate of $AgBrO_3$ is formed. From the stoichiometry of the reaction, deduce the final concentration of bromate ion. What is the concentration of Ag^+ remaining in the solution?

6.9 Lead iodate, $Pb(IO_3)_2$, is a sparingly soluble salt with a solubility product of 2.6×10^{-13}. To 35.0 ml of a 0.150 M $Pb(NO_3)_2$ solution, 15.0 ml of a solution of 0.800 M KIO_3 are added, and a precipitate of $Pb(IO_3)_2$ formed. What are the concentrations of Pb^{++} and IO_3^- left in the solution?

6.10 Formic acid, HCO_2H, loses one proton upon ionization and has a dissociation constant of 1.8×10^{-4} at 25°C. Calculate the concentrations of HCO_2H, H_3O^+, HCO_2^-, and OH^- in (a) a solution made by adding 1.00 mole of HCO_2H to enough pure water to make 1.00 liter of solution; (b) a solution made by adding 1.00×10^{-2} mole of HCO_2H to sufficient water to make 1.00 liter of solution. Indicate the approximations made, and show that they are justified. Obtain answers that are within 5% of the exact values.

6.11 Ammonia, NH_3, produces hydroxyl ions in aqueous solution according to the reaction

$$NH_3 + H_2O = NH_4^+ + OH^-,$$

which has an equilibrium constant of 1.80×10^{-5}. Calculate the concentrations of NH_3, NH_4^+, H_3O^+, and OH^- in (a) a solution prepared from 0.010 mole of ammonia and 1.00 liter of water; (b) a solution prepared from 1.00×10^{-4} mole of ammonia and enough water to make 1.00 liter of solution. Check the validity of all approximations, and refine the answers by successive approximation until you feel you are within 5% of the exact answer.

6.12 Calculate the concentrations of H_3O^+, HOAc, and OAc^-, and OH^- in a solution that is prepared from 0.150 mole of HCl, 0.100 mole HOAc, and enough water to make 1.00 liter of solution. The dissociation constant for HOAc is 1.85×10^{-5}, and HCl is totally dissociated in aqueous solution.

6.13 In dilute aqueous solution sulfuric acid can be regarded as totally dissociated to H_3O^+ and HSO_4^-. The bisulfate ion, HSO_4^-, is itself a weak acid with a dissociation constant of 1.20×10^{-2}. Calculate the concentration of H_3O^+, HSO_4^-, $SO_4^=$, and OH^- in a solution prepared by dissolving 0.100 mole of H_2SO_4 in enough water to make 1.00 liter of solution.

6.14 To 1 liter of a solution containing 0.150 M NH_4Cl there is added 0.200 mole of solid NaOH. What are the ionic and molecular species that are in major concentration when equilibrium is reached? Calculate the concentrations of NH_3, OH^-, and NH_4^+ at equilibrium if the dissociation constant for ammonia is 1.8×10^{-5}.

6.15 Consider aqueous solutions of a strong (totally dissociated) acid such as HCl. At high concentrations, the acid itself is the only important source of H_3O^+, but at concentrations near 10^{-7} M and below, the ionization of water contributes appreciably to the concentration of H_3O^+. Use the charge balance expression and the relation $[H_3O^+][OH^-] = K_w$ to derive an exact equation for the concentration of H_3O^+.

6.16 Calculate the hydrolysis constant for the reaction

$$HCO_2^- + H_2O = HCO_2H + OH^-,$$

and find the concentrations of H_3O^+, OH^-, HCO_2^-, and HCO_2H in a solution of 0.15 M HCO_2Na. The dissociation constant of formic acid is 1.8×10^{-4}.

6.17 The dissociation constant of HCN, hydrocyanic acid, is 4.8×10^{-10}. What is the concentration of H_3O^+, OH^-, and HCN in a solution prepared by dissolving 0.160 mole of NaCN in 450 ml of water at 25°C?

6.18 Calculate the solubility of FeS in a saturated solution of H_2S which has a concentration of H_3O^+ equal to $1.0 \times 10^{-3} M$. The equilibrium constant for the dissociation of H_2S by the reaction

$$H_2S + 2H_2O = 2H_3O^+ + S^=$$

is 1.1×10^{-21}.

6.19 Calculate the solubility of lead sulfate in a solution of $0.100 M$ H_3O^+ by taking account of the reaction

$$HSO_4^- + H_2O = H_3O^+ + SO_4^=.$$

The solubility product for $PbSO_4$ is 1.8×10^{-8}. Be sure to justify all approximations.

6.20 The solubility of $Mg(OH)_2$ is increased by the addition of ammonium salts due to the reaction

$$Mg(OH)_2(s) + 2NH_4^+ = 2NH_3 + 2H_2O + Mg^{++}.$$

Calculate the equilibrium constant for this reaction if the solubility product of $Mg(OH)_2$ is 5.5×10^{-12}, and the basic dissociation constant of ammonia is 1.8×10^{-5}. Find the solubility of $Mg(OH)_2$ in a solution that contained $1.00 M$ NH_4Cl before addition of $Mg(OH)_2$. Refine the answer to within 5% of the exact value.

6.21 A solution is prepared by dissolving 0.200 mole of sodium formate, HCO_2Na, and 0.250 moles of formic acid, HCO_2H, in approximately 200($\pm$50) ml of water. Calculate the concentrations of H_3O^+ and OH^-. The dissociation constant of formic acid is 1.8×10^{-4}.

6.22 Into 1.00 liter of a solution of $0.250 M$ HCl is placed 0.600 mole of solid sodium acetate. Assume that no volume change occurs, and calculate the concentration of OAc^-, HOAc, H_3O^+, and OH^-.

6.23 A solution of an unknown acid was titrated with base and the equivalence point reached when 36.12 ml of $0.100 M$ NaOH had been added. Then 18.06 ml of $0.100 M$ HCl were added to the solution and the measured pH was found with a pH meter to be 4.92. Calculate the dissociation constant of the unknown acid.

6.24 Consider a solution of carbonic acid, whose initial concentration is $0.04 M$ H_2CO_3. A certain amount of base is added until the pH of the solution reaches 5. The first and second ionization constants of carbonic acid are 4.2×10^{-7} and 4.8×10^{-11} respectively. Calculate the following concentration *ratios*: $[HCO_3^-]/[H_2CO_3]$, $[CO_3^=]/[HCO_3^-]$, $[CO_3^=]/[H_2CO_3]$. Calculate the fraction of the total carbonate material that is present as H_2CO_3 at pH 5. Similarly, calculate the fraction of the total carbonate material that is present as HCO_3^- and as $CO_3^=$ at this pH. Repeat the calculations for pH 7, pH 9, and pH 11, and plot for each species the fraction present as a function of pH.

6.25 From the second ionization constant of carbonic acid, calculate the equilibrium constant for the hydrolysis of carbonate ion to the bicarbonate ion, HCO_3^-. From this, compute the bicarbonate and hydroxide ion concentrations in a $0.050 M$ Na_2CO_3 solution. Is the hydrolysis of HCO_3^- to H_2CO_3 important in this instance? Why?

6.26 Find the concentrations of H_3O^+ and OH^- in a solution prepared by adding 0.250 mole of Na_2CO_3 and 0.300 mole of $NaHCO_3$ to 250 ml of water.

6.27 A carbonate buffer solution is prepared by dissolving 30.0 gm of Na_2CO_3 in 350 ml of water and adding 150 ml of 1.00 M HCl. Calculate the pH of the solution.

6.28 An unknown student takes an unknown weight of an unknown weak acid, dissolves it in an unknown amount of water, and titrates it with a strong base of unknown concentration. When he has added 10.00 ml of base he notices that the concentration of H_3O^+ is 1.0×10^{-5} M. He continues the titration until he reaches the equivalence point for removal of one proton. At this time his buret reads 22.22 ml. What is the dissociation constant of the acid?

6.29 A solution is prepared by adding 2.05 gm of sodium acetate, $NaC_2H_3O_2$, to 100 ml of 0.100 M HCl solution. What is the resulting concentration of H_3O^+? A subsequent addition of 6.00 ml of 0.100 M HCl is made. What is the new concentration of H_3O^+?

6.30 Cuprous ion forms an ammonia complex which dissociates in stages, according to the reactions

$$Cu(NH_3)_2^+ = Cu(NH_3)^+ + NH_3, \quad K_1 = 2 \times 10^{-5},$$

$$Cu(NH_3)^+ = Cu^+ + NH_3, \quad K_2 = 6 \times 10^{-7}.$$

Consider now a pure solution of $Cu(NH_3)_2Cl$. (a) Give the relation between the concentrations of NH_3, $Cu(NH_3)^+$, and Cu^+. (b) Use this relation to express the concentration of NH_3 as a function of the concentration of $Cu(NH_3)_2^+$ and various constants. (c) Calculate the concentration of ammonia in a solution of 0.01 M $Cu(NH_3)_2Cl$ to a first approximation. Are further approximations necessary to obtain 5% accuracy?

6.31 The three dissociation constants for the successive ionization of phosphoric acid, H_3PO_4, are $K_1 = 7.5 \times 10^{-3}$, $K_2 = 6.2 \times 10^{-8}$, and $K_3 = 1 \times 10^{-12}$. From the equilibrium constants, determine the principal species of phosphate material (H_3PO_4, $H_2PO_4^-$, $HPO_4^=$, or PO_4^{-3}) at the following values of the pH: 1, 5, 10, 14. What is the pH of a solution of equimolar H_3PO_4 and $H_2PO_4^-$? Is the fraction of phosphate material present as $HPO_4^=$ and PO_4^{-3} important in such a solution?

6.32 What is the concentration of H_3O^+ in a pure solution of: (a) NaH_2PO_4; (b) Na_2HPO_4; (c) 1 M Na_3PO_4?

6.33 What is the concentration of H_3O^+ in a solution prepared by adding to 300 ml of 0.500 M H_3PO_4 (a) 250 ml of 0.300 M NaOH; (b) 500 ml of 0.500 M NaOH; (c) 40 ml of 1.00 M NaOH?

6.34 A solution contains a mixture of two weak acids, 0.05 M HA and 0.01 M HB, which have dissociation constants of 2×10^{-5} and 6×10^{-5}, respectively. Find the charge balance equation for this system, and use it to derive an expression which gives the concentration of H_3O^+ as a function of the concentrations of HA and HB and various constants. What is the concentration of H_3O^+ in this solution?

CHAPTER 7

OXIDATION-REDUCTION REACTIONS

In Chapter 6 we pointed out that acid-base reactions form a large class of chemical processes that have in common the act of *proton transfer*. There is an equally large and important group of reactions that all involve *electron transfer* in either an obvious or subtle way. We are referring, of course, to oxidation-reduction reactions. For instance,

$$Zn + Cu^{++} = Zn^{++} + Cu \tag{7.1}$$

is an example of an oxidation-reduction reaction in which the feature of electron transfer is clear, while

$$2CO + O_2 = 2CO_2$$

is also an oxidation-reduction reaction, but one in which electron transfer is not so obvious.

One might ask how the term "oxidation" became generalized so as to apply to reactions in which electrons were transferred, regardless of whether oxygen was involved. Once it is recognized that in the oxidation

$$Zn + \tfrac{1}{2}O_2 = ZnO,$$

each zinc atom has lost two electrons and has become a zinc ion Zn^{++}, and that the same change occurs in

$$Zn + Cl_2 = ZnCl_2,$$
$$Zn + 2H^+(aq) = Zn^{++}(aq) + H_2,$$

then it seems logical to refer to all processes in which zinc loses electrons as oxidations. The same argument can be applied to the reactions of any substance, and so it is a useful generalization to say that a chemical substance is *oxidized when it loses electrons*.

The loss of electrons by one substance must be accompanied by the gain of electrons by some other reagent, and this latter process is called *reduction*. In reaction (7.1) the metallic zinc is oxidized to Zn^{++}, and Cu^{++} is reduced to metallic copper. It is common to call the substance that brings about the reduction of another a *reducing agent* or a *reductant*, and the substance responsible for oxidizing another is called an *oxidizing agent* or an *oxidant*. In reaction (7.1), zinc is a reductant (and is oxidized) and Cu^{++} is an oxidant (and is reduced).

7.1 OXIDATION STATES

The oxidation state concept is derived from the necessity to describe the changes brought about by oxidation-reduction reactions. For simple monatomic substances, it is convenient and direct to define the oxidation state or oxidation number as the atomic number of the atom minus the number of orbital electrons, or more simply, as the net charge on the atom. Thus the oxidation states of $S^=$, Cl^-, Cu^+, Co^{++}, and Fe^{+3} are -2, -1, $+1$, $+2$, and $+3$ respectively. The oxidation state of an element in any of its allotropic forms is always zero.

While there is a direct relation between oxidation state and the net charge on a monatomic species, the extension of the oxidation state concept to polyatomic species is less clear-cut. We might ask what the oxidation state of each atom in H_2O or NO_3^- is, for example. If we insist that the oxidation state is the actual charge on an atom in a molecule, then to assign oxidation states would at the very least require a detailed knowledge of the exact charge distribution in the molecule. This information is virtually never available. However, we can extend the oxidation state concept to polyatomic systems if we abandon the idea that the oxidation state is the *true* charge on an atom. We have only to decide *arbitrarily* that in a compound such as NO, the oxygen atom will be assigned an oxidation state of -2, just as it is in the compound ZnO. This is equivalent to saying that the oxygen atom in NO is arbitrarily assigned 10 of the 15 electrons in the molecule. The nitrogen atom must be assigned 5 electrons, 2 less than its atomic number, and consequently it has an oxidation state of $+2$.

The assignment of -2 as the oxidation state of the oxygen atom in NO was arbitrary, and we must not suppose that there is actually a charge of -2 on the oxygen atom, and a $+2$ charge on the nitrogen atom. Indeed, the experimental evidence is to the contrary, and indicates that the 15 electrons are nearly equally distributed around the two nuclei. However, even though the assignment of oxidation numbers in polyatomic molecules is an arbitrary procedure, and may have very little to do with the actual charge distribution in these species, it is still useful, as we shall see. Here then is the set of rules used to assign oxidation states in polyatomic molecules:

1. The oxidation state of all elements in any allotropic form is zero.

2. The oxidation state of oxygen is -2 in all its compounds except peroxides like H_2O_2 and Na_2O_2.

3. The oxidation state of hydrogen is $+1$ in all its compounds except those with the metals, where it is -1.

4. All other oxidation states are selected so as to make the algebraic sum of the oxidation states equal to the net charge on the molecule or ion.

It is also useful to remember that certain elements almost always display the same oxidation state, $+1$ for the alkali metals, $+2$ for the alkaline earth metals, and -1 for the halogens, except when they are combined with oxygen.

As an illustration of the application of these rules, let us deduce the oxidation states of Cl and N in the ions ClO^-, NO_2^-, and NO_3^-. In the case of ClO^-, we first assign oxygen an oxidation state of -2, and then deduce the value for chlorine by rule 3:

$$\text{oxidation state of O} + \text{oxidation state of Cl} = -1,$$
$$-2 + \text{oxidation state of Cl} = -1,$$
$$\text{oxidation state of Cl} = -1 + 2 = +1.$$

By a similar procedure we can decide that in NO_2^- the oxidation state of nitrogen is $+3$, and in NO_3^-, it is $+5$.

To see one of the uses of oxidation numbers, consider the reaction

$$ClO^- + NO_2^- = NO_3^- + Cl^-.$$

The net charge on the chlorine and nitrogen containing ions is the same in products and reactants; so if this reaction is an oxidation-reduction process, the feature of electron transfer is not at all obvious. In fact, one might look upon this reaction as an *oxygen atom* transfer from ClO^- to NO_2^-. However, by using oxidation numbers we can see that chlorine has been reduced from the $+1$ state in ClO^- to the -1 state in Cl^-, while nitrogen has been oxidized from the $+3$ state in NO_2^- to the $+5$ state in NO_3^-.

Contrast this example with the reaction

$$2CCl_4 + K_2CrO_4 = 2Cl_2CO + CrO_2Cl_2 + 2KCl.$$

Is this an oxidation-reduction reaction? The introduction of oxygen into a carbon compound certainly makes it look so. To be sure, let us calculate the oxidation state of chromium in K_2CrO_4 and CrO_2Cl_2. In the first instance we need consider only the ion $CrO_4^=$. Assignment of -2 to the four oxygen atoms means that chromium must have an oxidation number of $+6$, if the net charge on the ion is -2. In chromyl chloride, CrO_2Cl_2, assignment of -2 to each oxygen atom and -1 to each chlorine atom requires that the chromium ion again be in the $+6$ oxidation state. Therefore, chromium has not changed its oxidation state in the reaction. By following the convention that the oxidation state of chlorine is -1 except when it is combined directly with oxygen, we conclude that chlorine has not changed its oxidation state in the reaction. Finally, carbon has remained in the same oxidation state, $+4$ in CCl_4 and $+4$ in Cl_2CO. Thus the reaction is not an oxidation-reduction process.

These two examples show us that one of the important uses of the oxidation state concept is to provide an electron bookkeeping device that allows us to recognize an oxidation-reduction reaction. A second use of the concept is to provide a framework within which chemical similarities may be recognized and chemical properties correlated. For example, the acidic properties of the transition metal ions in the $+2$ state are quite similar, and the same may be said for the $+3$ ions. However, the $+3$ ions as a group are distinctly more acidic than the $+2$ ions, and this increase of acidity with oxidation number is found quite generally in the chemistry of other elements. In our study of the descriptive chemistry of the elements we shall find other examples of the correlation between chemical behavior and oxidation state. Finally, the oxidation state concept is useful in balancing the equations of oxidation-reduction reactions, as we shall see later in this chapter.

7.2 THE HALF-REACTION CONCEPT

A most remarkable feature of oxidation-reduction reactions is that they can be carried out with the reactants separated in space, and linked only by an electrical connection. Consider Fig. 7.1, an illustration of a galvanic cell which involves the reaction between metallic zinc and cupric ion:

$$Zn(s) + Cu^{++}(aq) \rightarrow Cu(s) + Zn^{++}(aq).$$

The cell consists of two beakers, one of which contains a solution of Cu^{++} and a copper rod, the other a Zn^{++} solution and a zinc rod. A connection is made between the two solutions by means of a "salt bridge," a tube containing a solution of an electrolyte, generally NH_4NO_3 or KCl. Flow of the solution from the salt bridge is prevented either by plugging the ends of the bridge with glass wool, or by using a salt dissolved in a gelatinous material as the bridge electrolyte. When the two metallic rods are connected through an ammeter, there is immediately evidence that a chemical reaction is occurring. The zinc rod starts to dissolve, and copper is deposited on the copper rod. The solution of Zn^{++} becomes more concentrated, and the solution of Cu^{++} becomes more dilute. The ammeter indicates that electrons are flowing from the zinc rod to

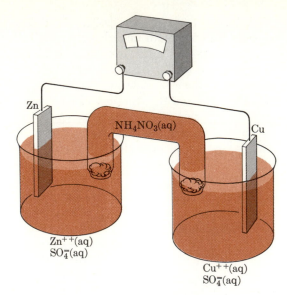

FIG. 7.1   A galvanic cell.

the copper rod. This activity continues as long as the electrical connection and the salt bridge are maintained, and visible amounts of reactants remain.

Now let us analyze what happens in each beaker more carefully. We note that electrons flow from the zinc rod through the external circuit, and that zinc ions are produced as the zinc rod dissolves. We can summarize these observations by writing

$$\text{Zn} = \text{Zn}^{++}(\text{aq}) + 2\text{e}^{-} \qquad \text{(at the zinc rod)}.$$

Also, we observe that electrons flow to the copper rod as cupric ions leave the solution and metallic copper is deposited. We can represent these occurrences by

$$2\text{e}^{-} + \text{Cu}^{++}(\text{aq}) = \text{Cu} \qquad \text{(at the copper rod)}.$$

In addition, we must examine the purpose of the salt bridge. Since zinc ions are produced as electrons leave the zinc electrode, we have a process which tends to produce a net positive charge in the left beaker. Similarly, the arrival of electrons at the copper electrode and their reaction with cupric ions tends to produce a net negative charge in the right beaker. The purpose of the salt bridge is to prevent any net charge accumulation in either beaker by allowing negative ions to leave the right beaker, diffuse through the bridge, and enter the left beaker. At the same time, there can be a diffusion of positive ions from left to right. If this diffusional exchange of ions did not occur, the net charge accumulating in the beakers would immediately stop the electron flow through the external circuit, and the oxidation-reduction reaction would stop. Thus, while the salt bridge does not participate chemically in the cell reaction, it is necessary if the cell is to operate.

The analysis of the operation of this galvanic cell quite naturally suggests that the overall oxidation-reduction reaction can be separated into two half-reactions:

$$
\begin{array}{ll}
\mathrm{Zn} = \mathrm{Zn}^{++} + 2e^- & \text{oxidation} \\
\underline{2e^- + \mathrm{Cu}^{++} = \mathrm{Cu}} & \text{reduction} \\
\mathrm{Zn} + \mathrm{Cu}^{++} = \mathrm{Zn}^{++} + \mathrm{Cu} &
\end{array}
$$

Many other oxidation-reduction reactions can be carried out successfully in galvanic cells, and it is natural to think of these cell processes as separated into the half-reactions which occur at the two electrodes. However, any oxidation-reduction reaction occurring under any circumstances can be *conceptually* separated into half-reactions. The benefits are these:

1. The half-reaction concept can aid greatly in the balancing of oxidation-reduction equations.

2. Half-reactions form the framework used to compare the strength of various oxidizing and reducing agents.

In the next two sections we shall examine these items in detail.

7.3 BALANCING OXIDATION-REDUCTION REACTIONS

The "half-reaction method" of balancing oxidation-reduction equations consists of four steps:

1. Identifying the species being oxidized or reduced.
2. Writing separate half-reactions for the oxidation and reduction processes.
3. Balancing these half-reactions with respect to atoms and electrical charge.
4. Combining the balanced half-reactions to form the overall net oxidation-reduction reaction.

We shall illustrate these steps by balancing the equation for the reaction

$$
\mathrm{H_2O_2} + \mathrm{I}^- \rightarrow \mathrm{I_2} + \mathrm{H_2O},
$$

which occurs in an acidic aqueous solution.

Use of oxidation numbers tells us immediately that the iodide ion is *oxidized* from the -1 state to elemental iodine, whose oxidation number is zero. Similarly, the oxidation state of oxygen in $\mathrm{H_2O_2}$ is -1, while in $\mathrm{H_2O}$ it is -2, so that hydrogen peroxide is being *reduced* to water. Therefore, we have

$$
\begin{array}{ll}
\mathrm{I}^- \rightarrow \mathrm{I_2}, & \text{oxidation,} \\
\mathrm{H_2O_2} \rightarrow \mathrm{H_2O}, & \text{reduction.}
\end{array}
$$

With the oxidation and reduction processes identified, and "skeleton" half-reactions written, we pass to the third step of the balancing procedure. We can balance the oxidation of iodide chemically by writing

$$2I^- \rightarrow I_2,$$

but this is still not a balanced equation, since the net charge on the reactants and products is not the same. However, the oxidation process amounts to removing one electron from each of two iodide ions, so we have

$$2I^- = I_2 + 2e^-,$$

as a half-reaction balanced with respect to atoms and net charge.

We can start balancing the reduction process by writing

$$H_2O_2 \rightarrow 2H_2O,$$

which balances with respect to oxygen atoms, but not with respect to hydrogen atoms. We must add some form of hydrogen to the left-hand side; the problem is to decide on the appropriate reagent. Examination of the overall reaction shows that hydrogen appears only in the +1 state in both reactants and products. Since hydrogen is not oxidized or reduced in the reaction, any hydrogen we introduce in balancing the half-reactions must be in the +1 state. Since the reaction occurs in acidic aqueous solution, protons are available, so we can write*

$$2H^+ + H_2O_2 \rightarrow 2H_2O,$$

which balances the half-reaction chemically, but not electrically, since the net charge on each side is not the same. We can remedy this by adding two electrons to the left-hand side to obtain

$$2e^- + 2H^+ + H_2O_2 = 2H_2O,$$

which is the balanced half-reaction for the reduction process.

Before we carry out step 4, we note that in *balancing* the separate half-reactions, we made use of the general requirements of atom and charge conservation, but did not use the oxidation number concept. Let us see whether the numbers of electrons introduced into these half-reactions on the basis of charge balance are consistent with the changes in oxidation numbers. In the oxidation of iodide ion to iodine, the oxidation number change is from -1 to 0 for each of two iodine atoms, and this is consistent with the two electrons appearing on the right-hand side of the half-reaction. In the reduction of hydrogen peroxide, each of two oxygen atoms is reduced from the -1 to the -2 state, which means

---

* Here we are reverting to the practice of writing the aqueous proton simply as H^+ instead of H_3O^+. This procedure simplifies the appearance of oxidation-reduction equations by decreasing the number of water molecules that must be written.

that two electrons are needed for each hydrogen peroxide molecule, as we have found already. Therefore the half-reactions we have written are consistent with the principles of atom and charge conservation and our conventions regarding oxidation states.

To write a balanced net reaction, we must combine the two half-reactions so that electrons do not appear as reactants or products. This can be done very simply in our present example; we merely add the two half-reactions as they stand:

$$2I^- = I_2 + 2e^-$$
$$2e^- + 2H^+ + H_2O_2 = 2H_2O$$
$$\overline{2H^+ + 2I^- + H_2O_2 = 2H_2O + I_2}$$

Since each half-reaction was balanced chemically and electrically, so is the net reaction.

The use of half-reactions permits us to balance equations using only the principles of atom and charge conservation, and to reserve oxidation numbers for use as a check on our work. As a more challenging example of the procedure, consider

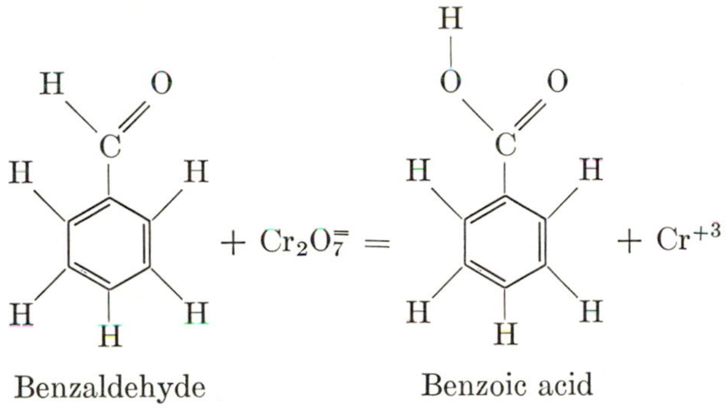

Benzaldehyde   Benzoic acid

It might be rather tedious to compute the oxidation numbers in benzaldehyde and benzoic acid, and we do not need to do so in order to balance the equation. We start with

$$C_6H_5CHO \rightarrow C_6H_5COOH$$

and balance chemically by writing

$$C_6H_5CHO + H_2O \rightarrow C_6H_5COOH + 2H^+,$$

which is a way of introducing the oxygen atom in the -2 state required on the left-hand side. The charge balance requirement results in two electrons appearing on the right-hand side, so the completed half-reaction is

$$C_6H_5CHO + H_2O = C_6H_5COOH + 2H^+ + 2e^-.$$

To check, we note that the *sum* of the oxidation numbers of seven carbon atoms in C_6H_5CHO must be $-(6-2) = -4$, since there are six hydrogen atoms in the $+1$ state and one oxygen atom in the -2 state, and the molecule is uncharged. Similarly, the *sum* of the oxidation numbers of the seven carbon atoms in C_6H_5COOH is $-(6-4) = -2$, since there are two oxygen atoms in the -2 state in benzoic acid. Therefore the net change in the oxidation numbers of the carbon atoms is $-2 + 4 = +2$, which means that the molecule loses two electrons as it is oxidized. This is consistent with the way in which our half-reaction is written.

To complete the example, we must balance

$$Cr_2O_7^= \rightarrow 2Cr^{+3}.$$

Introducing H^+ on the left and H_2O on the right gives

$$14H^+ + Cr_2O_7^= \rightarrow 2Cr^{+3} + 7H_2O.$$

To balance the half-reaction electrically we need 6 electrons on the left-hand side, so we get

$$6e^- + 14H^+ + Cr_2O_7^= = 2Cr^{+3} + 7H_2O.$$

Finally, to eliminate electrons from the overall reactions we must add the two half-reactions as follows:

$$3 \times [C_6H_5CHO + H_2O = C_6H_5COOH + 2H^+ + 2e^-]$$
$$\underline{1 \times [6e^- + 14H^+ + Cr_2O_7^= = 2Cr^{+3} + 7H_2O]}$$
$$3C_6H_5CHO + Cr_2O_7^= + 8H^+ = 3C_6H_5COOH + 2Cr^{+3} + 4H_2O$$

Our two examples have shown that in order to achieve material balance with respect to oxygen in the -2 state and hydrogen in the $+1$ state, we may introduce protons and water molecules as needed on either side of the half-reactions. This is true as long as the reaction being balanced is taking place in acidic aqueous solution. However, for reactions which occur in basic solution, the procedure is slightly different, as we shall see from our next example. We shall balance

$$ClO^- + CrO_2^- \rightarrow CrO_4^= + Cl^- \qquad \text{(basic solution)}.$$

The half-reaction involving chlorine is

$$ClO^- \rightarrow Cl^-.$$

Oxygen in the -2 state must appear in the products, and since we are dealing with a basic solution we might introduce it either as OH^- or H_2O. In order to avoid deciding which should be used, we first use oxidation numbers to decide how many electrons must appear, then use charge balance to decide where and how much OH^- must be used, and finally complete the balancing by adding H_2O where required by the atom conservation principle.

Thus, we say that the reduction of chlorine from the $+1$ state in ClO^- to the -1 state in Cl^- requires two electrons on the left-hand side of the equation:

$$2e^- + ClO^- \rightarrow Cl^-.$$

It is now clear that to achieve charge balance, we must add OH^- to the right-hand side:

$$2e^- + ClO^- \rightarrow Cl^- + 2OH^-.$$

Finally, chemical balance is reached by adding water to the left-hand side, which gives

$$2e^- + H_2O + ClO^- = Cl^- + 2OH^-.$$

It is true that we might have started with

$$ClO^- \rightarrow Cl^-,$$

and immediately achieved chemical balance by adding a water molecule to the left, and two hydroxide ions on the right:

$$ClO^- + H_2O \rightarrow Cl^- + 2OH^-.$$

Then by charge balance two electrons are required on the left, and the balanced half-reaction results, without the use of oxidation numbers. However, the immediate introduction of OH^- and H_2O is often a confusing procedure, and, as we have pointed out, it can be simplified by using oxidation numbers in the balancing process.

To finish the problem, we balance

$$CrO_2^- \rightarrow CrO_4^=,$$

by noting that the chromium atom changes from the $+3$ to the $+6$ state, so three electrons are required on the right:

$$CrO_2^- \rightarrow CrO_4^= + 3e^-.$$

Charge balance requires that $4OH^-$ must appear on the left to give

$$4OH^- + CrO_2^- \rightarrow CrO_4^= + 3e^-,$$

while material balance requires $2H_2O$ on the right:

$$4OH^- + CrO_2^- = CrO_4^= + 2H_2O + 3e^-.$$

To obtain the net reaction, we combine the half-reactions by writing

$$\begin{array}{r} 3 \times [2e^- + H_2O + ClO^- = Cl^- + 2OH^-] \\ 2 \times [4OH^- + CrO_2^- = CrO_4^= + 2H_2O + 3e^-] \\ \hline 2OH^- + 3ClO^- + 2CrO_2^- = 3Cl^- + 2CrO_4^= + H_2O \end{array}$$

As our final example, we choose a disproportionation reaction, where one substance is both oxidized and reduced:

$$P_4 + OH^- \rightarrow PH_3 + H_2PO_2^-.$$

The oxidation process is

$$P_4 \rightarrow 4H_2PO_2^-.$$

In the hypophosphite anion, $H_2PO_2^-$, phosphorus is in the $+1$ oxidation state, so that four electrons must be produced in the half-reaction as written:

$$P_4 \rightarrow 4H_2PO_2^- + 4e^-.$$

We need only add $8OH^-$ to the left-hand side to achieve charge and chemical balance:

$$8OH^- + P_4 = 4H_2PO_2^- + 4e^-.$$

The reduction reaction is

$$P_4 \rightarrow 4PH_3.$$

Since phosphorus changes from the 0 state to the -3 state, 12 electrons are needed on the left-hand side and 12 hydroxide ions must be added to the right-hand side of the equation. This gives us

$$12e^- + 12H_2O + P_4 = 4PH_3 + 12OH^-$$

when the reaction is balanced chemically. To obtain the net reaction we write

$$\begin{array}{r} 3 \times [8OH^- + P_4 = 4H_2PO_2^- + 4e^-] \\ 1 \times [12e^- + 12H_2O + P_4 = 4PH_3 + 12OH^-] \\ \hline 12OH^- + 4P_4 + 12H_2O = 12H_2PO_2^- + 4PH_3 \end{array}$$

The four examples we have discussed illustrate most of the difficulties that can arise in balancing oxidation-reduction equations.

7.4 GALVANIC CELLS

In Section 7.2 we used the qualitative features of galvanic cells to show the natural origin of the half-reaction concept. In this section we shall discuss electrochemical cells more thoroughly to see how they can be used to give us a quantitative comparison of the strengths of oxidizing and reducing agents.

First let us examine a few of the common types of electrodes that are used in galvanic cells. Very often the electrodes are metals that are "active" in the operation of the cell. That is, the metallic electrodes are dissolved or formed as the cell reaction proceeds. As an example, we already have the zinc and copper rod electrodes which are respectively consumed and formed as the reaction

$$Zn + Cu^{++} = Zn^{++} + Cu$$

runs from left to right.

Also common are inert or "sensing" electrodes which are left unchanged by the net cell reaction. For example, consider the cell shown in Fig. 7.2. In the left beaker there is a mixture of ferrous and ferric ion solutions and a platinum strip, while in the right beaker there is the familiar copper electrode in contact with a solution of cupric ion. As the cell operates, copper metal is oxidized and ferric ion is reduced:

$$Cu + 2Fe^{+3} = 2Fe^{++} + Cu^{++}.$$

Thus at the platinum electrode, ferric ions acquire electrons and become ferrous ions, while the electrode remains unchanged. In order to remain unchanged by the cell reaction, the electrode must be made of inert material; platinum and carbon are the two substances most commonly used.

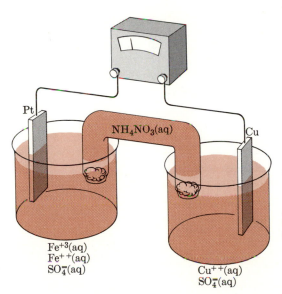

Pt

$NH_4NO_3(aq)$

Cu

$Fe^{+3}(aq)$
$Fe^{++}(aq)$
$SO_4^-(aq)$

$Cu^{++}(aq)$
$SO_4^-(aq)$

A galvanic cell. The half-cell on the left makes use of a platinum strip as an inert sensing electrode.

FIG. 7.2

A third common type is called a "gas electrode" and is actually quite closely related to the inert electrode just discussed. Figure 7.3 shows a hydrogen gas electrode operating in conjunction with a copper half-cell. The overall cell reaction in this case is

$$H_2(g) + Cu^{++}(aq) = Cu(s) + 2H^+(aq).$$

The hydrogen gas electrode is a piece of platinum whose surface is saturated by hydrogen gas at 1-atm pressure. The surface of the electrode serves as a place where hydrogen molecules can be converted to protons by the reaction

$$H_2(g) = 2H^+(aq) + 2e^-.$$

Depending on the direction of the overall cell reaction, the reverse process may also occur at the gas electrode. Thus the platinum metal itself is left unchanged, and serves to deliver or remove electrons as needed. In order to increase the rate at which oxidation or reduction can occur, the surface area of the gas electrode is increased by the deposition of finely divided platinum.

Now we return to the cell made up of the zinc and copper half-cells shown in Fig. 7.1, and this time imagine the electrodes connected to the terminals of a voltmeter. A few experiments would show that at a constant temperature, the voltage of the cell is a function of the *ratio* of the concentrations of the zinc and cupric ions. If the temperature is 25°C, and the concentrations are equal, the voltmeter reads 1.10 volts. If the zinc ion concentration is increased, or the concentration of cupric ion decreased, the voltage *decreases*, and vice versa.

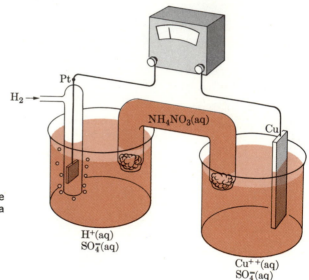

FIG. 7.3 A galvanic cell in which the half-cell on the left uses a hydrogen gas electrode.

Let us now imagine the copper–cupric ion half-cell replaced by a half-cell consisting of a silver wire immersed in a solution of silver nitrate. Once again experiments would show us that the voltage of the cell depends on the ratio of the ion concentrations, and when these are equal, the cell potential is 1.56 volts. This is substantially larger than the voltage produced by the zinc-copper cell operating under the same concentration conditions. Thus we see that the voltage of a galvanic cell is characteristic both of the chemical substances involved in the cell reaction and of their concentrations. To facilitate comparison of different galvanic cells, each should be characterized by a voltage measured under some set of standard conditions of concentration and temperature. The

standard conditions that have been chosen are 1-M concentration* for all dissolved materials, 1-atm pressure for all gases, and for solids, their most stable form at 25°C. The voltage measured under these conditions is called the **standard cell potential,** and is given the symbol $\Delta\varepsilon^0$.

The standard cell potential is a very useful and important quantity. In the first place, $\Delta\varepsilon^0$ in part determines the amount of work that the galvanic cell can do when it is operating under standard conditions. Imagine that the terminals of the cell are connected to an electric motor of 100% efficiency. Then, when a current i flows through the voltage difference $\Delta\varepsilon^0$ for a time t, the work performed is

$$\Delta\varepsilon^0 \times i \times t = \text{electrical work,}$$

and since the product of current and time is the total charge q passed,

$$\Delta\varepsilon^0 \times q = \text{electrical work.}$$

Thus the work which an electrochemical cell can do is given by the product of its voltage and the amount of charge it can pass. When a cell operates under standard conditions, its voltage $\Delta\varepsilon^0$ depends only on the chemical nature of the reactants and products. On the other hand, the amount of charge q that a given cell can deliver depends on the amount (not the concentration) of material available for the cell reaction. Therefore, of the factors that determine the ability of an electrochemical cell to do work, only $\Delta\varepsilon^0$ is directly related to the chemical nature of the reacting species.

The most important aspect of the standard cell potential is that it can be taken as a quantitative measure of the tendency of reactants *in their standard states* to form products *in their standard states*. In short, $\Delta\varepsilon^0$ represents the *driving force* of the chemical reaction. Figure 7.4 demonstrates the sense of this statement. A zinc-copper standard cell, whose $\Delta\varepsilon^0$ is 1.10 volts, is connected to an independent, variable voltage supply such that the variable voltage opposes the cell voltage. An ammeter indicates the direction of the flow of electrons.

When the variable voltage is less than 1.10 volts, the ammeter shows that electrons flow from the zinc electrode through the external circuit to the copper electrode. Therefore the spontaneous cell reaction under these conditions is

$$\text{Zn} + \text{Cu}^{++} \rightarrow \text{Cu} + \text{Zn}^{++},$$

just as it is when the electrodes are connected by a direct wire, or when metallic zinc is added to a solution of cupric ion.

If the variable voltage is increased, we find that when it reaches 1.10 volts, no current flows through the ammeter, and no net cell reaction occurs. The chemical "driving force" causing reactants to go to products is opposed by an

---

* Actually, not 1 molar but 1 molal. However, the difference between the two is small and can be neglected for our purposes.

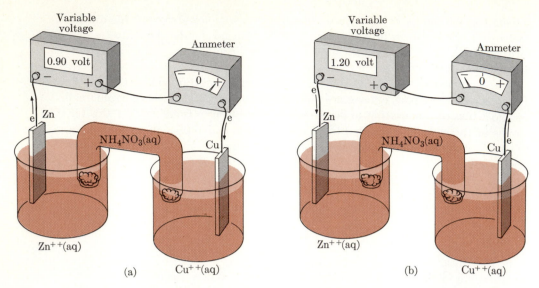

FIG. 7.4 The effect of opposing a cell with an external variable voltage. In (a) the external voltage is less than the cell voltage and electron flow is in the clockwise direction. In (b) the external voltage exceeds the cell voltage and the electron flow is reversed.

external electromotive force of equal magnitude. As the variable voltage is increased further and becomes greater than 1.10 volts, the ammeter shows that the direction of the electron flow is *from* the copper electrode *to* the zinc electrode. Therefore, copper is being converted to cupric ion, and zinc ion to zinc metal, so the cell reaction is

$$Cu + Zn^{++} \rightarrow Cu^{++} + Zn.$$

This is just the reverse of the spontaneous reaction of a short-circuited cell.

It seems reasonable to take the cell potential as a measure of the chemical driving force of the reaction, for when the cell is opposed by a numerically greater electromotive force, the spontaneous cell reaction is reversed, and an "electrolysis" occurs. However, we must be careful to remember that at best the *standard cell potential* measures the tendency of reactants in *their standard states* to form products in *their standard states*, and that the driving force of the reaction for any other states of the reactants and products will in general be different.

We have been emphasizing the significance of the magnitude of the standard cell potentials, and now we must introduce the convention concerning the sign of $\Delta \varepsilon^0$. If a reaction proceeds spontaneously from left to right as written, its $\Delta \varepsilon^0$ is given a positive sign, as in

$$Zn + Cu^{++} = Cu + Zn^{++}, \qquad \Delta \varepsilon^0 = +1.10 \text{ volts.}$$

If the spontaneous direction is from right to left, $\Delta\varepsilon^0$ is given a negative sign. Therefore we would have

$$Cu + Zn^{++} = Cu^{++} + Zn, \qquad \Delta\varepsilon^0 = -1.10 \text{ volts.}$$

Thus we can say that the more positive $\Delta\varepsilon^0$ is, the greater is the tendency of the reaction to proceed from left to right.

In the case of the copper-zinc cell, the value of $\Delta\varepsilon^0$ can be accepted as a measure of the tendency of zinc metal to lose electrons and become zinc ion, *and* of cupric ion to accept electrons and become copper metal. In other words, $\Delta\varepsilon^0$ is a simultaneous representation of the strength of zinc metal as a reductant, *and* of the strength of cupric ion as an oxidant. However, in order to compare the strengths of various oxidizing agents or reducing agents, it would be helpful if we could obtain a measure of the tendency of a given *half-reaction* to proceed. That is, we would like to have available *standard half-cell potentials*, rather than the values of $\Delta\varepsilon^0$.

We can properly regard any value of $\Delta\varepsilon^0$ as the sum of two half-cell potentials, one associated with each of the half-reactions in the cell. But because every galvanic cell involves *two* half-reactions, we can never measure absolute values of the individual half-cell potentials, only the sum of two of them. However, it is still possible to obtain numerical values for the half-cell potentials just by arbitrarily assigning one half-reaction a potential of zero. This procedure is completely analogous to choosing Greenwich, England, as the zero of longitude, for although only differences in longitude can be measured, once one point is *assigned* a definite longitude, all other points also assume definite numerical values. Accordingly, it has been decided to assign the hydrogen gas–hydrogen ion half-reaction,

$$H_2(1 \text{ atm}) = 2H^+(1\ M) + 2e^-,$$

a half-cell potential ε^0 of zero volts, when the reactants and products are in their standard states.

To assign half-cell potentials to other half-reactions we proceed in the following manner. We first measure the magnitude of the standard voltage generated when each half-cell is combined with the hydrogen half-cell. Thus, when the zinc–zinc ion half-cell operates with the standard hydrogen electrode, we note that the measured voltage is 0.76 volt, and that electrons flow from the zinc electrode to the hydrogen electrode. Hence the spontaneous cell reaction is

$$Zn(s) + 2H^+(1\ M) \rightarrow Zn^{++}(1\ M) + H_2(1 \text{ atm}).$$

Similarly, measurement of the voltage of the copper–cupric ion half-cell working with the hydrogen electrode shows 0.34 volt, and that electrons flow from the hydrogen electrode to the copper electrode. The direction of the spontaneous cell reaction must be

$$Cu^{++}(1\ M) + H_2(1\ atm) \rightarrow Cu + 2H^+(1\ M).$$

The information we now have allows us to say that the absolute *magnitude* of the zinc–zinc ion half-cell potential is 0.76 volt, and that the absolute *magnitude* of the copper–cupric ion half-cell potential is 0.34 volt. But should these voltages be given the same sign? The answer is no, for the following reasons. We wish, in the case of half-cell potentials, to use the idea that the more positive the potential, the greater the driving force is for the reaction to proceed from left to right. Therefore, if we adopt the *convention* that all half-reactions be written as reductions, as in

$$Zn^{++} + 2e^- = Zn,$$
$$2H^+ + 2e^- = H_2,$$
$$Cu^{++} + 2e^- = Cu,$$

then the magnitude and sign of the half-cell potentials must reflect the relative tendencies of the reactions to proceed from left to right. Now, the experiments mentioned above tell us that in the zinc-hydrogen cell, zinc metal spontaneously reduces hydrogen ion. Therefore, the half-cell potential of the hydrogen ion–hydrogen reaction must be more positive than that of the zinc ion–zinc reaction, since when the two reactions are combined, it is the former that proceeds from left to right. Consequently we have

$$Zn^{++} + 2e^- = Zn, \qquad \varepsilon^0 = -0.76\ volt$$

for the sign and magnitude of the zinc half-cell potential.

Similarly, experiments with the copper-hydrogen cell show that cupric ion is a better oxidant than hydrogen ion, or that the Cu^{++}, Cu half-reaction has a greater tendency to proceed from left to right than does the H^+, H_2 half-reaction. Accordingly, the half-cell potential of the cupric ion–copper reaction must be more positive than that of the hydrogen ion–hydrogen gas reaction, and by our measurements we have

$$Cu^{++} + 2e^- = Cu, \qquad \varepsilon^0 = +0.34\ volt.$$

Our table of half-cell potentials now reads

$$Zn^{++} + 2e^- = Zn, \qquad \varepsilon^0 = -0.76\ volt,$$
$$2H^+ + 2e^- = H_2, \qquad \varepsilon^0 = 0.00\ volt,$$
$$Cu^{++} + 2e^- = Cu, \qquad \varepsilon^0 = +0.34\ volt,$$

with increasing voltage indicating increasing tendency of the half-reaction to proceed from left to right.

We can check the consistency of our assignments by calculating the standard potential of the reaction

$$Zn + Cu^{++} = Zn^{++} + Cu$$

from the standard half-cell potentials

$$Zn^{++} + 2e^- = Zn, \qquad \varepsilon^0 = -0.76 \text{ volt},$$
$$Cu^{++} + 2e^- = Cu, \qquad \varepsilon^0 = +0.34 \text{ volt}.$$

We must combine these reactions such that zinc metal appears on the left-hand side of the equation, and electrons are eliminated. Therefore, we reverse the direction of the first half-reaction, and *accordingly reverse the sign of its half-cell potential:*

$$
\begin{array}{ll}
Zn = Zn^{++} + 2e^- & \varepsilon^0 = +0.76 \\
2e^- + Cu^{++} = Cu & \varepsilon^0 = +0.34 \\
\hline
Zn + Cu^{++} = Cu + Zn^{++} & \Delta\varepsilon^0 = \quad 1.10 \text{ volts}
\end{array}
$$

The resulting $\Delta\varepsilon^0$ is numerically equal to the experimentally measured cell potential. Furthermore, the positive sign of the calculated $\Delta\varepsilon^0$ tells us that the spontaneous direction of the reaction is from left to right as written; this is also found experimentally. Thus our sign convention is internally consistent.

Another example can illustrate the construction of a table of half-cell potentials, and lead us to an important point concerning the combination of half-reactions and half-cell potentials. When a standard cell is constructed from the hydrogen electrode and a silver wire electrode immersed in $1\,M\,Ag^+$, it generates 0.80 volt, and the hydrogen electrode is negative. Consequently, as the cell operates, electrons must be produced at the hydrogen electrode and consumed at the silver electrode. The spontaneous cell reaction must be

$$H_2 + 2Ag^+ \rightarrow 2Ag + 2H^+.$$

The direction of the spontaneous cell reaction tells us that silver ion is a better oxidant than hydrogen ion. Thus the half-reaction

$$Ag^+ + e^- = Ag$$

has greater tendency to proceed as written than does

$$2H^+ + 2e^- = H_2.$$

Therefore the half-cell potential for the silver ion–silver reaction must be $+0.80$ volt.

Let us now combine the Ag^+, Ag half-reaction with the Cu^{++}, Cu half-reaction to obtain the voltage of the corresponding galvanic cell. We have

$$Cu^{++} + 2e^- = Cu, \qquad \varepsilon^0 = +0.34 \text{ volt},$$
$$Ag^+ + e^- = Ag, \qquad \varepsilon^0 = +0.80 \text{ volt}.$$

To eliminate electrons in the overall reaction, the direction of the Cu^{++}, Cu half-reaction must be reversed, and the Ag^+, Ag half-reaction itself must be multiplied by two:

$$
\begin{array}{ll}
Cu = Cu^{++} + 2e^- & -0.34 \\
\underline{2e^- + 2Ag = 2Ag} & \underline{+0.80} \\
Cu + 2Ag^+ = Cu^{++} + 2Ag & \Delta\varepsilon^0 = 0.46 \text{ volt}
\end{array}
$$

This calculated voltage agrees with that obtained by direct measurement from the cell itself, and its sign correctly indicates that copper metal should reduce silver ion as the cell operates.

Note carefully that the sign of the Cu^{++}, Cu half-cell potential was changed when the direction of the half-reaction was reversed, but when the Ag^+, Ag half-reaction was multiplied by two, *the half-cell potential was not.* The reason for this is that the voltage is associated *only* with the driving *force* of a reaction, and is intimately connected with the *direction* in which a reaction proceeds. On the other hand, multiplying an equation by two is an operation which has to do only with the *amounts* of materials available for reaction, and not with the driving force or half-cell potential.

Now we can summarize the conventions regarding half-cell potentials:

1. The standard hydrogen electrode is assigned a potential of exactly zero volts.

2. When all half-reactions are written as reductions, that is, in the form

$$\text{Oxidant} + ne^- = \text{reductant,}$$

the reactions that proceed to the right more readily than the H^+, H_2 reaction are assigned a positive voltage, and those that proceed with a smaller driving force are given negative half-cell voltages.

3. The size (in the algebraic sense) of the half-cell potential is a quantitative measure of the tendency of the half-reaction to proceed from left to right.

4. If the direction in which a half-reaction is written is reversed, the sign of its half-cell potential is reversed. However, when a half-reaction is multiplied by a positive number, its voltage is unchanged.

We have in this discussion adopted the convention that tabulated half-reactions are written as reductions proceeding from left to right. This convention, long traditional in European countries, has been recommended and largely adopted for international use. The convention used in the United States until recently has been to write half-reactions as oxidations, as for example,

$$
\begin{array}{ll}
Zn = Zn^{++} + 2e^-, & \varepsilon^0 = 0.763, \\
2Cl^- = Cl_2 + 2e^-, & \varepsilon^0 = -1.36.
\end{array}
$$

Because the direction of writing the reaction has changed, the sign of ε^0 has been reversed. Half-reactions tabulated in this way will be encountered frequently in older books. Reactions written according to either convention con-

Table 7.1 Standard reduction potentials at 25°C

Half-reaction	$\mathcal{E}^0$ (volt)	Half-reaction	$\mathcal{E}^0$ (volt)
Acid solution		$Hg_2Cl_2 + 2e^- = 2Hg + Cl^-$	0.2676
$Li^+ + e^- = Li$	-3.045	$Cu^{++} + 2e^- = Cu$	0.337
$Ca^{++} + 2e^- = Ca$	-2.866	$Fe(CN)_6^{-3} + e^- = Fe(CN)_6^{-4}$	0.36
$Na^+ + e^- = Na$	-2.714	$Cu^+ + e^- = Cu$	0.521
$La^{+3} + 3e^- = La$	-2.52	$I_2(s) + 2e^- = 2I^-$	0.535
$Mg^{++} + 2e^- = Mg$	-2.36	$I_3^- + 2e^- = 3I^-$	0.536
$AlF_6^{-3} + 3e^- = Al + 6F^-$	-2.07	$PtCl_4^- + 2e^- = Pt + 4Cl^-$	0.73
$Al^{+3} + 3e^- = Al$	-1.66	$Fe^{+3} + e^- = Fe^{++}$	0.77
$SiF_6^- + 4e^- = Si + 6F^-$	-1.24	$Hg_2^{++} + 2e^- = 2Hg$	0.788
$V^{++} + 2e^- = V$	-1.19	$Ag^+ + e^- = Ag$	0.799
$Mn^{++} + 2e^- = Mn$	-1.18	$2Hg^{++} + 2e^- = Hg_2^{++}$	0.920
$Zn^{++} + 2e^- = Zn$	-0.763	$Br_2 + 2e^- = 2Br^-$	1.087
$Cr^{+3} + 3e^- = Cr$	-0.744	$IO_3^- + 6H^+ + 5e^- = \frac{1}{2}I_2 + 3H_2O$	1.19
$Fe^{++} + 2e^- = Fe$	-0.44	$O_2 + 2H^+ + 4e^- = 2H_2O(l)$	1.23
$Cr^{+3} + e^- = Cr^{++}$	-0.41	$Cr_2O_7^- + 14H^+ + 6e^- = 2Cr^{+3} + 7H_2O$	1.33
$PbSO_4 + 2e^- = Pb + SO_4^-$	-0.359	$Cl_2 + 2e^- = 2Cl^-$	1.36
$Co^{++} + 2e^- = Co$	-0.277	$PbO_2 + 4H^+ + 2e^- = Pb^{++} + 2H_2O$	1.45
$Ni^{++} + 2e^- = Ni$	-0.250	$Au^{+3} + 3e^- = Au$	1.50
$Pb^{++} + 2e^- = Pb$	-0.126	$MnO_4^- + 8H^+ + 5e^- = Mn^{++} + 4H_2O$	1.51
$D^+ + e^- = \frac{1}{2}D_2$	-0.0034	$O_3 + 2H^+ + 2e^- = O_2 + H_2O$	2.07
$H^+ + e^- = \frac{1}{2}H_2$	0 (definition)	$F_2 + 2e^- = 2F^-$	2.87
$Cu^{++} + e^- = Cu^+$	0.153	$H_4XeO_6 + 2H^+ + 2e^- = XeO_3 + 3H_2O$	3.0

Half-reaction	$\mathcal{E}^0$ (volt)
Basic solution	
$H_2AlO_3^- + H_2O + 3e^- = Al + 4OH^-$	-2.33
$CrO_2^- + 2H_2O + 3e^- = Cr + 4OH^-$	-1.27
$ZnO_2^- + 2H_2O + 2e^- = Zn + 4OH^-$	-1.21
$Sn(OH)_6^- + 2e^- = HSnO_2^- + H_2O + 3OH^-$	-0.93
$HSnO_2^- + H_2O + 2e^- = Sn + 3OH^-$	-0.91
$HPbO_2^- + H_2O + 2e^- = Pb + 3OH^-$	-0.54
$Co(OH)_3 + e^- = Co(OH)_2 + OH^-$	0.17
$IO_3^- + 3H_2O + 6e^- = I^- + 6OH^-$	0.26
$ClO_3^- + H_2O + 2e^- = ClO_2^- + 2OH^-$	0.33
$ClO_4^- + H_2O + 2e^- = ClO_3^- + 2OH^-$	0.36
$O_2 + H_2O + 4e^- = 4OH^-$	0.40
$HO_2^- + H_2O + 2e^- = 3OH^-$	0.88
$ClO^- + H_2O + 2e^- = Cl^- + 2OH^-$	0.89
$HXeO_4^- + 3H_2O + 6e^- = Xe + 7OH^-$	0.9

tain the same amount of information. We have adopted the international convention in anticipation of its increasingly common use. However, in using any table of half-cell potentials, one should at the outset ascertain the convention that is being employed.

Table 7.1 gives the half-cell potentials of a number of reactions. Such a table not only gives us a quantitative comparison of the strengths of oxidizing and reducing agents, it is a very compact way of storing chemical information. If the values of ε^0 for 50 half-cells are tabulated, it is possible to calculate from these the values of $\Delta\varepsilon^0$ for $(50 \times 49)/2$ reactions.

Example 7.1 By using Table 7.1, arrange the following substances in order of increasing strength as reductants: Zn, Pb, Al. Also, of Ag^+, Cl_2, O_3, which is the strongest oxidant, and which is the weakest oxidant?

To find the order of increasing reducing strength, we note that any reductant in Table 7.1 is stronger than all others below it. Therefore, reducing strength increases in the order Pb, Zn, Al.

Oxidizing agents appear on the left-hand side of the half-reactions in Table 7.1, and any oxidant is stronger than all those above it. Therefore O_3 is the strongest of the three oxidizing agents, and Ag^+ is the weakest.

7.5 THE NERNST EQUATION

Until now, we have been exclusively concerned with galvanic cells operating under standard concentration conditions, and have associated the sign and magnitude of $\Delta\varepsilon^0$ with the driving force for chemical reaction. We must be careful about this use of $\Delta\varepsilon^0$, however. Remember that $\Delta\varepsilon^0$ tells us only whether products in their standard states will be formed spontaneously from reactants in their standard states. For example, the fact that for the reaction

$$Co(s) + Ni^{++}(aq) = Co^{++}(aq) + Ni(s),$$

$\Delta\varepsilon^0 = 0.03$ volt tells us only that if *both* Ni^{++} and Co^{++} are present at 1-M concentrations, nickel metal will be formed, and cobalt will be converted to the ion. However, experiment shows that if the concentration of Ni^{++} is 0.01 M, and that of Co^{++} is 1 M, then the direction of the spontaneous reaction is *reversed*. Therefore, before we can predict the direction of spontaneous reaction for anything other than standard concentration conditions, we must learn how the voltage of a galvanic cell depends on concentration.

Experimental measurements of cell voltage as a function of reagent concentration show that for any general reaction

$$aA + bB = cC + dD,$$

the cell voltage $\Delta\varepsilon$ is given by

$$\Delta\varepsilon = \Delta\varepsilon^0 - \frac{0.059}{n} \log \frac{[C]^c[D]^d}{[A]^a[B]^b}.$$

Here $\Delta\mathcal{E}^0$ is the standard cell potential, n is the number of electrons transferred in the reaction as written, and the logarithm is taken to the base 10. The factor 0.059 is common to all cells operating at a temperature of 25°C. This expression, called the **Nernst equation,** is substantiated by a wealth of experimental data, and also can be derived from the more fundamental principles of thermo-dynamics, as we shall see in Chapter 8.

Let us apply the Nernst equation to a specific example, the reaction

$$Co + Ni^{++} = Co^{++} + Ni, \quad \Delta\mathcal{E}^0 = 0.03 \text{ volt.}$$

We have

$$\Delta\mathcal{E} = 0.03 - \frac{0.059}{2} \log \frac{[Co^{++}]}{[Ni^{++}]}, \quad (7.2)$$

where we have substituted 0.03 for $\Delta\mathcal{E}^0$, and since two electrons are transferred in the reaction as written, $n = 2$. Although metallic nickel and cobalt are involved in the net reaction, they are not included in the concentration term, since their concentrations are constant. Thus in writing the concentration term in the Nernst equation, we follow the same conventions as in writing equilib-rium-constant expressions.

If the reaction had been written

$$2Co + 2Ni^{++} = 2Co^{++} + 2Ni, \quad \Delta\mathcal{E}^0 = 0.03 \text{ volt,}$$

then n would equal 4, and the appropriate Nernst equation would be

$$\Delta\mathcal{E} = 0.03 - \frac{0.059}{4} \log \frac{[Co^{++}]^2}{[Ni^{++}]^2}$$

$$= 0.03 - \frac{0.059}{2} \log \frac{[Co^{++}]}{[Ni^{++}]}.$$

We see from this that the form of the Nernst equation is consistent with the idea that the voltage associated with a reaction is unaffected by multiplying that reaction by a positive number.

The Nernst equation shows that the cell voltage is related to the logarithm of the reagent concentrations. In the particularly simple situation represented by Eq. (7.2), when the concentration ratio $[Co^{++}]/[Ni^{++}]$ changes by a factor of 10, the cell voltage changes by an amount equal to 0.059/2, or 0.03 volt. If the reactant concentration is increased, or the product concentration decreased, the cell voltage becomes more positive. For example, if $[Ni^{++}] = 1\ M$ and $[Co^{++}] = 0.1\ M$, then

$$\Delta\mathcal{E} = 0.03 - \frac{0.059}{2} \log 0.1 = 0.03 + 0.03$$

$$= 0.06 \text{ volt.}$$

Therefore, if we continue to associate the magnitude of the cell voltage with the tendency for the reaction to proceed, we can decide that the driving force for

$$Co + Ni^{++}(1\ M) = Co^{++}(0.1\ M) + Ni$$

is greater than for

$$Co + Ni^{++}(1\ M) = Co^{++}(1\ M) + Ni.$$

We can also calculate $\Delta\mathcal{E}$ when $[Co^{++}] = 1\ M$ and $[Ni^{++}] = 0.01\ M$:

$$\Delta\mathcal{E} = 0.03 - \frac{0.059}{2}\log\frac{1}{0.01} = 0.03 - 0.059$$

$$= -0.03\ \text{volt}.$$

The negative value of $\Delta\mathcal{E}$ means that the reaction

$$Co + Ni^{++}(0.01\ M) = Co^{++}(1\ M) + Ni$$

proceeds spontaneously from right to left. This example shows that with a table of half-cell potentials and the Nernst equation, we should be able to predict the spontaneous direction of a reaction under any concentration conditions.

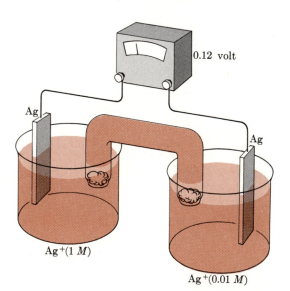

0.12 volt

Ag

Ag

Ag$^+$(1 M)

Ag$^+$(0.01 M)

FIG. 7.5   A silver-ion concentration cell.

The Nernst equation suggests that we should be able to generate a voltage by a concentration difference *alone*, even though the standard potential of a cell might be zero. As an example, consider the cell shown in Fig. 7.5. One half-cell consists of a 1 M Ag$^+$ solution and a silver wire electrode, while in the other half-cell a silver wire dips into 0.01 M Ag$^+$. Such an arrangement is called a **concentration cell.** Experiment shows that a voltage is generated, and that the electrode immersed in the more dilute solution is negative. Therefore, the half-reactions that occur must be

$$Ag(s) = Ag^+(0.01\ M) + e^- \qquad \text{(in the dilute solution)},$$
$$e^- + Ag^+(1\ M) = Ag(s) \qquad \text{(in the concentrated solution)},$$

and the net cell reaction is

$$Ag^+(1\ M) \rightarrow Ag^+(0.01\ M).$$

That is, the spontaneous action of the cell tends to equalize the two concentrations. The standard voltage of the cell reaction is zero, but the voltage generated in the present circumstances is

$$\Delta\mathcal{E} = 0 - \frac{0.059}{1} \log \frac{0.01}{1}$$
$$= 0.12 \text{ volt.}$$

The fact that there is a voltage, and thus a driving force for "reaction," should not be surprising. We know that if a concentrated solution is put in physical contact with a dilute solution, the two will mix *spontaneously* by diffusion to form a solution of uniform intermediate concentration. It is this natural tendency for the two solutions to mix that is measured by the voltage of this concentration cell.

Cell Potentials and Equilibrium Constants

It is informative to use the Nernst equation to predict what will happen to the voltage of a galvanic cell as its reactants are consumed and its products formed by the cell reaction. To be specific, let us imagine a nickel-cobalt cell whose voltage initially is 0.03 volt to be short circuited by a direct connection between the two electrodes. Then the reaction

$$Co + Ni^{++} = Co^{++} + Ni$$

proceeds spontaneously, and the concentration of Ni^{++} decreases while the concentration of Co^{++} increases. We can also imagine that a voltmeter is periodically inserted in the circuit to measure the cell potential as the reaction proceeds. The Nernst equation tells us that as the concentration of Ni^{++} diminishes and that of Co^{++} increases, the measured cell potential will decrease, and this is found experimentally.

Now, electrons will flow through the external circuit and the reaction will continue as long as there is a voltage difference between the two electrodes. However, as reactants are consumed and products formed, this voltage difference becomes steadily smaller and eventually becomes zero. At this point electron flow and net chemical reaction cease, and the concentrations of reactants and products remain constant indefinitely. These are conditions characteristic of chemical equilibrium, and indeed when $\Delta\varepsilon = 0$, concentrations of Co^{++} and Ni^{++} have been reached such that the reaction

$$Co(s) + Ni^{++}(aq) = Co^{++}(aq) + Ni(s)$$

is at equilibrium.

The same sort of analysis can be carried through for any reaction, so for the general process

$$aA + bB = cC + dD,$$

we can say that at equilibrium,

$$\Delta\varepsilon = 0 = \Delta\varepsilon^0 - \frac{0.059}{n} \log \left(\frac{[C]^c[D]^d}{[A]^a[B]^b}\right)_{eq}, \tag{7.3}$$

where the subscript on the concentration term indicates that all concentrations are those found at equilibrium.

The concentration term in Eq. (7.3) is equal to the equilibrium constant K for the reaction

$$\frac{[C]^c[D]^d}{[A]^a[B]^b} = K.$$

Therefore, Eq. (7.3) becomes

$$0 = \Delta\varepsilon^0 - \frac{0.059}{n} \log K,$$

$$\log K = \frac{n\Delta\varepsilon^0}{0.059},$$

$$K = 10^{n\Delta\varepsilon^0/0.059}. \tag{7.4}$$

Equation (7.4) is particularly interesting and important, because it is an exact relation between K and $\Delta\varepsilon^0$, the two factors that we have used to measure the tendency of reactions to proceed toward the products. We see that if $\Delta\varepsilon^0$ is positive, K will be greater than unity, and the larger $\Delta\varepsilon^0$ is, the larger is K. Our criteria that a large K, or a large positive $\Delta\varepsilon^0$, indicate a strong tendency for reactants to be converted to products are therefore consistent with each other.

We have taken a negative value of $\Delta\varepsilon^0$ to mean that reactants in their standard states will not proceed spontaneously to form products in their standard states. However, a negative $\Delta\varepsilon^0$ does *not* mean that no products whatsoever will be formed from reactants. As Eq. (7.4) shows, a negative $\Delta\varepsilon^0$ means only that the equilibrium constant for the reaction is less than unity, so that from reactants in their standard states, products in some concentration *less* than the standard value will be formed.

Example 7.2 Calculate $\Delta\mathcal{E}^0$ and K for the reaction

$$2Fe^{+3} + 3I^- = 2Fe^{++} + I_3^-.$$

From Table 7.1 we get

$$e^- + Fe^{+3} = Fe^{++}, \qquad \mathcal{E}^0 = 0.771,$$
$$2e^- + I_3^- = 3I^-, \qquad \mathcal{E}^0 = 0.536.$$

We combine these half-reactions by subtracting the second from twice the first to get

$$2Fe^{+3} + 3I^- = 2Fe^{++} + I_3^-.$$

To obtain $\Delta\mathcal{E}^0$, we subtract $\mathcal{E}^0$ of the second half-reaction from $\mathcal{E}^0$ of the first:

$$\Delta\mathcal{E}^0 = 0.771 - 0.536 = 0.235 \text{ volt.}$$

Therefore,

$$K = 10^{n\Delta\mathcal{E}^0/0.059} = 10^{(2)(0.235)/0.059} = 9.3 \times 10^7.$$

Thus a rather modest positive cell voltage corresponds to a large equilibrium constant.

Example 7.3 The $\Delta\mathcal{E}^0$ for the reaction

$$Fe + Zn^{++} = Zn + Fe^{++}$$

is -0.32 volt. What is the equilibrium concentration of Fe^{++} reached when a piece of iron is placed in a 1 M Zn^{++} solution? The equilibrium constant is

$$K = 10^{n\Delta\mathcal{E}^0/0.059} = 10^{2(-0.32)/0.059}$$
$$= 1.4 \times 10^{-11}.$$

Since

$$K = \frac{[Fe^{++}]}{[Zn^{++}]},$$

then when $[Zn^{++}] = 1\ M$, it follows that $[Fe^{++}] = 1.4 \times 10^{-11}\ M$.

These examples show that connecting $\Delta\mathcal{E}^0$ and K allows us to measure equilibrium constants that are very large or very small, by carrying out experiments in which all reagents have the convenient concentration of $1\ M$, or 1-atm pressure.

The value of the solubility product constant can be found by measurement of the voltage of a type of concentration cell. Suppose, for example, that a cell is constructed from a standard silver–silver ion half-cell, and another half-cell consisting of a silver wire immersed in a solution which contains $10^{-3}\ M$ Cl^- and is saturated with respect to solid silver chloride. This combination is a silver-ion concentration cell, since in the second half-cell, the silver-ion concentration is

$$[Ag^+] = \frac{K_{sp}}{[Cl^-]} = 10^3\ K_{sp},$$

where K_{sp} is the solubility product constant for silver chloride. Therefore the voltage generated by the cell will be

$$\Delta \varepsilon = 0 - \frac{0.059}{1} \log \frac{[Ag^+]_2}{[Ag^+]_1}$$

$$= 0 - \frac{0.059}{1} \log \frac{10^3 \, K_{sp}}{1}$$

$$= -3 \times 0.059 \log K_{sp},$$

since $\Delta \varepsilon^0 = 0$ for a concentration cell. For this particular cell the measured value of $\Delta \varepsilon$ is $+1.69$ volt, so

$$1.69 = -3 \times 0.059 \log K_{sp},$$
$$K_{sp} = 2.8 \times 10^{-10}.$$

Once again we see that measurement of cell voltages is a convenient way to obtain equilibrium constants that would be virtually impossible to evaluate by direct chemical analysis.

7.6 OXIDATION-REDUCTION TITRATIONS

Oxidation-reduction titrations are important in analytical chemistry, and a discussion of them provides considerable insight into the operation of galvanic cells as well as the use of the Nernst equation. As an example, let us consider the titration of ferrous ion by ceric ion:

$$Fe^{++} + Ce^{+4} \rightarrow Fe^{+3} + Ce^{+3}. \qquad (7.5)$$

This titration can be carried out with the apparatus shown in Fig. 7.6. The ceric ion is added from the buret to a beaker containing the ferrous ion. Into this beaker are placed a salt bridge and a platinum wire, so that it may be operated as a half-cell against the standard hydrogen electrode. Because the voltage of the standard hydrogen half-cell is zero by definition, the voltmeter gives the half-cell potential of the solution being titrated.

Either of the two half-reactions

$$Ce^{+4} + e^- = Ce^{+3}, \qquad (7.6a)$$
$$Fe^{+3} + e^- = Fe^{++} \qquad (7.6b)$$

can take place at the platinum electrode. Which of them actually determines the potential of the platinum wire? To answer, we first assume that after each amount of Ce^{+4} is added from the buret, reaction (7.5) takes place and rapidly reaches a position of equilibrium. Now, equilibrium between Fe^{++}, Fe^{+3}, Ce^{+3} and Ce^{+4} means that the half-cell potentials of reactions (7.6) are *identical*. Therefore we may regard the beaker either as a Fe^{+3}, Fe^{++} half-cell or as a Ce^{+4}, Ce^{+3} half-cell, whichever is more convenient.

Let us see how the half-cell potential changes as the reaction proceeds. If we think of the beaker as a Fe^{+3}, Fe^{++} half-cell, its voltage is given by

$$\mathcal{E} = \mathcal{E}^0_{Fe} - \frac{0.059}{1} \log \frac{[Fe^{++}]}{[Fe^{+3}]},$$

and this is also the voltage read by the voltmeter. At the start of the titration $[Fe^{+3}] = 0$, so the cell potential should be infinite. In fact, it never reaches this value, because trace impurities and the electrical operation of the cell itself make $[Fe^{+3}]$ nonzero, but even so, the initial voltage is large.

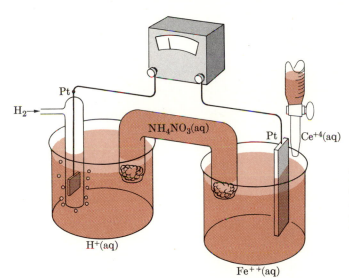

Apparatus for performing an oxidation-reduction titration by measuring the half-cell potential of the titrated solution.

FIG. 7.6

Suppose now that a fraction f of the ferrous ion has been titrated. If C_0 and V are the initial concentration of ferrous ion and the initial volume of the solution respectively, the *amount* of ferrous ion left is $C_0V(1-f)$, and the amount of ferric ion produced is C_0Vf. The resulting volume of the solution is $V + v$, where v is the volume of ceric solution added. Therefore the voltage is

$$\mathcal{E} = \mathcal{E}^0_{Fe} - 0.059 \log \frac{C_0V(1-f)/(V+v)}{C_0Vf/(V+v)}$$

$$= \mathcal{E}^0_{Fe} - 0.059 \log \frac{1-f}{f}. \tag{7.7}$$

That is, the cell voltage depends only on the fraction titrated. Accordingly, we can use Eq. (7.7) to construct the titration curve of $\mathcal{E}$ vs. f shown in Fig. 7.7, as long as $0 < f < 1$. Note that when f is equal to 0.5, $\mathcal{E} = \mathcal{E}^0_{Fe}$.

Just as for acid-base titrations, the equivalence point of an oxidation-reduction titration requires special treatment. The reason for this is that in our present example, when f equals unity, the concentration of ferrous ion is not really zero, but is actually some small value determined by the equilibrium between it and the other ions. However, even though some Fe^{++} remains unconverted at the equivalence point, there must be an exactly equivalent amount of Ce^{+4} present in solution, if f is truly unity. Therefore, at the equivalence point

$$[Ce^{+4}] = [Fe^{++}], \tag{7.8a}$$

$$[Ce^{+3}] = [Fe^{+3}]. \tag{7.8b}$$

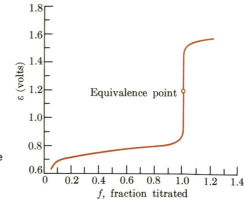

FIG. 7.7 Half-cell potential as a function of f, the fraction of Fe^{++} titrated with Ce^{+4}.

Since these two relations hold simultaneously only at the equivalence point, we can use them to find the corresponding potential. First, we remember that the half-cell potential can always be written in two ways:

$$\varepsilon = \varepsilon_{Fe}^0 - 0.059 \log \frac{[Fe^{++}]}{[Fe^{+3}]}$$

$$= \varepsilon_{Ce}^0 - 0.059 \log \frac{[Ce^{+3}]}{[Ce^{+4}]} .$$

Adding these expressions gives

$$2\varepsilon = \varepsilon_{Fe}^0 + \varepsilon_{Ce}^0 - 0.059 \log \frac{[Fe^{++}][Ce^{+3}]}{[Fe^{+3}][Ce^{+4}]} . \tag{7.9}$$

But Eqs. (7.8) tell us that

$$\frac{[Ce^{+4}]}{[Ce^{+3}]} = \frac{[Fe^{++}]}{[Fe^{+3}]} ,$$

$$\frac{[Fe^{++}][Ce^{+3}]}{[Fe^{+3}][Ce^{+4}]} = 1$$

at the equivalence point. Therefore, under the same circumstance Eq. (7.9) becomes

$$2\mathcal{E}_{ep} = \mathcal{E}^0_{Fe} + \mathcal{E}^0_{Ce} - 0.059 \log 1,$$
$$\mathcal{E}_{ep} = \tfrac{1}{2}(\mathcal{E}^0_{Fe} + \mathcal{E}^0_{Ce}). \qquad (7.10)$$

Thus, $\mathcal{E}_{ep}$, the half-cell potential at the equivalence point, is the average of the two standard half-cell potentials. Equation (7.10) is valid for any situation which involves two half-reactions with the *same* number of electrons.

To calculate the half-cell potential when f is greater than unity, it is convenient to use the expression written in terms of the ceric-cerous half-reactions:

$$\mathcal{E} = \mathcal{E}^0_{Ce} - 0.059 \log \frac{[Ce^{+3}]}{[Ce^{+4}]}. \qquad (7.11)$$

The reason for this is that after the equivalence point has been passed, the concentrations of both Ce^{+4} and Ce^{+3} are large and easy to calculate approximately. In fact, the amount of Ce^{+3} present is virtually constant, because essentially all the Fe^{++} has been titrated. On the other hand, it is not convenient to use the Nernst equation written in terms of the Fe^{++}, Fe^{+3} half-cell potential, because the concentration of Fe^{++} is very small and not constant, and must be carefully calculated using the equilibrium constant for the reaction and the concentrations of the other ions. Equation (7.11) allows a much more direct calculation of the half-cell potential.

Examination of the titration curve shows that the electrode potential changes very rapidly in the vicinity of the equivalence point. In fact, between $f = 0.999$ and $f = 1.001$, the voltage change is 0.48 volt. It would seem, then, that the equivalence point could be located with great precision by following the progress of the titration by a voltmeter. In fact, elaborations of this idea provide the most elegant and convenient ways of carrying out oxidation-reduction titrations.

From this titration curve we can see that in order to have a large change in voltage near the equivalence point, a *minimum* requirement is that the cell voltages at $f = 0.5$ and at $f = 1$ be as different as possible. Thus the quantity

$$\begin{aligned}
\mathcal{E}_1 - \mathcal{E}_{0.5} &= \tfrac{1}{2}(\mathcal{E}^0_{Fe} + \mathcal{E}^0_{Ce}) - \mathcal{E}^0_{Fe} \\
&= \tfrac{1}{2}(\mathcal{E}^0_{Ce} - \mathcal{E}^0_{Fe}) \\
&= \Delta\mathcal{E}^0/2
\end{aligned}$$

must be as large as possible. Now, $\Delta\mathcal{E}^0$ is just the standard potential associated with the titration reaction, so since

$$K = 10^{n\Delta\mathcal{E}^0/0.059},$$

our requirement for a large voltage change near the equivalence point is that the equilibrium constant for the titration reaction must be large. For the ferrous ion–ceric ion reaction we find K to be

$$K = 10^{0.84/0.059}$$
$$= 1.7 \times 10^{14},$$

which is large indeed.

When the two half-reactions involved in the titration reaction have different numbers of electrons, the expression for the potential at the equivalence point is somewhat different from Eq. (7.10). Let us illustrate this for the reaction

$$5Fe^{++} + MnO_4^- + 8H^+ = 5Fe^{+3} + Mn^{++} + 8H_2O,$$

which involves the following half-reactions:

$$Fe^{+3} + e^- = Fe^{++},$$
$$MnO_4^- + 8H^+ + 5e^- = Mn^{++} + 4H_2O.$$

As we noted before, the half-cell potential for the titration mixture can be written in two ways:

$$\varepsilon = \varepsilon_{Fe}^0 - 0.059 \log \frac{[Fe^{++}]}{[Fe^{+3}]} \tag{7.12}$$

$$= \varepsilon_{Mn}^0 - \frac{0.059}{5} \log \frac{[Mn^{++}]}{[MnO_4^-][H^+]^8}. \tag{7.13}$$

At the equivalence point we must have

$$5[MnO_4^-] = [Fe^{++}],$$
$$5[Mn^{++}] = [Fe^{+3}],$$

and therefore it follows that

$$\frac{[Fe^{++}]}{[Fe^{+3}]} = \frac{[MnO_4^-]}{[Mn^{++}]}.$$

Substitution of this into Eq. (7.12) gives

$$\varepsilon_{ep} = \varepsilon_{Fe}^0 - 0.059 \log \frac{[MnO_4^-]}{[Mn^{++}]}. \tag{7.14}$$

Now we multiply Eq. (7.13) by 5, add the result to Eq. (7.14), and get

$$\varepsilon_{ep} + 5\varepsilon_{ep} = \varepsilon_{Fe}^0 + 5\varepsilon_{Mn}^0 - 0.059 \log \frac{[Mn^{++}]}{[MnO_4^-]} \frac{[MnO_4^-]}{[Mn^{++}][H^+]^8},$$

$$6\varepsilon_{ep} = \varepsilon_{Fe}^0 + 5\varepsilon_{Mn}^0 + 0.059 \log [H^+]^8,$$

$$\varepsilon_{ep} = \frac{\varepsilon_{Fe}^0 + 5\varepsilon_{Mn}^0}{6} + \frac{0.059}{6} \log [H^+]^8.$$

This expression shows that when the numbers of electrons in the two half-reactions are different, the potential at the equivalence point is the *weighted* average of the two half-cell potentials. Also, we can see that the potential at the equivalence point may be determined by the concentrations of any species that appear in the net titration reaction.

Our exclusive concern has been with situations in which the spontaneous reaction occurring in an electrochemical cell provides a source of voltage and electrical power. The opposite case is that of *electrolysis*, where the application of an *external* source of voltage is used to carry out a chemical change. Electrolytic processes are of great importance in industry today, and as we shall see in Chapter 10, were influential in the development of ideas about the electrical nature of matter.

Perhaps the simplest of electrolytic processes occurs when two copper strips connected to opposite terminals of a voltage source are both dipped into an aqueous solution of copper sulfate. A current passes and at the copper strip connected to the negative terminal, more metallic copper is deposited, while at the other electrode copper metal is oxidized to Cu^{++}. The electrode at which reduction occurs is *always* called the cathode, while the anode is always the electrode at which oxidation takes place. Thus we have

$$Cu^{++} + 2e^- = Cu \quad \text{at the cathode,}$$
$$Cu = Cu^{++} + 2e^- \quad \text{at the anode.}$$

If the anode of such a cell is made out of impure copper (99.0%), it is possible to deposit at the cathode copper of a purity of 99.98%. Thus this and other electrolytic refining processes find considerable use in the preparation of large quantities of pure metals.

Electrolytic techniques make possible the recovery of the most active of the elements from their compounds. A glance at a table of standard electrode potentials shows that such ions as Na^+, Mg^{++}, and Al^{+3} are extremely difficult to reduce:

$$Na^+ + e^- = Na, \quad \mathcal{E}^0 = -2.714 \text{ volts,}$$
$$Mg^{++} + 2e^- = Mg, \quad \mathcal{E}^0 = -2.37 \text{ volts,}$$
$$Al^{+3} + 3e^- = Al, \quad \mathcal{E}^0 = -1.66 \text{ volts.}$$

In fact, there is no readily available chemical reagent that can reduce these ions to the metals in large quantities. As a result the commercial preparation of the active metals involves electrolytic reduction at the cathode, as for instance in the electrolysis of fused $MgCl_2$:

$$Mg^{++} + 2e^- \rightarrow Mg \quad \text{at the cathode,}$$
$$2Cl^- \rightarrow Cl_2 + 2e^- \quad \text{at the anode.}$$

Only about 6 volts need be applied to effect such a reaction, but even this voltage makes the cathode of an electrolytic cell an extremely powerful reducing agent.

The quantitative aspect of electrolysis is straightforward: the number of moles of material oxidized or reduced at an electrode is related by the stoichiometry of the electrode reaction to the amount of electricity passed through the cell. For example, 1 mole of electrons will reduce and deposit 1 mole of Ag^+, or 0.5 mole of Cu^{++}, or 0.33 mole of Al^{+3}, as suggested by the following electrode reactions:

$$Ag^+ + e^- = Ag,$$
$$Cu^{++} + 2e^- = Cu,$$
$$Al^{+3} + 3e^- = Al.$$

While it is convenient to "count" atoms by weighing substances, the number of moles of electrons delivered to an electrode is most easily measured in terms of total electrical charge. Since the charge of one electron is 1.6021×10^{-19} coulomb (coul), the charge on 1 mole of electrons is

$$1.6021 \times 10^{-19} \text{ coul/electron} \times 6.0225 \times 10^{23} \text{ electrons/mole} = 96,487 \text{ coul.}$$

This quantity of electricity is called the *faraday* and given the symbol $\mathfrak{F}$. Thus 1 faraday of electricity will reduce 1 mole of Ag^+ or 0.5 mole of Cu^{++} to the metal.

The number of coulombs of charge passed through a cell in an electrolysis can be calculated from the measured current and the length of time the current flows. Since

$$1 \text{ ampere (amp)} = 1 \text{ coul/sec,}$$

we have

$$\text{coulombs passed} = \text{current (in amp)} \times \text{time (in sec).}$$

With this in mind, we can calculate, for example, that if a solution of $CuSO_4$ is electrolyzed for 7.00 min with a current of 0.60 amp, the number of coulombs delivered is

$$\text{coulombs} = \text{amperes} \times \text{seconds} = 0.60 \times 7.00 \times 60$$
$$= 252 \text{ coul.}$$

This corresponds to

$$\frac{252 \text{ coul}}{96,487 \text{ coul/faraday}} = 2.61 \times 10^{-3} \text{ faraday.}$$

Consequently, 1.30×10^{-3} mole of copper metal is deposited in this electrolysis. In the commercial production of metals like magnesium the currents employed are approximately 50,000 amp. This current corresponds to about 0.5 faraday/sec, or to roughly 0.25 mole of magnesium metal deposited per second.

Even a casual examination of the natural world reveals an almost overwhelming number of examples of natural processes in which oxidation-reduction reactions play a central role. Indeed, photosynthesis, the basic process which sustains life on earth, is the light induced reduction of carbon dioxide and water to carbohydrate plant material, accompanied by the oxidation of some of the oxygen in these compounds to the elemental state. The metabolism of foods by animals is nearly the exact reverse of this process: oxidation of carbohydrates and other food materials to carbon dioxide and water. In detail, the fundamental life processes are executed by a very large number of oxidation-reduction reactions which interact with one another in a subtle manner. An indication of the nature of some of these biologically significant oxidation-reduction reactions will be found in Chapter 18. For the present, however, we shall investigate the application of the electrochemical principles to the apparently simpler phenomena of the corrosion of metals, and the storage and production of energy by galvanic cells.

Corrosion

In the United States, more than $\$10^7$ a year is lost to corrosion—the loss of material due to chemical attack. The mechanism of this loss is in some circumstances simply the solution process. However, in the case of the oxidative corrosion of metals, the mechanism involved is electrochemical in nature, and is very closely related to the galvanic cell phenomena discussed in Section 7.4.

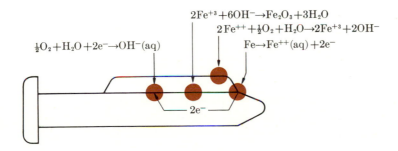

$$2Fe^{+3}+6OH^-\rightarrow Fe_2O_3+3H_2O$$
$$2\,Fe^{++}+\tfrac{1}{2}O_2+H_2O\rightarrow 2Fe^{+3}+2OH^-$$
$$Fe\rightarrow Fe^{++}(aq)+2e^-$$

$$\tfrac{1}{2}O_2+H_2O+2e^-\rightarrow OH^-(aq)$$

$$2e^-$$

Corrosion of a wet iron spike by atmospheric oxygen. Oxidation of iron takes place principally at the highly stressed areas near the point.

FIG. 7.8

To analyze the corrosion mechanism of iron, consider Fig. 7.8, where an iron spike is shown with its partially wet surface exposed to atmospheric oxygen. The iron spike is a highly imperfect solid: it consists of randomly oriented microcrystals which have imperfect lattices and which incorporate impurity atoms. The iron atoms near the boundaries of these microcrystals or grains are relatively weakly bound, and at some sites they may easily enter the aqueous phase as ions:

$$Fe(s) = Fe^{++}(aq) + 2e^-.$$

Thus certain of these grain boundaries serve as anodes, where iron is oxidized to ferrous ion. This process cannot long continue unless something is done to remove the electrons. Since iron is a good electrical conductor, these extra electrons can travel to sites on the iron surface which can facilitate a reduction reaction. If dissolved oxygen is in the aqueous phase, the reaction

$$\tfrac{1}{2}O_2 + H_2O + 2e^- = 2OH^-$$

can occur at microscopic cathodes distributed on the surface. The net result is the production of Fe^{++} and OH^-, with the disappearance of metallic iron and oxygen. Further direct oxidation of the ferrous ion is possible:

$$2Fe^{++} + \tfrac{1}{2}O_2 + H_2O = 2Fe^{+3} + 2OH^-.$$

Finally, if Fe^{+3} can diffuse to the region of the cathode where OH^- is in abundance, the reaction

$$2Fe^{+3} + 6OH^- = Fe_2O_3 + 3H_2O$$

can occur, and the insoluble solid iron oxide, or rust, is precipitated.

In ordinary iron objects, the cathode and anode regions are so close that the unaided eye cannot distinguish them, and rust apparently forms everywhere on a wet surface. However, even in these circumstances it is observed that corrosion is much more rapid if the aqueous phase contains dissolved electrolytes. The reason for this is quite straightforward: These dissolved electrolytes play the same role as does the salt bridge in a galvanic cell. The charge separation that would result from production of Fe^{++} at the anode regions and formation of OH^- at the cathode region would slow and eventually stop the oxidation-reduction reaction. By providing a uniform reservoir of ions of both electrical charges, the dissolved salts prevent excess charge from accumulating at either electrode and thereby speed the electrochemical corrosion.

There are a number of ways to inhibit or prevent corrosion. The most obvious is to cover a susceptible metal with a polymeric coating like paint, which is relatively impervious to moisture and oxygen. In some circumstances, the protective coating may be a noble metal such as gold, which cannot be spontaneously oxidized by air. More commonly, the coating is a metal which protects itself and the substrate by formation of an impervious oxide layer. In this way, such metals as zinc, tin, nickel, or chromium can prevent the corrosion of iron.

For some items such as large underground tanks or pipelines, overall coating or plating is either impractical or not fully effective. In these cases it is possible to prevent corrosion by using a sacrificial anode. A block of easily oxidized metal such as zinc or magnesium is placed in the earth and connected electrically to the object to be protected. The zinc block serves as an anode, and as it dissolves by the reaction

$$Zn(s) = Zn^{++}(aq) + 2e^-,$$

it supplies electrons to the iron object, and prevents any part of it from being oxidized. Instead, the whole object acts as a cathode, where oxygen is reduced to OH^-. When the sacrificial anode is consumed, it can be replaced, and the protection continued.

Batteries and Fuel Cells

Galvanic cells provide a compact, safe way of storing energy so that it can be delivered in a particularly useful manner: as an electrical current driven by a voltage difference. A large number of pairs of half-reactions have been used for practical applications, with the choice of reactions being influenced by such considerations as the availability and expense of materials, mechanical stability, operating temperature, total energy stored per unit weight, and safety factors.

The most common type of battery has been the *Leclanché "dry" cell*. A zinc can forms the anode, and is consumed as the cell operates. A carbon rod in contact with manganese dioxide serves as the cathode, and an electrolyte of zinc and ammonium chlorides in water mixed with enough starch to prevent spillage is used. As the cell operates, the reaction

$$Zn + 2MnO_2 + H_2O = Zn(OH)_2 + Mn_2O_3$$

occurs, and produces approximately 1.2 volts. This Leclanché cell is an example of a "primary" cell: one that can be used once, but cannot be restored by reversing the current flow. Another example of a primary battery is the so-called *mercury cell*, which uses the reaction

$$Zn + HgO = ZnO + Hg$$

carried out in a potassium hydroxide paste electrolyte. It also produces approximately 1.2 volts.

Secondary batteries *can* be conveniently recharged, or restored to nearly new condition by reversing the current flow. The best known example of this type of device is the lead storage cell, which involves the reaction

$$Pb(s) + PbO_2(s) + 4H^+(aq) + 2SO_4^=(aq) = 2PbSO_4(s) + 2H_2O.$$

In operation, metallic lead is oxidized to lead sulfate at the anode, and lead dioxide is reduced to lead sulfate at the cathode. Since sulfuric acid is consumed in this process, the specific gravity of the electrolyte decreases as the cell operates, and is a convenient indicator of the charge state of the battery. Each cell of a commercial lead storage battery generates slightly more than 2 volts, and the common automobile battery consists of six such cells connected in series. To recharge the battery, an opposing voltage somewhat greater than 12 volts is applied to the electrodes, and the cell reaction is reversed, with lead sulfate being converted lead and lead dioxide. The restoration of the electrodes in the recharge process is never perfect, and metallic needles and other mechanically unstable growths appear on the electrodes. Eventually these growths produce internal short circuits and the cell operation ceases.

The major virtues of the lead storage battery are its reliability, lifetime, and relative simplicity. However, its liquid electrolyte is a disadvantage, as is the great weight of lead which is required to deliver a substantial current. Consequently, other secondary batteries are in use and under development. The nickel-cadmium cell, which operates on the reaction

$$Cd + Ni_2O_3 + 3H_2O = Cd(OH)_2 + 2Ni(OH)_2$$

and produces 1.3 volts, is used extensively to power small electronic devices. The alkali metal-sulfur cells, which use reactions like

$$2Li + S = Li_2S,$$

produce relatively high voltages ($1.9 - 2.3$ volts) and have the major advantage that they are light in weight. Unfortunately, they operate only at elevated temperatures (above $350°C$) and are not suitable for applications that require long standby and rapid starts.

The cells mentioned so far involve the oxidation of metallic electrodes by a variety of oxidizing agents. While the metallic electrode is a convenient portable source of small amounts of energy, it is not suitable for very large scale energy production. For such purposes, it would be very advantageous to have a cell that produces electricity by the oxidation of a readily available gaseous fuel, such as natural gas (CH_4), carbon monoxide, or hydrogen. Some notable progress has been made in this direction, and a few practical fuel cells have been developed.

Fuel cells have a great intrinsic advantage in the high efficiency with which they can convert the energy released in combustion of a fuel into useful work. In an ordinary electric power plant, fuel (usually oil) is burned to produce the steam necessary to drive a turbine, which in turn runs an electric generator. The overall efficiency of converting the energy of combustion to useful work has been raised to 35–40 percent, and significantly higher efficiencies cannot be expected. In Chapter 8, we shall find that the percentage efficiency η of steam turbines and other such heat engines is intrinsically limited to values given by the expression

$$\eta \leq 100 \frac{T_h - T_c}{T_h},$$

where T_h is the Kelvin temperature at which steam enters the turbine, and T_c is the lower temperature at which steam leaves the turbine. In practice, T_h and T_c are approximately 800°K and 400°K, respectively, and the maximum efficiency expected is only roughly 50 percent. In contrast, a fuel cell is not a heat engine, and does not have this intrinsic efficiency limitation. Practical fuel cells convert the energy of fuel combustion to useful work with an efficiency of 75 percent, with the principal limitation being the heat dissipated due to the internal resistance of the cell itself.

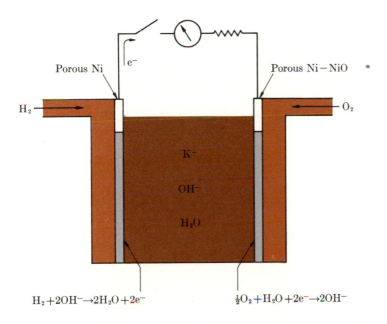

A schematic diagram of an H_2–O_2 fuel cell operating with an aqueous KOH electrolyte.   **FIG. 7.9**

Because of their very high efficiency and the nonpolluting nature of their operation, it would appear that fuel cells would be very desirable energy converters. Unfortunately, their application has been limited by the difficulty of finding suitable fuel-electrode-electrolyte combinations which allow rapid oxidation of the fuel. The most successful cell uses hydrogen as the fuel; a schematic diagram which illustrates its operation is shown in Fig. 7.9. Hydrogen at 40 atm pressure is forced into a porous nickel electrode, where it is oxidized to water in the presence of an aqueous potassium hydroxide electrolyte. The half-reaction is

$$H_2(g) + 2OH^-(aq) = 2H_2O + 2e^-.$$

The cathode of the cell is nickel covered with nickel oxide, which catalyzes the reduction of oxygen:

$$\tfrac{1}{2}O_2(g) + H_2O + 2e^- \rightarrow 2OH^-(aq).$$

Thus the overall cell reaction is the combustion of hydrogen and oxygen to liquid water. By using a very concentrated aqueous KOH electrolyte, the operating pressure of the cell can be lowered to nearly 1 atm, if the current demands are not great. Such low-pressure fuel cells have been used as power sources on some of the larger manned spacecraft.

While hydrocarbons have been oxidized to carbon dioxide and water at platinum electrodes, the cost of such cells precludes their widespread use. The most practical way to use hydrocarbons or coal in fuel cells at present is to provide a conversion stage in which hydrogen is generated by reaction with steam:

$$C(s) + H_2O = CO + H_2,$$
$$C_nH_{2n+2} + nH_2O = nCO + (2n + 1) H_2,$$
$$CO + H_2O = CO_2 + H_2.$$

The hydrogen produced in this manner can be used after purification to power a conventional fuel cell. More extensive applications of fuel cells await the development of large coal-to-hydrogen converters or inexpensive electrodes which permit the direct use of carbonaceous fuels.

7.9 CONCLUSION

The concepts discussed in this chapter will be used throughout our study of chemistry. For example, a classification scheme that organizes and simplifies much of the descriptive chemistry of the elements is based on oxidation numbers. Furthermore, many of the practical operations of analytical chemistry and of the chemical industry involve oxidation-reduction reactions. Perhaps most important, however, is the discovery that galvanic cell potentials provide a convenient way of assessing *quantitatively* the tendency of reactions to proceed as written. What is particularly significant is that through the use of *half-cell* potentials, we have a way of expressing the intrinsic ability of individual re-agents to perform as oxidants or reductants. Thus we are nearer to the goal of expressing and understanding what is meant by chemical reactivity. Another step toward this goal involves the study of thermodynamics, which is the subject of Chapter 8.

SUGGESTIONS FOR FURTHER READING

Barrow, G. M., *Physical Chemistry*, 3rd ed. New York: McGraw-Hill, 1973.

Butler, J. N., *Ionic Equilibrium*. Reading, Mass.: Addison-Wesley, 1964. Chapters 1 and 11.

Cragg, L. H., and R. P. Graham, *An Introduction to the Principles of Chemistry*. New York: Rinehart and Co., 1955. Chapters 16 and 17.

Daniels, F., and R. A. Alberty, *Physical Chemistry*, 3rd ed. New York: Wiley, 1966. Chapter 14.

Nyman, C. J., and R. E. Hamm, *Chemical Equilibrium*. Lexington, Mass.: Raytheon Education Co., 1968.

PROBLEMS

7.1 Complete and balance the following reactions which occur in acidic aqueous solution.

$$I_2 + H_2S = H^+ + I^- + S(s)$$
$$I^- + H_2SO_4 \text{ (hot, concentrated)} = I_2 + SO_2$$
$$Ag + NO_3^- = Ag^+ + NO$$
$$CuS + NO_3^- = Cu^{++} + SO_4^= + NO$$
$$S_2O_3^= + I_2 = I^- + S_4O_6^=$$
$$Zn + NO_3^- = Zn^{++} + NH_4^+$$
$$HS_2O_3^- = S + HSO_4^-$$
$$ClO_3^- + As_2S_3 = Cl^- + H_2AsO_4^- + SO_4^=$$
$$Cr_2O_7^= + C_2H_4O = C_2H_4O_2 + Cr^{+3}$$
$$MnO_4^= = MnO_2 + MnO_4^-$$

7.2 Complete and balance the following reactions which occur in basic aqueous solution.

$$Al + NO_3^- + OH^- = Al(OH)_4^- + NH_3$$
$$PbO_2 + Cl^- = ClO^- + Pb(OH)_3^-$$
$$N_2H_4 + Cu(OH)_2 = N_2 + Cu$$
$$Ag_2S + CN^- + O_2 = S + Ag(CN)_2^-$$
$$ClO^- + Fe(OH)_3 = Cl^- + FeO_4^=$$
$$HO_2^- + Cr(OH)_3^- = CrO_4^= + OH^-$$
$$Cu(NH_3)_4^{++} + S_2O_4^= = SO_3^= + Cu + NH_3$$
$$ClO_2 + OH^- = ClO_2^- + ClO_3^-$$
$$V + H_2O = HV_6O_{17}^{-3} + H_2$$
$$Mn(CN)_6^{-4} + O_2 = Mn(CN)_6^{-3}$$

7.3 Consult the table of standard electrode potentials, and select an oxidizing agent capable of transforming (a) Cl^- to Cl_2, (b) Pb to Pb^{++}, and (c) Fe^{++} to Fe^{+3}.

Similarly, select a reducing agent that can convert (d) Fe^{++} to Fe, (e) Ag^+ to Ag, and (f) Mn^{++} to Mn.

7.4 Which of the following oxidizing agents become stronger as the concentration of H^+ increases? Which are unchanged and which become weaker? (a) Cl_2, (b) $Cr_2O_7^=$, (c) Fe^{+3}, (d) MnO_4^-.

7.5 Compare the following standard electrode potentials for the ferrous–ferric ions and their cyanide complexes:

$$Fe^{+3} + e^- = Fe^{++}, \qquad \mathcal{E}^0 = 0.77 \text{ volt,}$$
$$Fe(CN)_6^{-3} + e^- = Fe(CN)_6^{-4}, \qquad \mathcal{E}^0 = 0.48 \text{ volt,}$$

On this basis, which ion, Fe^{++} or Fe^{+3}, is stabilized more by complexing with CN^-?

7.6 From the appropriate values of $\mathcal{E}^0$ drawn from Table 7.1, calculate $\Delta\mathcal{E}^0$ and the equilibrium constant for the reaction

$$Hg^{++} + Hg = Hg_2^{++}.$$

7.7 By use of appropriate half-cell potentials, calculate $\Delta\mathcal{E}^0$ and the equilibrium constant for the reaction

$$Fe^{+3} + I^- = Fe^{++} + \tfrac{1}{2}I_2.$$

State what you expect to happen when equal volumes of $2\,M$ Fe^{+3} and $2\,M$ I^- are mixed.

7.8 A half-cell (A) consisting of a strip of nickel dipping into a 1-M solution of Ni^{++}, and a half-cell (B) consisting of a strip of zinc dipping into a 1-M solution of Zn^{++} were successively connected with a standard hydrogen half-cell. The magnitudes of the individual half-cell potentials were then determined as

$$(A) \ \ Ni^{++} + 2e^- = Ni, \qquad |\mathcal{E}^0| = 0.25 \text{ volt,}$$
$$(B) \ \ Zn^{++} + 2e^- = Zn, \qquad |\mathcal{E}^0| = 0.77 \text{ volt.}$$

(a) When both the half-cells (A) and (B) were connected with the hydrogen half-cell, the metallic electrode (Ni or Zn) was found to be negative. What is the correct sign of the electrode potentials? (b) Of the substances Ni, Ni^{++}, Zn, Zn^{++}, which is the strongest oxidant? Which is the strongest reductant? (c) Will a noticeable reaction occur when metallic nickel is placed in a 1-M solution of Zn^{++}? metallic zinc is dipped into a 1-M solution of Ni^{++}? (d) Zinc forms a complex ion with hydroxide ion, $Zn(OH)_4^=$. If hydroxide ion were added to half-cell (B), would its electrode potential as written become more positive, less positive, or be unaffected? (e) If the half-cells (A) and (B) were connected together, which electrode would be negative? What would the cell voltage be?

7.9 An electrochemical cell is constructed of one half-cell in which a platinum wire dips into a solution containing 1-M Fe^{+3} and 1-M Fe^{+2}; the other half-cell consists of thallium metal immersed in 1-M Tl^+ solution. Given the following standard electrode potentials,

$$Tl^+ + e^- = Tl, \qquad \mathcal{E}^0 = -0.34,$$
$$Fe^{+3} + e^- = Fe^{++}, \qquad \mathcal{E}^0 = 0.77,$$

supply the desired information. (a) Which electrode is the negative terminal? (b) Which electrode is the cathode? (c) What is the cell voltage? (d) Write the reaction that proceeds from left to right as the cell operates spontaneously. (e) What is the equilibrium constant of this reaction? (f) How will the voltage of the cell be changed by decreasing the concentration of Tl^+?

7.10 A galvanic cell consists of a strip of cobalt metal, Co, dipping into $1\text{-}M$ Co^{++} solution, and another half-cell in which a piece of platinum dips into a $1\text{-}M$ solution of Cl^-. Chlorine gas at 1-atm pressure is bubbled into this solution. The observed cell voltage is 1.63 volts, and as the cell operates the cobalt electrode is negative. Given only that the standard potential for the chlorine–chloride ion half-cell is

$$\tfrac{1}{2}Cl_2 + e^- = Cl^-, \qquad \mathcal{E}^0 = 1.36 \text{ volts,}$$

supply the desired information. (a) What is the spontaneous cell reaction? (b) What is the standard potential of the cobalt electrode? (c) Would the cell voltage increase or decrease if the pressure of chlorine gas increased? (d) What would the cell voltage be if the concentration of Co^{++} were reduced to $0.01\ M$?

7.11 A cell consists of a standard Ag, Ag^+ half-cell ($1\text{-}M$ Ag^+) combined with another half-cell in which a silver wire dips into a solution of $1\text{-}M$ Br^- which is saturated and in contact with solid AgBr. The electrode of this latter cell is negative, and the cell generates 0.77 volt. What is the concentration of Ag^+ in equilibrium with $1\text{-}M$ Br^- and solid AgBr? What is the apparent solubility product of AgBr?

7.12 Two hydrogen–hydrogen ion half-cells are connected to make a single galvanic cell. In one of the half-cells the pH is 1.0, but the pH in the other half-cell is not known. The measured voltage delivered by the combination is 0.16 volt, and the electrode in the half-cell of known concentration is positive. Is the unknown concentration of H^+ greater or less than $0.1\ M$? What is the unknown concentration of H^+?

7.13 From the following standard electrode potentials,

$$Cu^{++} + 2e^- = Cu, \qquad \mathcal{E}^0 = 0.34,$$
$$Cu^{++} + \ e^- = Cu^+, \qquad \mathcal{E}^0 = 0.15,$$

calculate the equilibrium constant of the reaction

$$Cu + Cu^{++} = 2Cu^+.$$

Would you expect to be able to form appreciable amounts of Cu^+ by reaction of Cu with Cu^{++}? Consider that CuCl is a sparingly soluble salt with $K_{sp} = 3.2 \times 10^{-7}$. Calculate the equilibrium constant for the reaction

$$Cu + Cu^{++} + 2Cl^- = 2CuCl(s).$$

7.14 Consider the titration reaction

$$V(OH)_4^+ + Cr^{++} + 2H^+ = VO^{++} + Cr^{+3} + 3H_2O,$$

which involves the half-reactions

$$V(OH)_4^+ + 2H^+ + e^- = VO^{++} + 3H_2O, \qquad \mathcal{E}^0 = 1.00 \text{ volt,}$$
$$Cr^{+3} + e^- = Cr^{++}, \qquad \mathcal{E}^0 = -0.41 \text{ volt.}$$

Imagine that the titration is conducted by adding a solution of $V(OH)_4^+$ from a buret to a solution of Cr^{++} in a beaker which is connected by a salt bridge to a standard hydrogen electrode assembly. A platinum wire dips into the solution to be titrated, and a voltmeter measures the potential of the solution with respect to the standard hydrogen electrode. (a) Write an expression in terms of the concentrations of chromium species only that gives the potential of the titrated solution at any point during the titration. (b) Write a similar expression, but entirely in terms of the concentrations of vanadium species and hydrogen ion only. (c) What is the relation between these two expressions at any point in the titration? (d) What is the voltage reading when 0.91 of the initial Cr^{++} has been converted to Cr^{+3}? (e) What is the relation between Cr^{++} and $V(OH)_4^+$, and between Cr^{+3} and VO^{++} at the equivalence point of the titration? (f) Derive an expression that shows how the voltage at the equivalence point depends on the two standard potentials and the concentration of H^+.

7.15 Electrolytic cells containing as electrolytes zinc sulfate, silver nitrate, and copper sulfate were connected in series. A steady current of 1.50 amp was passed through them until 1.45 gm of silver were deposited at the cathode of the second cell. How long did the current flow? What weights of copper and of zinc were deposited?

7.16 In the electrolysis of sodium sulfate, the reaction that occurs at the anode can be written

$$2H_2O \rightarrow 4H^+ + O_2 + 4e^-.$$

If a steady current of 2.40 amp is passed through aqueous sodium sulfate for 1 hr, what volume of oxygen measured at 25°C and 1-atm pressure is evolved?

CHAPTER 8

CHEMICAL THERMODYNAMICS

The previous three chapters have been concerned with the quantitative description of reacting systems. We have found two related ways of expressing the tendency of reactants to be converted to products: by use of the equilibrium constant K for the reaction, or by means of its standard cell potential $\Delta\mathcal{E}^0$. While we can describe the extent to which a reaction proceeds, as yet we have no insight into why some reactions have large equilibrium constants, while those of other reactions are small. A study of chemical thermodynamics will lead us to an understanding of chemical reactivity by showing how the equilibrium constant of a reaction is related to the properties of individual reactants and products. The role of thermodynamics in understanding chemistry can be illustrated by the following diagram:

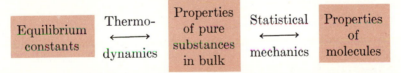

Note that thermodynamics only relates the properties of bulk matter to its behavior in physical and chemical processes. Its great strength is that it accomplishes this *without* making any assumptions about the molecular structure of matter. Because thermodynamics deals only with the macroscopic, observable properties of matter, without assumptions of its atomic nature, it is a subject of very general applicability, and immense reliability.

Thermodynamic reasoning is based on three laws. Two of these, the ones of most immediate application to our experience, are:

The energy of the universe is constant.

The entropy of the universe is increasing.

These laws are not derived. They are deduced from our experiments with the behavior of matter in bulk, and summarize the universal features of all our experience. Their generality has been demonstrated repeatedly, and we expect conclusions based on them to hold true in whatever new experiments we perform.

In order to use these laws, we must know what energy and entropy are—how they are measured and related to other properties of matter. Once we have accomplished this, we will be able to show how a number of things we have regarded as isolated empirical facts can actually be derived from these more fundamental laws of thermodynamics. For example, we will be able to prove that for a general reaction between ideal reagents

$$aA + bB = cC + dD,$$

there *should* be an equilibrium constant of the form

$$K = \frac{[C]^c[D]^d}{[A]^a[B]^b}.\tag{8.1}$$

That is, the existence of Eq. (8.1) is not just an isolated experimental fact; it is a consequence of the laws of thermodynamics and the properties of ideal gases and solutions. Furthermore, we shall find that we can associate a quantity with each compound and element, called its standard free energy, and that the equilibrium constant of any reaction can be expressed in terms of the free energies of reactants and products. Thus thermodynamics shows how the value of any equilibrium constant is related to the properties of individual pure reactants and products. This application alone makes thermodynamics an immensely helpful subject to the chemist.

8.1 SYSTEMS, STATES, AND STATE FUNCTIONS

In performing a controlled experiment, we select the part of the universe of interest to us, and attempt to isolate it from any uncontrolled disturbances. This object, whose properties we wish to study, is called the **system.** All other parts of the universe, whose properties are not of immediate interest, are called the **surroundings.** The surroundings may influence the properties of the system by, for example, determining its temperature or pressure, but in a carefully designed experiment these influences will be controlled and measurable.

Thermodynamics is concerned with the **equilibrium states** of systems. An equilibrium state is one in which the macroscopic properties of the system, such as its temperature, density, and chemical composition, are well defined and do not change with time. Thus thermodynamics is not concerned with the rate at which chemical or physical processes occur, nor does it attempt to describe systems while these changes are going on. Thermodynamic reasoning can be used to tell us whether it is possible, *in principle*, to go from one particular state of the reactants to some particular state of the products of a reaction, but it cannot tell us whether that change can be accomplished in a finite time. This information may seem limited but it is still very valuable. If application of thermodynamics shows that a particular reaction is impossible, there is no point to attempting to make it proceed. If thermodynamics shows the reaction is possible, in principle, then it may be worth the effort to accomplish it in practice. A notable example of this use of thermodynamics occurred in the efforts to convert graphite to diamond. Many attempts to accomplish this conversion in the laboratory failed, but thermodynamics showed the reaction was possible under certain conditions of high temperature and pressure. This assurance encouraged researchers to continue their efforts, which eventually were successful.

The description of thermodynamic systems is made by giving the values of certain quantities called **state functions.** A state function is a property of a system which has some definite value for each state, and which is independent of the manner in which the state is reached. Pressure, volume, and temperature are state functions, and there are five others which are important in thermodynamic arguments. State functions have two very important properties. First, assigning values to a few state functions (usually two or three) automatically fixes the values of all others. Second, when the state of a system is changed, the *changes* of the state functions depend *only on the initial and final states of the system*, and not on how the change is accomplished.

As an illustration of the first property of state functions, consider the consequence of assigning values to the volume V and the temperature T of one mole of an ideal gas. We know that then the pressure must assume the value $P = RT/V$. Thus the value of one state function is automatically determined by specifying the values of the volume and temperature. All other state functions also assume definite values, although the algebraic relation between them and volume and temperature may be complicated.

To demonstrate the second property of state functions, we need only consider a change in the state of an ideal gas from $P_1 = 1$ atm, $V_1 = 22.4$ liters, $T_1 = 273°$K to a final state in which $P_2 = 10$ atm, $V_2 = 4.48$ liters, and $T_2 = 546°$K. Then we say that the pressure change* ΔP is given by

$$\Delta P = P_2 - P_1 = 9 \text{ atm},$$

---

* The symbol Δ always stands for the operation of subtracting the *initial* value of a quantity from its *final* value. Thus $\Delta P = P_f - P_i = P_2 - P_1$.

and the volume change is

$$\Delta V = V_2 - V_1 = 17.9 \text{ liters,}$$

while the temperature change is

$$\Delta T = T_2 - T_1 = 273°\text{K.}$$

That is, the change in each of these state functions depends only on their values in the initial and final states of the system, and *not* on *how* the change was accomplished. It does not matter that during the change the pressure might have risen to 100 atm and the volume decreased to 0.224 liter. Changes in state functions are determined only by the initial and final states of the system, and not by the path taken between them.

This property of state functions is by no means trivial, even though it may seem obvious. Quantities whose values are not independent of *how* a change occurs are not state functions. For example, the angular separation between two points on the earth is a fixed constant which depends on the coordinates of the two points. On the other hand, the distance one covers in traveling between the points depends on the route one takes. Thus separation is a state function, but distance traveled is not. State functions are important in thermodynamics because the subject deals only with equilibrium states, and not with how a change in state occurs. Therefore thermodynamic decisions as to whether a particular change is possible must be based on the accompanying changes of state functions, for only these are independent of the way changes occur.

8.2 WORK AND HEAT

In mechanics, work is defined as the product of a force times a displacement. That is

$$\text{mechanical work} \equiv \text{force} \times \text{distance,}$$
$$w = f \times r,$$

where f is a constant force applied in the direction of the displacement r. Work is the means by which the energy of a mechanical system is *changed*. Thus, if we raise a mass m to a height h against the gravitational acceleration g, we apply a force mg over a distance h and do work

$$w = mg \times h$$

on the mass. We also say that we have changed the (potential) energy of the mass from an arbitrary amount taken as zero at the surface of the earth to a new value mgh at the height h.

If we apply a constant acceleration a to a free particle of mass m over a distance $r_2 - r_1$, the work done on the particle is $w = ma(r_2 - r_1)$. But

$$r_2 - r_1 = \left(\frac{v_2 + v_1}{2}\right) t,$$

where $(v_2 + v_1)/2$ is the average velocity over the distance $r_2 - r_1$, and t is the time taken to travel this distance. Also, we have

$$v_2 - v_1 = at,$$

so we get

$$w = ma(r_2 - r_1) = m \frac{(v_2 - v_1)}{t} \times \left(\frac{v_2 + v_1}{2}\right) t$$

$$= \frac{mv_2^2}{2} - \frac{mv_1^2}{2}. \tag{8.2}$$

The right-hand side of Eq. (8.2) is just the final kinetic energy of the particle minus the initial kinetic energy. Once again, the work done on a simple mechanical system is equal to the *change* in its energy.

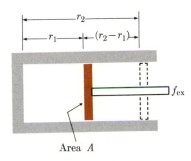

Expansion of a gas against an external force f_{ex}.

FIG. 8.1

A particularly important form of work is that associated with a pressure-volume change. Consider Fig. 8.1, which shows a gas confined by a piston expanding against a constant external force f_{ex}. We can again calculate the work as the product of the force and the displacement. However, in this and in all future applications, we want the symbol w to assume a special significance: w is the *work done on the system* by the surroundings. When a gas expands against an external force f_{ex}, it actually does work *on* the surroundings, so w, the work done on the gas, should be a negative quantity for this process. We therefore should write

$$w = -f_{ex}(r_2 - r_1)$$

for the work *done on* the gas during the expansion. Note that since r_2 is greater than r_1, and f_{ex} is always taken to be positive, w is in fact negative for the expansion. This is consistent with our choice of meaning for the symbol w.

To find an expression for w in terms of pressure and volume, we can introduce the area A of the piston to give

$$w = -\frac{f_{ex}}{A} \times A(r_2 - r_1).$$
(8.3)

But $A(r_2 - r_1) = \Delta V$, the volume change for the gas. Also f_{ex}/A is the force per unit area, or the external pressure, against which the gas expands. Thus Eq. (8.3) becomes

$$w = -P_{ex}\,\Delta V \quad \text{if pressure is constant.}$$
(8.4)

For a more general pressure-volume change in which pressure is not constant, we can calculate the work by first saying that an infinitesimal volume change dV produces an infinitesimal amount of work dw. Thus

$$dw = -P_{ex}\,dV.$$

During this infinitesimal volume change, the pressure remains virtually constant at P_{ex}. The work done in a finite displacement is the sum of such infinitesimals, or

$$w = -\int_{V_1}^{V_2} P_{ex}\,dV.$$
(8.5)

This is a general formula which allows us to compute the work, *if we know how P_{ex} depends on V*. If the external pressure is constant throughout the expansion, we can take it outside the integral sign and write

$$w = -P_{ex}\int_{V_1}^{V_2} dV = -P_{ex}(V_2 - V_1) = -P_{ex}\,\Delta V,$$

and we recover Eq. (8.4).

Note that it is the *external* pressure that is used in calculating the work. No matter what the gas pressure is, a volume change does no work unless the system is linked to the surroundings by an *external* force represented by P_{ex}. If this external force is zero, there is no mechanical link between the system and its surroundings, and no mechanical work can be done on or by the system.

Consideration of Eq. (8.5) shows that the work done in a process depends on *how* the change from V_1 to V_2 is accomplished. We can see this more clearly by referring to Fig. 8.2. There are two particularly simple paths by which a system may change its state from P_1, V_1, to P_2, V_2. In part (a), we first change the volume from V_1 to V_2 at a constant pressure P_1. Then we change the pressure from P_1 to P_2, keeping the volume constant. In part (b) we simply reverse the order of changes. In Fig. 8.2 the work done in following each of the paths is represented as the area under each of the curves followed from the initial to the final state. It is clear from this drawing that the work done depends on the path followed, even though the initial and final states are the same. Consequently, we must conclude that *work is not a state function*, for its value depends on the path taken between states.

Work is the only means by which energy can be transferred to and from the simple hypothetical systems of mechanics. However, we must recognize that there is another way in which energy can be exchanged with systems of the real world. If a temperature difference exists between a system and its surroundings, energy may be transferred by "heat flow"—radiation or conduction.

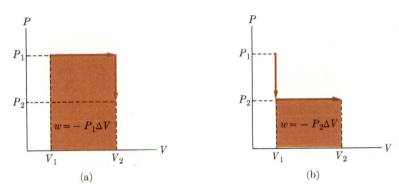

Work done in going from the initial to the final state depends on the path followed.

FIG. 8.2

The concept of heat is clouded by an historically based tendency to think of heat as "something" that "flows." In truth, heat is not a substance. It is, like work, a method by which systems exchange energy. The proof of this was given by James Joule who showed that the same change in state (i.e., a certain rise in temperature) can be accomplished either by doing work on a body, or by heating it. Furthermore, the amount of heat, measured in calories, and the amount of work, measured in joules, necessary to effect any given change always stand in a fixed ratio. That is, one calorie of heat always produces the *same* change in the state of a system as do 4.18 joules of work. Thus heat and work are both methods of changing the energy of a system, and 1 calorie = 4.18 joules. To distinguish between them, we need only say that work is energy transferred by virtue of a mechanical link between systems, and that heat is energy transferred due to a temperature difference.

Our discussion of the nature of heat allows one more conclusion. Heat is not a state function. This must be so, for as Joule showed, the same change in state may be brought about by using either heat or work. That means that the amount of heat used to make the state change depends on how the change in state is made. For the path in which work alone is used, the heat used is zero. For the path in which heat alone is used, the heat used is, of course, not zero. Thus the heat used depends on the path between states, and heat is not a state function.

8.3 THE FIRST LAW OF THERMODYNAMICS

Our discussion has suggested that there is a very important difference between heat and work on one hand, and energy on the other. Heat and work refer to *processes*—events in which displacements occur or temperature changes. In contrast, energy is a *property* that can be associated with a single equilibrium state of a system. It appears then that energy is a state function.

This can be seen most clearly by thinking about the energy of simple mechanical systems, for example, a free particle of mass m moving in a vacuum with velocity v. We know that such a system has a kinetic energy $\frac{1}{2}mv^2$. If it is at a height h above the surface of the earth, we say its potential energy relative to the earth's surface is mgh. We see that the kinetic and potential energies of this simple system are functions of its state, that is, its velocity and position. Thus the energy of a simple mechanical system is definitely a state function.

Any macroscopic amount of a chemical substance can be regarded as a collection of simple mechanical systems. Is it possible to associate an internal energy with a chemical system, and is this energy a state function? To answer this, let us consider two different states of a system, and two different paths, a and b, which connect them. If we say that in going from state 1 to state 2 along path a, we must put into the system an amount of energy ΔE_a, and if by making the same state change by following path b, we must put into the system an energy ΔE_b, then if internal energy is a state function,

$$\Delta E_a = \Delta E_b.$$

This must be true, since the change in a state function is independent of the path taken between states.

But suppose that energy is *not* a state function, and that for the sake of argument $\Delta E_b > \Delta E_a$. What would be the consequences? We could take the system from 1 to 2 along path a; this requires us to put in an amount of energy ΔE_a. Then we could *return* the system to state 1 by path b. This would allow us to *extract* an amount of energy ΔE_b. As a result of these operations, we would obtain a net amount of energy $\Delta E_b - \Delta E_a$, and the system would be unchanged. There would be nothing to prevent us from repeating the process and obtaining more energy.

As attractive as this creation of energy is, all attempts to achieve it have failed. The failure has been so well documented that it is accepted as a general truth and expressed as the law of conservation of energy: energy may be neither created nor destroyed, only transferred or changed from one form to another. As a consequence of the law of conservation of energy, we must conclude that $\Delta E_a = \Delta E_b$ for our example, and that the energy change must be independent of the path taken between states. In other words, internal energy is a state function, and the basis for this conclusion is the overwhelming experimental evidence that energy is conserved.

Now consider the effect of adding energy to a system as an amount of heat q. If energy is conserved, and if the system does no work, q must appear as a change in the internal energy of the system ΔE. Thus

$$\Delta E = q \qquad \text{(no work done)}.$$

If, in a separate experiment, we do work on a system but do not allow it to transfer heat to or from its surroundings, then the work done must appear as an internal energy change of the system. In this case

$$\Delta E = w \qquad \text{(no heat transferred)}.$$

In general, we can expect to find processes in which heat is added to, and work is done on, a system. The foregoing examples lead us to the expression

$$\Delta E = q + w, \tag{8.6}$$

$$\frac{\text{internal energy}}{\text{change}} = \frac{\text{heat added}}{\text{to system}} + \frac{\text{work done}}{\text{on system}}.$$

Equation (8.6) is a mathematical statement of the **first law of thermodynamics**. We can say, then, that the first law of thermodynamics is just the law of conservation of energy, in which specific account of heat effects has been taken.

Note carefully that *both* the heat *added to* a system and the work *done on* the system are assigned *positive* symbols. This is a convention which is now followed by most textbooks on thermodynamics and physical chemistry. In the past, however, many books followed the convention that w was the work done *by* the system. If this definition of w is made, then the first law of thermodynamics must be written

$$\Delta E = q - w \qquad \text{(older convention)},$$

and the pressure volume work is given by

$$w = \int P \, dV \qquad \text{(older convention)}.$$

Therefore, when another book is consulted, it is important to ascertain which convention is being followed, in order to avoid confusion.

We have remarked that thermodynamics deals only with macroscopic properties of matter, and does *not* use the results of the atomic theory in any way. However, in order to understand the significance of the thermodynamic state functions most fully, it is often helpful to use the results of the atomic and kinetic theories. Thus we may ask for an explanation, in terms of atomic properties, of what internal energy is. The internal energy of a system results from the kinetic energies of its molecules, the potential energy associated with forces between molecules, and the kinetic and potential energies of the electrons and nuclei in molecules. This may not be a complete list of contributions to the internal energy, and in fact we should add to it the energy associated with the existence of the mass of the system. When the internal energy of a system changes, some or all of these contributing energies change. The virtue of thermodynamics is that it shows us how to use the internal energy concept, without requiring that we analyze the individual contributions to the internal energy of a system.

Measurement of ΔE

Suppose we have a chemical process in which reactants at 25°C are completely converted to products at the same temperature. This is a change in the state of a chemical system, and there must be a definite value of ΔE associated with it. The value of ΔE is of interest, for it tells how the internal energies of reactants and products differ. It is a quantitative comparison between the mechanical stabilities of reactants and products. How can we measure the ΔE of a chemical reaction?

To answer, we need only refer to Eq. (8.6),

$$\Delta E = q + w, \tag{8.6}$$

and recognize that when a chemical reaction occurs, under ordinary circumstances the only way the system can do work is by a pressure-volume change. Thus

$$w = -\int_{V_1}^{V_2} P \, dV, \qquad \Delta E = q + \int_{V_1}^{V_2} P \, dV. \tag{8.7}$$

But if the reaction were run in a closed container so that the volume of the system was constant at V_1, we would have

$$\Delta E = q + \int_{V_1}^{V_2} P \, dV$$
$$= q + 0 \qquad \text{(constant } V\text{)}$$
$$= q_V. \tag{8.8}$$

We see that ΔE is numerically equal to the heat absorbed by the system when the process occurs at constant volume. The subscript in Eq. (8.8) emphasizes this point.

To measure ΔE, we need only carry out a reaction at constant volume and measure the heat evolved or absorbed. If heat is evolved, q_V is a negative number, and the internal energy of the products is lower than that of the reactants. Reactions in which heat is evolved are said to be *exothermic*. If heat is absorbed by the system during the reaction, q_V is positive, ΔE is positive, and the products have a greater internal energy than the reactants. Reactions in which heat is absorbed by the system are said to be *endothermic*.

Enthalpy

Commonly, chemical reactions are run not at constant volume, but at a *constant pressure* of 1 atm. Consequently, the heat absorbed under these conditions is not equal to ΔE or q_V. In order to discuss thermal effects for reactions run at constant pressure, it is convenient to define a new function of state by the equation

$$H \equiv E + PV. \tag{8.9}$$

The *enthalpy* H, defined by Eq. (8.9), is definitely a state function, since its value depends only on the values of E, P, and V. Note also that enthalpy must have the units of energy.

A change in enthalpy can be expressed as

$$\Delta H = \Delta E + \Delta(PV)$$
$$= q + w + \Delta(PV). \tag{8.10}$$

Let us restrict our attention to changes that occur only at constant pressure. For such changes,

$$\left. \begin{array}{l} w = -P\,\Delta V \\ \Delta(PV) = P\,\Delta V \end{array} \right\} \quad \text{constant pressure only.}$$

Using these relations in Eq. (8.10) gives us

$$\Delta H = q - P\,\Delta V + P\,\Delta V$$
$$= q_P. \tag{8.11}$$

Thus the enthalpy change is equal to the heat absorbed q_P when a reaction is carried out at constant pressure. For an exothermic process ΔH is negative, and for an endothermic process ΔH is positive.

How different are ΔH and ΔE? We have

$$\Delta H = \Delta E + \Delta(PV). \tag{8.12}$$

For reactions in which *only* liquids and solids are involved, very little volume change occurs, because the densities of all condensed substances containing the same atoms are similar. If the reactions are run at the relatively low pressure of 1 atm, $\Delta(PV)$ is very small, so we have

$$\Delta H \simeq \Delta E \quad \text{(reactions involving only solids and liquids).}$$

On the other hand, if gases are produced or consumed during the reaction, ΔH and ΔE can be quite different. Since for ideal gases,

$$PV = nRT,$$

it follows that at a constant temperature,

$$\Delta(PV) = \Delta nRT,$$

where Δn is the change in the number of moles of gas due to chemical reaction. Thus we obtain from Eq. (8.12),

$$\Delta H = \Delta E + \Delta nRT \qquad \text{(constant } T\text{).} \qquad (8.13)$$

When ΔH and ΔE are expressed in units of calories, we must use $R = 1.987$ cal/mole-deg in Eq. (8.13).

Example 8.1 When 1 mole of ice melts at 0°C and a constant pressure of 1 atm, 1440 cal of heat are absorbed by the system. The molar volumes of ice and water are 0.0196 and 0.0180 liter, respectively. Calculate ΔH and ΔE.

Since $\Delta H = q_P$ we have

$$\Delta H = 1440 \text{ cal} = 4770 \text{ J.}$$

To find ΔE by Eq. (8.12), we must evaluate $\Delta(PV)$.

Since $P = 1$ atm, we have

$$\Delta(PV) = P\,\Delta V = P(V_2 - V_1) = (1)(0.0180 - 0.0196)$$
$$= -1.6 \times 10^{-3} \text{ liter-atm} = -0.039 \text{ cal} = 0.16 \text{ J.}$$

Since $\Delta H = 1440$ cal, the difference between ΔH and ΔE is negligible, and we can say that $\Delta E = 1440$ cal or 4770 J.

Example 8.2 For the reaction

$$C(\text{graphite}) + \tfrac{1}{2}O_2 = CO$$

at 298°K and 1 atm, $\Delta H = -26,416$ cal. What is ΔE, if the molar volume of graphite is 0.0053 liter?

We see that the net change in the number of moles of gas is $\Delta n = +\tfrac{1}{2}$. Thus (ΔV) due to the net production of gas is $\tfrac{1}{2} \times 22.4 = 11.2$ liters. This is much greater than the volume decrease caused by the disappearance of solid graphite; so we can neglect the latter and say that

$$\Delta H = \Delta E + \Delta nRT,$$
$$-26,416 = \Delta E + \tfrac{1}{2}(1.987)(298),$$
$$\Delta E = -26,416 - 296 = -26,712 \text{ cal} = -111,763 \text{ J.}$$

We now realize that the ΔH associated with any change in state can, in principle, be found either directly as the heat absorbed by the system at constant pressure, or indirectly from a measured q_V and use of Eq. (8.12). A quantity useful in discussing chemical reactions is the **standard enthalpy change** ΔH^0. This is the enthalpy change of the system when the reactants in their standard states are converted to the products in their standard states. The standard state of a substance is its most stable form at 1-atm pressure and at a temperature which is usually specified as 298°K. Thus we can write

$$C(graphite) + O_2(g) = CO_2(g),$$
$$\Delta H^0_{298} = -94.05 \text{ kcal} = -393.5 \text{ kJ}.$$

This means that when one mole of carbon is completely converted to one mole of carbon dioxide, with reactants and products at 1-atm pressure and 298°K, 94.05 kcal of heat are evolved, and the standard enthalpy change for the reaction is -94.05 kcal.

The combustion of carbon to carbon dioxide can be carried out quantitatively in a calorimeter, and its accompanying ΔH measured conveniently. The same is true for the reaction

$$CO(g) + \tfrac{1}{2}O_2(g) = CO_2(g),$$
$$\Delta H^0_{298} = -67.63 \text{ kcal} = -283.0 \text{ kJ}.$$

In contrast, the combustion of carbon to *carbon monoxide*,

$$C(graphite) + \tfrac{1}{2}O_2(g) = CO(g), \qquad (8.14)$$

is difficult to carry out quantitatively. Unless an excess of oxygen is used, the combustion of carbon is incomplete, but if an excess of oxygen is used, some of the carbon monoxide is oxidized further to carbon dioxide. However, using the fact that enthalpy is a state function, we can make the direct measurement of ΔH for reaction (8.14) unnecessary.

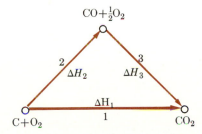

Alternative paths for the conversion of carbon and oxygen to carbon dioxide. Since enthalpy is a function of state, ΔH_1 must equal the sum of ΔH_2 and ΔH_3.

FIG. 8.3

The way to accomplish this is to realize that there are two paths which, in principle, can be used to convert graphite and oxygen to carbon dioxide. These paths are illustrated in Fig. 8.3. We might carry out the reaction directly by step 1, for which the enthalpy ΔH_1 is known. Alternatively we could proceed from reactants to products by steps 2 and 3, and with these steps are associated the enthalpy changes ΔH_2 and ΔH_3. Since H is a state function, ΔH for the conversion of carbon to carbon dioxide is *independent of the reaction path*. This means that

$$\Delta H_1 = \Delta H_2 + \Delta H_3.$$

Both ΔH_1 and ΔH_3 have been measured; therefore

$$\Delta H_1 = -94.05 \text{ kcal},$$
$$\Delta H_3 = -67.63 \text{ kcal},$$
$$\Delta H_2 = \Delta H_1 - \Delta H_3 = -26.42 \text{ kcal}.$$

This is the desired standard enthalpy change for the conversion of carbon to carbon monoxide:

$$C(s) + \tfrac{1}{2}O_2(g) = CO(g), \qquad \Delta H^0_{298} = -26.42 \text{ kcal}.$$

The argument we have just used is a specific example of *Hess' law of constant heat summation:* the heat evolved or absorbed at constant pressure for any chemical change is the same regardless of the path by which the change occurs. Our use of Hess' law is equivalent to the following procedure: We algebraically combine the chemical reactions whose enthalpy we know so as to obtain the desired reaction. To obtain the enthalpy of the reaction, we algebraically combine the known values of ΔH in the same way as the reactions. Thus

$$
\begin{array}{ll}
C(s) + O_2(g) = CO_2(g) & \Delta H^0 = -94.05 \text{ kcal} \\
-[CO(g) + \tfrac{1}{2}O_2(g) = CO_2(g)] & -[\Delta H^0 = -67.63] \text{ kcal} \\
\hline
C(s) + \tfrac{1}{2}O_2(g) = CO(g) & \Delta H^0 = -26.42 \text{ kcal}
\end{array}
$$

A slightly more involved calculation is needed to determine the ΔH^0 of

$$C(s) + 2H_2(g) = CH_4(g)$$

from the measured values of ΔH^0 for

(a) $\quad C(s) + O_2(g) = CO_2(g),$ $\qquad\qquad \Delta H^0 = -94.1 \text{ kcal},$

(b) $\quad H_2(g) + \tfrac{1}{2}O_2(g) = H_2O(l),$ $\qquad\qquad \Delta H^0 = -68.3 \text{ kcal},$

(c) $\ CH_4(g) + 2O_2(g) = CO_2(g) + 2H_2O(l), \ \ \Delta H^0 = -212.8 \text{ kcal}.$

To obtain the desired reaction, we must multiply equation (b) by 2, add equation (a) to it, and subtract equation (c). The values of ΔH^0 must be combined in exactly the same way. We have

$$
\begin{array}{ll}
\text{C(s)} + \text{O}_2(\text{g}) = \text{CO}_2(\text{g}) & \Delta H^0 = -\ 94.1 \text{ kcal}\\
2 \times [\text{H}_2(\text{g}) + \tfrac{1}{2}\text{O}_2(\text{g}) = \text{H}_2\text{O(l)}] & 2 \times [\Delta H^0 = -\ 68.3] \text{ kcal}\\
-\ [\text{CH}_4(\text{g}) + 2\text{O}_2(\text{g}) = \text{CO}_2(\text{g}) + 2\text{H}_2\text{O(l)}] & -\ [\Delta H^0 = -212.8] \text{ kcal}\\
\hline
\text{C(s)} + 2\text{H}_2(\text{g}) = \text{CH}_4(\text{g}) & \Delta H^0 = -\ 17.9 \text{ kcal}
\end{array}
$$

To summarize: a ΔH is associated with each reaction. When any number of reactions are algebraically combined to yield a net reaction, the values of ΔH are combined in exactly the same way to give the ΔH of the net reaction.

The use of Hess' law permits us to avoid performing many difficult calorimetric experiments. A particularly efficient way to tabulate known thermochemical information is by recording the **enthalpy of formation** of compounds. The enthalpy of formation is the ΔH of the reaction in which a pure compound is formed from its elements, with all substances in their standard states. Thus for the reactions

$$
\begin{array}{lll}
\text{C(s)} + \tfrac{1}{2}\text{O}_2(\text{g}) = \text{CO(g)}, & \Delta H^0 = \Delta H_f^0(\text{CO}) & = -26.4 \text{ kcal},\\
\text{H}_2(\text{g}) + \tfrac{1}{2}\text{O}_2(\text{g}) = \text{H}_2\text{O(l)}, & \Delta H^0 = \Delta H_f^0(\text{H}_2\text{O, l}) & = -68.3 \text{ kcal},\\
\text{H}_2(\text{g}) + \text{O}_2(\text{g}) + \text{C(s)} = \text{HCOOH(l)}, & \Delta H^0 = \Delta H_f^0(\text{HCOOH}) & = -97.8 \text{ kcal},
\end{array}
$$

the enthalpy changes are the enthalpies of formation of carbon monoxide, liquid water, and formic acid, respectively. The enthalpies of formation of *elements* in their standard states are zero, by definition.

To see why enthalpies of formation are useful, let us try to calculate the ΔH^0 of

$$
\text{HCOOH(l)} = \text{CO(g)} + \text{H}_2\text{O(l)}, \qquad \Delta H^0 = ?
$$

We can use thermochemical information available to us if we imagine that this reaction is conducted along a path in which formic acid is first decomposed to the elements C, H_2, and O_2, and these elements are then used to form CO and H_2O. This path is illustrated in Fig. 8.4.

Since ΔH^0 for the net reaction is independent of path, we have from Fig. 8.4,

$$
\Delta H_1 = \Delta H_2 + \Delta H_3.
$$

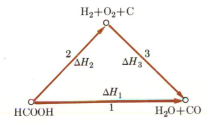

Alternative paths for the conversion of formic acid to carbon monoxide and water.

FIG. 8.4

Table 8.1 Enthalpies cf formation, ΔH_f^0 (kcal/mole) at 298°K

INORGANIC COMPOUNDS

$H_2O(g)$	−57.79	$CO(g)$	− 26.41
$H_2O(l)$	−68.32	$CO_2(g)$	− 94.05
$H_2O_2(g)$	−32.53	$CaO(s)$	−151.8
$O_3(g)$	34.0	$Ca(OH)_2(s)$	−235.6
$HCl(g)$	−22.06	$CaCO_3(s)$	−288.4
$SO_2(g)$	−70.96	$BaO(s)$	−133.5
$SO_3(g)$	−94.45	$BaCO_3(s)$	−290.8
$H_2S(g)$	− 4.81	$BaSO_4(s)$	−345.3
$N_2O(g)$	19.49	$Fe_2O_3(s)$	−196.5
$NO(g)$	21.60	$Al_2O_3(s)$	−399.1
$NO_2(g)$	8.09	$CuO(s)$	− 37.6
$NH_3(g)$	−11.04	ZnO	− 83.2

ORGANIC COMPOUNDS
Gases

Methane, CH_4	−17.89	Ethylene, C_2H_4	12.50
Ethane, C_2H_6	−20.24	Propylene, C_3H_6	4.88
Propane, C_3H_8	−24.82	l-butene, C_4H_8	0.28
n-butane, C_4H_{10}	−29.81	cis-2-butene, C_4H_8	− 1.36
Isobutane, C_4H_{10}	−31.45	trans-2-butene, C_4H_8	− 2.40
Acetylene, C_2H_2	54.19	Isobutene, C_4H_8	− 3.34

Liquids

Methanol, CH_3OH	−57.02	Acetic acid, CH_3COOH	−116.4
Ethanol, C_2H_5OH	−66.35	Benzene	11.72

GASEOUS ATOMS

H	52.1	C	171.7
O	59.1	N	112.5
Cl	29.0	Br	26.7

But ΔH_3 is the sum of the enthalpies of formation of CO and of H_2O:

$$\Delta H_3 = \Delta H_f^0(CO) + \Delta H_f^0(H_2O, l).$$

Also, ΔH_2 is the negative of the enthalpy of formation of formic acid, since step 2 is just the *reverse* of the formation of formic acid from its elements. Thus

$$\Delta H_2 = -\Delta H_f^0(HCOOH),$$

and

$$\Delta H_1 = \Delta H_f^0(CO) + \Delta H_f^0(H_2O, l) - \Delta H_f^0(HCOOH)$$
$$= +3.1 \text{ kcal}. \tag{8.15}$$

We can see from this example that the enthalpy of a reaction can be calculated from the enthalpies of formation of reactants and products. The general expression, of which Eq. (8.15) is a specific example, is

$$\Delta H = \sum \Delta H_f(\text{products}) - \sum \Delta H_f(\text{reactants}). \tag{8.15a}$$

The equation is a consequence of the fact that any reaction can proceed, in principle, by a path in which the first step is decomposition of reactants to the elements, and the second step is formation of the products from these elements. To the overall ΔH of reaction, the first step contributes $-\Delta H_f$ for each reactant, and the second contributes ΔH_f for each product.

Besides being a useful quantity in thermochemical calculations, the enthalpy of formation is a quantitative expression of the mechanical stability of a compound with respect to its elements. Table 8.1 gives the enthalpy of formation of a few common compounds. When ΔH_f^0 is positive, the compound is energetically less stable than its elements, and when ΔH_f^0 is negative, the compound has lower energy and is more stable than its elements.

Heat Capacity

The heat capacity of a substance is the amount of heat required to raise *one mole* of material one celsius degree. Because heat is *not* a state function, the amount required to produce a given change in state depends on the path followed. Therefore two types of heat capacity are used: C_P for changes at constant pressure, and C_V for changes at constant volume. The mathematical definitions are

$$C_P = \frac{dq_P}{dT} = \frac{dH}{dT}, \tag{8.16}$$

$$C_V = \frac{dq_V}{dT} = \frac{dE}{dT}. \tag{8.17}$$

The amount of heat needed to change the temperature of n moles of material from T_1 to T_2 is therefore

$$q_P = n \int_{T_1}^{T_2} C_P \, dT$$

$$= nC_P \int_{T_1}^{T_2} dT = nC_P \, \Delta T \qquad \text{if} \qquad C_P \text{ is a constant}; \tag{8.18}$$

$$q_V = n \int_{T_1}^{T_2} C_V \, dT$$

$$= nC_V \int_{T_1}^{T_2} dT = nC_V \, \Delta T \qquad \text{if} \qquad C_V \text{ is a constant}. \tag{8.19}$$

The difference between C_P and C_V can be found very simply. For one mole of material,

$$H = E + PV, \qquad \frac{dH}{dT} = \frac{dE}{dT} + \frac{d(PV)}{dT}, \qquad C_P = C_V + \frac{d(PV)}{dT}.$$

Table 8.2 Molar heat capacities at constant pressure, C_P (cal/mole-deg)

H_2	6.90	CO_2	8.96
O_2	7.05	CH_4	8.60
N_2	6.94	C_2H_6	12.71
CO	6.97	NH_3	8.63
Cl_2	8.14	$H_2O(g)$	5.92

For solids and liquids, $d(PV)/dT$ is generally small, so $C_P \cong C_V$. For ideal gases, $PV = RT$, and

$$\frac{d(PV)}{dT} = \frac{d(RT)}{dT} = R, \qquad C_P = C_V + R.$$

The gas constant $R \cong 2$ cal/mole-deg, and by referring to the heat capacities listed in Table 8.2 we can see that R, the difference between C_P and C_V, is an appreciable fraction of the heat capacity.

Temperature dependence of ΔH

We have been concerned with the values of ΔH for reactions at one temperature only. It is conceivable that ΔH for a reaction is a function of temperature. In this section we will show how application of the fact that ΔH is independent of reaction path supplies us with an expression for the temperature dependence of ΔH.

Consider the general reaction

$$a\text{A} + b\text{B} = c\text{C} + d\text{D}.$$

The conversion of reactants to products at a temperature T_1 can be carried out by either of the two paths shown in Fig. 8.5. Let us suppose we know ΔH_1, the enthalpy change when reactants and products are at temperature T_1. We wish to find ΔH_2, the enthalpy change when the reaction is run at temperature T_2.

FIG. 8.5   Alternative paths for converting reactants to products.

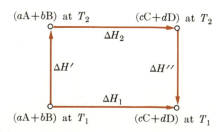

Referring to Fig. 8.5, we see that since ΔH is independent of path,

$$\Delta H_1 = \Delta H' + \Delta H_2 + \Delta H'',$$

where $\Delta H'$ is the enthalpy change associated with changing the temperature of the *reactants* at constant pressure from T_1 to T_2, and $\Delta H''$ is the enthalpy change which results from changing the temperature of the *products* from T_2 to T_1 at a constant pressure. The *total* heat capacity of the reactants is

$$C_P(\text{reactants}) = aC_P(\text{A}) + bC_P(\text{B}),$$

so for $\Delta H'$ we have

$$\Delta H' = \int_{T_1}^{T_2} C_P(\text{reactants}) \, dT.$$

Similarly,

$$C_P(\text{products}) = cC_P(\text{C}) + dC_P(\text{D}),$$

and therefore

$$\Delta H'' = \int_{T_2}^{T_1} C_P(\text{products}) \, dT.$$

Now ΔH_2 is the only unknown, so

$$\Delta H_2 = \Delta H_1 - \Delta H'' - \Delta H'$$

$$= \Delta H_1 - \int_{T_2}^{T_1} C_P(\text{products}) \, dT - \int_{T_1}^{T_2} C_P(\text{reactants}) \, dT.$$

We can change the sign of the second term on the right-hand side of this equation if we reverse the limits of integration. Thus

$$\Delta H_2 = \Delta H_1 + \int_{T_1}^{T_2} C_P(\text{products}) \, dT - \int_{T_1}^{T_2} C_P(\text{reactants}) \, dT.$$

This expression can be made more compact if we define

$$\Delta C_P = C_P(\text{products}) - C_P(\text{reactants})$$
$$= cC_P(\text{C}) + dC_P(\text{D}) - aC_P(\text{A}) - bC_P(\text{B}).$$

The integrals can be combined to give

$$\Delta H_2 = \Delta H_1 + \int_{T_1}^{T_2} \Delta C_P \, dT. \tag{8.20}$$

We see now that the difference in ΔH at the two temperatures depends on the *difference* of the heat capacities of the products and reactants. Often this heat-capacity difference is very small, and ΔH is virtually independent of temperature, particularly over small temperature ranges.

Example 8.3 Find ΔH^0 at 398°K for the reaction

$$CO + \tfrac{1}{2}O_2 = CO_2, \qquad \Delta H_{298}^0 = -67{,}640 \text{ cal} = -283.0 \text{ kJ}.$$

From Table 8.2 we find

$$C_P(CO) = 6.97, \qquad C_P(O_2) = 7.05, \qquad \text{and} \qquad C_P(CO_2) = 8.96,$$

all in cal/mole-deg. Therefore

$$\Delta C_P = 8.96 - 6.97 - 7.05/2 = -1.53 \text{ cal/deg},$$
$$\Delta H_{398}^0 = \Delta H_{298}^0 + \Delta C_P(398 - 298)$$
$$= -67{,}640 - 153 = -67{,}790 \text{ cal} = -283.6 \text{ kJ}.$$

Thus the ΔH of reaction is only slightly more negative at this higher temperature.

8.5 CRITERIA FOR SPONTANEOUS CHANGE

We have derived several relations from the first law of thermodynamics which help us make efficient use of calorimetric data. However, we have not yet achieved our major purpose—to learn to use the properties of individual substances to predict the extent to which chemical reactions proceed. It is true that we can associate an enthalpy of formation with each compound and with them calculate the ΔH of a reaction. But the value of ΔH alone is not a sufficient criterion to decide whether a reaction will proceed spontaneously from reactants to products. Granted that there are many exothermic reactions that have large equilibrium constants, it is nevertheless an experimental fact that endothermic reactions can also proceed almost to completion.

There are also physical processes that have a preferred spontaneous direction which cannot be rationalized on the basis of the first law of thermodynamics alone. An ideal gas expands spontaneously into an evacuated container. It does not do this in order to lower its energy, for experiments show that the energy of an ideal gas is independent of its volume. The reverse process, a spontaneous collection or compression of the gaseous molecules, is allowed by the first law of thermodynamics, but it never occurs. As another example, consider that we always observe that heat flows from a hot to a cold body. This process *and its reverse* both obey the law of conservation of energy. However, heat never does flow spontaneously from a cold to a hot body. It is clear, then, that the first law of thermodynamics alone does not explain the directions of spontaneous physical or chemical processes.

This conclusion should not be a surprise. In previous chapters we have remarked that the tendency of molecules to seek a state of minimum energy is insufficient to explain the occurrence of many chemical and physical changes. We had to recognize an additional tendency toward maximum molecular chaos. In this section we will be concerned with the thermodynamic description of this tendency toward molecular chaos, and we will see that the spontaneous direction of physical and chemical processes can be found by application of the second law of thermodynamics. First, some remarks about spontaneous and reversible processes are in order.

Reversibility and Spontaneity

A reversible process is one which is carried out so that the state functions of a system never differ by more than an infinitesimal amount from one moment to another. Since the thermodynamic functions change infinitely slowly in reversible processes, they are sometimes said to be *quasistatic processes*. Another characteristic of a reversible change is that the state functions of a system, like pressure and temperature, never differ from those of the surroundings by more than an infinitesimal amount. For example, to carry out an expansion reversibly we must have

$$P_{int} = P_{ex} + dP,$$

and for a reversible compression,

$$P_{int} = P_{ex} - dP,$$

where P_{int} is the pressure of the system. Since no more than an infinitesimal pressure difference exists between the system and surroundings, the net acceleration acting on the system is infinitesimally small, and any change will occur quasistatically. Likewise, for a reversible temperature change we must have $T_{ex} = T_{int} \pm dT$, if the heating or cooling is to take place infinitely slowly.

A spontaneous or irreversible mechanical change takes place at a finite rate. If the process involves a change in pressure or temperature, these variables differ by a finite amount between the system and its surroundings. Thus there is an important practical difference between a reversible and an irreversible process. The *direction* of a reversible process can be reversed at any time, just by making an infinitesimal change in the surroundings. That is, a reversible *compression* can be turned into a reversible *expansion* just by decreasing the externally applied pressure by an infinitesimal amount. On the other hand, an irreversible process cannot be stopped or reversed by an infinitesimal change in external conditions, for any such change cannot overcome the finite differences in pressure, temperature, or other thermodynamic functions which are responsible for the irreversible process.

Another of the important differences between reversible and irreversible processes is that the work done on a system in a reversible process is less than in the corresponding irreversible process between the same two states. In the case of a *reversible* compression, P_{ex} and P_{int} differ by only an infinitesimal amount, so we can write

$$w_{rev} = -\int P_{ex}\, dV = -\int (P_{int} + dP)\, dV \cong -\int P_{int}\, dV,$$

since the product of infinitesimals can be neglected. For an irreversible compression, $P_{ex} > P_{int}$, so we have

$$w_{irrev} = -\int_{V_1}^{V_2} P_{ex}\, dV > -\int_{V_1}^{V_2} P_{int}\, dV = w_{rev}.$$

Note that the choice of the direction of the inequality results from realizing that for a compression, $V_2 < V_1$; so the integrals (and the work) are positive. Since $P_{ex} > P_{int}$, the inequality must have the direction indicated. Thus we are able to conclude

$$w_{rev} < w_{irrev}. \tag{8.21}$$

As an illustration of Eq. (8.21), consider the reversible, isothermal compression of an ideal gas. Since for a reversible process $P_{int} = P_{ex}$, and $P = nRT/V$ for an ideal gas, we obtain

$$w_{rev} = -\int_{V_1}^{V_2} P_{ex}\, dV = -\int_{V_1}^{V_2} \frac{nRT}{V}\, dV = -nRT \int_{V_1}^{V_2} \frac{dV}{V},$$

$$w_{rev} = -nRT \ln \frac{V_2}{V_1}. \tag{8.22}$$

Thus the work done on the gas in the reversible compression is the area under the $P - V$ isotherm between V_1 and V_2, as shown in Fig. 8.6. Since $V_2 < V_1$, $w_{rev} > 0$, as must be true if work is done on the gas.

Let us choose the corresponding *irreversible* compression to be one in which the external pressure is suddenly increased from $P_{ex} = P_1 = nRT/V_1$ to $P_{ex} = P_2 = nRT/V_2$, *without an appreciable change in the volume of the system.* The compression from V_1 to V_2 then occurs with a constant external pressure $P_{ex} = P_2 = nRT/V_2$. This path is also illustrated in Fig. 8.6. The work done by the gas in this irreversible compression is

$$w_{irrev} = -\int_{V_1}^{V_2} P_{ex}\, dV = -P_2(V_2 - V_1) > 0.$$

The graphical representation of this work in Fig. 8.6 makes it clear that $w_{rev} < w_{irrev}$, as was concluded above for the general case.

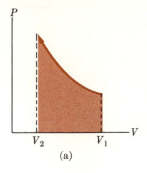 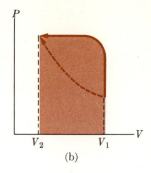

    (a)   (b)

Work done in the isothermal compression of an ideal gas: (a) reversible path; (b) irreversible path, for which P_{ex} is plotted. **FIG. 8.6**

    Now let us compare the work done in the reversible and irreversible *expansions* of a gas. For the reversible case, $P_{ex} = P_{int}$, and

$$w_{rev} = -\int_{V_1}^{V_2} P_{int}\, dV.$$

For the irreversible expansion, $P_{ex} < P_{int}$, and $V_2 > V_1$, so

$$w_{irrev} = -\int_{V_1}^{V_2} P_{ex}\, dV > -\int_{V_1}^{V_2} P_{int}\, dV = w_{rev}.$$

The direction of the inequality comes about because the integrals with their negative signs are negative quantities, and since $P_{ex} < P_{int}$, the integral on the left is less negative (greater) than the one on the right. Thus we again conclude

$$w_{rev} < w_{irrev},$$

which is algebraically the same result as we obtained earlier, but now we are saying that for an expansion, w_{rev} is more negative than w_{irrev}. Figure 8.7 illustrates this point.

    In the application of the second law of thermodynamics, the distinction between q_{irrev} and q_{rev} is important. We can deduce the relation between these two quantities from the inequality (8.21). Imagine the same change state, once carried out reversibly, and again carried out irreversibly. We can write

$$q_{rev} = \Delta E - w_{rev}, \qquad q_{irrev} = \Delta E - w_{irrev}.$$

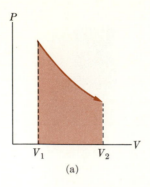 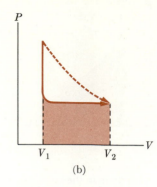

(a)   (b)

FIG. 8.7   Comparison of work done by a gas in isothermal expansions: (a) reversible path; (b) irreversible path, for which P_{ex} is plotted.

Since the change in state is the same, subtraction of the second of these equations from the first gives

$$q_{rev} - q_{irrev} = w_{irrev} - w_{rev}.$$

But by (8.21), $w_{rev} < w_{irrev}$, so

$$q_{rev} - q_{irrev} > 0, \qquad q_{rev} > q_{irrev}. \qquad (8.23)$$

Thus there is a general relation between the heat absorbed by a system in a reversible process, and that in the corresponding irreversible process. This inequality will be used to deduce the criterion for spontaneous change.

8.6 ENTROPY AND THE SECOND LAW

Just as the first law of thermodynamics is a general statement about the behavior of the state function, energy, the second law tells us the general behavior of another state function called **entropy.** The entropy change of a system for any change in state is defined by

$$\Delta S \equiv \int_1^2 \frac{dq_{rev}}{T}. \qquad (8.24)$$

In words, Eq. (8.24) says: take the system from state 1 to state 2 by a *reversible* path. To compute the entropy change of the system, divide each infinitesimal amount of heat by the temperature T at which it is absorbed by the system, and add all these quantities.

Entropy changes must *always* be computed by taking the system from the initial state to the final state by means of a *reversible* path. However, entropy is a state function, and thus ΔS is independent of the path. Although these two statements sound contradictory, they are not, since

$$\frac{dq_{rev}}{T} \neq \frac{dq_{irrev}}{T}.$$

The situation here is similar to that encountered in the calculation of ΔH; ΔH is independent of the path, but it is only equal to q when a process is carried out at constant pressure. The entropy change is also independent of the path, but it is equal to $\int dq/T$ only when the process is carried out reversibly. It is $\int dq/T$ which depends on how the process is done, and not ΔS of the system.

The formal statement of the second law of thermodynamics is:

The entropy S is a function of state. In a reversible process, the entropy of the universe is constant. In an irreversible process, the entropy of the universe increases.

As we have remarked, the thermodynamic laws are not derived mathematically, but are general expressions of experimental findings. To "prove" the first law of thermodynamics, that energy is a state function, we showed that to deny its validity would be to say that creation of energy is possible, and all other experience tells us this is not true. To "prove" the second law of thermodynamics, we will demonstrate that to deny it implies that gases can spontaneously compress themselves, and that heat can flow spontaneously from cold to hot regions.

Entropy Calculations

We shall demonstrate the use of Eq. (8.24) by calculating the entropy change that accompanies the isothermal expansion of an ideal gas. For an isothermal process Eq. (8.24) becomes

$$\Delta S = \int_1^2 \frac{dq_{rev}}{T}$$
$$= \frac{1}{T} \int_1^2 dq_{rev}$$
$$= \frac{q_{rev}}{T},$$

where q_{rev} is the heat added to the system in going from state 1 to state 2. We now wish to express ΔS in terms of the initial and final volumes of the gas. To do this we make use of the experimental fact that the internal energy of an ideal gas depends *only* on its temperature. Before we use this fact, we might remark that it is consistent with the gas-kinetic theory discussed in Chapter 2, where we wrote for an ideal gas,

$$E_{\text{trans}} = \tfrac{3}{2}RT.$$

Thus, if temperature is constant, the translational energy and the total internal energy of an ideal gas are constant. For an *isothermal* expansion, then,

$$\Delta E = 0 = q + w.$$

If the expansion is both reversible and isothermal, we have by Eq. (8.22)

$$q_{\text{rev}} = -w_{\text{rev}} = nRT \ln \frac{V_2}{V_1}.$$

Therefore

$$\Delta S = \frac{q_{\text{rev}}}{T} = nR \ln \frac{V_2}{V_1}. \qquad (8.25)$$

Note that if $V_2 > V_1$, the gas has expanded and its entropy has increased. If $V_2 < V_1$, we have a compression and a decrease in the entropy of the gas.

Are these last two statements consistent with the second law of thermodynamics? The restriction imposed by the second law is that the entropy change of the *universe* (system and surroundings) must be zero for a reversible process. Therefore, to see whether the second law is obeyed, we must find the entropy change of the *surroundings* as well as that of the system.

In a reversible expansion, the gas absorbs an amount of heat q_{rev}, so its entropy change is

$$\Delta S_{\text{gas}} = \frac{q_{\text{rev}}}{T}.$$

The surroundings, a thermostat at the temperature T, *lose* an amount of heat equal in *magnitude* to q_{rev}. Thus the entropy change of the surroundings is

$$\Delta S_{\text{surr}} = -\frac{q_{\text{rev}}}{T},$$

where we are to take q_{rev} as a positive number, the heat absorbed by the system. The total entropy change is

$$\Delta S = \Delta S_{\text{gas}} + \Delta S_{\text{surr}} = \frac{q_{\text{rev}}}{T} - \frac{q_{\text{rev}}}{T} = 0.$$

As the second law requires, $\Delta S = 0$ for a reversible process.

Consider now the irreversible isothermal expansion from V_1 to V_2. Since S is a state function, ΔS for the gas is independent of the path, so we can say

$$\Delta S_{\text{gas}} = nR \ln \frac{V_2}{V_1}.$$

How then does the irreversible expansion differ from the reversible process? Suppose that the expansion is carried out against zero external force. Then $w = 0$, and since $\Delta E = 0$ for an isothermal process on an ideal gas, $q = 0$. Therefore no heat is lost by the surroundings, which means that the entropy change of the surroundings is zero. For this irreversible expansion,

$$\Delta S = \Delta S_{gas} + \Delta S_{surr}$$

$$= nR \ln \frac{V_2}{V_1} + 0,$$

which is greater than zero. The entropy of the universe increases, as the second law states.

Now we can examine the possibility of a *spontaneous compression* of an ideal gas from V_1 to V_s. The entropy change of the gas would be

$$\Delta S_{gas} = nR \ln \frac{V_s}{V_1},$$

which, because $V_s < V_1$, is negative. If the compression is to occur spontaneously with no outside influence, then surely $\Delta S_{surr} = 0$, since the surroundings change in no way. Thus for the total entropy change, we obtain

$$\Delta S = nR \ln \frac{V_s}{V_1} + 0 < 0.$$

Since the total entropy change of the universe is negative, this spontaneous compression of a gas is impossible according to the second law. To put it another way: to deny the validity of the second law would be to say that spontaneous compressions of a gas are possible, when in fact they are never observed.

Let us consider one other application of the entropy criterion for spontaneous change. Two blocks of material at different temperatures T_h and T_c are brought together momentarily. The cold block absorbs a small amount of heat dq, and the hot one loses the same amount. The amount of heat transferred is so small that the temperatures of the two blocks change by a negligible amount. Does this spontaneous heat flow produce an increase in entropy?

We can find the entropy changes by realizing that if the cold body received the heat dq reversibly, its entropy change would be

$$dS_c = \frac{dq}{T_c}.$$

Likewise, if the hot body were to *lose* the heat dq reversibly, its entropy change would be

$$dS_h = -\frac{dq}{T_h},$$

where we are still thinking of dq as a positive number, the heat absorbed by the cold block. These entropy changes, calculated with the assumptions of reversible transfer of heat, are equal to the entropy changes experienced by the blocks when they are brought together momentarily and undergo an irreversible transfer of heat dq. Thus the total entropy change of the blocks is

$$dS = dS_c + dS_h$$
$$= \frac{dq}{T_c} - \frac{dq}{T_h} > 0.$$

Since $T_h > T_c$, the total entropy change is greater than zero, as the second law requires. If heat dq had passed *from* the cold block *to* the hot one, the entropy change would have been

$$dS = dS_c + dS_h$$
$$= -\frac{dq}{T_c} + \frac{dq}{T_h} < 0.$$

Thus the unaided flow of heat from a cold to a hot body violates the second law of thermodynamics.

These two examples suggest how we might use the second law to find the direction of spontaneous changes. We should compute the entropy change associated with a contemplated process. If the entropy change of the system and its surroundings is negative, the process will not occur. If the total entropy change is positive, the process will occur spontaneously. This is a possible way of using the second law, but we shall find that there are other more efficient procedures.

Temperature Dependence of Entropy

Let us calculate the entropy change which accompanies a finite temperature change. We must imagine a reversible process in which the temperature of the surroundings is never more than infinitesimally different from the temperature of the system. Then, in the expression

$$\Delta S = \int \frac{dq_{\text{rev}}}{T},$$

we can replace dq_{rev} by

$$dq_{rev} = nC_P \, dT,$$

or

$$dq_{rev} = nC_V \, dT,$$

depending on whether the process takes place at constant pressure or at constant volume. The results are

$$\Delta S = \int_{T_1}^{T_2} \frac{nC_P}{T} \, dT \qquad (8.26)$$

$$= \int_{T_1}^{T_2} \frac{nC_V \, dT}{T}. \qquad (8.27)$$

If the temperature interval is small, C_P or C_V can be regarded as constant, and we obtain

$$\Delta S = nC_P \ln \frac{T_2}{T_1}, \qquad \Delta S = nC_V \ln \frac{T_2}{T_1}.$$

More frequently C_P and C_V cannot be taken as constants, and the exact form of their temperature dependence must be known before Eqs. (8.26) and (8.27) can be integrated.

8.7 MOLECULAR INTERPRETATION OF ENTROPY

While thermodynamics makes no assumptions about the structure of matter, our understanding of the thermodynamic functions can be deepened if we try to interpret them in terms of molecular properties. We have seen that gas pressure arises from molecular collisions with the walls of a vessel, that temperature is a parameter which expresses the average kinetic energy of molecules, and that internal energy consists of the kinetic and potential energies of all atoms, molecules, electrons, and nuclei in a system. What molecular property does entropy reflect?

To answer this question we must first recognize that there are two ways of describing the state of a thermodynamic system: the *macroscopic* description provided by the values of state functions like P, V, and T, and the *microscopic* description which would involve giving the position and velocity of every atom in the system. The *complete* microscopic description is never used for thermodynamic systems, since just to write down the $3 \times 6 \times 10^{23}$ positional coordinates and the $3 \times 6 \times 10^{23}$ velocity components of a mole of monatomic material would require a pile of $8'' \times 11''$ paper 10 light years high. Moreover, this single microscopic description would be valid only for an instant, since atomic positions and velocities are always changing rapidly. Thus, as we observe any thermodynamic system in an equilibrium *macroscopic* state, its *microscopic* state is changing at an enormous rate.

Despite this molecular activity, the properties of a macroscopic state remain constant. This must mean that there are very many microscopic states consistent with any macroscopic state. *Entropy is a measure of the number of microscopic states associated with a particular macroscopic state.*

To explore this point more thoroughly, let us use a deck of cards as an analog to a thermodynamic system. There are two distinct macroscopic states of the deck: either it is "ordered," with the cards in some standard sequence, or it is "disordered," with the cards in a random sequence. The microscopic state of the deck can be specified by giving the exact order in which the cards are arranged. We can see that there is only one microscopic state which corresponds to the "ordered" macroscopic state. On the other hand, there are many microscopic states associated with the "disordered" macroscopic state, because there are many random sequences of the cards. Since entropy measures, and increases with, the number of microscopic states of the system, we can say that the disordered state has higher entropy than the ordered state.

By using this analysis, we can see why a deck of cards moves from an ordered macroscopic state to a disordered state as the cards are shuffled. Since there are more microscopic states associated with the disordered macroscopic state, it is simply more probable for the deck to end up in the more disordered condition. If we apply this reasoning to the behavior of thermodynamic systems, we can see that entropy has a natural tendency to increase because this corresponds to the movement of systems from conditions of low probability to states of greater probability.

It is now possible to understand why a gas expands spontaneously into a vacuum. In the larger volume, each molecule has more positions available than in the smaller volume. Consequently, in the larger volume, the gas has more microscopic states associated with it than it had in the smaller volume. The gas is found to fill the container because that is the most probable condition for it to be in. Apparently it is not impossible for the gas molecules to come together spontaneously, but it is overwhelmingly improbable for them to do so.

There is another conceptual matter we must discuss. We have noted that there is a general tendency of systems to move toward a state of molecular chaos. Why is it that these more disordered states are more probable than ordered states? The answer lies in what we really mean by disorder. A disordered system is one for which we have a relatively small amount of information about the exact microscopic state. The reason we lack this detailed knowledge is that the system has many microscopic states available to it, and the best we can do is guess that it is in some one of them at any instant. If only a few microscopic states were possible, we might be able to make an accurate guess of which one the system was in, and thus make a detailed description of the positions and velocities of the molecules. Thus a "disordered" system is one that has a relatively large number of microscopic states available to it, and that is why a disordered state is more probable than an ordered state.

The values of ΔS associated with phase changes provide a simple illustration of the connection between entropy and molecular chaos. When a solid melts reversibly at constant pressure, it absorbs an amount of heat equal to ΔH_f, the enthalpy of fusion. Thus the entropy change upon fusion is

$$\Delta S = \frac{q_{\text{rev}}}{T} = \frac{\Delta H_f}{T_f},$$

which is always positive. For the ice-water transition, $\Delta H_f = 1440$ cal, $T_f = 273°$K, and $\Delta S = 5.28$ cal/deg, or 5.28 entropy units (eu). We also know that in a liquid the molecules are in a more disordered state than in a crystalline solid, and this is consistent with the increase of entropy upon melting.

When a liquid is converted to a vapor at constant temperature, it absorbs heat, and therefore its entropy increases. We can also recognize that there is a corresponding increase in molecular chaos as a result of the evaporation. If the evaporation is carried out reversibly at the boiling temperature T_b to form the vapor at 1-atm pressure, we have

$$\Delta S = \frac{q_{\text{rev}}}{T} = \frac{\Delta H_{\text{vap}}}{T_b}.$$

For one mole of diethyl ether, $\Delta H_{\text{vap}} = 6500$ cal, $T_b = 308°$K, and $\Delta S = 21.1$ eu. When similar calculations are made for a variety of other liquids, it is found that $\Delta S \cong 21$ eu in virtually all cases. This means that the increase in molecular disorder upon evaporation is nearly the same for all liquids.

8.8 ABSOLUTE ENTROPIES AND THE THIRD LAW

In our use of the enthalpy, we found it helpful to select a certain state of matter and assign to it a definite enthalpy of formation. Our choice, that the enthalpy of formation of all elements in their standard states is zero, is based on convenience alone. Any other state of the elements might have been assigned zero enthalpy. In the case of entropy, the situation is somewhat different, for the association of entropy with the number of microscopic states available to a system suggests a natural choice for the entropy zero. In a perfect crystal at absolute zero there is only one possible microscopic state: each atom must be at a crystal lattice point, and must have a minimum energy. Thus we can say that this is a state of perfect order, or of zero entropy. This important decision is expressed in the **third law of thermodynamics**:

The entropy of perfect crystals of all pure elements and compounds is zero at the absolute zero of temperature.

The third law allows us to assign an absolute entropy to each element and compound at any temperature. From Eq. (8.26) we have for one mole of material,

$$S_T - S_0 = \int_0^T \frac{C_P\, dT}{T},$$

$$S_T = \int_0^T \frac{C_P\, dT}{T}, \tag{8.28}$$

since $S_0 = 0$, according to the third law of thermodynamics. The heat capacity of substances depends on temperature, so in order to find the entropy at 298°K of a material like diamond, we would have to measure C_P as a function of temperature from 0°K to 298°K. Then we could evaluate the integral in Eq. (8.28) graphically by plotting C_P/T as a function of T, and measuring the area under the curve. Such a plot is shown in Fig. 8.8.

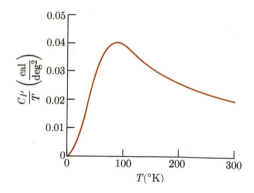

FIG. 8.8 C_P/T as a function of temperature for copper metal. The area under the curve is equal to the absolute molar entropy of copper at 298°K.

Suppose that we are interested in the standard absolute entropy S_{298}^0 of a substance that melts at some temperature T_f less than 298°K. Then the entropy associated with this phase transition must be included in the calculation of the absolute entropy. To do this, we modify Eq. (8.28) to give

$$S_{298}^0 = \int_0^{T_f} \frac{C_P}{T}\, dT + \frac{\Delta H_f}{T_f} + \int_{T_f}^{298} \frac{C_P'}{T}\, dT,$$

where C_P and C_P' are the heat capacities of the solid and liquid, respectively, and ΔH_f is the enthalpy of fusion. If any other phase changes, such as vaporization, occur between 0°K and 298°K, their contribution to the entropy must be included in a similar fashion.

Table 8.3 Absolute entropies, S^0 (cal/mole-deg) at 298°K

Solid elements		Solid compounds		Liquids	
Ag	10.20	BaO	16.8	Br_2	36.4
B	1.7	$BaCO_3$	26.8	H_2O	16.73
Ba	15.1	$BaSO_4$	31.6	Hg	18.17
C (graph)	1.37	CaO	9.5		
C (diam)	0.6	$Ca(OH)_2$	17.4		
Ca	9.95	$CaCO_3$	22.2		
Cu	7.97	CuO	10.4		
Fe	6.49	Fe_2O_3	21.5		
S (rh)	7.62	ZnO	10.5		
Zn	9.95	ZnS	13.8		

Monatomic gases		Diatomic gases		Polyatomic gases	
He	30.13	H_2	31.21	H_2O	45.1
Ne	34.95	D_2	34.6	CO_2	51.1
Ar	36.98	F_2	48.6	SO_2	59.4
Kr	39.19	Cl_2	53.3	H_2S	49.1
Xe	40.53	Br_2	58.6	NO_2	57.5
H	27.39	CO	47.3	N_2O	52.6
F	37.92	NO	50.3	NH_3	46.0
Cl	39.46	N_2	45.7	O_3	56.8
Br	41.80	O_2	49.0		
I	43.18	HF	41.5		
N	36.61	HCl	44.6		
C	37.76	HBr	47.4		
O	38.47	HI	49.3		

Organic compounds

Gases

Methane, CH_4	44.5	Ethylene, C_2H_4	52.45
Ethane, C_2H_6	54.8	Propylene, C_3H_6	63.80
Propane, C_3H_8	64.5	1-butene, C_4H_8	73.48
n-butane, C_4H_{10}	74.10	cis-2-butene, C_4H_8	71.9
Isobutane, C_4H_{10}	70.42	trans-2-butene, C_4H_8	70.9
Acetylene, C_2H_2	49.99	Isobutene, C_4H_8	70.2

Liquids

Methanol, CH_3OH	30.3	Acetic acid, CH_3COOH	38.2
Ethanol, C_2H_5OH	38.4	Benzene, C_6H_6	48.5

Table 8.3 gives the absolute entropies of a few elements and compounds. Note that the substances of similar molecular structure have nearly the same entropy. For example, among the solids the substances having the lowest entropies are hard rigid crystals with light atoms. This occurs because the entropy or disorder of a crystal is related to the amplitudes of vibration of its atoms about their lattice points. In soft crystals composed of heavy atoms the amplitude of vibration is relatively large, and in effect each atom moves in a

larger volume and has more freedom than is available in a more rigid crystal. Just as in the case of gases, this larger volume available to atoms means larger entropy.

We can see also that the entropies of all monatomic gases are nearly the same, and tend to increase with atomic mass. The entropies of diatomic gases are greater than those of the monatomic gases, and the entropies of the triatomics are higher still. In general, as molecular complexity increases, entropy increases, for in a complex molecule the atoms can vibrate about their equilibrium positions, and just as in solids, this motion contributes to the number of possible microscopic states and to the entropy. It is also true that the ability of a polyatomic molecule to rotate about its center of mass contributes to the entropy, and this contribution becomes larger as the molecule becomes more complex.

The relation between molecular structure and entropy can be made quantitative; it is possible to calculate the entropy of a substance from the values of certain of the mechanical properties of its molecules. This is the subject material of **statistical mechanics,** which was mentioned at the beginning of this chapter. We shall discuss some of the quantitative aspects of this procedure in Section 8.14. For the present, however, we shall find it useful to remember the qualitative relation between entropy and molecular complexity.

It is possible to calculate the entropy changes that accompany chemical reactions by using the table of absolute entropies. For the general reaction,

$$aA + bB = cC + dD,$$
$$\Delta S^0 = cS^0(C) + dS^0(D) - aS^0(A) - bS^0(B),$$
$$= \sum S^0 \text{ (products)} - \sum S^0 \text{ (reactants)}.$$

This procedure is of course, similar to the procedure used to find ΔH^0 from enthalpies of formation.

Example 8.4 Calculate the ΔS^0 for each of the following reactions:

(a) $\frac{1}{2}N_2(g) + \frac{1}{2}O_2(g) = NO(g)$,

(b) $Ca(s) + \frac{1}{2}O_2(g) = CaO(s)$,

(c) $\frac{1}{2}H_2 = H$.

For reaction (a), we have

$$\Delta S^0 = S^0(NO) - \frac{1}{2}S^0(N_2) - \frac{1}{2}S^0(O_2)$$
$$= 50.3 - \frac{1}{2}(45.7) - \frac{1}{2}(49.0)$$
$$= 3.0 \text{ eu}.$$

The entropy change is small, since reactants and products have similar structures.

A similar procedure for reaction (b) gives

$$\Delta S^0 = S^0(\text{CaO}) - S^0(\text{Ca}) - \tfrac{1}{2}S^0(\text{O}_2)$$
$$= -24.9.$$

There is a substantial decrease in entropy here because the oxygen is being converted from a form in which there is considerable disorder (O_2), to a form in which the atoms are well localized at crystal lattice points. For reaction (c),

$$\Delta S^0 = S^0(\text{H}) - \tfrac{1}{2}S^0(\text{H}_2)$$
$$= 11.9.$$

Now we have an increase in entropy, because when one hydrogen atom is dissociated from another, there are more microscopic states available to them than when they are linked.

Our criteria for reversible and irreversible processes are

$$\Delta S = 0, \quad \text{reversible process,}$$
$$\Delta S > 0, \quad \text{irreversible process,}$$

and we have now seen how to apply them to the processes of gas expansion and heat flow. While these relations allow us to decide whether a contemplated process will be reversible or irreversible, they are not always very convenient to use. In particular, the entropy change referred to is that of the *system and its surroundings*. If the criterion of spontaneity were expressed in terms of the properties of the *system alone*, it would be much easier to use.

To accomplish this, all we need do is define a new function of state called the **Gibbs free energy,** G:

$$G \equiv H - TS.$$

To find the criterion of spontaneity in terms of G, we first write down its differential

$$dG = dH - T\,dS - S\,dT.$$

Let us restrict our arguments to the conditions most common in chemical processes: constant temperature and pressure. In these circumstances

$$dT = 0,$$
$$dH = dq,$$
$$dG = dq - T\,dS \qquad (\text{constant } P, T).$$

But by the definition of entropy, $T\,dS = dq_{rev}$; therefore

$$dG = dq - dq_{rev} \qquad \text{(constant } P,\ T).$$

There are now two possibilities. If a process is reversible, $dq = dq_{rev}$, so

$$dG = 0 \qquad \text{(reversible process, constant } P,\ T). \tag{8.29a}$$

If a process is irreversible, $q < q_{rev}$, and $dq < dq_{rev}$, by our arguments in Section 8.5. Therefore,

$$dG = dq - dq_{rev},$$
$$dG < 0 \qquad \text{(irreversible process, constant } P,\ T). \tag{8.29b}$$

Equations (8.29a) and (8.29b) apply to infinitesimal changes. For finite changes they become

$$\Delta G = 0, \qquad \text{reversible process,} \tag{8.30a}$$
$$\Delta G < 0, \qquad \text{irreversible process.} \tag{8.30b}$$

To decide whether a given process will be spontaneous when carried out at constant temperature and pressure, we have only to calculate the ΔG of *the system alone*. If ΔG is negative, the process will be spontaneous. If ΔG is zero, the initial and final states can exist at equilibrium with each other, with no net change. If ΔG is positive, the process will not occur spontaneously, but its reverse will.

Let us test this criterion on a simple phase change, the evaporation of water to form vapor at *1-atm pressure*. The free energy change is given by

$$G = H - TS,$$
$$\Delta G = \Delta H - T\,\Delta S, \tag{8.31}$$

for a process at constant temperature. For the reaction

$$H_2O(l) = H_2O(g) \qquad (P = 1 \text{ atm}),$$

$\Delta H = 9710$ cal, and $\Delta S = 26$ eu. We then have

$$\Delta G = 9710 - 26T. \tag{8.32}$$

Let us find the temperature that makes $\Delta G = 0$, for when this is true, liquid water and water vapor at 1-atm pressure will be in equilibrium with each other. To make $\Delta G = 0$, we must have

$$0 = 9710 - 26T,$$
$$T = 373°K.$$

This temperature is, of course, the normal boiling point of water, the temperature at which the liquid and vapor at 1-atm pressure are in equilibrium.

To find the conditions under which the evaporation of water to the vapor at 1 atm is spontaneous, we return to Eq. (8.32) and require that $\Delta G < 0$. This will occur if

$$T > \frac{9710}{26} > 373°\text{K}.$$

Therefore, at temperatures greater than 100°C, the formation of water vapor at 1-atm pressure is spontaneous. Thus the conclusion reached by thermodynamic argument is consistent with our intuitive expectation that water heated above its boiling temperature will evaporate irreversibly to vapor at 1-atm pressure.

Finally, we note from Eq. (8.32) that if $T < 373°\text{K}$, then ΔG is positive. This means that at temperatures below the normal boiling point, the evaporation of water to form the vapor at *1-atm pressure* will not occur. On the other hand, for the reverse process, water vapor at 1 atm going to liquid water, ΔG is negative when $T < 373°\text{K}$. Thus the condensation of a *supersaturated* vapor is a spontaneous irreversible process.

8.10 FREE ENERGY AND EQUILIBRIUM CONSTANTS

To decide whether a certain change of state is spontaneous, we have only to evaluate the accompanying free-energy change and apply Eq. (8.30). Note, however, that entropy and, consequently, free energy depend on pressure. Therefore we must be careful to specify the pressure or, in general, the concentration conditions for which a free-energy change is evaluated. Hence, it is convenient to tabulate the **standard free-energy change** ΔG^0 of a process, where ΔG^0 is the free-energy change that accompanies the conversion of reactants in their standard states to products in their standard states.

In discussing thermochemical problems, we associated a standard enthalpy of formation with each compound in its standard state. In a similar manner we can define a **standard free energy of formation** ΔG_f^0 as the free-energy change that occurs when one mole of a compound in its standard state is formed from its elements in their standard states. It is not difficult to obtain values of ΔG_f^0, since this quantity is related to ΔH_f^0 and ΔS_f^0 by

$$\Delta G_f^0 = \Delta H_f^0 - T\,\Delta S_f^0,$$

where all the thermodynamic quantities are evaluated at a single temperature T. Table 8.4 gives the values of ΔG_f^0 at 298°K for several compounds. The standard free energy of formation of all *elements* is *defined* to be zero.

Once we have the value of ΔG_f^0 for each compound, we can compute the standard free-energy change for any reaction

$$a\text{A} + b\text{B} = c\text{C} + d\text{D}$$

Table 8.4 Free energy of formation, ΔG_f^0 (kcal/mole) at 298°K

GASES		SOLIDS	
H_2O	−54.64	BaO	−126.3
H_2O_2	−24.7	$BaSO_4$	−350.2
O_3	39.06	$BaCO_3$	−272.2
HCl	−22.77	CaO	−144.4
SO_2	−71.79	$CaCO_3$	−269.8
SO_3	−88.52	$Ca(OH)_2$	−214.3
H_2S	− 7.89	Fe_2O_3	−177.1
N_2O	24.9	Al_2O_3	−376.8
NO	20.72	CuO	− 30.4
NO_2	12.39	Cu_2O	− 34.98
NH_3	− 3.97	SiO_2	−192.4
CO	−32.81	ZnO	− 76.05
CO_2	−94.26	PbO_2	− 52.34

ORGANIC COMPOUNDS

Gases

Methane, CH_3	−12.14	Ethylene, C_2H_4	16.28
Ethane, C_2H_6	− 7.86	Propylene, C_3H_6	14.90
Propane, C_3H_8	− 5.61	1-butene, C_4H_8	17.09
n-butane, C_4H_{10}	− 3.75	*cis*-2-butene, C_4H_8	15.74
Isobutane, C_4H_{10}	− 4.3	*trans*-2-butene, C_4H_8	15.05
Acetylene, C_2H_2	50.00	Isobutene, C_4H_8	13.88

Liquids

Methanol, CH_3OH	−39.73	Acetic acid, CH_3COOH	−93.8
Ethanol, C_2H_5OH	−41.77	Benzene, C_6H_6	29.76

GASEOUS ATOMS

H	48.57	I	16.77
F	14.2	C	160.84
Cl	25.19	N	81.47
Br	19.69	O	54.99

by the expression

$$\Delta G^0 = c\,\Delta G_f^0(C) + d\,\Delta G_f^0(D) - a\,\Delta G_f^0(A) - b\,\Delta G_f^0(B).$$

In general,

$$\Delta G^0 = \sum \Delta G_f^0(\text{products}) - \sum \Delta G_f^0(\text{reactants}). \qquad (8.33)$$

If ΔG^0 for a chemical reaction is negative, the reactants *in their standard states* will be converted spontaneously to products *in their standard states*. If ΔG^0 is positive, this conversion will not be spontaneous; however, the corresponding reverse reaction will be.

Just because ΔG^0 for a reaction is positive does not mean that no products are formed from reactants in their standard states. *Some* products can be formed, but not in concentrations as great as that of the standard state. Our problem now is to find out how the magnitude of ΔG^0 is related to the actual amounts of reactants and products present when a reaction reaches equilibrium.

To accomplish this, we must have an expression for the dependence of free energy on pressure. From the definition of free energy,

$$G = H - TS = E + PV - TS,$$

we obtain

$$dG = dE + P\, dV + V\, dP - T\, dS - S\, dT.$$

But for a situation in which only pressure-volume work can be done, $dE = dq - P\, dV$, so

$$dG = dq + V\, dP - T\, dS - S\, dT.$$

Equating $T\, dS$ and dq then gives us

$$dG = V\, dP - S\, dT.$$

For a pressure change at constant temperature,

$$dG = V\, dP. \qquad (8.34)$$

In the following discussion, quantities that apply to one mole of material will be denoted by a bar superscript, as in $\overline{V} = RT/P$. Then for one mole of an ideal gas, Eq. (8.34) becomes

$$d\overline{G} = \frac{RT}{P}\, dP.$$

Let us integrate this expression, taking as one limit of pressure $P^0 = 1$ atm, the standard pressure. Then the corresponding limit for $\overline{G}$ will be $\overline{G}^0$, the standard free energy of one mole of the ideal gas. We get

$$\int_{\overline{G}^0}^{\overline{G}} d\overline{G} = \int_{P^0}^{P} \frac{RT}{P}\, dP,$$

$$\overline{G} - \overline{G}^0 = RT \ln \frac{P}{P^0} = RT \ln P,$$

where $\overline{G}$ is the molar free energy at any pressure P (in atm) and $\overline{G}^0$ is the standard free energy. If, instead of 1 mole, n moles are considered, we get

$$n\overline{G} = n\overline{G}^0 + nRT \ln P. \qquad (8.35)$$

Equation (8.35) is just what we need to relate ΔG^0 to the equilibrium constant. The next step is to calculate ΔG for the general reaction between ideal gases,

$$a\mathrm{A}(P_\mathrm{A}) + b\mathrm{B}(P_\mathrm{B}) = c\mathrm{C}(P_\mathrm{C}) + d\mathrm{D}(P_\mathrm{D}),$$

where P_A, P_B, etc., are the pressures of the reactants and products. We have

$$\Delta G = \sum G(\text{products}) - \sum G(\text{reactants})$$
$$= c\overline{G}(\mathrm{C}) + d\overline{G}(\mathrm{D}) - a\overline{G}(\mathrm{A}) - b\overline{G}(\mathrm{B}).$$

Use of Eq. (8.35) gives

$$\Delta G = [c\overline{G}^0(\mathrm{C}) + d\overline{G}^0(\mathrm{D}) - a\overline{G}^0(\mathrm{A}) - b\overline{G}^0(\mathrm{B})] + cRT \ln P_\mathrm{C}$$
$$+ d\,RT \ln P_\mathrm{D} - a\,RT \ln P_\mathrm{A} - b\,RT \ln P_\mathrm{B}.$$

The bracketed terms are equal to ΔG^0, and the remaining terms can be combined to give

$$\Delta G = \Delta G^0 + RT \ln \frac{(P_\mathrm{C})^c (P_\mathrm{D})^d}{(P_\mathrm{A})^a (P_\mathrm{B})^b}. \tag{8.36}$$

This is an important equation. It relates the free-energy change for any ideal gas reaction involving *arbitrary* pressures of reactants and products to the standard free-energy change and the pressures of the reagents.

Suppose that the pressures in Eq. (8.36) are those that exist when reactants and products are in equilibrium with each other. Then $\Delta G = 0$, since initial and final states are in equilibrium, and

$$0 = \Delta G^0 + RT \ln \left[\frac{(P_\mathrm{C})^c (P_\mathrm{D})^d}{(P_\mathrm{A})^a (P_\mathrm{B})^b}\right]_{\mathrm{eq}}.$$

Because the pressures are those that exist at equilibrium, the term in brackets is equal to the equilibrium constant K, so

$$\Delta G^0 = -RT \ln K. \tag{8.37}$$

Equation (8.37) is the quantitative relation between the standard free-energy change and the equilibrium constant that we have been seeking.

The importance of Eq. (8.37) cannot be overstated. In the first place, it constitutes a proof that there is such a thing as the equilibrium constant. That is, since G is a state function, ΔG^0 must be a *fixed constant* whose value depends only on the temperature and the nature of the reactants and products in their standard states. Therefore Eq. (8.37) says that at a fixed temperature, the concentration ratio

$$\frac{(P_\mathrm{C})^c (P_\mathrm{D})^d}{(P_\mathrm{A})^a (P_\mathrm{B})^b} = K$$

is a fixed constant at equilibrium.

The second important feature of Eq. (8.37) is that it supplies the bridge between properties of individual substances and the extent to which reactions proceed. The standard free-energy change can be calculated from the values of ΔG_f^0 for reactants and products, and these quantities can be obtained from the values of ΔH_f^0 and S^0. Therefore Eq. (8.37) is the final step in the calculation of chemical reactivity from the "thermal" properties, ΔH_f^0 and S^0, of pure substances.

Finally, Eq. (8.37) permits us to make a more thorough interpretation of the meaning of the sign of ΔG^0. The use of antilogarithms gives

$$K = e^{-\Delta G^0/RT} = 10^{-\Delta G^0/2.3RT}. \qquad (8.38)$$

We can see now that if $\Delta G^0 < 0$, the exponent will be positive, K will be greater than unity, and will increase as ΔG^0 becomes more negative. Thus reactions with the largest negative values of ΔG^0 will tend to proceed to completion to the greatest extent. Conversely, if $\Delta G^0 > 0$, K will be less than unity, and although some products will be present at equilibrium, most material will be in the form of reactants. The special and rare case in which $\Delta G^0 = 0$ corresponds to an equilibrium constant of unity.

Writing Eq. (8.38) in a slightly expanded form gives more insight into the "driving forces" of chemical reactions. We use

$$\Delta G^0 = \Delta H^0 - T\,\Delta S^0, \qquad (8.39)$$

which with Eq. (8.38) gives us

$$K = e^{\Delta S^0/R}e^{-\Delta H^0/RT} \qquad (8.40a)$$
$$= 10^{\Delta S^0/2.3R}10^{-\Delta H^0/2.3RT}. \qquad (8.40b)$$

We see that the larger ΔS^0 is, the larger K is. Thus the tendency toward maximum molecular chaos directly influences the magnitude of the equilibrium constant. It is also clear that the more negative ΔH^0 is, the larger K is. In this way, the tendency of atoms to seek the state of lowest energy helps determine the equilibrium constant.

Since the standard state for gases was chosen as 1 atm, the concentration unit to be used in the equilibrium constants of gaseous reactions is the atmosphere. However, application of the general relation Eq. (8.37) is not restricted to gases. To deal with reactions in solution, the free energy of any "ideal" solute can be expressed as

$$\overline{G} = \overline{G}^0 + RT \ln \frac{C}{C^0},$$

where C is any appropriate concentration unit, and C^0 is the standard concentration, for example 1 M. Using this equation, a derivation can be carried through to yield Eq. (8.37) with K expressed in concentration units. Thus, provided care is taken to specify correct concentration units, Eq. (8.37) can be applied to any reaction taking place under ideal conditions in solution or in the gas phase.

Example 8.5 Calculate the standard free energy of formation for ozone at 298°K from the values of ΔH_f^0 and S^0 given in Tables 8.1 and 8.3. The formation reaction is

$$\tfrac{3}{2}O_2(g) = O_3(g).$$

We will calculate $\Delta G_f^0(O_3)$ by the expression

$$\Delta G_f^0 = \Delta H_f^0 - T\,\Delta S_f^0.$$

The value of ΔH_f^0 obtained directly from Table 8.1 is $\Delta H_f^0(O_3) = 34.0$ kcal. The entropy of formation must be calculated from

$$\Delta S_f^0 = S^0(O_3) - \tfrac{3}{2}S^0(O_2)$$
$$= 56.8 - \tfrac{3}{2}(49.0),$$
$$\Delta S_f^0(O_3) = -16.7 \text{ cal/mole-deg.}$$

Therefore

$$\Delta G_f^0 = 34,000 - (298)(-16.7)$$
$$= 34,000 + 4980$$
$$= 39,000 \text{ cal/mole.}$$

Example 8.6 Calculate ΔG^0 and K for the reaction

$$NO + O_3 = NO_2 + O_2.$$

Is the size of the equilibrium constant principally a consequence of the ΔH^0 or of the ΔS^0 for this reaction? For ΔG^0 we have from Table 8.4,

$$\Delta G^0 = \Delta G_f^0(NO_2) + \Delta G_f^0(O_2) - \Delta G_f^0(NO) - \Delta G_f^0(O_3)$$
$$= 12.39 + 0 - 20.7 - 39.0$$
$$= -47.3 \text{ kcal.}$$

By Eq. (8.38)

$$K = 10^{-\Delta G^0/2.3RT} = 10^{+(47,300)/1360} = 5 \times 10^{34}.$$

The equilibrium constant greatly favors the products. By Eq. (8.40a)

$$K = e^{\Delta S^0/R}e^{-\Delta H^0/RT}.$$

We can evaluate the contribution of enthalpy and entropy to the equilibrium separately:

$$\Delta H^0 = \Delta H_f^0(NO_2) + \Delta H_f^0(O_2) - \Delta H_f^0(NO) - \Delta H_f^0(O_3)$$
$$= 8.09 + 0 - 21.60 - 34.0$$
$$= -47.5 \text{ kcal.}$$

For ΔS^0 we have

$$\Delta S^0 = S^0(NO_2) + S^0(O_2) - S^0(NO) - S^0(O_3)$$
$$= 57.5 + 49.0 - 50.3 - 56.8$$
$$= -0.6 \text{ cal/deg.}$$

Thus by Eq. (8.40b)

$$K = 10^{+\Delta S^0/2.3R}10^{-\Delta H^0/2.3RT}$$
$$= 10^{-0.13}10^{+34.8}$$

The entropy change for the reaction is very small, since the geometrical molecular structures of products and reactants are very similar. Thus the real "driving force" for this reaction is the fact that the products are energetically more stable than the reactants.

8.11 ELECTROCHEMICAL CELLS

We have already at our disposal one method of obtaining standard free-energy changes for chemical reactions: the use of Eq. (8.39). In this section we will show that measurement of the standard cell potential $\Delta \varepsilon^0$ can give us ΔG^0 for a reaction. This should not be surprising, since we know that ΔG^0 and the standard cell potential are both related to the equilibrium constant; they certainly should be related to each other.

To establish the relation between ΔG^0 and $\Delta \varepsilon^0$, we must first find the connection between free energy and electrical work. From the definition of free energy we get

$$dG = dH - T\,dS - S\,dT$$
$$= dq + dw + P\,dV + V\,dP - T\,dS - S\,dT.$$

Once again we restrict our argument to a reversible process at constant temperature and pressure, and let $dq = T\,dS$. This gives us

$$dG = dw + P\,dV.$$

The term $P\,dV$ is the work done by volume changes, while dw represents *all* the work done by the system. If the system is an electrochemical cell, dw includes pressure-volume work *and* the electrical work, so we can say

$$w = w_{PV} + w_{\text{elec}},$$
$$dG = dw + P\,dV$$
$$= dw_{PV} + dw_{\text{elec}} + P\,dV$$
$$= dw_{\text{elec}},$$
$$\Delta G = +w_{\text{elec}} \quad (P, T \text{ constant}). \qquad (8.41)$$

Thus ΔG for a process is the reversible electrical work done on the system.

Our second step is to express the electrical work as the product of an amount of charge times the voltage difference through which it is transferred. For n moles of electrons, this is given by

$$w_{\text{elec}} = -n \times \mathfrak{F} \times \Delta\mathcal{E},$$

where $\mathfrak{F}$ is the faraday, the charge on one mole of electrons. The minus sign appears here for the same reason it appears in the expression for the pressure-volume work. If the system is capable of producing a potential difference $\Delta\mathcal{E}$ between electrodes, it does work *on* the surroundings as the charge n passes. Hence w, the work done *on* the system must be negative, and the minus sign accomplishes this. Therefore the relation between ΔG and $\Delta\mathcal{E}$ is

$$\Delta G = -n\mathfrak{F}\,\Delta\mathcal{E}. \tag{8.42}$$

If the change is between standard states, then

$$\Delta G^0 = -n\mathfrak{F}\,\Delta\mathcal{E}^0. \tag{8.43}$$

Equation (8.43) shows that measurement of the standard cell potential can give us the value of ΔG^0 directly. This measurement is in fact one of the most convenient ways to obtain free energies of formation. If ΔG_f^0 is known for all substances in the cell reaction except one, measurement of $\Delta\mathcal{E}^0$ gives ΔG^0, and then use of Eq. (8.33) allows calculation of the unknown ΔG_f^0.

In Chapter 7, we introduced the Nernst equation,

$$\Delta\mathcal{E} = \Delta\mathcal{E}^0 - \frac{0.059}{n}\log Q,$$

where Q is the quotient formed by the product of the concentrations of the products, each raised to its stoichiometric coefficient, divided by a similar product of reactant concentrations. We can now derive the Nernst equation from the more general relation Eq. (8.36). First we rewrite Eq. (8.36) in terms of concentrations:

$$\Delta G = \Delta G^0 + RT \ln \frac{[C_{\text{C}}]^c[C_{\text{D}}]^d}{[C_{\text{A}}]^a[C_{\text{B}}]^b}$$

or

$$\Delta G = \Delta G^0 + RT \ln Q.$$

We introduce Eqs. (8.42) and (8.43) to give

$$\Delta\mathcal{E} = \Delta\mathcal{E}^0 - \frac{RT}{n\mathfrak{F}}\ln Q.$$

Converting to base-ten logarithms results in

$$\Delta\mathcal{E} = \Delta\mathcal{E}^0 - \frac{2.3\,RT}{n\mathfrak{F}}\log Q.$$

To evaluate the constant terms we let

$$R = 1.98 \text{ cal/mole-deg,}$$
$$T = 298°\text{K,}$$
$$\mathcal{F} = 23{,}061 \text{ cal/volt,}^*$$

and find that

$$\Delta\mathcal{E} = \Delta\mathcal{E}^0 - \frac{0.059}{n} \log Q,$$

which is the usual form of the Nernst equation.

Equation (8.42) shows that the voltmeter connected to an electrochemical cell is really a free-energy meter. A cell reaction proceeds so long as $\Delta\mathcal{E}$ or ΔG is not zero. When $\Delta\mathcal{E} = 0$, the cell has "run down" or reached equilibrium; that is, the reactants and products have reached concentrations at which their free energies are the same.

8.12 TEMPERATURE DEPENDENCE OF EQUILIBRIA

While Le Chatelier's principle provides a qualitative guide for predicting how equilibria are affected by changes in temperature, we can obtain a quantitative relation between K and T by using the thermodynamic concepts available to us. To derive this expression, we combine two fundamental relations, Eqs. (8.37) and (8.39),

$$\Delta G^0 = -RT \ln K, \tag{8.37}$$
$$\Delta G^0 = \Delta H^0 - T \Delta S^0, \tag{8.39}$$

to give

$$\ln K = -\frac{\Delta H^0}{RT} + \frac{\Delta S^0}{R}. \tag{8.44}$$

This equation says that if ΔH^0 and ΔS^0 are constants independent of temperature, $\ln K$ is a linear function of $1/T$. But are ΔH^0 and ΔS^0 independent of temperature? Equation (8.20) shows that if the difference in the heat capacities of reactants and products is very small, then ΔH^0 is essentially independent of temperature. Likewise, since ΔS^0 at any temperature T can be expressed as

$$\Delta S^0 = \Delta S^0_{298} + \int_{298}^{T} \frac{\Delta C_P}{T} \, dT,$$

---

* This value is derived by writing

$$\mathcal{F} = 96{,}487 \text{ coul} = 96{,}487 \text{ volt-coul/volt}$$
$$= 96{,}487 \text{ joules/volt,}$$
$$1 \text{ cal} = 4.1840 \text{ joules,}$$
$$\mathcal{F} = 96{,}487/4.1840 = 23{,}061 \text{ cal/volt.}$$

if $\Delta C_P \cong 0$, ΔS^0 is also independent of temperature. If we accept this approximation, then Eq. (8.44) teaches us that for an exothermic reaction, K decreases as T increases, and for an endothermic reaction, K increases as T increases. These facts are consistent with the qualitative conclusions based on Le Chatelier's principle.

Not only does the sign of ΔH^0 indicate the direction in which K changes; for a given temperature variation, the magnitude of ΔH^0 determines how rapidly K changes as a function of temperature. According to Eq. (8.44), plotting $\ln K$ as a function of $1/T$ should give a straight line whose slope is $-\Delta H^0/R$. Thus the more negative ΔH^0 is, the faster $\ln K$ should decrease as T increases, and vice versa. Figure 8.9 demonstrates the validity of these conclusions.

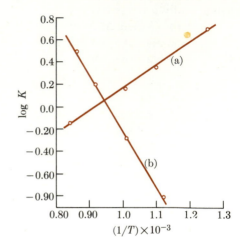

FIG. 8.9

Log K versus $1/T$ for two reactions. (a) $CO_2(g) + H_2(g) = CO(g) + H_2O(g)$, $\Delta H = -9.1$ kcal; (b) $SO_3(g) = SO_2(g) + \frac{1}{2}O_2(g)$, $\Delta H = 22$ kcal.

Another form of Eq. (8.44) that is particularly useful for numerical calculations can be obtained by writing

$$\ln K_1 = -\frac{\Delta H^0}{RT_1} + \frac{\Delta S^0}{R}, \qquad \ln K_2 = -\frac{\Delta H^0}{RT_2} + \frac{\Delta S^0}{R},$$

and then subtracting the first of these expressions from the second. The result is

$$\ln \frac{K_2}{K_1} = -\frac{\Delta H^0}{R}\left(\frac{1}{T_2} - \frac{1}{T_1}\right). \tag{8.45}$$

Equation (8.45) shows that if we know the value of ΔH^0 and of the equilibrium constant at one temperature, we can calculate K at any other temperature. Also, if we measure K at two temperatures, we can calculate ΔH^0 by Eq. (8.45). Thus it is possible to obtain ΔH^0 for a reaction without ever doing a calorimetric experiment.

Example 8.7 For the reaction

$$NO(g) + \tfrac{1}{2}O_2(g) = NO_2(g),$$

$\Delta G^0 = -8.33$ kcal and $\Delta H^0 = -13.5$ kcal at 298°K. Calculate the equilibrium constant at 298°K and at 598°K.

By Eq. (8.38),

$$K = 10^{-\Delta G^0/(2.3RT)}$$

$$= 10^{+8330/(2.3)(1.99)(298)}$$

$$= 1.28 \times 10^6.$$

To find the equilibrium constant at 598°K, we cannot use Eq. (8.38), since we do not know ΔG^0 at this temperature. However, we can use Eq. (8.45):

$$\ln \frac{K_2}{K_1} = 2.3 \log \frac{K_2}{K_1} = -\frac{\Delta H^0}{R}\left(\frac{1}{T_2} - \frac{1}{T_1}\right),$$

$$2.3(\log K_2 - 6.12) = \frac{13{,}500}{1.99}\left(\frac{1}{598} - \frac{1}{298}\right),$$

$$\log K_2 = 1.16,$$

$$K_2 = 14.4.$$

The equilibrium constant of this exothermic reaction is smaller at the higher temperature, which is consistent with Le Chatelier's principle.

Example 8.8 For the reaction

$$\tfrac{1}{2}N_2(g) + \tfrac{3}{2}H_2(g) = NH_3(g),$$

$K = 1.3 \times 10^{-2}$ at 673°K, and $K = 3.8 \times 10^{-3}$ at 773°K. What is ΔH^0 for this reaction in this temperature range? We simply substitute the values of K and T in Eq. (8.45) to give

$$-2.3 \log \frac{3.8 \times 10^{-3}}{1.3 \times 10^{-2}} = \frac{\Delta H^0}{1.99}\left(\frac{1}{773} - \frac{1}{673}\right),$$

$$\Delta H^0 = -12.7 \text{ kcal.}$$

8.13 COLLIGATIVE PROPERTIES

The equations that we have developed will be very useful in our subsequent discussions of the descriptive chemistry of the elements. However, it is worth noting that Eq. (8.45) in particular can be used to analyze the physical processes of boiling-point elevation and freezing-point depression by dissolved substances. By using Eq. (8.45), we can find out how the empirical constants for boiling-point elevation and freezing-point depression are related to more fundamental properties of the solvent.

Let us first consider the boiling-point elevation phenomenon in an ideal solution of a volatile solvent and a nonvolatile solute. The equilibrium is between liquid solvent at a mole fraction x_1, and its vapor at a pressure of 1 atm. The corresponding reaction is

$$\text{liquid (concentration } x_1) = \text{vapor (1 atm)}.$$

The equilibrium constant for this reaction is

$$K = \frac{[\text{vapor}]}{[\text{liquid}]} = \frac{1 \text{ atm}}{[x_1]} = \frac{1}{[x_1]}.$$

Since we are always interested in the normal boiling point of the solution, the vapor concentration is always 1 atm.

Depending on the concentration x_1, the solution boils at different temperatures. To connect x_1 and the boiling temperature, we must find the temperature dependence of the equilibrium constant. Our general expression is

$$\ln \frac{K_2}{K_1} = -\frac{\Delta H^0}{R} \left(\frac{1}{T_2} - \frac{1}{T_1} \right). \tag{8.45}$$

Let us choose T_1 as the boiling point T_b of the pure solvent; thus $x_1 = 1$, and $K_1 = 1$ at $T_1 = T_b$. For K_2 we substitute $1/x_1$, the value of the equilibrium constant at the arbitrary temperature $T_2 = T$. Also ΔH^0 must be the enthalpy of vaporization. With these substitutions, Eq. (8.45) becomes

$$\ln \frac{1}{x_1} = -\frac{\Delta H_{\text{vap}}}{R} \left(\frac{1}{T} - \frac{1}{T_b} \right),$$

$$\ln x_1 = \frac{\Delta H_{\text{vap}}}{R} \left(\frac{T_b - T}{TT_b} \right).$$

The boiling-point elevation ΔT is just

$$\Delta T = T - T_b,$$

and if the solution is dilute, ΔT is small and $T \simeq T_b$. Therefore we can set TT_b equal to T_b^2, and get

$$\ln x_1 = -\frac{\Delta H_{\text{vap}}}{RT_b^2} \Delta T.$$

We can simplify this expression further. If we have a two-component mixture, then $x_1 = 1 - x_2$, and

$$\ln x_1 = \ln (1 - x_2) \cong -x_2,$$

where the last equality holds if x_2 is small. Consequently we can write

$$-x_2 = -\frac{\Delta H_{vap}}{RT_b^2}\Delta T, \qquad \Delta T = \frac{RT_b^2}{\Delta H_{vap}}x_2.$$

Let us recall that in Chapter 4 we showed that for *dilute* solutions, x_2 was related to the molality m by

$$x_2 \cong \frac{M_1}{1000}m,$$

where M_1 is the molecular weight of the solvent. Using this relation gives

$$\Delta T = \left(\frac{RT_b^2 M_1}{1000\,\Delta H_{vap}}\right)m \tag{8.46}$$

$$= K_b m.$$

Equation (8.46) provides us with an explicit expression of the boiling-point elevation constant K_b in term of T_b, M_1, and ΔH_{vap}. All of these are properties of the solvent alone, so K_b should be applicable to any ideal solution of a particular solvent.

The same type of analysis can be applied to the freezing-point depression phenomenon. In this case, the equilibrium is between a pure solid solvent and the same material as a liquid of concentration x_1. The reaction we consider is

$$\text{solid (pure)} = \text{liquid (concentration } x_1\text{)}.$$

Since the concentration of the pure solid is contant, the equilibrium constant is simply

$$K = [x_1].$$

Once again we are interested in the temperature at which equilibrium is reached for solutions of various concentration, and therefore we must find the relation between K and T. We start with Eq. (8.45) and choose $T_1 = T_{fus}$, the freezing point of the pure solvent. Therefore at T_1, $x_1 = 1$ and $K_1 = 1$. The equilibrium constant at the arbitrary temperature $T = T_2$ has the value $K_2 = K = x_1$, and ΔH^0 is the enthalpy of fusion. Thus

$$\ln x_1 = -\frac{\Delta H_{fus}^0}{R}\left(\frac{1}{T} - \frac{1}{T_{fus}}\right)$$

$$= -\frac{\Delta H_{fus}^0}{R}\left(\frac{T_{fus} - T}{TT_{fus}}\right).$$

We realize that the freezing-point depression ΔT is equal to $T_{fus} - T$, and that $TT_{fus} \cong T_{fus}^2$ for small ΔT. As a result we get

$$\ln x_1 = -\frac{\Delta H_{fus}^0}{R} \frac{\Delta T}{T_{fus}^2}.$$

Setting $\ln x_1 = \ln (1 - x_2) \cong -x_2$ as before gives us for ΔT

$$\Delta T = \frac{RT_{fus}^2}{\Delta H_{fus}^0} x_2.$$

Conversion of x_2 to molality results in

$$\Delta T = \left(\frac{RT_{fus}^2 M_1}{1000 \, \Delta H_{fus}^0}\right) m \qquad (8.47)$$

$$= K_{fm}.$$

Equation (8.47) shows that the freezing-point depression constant K_f is a function only of the properties of the solvent itself.

We turn now to the phenomenon of osmotic pressure, which has been described in Section 4.4. As pointed out there, we are dealing with an equilibrium between a pure solvent and the same solvent in a solution to which some external pressure has been applied. This equilibrium will be reached only when the molar free energy of the solvent *in solution* is the same as that of the pure solvent. The molar free energy of an ideal solvent of mole fraction x_1 is

$$\overline{G} = \overline{G}^0 + RT \ln x_1,$$

where G^0 is the free energy of the pure solvent. Since x_1 is less than unity, $\ln x_1$ is negative, and the solvent in solution has a smaller free energy than the pure solvent.

If an external pressure is applied to the solution, the free energy of the solvent may be increased to the point at which it is equal to the free energy of the pure solvent. According to Eq. (8.34), the effect of pressure on free energy is given by

$$dG = V \, dP$$

for a process at constant temperature. Since the osmotic pressure π is the pressure on the solution in excess of that exerted on the pure solvent, the increase of the molar free energy due to this pressure is

$$\Delta \overline{G} = \overline{V} \int_0^\pi dP = \pi \overline{V}.$$

We have assumed that the solvent is virtually incompressible, so that the molar volume $\overline{V}$ is independent of pressure.

The combined effects of dilution and of external pressure on the free energy of the solvent are given by

$$\overline{G} = \overline{G}^0 + RT \ln x_1 + \pi \overline{V}.$$

When the solvent in solution is in equilibrium with the pure solution, $\overline{G} = \overline{G}^0$, and

$$\pi \overline{V} = -RT \ln x_1.$$

Substituting $x_1 = 1 - x_2$ and expanding the logarithm as before gives

$$\pi \overline{V} = RT x_2.$$

If the solution is dilute, $x_2 \cong n_2/n_1$, and $\overline{V} = V/n_1$, where V is the volume of the solution and n_1 and n_2 are the numbers of moles of solvent and solute, respectively. Therefore

$$\pi \frac{V}{n_1} = RT \frac{n_2}{n_1},$$

$$\pi = \frac{RT}{V} n_2 = cRT, \tag{8.48}$$

where c is the concentration of solute in moles per liter. Thus we have arrived at the expression used in Section 4.4 to relate the osmotic pressure π to the concentration and temperature of the solution.

8.14 HEAT ENGINES

Early work in thermodynamics was very much concerned with the operation and efficiency of devices for converting heat into useful work. Indeed, there are two common statements of the second law of thermodynamics which have to do with the existence of natural limitations on the conversion of heat to work. In this section, we shall use our understanding of the second law to deduce the limiting efficiency with which heat can be converted to work in cyclical or repetitive processes such as occur in practical engines. To do this, we shall analyze the behavior of an idealized device called the Carnot heat engine.

The first component of a Carnot heat engine is a source of heat which is maintained at a constant high temperature T_h. In an actual engine, this source of heat might be a combustion chamber or nuclear reactor. The second component of the Carnot engine is a heat sink, maintained at a low temperature T_c, which receives any heat which might be discarded by the engine as it operates. In a real working engine this heat sink might be the atmosphere or a cooling bath. The engine operates by carrying a working substance through a cycle, or by a sequence of changes which converts heat into work and returns the working substance back to its original state. The working material is most

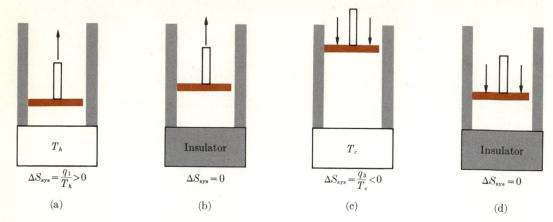

FIG. 8.10 The operation of a Carnot heat engine: (a) Isothermal expansion at T_h. (b) Adiabatic expansion. (c) Isothermal compression at T_c. (d) Adiabatic compression to original state.

commonly a gas, but it may be any substance. For convenience, we shall refer to the working substance as a gas.

The cycle used in the Carnot engine is shown in Fig. 8.10. There are four steps, *each of which is carried out reversibly*:

1. The working gas absorbs q_1 units of heat from the hot reservoir; while doing so, the gas expands isothermally and reversibly and does work. The value of q_1 is positive, but w_1, the work done *on* the gas, is negative.

2. The working substance is thermally isolated from the surroundings. The gas expands, and does work. However, $q_2 = 0$, since the gas is thermally isolated. Such a process is called an *adiabatic* expansion. Since the gas does work without receiving heat, its internal energy and temperature decrease. The adiabatic expansion continues until the temperature of the gas has dropped from T_h to T_c, the temperature of the heat sink, or cold reservoir.

3. The gas is brought into thermal contact with the cold reservoir and then compressed isothermally. This process deposits heat into the cold reservoir, so q_3, *the heat absorbed by the gas*, is a negative number. Work is done on the gas as it is compressed, so w_3 is a positive number.

4. In the final step, the gas is again thermally insulated and compressed adiabatically. Therefore, $q_4 = 0$, and w_4 is a positive number. During the compression, the gas temperature rises to T_h.

At the end of the cycle the gas has returned to its initial state. Therefore, $\Delta S_{\text{gas}} = 0$, $\Delta E_{\text{gas}} = 0$, and so

$$q_1 + q_3 = -w_1 - w_2 - w_3 - w_4 = -w. \tag{8.49}$$

Here w is the net work done *on* the gas, so $-w$ is the net work done *by* the gas. Also $q_1 + q_3$ is the net heat added *to* the gas, so Eq. (8.49) is a statement of the conservation of energy.

To find the limitations on the net work done by the gas, we must consider the entropy changes that occur during one cycle. According to the second law of thermodynamics, the total entropy change of the gas and the two reservoirs must be zero, because the whole cycle is carried out reversibly. That is,

$$\Delta S_{gas} + \Delta S_{res} = 0.$$

But $\Delta S_{gas} = 0$, because the final state of the gas is the same as the initial state, and therefore $\Delta S_{res} = 0$. Thus we can write the entropy change of the reservoirs as

$$\Delta S_{res} = -\frac{q_1}{T_h} - \frac{q_3}{T_c} = 0.$$

The minus signs appear here, because when q_1 (or q_3) is absorbed by the gas, the same amount is *lost* by the reservoirs. Now we use

$$q_1 + q_3 = -w$$

to find that

$$-\frac{q_1}{T_h} + \frac{q_1 + w}{T_c} = 0$$

or

$$-\frac{w}{q_1} = 1 - \frac{T_c}{T_h} \equiv \eta. \qquad (8.50)$$

The quantity $-w/q_1$ is just the net work done on the outside world, divided by the heat extracted from the high temperature source. Therefore, it is the efficiency η of the reversible Carnot engine.

Equation (8.50) shows that in order for the efficiency η to approach unity, the temperature ratio T_c/T_h must approach zero. It can be proved that no other heat engine can exceed the efficiency of the Carnot engine, and thus Eq. (8.50) provides the limiting value of the efficiency of any heat engine operating between the two temperatures T_h and T_c. Clearly, it is advantageous to make T_h as large, and T_c as small as possible, but there are practical limitations imposed by nature of heat sources, the properties of working materials, and the availability of cold reservoirs. In commercial steam heating plants, T_h and T_c can be maintained at values such that the efficiency is 45 percent even with friction and heat losses included. In contrast, the automobile engine operates at an efficiency of only 15 percent, partly because of a rather high exhaust manifold temperature.

In Section 8.7 we advanced the qualitative, intuitive argument that entropy was related to our lack of detailed knowledge or information about the microscopic state of a system. In this section we shall make this idea somewhat more precise, and thereby indicate how entropy can be *calculated* from the *mechanical properties* of the molecules which make up the system. To begin, we must specify what we mean by information.

The various sources of information such as clocks, radio dials, stop lights, etc., which we encounter in our lives can be thought of as devices which send messages to us. How much information is in each message? A little reflection suggests that the amount of information in any one message decreases as the probability that that particular message might be sent increases. For example, a railroad gate has two possible messages: if it is up, no train is coming in the immediate future; if it is down, a train will be along sometime soon. The message that the gate is closed may be very interesting, but does not contain much information, since there is only one other possibility. The message that the gate is open contains even less information, since this is the usual or more probable situation. Contrast this situation with the information conveyed by a 24-hour digital clock that gives the time to the nearest minute. If we have absolutely no information concerning the time, we know that there are $24 \times 60 = 1440$ messages which might be sent, and consequently the information in any one of them is quite large. If we know, by virtue of some previous experience, that the time is 12 o'clock plus or minus 15 minutes, then the amount of information in a message is correspondingly less. We assert, then, that if p_i is the probability that a message will be sent, the information I_i in the message is a function $I_i(p_i)$ of p_i, and *increases as p_i decreases*.

Suppose two messages i and j are sent quite independently of each other. It is reasonable to require that the total information received, I_{ij}, is the sum of the information contained in the separate messages:

$$I_{ij}(p_{ij}) = I_i(p_i) + I_j(p_j).$$

Here p_{ij} is the probability that the two messages i and j will be sent. If messages i and j are truly independent, the probability of both being sent is just the probability of one times the probability of the other. Thus

$$p_{ij} = p_i \times p_j.$$

We see, then, that information is a function of the probability of a message, that *independent* information is additive, and that *independent* probabilities are multiplicative. How then does information depend on probability? The only function that satisfies these conditions is the logarithm, and so we must have

$$I_i = -k \ln p_i = k \ln (1/p_i), \tag{8.51}$$

where k is, for the moment, an arbitrary positive constant. Since $p_i \leq 1$, the minus sign is inserted so that I_i will be a positive quantity. We see from the definition that

$$I_i + I_j = -k(\ln p_i + \ln p_j) = k \ln p_i p_j,$$

and since $p_{ij} = p_i \times p_j$ for independent probabilities, it follows that

$$I_i + I_j = -k \ln p_{ij} = I_{ij}.$$

This shows that, as we required, information is indeed additive if the probabilities of the messages are independent.

Now we can start to apply this analysis to the problem of the entropy of thermodynamic systems. The microscopic state of a system of independent molecules such as an ideal gas can be regarded as a message. This message would consist, if we were to use quantum mechanics, of all the quantum numbers which specify the motion of all the atoms. If we were to use classical mechanics, it would consist of a list of the positions and momenta of all the atoms. Clearly, with as many as a mole of atoms, the number of different microscopic states possible is enormous. The probability that any one of them will occur is, therefore, very very small, and the information contained in the message that a particular one has occurred is very great. Since we never do know the microstate of a thermodynamic system, we can regard $-k \ln p_j$ as a piece of *missing* information about the system. Of course, other microstates occur, and the missing information associated with them may be different from $-k \ln p_j$. The *average* missing information would be

$$\overline{I} = \overline{-k \ln p_i} = -k \sum_i p_i \ln p_i.$$

That is, we take the missing information associated with each microstate, weight it with the probability that the microstate occurs, and add these quantities together.

We now make a major assertion: the entropy of a system is proportional to the average missing information associated with it, and is given by

$$S = -k \sum_i p_i \ln p_i$$

where k has the value of Boltzmann's constant, 1.380×10^{-16} erg/degree. Clearly, in order to test this assertion we would have to calculate the entropy of a system and compare it to an experimental measurement. To do this calculation, we must know the values of p_i. This is achieved by saying that in the absence of any information concerning the microscopic state of the system, each microscopic state has the same probability of occurring. Therefore, if there is a total of Ω microscopic states available to the system, the probability of each is $1/\Omega$. We can then deduce that

$$S = -k \sum_i p_i \ln p_i = -k \sum_i p_i \ln \frac{1}{\Omega}$$

$$= k \ln \Omega \sum_i p_i = (k \ln \Omega)\Omega \times \frac{1}{\Omega},$$

$$S = k \ln \Omega, \tag{8.52}$$

since there are Ω microstates, each with probability $1/\Omega$. We have now arrived at a fundamental equation of statistical thermodynamics: the entropy is Boltzmann's constant times the natural logarithm of the number of microscopic states available to the system. When Ω is calculated correctly, this equation is in exact agreement with experiment. We have found in Eq. (8.52) the origin of our statement that entropy is a measure of the number of microscopic states associated with a macroscopic state.

It appears that it was necessary to make two postulates in order to derive Eq. (8.52). The first was that entropy was equal to the average missing information associated with a system, and the second was that the probability of each microstate was the same. Actually, it is possible to show that the quantity

$$\sum_i p_i \ln p_i$$

has its maximum value when all the p_i are equal. Thus we can condense our postulates and say that entropy is proportional to the maximum missing information associated with a system.

The actual evaluation of Ω has been carried out for the ideal gas, the ideal solid, and other systems of chemical interest. Such calculations are moderately involved, and will not be carried out here. We can, however, show how Ω for an ideal gas depends on the state parameters V and T, and use this result to calculate ΔS for an isothermal expansion and a temperature change at constant volume.

Consider an isothermal expansion of an ideal gas. The initial and final entropies S_i and S_f are different, and so the microstates Ω_i and Ω_j associated with the initial and final macroscopic states are different. Thus

$$\Delta S = S_f - S_i = k \ln \Omega_f - k \ln \Omega_i$$

$$= k \ln \frac{\Omega_f}{\Omega_i},$$

and so we have to calculate only the *ratio* Ω_f/Ω_i rather than their absolute values. Suppose that there were only one molecule in the gas. The number of microstates available to this molecule should be proportional to the number of places the molecule can be, and hence to the volume of the container. This is also true for *each* molecule in a system of N molecules. The number of microstates available to the whole system is the number of states available to molecule 1 times the states available to molecule 2, and so on. Thus for an ideal gas of N molecules, we have

$$\Omega \propto V^N.$$

Consequently, for an isothermal expansion,

$$\Omega_f/\Omega_i = (V_f/V_i)^N, \tag{8.53}$$

and therefore

$$\Delta S = k \ln (\Omega_f/\Omega_i) = k \ln (V_f/V_i)^N$$
$$= Nk \ln (V_f/V_i)$$

or

$$\Delta S = nN_0 k \ln (V_f/V_i) = nR \ln (V_f/V_i) \tag{8.54}$$

if the number of molecules N is expressed as the number of moles n times Avogadro's number N_0. Equation (8.54) is the same as the result we obtained earlier by using thermodynamic reasoning, and helps to confirm the idea that our probabilistic or statistical interpretation is correct.

Our analysis reenforces the explanation given in Section 8.7 of why a gas expands spontaneously from a small volume into a large volume. The value of Ω_s for the gas in the small volume is smaller than the value of Ω_l for the gas in the larger volume. We can say, therefore, that because the gas in the large volume can exist in that state in many many more ways than it can exist in the small volume, it is overwhelmingly more probable for the gas to occupy the large volume. To see how enormously more probable the occupation of the larger volume is, consider the situation in which one mole of an ideal gas is given the opportunity to double its volume. By Eq. (8.53), for one mole of gas,

$$\Omega_l/\Omega_s = (V_l/V_s)^{N_0} = 2^{N_0} = 2^{(6 \times 10^{23})}.$$

Thus the chance that the gas will remain in the smaller volume is less than about one in $10^{10^{23}}$, which is rather negligible. We can attribute the "spontaneous" free expansion of a gas to the fact that the molecules are just doing what is overwhelmingly probable for them to do.

We shall now give a heuristic demonstration of how statistical considerations can be used to calculate the entropy change of an ideal monatomic gas which is heated at constant volume. As energy is added to the gas, the average momentum of the molecules increases. Since the range of momenta in which we would find molecules increases with increasing energy of the gas, we can deduce that the number of microscopic momentum states increases with increasing temperature.

For one molecule, the number of momentum states should increase proportionally to $\bar{p}^3$, the cube of the average linear momentum, since each molecule has three independent momenta along the three coordinate axes. This dependence of the number of microstates on $\bar{p}^3$ is completely analogous to the dependence of the number of microstates on $L^3 = V$. For N molecules, the number of microstates is proportional to $\bar{p}^{3N}$. Since $p \propto \bar{\epsilon}^{1/2} \propto T^{1/2}$, where $\bar{\epsilon}$ is the average molecule energy, we have

$$\Omega \propto T^{3N/2}.$$

Thus for a temperature increase at constant volume,

$$\Delta S = k \ln (\Omega_f/\Omega_i), \qquad \Omega_f/\Omega_i = (T_f/T_i)^{3N/2},$$
$$\Delta S = k \ln (T_f/T_i)^{3N/2}$$
$$= \tfrac{3}{2}Nk \ln (T_f/T_i),$$
$$\Delta S = n\tfrac{3}{2}R \ln (T_f/T_i).$$

But for the monatomic gas we are treating

$$C_V = \tfrac{3}{2}R,$$

so

$$\Delta S = nC_V \ln \frac{T_f}{T_i},$$

which is identical with the result obtained by thermodynamic reasoning.

Our demonstrations of how ΔS can be calculated for a volume change or temperature change of an ideal monatomic gas suggest another way of stating the relationship of entropy to molecular properties. We can say that since S increases as $Nk \ln V$, entropy increases as the logarithm of the volume available to molecules in *coordinate space*. Also, since S increases as $Nk \ln (\bar{p})^3$, we can say that entropy increases as the logarithm of volume available to molecules in *momentum space*—a three-dimensional space in which the coordinates are the Cartesian components of momentum. The concept of available volume in coordinate space and momentum space is very important in understanding how entropy depends on the mechanical properties of molecules. The combined six-dimensional space made up of coordinate space and momentum space is called *phase space*. Thus it is true that as V or T increases, the available volume in phase space increases, the number of microstates of the system increases, our lack of knowledge of the detailed microstate increases, and the entropy of the system increases.

Having encountered the statistical interpretation of why entropy increases with increasing volume and temperature, we now analyze, in a similar manner, the problem of the dissociation of diatomic molecules. We ask for the entropy increase associated with

$$AB(g) \rightarrow A(g) + B(g)$$

for an ideal gas at constant temperature. We shall continue to use arguments based on the available number of microstates, or the available volume in phase space, and to assume that classical mechanics rather than quantum mechanics applies.

If the dissociation occurs at constant temperature, the number of momentum microstates, or the available volume in momentum space, is the same for products and reactants. Each atom, whether bound or free, has three independent Cartesian momentum coordinates and, at the same temperature, has the same momentum states available to it. Thus to find the entropy change upon dissociation, we need to consider only the change in available positional microstates or the change in coordinate space available to the atoms.

We treat the products first. Since the number of positional microstates for one atom is proportional to V, the value of Ω_p for N_A A-atoms and N_B B-atoms is proportional to $V^{N_A}V^{N_B} = V^{2N}$, since all the $2N$ product atoms move independently of one another.

For the reactant molecules, the positions of the bonded pairs of atoms are not independent, but are correlated by the chemical bond force. We can regard the motion of one of the atoms (say A) in the molecule as independent, however, since it can go anywhere in the container, dragging the other atom B along. The number of microstates available to atom A is proportional to V.

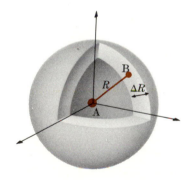

The volume available to atom B when atom A is held fixed is a spherical shell of radius R equal to the bond length, and thickness ΔR equal to the amplitude of bond oscillation.

FIG. 8.11

We now seek the positions available to the second atom in the diatomic molecule which are independent of the position of the first atom. As Fig. 8.11 shows the volume independently available to atom B lies in a spherical shell about atom A. The radius of the shell is R, the bond length of the diatomic molecule, and the thickness is ΔR, the amplitude of oscillation of the bonded atoms. The associated volume available to the second atom is, therefore, $4\pi R^2 \Delta R$. The number of microstates available to one molecule is proportional to the product of V and $4\pi R^2 \Delta R$, and for N molecules we have

$$\Omega_r \propto (V \times 4\pi R^2 \Delta R)^N.$$

The entropy change for the dissociation reaction can now be given as

$$\Delta S = k \ln \frac{\Omega_p}{\Omega_r} = k \ln \frac{(V)^{2N}}{(V 4\pi R^2 \, \Delta R)^N}$$

$$= Nk \ln \frac{V}{4\pi R^2 \, \Delta R}.$$

Thus the entropy change is proportional to the logarithm of a simple volume ratio: V, the volume an atom has when it is free, divided by $4\pi R^2 \, \Delta R$, the independent volume that the atom has available to it when it is bonded. It is clear that since $V \gg 4\pi R^2 \, \Delta R$, the entropy of dissociation is positive. We see that the larger R is, the larger is the volume available to the atom in the molecule, and the smaller is the entropy of dissociation. Also, the larger the vibrational amplitude ΔR, the smaller the entropy of dissociation. Further analysis shows that this vibrational amplitude can be expressed as $(2\pi kT/f)^{1/2}$, where f is the spring or Hooke's law constant for changing the bond distance slightly from its most favorable position. Therefore, the stiffer the bond, the smaller ΔR, and the larger the entropy gained upon dissociation.

Although our analysis applies directly to the dissociation of diatomic molecules, the physical idea that entropy increases with an increase in the freedom or volume available to atoms for *independent* motion is of general validity. The positive entropy changes associated with melting of a crystal, evaporation of a liquid, and dissociation of diatomic and polyatomic molecules are immediate consequences of this principle.

8.15 CONCLUSION

Thermodynamics is a subject of great scope and enormous utility. It proceeds from certain experimentally determined laws that concern the behavior of systems in general, and shows how to make conclusions or predictions about specific changes that a particular system may undergo. From the first law of thermodynamics, we learn how to use in a systematic and efficient way the measured energy changes that accompany chemical reactions. By measuring the ΔH of *relatively few* reactions we can tabulate enthalpies of formation of compounds and from these calculate ΔH for *any* reaction that involves compounds in the table. By doing this we avoid having to perform many calorimetric experiments, some of which may be very difficult or impossible. From the second law of thermodynamics, we learn the criteria for the spontaneity of any chemical or physical process. By developing these criteria we discover the utility of ΔG_f^0, the standard free energy of formation. This quantity gives us a measure of the intrinsic stability of a substance with respect to chemical change. If ΔG_f^0 is very negative, the compound is more stable than its elements, and will in general tend to be formed rather than consumed by any chemical reaction in which it is involved. In contrast, a compound with a large positive ΔG_f^0 represents a relatively unstable arrangement of atoms, and will have a tendency to be converted to other, more stable arrangements—either elements or other compounds of them. Thus in ΔG_f^0 we have a quantitative expression of the intrinsic chemical reactivity of a substance.

We must be careful to recognize the limitations of thermodynamics. While we can measure values of ΔG_f^0 for various compounds and use these to predict equilibrium constants, thermodynamics does not contain explanations of *why* one molecule is more stable than another. Such explanations are drawn from the quantum theory of molecular structure, an interesting and complex subject that is still growing and being refined. But even the proper predictions of thermodynamics are sometimes of a conditional nature. We may calculate that the equilibrium constant of a particular reaction is large, but thermodynamics offers no guarantee that this reaction will proceed fast enough to be observed. Thus thermodynamics provides a way of predicting what is possible in principle; it tells us what *can* happen rather than what is certain to happen. Despite this limitation, thermodynamics is of substantial help in organizing and understanding chemical phenomena, as we shall see in subsequent chapters.

SUGGESTIONS FOR FURTHER READING

Allen, J. A., *Energy Changes in Chemistry*. Boston: Allyn and Bacon, 1966.

Hargreaves, G., *Elementary Chemical Thermodynamics*. London: Butterworths, 1961.

Harvey, K. B., and G. B. Porter, *Introduction to Physical Inorganic Chemistry*. Reading, Mass.: Addison-Wesley, 1963. Chapter 8.

Mahan, B. H., *Elementary Chemical Thermodynamics*. New York: W. A. Benjamin, 1963.

Nash, L. K., *Elements of Chemical Thermodynamics*. Reading, Mass.: Addison-Wesley, 1962.

Van Ness, H. C., *Understanding Thermodynamics*. New York: McGraw-Hill, 1969.

PROBLEMS

8.1 Calculate the enthalpy of formation of $Ca(OH)_2(s)$ from the following data:

$$H_2(g) + \tfrac{1}{2}O_2(g) = H_2O(l), \qquad \Delta H = -68.3 \text{ kcal};$$
$$CaO(s) + H_2O(l) = Ca(OH)_2(s), \qquad \Delta H = -15.3 \text{ kcal};$$
$$Ca(s) + \tfrac{1}{2}O_2(g) = CaO(s), \qquad \Delta H = -151.8 \text{ kcal}.$$

8.2 From the data contained in Table 8.1, calculate ΔH for each of the following reactions:

$$Fe_2O_3(s) + 3CO(g) = 3CO_2(g) + 2Fe(s),$$
$$2NO_2(g) = 2NO(g) + O_2(g),$$
$$N(g) + NO(g) = N_2(g) + O(g).$$

8.3 A sample of solid naphthalene, $C_{10}H_8$, weighing 0.600 gm is burned to $CO_2(g)$ and $H_2O(l)$ in a constant-volume calorimeter at $T = 298°K$. In this experiment, the observed temperature rise of the calorimeter and its contents is 2.270°C. In a separate experiment, the total heat capacity of the calorimeter was found to be 2556 cal/deg. What is ΔE for the combustion of one mole of naphthalene? What is ΔH for this reaction? By using the tabulated values for $\Delta H_f^0(CO_2)$ and $\Delta H_f^0(H_2O, l)$, calculate the enthalpy of formation of naphthalene.

8.4 One mole of an ideal gas at 300°K expands isothermally and reversibly from 5 to 20 liters. By remembering that for an ideal gas, E is constant at constant temperature, calculate the work done and the heat absorbed by the gas. What is ΔH for this process?

8.5 Ethylene, C_2H_4, and propylene, C_3H_6, can be hydrogenated according to the reactions

$$C_2H_4(g) + H_2(g) = C_2H_6(g),$$
$$C_3H_6(g) + H_2(g) = C_3H_8(g),$$

to yield ethane, C_2H_6, and propane, C_3H_8, respectively. From data contained in Table 8.1, calculate ΔH for these reactions. Do these answers suggest that ΔH for any reaction of the type

$$C_nH_{2n} + H_2 = C_nH_{2n+2}$$

might be approximately equal to a constant? Test this idea with another such reaction involving compounds in Table 8.1.

8.6 Calculate ΔS^0 for the following chemical reactions, all carried out at 298°K.

$$Ca(s) + \tfrac{1}{2}O_2(g) = CaO(s),$$
$$CaCO_3(s) = CaO(s) + CO_2(g),$$
$$H_2(g) = 2H(g),$$
$$N_2(g) + O_2(g) = 2NO(g).$$

Explain the sign and/or magnitude of ΔS^0 for each reaction by qualitatively evaluating the change in molecular chaos or disorder that accompanies the reaction. State whether or not the attendant entropy change tends to favor production of the products of the reaction.

8.7 (a) One mole of an ideal gas expands reversibly from a volume of 2 to 20 liters. Calculate the entropy change of the system and of the surroundings. (b) The same isothermal expansion takes place irreversibly such that no work is done on or by the ideal gas. Calculate the entropy change of the system and of its surroundings. (c) Using the answers you have accumulated, show numerically that the spontaneous contraction of an ideal gas in an isolated system would violate the second law of thermodynamics.

8.8 Compute the entropy of vaporization of the following liquids at their normal boiling points.

	T_b(°K)	ΔH_{vap} (kcal)		T_b(°K)	ΔH_{vap} (kcal)
Cl_2	238.5	4.87	$PbCl_2$	1145	24.8
C_6H_6	353	7.35	H_2O	373	9.72
$CHCl_3$	334	7.02	C_2H_5OH	351	9.22

The fact that for most liquids, ΔS_{vap} at the boiling point is nearly always 21 eu is called Trouton's rule. In liquid water and ethyl alcohol, molecules are linked by relatively strong hydrogen bonds between the hydrogen of one molecule and the oxygen atom of another. Is this "ordering" effect consistent with the way in which water and ethanol deviate from Trouton's rule? Explain.

8.9 When a mole of water supercooled to −10°C freezes isothermally, what is its entropy change? The process as described is irreversible, so in order to calculate ΔS, a reversible path between initial and final states must be found. One such path is

$$H_2O(l), -10°C \rightarrow H_2O(l), 0°C,$$
$$H_2O(l), 0°C \rightarrow H_2O(s), 0°C,$$
$$H_2O(s), 0°C \rightarrow H_2O(s), -10°C.$$

The molar enthalpy of fusion of ice at 0°C is 1440 cal, the molar heat capacity of ice is 9.0 cal/mole-deg, and the molar heat capacity of water is 18.0 cal/mole-deg. Use these data to compute ΔS for the water when it freezes at −10°C.

The enthalpy of fusion of ice at −10°C is 1350 cal/mole. Find the entropy change of the surroundings when 1 mole of water freezes at −10°C. What is the total entropy change of the system and surroundings for this process? Is the process irreversible according to the second law of thermodynamics?

8.10 From the data given below, calculate the absolute entropy of solid silver at 300°K by plotting C_P/T as a function of T and determining the area under the curve by counting squares on the graph paper.

$T(°K)$	15	20	30	40	50	70
C_P(cal/mole-deg)	0.16	0.41	1.14	2.01	2.78	3.90
$T(°K)$	90	130	170	210	250	300
C_P(cal/mole-deg)	4.57	5.29	5.64	5.84	5.91	6.09

8.11 From each of the following pairs of substances, choose the one which you would expect to have the greater absolute entropy. Except where specified, assume one mole of each material at the same temperature and pressure.

(a) C(graph), Ag(s) (b) B(298°K), B(398°K)
(c) $Br_2(g)$, 2Br(g) (d) Ar(1 atm), Ar(0.1 atm)

8.12 At the normal boiling temperature of water, $\Delta H_{vap} = 9.72$ kcal/mole. By assuming that the volume of 1 mole of liquid water is negligible, and that water vapor is an ideal gas, calculate q, w, ΔE, ΔS, and ΔG, for the reversible vaporization of 1 mole of water at a constant pressure of 1 atm and at a temperature of 373°K.

8.13 Calculate ΔG^0 and K at 25°C for the reaction

$$NO(g) + \tfrac{1}{2}O_2(g) = NO_2(g)$$

from data tabulated in this chapter. Which factor, enthalpy or entropy, makes K greater than unity and thereby provides the principal driving force for the reaction?

8.14 Tabulated below are ΔH_f^0, ΔG_f^0, and S^0 for four substances:

	ΔH_{298}^0(kcal)	ΔG_{298}^0(kcal)	S_{298}^0(eu)
CO	−26.42	−32.79	47.3
CO_2(g)	−94.05	−94.24	51.1
H_2O(g)	−57.80	−54.64	–
H_2(g)	–	–	31.2

From these data only, calculate ΔH^0, ΔG^0, and ΔS^0 for the reaction

$$H_2O(g) + CO(g) = H_2(g) + CO_2(g).$$

If these substances are perfect gases, what is ΔE_{298}^0 for this reaction? Finally, compute the absolute entropy of $H_2O(g)$ at 298°K.

8.15 For the reaction

$$SO_2(g) + \tfrac{1}{2}O_2(g) = SO_3(g),$$

calculate ΔG^0 and ΔH^0 from data tabulated in this chapter. Compute the equilibrium constant at 298°K and at 600°K by assuming that ΔH is independent of temperature.

8.16 For the reaction

$$\tfrac{1}{2}N_2(g) + \tfrac{1}{2}O_2(g) = NO(g),$$

the equilibrium constant is 1.11×10^{-2} at 1800°K and 2.02×10^{-2} at 2000°K. Calculate ΔG_{2000}^0, the standard free-energy change at 2000°K, and compare it with ΔG_{298}^0 which you can obtain from Table 8.4. From the two values of the equilibrium constant, compute ΔH for the reaction.

8.17 The equilibrium vapor pressure of water over $BaCl_2 \cdot H_2O$ is 2.5 mm at 25°C. What is ΔG for the process

$$BaCl_2 \cdot H_2O(s) \rightarrow BaCl_2(s) + H_2O(g),$$

where the water vapor is *imagined* to be a 1-atm pressure? What is ΔG for the process if water vapor is produced at 2.5 mm?

8.18 For the reaction

$$BaSO_4(s) = Ba^{++}(aq) + SO_4^{=}(aq),$$

$\Delta H = 5800$ cal. Does barium sulfate become more or less soluble in water as temperature increases? At 25°C the solubility product of barium sulfate is 1.1×10^{-10}. What is its value at 90°C, given that ΔH for the reaction is constant?

8.19 For the reaction

$$N_2O_4(g) = 2NO_2(g),$$

the following values of $\log_{10}K$ have been obtained at different temperatures:

$\log_{10}K$	−1.45	−1.02	−0.587	−0.036	0.379	0.903
$T(°K)$	282	298	306	325	343	362

Plot $\log K$ as a function of $1/T$, and determine ΔH for the reaction in this manner. Is the slope of the line obtained from these data equal to $-\Delta H/R$ or to $-\Delta H/2.3R$?

8.20 For water, $\Delta H_{fus} = 1.44$ kcal/mole, and $\Delta H_{vap} = 9.72$ kcal/mole. From these data, calculate the molal freezing-point depression constant and the molal boiling-point elevation constant. Do they compare well with the directly measured values of 1.86 and 0.52, respectively?

8.21 In the Daniell cell, the reaction

$$Zn(s) + Cu^{++}(aq) = Zn^{++}(aq) + Cu(s)$$

occurs spontaneously, and the cell delivers 1.10 volts when all substances are at 1-M concentration. What is ΔG^0 for this reaction, expressed in units of (i) joules and (ii) calories? Calculate the equilibrium constant of the reaction from (a) the value of $\Delta \mathcal{E}^0$ and (b) the value of ΔG^0.

8.22 Consider the following half-reactions and their standard electrode potentials:

1. $Fe^{++} + 2e^- = Fe,$ $\mathcal{E}^0 = -0.440$ volt;
2. $Fe^{+3} + e^- = Fe^{++},$ $\mathcal{E}^0 = 0.771$ volt;
3. $Fe^{+3} + 3e^- = Fe,$ $\mathcal{E}^0 = -0.036$ volt.

Although half-reaction 3 is the sum of half-reactions 1 and 2, the electrode potential of half-reaction 3 is *not* equal to the sum of the first two electrode potentials. In general, it is not possible to combine $\mathcal{E}^0$'s for half-reactions directly to obtain the $\mathcal{E}^0$ of a third *half-reaction*. Strictly, only free energies, and not electrode potentials, may be combined in these instances. With this in mind, execute the following: (a) Find ΔG^0 for half-reactions 1 and 2 by using $\Delta G^0 = -n\mathcal{F}\mathcal{E}^0$. (b) Add these quantities in order to obtain ΔG^0 for reaction 3. (c) From the result of part (b), calculate $\mathcal{E}^0$ for reaction 3, and compare it with the value given. (d) Use this type of analysis to decide why it is legitimate to calculate $\Delta \mathcal{E}^0$ for a net reaction in which electrons do not appear, simply by algebraic combination of the $\mathcal{E}^0$'s of two half-reactions.

CHAPTER 9

CHEMICAL KINETICS

A chemical reaction has two general characteristics of foremost importance: the position of equilibrium and the reaction rate. In considering chemical equilibrium, we are concerned only with the relative stabilities of products and reactants and their relative concentrations at equilibrium and not with the pathway from the initial to final state. On the other hand, in treating reaction rates, we are concerned not only with how fast reactants are converted to products, but also with the sequence of physical and chemical processes by which this conversion occurs. Indeed, reaction rates are studied in order to obtain a detailed picture of what molecules do to each other when they react.

The behavior of a chemical system may be determined either by equilibrium effects or by reaction rates. For example, when a small amount of aqueous silver ion is added to an equimolar solution of chloride and iodide ions, we immediately obtain a small amount of solid silver iodide, and no silver chloride. This occurs because the reaction

$$Ag^+ + I^- = AgI$$

has a much larger equilibrium constant than the reaction

$$Ag^+ + Cl^- = AgCl,$$

so that silver iodide is the product favored by the equilibrium constants. In contrast, consider the two ways in which ethyl alcohol, CH_3CH_2OH, may be dehydrated:

$$CH_3CH_2OH \xrightarrow{H_2SO_4} CH_2CH_2 + H_2O \qquad \text{(concentrated acid, 170°C)},$$

$$2CH_3CH_2OH \xrightarrow{H_2SO_4} CH_3CH_2OCH_2CH_3 + H_2O \qquad \text{(dilute acid, 140°C)}.$$

Either ethylene or diethyl ether can be the principal product of the dehydration, depending on the temperature and acid concentration. This selectivity is *not* due to a change in equilibrium constants. Rather, ethylene is obtained at the higher temperature because under these conditions it is *formed faster* than diethyl ether. At the lower temperature, the situation is reversed; the rate of formation of diethyl ether is greater than that of ethylene. In general, when any two compounds are mixed, there may be a large number of reactions which are possible, but the reaction or reactions which are actually observed are the ones which proceed fastest.

Our example shows that it is possible to influence the products of chemical change by controlling factors which affect reaction rates. Naturally the rates of reactions are in large measure determined by the nature of the reactants, but there are other factors, more at our disposal, which are also influential. As our example suggests, the first of these is concentration—both of the reactants themselves, and of other added reagents called catalysts. The latter affect rates even though they may not be involved in the stoichiometry of the overall reaction. Temperature also is an important parameter; the rates of some reactions increase profoundly when temperature is raised, while the rates of others are almost insensitive to temperature changes. Two additional factors apply only to reactions which take place at the boundary surface between two phases. These reactions are classified as heterogeneous, and include the combustion of solid particles, the dissolution of metals in acid, and the evaporation of condensed materials. In such systems, reaction rates increase as the available surface area increases, and may be further augmented by agitation, which speeds the transport of fresh reagents to the phase boundary. We now turn to a detailed discussion of each of these rate-controlling factors.

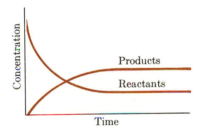

Time dependence of the concentrations of reactants and products in a chemical reaction.

FIG. 9.1

9.1 CONCENTRATION EFFECTS

Figure 9.1 shows the usual behavior of the concentration of a reactant and a product as a chemical reaction progresses. The concentrations at first change rapidly and then more slowly approach the limiting concentrations found when the reaction comes to equilibrium. Not only does the concentration of a reactant diminish but the rate of change of its concentration (the slope of the concentration-time curve) also decreases as the reaction progresses. Since this rate of change of concentration is not constant, it is best expressed as a time derivative, dc/dt, which gives the change in concentration per unit time, at a particular instant or at a particular concentration of reactants. We shall use such time derivatives of concentration to express reaction rates quantitatively.

Differential Rate Laws

In general, as reaction products are formed, they react with each other and re-form reactants. Therefore the *net* rate at which the reaction proceeds from left to right is

$$\text{net reaction rate} = \text{forward rate} - \text{reverse rate}.$$

When the reaction reaches equilibrium, the net reaction rate is zero, and the forward rate equals the reverse rate. When the reaction mixture is far from its equilibrium composition, either the forward or reverse rate is dominant, depending on whether reactants or products are in excess of the equilibrium value. To simplify our initial discussions, we shall limit ourselves to cases in which only the reaction rate in the forward direction is important. This is the situation when the reactants are first brought together and the mixture is far from equilibrium.

With this restriction in mind, let us consider the reaction

$$NO + O_3 \rightarrow NO_2 + O_2,$$

where the arrow means that only the reaction from left to right is important. What algebraic relations connect the various derivatives $d[NO]/dt$, $d[O_3]/dt$, $d[NO_2]/dt$, and $d[O_2]/dt$? The stoichiometry of the reaction shows that the concentrations of nitric oxide and ozone must decrease at the same rate, which in turn is exactly the rate at which the concentrations of nitrogen dioxide and oxygen increase. Since the concentrations of nitric oxide and ozone are diminishing, $d[NO]/dt$ and $d[O_3]/dt$ are *negative numbers*, while $d[NO_2]/dt$ and $d[O_2]/dt$ are positive. Thus we have

$$-\frac{d[NO]}{dt} = -\frac{d[O_3]}{dt} = \frac{d[NO_2]}{dt} = \frac{d[O_2]}{dt} = \text{reaction rate.}$$

As another illustration of the relations between concentration derivatives, consider the reaction

$$2HI(g) \rightarrow H_2(g) + I_2(g).$$

Since 2 moles of HI disappear for each mole of H_2 formed, the rate of change of the HI concentration must be twice the rate of change of the H_2 concentration. Remembering that $d[HI]/dt$ is a negative number, we can write

$$-\frac{1}{2}\frac{d[HI]}{dt} = \frac{d[H_2]}{dt} = \frac{d[I_2]}{dt}.$$ (9.1)

This shows that the concentrations of the various reactants and products may change at different rates. What, then, is the rate of this reaction? Is it the rate at which the concentration of HI changes, or the rate at which the concentration of H_2 changes?

We can solve this problem by applying our arguments to the general reaction

$$aA + bB \rightarrow cC + dD.$$

A little reflection shows that the relation between the various derivatives is

$$-\frac{1}{a}\frac{d[A]}{dt} = -\frac{1}{b}\frac{d[B]}{dt} = \frac{1}{c}\frac{d[C]}{dt} = \frac{1}{d}\frac{d[D]}{dt}.$$

Since all these quantities are equal, it is natural to take the formal definition of "the rate of reaction" as the time derivative of a concentration divided by the appropriate stoichiometric coefficient and converted to a positive number. Thus the rate of the HI decomposition reaction is equal to any of the expressions in Eq. (9.1). Application of these ideas to the reaction

$$30CH_3OH + B_{10}H_{14} \rightarrow 10B(OCH_3)_3 + 22H_2$$

gives

$$\text{rate of reaction} \equiv -\frac{1}{30}\frac{d[CH_3OH]}{dt} = -\frac{d[B_{10}H_{14}]}{dt}$$
$$= \frac{1}{10}\frac{d[B(OCH_3)_3]}{dt} = \frac{1}{22}\frac{d[H_2]}{dt}.$$

Any of the derivatives of concentration, appropriately modified, may be used to express the rate of reaction.

The mathematical expression which shows how the rate of reaction depends on concentration is called the **differential rate law**. In many instances, it is possible to express the differential rate law as a product of reagent concentrations, each raised to some power. Accordingly, for the reaction

$$3A + 2B \rightarrow C + D,$$

the differential rate law may have the form

$$-\frac{1}{3}\frac{d[A]}{dt} = \frac{d[C]}{dt} = k[A]^n[B]^m.$$

The exponents n or m are generally integers or half-integers; n is called the **order** of the reaction with respect to A, and m is the order of the reaction with respect to B. The sum $n + m$ is called the overall order of the reaction. It is important to realize that n and m are *not necessarily equal to the stoichiometric coefficients of* A *and* B *in the net reaction*. The order with respect to each reagent must be found *experimentally* and cannot be predicted or deduced from the equation for the reaction. For example, experiments show that the differential rate law for the reaction $H_2 + I_2 \rightarrow 2HI$ is

$$-\frac{d[H_2]}{dt} = k[H_2][I_2].$$

Since each concentration is raised to the first power, we say that this reaction is of first order with respect to H_2, first order with respect to I_2, and second order overall. In contrast, the differential rate law for the apparently similar reaction $H_2 + Br_2 \rightarrow 2HBr$ is

$$-\frac{d[H_2]}{dt} = k'[H_2][Br_2]^{1/2},$$

that is, first order with respect to hydrogen, one-half order with respect to bromine, and three-halves order overall. Although these two reactions have the same stoichiometry, and despite the similarities of iodine and bromine, the rate laws of the reactions are different.

The constant k which appears in the differential rate law is called the **rate constant** or more formally, the specific-reaction rate constant, since it is numerically equal to the rate the reaction would have if all concentrations were set equal to unity. Each reaction is characterized by its own rate constant whose value is determined by the nature of the reactants and the temperature. From the numerical value of the rate constant, we can calculate the rate of a reaction under particular concentration conditions. In essence, the rate constant is a numerical expression of the effect of the nature of the reactants and the temperature on reaction rate. Consequently, one of the goals of theoretical chemistry is to be able to understand, or better, to predict the values of rate constants from a knowledge of the electronic structure of reactants and products.

The Integrated Rate Laws

The differential rate laws show how the rates of reaction depend on the concentrations of reagents. It is also useful to know how the concentrations depend on time; this information can be obtained from the differential rate law by integration. The decomposition of dinitrogen pentoxide provides us with our first example:

$$N_2O_5 \rightarrow 2NO_2 + \tfrac{1}{2}O_2.$$

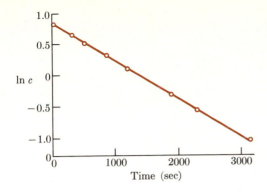

FIG. 9.2

The natural logarithm of the concentration of N_2O_5, plotted as a function of time. The slope of the line is the negative of the first-order rate constant.

Experiments show that the reaction is first order with respect to the concentration of N_2O_5,

$$-\frac{d[N_2O_5]}{dt} = k[N_2O_5].$$

If we let c stand for the concentration of N_2O_5, the differential rate law is

$$-\frac{dc}{dt} = kc,$$

which we can rearrange to

$$-\frac{dc}{c} = k\,dt.$$

The left-hand side of the equation is a function of c only, and apart from a constant, the right-hand side contains only the differential of time. Therefore we can integrate both sides, taking as limits c_0, which is the concentration at $t = 0$, and c, the concentration at time t. Thus

$$-\int_{c_0}^{c} \frac{dc}{c} = k\int_{0}^{t} dt$$

$$-\ln c\big|_{c_0}^{c} = kt\big|_{0}^{t}$$

$$-\ln \frac{c}{c_0} = kt.$$

This equation shows that for a first-order reaction, the logarithm of the reactant concentration decreases linearly as time increases. It suggests that if we plot $\ln c$ as a function of t, we will obtain a straight line whose slope is $-k$. Figure 9.2 shows that the data for the decomposition of N_2O_5 do indeed yield a straight line when plotted this way.

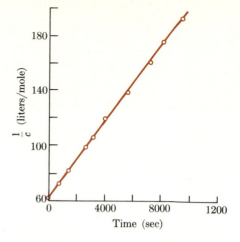

FIG. 9.3

The reciprocal of the concentration of butadiene, plotted as a function of time. The slope of the line equals the rate constant for the dimerization reaction.

If a reaction is second order, the time dependence of the reactant concentration is different. Our example for this case is the dimerization of butadiene, C_4H_6:

$$C_4H_6(g) \rightarrow \tfrac{1}{2}C_8H_{12}(g),$$

which follows the second-order differential rate law

$$-\frac{d[C_4H_6]}{dt} = k[C_4H_6]^2.$$

If we substitute c for the concentration of C_4H_6, we can rearrange the equation to read

$$-\frac{dc}{c^2} = k\,dt,$$

and once again each side can be integrated between the limits c_0, c and 0, t:

$$-\int_{c_0}^{c} \frac{dc}{c^2} = k\int_0^t dt,$$

$$\frac{1}{c}\bigg|_{c_0}^{c} = kt\big|_0^t,$$

$$\frac{1}{c} - \frac{1}{c_0} = kt.$$

Thus for a second-order reaction, the reciprocal of the reactant concentration is a linear function of time, so that a plot of $1/c$ as a function of t should be a straight line with a slope k and an intercept $1/c_0$. Figure 9.3 shows how well the dimerization of C_4H_6 conforms to this prediction.

These are but two examples of how the differential rate law can be converted to an expression which gives the time dependence of concentration. Other more complicated differential rate laws can be integrated, but the process is sometimes difficult and often produces cumbersome algebraic expressions. This is not a serious drawback, since virtually all the interesting information about reaction rates can be obtained by using the differential rate law expressions.

Experimental Determination of Rate Laws

The order of a reaction and its rate constant must be found experimentally, and the differential and integral rate laws which we have discussed suggest how this can be done. The differential rate law for the reaction

$$aA + bB \rightarrow cC$$

is often of the form

$$-\frac{1}{a}\frac{d[A]}{dt} = k[A]^n[B]^m.$$

Our first object is to determine the exponents n and m. One way of determining the order with respect to A is to make up a series of mixtures which contain the same concentration of B but different concentrations of A. Then the *initial rate* of reaction is found for each of these mixtures by measuring the change in concentration of one of the reactants or products which occurs in the first small time interval after the reagents are mixed. For instance, $\Delta[A]/\Delta t$ is a good approximation to $d[A]/dt$ if the time interval is short enough, and from it we can calculate the rate of reaction. Since in the series of experiments the only variable is the initial concentration of A, the experimentally determined initial rates should vary as the initial concentration of A, raised to the power n. If the reaction rate doubles when A is doubled, the rate depends on the first power of the concentration of A, and n is equal to one. If the reaction rate increases by a factor of four when the concentration of A is doubled, n must equal 2, and the reaction is second order. Once the order with respect to A is determined, the procedure can be repeated. This time we hold the concentration of A constant, vary the concentration of B, and deduce the order of the reaction with respect to B. When the order with respect to each reagent is known, we can calculate k by dividing the measured rate of reaction by the concentrations of the reactants, each raised to the appropriate power:

$$k = \frac{-(1/a)(d[A]/dt)}{[A]^n[B]^m}.$$

The success of this procedure depends on our ability to accurately evaluate the initial rate. Since it is difficult to measure accurately the small change in concentration $\Delta[A]$ which occurs in a small time interval Δt, the method we have outlined is not always satisfactory. As an alternative, we can make use of the integrated rate laws to find both the order and rate constant of a reaction. The only integrated rate laws we are familiar with pertain to reactions in which there is only one reactant, as in

$$N_2O_5 \rightarrow 2NO_2 + \tfrac{1}{2}O_2$$

and

$$2HI \rightarrow H_2 + I_2,$$

where the differential rate law has the form

$$-\frac{1}{a}\frac{d[A]}{dt} = k[A]^n.$$

To deal with such a reaction, we must determine the concentration of A at various times as the reaction proceeds. Then we prepare two graphs: one a plot of $\ln[A]/[A_0]$ as a function of time, the other a plot of $1/[A]$ as a function of time. If the logarithmic graph is linear and the reciprocal plot curved, we can conclude that the reaction is a first-order process like the decomposition of N_2O_5. Moreover, since the general integrated rate law for a first-order process is $\ln c/c_0 = -kt$, the slope of the logarithmic graph can tell us the rate constant. On the other hand, if the logarithmic plot is curved and the reciprocal plot linear, the reaction has a second-order rate law. The reaction rate constant can be calculated from the slope of the reciprocal plot, since the general equation of the line is $1/c = 1/c_0 + kt$.

It is possible that neither $\ln c$ nor $1/c$ will be linear functions of time; this merely shows that the reaction does not follow the simple rate laws we have discussed. The procedures used to identify which of the more complicated rate laws does apply are analogous to the methods we have just outlined and only involve the use of slightly more complicated algebraic expressions.

9.2 REACTION MECHANISMS

We now attack the problem of obtaining a complete chemical description of how reactant molecules are converted to products. In some reactions this conversion occurs in one step; two reactant molecules collide and as a result form the observed product molecules. An example of such a one-step conversion of reactants to products is

$$NO + O_3 \rightarrow NO_2 + O_2.$$

On the other hand, most chemical reactions do not follow such a simple path from reactants to products. For example, the reaction

$$H_2O_2 + 2Br^- + 2H^+ \rightarrow Br_2 + 2H_2O$$

is *not* the result of the simultaneous collision of two hydrogen ions, two bromide ions, and a hydrogen peroxide molecule. The chance of having five species come to the same place at the same time is very small, so small that such a process could never produce products at the rate which is experimentally observed. The actual reaction path consists of two successive processes, neither of which involves the collision of more than three particles:

$$Br^- + H^+ + H_2O_2 \rightarrow HOBr + H_2O,$$
$$H^+ + HOBr + Br^- \rightarrow H_2O + Br_2.$$

Other reactions may involve a great many steps. For example, the decomposition of N_2O_5 follows the path

$$N_2O_5 + N_2O_5 \rightarrow N_2O_5^* + N_2O_5,$$
$$N_2O_5^* \rightarrow NO_2 + NO_3,$$
$$NO_2 + NO_3 \rightarrow NO + NO_2 + O_2,$$
$$NO + NO_3 \rightarrow 2NO_2,$$

where $N_2O_5^*$ stands for an energized molecule capable of dissociating. Each one of the steps is called an **elementary process,** since each is a simple event in which some kind of transformation occurs. The collection of elementary processes by which an overall reaction occurs is called a **reaction mechanism.** The mechanism of a reaction must be determined experimentally; to understand how this is done we must first discuss the three types of elementary process.

Elementary Processes

Elementary processes are classified according to the number of molecules which they involve. An event in which only one reactant molecule participates is called a **unimolecular process.** The decomposition or rearrangement of an energized molecule is a unimolecular elementary process:

$$O_3^* \rightarrow O_2 + O,$$

$$\begin{matrix} CH_2^* \\ \diagup \quad \diagdown \\ CH_2 \!-\! CH_2 \end{matrix} \rightarrow CH_3CH\!=\!CH_2$$

A **bimolecular process** always involves two reacting molecules. For example,

$$NO + O_3 \rightarrow NO_2 + O_2,$$
$$Cl + CH_4 \rightarrow CH_3 + HCl,$$

and

$$Ar + O_3 \rightarrow Ar + O_3^*$$

are all bimolecular processes. No chemical change occurs in the last process, but the collision between Ar and O_3 does supply the ozone molecule with excess internal energy which can eventually cause it to dissociate, as we have noted above.

Elementary processes in which three particles participate are said to be **termolecular.** Most termolecular processes involve the association or combination of two particles which is made possible by a third particle whose role is to remove the excess energy produced when a chemical bond forms. As examples we have

$$O + O_2 + N_2 \rightarrow O_3 + N_2$$

and

$$O + NO + N_2 \rightarrow NO_2 + N_2.$$

By the conservation of energy, an ozone molecule formed by association of an oxygen atom and molecule has enough energy to redissociate. Only if some of this energy is removed by a third particle, the nitrogen molecule, can a stable product be formed. Elementary processes with molecularity greater than three are not known, since collisions in which more than three particles come together simultaneously are very rare.

We have emphasized that the order of a reaction cannot, in general, be predicted from the stoichiometry of the overall reaction. However, the order of an *elementary process* is predictable. For example, consider the general bimolecular elementary process

$$A + B \rightarrow C + D.$$

Now in order for a molecule of A and a molecule of B to react, they must at the very least collide with each other. The rate at which collisions between A- and B-molecules occur is directly proportional to the concentrations of A and of B. Therefore any bimolecular elementary process must follow the second-order rate law

$$-\frac{d[A]}{dt} = k[A][B].$$

A similar argument applies to the problem of finding three particles in collision and leads us to conclude that a termolecular elementary process,

$$A + B + C \rightarrow D + E,$$

follows a rate law which is overall third order, and first order with respect to each reactant:

$$-\frac{d[A]}{dt} = k[A][B][C].$$

Finally, let us consider the unimolecular process. Here we deal with a collection of molecules, each rearranging or decomposing independently of the others. It seems clear that as the number of molecules increases, the number that decompose in a given time interval will also increase. Thus the rate of reaction will be proportional to the first power of the concentration, and the unimolecular process

$$A^* \rightarrow B$$

will follow a first-order rate law:

$$-\frac{d[\text{A}^*]}{dt} = k[\text{A}^*].$$

Our general conclusion is that *for elementary processes the molecularity and order are the same;* a unimolecular process is first order, a bimolecular process is always second order, and a termolecular process is third order. However, it is important to realize that *the converse is not true:* not all first-order reactions are unimolecular elementary processes, second-order reactions are not all bimolecular, and third-order reactions are not necessarily termolecular. We have had one illustration of this point already; the decomposition of N_2O_5 is a first-order reaction, but it proceeds by a complex mechanism which consists of *both* unimolecular and bimolecular elementary processes. We will find an explanation for this by examining the relation between reaction mechanisms and rate laws.

Mechanisms and Rate Laws

We now have to find how the experimentally observed order and rate of an overall reaction are related to the order and rates of the elementary processes which comprise its mechanism. Fortunately this question has an answer which is simple and direct for most reactions. Consider the hypothetical reaction,

$$3\text{A} + 2\text{B} \rightarrow \text{C} + \text{D},$$

which we will assume follows the mechanism

$$
\begin{array}{c}
\text{A} + \text{B} \rightarrow \text{E} + \text{F} \\
\text{A} + \text{E} \rightarrow \text{H} \\
\text{A} + \text{F} \rightarrow \text{G} \\
\underline{\text{H} + \text{G} + \text{B} \rightarrow \text{C} + \text{D}} \\
3\text{A} + 2\text{B} \rightarrow \text{C} + \text{D}
\end{array}
$$

The products C and D are the result of a sequence of four elementary processes, and it is indisputable that the products can be formed *no faster than the rate of the slowest step* in this sequence. Therefore, if one of the steps is much slower than all the others, the rate of the overall reaction will be limited by, and be exactly equal to, the rate of this slow step. Consequently, the slowest elementary process in a sequence is called the *rate-determining step.* Suppose that the first step in the above mechanism is the slowest, and that its rate constant is k_1. Since it is a bimolecular elementary process, its rate is overall second order: first order with respect to A and first order with respect to B. As a result the observed rate law for the overall reaction will be

$$-\frac{1}{3}\frac{d[\text{A}]}{dt} = k_1[\text{A}][\text{B}].$$

Thus a complicated mechanism can result in a very simple rate law. In this argument we also find the explanation of why all reactions which follow second-order rate laws are not necessarily bimolecular elementary processes. The overall reaction,

$$3A + 2B \rightarrow C + D$$

is not an elementary process, it is complex; yet it follows a second-order rate law, because its slowest step is a bimolecular process.

To turn from the abstract to the concrete, let us examine a few complex reactions which follow simple rate laws. For instance the reaction

$$2NO_2 + F_2 \rightarrow 2NO_2F$$

follows a second-order rate law:

$$-\frac{1}{2}\frac{d[NO_2]}{dt} = k_{exp}[NO_2][F_2].$$

The rate law indicates that both NO_2 and F_2 are involved in the rate-determining step, but stoichiometry shows that any reaction between NO_2 and F_2 must produce something besides NO_2F. These two facts suggest that the most likely mechanism for the reaction is

$$1. \quad NO_2 + F_2 \xrightarrow{k_1} NO_2F + F \qquad \text{(slow)},$$
$$2. \quad F + NO_2 \xrightarrow{k_2} NO_2F \qquad \text{(fast)}.$$

The first bimolecular process is the rate-determining step. Its rate law, and thus that of the overall reaction, is second order. Since the overall reaction proceeds at exactly the rate of reaction 1, k_{exp} must equal k_1.

We have already said that the mechanism of the reaction

$$2Br^- + 2H^+ + H_2O_2 \rightarrow Br_2 + 2H_2O$$

is

$$1. \quad H^+ + Br^- + H_2O_2 \xrightarrow{k_1} HOBr + H_2O,$$
$$2. \quad HOBr + H^+ + Br^- \xrightarrow{k_2} Br_2 + H_2O.$$

How was this conclusion reached? The most important clue to the mechanism of a reaction is the rate law, which in this case is

$$\frac{d[Br_2]}{dt} = k_{exp}[H_2O_2][H^+][Br^-].$$

This tells us that only H_2O_2, H^+, and Br^- are involved in the rate-determining step. To decide what the products of the rate-determining step are, we must use imagination, and be guided by our knowledge of descriptive chemistry and the principles of atom and charge conservation. Stoichiometry shows that H_2O and $HOBr$ are at least possible products of a reaction between H^+, Br^-, and H_2O_2. Furthermore, HOBr is a known chemical species, although it is quite unstable. Consideration of the molecular structure of the reactants shows that HOBr and H_2O can be formed from the reactants without serious departures from normal molecular geometry. That is, we might picture the slow step as

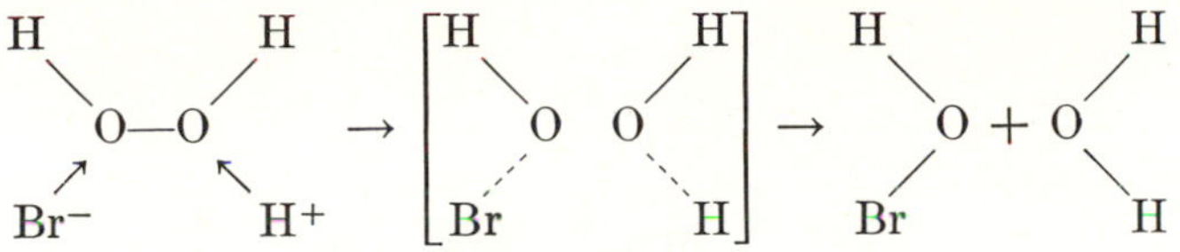

The species in brackets represents the intermediate situation in which the O—O bond is breaking while the Br—O and H—O bonds form. Such unstable structures are called **activated complexes,** and last for only about 10^{-13} sec.

To justify the final step of the mechanism, we must use some supplementary knowledge of descriptive chemistry. It is possible to prepare neutral or slightly alkaline mixtures of HOBr and Br^-, and when these solutions are acidified, bromine is formed very rapidly. This is independent evidence that the reaction

$$HOBr + H^+ + Br^- \rightarrow H_2O + Br_2$$

is very fast. In other words, the second step of our mechanism is consistent with our chemical experience.

The mechanism of a reaction may change if the conditions under which it is run are altered. The reaction between carbon monoxide and nitrogen dioxide,

$$NO_2 + CO \rightarrow CO_2 + NO,$$

follows the rate law

$$\frac{d[CO_2]}{dt} = k[NO_2][CO]$$

at temperatures above approximately 500°K. The reaction mechanism is a single elementary process in which an oxygen atom is transferred:

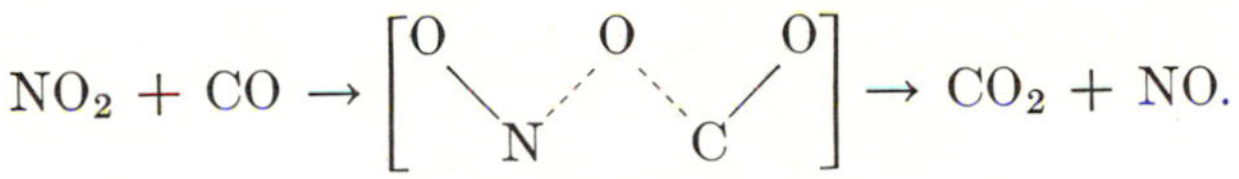

At lower temperatures the rate law changes to

$$\frac{d[CO_2]}{dt} = k'[NO_2]^2,$$

which does not involve the concentration of carbon monoxide at all. The explanation is that the low-temperature mechanism is

$$NO_2 + NO_2 \rightarrow NO_3 + NO \quad \text{(slow)},$$
$$NO_3 + CO \rightarrow NO_2 + CO_2 \quad \text{(fast)}.$$

The first of these reactions is the slower, rate-determining step, and accordingly the rate is independent of the carbon monoxide concentration, as long as some is present. At high temperatures, this pair of reactions is slower than the direct reaction between NO_2 and CO, but the reverse is true at lower temperatures. This is the reason that the mechanism changes; the reaction goes by the fastest path available.

As another example of a complex reaction mechanism, we select the reaction between gaseous hydrogen and bromine:

$$H_2(g) + Br_2(g) \rightarrow 2HBr(g).$$

This reaction is interesting because it follows a three-halves-order rate law:

$$\frac{1}{2} \frac{d[HBr]}{dt} = k[H_2][Br_2]^{1/2},$$

and our analysis so far does not suggest how half-integer rate laws come about. Considerable experimentation has shown the mechanism to be

$$\left.\begin{array}{l} Br_2 + M \xrightarrow{k_1} 2Br + M \\ 2Br + M \xrightarrow{k_{-1}} Br_2 + M \end{array}\right\} \text{ (fast equilibrium),}$$

$$Br + H_2 \xrightarrow{k_2} HBr + H \quad \text{(slow)},$$

$$H + Br_2 \xrightarrow{k_3} HBr + Br \quad \text{(fast)}.$$

The first two elementary processes result in a rapidly established equilibrium between molecular bromine and its atoms. The symbol M stands for any molecule capable of both colliding with Br_2 so as to cause its dissociation and removing the excess energy from a pair of atoms so that they may combine. The third and fourth processes convert hydrogen and bromine to hydrogen bromide, *without a net consumption of bromine atoms*. The rate-determining step is the reaction between a bromine atom and a hydrogen molecule; hence the rate of the reaction is given by

$$\frac{1}{2} \frac{d[HBr]}{dt} = k_2[H_2][Br]. \tag{9.2}$$

To find the rate law expressed in terms of the bromine-molecule concentration, we make use of the equilibrium relation between atomic and molecular bromine:

$$Br_2 = 2Br, \qquad \frac{[Br]^2}{[Br_2]} = K_{eq}, \qquad [Br] = \sqrt{K_{eq}[Br_2]}\,.$$

Substituting this into Eq. (9.2), we find

$$\frac{1}{2} \frac{d[HBr]}{dt} = k_2 K_{eq}^{1/2}[H_2][Br_2]^{1/2},$$

which has the same form as the experimentally determined rate law. Furthermore, we see that the experimental rate constant is actually the product of a rate constant k_2 and the square root of an equilibrium constant, $K_{eq}^{1/2}$. This shows how important the determination of reaction mechanisms is, for in order to understand or interpret the size of an experimentally determined rate constant k, we must know whether it is equal to the rate constant of an elementary process, or whether it is actually some algebraic combination of rate constants and equilibrium constants.

While we have stressed the importance of the rate law in the determination of reaction mechanisms, the rate law alone often does not allow us to make a unique choice when several mechanisms are possible. An outstanding example of this fact is provided by the reaction between nitric oxide and oxygen:

$$2NO + O_2 \rightarrow 2NO_2.$$

The rate law is

$$-\frac{d[O_2]}{dt} = k[NO]^2[O_2].$$

Two possible mechanisms, both consistent with the rate law, are

$$NO + NO \overset{K}{\rightleftharpoons} N_2O_2 \quad \text{(fast, at equilibrium)},$$
$$N_2O_2 + O_2 \overset{k}{\longrightarrow} 2NO_2 \quad \text{(slow)},$$
$$-\frac{d[O_2]}{dt} = k[N_2O_2][O_2] = kK[NO]^2[O_2];$$

and

$$NO + O_2 \overset{K'}{\rightleftharpoons} OONO \quad \text{(fast, at equilibrium)},$$
$$NO + OONO \overset{k'}{\longrightarrow} 2NO_2 \quad \text{(slow)},$$
$$-\frac{d[O_2]}{dt} = k'[OONO][NO] = k'K'[NO]^2[O_2].$$

The difficulty is that the rate law really tells us only the atomic composition of the activated complex, and for both of these mechanisms, it is the same, O_4N_2. A definite choice between the two mechanisms will be possible only when the structure of the intermediate is discovered through use of molecular spectroscopy. In general, not only the rate law, but information from every possible source must be used to select a reaction mechanism.

The Steady-State Approximation

The mechanisms that we have discussed so far have been of two types. The simplest situation is one in which the first step is slow and rate-determining, and is followed by very rapid subsequent reactions. The other situation, also quite simple, occurs when the first step of the mechanism is a rapid equilibrium which produces an intermediate which reacts slowly in the rate-determining step. We must anticipate the occurrence of the intermediate case, in which all steps of the mechanism proceed at comparable rates. The exact deduction of the rate law for this situation can be quite complicated. Fortunately, however, the steady-state approximation provides us with a simple means of finding the rate law under most conditions.

Consider the following general mechanism, which applies to many thermal decompositions and isomerizations:

$$A + M \underset{k_{-1}}{\overset{k_1}{\rightleftarrows}} A^* + M,$$

$$A^* \xrightarrow{k_2} B + C.$$

In the first step, the molecule of interest A undergoes a collision with any molecule M, and the result is that A^*, a molecule with a considerable amount of internal energy, is produced. The reverse process in which A^* is deactivated by collision with M can also occur. Finally, if A^* is left alone, it decomposes to products B and C. We know that if the first step is slow and rate-determining, the rate law is

$$\frac{d[B]}{dt} = k_1[A][M]. \tag{9.3}$$

On the other hand, if k_{-1} is large, and thus A^* is in rapid equilibrium with A, we have

$$\frac{[A^*]}{[A]} = \frac{k_1}{k_{-1}},$$

$$\frac{d[B]}{dt} = k_2[A^*] = \frac{k_1 k_2}{k_{-1}}[A]. \tag{9.4}$$

We will use the steady-state approximation to find the general rate law for this system.

The excited molecule A^* begins to be formed when the A-molecules are heated in order to start the reaction. At first the concentration of A^* may increase fairly rapidly, but as its concentration builds up, it starts to be de-activated and to decompose to products. Thus we can anticipate reaching a condition in which the rate at which A^* is created is just balanced by the rate at which it is destroyed. At this point, the concentration of A^* will be finite, and very nearly constant in time. We can find this *steady-state concentration* by writing

$$\text{rate of production of } A^* = \text{rate of destruction of } A^*$$

$$k_1[A][M] = k_{-1}[A^*][M] + k_2[A^*],$$

$$[A^*] = \frac{k_1[A][M]}{k_{-1}[M] + k_2}. \tag{9.5}$$

Now, since the rate of reaction is

$$\frac{d[B]}{dt} = k_2[A^*],$$

we can substitute the steady-state expression for A^* and get

$$\frac{d[B]}{dt} = \frac{k_1 k_2 [A][M]}{k_{-1}[M] + k_2} \tag{9.6}$$

for the general form of the rate law.

We shall now find the conditions under which the general expression, Eq. (9.6), reduces to either of the two simpler rate laws, Eq. (9.3) and Eq. (9.4). Suppose we work at gas pressures low enough so that $k_{-1}[M] \ll k_2$. Physically this means that virtually every A^* formed will proceed to products, and that reaction 1 is the rate-determining step. From Eq. (9.6) we can write

$$\frac{d[B]}{dt} = \frac{k_1 k_2 [A][M]}{k_{-1}[M] + k_2} \cong \frac{k_1 k_2 [A][M]}{k_2} \cong k_1 [A][M] \qquad \text{(if } k_{-1}[M] \ll k_2\text{)},$$

which is the rate law we expect if step 1 is rate-determining. The opposite situation is $k_{-1}[M] \gg k_2$, which we can achieve by making the pressure quite high. Physically this means that few of the A^* decompose to products, and that A^* is essentially in equilibrium with A. In the denominator of Eq. (9.6) we can neglect k_2, and get

$$\frac{d[B]}{dt} = \frac{k_1 k_2 [A][M]}{k_{-1}[M] + k_2} \cong \frac{k_1 k_2 [A][M]}{k_{-1}[M]}$$

$$= \frac{k_1}{k_{-1}} k_2 [A],$$

which is just what we expect if A^* is in near equilibrium with A. Thus the general expression for the rate, Eq. (9.6), includes as special cases the simple situation where either the first or second step of the mechanism is rate-determining.

The steady-state approximation consists, as we have seen, of selecting an intermediate in the reaction mechanism, and calculating its concentration by assuming that it is destroyed as rapidly as it is formed. This procedure cannot be strictly accurate for all times during the reaction, since it implies that the concentration of the intermediate is constant. This is not true at the beginning of the reaction, when the concentration of the intermediate rises from zero toward its steady-state value. Nevertheless, when the concentration of the intermediate is small, the approximation is sufficiently accurate to be of very great use and importance in the analysis of mechanisms.

As a practical example of the application of the steady-state approximation, we can consider the decomposition of gaseous N_2O_5:

$$2N_2O_5 \rightarrow 4NO_2 + O_2,$$

which has the rate law

$$\frac{d[O_2]}{dt} = k[N_2O_5].$$

A great deal of evidence shows that the mechanism is

$$N_2O_5 \underset{k_{-1}}{\overset{k_1}{\rightleftarrows}} NO_2 + NO_3,$$

$$NO_3 + NO_2 \overset{k_2}{\longrightarrow} NO + NO_2 + O_2,$$

$$NO_3 + NO \overset{k_3}{\longrightarrow} 2NO_2.$$

The molecule NO_3 is an intermediate whose concentration is small and can be calculated by the steady-state approximation. The same can be said for NO. In the latter case we have

rate of production of NO = rate of destruction of NO,

$$k_2[NO_2][NO_3] = k_3[NO][NO_3],$$

$$[NO] = (k_2/k_3)[NO_2]. \tag{9.7}$$

For NO_3 we proceed as follows:

rate of production of NO_3 = rate of destruction of NO_3,

$$k_1[N_2O_5] = (k_{-1}[NO_2] + k_2[NO_2] + k_3[NO])[NO_3].$$

If we substitute Eq. (9.7) for [NO] and solve for $[NO_3]$ we get

$$[NO_3] = \frac{k_1[N_2O_5]}{k_{-1}[NO_2] + 2k_2[NO_2]}$$

for the steady-state concentration of NO_3. Now, the rate of production of oxygen, which is also the rate of reaction, is

$$\frac{d[O_2]}{dt} = k_2[NO_2][NO_3].$$

With the steady-state approximation for $[NO_3]$ this becomes

$$\frac{d[O_2]}{dt} = \frac{k_1 k_2[N_2O_5]}{k_{-1} + 2k_2},$$

which is of the same form as the experimental rate law. Note that the relation between the experimental rate constant k and the rate constants for the individual steps is

$$k = \frac{k_1 k_2}{k_{-1} + 2k_2}.$$

Question. The steady-state approximation is applied to a reaction intermediate, but never to a reactant or product. Why?

Chain Reactions

We have encountered reaction mechanisms which involved intermediates that were created in one step, and consumed in another step to give the reaction products. However, there are a very large number of so-called *chain reactions* of the type exemplified by

$$H_2(g) + Cl_2(g) \rightarrow 2HCl(g)$$

which has the mechanism

$$Cl_2 + \text{light} \longrightarrow 2Cl$$
$$Cl + H_2 \xrightarrow{k_2} HCl + H,$$
$$H + Cl_2 \xrightarrow{k_3} HCl + Cl,$$
$$2Cl + M \xrightarrow{k_4} Cl_2 + M.$$

We see that in step 1, the reactive intermediate Cl is produced, and this in turn reacts by step 2 to produce a product HCl molecule. However, the hydrogen atom also produced by step 2 can react with Cl_2 to give another HCl molecule and a chlorine atom. Thus the net result of steps 2 and 3 is formation of two molecules of HCl, *without* consumption of the intermediate chlorine atom. This mechanism is analogous to that encountered for the H_2-Br_2 reaction. The possibility that steps 2 and 3 can be repeated indefinitely is responsible for the name "chain reaction." Step 1, which first produces the chain carrier Cl, is called an *initiation reaction*, while step 4 is a *chain termination reaction*. Steps 2 and 3 are said to be *chain propagation reactions*.

Chain reactions occur in flames, explosions, and atmospheric and life processes, and are important in the production of synthetic polymers. Examples of the latter are found in the polymerization of ethylene (CH_2CH_2) and vinyl chloride (CH_2CHCl) to make polyethylene and polyvinyl chloride, respectively. The direct reaction

$$2n(CH_2CHCl) \rightarrow (-CH_2CHCH_2CH-)_n$$
$$| |$$
$$Cl Cl$$

is very slow. If, however, a small amount of a substance which will produce atoms or free radicals is present, the polymerization proceeds via a chain reaction. Benzoyl peroxide is a convenient initiator, for it decomposes to benzoyl radicals ($C_6H_5CO_2 \cdot$) and benzyl radicals ($C_6H_5 \cdot$):

$$C_6H_5\overset{\overset{\displaystyle O}{\|}}{C}\!-\!O\!-\!O\!-\!\overset{\overset{\displaystyle O}{\|}}{C}\!-\!C_6H_5 \rightarrow C_6H_5\overset{\overset{\displaystyle O}{\|}}{C}\!-\!O \cdot + C_6H_5 \cdot + CO_2.$$

These radicals ($R \cdot$) then react with vinyl chloride, and a radical chain is propagated:

$$R \cdot + \underset{\underset{\displaystyle Cl}{|}}{CH_2\!\!=\!\!CH} \rightarrow \underset{\underset{\displaystyle Cl}{|}}{RCH_2\!-\!CH} \cdot,$$

$$\underset{\underset{\displaystyle Cl}{|}}{RCH_2\!-\!CH} \cdot + \underset{\underset{\displaystyle Cl}{|}}{CH_2\!\!=\!\!CH} \rightarrow \underset{\underset{\displaystyle Cl}{|}\,\,\underset{\displaystyle Cl}{|}}{RCH_2CHCH_2CH} \cdot,$$

$$R(CH_2CHCl)_n \cdot + CH_2CHCl \rightarrow R(CH_2CHCl)_{n+1} \cdot.$$

The chain termination step occurs when two such polymer radicals combine with each other.

The number of intermediates or chain carriers in a reaction mixture is usually determined by the relative rates of the initiation and termination steps. A good example of this is the H_2-Br_2 reaction, which we discussed earlier. The relative rates of dissociation of bromine and recombination of bromine atoms determine the bromine atom concentration

$$\left.\begin{aligned} Br_2 + M \xrightarrow{k_1} 2Br + M\\ M + 2Br \xrightarrow{k_{-1}} Br_2 + M \end{aligned}\right\} \text{ fast equilibrium,}$$

$$[Br] = (k_1/k_{-1})^{1/2}[Br_2]^{1/2},$$

and the chain is propagated by

$$Br + H_2 \rightarrow HBr + H,$$
$$H + Br_2 \rightarrow HBr + Br,$$

which produce as many atoms as they consume. There are, however, chain reactions in which some steps produce more radicals or chain carriers than they consume. Such steps are called *chain-branching reactions*, examples of which occur in the hydrogen-oxygen reaction:

$$\begin{aligned} O_2 + M &\rightarrow 2O &&\text{initiation}\\ O + H_2 &\rightarrow OH + H &&\text{branching}\\ H + O_2 &\rightarrow OH + O &&\text{branching}\\ OH + H_2 &\rightarrow H_2O + H &&\text{propagation}\\ \left.\begin{aligned} H + O_2 + M &\rightarrow HO_2 + M\\ 2HO_2 &\rightarrow H_2O_2 + O_2 \end{aligned}\right\} &&\text{termination} \end{aligned}$$

If the rate of the branching reaction is greater than the rate of the termination reactions, the concentration of chain carriers grows steadily, and hence the rate of propagation increases essentially without limit. This nearly unbounded reaction rate is what produces an explosion. The slow, controlled reaction can be maintained if the termination step is made fast enough so that branching reactions cannot produce ever-increasing numbers of chain carriers, or if the rate at which reactants are fed into the reaction mixture is limited.

To summarize our discussion of concentration effects and reaction mechanisms, we note that the investigation of a reaction rate involves the following steps:

1. The rate law is determined by studying the effect of concentration of reactants on the rate of reaction.

2. The rate law, together with imagination, general chemical experience, and the principles of stoichiometry and molecular structure, is used to deduce a mechanism for the reaction.

3. The mechanism is used to show that the measured rate constant is either the rate constant for one of the elementary processes, or is an algebraic combination of elementary rate constants and equilibrium constants.

4. The temperature dependence of the rate constant is determined. This information permits us to interpret the magnitude of the rate constant in terms of the nature of the reacting molecules.

The last step has not yet been discussed. We will turn to it after a brief analysis of the relation between reaction rates and chemical equilibria.

9.3 REACTION RATES AND EQUILIBRIA

We noted in Chapter 5 that in a state of chemical equilibrium, the rates of the forward reaction and its reverse are exactly equal. This principle allows us to establish a relation between equilibrium constants and rate constants. Let us first consider the reaction

$$CO + NO_2 \xrightarrow{k_1} CO_2 + NO$$

and its reverse

$$NO + CO_2 \xrightarrow{k_{-1}} NO_2 + CO.$$

Both of these reactions are elementary processes and, at temperatures above $500°K$, are the only reactions responsible for the interconversion of CO_2, NO, NO_2, and CO. When a mixture of these molecules reaches chemical equilibrium, the rates of the two reactions must be equal. At equilibrium, then,

$$k_1[CO]_e[NO_2]_e = k_{-1}[CO_2]_e[NO]_e,$$

where the subscript e indicates that the concentrations are those found at chemical equilibrium. We can rearrange this expression to

$$\frac{k_1}{k_{-1}} = \frac{[CO_2]_e[NO]_e}{[CO]_e[NO_2]_e}.$$

The quotient of the equilibrium concentrations is equal to the equilibrium constant, so this equation shows us that

$$\frac{k_1}{k_{-1}} = K_{eq}.$$

This is the general relation which connects the equilibrium constant and the rate constants for the forward and reverse of any *elementary process*.

It is not difficult to extend this argument to reactions which proceed by a multistep mechanism. Our example now is

$$NO_2 + F_2 \xrightarrow{k_1} NO_2F + F,$$
$$F + NO_2 \xrightarrow{k_2} NO_2F.$$

The condition for equilibrium in such a system is that *each elementary process and its reverse* proceed at the same rate. The reverse reactions of our mechanism are

$$F + NO_2F \xrightarrow{k_{-1}} NO_2 + F_2,$$
$$NO_2F \xrightarrow{k_{-2}} NO_2 + F.$$

The equilibrium condition requires that

$$k_1[NO_2]_e[F_2]_e = k_{-1}[NO_2F]_e[F]_e,$$
$$k_2[NO_2]_e[F]_e = k_{-2}[NO_2F]_e.$$

Now we combine these two expressions in a way which will eliminate the concentration of fluorine atoms. Multiplication of the left- and right-hand sides respectively gives

$$k_1k_2[NO_2]_e^2[F_2]_e[F]_e = k_{-1}k_{-2}[NO_2F]_e^2[F]_e.$$

Cancellation of the fluorine atom concentration and rearrangement leave us with

$$\frac{k_1k_2}{k_{-1}k_{-2}} = \frac{[NO_2F]_e^2}{[NO_2]_e^2[F_2]_e}.$$

Once again, the concentration quotient is equal to the equilibrium constant, so we conclude that

$$\frac{k_1k_2}{k_{-1}k_{-2}} = K_{eq}.$$

The principle that at equilibrium, each elementary process and its reverse proceed at the same rate establishes the connection between the equilibrium constant of a reaction and the rate constants of its elementary processes. This idea that at equilibrium, each elementary process is exactly balanced by its reverse reaction is called the **principle of detailed balancing,** or sometimes the **principle of microscopic reversibility.**

9.4 COLLISION THEORY OF GASEOUS REACTIONS

Now that the connection between reaction mechanism, rate constants, and equilibrium constants has been found, we can begin a theoretical analysis of the factors which determine the magnitude of the specific rate constant. We will treat gaseous bimolecular reactions, for which the theory is best established.

It is a fundamental idea of the collision theory of reactions that a *minimum* condition for two molecules A and B to react is that their centers of mass must come to within a certain critical distance of each other. We will call this distance ρ. Its exact value depends on the nature of the molecules which react, but we would expect that, in general, ρ is not much larger than the length of a chemical bond, that is, about 2 or 3 A. In Section 2.6 we found the expression for the number of collisions experienced by one molecule in one second:

$$\text{collisions/molecule-sec} = \pi \rho^2 \bar{c} n,$$

where n is the number of molecules per cubic centimeter, and $\bar{c}$ is the average relative speed of the molecules in centimeters per second. We can adapt this expression to calculate the total rate at which A-molecules collide with B-molecules. If the concentration of B in molecules/cm^3 is n_B, then the number of collisions that *one* A-molecule makes per second with B-molecules is $\pi \rho^2 \bar{c} n_B$. If the concentration of A-molecules is n_A, then the total number of A-B collisions/sec-cm^3 is

$$\text{collisions/}cm^3\text{-sec} = \pi \rho^2 \bar{c} n_A n_B. \tag{9.8}$$

This is the total collision rate, and if molecules reacted upon every collision, it would be equal to the chemical reaction rate. Although we have expressed this rate as a number of collisions/cm^3-sec, we see that the combined units on the right-hand side of Eq. (9.8) are concentration/sec:

$$\pi \rho^2 \bar{c} n_A n_B \left(\frac{cm^2}{\text{molecule}} \right) \left(\frac{cm}{\text{sec}} \right) \left(\frac{\text{molecules}}{cm^3} \right)^2 = \pi \rho^2 \bar{c} n_A n_B \frac{\text{molecules}}{cm^3\text{-sec}}.$$

Strictly, the units are molecules that react (collide) per second per cubic centimeter.

Because Eq. (9.8) apparently represents a maximum possible reaction rate, it is interesting to evaluate it for a typical situation. If we take both gases to be at 1-atm pressure and 0°C, then $n_A = n_B = 2.8 \times 10^{19}$ molecules/cm^3.

Common values of ρ and $\bar{c}$ are 3×10^{-8} cm and 5×10^4 cm/sec, respectively. Therefore

$$\text{collision rate} = (3.14)(3 \times 10^{-8})^2(5 \times 10^4)(2.8 \times 10^{19})^2$$
$$= 1.1 \times 10^{29} \text{ molecules/cm}^3\text{-sec}$$
$$= 1.8 \times 10^8 \text{ moles/liter-sec.}$$

The total collision rate corresponds to an enormous reaction rate. If the two reacting gases, each at 1 atm, could be mixed, they would be almost entirely consumed by reaction in 10^{-9} sec. A few reactions actually are nearly this fast; among them are

$$N + NO \rightarrow N_2 + O,$$
$$O + NO_2 \rightarrow NO + O_2.$$

On the other hand, there are many common reactions used in laboratory and industrial preparative chemistry whose rates are 10^{-2} or 10^{-3} moles/liter-sec, or 10^{-10} to 10^{-11} times as fast as the total collision rate. There are other reactions that proceed even more slowly. Therefore, there must be criteria for reaction, other than those considered so far, that are responsible for this enormous variation in reaction rate.

The clue that led to the discovery of the most important criterion for chemical reaction is the fact that reaction rates are in general very sensitive to temperature. Although the rates of *some* reactions are virtually independent of temperature, a temperature change of 10° increases the rate of most bimolecular reactions by factors which commonly lie between 1.5 and 5. The total collision-rate formula, Eq. (9.8), does not account for this behavior, for it suggests that the only way in which temperature affects reaction rate is through the mean speed $\bar{c}$, which is proportional to $T^{1/2}$. Thus

$$\text{total collision rate} \propto \bar{c} \propto T^{1/2}.$$

If T is initially 300°K, increasing it to 310°K increases the collision rate by a factor of

$$\left(\frac{310}{300}\right)^{1/2} = 1.015.$$

We see that the average molecular speed is rather insensitive to temperature, and its variation cannot account for the temperature dependence of reaction rates.

What feature of a gas is sensitive to temperature? If we consult Fig. 9.4, the answer is clear. In Fig. 9.4, we find the Maxwell-Boltzmann *energy* distribution function plotted for two temperatures. The area under either one of these curves corresponding to energies equal to or greater than the value E_a is equal to the fraction of molecules that collide with kinetic energy of relative motion equal to or greater than E_a. As the temperature changes, the area under the distribution curve for energies greater than E_a changes. If E_a lies in the tail of the distribution curve, the area may change by a large factor as temperature changes. Thus, if we suppose that the only molecules which react are those that collide with energy greater than a certain minimum, we can explain why chemical reaction rates are generally smaller and more temperature sensitive than total collision rates. The minimum energy of relative motion necessary for reaction is called the **activation energy,** and is a factor of foremost importance in determining the magnitude of the reaction rate.

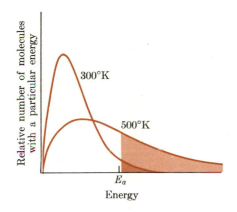

The distribution of molecular kinetic energies at two temperatures. The number of molecules with energy E_a or greater is proportional to the shaded area for each temperature.

FIG. 9.4

In any collision that leads to reaction, some chemical bonds are broken and some new bonds may be formed. During any individual reactive collision, the *total energy* of the colliding particles *remains constant*, but this total energy can be interconverted between kinetic energy and potential energy of the participating atoms. The origin of the activation-energy requirement is most easily explained if we assume that between the atomic arrangement we call products and that called reactants, there is an atomic arrangement which has a potential energy greater than that of reactants or products. In order to pass from reactants to products, a colliding pair of molecules must possess a total energy at least equal in magnitude to the potential energy of this intermediate atomic configuration.

Let us consider a specific example. The reaction

$$Br + H_2 \rightarrow HBr + H$$

is of the simplest type, since it involves only three atoms. For simplicity, we shall assume that the reactants collide such that the nuclei of the three atoms lie on the same straight line at all times. The potential energy of this linear system of atoms is plotted in Fig. 9.5 as a function of a "reaction coordinate" that represents the progress of the three atoms from the form of reactants to that of products. When Br and H_2 are many angstroms from each other, the reaction coordinate is simply the distance between their centers of mass. In this region, the potential energy is essentially constant. As the bromine atom nears the hydrogen molecule, there may be at first a slight lowering of the potential energy due to van der Waals attractive forces, but as the bromine atom moves still closer, the potential energy of the system rises.

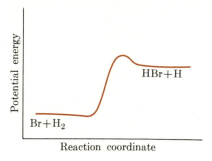

FIG. 9.5 The potential energy as a function of the reaction coordinate for the linear system of atoms in the reaction $Br + H_2 \rightarrow HBr + H$.

In the region where all three atoms are close, the reaction coordinate represents a simultaneous shrinking of the Br—H distance and expansion of the H—H distance. Theoretical calculations suggest that the maximum potential energy is reached when the Br—H and H—H distances are comparable, both approximately 1.5 A. In this situation the central hydrogen atom is partially bonded to both terminal atoms. This configuration of maximum potential energy is the activated complex. The activated complex decays as the external hydrogen atom moves away and the molecule of HBr is formed. To the right of the potential-energy maximum, the reaction coordinate is the distance between the centers of mass of HBr and H, and the potential energy along this coordinate is substantially constant.

We can now ask how the potential-energy profile is changed when the three atoms do not all lie on a straight line. In the particular case we are discussing, the potential energy is unaffected for configurations that are removed from the potential-energy maximum, but the height of the maximum increases as the activated complex becomes bent, as shown in Fig. 9.6. Therefore, while it is possible for collisions in which the three atoms are not collinear to lead to reaction, the energy requirement for such collisions is greater than for those in which the atoms are collinear.

Similar considerations apply to other, more complicated reaction processes. The reaction

$$CO + NO_2 \rightarrow NO + CO_2$$

involves an activated complex for which the geometry of lowest energy is zigzag:

$$OC + O \overset{N}{\diagdown} O \rightarrow [O{-}C \cdots O \overset{N}{\diagdown} O] \rightarrow OCO + NO.$$

If the collision occurs in a way that forces the geometry of the activated complex to depart seriously from this most favorable arrangement, the energy barrier between reactants and products is higher and fewer collisions can meet this requirement. In the extreme case it is difficult to see how a collision with the orientation

$$O{-}C \quad N \overset{\diagup O}{\underset{\diagdown O}{}}$$

can lead to reaction at all, since the atoms that must eventually become bonded are separated from each other.

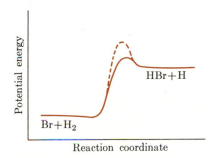

The potential energy as a function of the reaction coordinate for the reaction $Br + H_2 \rightarrow HBr + H$. The solid line represents the linear system of atoms, and the dashed line a nonlinear configuration of atoms.

FIG. 9.6

Our discussion has revealed two related factors that influence reaction rates. Not only must molecules collide, they must collide with a restricted range of relative orientations, and with enough initial kinetic energy of relative motion to be able to pass over the potential-energy barrier to products. The energy and orientation factors are related because it is the orientation of the colliding molecules that determines, to a certain extent, what the energy requirement is. When the energy and orientation criteria are applied, the theoretical expression

Table 9.1 Dependence of exponential factor $e^{-E_a/RT}$ on E_a and T

$T(°K)$ \ E_a	10 kcal	20 kcal	30 kcal
298	4.6×10^{-8}	2.1×10^{-15}	9.9×10^{-23}
400	3.4×10^{-6}	1.2×10^{-11}	4.0×10^{-17}
600	2.3×10^{-4}	5.2×10^{-8}	1.2×10^{-11}
800	1.8×10^{-3}	3.4×10^{-6}	6.4×10^{-9}

for the rate of a bimolecular gas reaction becomes

$$\text{rate} = p \left(\frac{8\pi kT}{\mu} \right)^{1/2} \rho^2 e^{-E_a/RT} n_A n_B, \tag{9.9}$$

where μ is a combination of molecular masses given by

$$\mu = \frac{m_A m_B}{m_A + m_B},$$

k is Boltzmann's constant, and R is the ideal gas constant in units of cal/mole-deg. The quantity p is called the *steric factor* and is related to the orientation requirement. Its value depends on the complexity of the reacting molecules and on how sensitive the height of the potential-energy barrier is to distortions of the activated complex. Methods are available for estimating p from the mechanical and geometric properties of the activated complex. In general, p is approximately 10^{-1} for reactions between atoms and simple molecules, but may be as small as 10^{-5} for a reaction between two complicated molecules.

The factor $e^{-E_a/RT}$ arises from the energy requirement, since E_a is the activation energy, the minimum energy required to form the activated complex from the reactants. Table 9.1 gives the value of this exponential factor for several choices of temperature and activation energy. It is clear from these numbers that the activation-energy requirement can have a profound influence on the reaction rate. Table 9.2 lists a few reactions and their experimentally deter-

Table 9.2 Activation energies of some bimolecular reactions

Reaction	E_a(kcal/mole)
$NO + O_3 \rightarrow NO_2 + O_2$	2.5
$H + D_2 \rightarrow HD + D$	8
$Br + H_2 \rightarrow HBr + H$	17.6
$Cl + H_2 \rightarrow HCl + H$	5.5
$H + Br_2 \rightarrow HBr + Br$	1.2

mined activation energies. Differences in the activation energies are largely responsible for the large range of the magnitudes of chemical reaction rates.

Figure 9.7 shows how the activation energy for a reaction and its reverse are related to the overall energy change for the reaction. Any reaction and its reverse have the same activated complex, so if we call E_{af} the activation energy for the forward reaction, and E_{ar} the activation energy for the reverse process, we get

$$\Delta E = E_{af} - E_{ar}$$

for the overall energy change for the forward reaction.

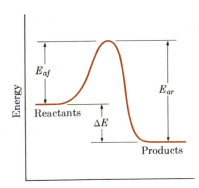

The relation between the activation energy of a forward reaction E_{af}, that of the reverse reaction E_{ar}, and ΔE, the net internal energy change.

FIG. 9.7

Let us return to Eq. (9.9) and set the rate of the bimolecular reaction equal to $k'n_A n_B$, where k' is the bimolecular rate constant. This gives us

$$k'n_A n_B = p\left(\frac{8\pi kT}{\mu}\right)^{1/2}\rho^2 e^{-E_a/RT}n_A n_B.$$

Canceling the concentrations leads to

$$k' = p\left(\frac{8\pi kT}{\mu}\right)^{1/2}\rho^2 e^{-E_a/RT} \tag{9.10}$$

as the expression for the bimolecular rate constant. Equation (9.10) shows that the rate constant is determined by the temperature and by the nature of the reactants through the factors p, ρ, μ, and E_a. Now that we have identified the principal factors that determine a bimolecular rate constant, we might ask whether it is possible to predict reaction rates theoretically. In principle this is possible; in practice it is not often done satisfactorily. To predict a reaction rate, one must know all the mechanical properties of the activated complex.

There are some methods that can be used to estimate geometric properties of activated complexes, but it is virtually impossible to predict activation energies accurately. Since reaction rates are so sensitive to the value of the activation energy, it is not possible at the present time to make very useful predictions of the magnitudes of many reaction rates.

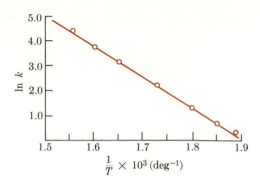

FIG. 9.8   The natural logarithm of the rate constant as a function of reciprocal temperature for the dimerization of butadiene. The slope of the line equals $-E_a/R$.

9.5 TEMPERATURE EFFECTS

Inasmuch as it is usually not possible to predict the activation energies of chemical reactions, these quantities must be obtained experimentally. To see how this is done, let us first note that in Eq. (9.10), the factors other than $e^{-E_a/RT}$ are rather insensitive to temperature. In particular, we demonstrated in Section 9.4 that $T^{1/2}$ changes by only a very small factor when T is changed by 10 K°. On the other hand, reference to Table 9.1 shows how rapidly the factor $e^{-E_a/RT}$ varies with temperature. With these ideas in mind, we can rewrite Eq. (9.10) in the form

$$k = Ae^{-E_a/RT} \tag{9.11}$$

and regard the pre-exponential factor A as virtually independent of temperature. Taking the natural logarithm of Eq. (9.11) gives

$$\ln k = \ln A - \frac{E_a}{RT} \tag{9.12}$$

$$\ln k \cong -\frac{E_a}{RT} + \text{constant}. \tag{9.13}$$

Equation (9.13) suggests that a plot of $\ln k$ as a function of $1/T$ should be a straight line, and as Fig. 9.8 demonstrates, this is found experimentally. Our analysis shows that the slope of this plot is $-E_a/R$, and experimental activation energies are obtained by measuring k at several temperatures, plotting the data as in Fig. 9.8, and calculating E_a from the slope.

If the activation energy and the value of the rate constant at one temperature are known, it is possible to calculate the rate constant for any other temperature. To show this, we need only write Eq. (9.12) for two temperatures:

$$\ln k_2 = \ln A - \frac{E_a}{RT_2},$$

$$\ln k_1 = \ln A - \frac{E_a}{RT_1}.$$

Subtracting the second of these from the first gives

$$\ln \frac{k_2}{k_1} = -\frac{E_a}{R}\left(\frac{1}{T_2} - \frac{1}{T_1}\right). \tag{9.14}$$

Thus if T_1, k_1, and E_a are known, k_2 can be calculated for any choice of T_2.

Example 9.1 For the reaction,

$$C_2H_5I + OH^- \rightarrow C_2H_5OH + I^-,$$

$k = 5.03 \times 10^{-2}\ M^{-1} \cdot \sec^{-1}$ at 289°K   and $k = 6.71\ M^{-1} \cdot \sec^{-1}$ at 333°K.

What is the activation energy of the reaction? What is its rate constant at 305°K?
By Eq. (9.14),

$$E_a = R\left(\frac{T_1 T_2}{T_2 - T_1}\right)\ln \frac{k_2}{k_1}$$

$$= 1.99 \frac{(289)(333)}{44} 2.3 \log \frac{6.71}{5.03 \times 10^{-2}}$$

$$= 21,200\ \text{cal} = 21.2\ \text{kcal}.$$

To find the rate constant at 305°K, we use E_a and the rate constant at 333°K to get

$$2.3 \log k_2 = 2.3 \log k_1 - \frac{E_a}{R}\left(\frac{T_1 - T_2}{T_1 T_2}\right)$$

$$= 2.3 \log 6.71 - \frac{21,300}{1.99}\left[\frac{333 - 305}{(333)(305)}\right],$$

$$\log k_2 = 0.826 - 1.28 = -0.45,$$

$$k_2 = 0.35\ M^{-1} \cdot \sec^{-1}.$$

The experimentally measured value is 0.37 $M^{-1} \cdot \sec^{-1}$.

Our discussion of the effect of temperature on reaction rate has dealt exclusively with bimolecular elementary reactions. Now let us consider some other cases. A unimolecular rearrangement reaction always involves collisional processes that produce and destroy energized species. For the rearrangement of cyclopropane to propylene, the important steps are

$$
\begin{array}{c}
\text{CH}_2 \\
\diagdown \\
\text{H}_2\text{C} \text{——} \text{CH}_2
\end{array}
+ \text{M}
\;\underset{k_2}{\overset{k_1}{\rightleftarrows}}\;
\left[
\begin{array}{c}
\text{CH}_2 \\
\diagdown \\
\text{H}_2\text{C} \text{——} \text{CH}_2
\end{array}
\right]^*
+ \text{M},
$$

$$
\left[
\begin{array}{c}
\text{CH}_2 \\
\diagdown \\
\text{H}_2\text{C} \text{——} \text{CH}_2
\end{array}
\right]^*
\;\overset{k_3}{\longrightarrow}\; \text{CH}_3\text{CH}{=}\text{CH}_2,
$$

where M is any molecule, including cyclopropane itself. Only the cyclopropane molecules with considerable internal energy of vibration can rearrange to propylene. The role of the collisions with the molecules M is to maintain a certain concentration of the cyclopropane molecules in this energized condition, and the ratio of k_1 to k_2 is the equilibrium constant for the formation of energized molecules:

$$
\frac{[\text{cyclopropane}^*]}{[\text{cyclopropane}]} = \frac{k_1}{k_2}.
$$

General thermodynamic considerations suggest that this equilibrium constant should be proportional to $e^{-E_a/RT}$, where E_a is the minimum energy a cyclopropane molecule must have in order to be considered activated; that is, in order to be able to react by step 3. The decomposition of the activated molecules occurs with some rate constant k_3 characteristic of the structure of the molecule, and k_3 is not strongly temperature dependent. Thus for the temperature dependence of the overall first-order reaction we get

$$
k = \frac{k_1}{k_2} k_3 = \text{constant} \times e^{-E_a/RT},
$$

where k is the experimental rate constant defined by

$$
-\frac{d[\text{cyclopropane}]}{dt} = k[\text{cyclopropane}].
$$

Thus for reactions that involve a unimolecular decomposition step we again obtain an exponential relation between the rate constant and temperature.

It is not always true that the rate of a reaction increases as temperature increases. For example, the recombination of iodine atoms,

$$
\text{I} + \text{I} + \text{Ar} \rightarrow \text{I}_2 + \text{Ar},
$$

is faster at lower temperatures than at higher temperatures. The same is true for other atomic recombinations. The explanation of this behavior lies in the mechanism of the reaction:

$$I + Ar + Ar \underset{k_{-1}}{\overset{k_1}{\rightleftharpoons}} IAr + Ar, \qquad \text{fast equilibrium,}$$

$$IAr + I \xrightarrow{k_2} I_2 + Ar, \qquad \text{slow.}$$

The weakly bound complex IAr is in equilibrium with iodine and argon atoms. Thus

$$[IAr] = (k_1/k_{-1})[Ar][I] \qquad \text{and} \qquad \frac{d[I_2]}{dt} = k_2[I][IAr]$$

give us

$$\frac{d[I_2]}{dt} = \frac{k_1}{k_{-1}} k_2[I]^2[Ar].$$

Since k_1/k_{-1} is the equilibrium constant for what must be an exothermic association reaction, it must decrease as temperature increases. Now k_2 is the rate constant for a bimolecular process, and if there were any activation energy required for reaction 2, k_2 would increase with temperature. However, for virtually all atom recombinations there is no activation energy for step 2, and the overall rate of recombination follows the behavior of k_1/k_{-1}, and decreases as temperature increases.

To summarize our discussion of the effect of temperature on reaction rates, we can remark that the rate constants of bimolecular elementary reactions generally increase with increasing temperature at a rate determined by the activation energy of the reaction. However, the behavior of the overall rate constant of a reaction that follows a complex mechanism is difficult to predict. Depending on the nature of the mechanism, the overall rate constant may increase, decrease, or stay virtually constant as the temperature is changed.

9.6 RATES OF REACTIONS IN SOLUTION

Reactions in the gas phase occur by processes in which no more than three molecules collide at the same time. The situation is quite different for reactions which take place in solution. Any reactant molecules brought together in solution not only collide with each other, but are constantly subjected to forces due to their several neighboring solvent molecules. It would appear, then, that any reaction in solution is a complicated event in which the behavior of not only the reactants but 10 or 20 surrounding solvent molecules must be considered. Despite this apparent complexity, we can achieve a good understanding of solution reactions by analyzing three general factors which affect the reaction rate. They are:

1. The rate at which initially separated reactant molecules come together and become neighbors. This is called the rate of encounters.

2. The time that two reactants spend as neighbors before moving away from each other. This is called the duration of an encounter. During this time the two reactants may collide or vibrate against each other hundreds of times.

3. The requirements of energy and orientation which two neighboring reactant molecules must satisfy in order to react.

The first and third factors are similar to the concepts which entered our discussion of gas-phase reactions, while the second involves a phenomenon which occurs only in dense phases. Any one of these three factors may determine the rate at which a reaction occurs, and we shall now consider each of them in more detail.

There are several reactions in which there are no activation-energy or orientation requirements; the molecules react as soon as they become neighbors. Consequently the reaction rate is limited only by our first factor—the rate at which encounters occur. Two examples of such reactions are

$$I + I \rightarrow I_2 \qquad \text{(in CCl}_4 \text{ solution),}$$
$$H_3O^+ + HSO_4^- \rightarrow H_2SO_4 + H_2O \quad \text{(in H}_2O \text{ solution).}$$

Since the reactants are of very simple structure, and since the products are energetically much more stable than the reactants, it is not surprising that there are no orientation or energy restrictions for reaction. The reactant molecules move together through the liquid by diffusion. Since the rate of the reaction is determined only by the rate at which reactant molecules can diffuse together, these reactions are said to be *diffusion controlled*. The rate at which diffusion occurs depends on the nature of the solvent. If the solvent molecules are large and exert strong forces on one another, they will impede the motion of reactant molecules and diminish the rates of diffusion-controlled reactions. Therefore, diffusion-controlled reactions are fastest in solvents of low viscosity, where the solvent molecules are easily pushed aside by the diffusion reactants.

Diffusion-controlled reactions are very rapid, and consequently it is difficult to measure their rate constants. However, considerable effort and ingenuity have been applied to this problem, and Table 9.3 contains only a few of the measured rate constants for diffusion-controlled reactions. The neutralization reaction between H_3O^+ and OH^- is the fastest of all reactions that take place in aqueous solution, because H_3O^+ and OH^- diffuse through water faster than any other ions. The rate constant for the corresponding reaction between H_3O^+ and F^- is smaller, since F^- diffuses through water more slowly than does OH^-. Comparison of the rate constants of the last three reactions is interesting because it shows how the charge on the reacting species affects the reaction rate. We would expect the rate at which two ions of opposite charge diffuse together to be greater than the rate of encounters of an ion and a neutral species, or of two ions of the same charge. The data substantiate this expecta-

Table 9.3 Rate constants for diffusion controlled reactions

Reaction	k(liters/mole-sec)
$H_3O^+ + OH^- \rightarrow 2H_2O$	1.4×10^{11}
$H_3O^+ + F^- \rightarrow HF + H_2O$	1×10^{11}
$H_3O^+ + HS^- \rightarrow H_2S + H_2O$	7.5×10^{10}
$H_3O^+ + N(CH_3)_3 \rightarrow H_2O + HN(CH_3)_3^+$	2.6×10^{10}
$H_3O^+ + CuOH^+ \rightarrow Cu^{++}(aq) + H_2O$	1×10^{10}
$H_3O^+ + (NH_3)_5CoOH^{+2} \rightarrow (NH_3)_5CoH_2O^{+3} + H_2O$	4.8×10^9

tion, since the rate of these neutralizations decreases as the like charge on the reactants increases.

The second of the rate-controlling factors also involves the effect of the solvent on the motion of reactant molecules. Consider, for example, an iodine molecule which has somehow received enough energy to dissociate into atoms. If this molecule is in the gas phase, this dissociation occurs immediately, and within 10^{-11} sec the two atoms may be separated by several hundred angstroms. However, if the dissociating molecule is in a solution, it is surrounded by solvent molecules which impede the separation of the atoms, and which thereby may prevent the dissociation. In essence, the iodine molecule is held in a "cage" of solvent molecules, and the action of the solvent which prevents dissociation is known as the "cage effect." In our example, the cage effect diminishes the net rate of dissociation of iodine molecules. However, there are situations in which the cage effect can work to increase the reaction rate. Suppose two molecules diffuse together but find upon their first collision that they do not satisfy the energy or orientation requirements for reaction. Then, instead of immediately separating, they are held as neighbors by the solvent cage. The duration of such an encounter may be 10^{-10} sec, and during this time the molecules may acquire the energy or orientation required to react. This argument shows one essential difference between reactions in the gas and liquid phases. In the gas phase, molecules *collide* and, if no reaction occurs, separate immediately. In solution, molecules make an *encounter* during which they may collide several hundred times. If no reaction occurs, they eventually separate.

The third of the rate-controlling factors, the energy-orientation requirement, is primarily determined by the nature of the reacting species. Even here, however, the solvent may play a role. A good example is the reaction

$$(CH_3)_3CCl + OH^- \rightarrow (CH_3)_3COH + Cl^-,$$

which proceeds by the mechanism

$$(CH_3)_3CCl \rightarrow (CH_3)_3C^+ + Cl^- \quad \text{(slow)},$$
$$(CH_3)_3C^+ + OH^- \rightarrow (CH_3)_3COH \quad \text{(fast)}.$$

Of the reactants, only $(CH_3)_3CCl$ is involved in the rate-determining step, and consequently the reaction is first order with respect to $(CH_3)_3CCl$, and does not depend on the concentration of OH^-. This reaction can be studied in aqueous solution and in solvents which are mixtures of water and organic liquids. The result is that the rate of the reaction depends profoundly on the nature of the solvent and increases with increasing water concentration. The explanation of this behavior is that the rate-determining step involves the production of a pair of ions, and this step would be expected to proceed more easily or more frequently in a solvent in which ions are more stable. In other words, a polar solvent to which ions are strongly attracted should increase the rate of ionization. This is precisely what is found: the reaction is 10^4 times faster in a 90% water–10% acetone mixture than in 10% water–90% acetone. In general we can expect a reaction to be fastest in a solvent in which the activated complex is most stable. If, as in the above example, the activated complex is highly polar, the reaction should be fastest in a solvent of polar molecules. On the other hand, a reaction in which polar or ionic molecules form a nonpolar or neutral activated complex will be fastest in a nonpolar solvent.

9.7 CATALYSIS

We observed earlier that there are many reactions which, while having large equilibrium constants, proceed at extremely small rates. In order to take advantage of these reactions, particularly for industrial processes, it is important to find ways to increase their rates. This is the general problem of catalysis. According to the usual formal definition, a catalyst is a substance which increases the speed of a chemical reaction without itself undergoing change. In practice this definition proves too restrictive. There are many instances in which a substance not required in the overall stoichiometry will increase the reaction rate and be changed at the same time. A simple example occurs in the hydrolysis of an ester; the overall equation is

$$CH_3COOC_2H_5 + H_2O \rightarrow C_2H_5OH + CH_3COOH.$$

Hydroxide ion is not required by the stoichiometry of this reaction, but its addition does increase the reaction rate. However, one of the reaction products is an acid, and any added hydroxide ion is consumed as the reaction proceeds. Even so, we call the hydroxide ion a catalyst, for according to common usage, a catalyst is any reagent which can increase the rate of reaction while not being actually *required* for the stoichiometry.

How do catalysts increase reaction rates? The general answer is that they provide new faster paths by which a reaction can proceed. This can be done in a variety of ways. Consider, for instance, the reaction between ceric and thallous ions:

$$2Ce^{+4} + Tl^+ \rightarrow 2Ce^{+3} + Tl^{+3}.$$

This reaction is slow because in order to remove two electrons from Tl^+ simultaneously, a three-body collision between ions of the same charge is required. However, the reaction is catalyzed by Mn^{++}, which acts in the following way:

$$Ce^{+4} + Mn^{++} \rightarrow Ce^{+3} + Mn^{+3},$$
$$Ce^{+4} + Mn^{+3} \rightarrow Ce^{+3} + Mn^{+4},$$
$$Mn^{+4} + Tl^+ \rightarrow Tl^{+3} + Mn^{++}.$$

That is, the presence of Mn^{++} permits a new mechanism or reaction path, whereby the single slow termolecular process is replaced by three faster bimolecular reactions.

Catalysts can also act by modifying the electronic structure of the reactants. For example, the conversion of an alcohol to an organic halide is catalyzed by hydrogen ion. The reaction is

$$Br^- + C_2H_5OH \rightarrow C_2H_5Br + OH^-.$$

The role of the hydrogen ion seems to be to facilitate the ejection of the OH group in the following way:

$$H^+ + C_2H_5OH = C_2H_5\overset{\displaystyle H\,+}{\overset{\displaystyle |}{O}}-H \quad \text{(rapid equilibrium)},$$

$$Br^- + \overset{\displaystyle CH_3}{\underset{\displaystyle H}{\diagdown}}\overset{\displaystyle H\,+}{\underset{\displaystyle |}{C}}-O-H \rightarrow Br \cdots \overset{\displaystyle CH_3}{\underset{\displaystyle H}{C}} \cdots OH_2 \rightarrow CH_3CH_2Br + H_2O \quad \text{(slow).}$$

The presence of the proton on the hydroxyl group of the alcohol apparently lowers the activation energy of the second slow step.

The reactions we have discussed are examples of *homogeneous catalysis*, since the catalytic processes occur in one phase. Phase boundaries or surfaces also can increase reaction rates; this is called *heterogeneous catalysis*. One of the most outstanding examples of heterogeneous catalysis is the hydrogenation of unsaturated organic compounds. The reaction

$$H_2 + C_2H_4 \rightarrow C_2H_6$$

is immeasurably slow at moderate temperatures in the gas phase, but occurs readily at the surface of metals such as nickel, platinum, and palladium. Separate experiments show that these metals can "dissolve" or absorb large quantities of hydrogen, apparently by incorporating it in the metallic lattice as hydrogen atoms. We might represent this process by

$$\tfrac{1}{2}H_2(g) + M = M \cdot H$$

where M represents the metal, and $M \cdot H$ stands for the "pool" of atomic hydrogen in the metallic lattice. Hydrogen atoms are known to be more reactive than hydrogen molecules, and thus the hydrogenation of C_2H_4 may take place by

$$2M \cdot H + C_2H_4 \xrightarrow{\text{at surface}} C_2H_6.$$

In effect, the metal provides a new reaction path of low activation energy by dissociating the molecular hydrogen to atoms.

Metal oxides are often effective catalysts for oxidation reactions. Two examples are

$$CO + \tfrac{1}{2}O_2 \xrightarrow{Cu_2O} CO_2, \qquad SO_2 + \tfrac{1}{2}O_2 \xrightarrow{V_2O_5} SO_3.$$

An outstanding feature of these metal oxide catalysts is their specificity—an oxide particularly effective as a catalyst for the oxidation of carbon monoxide may have no effect on the rate of sulfur dioxide oxidation. Presumably this is due to differences in the interaction or bonding which can occur between the surface and the various substances to be oxidized. However, since the nature of surfaces and their interactions with gaseous molecules is one of the most complicated problems in chemistry, the detailed explanation of much of surface catalysis has not yet been found.

Enzyme Catalysis

In living systems, a great many very complicated molecular transformations are catalysed by large protein molecules called enzymes. These catalysts can be quite specific: for example, the enzyme urease catalyzes the hydrolysis of urea, $(NH_2)_2 CO$, by the reaction

$$(NH_2)_2 CO + 2H_2O \xrightarrow{\text{urease}} 2NH_4^+ + CO_3^=,$$

but has no effect on the hydrolysis rate of any other molecule, even those whose molecular structure is very similar to that of urea. This specificity is a very important feature of enzyme action. It provides a mechanism for allowing the highly selected reactions which are necessary for the function of living cells to occur rapidly at moderate temperatures. The specificity of enzymes varies, however. The enzyme α-chymotrypsin, which is secreted by the human pancreas, can catalyze the hydrolysis of esters, amides, and polypetides (see Chapters 17 and 18), and is used by the body to speed the digestion of small protein molecules.

Enzymes function by forming an association or complex with the molecule (called the substrate) whose transformation they catalyze. This enzyme-substrate complex may dissociate back to reactant substrate and free enzyme, or to free enzyme and product molecules. Thus we have the general mechanism

$$E + S \underset{k_{-1}}{\overset{k_1}{\rightleftharpoons}} ES,$$

$$ES \overset{k_2}{\longrightarrow} E + P,$$

where E stands for the enzyme molecule, S for the substrate, ES for the enzyme-substrate complex, and P for the product molecules. This simple three-step mechanism does not deal with the detailed questions of how the enzyme and substrate are bound, and what atomic motions lead to product formation. Such questions are currently of very great interest to biochemists, and in some cases, a very detailed picture of enzyme action can be built up. Nevertheless, the general behavior of most enzyme-catalyzed reactions can be understood in terms of the simple three-step mechanism.

Let us find the rate law given by this mechanism. To conform to common usage in biochemistry, we write V for the rate at which products appear. Then

$$V = k_2(ES), \tag{9.15}$$

and to proceed we must find an expression for the concentration of enzyme-substrate complexes. In the steady-state, when the rates of formation and destruction of ES just balance, we have

$$k_1[E][S] = (k_{-1} + k_2)[ES]. \tag{9.16}$$

This equation might be solved for the concentration of the enzyme-substrate complex [ES], but it contains the concentration of free enzyme [E], which is unknown. There is a way around this difficulty, however. We can write the material balance equation for the enzyme as

$$[E_0] = [E] + [ES],$$

where $[E_0]$ is the total concentration of enzyme material. Solving this equation for [E], and substituting the result in Eq. (9.16), we get

$$k_1[S]([E_0] - [ES]) = (k_{-1} + k_2)[ES],$$

$$[ES] = \frac{k_1[E_0][S]}{k_{-1} + k_2 + k_1[S]}$$

for the concentration of the enzyme-substrate complex. Substituting this in Eq. (9.15) we find that the rate of reaction is given by

$$V = \frac{k_1 k_2[E_0][S]}{k_{-1} + k_2 + k_1[S]}.$$

Now we divide numerator and denominator by k_1, and make the definition

$$K_m \equiv \frac{k_{-1} + k_2}{k_1}.$$

The result is

$$V = \frac{k_2[E_0][S]}{K_m + [S]}. \tag{9.17}$$

This expression is known as the Michaelis-Menten equation, after the first two scientists who used it to represent the kinetics of enzyme-catalyzed reactions. By taking the reciprocal of both sides, we obtain another common form of the Michaelis-Menten rate law:

$$\frac{1}{V} = \frac{1}{k_2[E_0]} + \frac{K_m}{k_2[E_0][S]}. \tag{9.18}$$

Thus the reciprocal of the reaction rate is a linear function of the reciprocal of the substrate concentration when the total amount of enzyme is held constant.

Let us analyze the dependence of the reaction velocity V on the concentration of the substrate molecule S given by Eq. (9.17). At sufficiently low concentration of substrate, we can satisfy the inequality

$$[S] \ll K_m,$$

and therefore the concentration of substrate can be neglected compared to K_m in the denominator of Eq. (9.17). Thus we get

$$V = \frac{k_2}{K_m} [E_0][S], \qquad [S] \ll K_m,$$

and we see that at low substrate concentrations, the rate is first order with respect to both substrate and enzyme.

Now consider the contrasting situation in which the substrate concentration is so high that

$$[S] \gg K_m.$$

In this case, we can neglect K_m compared to $[S]$ in the denominator of Eq. (9.17). The result is that the concentration of the substrate cancels from the rate law expression, and we get

$$V = \frac{k_2}{K_m} [E_0], \qquad [S] \gg K_m.$$

Thus the rate law is now first order with respect to the enzyme, but zero order with respect to substrate.

What is the physical reason why the reaction rate changes from first to zero with respect to substrate as the substrate concentration is increased? Each enzyme molecule has one or more "active sites" at which the substrate must be bound in order for the catalyzed transformation to products to occur. When the concentration of substrate is low, most of these active sites are unoccupied at any time. Increasing the substrate concentration increases the number of these active sites which are occupied and the rate of reaction therefore increases. When very high substrate concentrations are reached, virtually all the active sites of the enzyme are occupied at any time, and further increases of the substrate concentration cannot increase the number of enzyme-substrate complexes. The rate is therefore unaffected by changes in the substrate concentration. Examination of the expression for the enzyme-substrate complex concentration,

$$[ES] = \frac{[E_0][S]}{K_m + [S]},$$

confirms this picture. At low substrate concentrations we have

$$[ES] \cong \frac{[E_0][S]}{K_m}, \qquad [S] \ll K_m,$$

so that [ES] increases linearly with [S]. At high substrate concentrations, we find

$$[ES] \cong [E_0], \qquad [S] \gg K_m,$$

which means that all active sites on the enzyme are occupied by substrate molecules.

One way of affecting the pattern of cell functions is to inhibit the action of specific enzymes with chemical agents. This is in fact the basis for the action of several chemotherapeutic agents. Some molecules which act as inhibitors of an enzyme are chemically and structurally very similar to the normal substrate molecule and can occupy the active sites on the enzyme, but do not react to give the normal products. Such agents are known as competitive inhibitors, since they compete with the substrate for the active sites on the enzyme.

Let us determine the form of the rate law for an enzyme-catalyzed reaction subject to competitive inhibition. The basic rate law is still

$$V = k_2[ES],$$

but now we must find how the inhibitor affects the concentration of the enzyme-substrate complex. Let the inhibitor molecule be designated by I. For the formation and destruction of an enzyme-inhibitor complex EI by the reactions

$$E + I = EI$$

there is an equilibrium constant defined by

$$K = \frac{[EI]}{[E][I]} .$$

The material balance equation for the enzyme is

$$[E_0] = [E] + [EI] + [ES]$$
$$= [E](1 + K[I]) + [ES].$$

Solving this for the enzyme concentration [E], and substituting the result into the steady state expression for [ES],

$$k_1[E][S] = (k_{-1} + k_2)[ES],$$

we can find the concentration of the enzyme-substrate complex,

$$[ES] = \frac{[E_0][S]}{K_m(1 + K[I]) + [S]} .$$

The rate of reaction is therefore

$$V = \frac{k_2[E_0][S]}{K_m(1 + K[I]) + [S]} . \tag{9.19}$$

We can see from this that if $K[I]$ is large compared with unity, the rate of reaction is diminished. Since

$$K[I] = \frac{[EI]}{[E]} ,$$

this condition occurs when an appreciable fraction of the active sites on the enzyme are occupied by inhibitor molecules.

9.8 CONCLUSION

The chemical reactions we encounter range from the almost instantaneous to the imperceptibly slow. It is possible, nevertheless, to understand the existence of this great variety of rates in terms of reaction mechanisms and the nature of elementary processes. In analyzing reaction rates, we have found that very many chemical reactions proceed by a series of steps. For any chemical reaction, there may be more than one possible sequence of steps leading from reactants to products, and consequently, the reaction mechanism must be determined experimentally by studying the concentration dependence of the reaction rate. A marked change in the temperature, introduction of a catalyst, or some other change in the conditions of the reaction may cause the mechanism of a reaction to change. In some circumstances, reactions proceed by two or more mechanisms simultaneously. In any case, a reaction proceeds by the fastest mechanism available to it.

The elementary processes of a mechanism are events which involve from one to three molecules. In an elementary process, energy may be transferred from one molecule to another, bonds may be broken or formed, or electrons may be transferred. In order for a bimolecular reaction, in which some bonds are broken and others are made, to occur, the molecules must collide with a certain orientation and with a certain minimum energy. Because of the orientation and energy requirements, the reaction rate is generally less than the total collision rate. While it is possible to predict the influence of the orientation requirement from the structures of the reacting molecules, it is considerably more difficult to predict what the activation energy of a reaction will be. As the techniques of quantum-mechanical calculation improve, it will prove possible to calculate activation energies, and then we will be able to predict the rates of many elementary processes. In the meantime, however, analysis of experimental rate studies provides us with a description of what molecules do to each other when they react.

SUGGESTIONS FOR FURTHER READING

Daniels, F., and R. A. Alberty, *Physical Chemistry*, 3rd ed. New York: Wiley, 1966.

Frost, A. A., and R. G. Pearson, *Kinetics and Mechanism*, 2nd ed. New York: Wiley, 1961.

Harris, G. M., *Chemical Kinetics*. Boston: D. C. Heath and Co., 1966.

Harvey, K. B., and G. B. Porter, *Introduction to Physical Inorganic Chemistry*. Reading, Mass.: Addison-Wesley, 1963. Chapter 10.

King, E. L., *How Chemical Reactions Occur*. New York: Benjamin, 1963.

Sykes, A. G., *Kinetics of Inorganic Reactions*. Oxford: Pergamon Press, 1966.

PROBLEMS

9.1 If concentrations are measured in moles per liter, and time in seconds, what are the units of the rate constant for (a) a first-order reaction (b) a second-order reaction and (c) a third-order reaction?

9.2 For the reaction between gaseous chlorine and nitric oxide,

$$2NO + Cl_2 \rightarrow 2NOCl,$$

it is found that doubling the concentration of both reactants increases the rate by a factor of eight, but doubling the chlorine concentration alone only doubles the rate. What is the order of the reaction with respect to nitric oxide and chlorine?

9.3 The following terms are grouped in pairs which are sometimes confused with each other. Give a brief definition and explanation of each term that distinguishes one member of the pair from the other.

reaction rate	reaction rate constant
order	molecularity
activation energy	activated complex

9.4 The reaction $I^- + OCl^- \rightarrow Cl^- + OI^-$ follows the rate law $d[OI^-]/dt = k'[I^-][OCl^-]$, but k' proves to be a function of the hydroxide-ion concentration. For hydroxide concentrations of $1.00\ M$, $0.50\ M$, and $0.25\ M$, k' is equal to 61, 120, and 230 liters/mole-sec, respectively, at 25°C. What is the order of this reaction with respect to hydroxide ion?

The mechanism of this reaction is

$$OCl^- + H_2O\ =\ HOCl + OH^- \qquad \text{(fast equilibrium)},$$
$$HOCl + I^- \rightarrow HOI + Cl^- \qquad \text{(slow)},$$
$$HOI + OH^- \rightarrow H_2O + OI^- \qquad \text{(fast)}.$$

Show that this is consistent with the rate law for the reaction.

9.5 The reaction between carbon monoxide and chlorine to form phosgene (Cl_2CO),

$$Cl_2 + CO\ \rightarrow\ Cl_2CO.$$

has the rate law

$$\frac{d[Cl_2CO]}{dt}\ =\ k[Cl_2]^{3/2}[CO].$$

Show that the following mechanism is consistent with this rate law:

$$Cl_2 + M\ \overset{K}{=}\ 2Cl + M, \qquad \text{(fast equilibrium)},$$
$$Cl + CO + M\ =\ ClCO + M, \qquad \text{(fast equilibrium)},$$
$$ClCO + Cl_2 \rightarrow Cl_2CO + Cl, \qquad \text{(slow)}.$$

9.6 In acid solution the rate of the reaction

$$NH_4^+ + HNO_2 \rightarrow N_2 + 2H_2O + H^+$$

is consistent with the mechanism

$$HNO_2 + H^+\ =\ H_2O + NO^+ \quad \text{(rapid equilibrium)},$$
$$NH_4^+\ =\ NH_3 + H^+ \quad \text{(rapid equilibrium)},$$
$$NO^+ + NH_3 \rightarrow NH_3NO^+ \qquad \text{(slow)},$$
$$NH_3NO^+ \rightarrow H_2O + H^+ + N_2 \quad \text{(fast)}.$$

Write the rate law which is consistent with this mechanism by expressing $d[NH_4^+]/dt$ as a function of $[NH_4^+]$, $[HNO_2]$, and $[H^+]$.

9.7 Consider the set of reactions

$$A + B \underset{k_{-1}}{\overset{k_1}{\rightleftarrows}} C + D,$$
$$C + E \xrightarrow{k_2} F.$$

What relationships between the magnitudes of k_1, k_{-1}, and k_2 and reagent concentrations will lead to the following rate laws?

a) $\dfrac{d[F]}{dt} = k\,\dfrac{[A][B][E]}{[D]}$

b) $\dfrac{d[F]}{dt} = k'[A][B]$

9.8 The following data give the concentration of gaseous butadiene as a function of time at 500°K. Plot them as $\ln c$ vs. t and as $1/c$ vs. t. Determine the order of the reaction, and calculate the rate constant.

t (sec)	C (moles/liter)	t (sec)	C (moles/liter)
195	1.62×10^{-2}	4140	0.89×10^{-2}
604	1.47×10^{-2}	4655	0.80×10^{-2}
1246	1.29×10^{-2}	6210	0.68×10^{-2}
2180	1.10×10^{-2}	8135	0.57×10^{-2}

9.9 The following data give the pressure of gaseous N_2O_5 as a function of time at 45°C. Plot them first as $1/P$ vs. t, and then as $\ln P$ vs. t. Ascertain the order of the reaction, and calculate the rate constant.

t (sec)	P (mm)	t (sec)	P (mm)
0	348	3600	58
600	247	4800	33
1200	185	6000	18
2400	105	7200	10

9.10 The following data give the temperature dependence of the rate constant for the reaction $N_2O_5 \rightarrow 2NO_2 + \frac{1}{2}O_2$. Plot them and calculate the activation energy of the reaction.

T (°K)	k (sec^{-1})	T (°K)	k (sec^{-1})
338	4.87×10^{-3}	308	1.35×10^{-4}
328	1.50×10^{-3}	298	3.46×10^{-5}
318	4.98×10^{-4}	273	7.87×10^{-7}

9.11 It is often stated that near room temperature, a reaction rate doubles if the temperature increases by 10°. Calculate the activation energy of a reaction that obeys this rule exactly. Would you expect to find this rule violated frequently?

9.12 An endothermic reaction has a positive internal energy change ΔE. In such a case, what is the minimum value that the activation energy can have? (Refer to Fig. 9.7).

9.13 An electronically excited atom can either fluoresce or lose its energy by collision with some other molecule. For example,

$$Hg^* \xrightarrow{k_1} Hg + h\nu,$$

$$Hg^* + Ar \xrightarrow{k_2} Hg + Ar.$$

These reactions are elementary processes. What is the rate law of each? What is the expression for the fraction of atoms lost by fluorescence at a given pressure of Ar?

9.14 Nitramide, O_2NNH_2, decomposes slowly in aqueous solution according to the reaction

$$O_2NNH_2 \rightarrow N_2O + H_2O.$$

The experimental rate law is

$$\frac{d[N_2O]}{dt} = k \frac{[O_2NNH_2]}{[H^+]}.$$

a) Which of the following mechanisms seems most appropriate?

1. $\qquad O_2NNH_2 \xrightarrow{k_1} N_2O + H_2O \qquad$ (slow)

2. $O_2NNH_2 + H^+ \underset{k_{-2}}{\overset{k_2}{\rightleftharpoons}} O_2NNH_3^+ \qquad$ (fast equilibrium)

$\qquad O_2NNH_3^+ \xrightarrow{k_3} N_2O + H_3O^+ \qquad$ (slow)

3. $\qquad O_2NNH_2 \underset{k_{-4}}{\overset{k_4}{\rightleftharpoons}} O_2NNH^- + H^+ \quad$ (fast equilibrium)

$\qquad O_2NNH^- \xrightarrow{k_5} N_2O + OH^- \qquad$ (slow)

$\qquad H^+ + OH^- \xrightarrow{k_6} H_2O \qquad\qquad$ (fast)

b) What is the algebraic relation between the k in the experimental rate law and the rate constants in the mechanism you chose? (c) What is the algebraic relation between the equilibrium constant for the overall reaction and the rate constants for the elementary process and their reverse reactions?

CHAPTER 10

THE ELECTRONIC STRUCTURE OF ATOMS

The strength of a science is that its conclusions are derived by logical arguments from facts that are the results of controlled experiments. Science has produced a picture of the microscopic structure of the atom, but it is a picture so detailed and so subtle of something which is so removed from our immediate experience that it is difficult to see how many of its features were constructed. This is so because many experiments have contributed to our ideas about the atom; even now as more experiments are done, the picture is being refined and revised. Yet among all the experiments used to form the theory of atomic structure, there stand a few which have been most influential in shaping its major features. In this chapter we shall examine these experiments and see how they contributed to the development of the atomic theory. Then, equipped with this logical background, we shall discuss the detailed features of the atomic theory itself. We shall examine the development of the theory of atomic structure in an historical context, for the order in which the most significant experiments were done is, surprisingly, the order in which the logic of the theory of atomic structure is most clear. There were essentially three great steps: the discovery of the electrical nature of matter and the nature of electricity itself (1900), the discovery that the atom consists of a nucleus surrounded by electrons (1911), and the discovery of the mechanical laws which govern the behavior of electrons in atoms (1925).

10.1 ELECTRICAL NATURE OF MATTER

The first important clues to the nature of electricity and the electrical structure of atoms came in 1833 as a result of Faraday's investigations of electrolysis. His findings can be summarized by two statements:

1. The weight of a given material deposited at an electrode by a given amount of electricity is always the same.

2. The weights of various materials deposited, evolved, or dissolved at an electrode by a fixed amount of electricity are proportional to the equivalent weights of these substances.

The second of these laws is particularly revealing, if we remember that the equivalent weight of any substance contains the same number of molecules, or an *integral multiple* thereof. Then we see that the laws of electrolysis are analogous to the laws of chemical combination which originally suggested the existence of atoms. If a fixed number of atoms reacts only with a certain fixed amount of electricity, it seems reasonable to suppose that electricity itself is composed of particles. Accordingly, an elementary electrode process must involve one molecule combining with or losing a small integral number of these electrical particles. Although Faraday did not realize this implication of his work, he did sense the relation between electricity and chemical bonding, for in his writings we find: "I have such conviction that the power which governs electrodecomposition and ordinary chemical attractions is the same."

The implications of Faraday's experiments were recognized in 1874 by G. J. Stoney, who first suggested the name *electron* for the fundamental electrical particle. However, it was not until 1897 that any firm experimental evidence for the existence and properties of the electron was found. The source of the decisive information was the investigation of the electrical conductivity of gases at low pressure. Gases are ordinarily electrical insulators, but when subjected to high voltages at pressures below 0.01 atm they "break down," and electrical conduction accompanied by light emission ensues. When the gas pressure is lowered to 10^{-4} atm, the electrical conduction persists, the luminosity of the gas decreases, and if the voltages involved are high enough (5000 to 10,000 volts), the glass container begins to glow or fluoresce faintly. By 1890 various experimenters had shown that this fluorescence is the result of bombardment of glass by "rays" which, originating at the cathode or negative electrode, travel in straight lines until they strike either the positive electrode or the walls of the tube. Other experiments showed that these "cathode rays" could be deflected by a magnetic field, just as a wire is which carries an electric current.

Experiments of J. J. Thomson

In 1897 J. J. Thomson demonstrated that when the cathode rays were deflected onto an electrode of an electrometer, the instrument acquired a negative charge. Furthermore, he was the first to show that the rays could be deflected by the application of an electric field that caused them to move away from the negative electrode. All of these results were found, irrespective of the gas present or the materials used to construct the discharge tube. Thomson provides us with a succinct summary and assessment of the results:

"As cathode rays carry a charge of negative electricity, are deflected by an electrostatic force as if they were negatively electrified, and are acted on by a magnetic force in just the way in which this force would act on a negatively electrified body moving along the path of the rays, I can see no escape from the conclusion that they are charges of negative electricity carried by particles of matter."

What was the nature of these particles? The fact that they were found regardless of what gas was used in the discharge tube *suggested* that they were not one particular type of electrified atom, but rather a universal fragment found in all atoms. The ratio of charge to mass of various ions had been obtained from electrolysis experiments, and Thomson recognized that a determination of the charge-to-mass ratio for the cathode-ray particle would help identify it as either an ion or some other charged fragment. Accordingly, he determined the charge-to-mass ratio (e/m) by two different methods.

In his first determination Thomson bombarded an electrode with cathode rays and measured both the current delivered to the electrode and the temperature rise produced by the bombardment. From the temperature rise and the heat capacity of the electrode he calculated the energy, W, which the cathode-ray particles delivered; this he took as equal to the kinetic energy of the particles:

$$W = \frac{N \cdot mv^2}{2}.$$

Here N is the number of particles of mass m and velocity v which arrived at the electrode during the experiment. Since $mv^2/2$ is the kinetic energy of one particle, $Nmv^2/2$ is the total kinetic energy of the particles which struck the electrode. The total charge, Q, collected at the electrode during the experiment is related to N and e, the charge on each particle, by

$$Q = Ne.$$

Combining these two equations gives

$$\frac{Q}{W} = \frac{2}{v^2}\left(\frac{e}{m}\right). \tag{10.1}$$

As we have said, Thomson could measure Q and W; to calculate e/m he needed only to measure the velocity of the particles. He accomplished this by measuring their deflection by a magnetic field of known strength, H. In a magnetic field, particles of charge e and mass m moving with velocity v travel in a circular path of radius r; the relation between these quantities is

$$v = \frac{erH}{m}.$$

Combining this with Eq. (10.1) gives

$$\frac{e}{m} = \frac{2W}{r^2 H^2 Q}.$$

All the quantities on the right-hand side of this equation can be measured, since the radius of curvature r produced by the known magnetic field H can be determined from the fluorescence produced by the particle beam. The value of e/m so obtained by Thomson is in acceptable agreement with the best modern determinations of the charge-to-mass ratio of the electron.

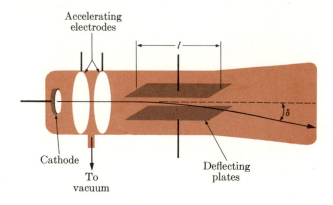

FIG. 10.1 Schematic representation of Thomson's apparatus for measuring e/m. Coils for producing a magnetic field perpendicular to the page are not shown.

When a quantity such as e/m is determined for the first time, it is quite proper, indeed imperative, to ask whether the experiment performed really measures the desired quantity, and not some other unsuspected experimental artifact. One way to answer this question is to repeat the determination by a second experimental method which is as different from the first as possible. Agreement between the two methods suggests, but does not prove, the validity of the result. Thomson's second procedure for determining e/m involved the apparatus shown in Fig. 10.1. A beam of cathode-ray particles passed through a region in which they could be subjected to electric and magnetic fields. Either field applied alone could deflect the beam from its horizontal trajectory, but the direction of the magnetic deflection was opposite to that produced by the electric field. Thus, if the electric field was applied and held constant, the magnitude of the magnetic field could be adjusted so as to return the beam to its original horizontal trajectory. In this condition, the force on the particles due to the magnetic field, *Hev*, was equal to the force due to the electric field,

eE. Thus

$$Hev = eE$$

and so

$$v = \frac{E}{H}.$$ (10.2)

The velocity of the particles could be calculated from measurements of E and H.

The second step of the experiment was to remove the magnetic field and measure the deflection of the beam produced by the electric field alone. As the particles pass between the plates, the electric force eE produces a deflection δ which, as Fig. 10.1 shows, can be calculated by the method of similar triangles from the displacement of the spot observed at the end of the tube. The electric force eE is related, by Newton's second law, to an acceleration, a:

$$eE = \text{force} = \text{mass} \times \text{acceleration} = ma,$$

$$a = \frac{eE}{m}.$$ (10.3)

The deflection δ can be related to a and t, the time the particles spend between the plates, by

$$\delta = \tfrac{1}{2}at^2,$$ (10.4)

which is a well-known result of elementary mechanics. Finally, t can be expressed in terms of the length of the plates l and the velocity of the particles v:

$$t = \frac{l}{v}.$$ (10.5)

If we now combine Eqs. (10.3), (10.4), and (10.5), we obtain

$$\delta = \frac{1}{2} \frac{eE}{m} \left(\frac{l}{v}\right)^2.$$

But v is given by Eq. (10.2); introducing this relation and rearranging the expression gives us

$$\frac{e}{m} = \frac{2\delta}{l^2} \frac{E}{H^2}.$$

Everything on the right-hand side of this equation can be measured experimentally, and thus e/m can be found. The currently accepted value for e/m is 1.76×10^8 coul/gm, or 5.27×10^{17} electrostatic units per gram (esu/gm).

The significance of e/m for cathode rays became apparent when its value was compared with the charge-to-mass ratios of ions, which had been obtained from electrolysis experiments. The charge-to-mass ratio of the cathode ray was over 1000 times larger than that of any ion. Furthermore, while the charge-to-mass ratios of various ions were different, e/m for cathode rays was a constant independent of the gas used in the discharge tube. These facts led Thomson to conclude that cathode rays were not electrified atoms, but corpuscular fragments of atoms; in our modern terminology, electrons.

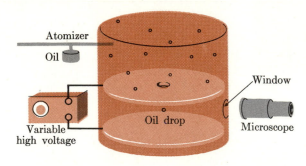

FIG. 10.2   Schematic diagram of Millikan's apparatus for measuring the fundamental unit of charge.

Millikan's Contribution

The ultimate demonstration of the particulate nature of electricity came from the famous oil-drop experiment of R. A. Millikan. Using the apparatus shown in Fig. 10.2, Millikan proved that all electric charges are multiples of one definite elementary unit whose value is 1.6×10^{-19} coul or 4.80×10^{-10} esu. To perform the experiment, spherical oil drops from the atomizer are led into the observation chamber. There they become charged by a collision with gaseous ions produced by the action of radium or x-rays on air. A charged oil drop is recognized by its response to an electric field, and its motion observed through the microscope. When the electric field is zero, the drop is subject only to the force of gravity and falls; because of air resistance the drop does not continually accelerate, but reaches a constant velocity given by

$$v = \frac{mg}{6\pi\eta r} = \frac{\text{gravitation force}}{\text{viscous resistance}},$$

where g is the acceleration of gravity, m and r are the mass and radius of the drop, and η is the viscosity of air. This equation, together with the expression

$$\text{density} = m/\tfrac{4}{3}\pi r^3,$$

which relates the known density of the oil drop to its mass and its radius, allows the calculation of m and r from the measured velocity and density.

If the same drop contains an amount of charge q, and is subjected to a field E, the electrical force causing an upward motion of the drop is qE. Due to the action of gravity, the net force on the drop is $qE - mg$, so its velocity in the upward direction is

$$v' = \frac{qE - mg}{6\pi\eta r}.$$

Since v' and E are measurable, and m, g, η, and r are known, q can be calculated. Millikan found that q was always an integral multiple of 4.8×10^{-10} esu. This result shows that electricity is particulate, and the fundamental unit of charge is 4.8×10^{-10} esu. The assumption that this fundamental unit is equal to the charge on the electron, together with the measured value of e/m, gives 9.1×10^{-28} gm for the mass of the electron.

The experiments of Millikan and Thomson have been discussed in detail, for they show how extremely important fundamental quantities can be determined by the use of fairly simple apparatus and the most elementary laws of physics. They are, without question, two of the greatest experiments in all of physical science.

10.2 THE STRUCTURE OF THE ATOM

While the nature of electricity was being established, scientists began to formulate a detailed picture of the atom. It was not difficult to estimate the atomic size, for the molar volume of a solid expressed in cm^3/mole, divided by Avogadro's number, gives the atomic volume as roughly 10^{-24} cm^3. Taking the cube root of the volume shows that the characteristic size of an atom is approximately 10^{-8} cm. But Thomson's experiments demonstrated that small as an atom was, it contained even smaller particles of negative electricity. Since atoms were ordinarily electrically neutral, it was clear they must also contain positive electricity. Furthermore, since electrons were so light, it seemed proper to associate most of the mass of an atom with its positive electricity. If the positive electricity contained most of the atomic mass, it was reasonable that it should occupy most of the atomic volume. Consequently, Thomson proposed that an atom was a uniform sphere of positive electricity of about 10^{-8}-cm radius, with the electrons embedded in this sphere in a way which would give the most stable electrostatic arrangement. Thomson tried to relate the relative stabilities of various numbers of charges in the atom to the periodic chemical properties of the elements, and even developed a theory of chemical bonding. Appealing as this simple model was, and despite its occasional successes, it had to be abandoned in 1911, when E. R. Rutherford showed that it was completely inconsistent with his observations of the scattering of α-particles by thin metal foils.

The Rutherford Scattering Experiment

The α-particle scattering experiment, perhaps the most influential single experiment used in the development of the theory of atomic structure, is shown schematically in Fig. 10.3. A narrow parallel beam of α-particles impinges on a thin metal foil (10^4 atoms thick), and the angular distribution of the scattered particles is obtained by counting the scintillations, or light flashes, produced on a zinc sulfide screen. The significant qualitative result of the experiment is that while most of the α-particles pass through the foil either undeflected or deflected by only small angles, a few particles are scattered at large angles, up to 180°.

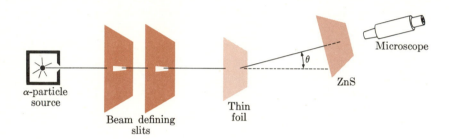

FIG. 10.3   Schematic diagram of Rutherford's α-particle scattering experiment. The region traversed by α-particles is evacuated.

At the time the experiment was first performed, Rutherford knew that α-particles were doubly ionized helium atoms with an atomic mass of 4; moreover, their velocities had been measured by the method of magnetic deflection discussed earlier. Consequently, Rutherford knew that the kinetic energy of the α-particles was very large, and he realized that in order to produce a large deflection of such an energetic particle, the atom must be the seat of an enormous electrical force. It was also clear that this force had to be exerted by a body of considerable mass, for a light body such as the electron would be swept aside by the heavier α-particle. Finally, the fact that only a *few* α-particles received large deflections suggested that the large electrical force was confined to very small regions of space which were missed by most of the α-particles. In other words, instead of being a sphere of uniform mass and charge density as Thomson had proposed, the atom was highly nonuniform. While the electrons might occupy the volume associated with the $\sim 10^{-8}$-cm dimension of the atom, the positive electricity had to be concentrated in a tiny but weighty "nucleus."

By assuming that the force between the nucleus and the α-particle was given by Coulomb's law, Rutherford showed that the trajectory of the α-particle deflected by an atom should be a hyperbola. As shown by Fig. 10.4, the deflection angle θ, which is the external angle between the asymptotes of the hyperbola, depends on the aiming error, or impact parameter b. The mathematical analysis shows that

$$\tan \tfrac{1}{2}\theta = \frac{zZe^2}{mv^2b},$$

where z and Z are the atomic numbers* of the α-particles and the nucleus, e is the magnitude of the electronic charge, and m and v are the mass and velocity of the α-particle. Thus we see that when $b = 0$, $\theta = 180°$, which is just what we expect for a head-on collision. In a given scattering experiment, z, Z, m, and v are constants, and since a relatively wide beam of α-particles is used, all values of b occur, and scattering is seen at all angles.

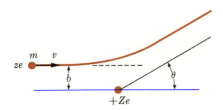

Trajectory of an α-particle passing near a nucleus of charge Ze. The α-particle has veloc-   **FIG. 10.4**
ity v, mass m, charge ze, and impact parameter or aiming error b.

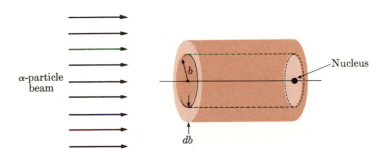

The probability of an α-particle passing between b and $b + db$ is proportional to the area   **FIG. 10.5**
of the ring, $2\pi b\, db$.

As shown in Fig. 10.5, the probability that the impact parameter b will lie in the range b, $b + db$ is proportional to the area of a ring of radius b and width db. This area is equal to $2\pi b\, db$, the product of the circumference of the ring, $2\pi b$, and its width, db. The area increases as b increases, and consequently large

* The atomic number is the number of fundamental units (4.8×10^{-10} esu) of positive charge on the nucleus.

values of b are more probable than small values. Thus most of the scattering will be in the forward direction (b large, θ small), and few particles will be deflected through large angles. This is just a sophisticated way of saying that there are more ways of missing a small target than of hitting it.

By using the probability of the various values of b, Rutherford deduced that the fraction, $f(\theta)$, of the initial α-particles scattered through an angle θ is given by

$$f(\theta) = 2\pi t\rho \left(\frac{zZe^2}{2mv^2}\right)^2 \frac{\sin\theta}{\sin^4(\theta/2)}, \tag{10.6}$$

where t is the thickness of the foil and ρ is its density expressed in atoms/cm^3. This expression for the angular distribution of particles is valid *only if the force between the nucleus and α-particles is given by Coulomb's law*. Geiger and Marsden, working in Rutherford's laboratory, showed that the experimental distribution of scattered particles followed Eq. (10.6) within experimental error. Furthermore, the quantity ($zZe^2/2mv^2$) could be evaluated from the experiments, and since z, e, m, and v were known, the atomic number Z of the scattering nucleus could be evaluated. For the gold nucleus, Rutherford calculated Z to be 100 ± 20; this is a reasonable approximation to 79, which we know is the correct value. Thus the α-particle scattering experiment was one of the first ways of estimating the atomic number of an atom, and as the experiment was refined, it became possible to measure Z exactly.

We have pointed out that the nucleus is small compared with 10^{-8} cm, but just how small is it? The scattering experiment can tell us. When an α-particle is deflected through 180°, it has made a head-on collision with a nucleus. In such a collision the α-particle approaches the nucleus until the Coulomb potential energy of repulsion, zZe^2/r, becomes equal to its initial kinetic energy, $\frac{1}{2}mv^2$. Thus the equation

$$\tfrac{1}{2}mv^2 = \frac{zZe^2}{r_{\min}}$$

allows us to calculate $r_{\min}$, the distance of closest approach, if all other factors are known.

For α-particles obtained from the disintegration of radium,

$$v = 1.6 \times 10^9 \text{ cm/sec,}$$
$$e = 4.8 \times 10^{-10} \text{ esu,}$$
$$m = 6.68 \times 10^{-24} \text{ gm,}$$

and if the scattering nucleus is copper,

$$Z = 29.$$

Thus

$$r_{\min} = \frac{zZe^2}{(1/2)mv^2} = \frac{2 \times 29 \times (4.8 \times 10^{-10})^2}{(1/2)(6.68 \times 10^{-24}) \times (1.6 \times 10^9)^2}$$
$$= 1.6 \times 10^{-12} \text{ cm.}$$

Since particles can come to within nearly 10^{-12} cm of the nucleus and still be scattered according to Coulomb's law, the nucleus itself must be smaller than 10^{-12} cm. Other experiments with faster α-particles and lighter nuclei (Z smaller, r_{min} smaller) show that the Coulomb scattering law is *not* obeyed if the α-particles come closer to the nucleus than about 0.8×10^{-12} cm; this does indeed imply that the positive charge on the nucleus occupies a sphere of approximately 10^{-12} cm radius. Thus the α-particle scattering experiment not only provided a qualitative indication of the existence of the nucleus, but also produced quantitative measurement of the nuclear charge and size.

10.3 ORIGINS OF THE QUANTUM THEORY

There was a serious difficulty with Rutherford's model of the atom: According to all the principles of physics known in 1911, the nuclear atom should have been unstable. If the electrons were stationary, there was nothing to keep them from being drawn into the nucleus; if they were in circular motion, the well-documented laws of electromagnetics predicted that the atom should radiate light until all electronic motion ceased. Only two years after Rutherford's proposal, Niels Bohr attempted to resolve this apparent paradox by analyzing atomic structure in terms of the quantum theory of energy which had been advanced by Max Planck in 1900. Before discussing Bohr's ideas about the behavior of electrons in atoms, let us examine the experiments which led to the development of the principles which Bohr used.

Classical Theory of Radiation

Before 1900, it was generally accepted that light was electromagnetic wave motion. That is, all experiments with light could be understood if it was pictured as oscillating electric and magnetic fields which were propagated through space. In Section 3.3 we discussed the electromagnetic wave theory and one of its most successful applications, the diffraction of x-rays. For our purposes now we need to call attention to only one more feature of classical radiation theory. According to electromagnetic theory, the energy contained in, or carried by, an electromagnetic wave is proportional to the squares of the maximum amplitudes of the electric and magnetic waves:

$$\text{energy} \propto (E_{max}^2 + H_{max}^2) \propto \text{light intensity.}$$

The important feature of this equation is that the energy of a wave depends *only on its amplitude*, and not on its frequency or wavelength.

The electromagnetic wave theory was eminently successful in explaining optical phenomena such as diffraction and scattering, which occur when waves encounter particles whose size is roughly the same as the wavelength. Yet despite many reassuring successes, the classical wave theory of light could not explain the nature of the radiation from a heated solid body. Experiments demonstrated that this radiation was distributed at different frequencies according to the curves shown in Fig. 10.6. As the temperature of the radiating body is increased, the frequency at which most of the light is emitted becomes higher. This corresponds to the body passing through the stages of red, yellow, and white heat as its temperature is raised. The distribution of radiated frequencies predicted by the wave theory is shown by the dashed line in Fig. 10.6, and it obviously disagrees with the experimental findings.

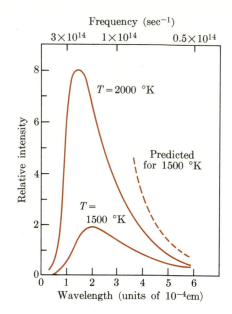

FIG. 10.6   The relative intensity of radiation from a heated solid as a function of frequency or wavelength. The dashed line represents prediction of the classical theory of matter.

In 1900 Planck resolved this discrepancy, but only by making an extreme departure from usual laws of physics. Planck had to assume that a mechanical system *cannot* have any arbitrary energy, but only certain selected energy values. Let us see how Planck applied this assumption. An electromagnetic wave of frequency ν was thought to be radiated from the surface of a solid by a group of atoms oscillating with the same frequency. Planck's assumption was that this group of atoms, the oscillator, could not have an arbitrary energy, but had to have an energy $\epsilon = nh\nu$, where n is a positive integer, ν is the oscillator frequency, and h is a constant to be determined.* This expression is known as

---

* Subsequently, spectroscopic measurements have shown that without question, the energy of a molecular oscillator is quantized. The allowed energy levels are given by $\epsilon = (n + \frac{1}{2})h\nu$, which is very nearly what Planck assumed.

Planck's quantum hypothesis, since it proposes that a system has discrete bits, or quanta, of energy. When such an oscillator radiates it must lose energy; thus, if the oscillator is to radiate, n, the quantum number of the oscillator, must be greater than zero. How does this explain why the high-frequency radiation from a body is so feeble?

Planck assumed that the oscillators were in equilibrium with each other, and consequently that their energies were distributed according to the Boltzmann distribution law. That is, the relative chance of finding an oscillator with energy $nh\nu$ was given by $e^{-nh\nu/kT}$. Now this expression shows that the chance of finding an oscillator of high frequency which has enough energy to radiate $(n > 0)$ is very small, since as ν increases, $e^{-nh\nu/kT}$ decreases. This explains why there is very little radiation at high frequencies: at equilibrium, the high-frequency oscillators rarely possess the minimum energy, $h\nu$, which they need in order to radiate. Hence the assumption that the energy of an oscillator cannot have continuous values leads to excellent agreement between theory and experiment. It should also be noted that the Planck quantum hypothesis was used by Einstein to explain the temperature dependence of the heat capacities of solids, as we discussed in Section 3.5. The success of the Einstein theory also substantiates the idea of quantized energy levels for oscillators.

The existence of separated energy "levels" is a concept that is difficult to accept, for it is contrary to all ordinary experience with macroscopic physical systems. Therefore, it is not surprising that scientists, including Planck, were initially suspicious of the quantum hypothesis. It had been *designed* to explain radiation from heated bodies; it could not be accepted as a general principle until it had been tested by other applications. One consequence of the quantum hypothesis which was tested almost immediately concerned the nature of light. If an oscillator could radiate only by a discrete act in which its energy changed from $nh\nu$ to $(n-1)h\nu$, then was it not reasonable that the light itself was composed of discrete entities of energy $h\nu$? This idea found application and support in Einstein's explanation of the photoelectric effect.

The Photoelectric Effect

By 1902 it was known that light impinging on a clean metallic surface in vacuum caused the surface to emit electrons. The existence of this photoelectric effect was not surprising; it was to be expected from the classical theory of light that the energy of the electromagnetic wave could be used to eject an electron from the metal. However, the wave picture of light was completely incapable of explaining the details of the experiment. In the first place, no electrons were emitted unless the frequency of the light was greater than some critical value ν_0, as shown in Fig. 10.7(a). Second, the electrons emitted had kinetic energies which increased as the frequency of the light increased, as shown in Fig. 10.7(b). Finally, increasing the light intensity did not change the energy of the electrons, but did increase the number emitted per unit time. According to the wave theory the energy of light is independent of its frequency; hence the wave theory could explain neither the frequency dependence of the kinetic energy nor the existence of a photoelectric threshold frequency, ν_0. Furthermore, the wave theory predicted that the energy of the electrons should increase as the light intensity increased, and this was in conflict with the experimental results.

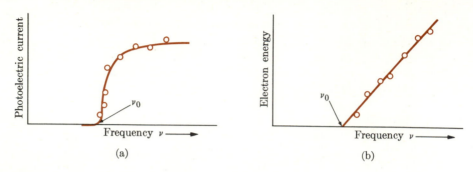

FIG. 10.7 The photoelectric effect: (a) emitted current as a function of frequency; (b) maximum kinetic energy of electrons as a function of frequency.

In 1905, Einstein pointed out that the photoelectric effect could be explained if light consisted of discrete particles or photons of energy $h\nu$. He proposed that a photon of frequency ν and energy $h\nu$ impinges on the metallic surface and gives up its energy to an electron. A certain amount, ϵ, of this energy is used to overcome the attractive forces between the electron and the metal; the rest is available to the ejected electron and appears as kinetic energy, $\frac{1}{2}mv^2$. The law of conservation of energy yields

$$h\nu = \epsilon + \tfrac{1}{2}mv^2.$$

It is clear that ϵ represents a minimum energy that the photon must have to eject the electron. If we express ϵ in terms of a frequency, that is if we write $\epsilon = h\nu_0$, then our equation becomes

$$h\nu = h\nu_0 + \tfrac{1}{2}mv^2,$$
$$\tfrac{1}{2}mv^2 = h\nu - h\nu_0.$$

Thus, if the energy of the ejected electrons is plotted as a function of frequency, there should result a straight line whose slope is equal to Planck's constant h, and whose intercept is $h\nu_0$. We have already seen in Fig. 10.7(b) that this is indeed found. The additional fact that the number of photoelectrons increases with the intensity of light indicates that we should associate light intensity with the number of photons arriving at a point per unit time.

The success of the photon theory was impressive, but by no means did it clarify the nature of radiation. Is light really composed of particles or of waves? There is support for both pictures. We will reserve a discussion of this problem until later; for the moment we need only note that by 1905 the association between energy and *frequency* of radiation was clear, and this, together with Rutherford's picture of the atom, allowed Niels Bohr to propose in 1913 a detailed model of the behavior of electrons in atoms.

The Bohr Atom

The work of Bohr was the first application of the quantum hypothesis to atomic structure which was in any way successful. Bear in mind, however, that Bohr's theory was incorrect; it was abandoned after twelve years in favor of our present quantum theory of atomic structure. Nevertheless, there was enough substance in Bohr's ideas to enable him to explain why only *certain* frequencies of light were radiated by atoms, and in some cases to predict the values of these frequencies. Furthermore, Bohr's proposals greatly helped Moseley to understand his measurements of the frequencies of emitted x-rays and to use them to determine atomic numbers. Thus, although it was eventually abandoned, this early theory was an important step in the understanding of atomic structure.

High voltage

Slit

Photographic
plate

Schematic diagram of the essential elements of a spectrograph and light source.

FIG. 10.8

The first success of the Bohr theory was in explaining the light emission or spectra of atoms. An apparatus for the measurement of atomic spectra is shown in Fig. 10.8. The light source is an electrical discharge through the gas to be investigated. In the case of hydrogen, the bombardment of the hydrogen molecules with electrons results in the production of hydrogen atoms. Some of these atoms acquire excess internal energy which they radiate as visible, ultra-violet, and infrared light. The light from the discharge tube passes through a slit and a prism which disperses the radiation into its various frequencies. These appear as lines (images of the slit) at different positions on the photographic plate. Such devices, called spectrographs, were available after 1859, and by 1885 Balmer recognized that the frequencies emitted by the hydrogen atom could be expressed by the formula

$$\nu = \left(\frac{1}{4} - \frac{1}{n^2}\right) \times 3.29 \times 10^{15} \text{ cycles/sec,}$$

where n is an integer greater than or equal to three. The simplicity of this formula was intriguing, and other empirical relations among the frequencies emitted by other atoms were sought, but none were found that involved integers in such a simple way.

Bohr developed a model of the hydrogen atom which allowed him to explain why the frequencies emitted obeyed such a simple law. His reasoning involved the following postulates:

1. The electron in an atom has only certain definite stationary states of motion allowed to it; each of these stationary states has a definite, fixed energy.

2. When an atom is in one of these states it does not radiate; but when changing from a high-energy state to a state of lower energy the atom emits a quantum of radiation whose energy $h\nu$ is equal to the difference in the energy of the two states.

3. In any of these states the electron moves in a circular orbit about the nucleus.

4. The states of allowed electronic motion are those in which the angular momentum of the electron is an integral multiple of $h/2\pi$.

Of these four postulates, the first two are correct and are retained in the modern quantum theory. The fourth postulate is partially correct; the angular momentum of an electron is fixed, but not in quite the way Bohr proposed. The third postulate is entirely incorrect, and does not appear in modern quantum theory.

The derivation of the expression giving the energies of the allowed states of an atom is very simple. First, mechanical stability of the electron orbit requires that the Coulomb force between the electron and nucleus be balanced by the centrifugal force due to the circular motion:

$$\text{Coulomb force} = \text{centrifugal force}$$

$$\frac{Ze^2}{r^2} = \frac{mv^2}{r}.$$

Here m and v are the mass and velocity of the electron, Z is the number of units of elementary charge e on the atomic nucleus, and r is the electron-nucleus separation. Canceling one power of r gives us

$$\frac{Ze^2}{r} = mv^2. \tag{10.7}$$

Bohr's postulate for the angular momentum, mvr, was

$$mvr = n\frac{h}{2\pi}, \qquad n = 1, 2, 3 \ldots \tag{10.8}$$

where h is Planck's constant, 6.626×10^{-27} erg-sec. That is, the angular momentum had to be an integral multiple of $h/2\pi$. Eliminating v between Eqs. (10.7) and (10.8) gives

$$r = \frac{n^2 h^2}{(2\pi)^2 m Z e^2}, \qquad n = 1, 2, 3 \ldots \qquad (10.9)$$

It would appear from this that only certain orbits whose radii are given by Eq. (10.9) are allowed to the electron.

Now let us consider the total energy E of the electron, which is the sum of the kinetic energy, $mv^2/2$, and potential energy, $-Ze^2/r$:

$$E = \frac{1}{2} mv^2 - \frac{Ze^2}{r}.$$

But by Eq. (10.7) we can write

$$E = \frac{1}{2} \frac{Ze^2}{r} - \frac{Ze^2}{r} = -\frac{1}{2} \frac{Ze^2}{r}.$$

Substitution for r, using Eq. (10.9), gives us

$$E = -\frac{2\pi^2 m Z^2 e^4}{n^2 h^2}, \qquad n = 1, 2, 3 \ldots$$

This expression shows that the consequence of the postulates is that only certain energies are allowed to the atom. Figure 10.9 indicates how these energies depend on n for the simplest case of the hydrogen atom ($Z = 1$).

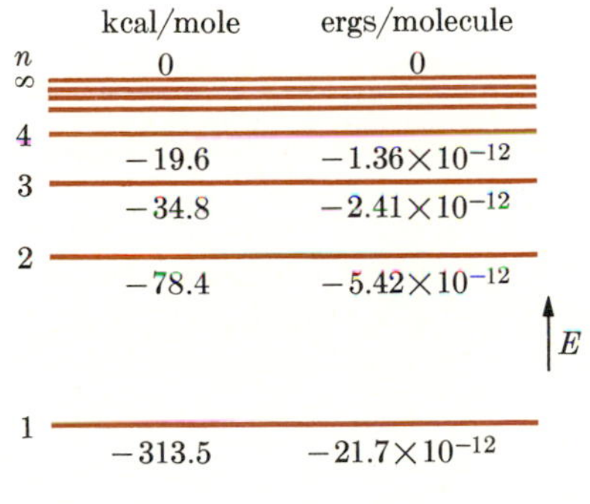

	kcal/mole	ergs/molecule
n	0	0
∞		
4		
3	-19.6	-1.36×10^{-12}
	-34.8	-2.41×10^{-12}
2		
	-78.4	-5.42×10^{-12}
		$\big\downarrow E$
1		
	-313.5	-21.7×10^{-12}

An energy-level diagram for the hydrogen atom. The spacing between energy levels is not drawn to scale.

FIG. 10.9

The energies are negative only because the energy of the electron in the atom is less than the energy of a free electron, which is taken as zero. The lowest energy level of the atom corresponds to $n = 1$, and as the quantum number increases, E becomes less negative. When $n = \infty$, $E = 0$, which corresponds to an ionized atom: the electron and nucleus are infinitely separated, and at rest.

According to Bohr's second postulate, the energy of any photon radiated by the atom should be equal to the difference in the energy of two levels. To ensure that the energy of the photon is positive, we take the absolute value of the energy difference and write

$$h\nu = |E_f - E_i| = \frac{2\pi^2 m Z^2 e^4}{h^2}\left(\frac{1}{n_f^2} - \frac{1}{n_i^2}\right), \qquad n_i > n_f, \qquad (10.10)$$

or

$$\nu = \frac{2\pi^2 m Z^2 e^4}{h^3}\left(\frac{1}{n_f^2} - \frac{1}{n_i^2}\right).$$

If n_f is set equal to 2, and the constant term evaluated, this expression is in numerical agreement with the formula which Balmer had found from the experimental hydrogen-atom spectrum. In other words, the Bohr expression was in agreement with the known experimental spectrum of the hydrogen atom.

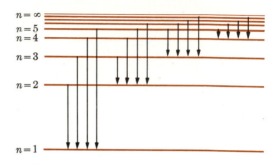

FIG. 10.10   Some of the predicted transitions between energy states of the hydrogen atom.

Furthermore, if n_f, the quantum number of the final state, is set equal to 1, Bohr's formula predicts a set of spectral lines for which $n_i \geq 2$, $n_f = 1$. Similarly, if $n_f = 3$, there should be a series of lines for which $n_i \geq 4$. These predicted transitions between states of different n are shown in Fig. 10.10, and *subsequent* to Bohr's work, all the predicted spectral lines have been found at the expected frequencies. Further application of Bohr's formula was made to other one-electron atoms, such as He^+ and Li^{++}. In each case of this kind, Bohr's prediction of the spectrum was correct.

X-ray Spectra and Atomic Number

The visible spectra of atoms arise from changes in the energy of the most weakly bound atomic electrons. However, under conditions of extreme electrical excitation, atoms can emit x-rays, high-energy radiations which result from energy changes of the electrons closest to the nucleus. Even though Bohr's theory was designed to explain the visible spectra due to the outer, or "valence," electrons, it was of great help in the first interpretations of x-ray spectra.

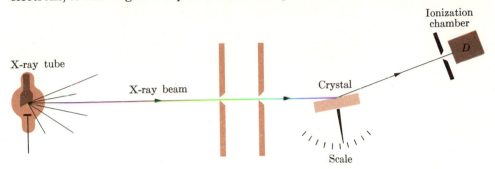

Schematic diagram of a spectrograph for measuring the wavelengths of x-rays emitted from the anode of an x-ray tube.　　**FIG. 10.11**

Figure 10.11 shows a simplified x-ray spectrograph. When a metallic anode is bombarded with electrons of very high velocity, x-rays which are characteristic of the anode material are emitted. As explained in Chapter 3, the wavelength of an x-ray can be found by measuring the angle θ through which the x-ray is diffracted by a crystal of known interatomic spacing d, and then using the Bragg law equation, $n\lambda = 2 \, d \sin \theta$. By employing various elements as his anode, Moseley observed in 1912 that each element emitted a different characteristic x-ray spectrum, and that the frequencies emitted increased regularly as the atomic weight increased. Using the Rutherford-Bohr model of the atom as a guide, Moseley realized that the x-ray frequency should be characteristic of the charge on the nucleus of the emitting atom. Accordingly, he found that he could fit all his observed frequencies by an empirical formula of the form

$$\nu = c(Z - b)^2,$$

where c and b were universal constants, valid for all elements, and Z was an *integer* whose value increased regularly by one unit for successive elements, taken in the order in which they appeared in the periodic table. Moseley then correctly concluded that the Z in his empirical formula was the atomic number, or the charge on the nucleus of the emitter. According to the theory of Bohr, the frequency emitted by a one-electron atom should be proportional to Z^2, the square of the charge on the nucleus, as Eq. 10.10 shows. Moseley suggested that the reason the x-ray frequencies were proportional to $(Z - b)^2$ and not Z^2 was that the electrons in the atom tended to shield one another from the nucleus. Thus, so far as any single electron was concerned, the effective nuclear charge is not Z, but $Z - b$. This interpretation is quite in accord with our present ideas of the origin of x-ray spectra and the behavior of electrons in atoms.

If entered in the periodic table strictly in order of their atomic weights, the elements Ni–Co, Ar–K, and Te–I appear in an order inconsistent with their chemical properties. Moseley found that when the elements are arranged in the order of their atomic numbers, these discrepancies are removed. In short, he showed that nuclear charge, and not nuclear mass, is most fundamental to chemical properties.

10.4 QUANTUM MECHANICS

It may seem surprising that the Bohr theory, initially so successful, had to be abandoned after only twelve years. Yet despite its successes a theory must be refined or rejected if it cannot explain all the relevant experimental facts. Even after the most searching refinements, the Bohr theory could not explain the details of the spectra of atoms with many electrons, nor could it provide a satisfactory picture of chemical bonding. These and other failures made it clear that Bohr's ideas could only be steppingstones or approximations to a universally applicable atomic theory.

There were two particularly objectionable features of theoretical physics in the early 1920's. One was the conflict between the wave and photon models of light. The other was that the idea of quantized energy had to be *imposed* on Newtonian mechanics, almost as an afterthought. It seemed necessary to set up a new mechanics which would relieve the wave-particle conflict, and which would introduce quantized energy as a consequence of some more basic principle.

Wave-Particle Duality

The first step in the development of the new quantum mechanics was taken by Louis de Broglie in 1924. His reasoning was somewhat as follows: Electromagnetic radiation had been thought of as a wave phenomenon for some time, yet the work of Einstein had shown that in certain experiments these "waves" had the properties of particles, or photons. Could the converse be true? Would things ordinarily called particles show the properties of waves in some experiments? The phenomena associated with wave behavior are diffraction and interference, and as we have mentioned, the appearance of these effects depends on how the length of a wave compares with the dimensions of the object it strikes. The task was to estimate the wavelength of the waves associated with particles. Starting with the Einstein relation between the energy and frequency of a photon, de Broglie wrote

$$h\nu = E,$$
$$\nu = c/\lambda,$$
$$hc/\lambda = E. \qquad (10.11)$$

From the theory of relativity he drew a relation between the momentum of light p, its velocity c, and its energy, $E = cp$. Combining this with Eq. (10.11) gives

$$h/\lambda = p, \qquad (10.12)$$

which we can interpret as the relation between the momentum of a photon and its wavelength. De Broglie suggested that this equation could be used to calculate the wavelength associated with any *particle* whose momentum was $p = mv$.

Table 10.1 gives the results of calculations using Eq. (10.12). The greater the mass and velocity of the particle, the shorter is its wavelength. The wavelength associated with any macroscopic particle is smaller than the dimensions of any physical system; thus diffraction or any other wave phenomena can never be observed with baseballs or even dust particles. On the other hand, electrons and even atoms can have such small momenta that their wavelengths are of the same dimension as the interatomic spacing in crystals. Therefore, when a beam of electrons impinges on a crystal, diffraction should be observed, as indeed it was, first in 1927, three years after de Broglie had advanced his ideas.

Table 10.1 Wavelengths of particles

Particle	Mass (gm)	Velocity (cm/sec)	Wavelength (A)
Electron at 300°K	9.1×10^{-28}	1.2×10^{7}	61
1-volt electron	9.1×10^{-28}	5.9×10^{7}	12.3
100-volt electron	9.1×10^{-28}	5.9×10^{8}	1.2
He atom at 300°K	6.6×10^{-24}	1.4×10^{5}	0.72
Xe atom at 300°K	2.2×10^{-22}	2.4×10^{4}	0.12

Instead of relieving the wave-particle conflict, de Broglie's proposal seemed to deepen the problem. Yet it was a progressive step, for the generalization of de Broglie's ideas produced a totally successful quantum mechanics. Today the almost universally accepted interpretation of the wave-particle conflict is that it is not really a conflict at all. In speaking of the behavior of atomic systems, we use the words of a language designed to describe the macroscopic world, and we have no right to suppose that only one of these words, wave or particle, will always characterize all properties of things which are not part of our macroscopic world. Therefore, we just accept the fact that whatever electrons and photons are, they have a dual nature; in some experiments their wave properties will be most obvious, and in others they will behave like particles.

The Uncertainty Principle

The terms position and velocity are used to describe the behavior of macroscopic particles. Is there any restriction to their application to subatomic "particles" which have wave properties? To see that there is, let us consider the problem of determining the position of an electron. If we use light to locate the electron, the general principles of optics tell us that we cannot resolve or locate the electron much more accurately than $\pm\lambda$, the wavelength of light used. Naturally we would try to make λ as small as possible, and so in principle locate the electron to any required degree of accuracy. But can we determine the momentum of the electron at the *same time as we determine its position?* The answer is no, for in determining the position of the electron we inevitably change its momentum by an unknown amount. To understand that this is so, we need only recognize that in order to locate an electron with a photon, there must be a collision between the two. A photon of wavelength λ has a momentum $p = h/\lambda$, and in the electron-photon collision, some unknown fraction of the momentum of the photon will be transferred to the electron. Thus the result of locating the electron to within a distance $\Delta x \approx \pm\lambda$ is to produce an uncertainty in its momentum which is roughly $\Delta p \approx h/\lambda$. The product of these two uncertainties is

$$\Delta p\, \Delta x \cong \frac{h}{\lambda}\lambda = h. \tag{10.13}$$

This is a crude derivation of Heisenberg's *Uncertainty Principle*, which states that there is a limit to the precision to which the position and momentum of a particle may be determined *simultaneously*. A more involved argument gives the precise form of the Uncertainty Principle as $\Delta p\, \Delta x \geq h/4\pi$.

A simultaneous and exact determination of position and momentum is just what is required to describe a trajectory, and thus the Uncertainty Principle tells us that there is a limit to the accuracy with which a particle trajectory can be known. Let us see how much the Uncertainty Principle allows us to say about the trajectories of electrons in atoms. In order to have a good idea where the electron is, we might wish to locate it to within 0.05 A, or 5×10^{-10} cm. According to the Uncertainty Principle, any such measurement of the electron position would have associated with it an uncertainty in momentum given by

$$\Delta p = \frac{h}{4\pi\, \Delta x} = \frac{6 \times 10^{-27}}{6 \times 10^{-9}} \cong 1 \times 10^{-18} \text{ gm-cm/sec.}$$

Since the mass of an electron is 9×10^{-28} gm, the uncertainty in the electron velocity is

$$\Delta v = \frac{\Delta p}{m} \cong \frac{1 \times 10^{-18}}{9 \times 10^{-28}} \cong 10^{9} \text{ cm/sec.}$$

According to this crude calculation, the uncertainty in the electron velocity would be nearly as large as the velocity of light, or as great as or even greater than what we might expect the actual electron velocity to be. In short, we have to say that the electron velocity is so uncertain that no possibility of specifying a trajectory exists. Here we find another failure of the Bohr theory. Its sharply defined electron trajectories can have no real meaning, for in view of the Uncertainty Principle, their existence can never be demonstrated experimentally.

The Schrödinger Equation

The de Broglie wave relation is the basis for predicting the behavior of freely moving particles. Shortly after it was proposed, Erwin Schrödinger demonstrated that the de Broglie expression could be generalized so as to apply to bound particles such as electrons in atoms. The heart of Schrödinger's theory is that the allowed energies of physical systems can be found by solving an equation which so resembles the equations of classical wave theory that it is called the wave equation. For the motion of one particle in one (the x) direction, the Schrödinger wave equation is

$$-\frac{h^2}{8\pi^2 m}\frac{d^2\psi}{dx^2} + V\psi = E\psi. \qquad (10.14)$$

The "knowns" in this equation are m, the mass of the particle, and V, its potential energy expressed as a function of x. The "unknowns" to be found by solving the equation are E, the quantized or allowed energies of the particle, and ψ, which is called the *wave function*. The quantity $d^2\psi/dx^2$ represents the rate of change of $d\psi/dx$, the rate of change of ψ. When this equation is applied to real systems such as the hydrogen atom, it is found that it cannot be solved unless E takes on *certain* values which are related by integers. Thus quantized energy and quantum numbers are an automatic consequence of the Schrödinger theory, and do not have to be tacked on to Newtonian mechanics as was done by Bohr.

What is ψ? By itself, it has no physical meaning. However, the square of the absolute value of ψ, $|\psi|^2$, does have an important physical interpretation. It is a mathematical expression of how the *probability* of finding a particle varies from place to place. Thus the exact trajectories of Newtonian mechanics and the Bohr theory do not appear in the results of the Schrödinger quantum mechanics; this, according to the Uncertainty Principle, is as it should be.

The Particle in a Box

As an example of the quantum mechanical description of matter, we shall solve the simplest of problems, the motion of a particle confined by impenetrable walls—the so-called particle in a box. This example will allow us to examine the properties of a simple wave function, and to see how quantized energies come about. In addition, the results will help us to understand qualitatively many types of more complicated quantum-mechanical problems.

A particle in any real box moves in three dimensions. To analyze the mechanics of its motion, however, it is often sufficient to deal only with one dimension (say the x-coordinate), since motion in the other directions is in principle no different. We consider then, a particle of mass m moving with a positive total energy E along the x-coordinate. There is an impenetrable wall at $x = 0$, and another at $x = L$. For $0 \le x \le L$, the potential energy is zero, and outside these limits it is taken to be infinite, due to the presence of the impenetrable walls.

Starting with the one-dimensional Schrödinger equation, and letting the symbol $\hbar$ stand for $h/2\pi$, we have

$$-\frac{\hbar^2}{2m}\frac{d^2\psi}{dx^2} + V\psi = E\psi.$$

We note that since V is zero for $0 \le x \le L$, for this region we can write

$$\frac{d^2\psi}{dx^2} = -\frac{2mE}{\hbar^2}\psi. \tag{10.15}$$

From the qualitative features of this equation we can get a picture of what our wave function ψ will look like. The second derivative $d^2\psi/dx^2$ is the curvature of the wave function. Equation (10.15) says, therefore, that since m, E, and $\hbar$ are positive quantities, whenever ψ is positive, its curvature is negative, or ψ is concave downward. Similarly, whenever ψ is negative, its curvature is positive, and ψ is concave upward. Whenever ψ is zero, its curvature is zero.

If we attempt to sketch a function ψ which has these curvature properties, we find that it begins to look like a wave. There are many functions that have this general appearance, and the simplest of them, the sine function, is in fact the solution of the Schrödinger equation for the particle in the box. To verify this, we assume ψ is equal to $A \sin bx$, where A and b are constants. Then we differentiate twice:

$$\psi = A \sin bx,$$

$$\frac{d\psi}{dx} = bA \cos bx,$$

$$\frac{d^2\psi}{dx^2} = -b^2 A \sin bx,$$

$$\frac{d^2\psi}{dx^2} = -b^2\psi.$$

We see that this last equation has exactly the same form as Eq. (10.15), and would be identical to it if b^2 were equal to $2mE/\hbar^2$. Thus the function that satisfies Eq. (10.15) is

$$\psi = A \sin\left(\frac{2mE}{\hbar^2}\right)^{1/2} x. \tag{10.16}$$

Up to this point we have not made any use of the fact that the walls are actually located at $x = 0$ and $x = L$. Thus the wave function we have found applies to a *free* particle, and not yet to one confined to a box. Note that the energy E of this free particle may be any positive value we please. That is, there is not yet any sign of quantized energies, or energy levels. This is an important observation, since it is part of a demonstration that quantized energy levels occur only when we *confine* a particle by potential energy barriers, or when we make its motion *periodic* in some manner.

We now consider the consequences of the walls of the box. If the walls are impenetrable, and if the square of the wave function represents the probability of finding a particle at a point, then it is reasonable to suppose that the wave function vanishes at the walls. More precisely, the wave function vanishes within the walls, and a general property that ψ is a continuous function then requires that ψ vanish at the walls. We have then the conditions

$$\psi(x = 0) = 0, \qquad \psi(x = L) = 0$$

to impose on our free-particle wave function. These two requirements are called the *boundary conditions* for the problem.

The first boundary condition, $\psi(0) = 0$ is satisfied automatically, since setting $x = 0$ in Eq. (10.16) gives $\psi = 0 = \sin (0)$. The second boundary condition can be satisfied only if E has certain values. We can deduce these values by noting that $\sin n\pi = 0$, where n is an integer. Thus if E is such that

$$\left(\frac{2mE}{\hbar^2}\right)^{1/2} L = n\pi, \qquad n = 1, 2, 3 \ldots,$$

the second boundary condition will be satisfied. Calling the values of E that satisfy this relation E_n, we square and transpose to get

$$E_n = \frac{n^2 h^2}{8mL^2}, \qquad n = 1, 2, 3 \ldots \tag{10.17}$$

These are the allowed or *quantized* values of the energy. The corresponding wave functions are

$$\psi_n = A \sin \left(\frac{2mE_n}{\hbar^2}\right)^{1/2} x \tag{10.18a}$$

or

$$\psi_n = A \sin \frac{n\pi x}{L}. \tag{10.18b}$$

It is of interest to note that the quantized energy levels can be obtained by asserting that our wave function must have the form of a standing wave between $x = 0$ and $x = L$. A standing wave has zero amplitude at the walls, and in order for this to be true, the distance L must be an integral multiple of half the wave length:

$$L = \frac{n\lambda}{2}, \qquad n = 1, 2, 3, \ldots$$

We now use the de Broglie relation $\lambda = h/p$, where p is the momentum mv:

$$L = \frac{n}{2}\frac{h}{mv}, \qquad mv = \frac{nh}{2L},$$

$$\frac{1}{2}mv^2 = \frac{(mv)^2}{2m} = \frac{1}{2m}\left(\frac{n^2h^2}{4L^2}\right) = \frac{n^2h^2}{8mL^2}.$$

Since all the energy is kinetic, $E_n = \frac{1}{2}mv^2$ and

$$E_n = \frac{n^2h^2}{8mL^2},$$

which is the correct expression for the allowed energy levels. Also, the mathematical representation of a standing wave of amplitude A between $x = 0$ and $x = L$ is

$$\psi_n = A\sin\frac{n\pi x}{L},$$

which is the correct wave function. Only in the simple cases where the potential energy is constant can the allowed energy levels and wave functions be deduced from the de Broglie relation.

Our wave function still contains the undetermined constant A. We can evaluate A by using the fact that $\psi_n^2(x)\,dx$ is the probability of finding the particle in state n in an interval dx at x. Therefore the sum (integral) of all such probabilities from $x = 0$ to $x = L$ must equal 1, since it represents the probability of finding the particle anywhere between 0 and L. Thus we must have

$$\int_0^L \psi_n^2\,dx = 1 \tag{10.19}$$

if ψ_n is to be a proper wave function. We can force this to be true by adjusting the value of the constant A in our wave function. That is, we substitute Eq. (10.18b) for ψ_n to get

$$A^2 \int_0^L \sin^2\left(\frac{n\pi x}{L}\right)dx = 1.$$

The value of the integral is $L/2$, so

$$A^2\frac{L}{2} = 1, \qquad A = \left(\frac{2}{L}\right)^{1/2}.$$

This then is the value that A must have if the wave function is to be used to calculate a probability of finding a particle. The procedure we have used to find A is called *normalization*, and a wave function that obeys Eq. (10.19) is said to be normalized to 1.

Our final results for the particle in a one-dimensional box are

$$E_n = \frac{n^2 h^2}{8mL^2}, \qquad \psi_n = \left(\frac{2}{L}\right)^{1/2} \sin\left(\frac{n\pi x}{L}\right).$$

These energy levels, the wave functions, and their squares are plotted in Fig. 10.12.

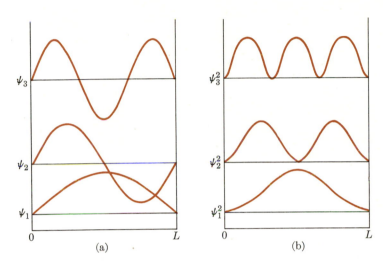

(a) (b)

(a) The wave functions for a particle in a box, and (b) their squares or the probability density of the particle as a function of position. The height at which the zero line for each state lies is proportional to E_n, the energy of the state.

FIG. 10.12

There are several properties of the energy levels and wave functions of the particle in a box which should be very carefully noted, since they appear qualitatively in solutions of more complicated problems.

1. Quantized energy levels appeared *only* when we confined the particle with the potential barriers. We can expect quantized energy levels whenever particle motion is confined or is periodic, as in a rotating molecule.

2. Equation (10.17) shows that the spacing between energy levels *increases* as the mass of the particle *decreases*, and as the space to which the particle is confined *decreases*. We can expect, in general, that effects of spacing of energy levels will be more prominent for systems of small mass confined to small regions of space. This is the qualitative reason why electrons confined to atoms have much more widely-spaced energy levels than atoms moving in a large box. It is also the reason why the motion of macroscopic systems

Table 10.2 Characteristic masses, lengths, and energy-level spacings

System	Characteristic mass (gm)	Characteristic length (cm)	Energy-level spacing (kcal/mole)
Nucleons in nucleus	10^{-24}	10^{-13}	$\sim 10^7$
Atom in solid	10^{-23}	10^{-9}	~ 0.3
Electron in atom	10^{-27}	10^{-8}	~ 100
Atom in a box	10^{-23}	10	$\sim 10^{-9}$

does not show quantum effects. Table 10.2 compares the energy-level spacing of several systems and shows how it is correlated with mass and degree of confinement.

3. The wave functions may have regions in which they are positive, and other regions in which they are negative. The sign of the wave function in various regions of space will prove to be important in our subsequent discussions of chemical bonding. Between the positive and negative regions, the wave functions pass through zero. These points are called the *nodes* of the wave function. In general, for wave functions of a given type, the one with the greater number of nodes will have the higher energy. The locations of the nodes of the electronic wave functions in molecules are very important in determining the bonding properties of electrons.

10.5 THE HYDROGEN ATOM

A complete theoretical treatment of the hydrogen atom using the Schrödinger equation has been accomplished, and the results agree with experimental information in every detail. Besides being an important test of quantum mechanics, the theoretical treatment of the hydrogen atom has another importance: the information derived from this simplest of all atomic systems is used to discuss and predict electron behavior in more complicated atoms and molecules. Thus in order to understand the periodicity of atomic properties and the nature of chemical bonding, it is necessary to thoroughly understand the behavior of the electron in the hydrogen atom.

In the old quantum theory of Bohr, it was necessary to postulate or assume the existence of quantum numbers. This is not so in modern quantum mechanics. All we need assume is the much more general principle that the Schrödinger equation correctly describes the behavior of any atomic system. When the Schrödinger equation is applied to the hydrogen atom, the quantum numbers appear as an automatic consequence of the mathematics, just as was true for the particle in a box. For the hydrogen atom, there are four of these quantum numbers which specify the allowed energies and general behavior of the atomic electron. These are listed on page 423 in decreasing order of importance.

1. The Principal Quantum Number n

This is a number which can assume any positive integral value, excluding only zero. As its name implies, it is most important, for its value determines the energy of the hydrogen atom (or any other one-electron atom of nuclear charge Z) by the formula

$$E = -\frac{2\pi^2 m e^4 Z^2}{n^2 h^2},\qquad (10.20)$$

where m and e are the electronic mass and charge. Equation (10.20), obtained by solution of the Schrödinger equation, is the same expression as Bohr had obtained earlier from his incorrect postulates.

2. The Angular-Momentum Quantum Number l

As its name implies, the value of l determines the angular momentum of the electron, with higher values of l corresponding to greater angular momentum. Now, if an electron has angular momentum, it has kinetic energy of angular motion, and the amount of this angular kinetic energy is limited by the total energy of the electron. Thus it is not surprising that the theory restricts the allowed values of l according to the value of n. Theory and experiment both show that l may assume all integral values from 0 to $n-1$ inclusive: that is, $0, 1, \ldots, n-2, n-1$.

3. The Magnetic Quantum Number m_l

An electron with angular momentum can be thought of as an electric current circulating in a loop, and consequently a magnetic field due to this current is expected and observed. The observed magnetism is determined by the value of m_l. Since this magnetism has its eventual source in the angular momentum of the electron, it is reasonable that the values allowed to m_l depend on the value of l, the angular-momentum quantum number. Theory and experiment both show that m_l can assume all integral values between $-l$ and $+l$, including zero. That is, m_l can equal $-l, -l+1, \ldots, 0, 1, \ldots, l-1, l$.

4. The Spin Quantum Number m_s

Besides the magnetic effect produced by its angular motion, the electron itself has an intrinsic magnetic property. A charged particle spinning about its own axis also behaves like a small magnet; hence we say that the electron has a spin. The quantum number associated with this spin has only two possible values: $+\frac{1}{2}$ and $-\frac{1}{2}$.

Since the value of n restricts the possible values of l, and the value of l in turn restricts the allowed values of m_l, only certain combinations of the quantum numbers are possible. For example, let us consider the lowest energy state, or ground state, of the hydrogen atom, for which $n = 1$. Since l is restricted to integer values between $n - 1$ and zero, it has only one possible value: $l = 0$ if $n = 1$. The value of l determines the allowed values of m_l; since only integers between $+l$ and $-l$ are permitted, only $m_l = 0$ is possible if $l = 0$. Finally, irrespective of the three other quantum numbers, m_s may be equal to $+\frac{1}{2}$ or $-\frac{1}{2}$. Thus we find that there are two ways in which the hydrogen atom can be in its ground state, and these correspond to the choices of 1, 0, 0, $+\frac{1}{2}$ and 1, 0, 0, $-\frac{1}{2}$ for n, l, m_l, and m_s, respectively.

Table 10.3 Quantum numbers and orbitals

n	l	Orbital	m_l	m_s	Number of combinations
1	0	1s	0	$+\frac{1}{2}, -\frac{1}{2}$	2
2	0	2s	0	$+\frac{1}{2}, -\frac{1}{2}$	2 ⎫ 8
2	1	2p	+1, 0, −1	$+\frac{1}{2}, -\frac{1}{2}$	6 ⎭
3	0	3s	0	$+\frac{1}{2}, -\frac{1}{2}$	2 ⎫
3	1	3p	+1, 0, −1	$+\frac{1}{2}, -\frac{1}{2}$	6 ⎬ 18
3	2	3d	+2, +1, 0, −1, −2	$+\frac{1}{2}, -\frac{1}{2}$	10 ⎭
4	0	4s	0	$+\frac{1}{2}, -\frac{1}{2}$	2 ⎫
4	1	4p	+1, 0, −1	$+\frac{1}{2}, -\frac{1}{2}$	6 ⎪ 32
4	2	4d	+2, +1, 0, −1, −2	$+\frac{1}{2}, -\frac{1}{2}$	10 ⎪
4	3	4f	+3, +2, +1, 0, −1, −2, −3	$+\frac{1}{2}, -\frac{1}{2}$	14 ⎭

Other combinations of the quantum numbers correspond to the *excited* electronic states of the hydrogen atom. If the electron is excited to the energy corresponding to $n = 2$, its angular-momentum quantum number may be equal either to $n - 1 = 1$, or to $n - 2 = 0$. If $l = 0$, the only allowed value of m_l is zero, and as before, m_s may be $+\frac{1}{2}$ or $-\frac{1}{2}$. If $l = 1$, m_l can assume any one of the values -1, 0, 1, and for each of these three values m_s can be $+\frac{1}{2}$ or $-\frac{1}{2}$. These possibilities are enumerated in Table 10.3, which shows that there are eight different ways in which an electron in the hydrogen atom can be in the $n = 2$ state. All of these eight combinations of quantum numbers correspond to the same energy. When the electron is excited to the $n = 3$ state, l can be 0, 1, or 2, and accordingly, a greater number of combinations of quantum numbers, 18 in all, is allowed. In general, the number of possible combinations of quantum numbers all with the same value of n is $2n^2$.

Each set of quantum numbers is associated with a different type of electronic motion, and now we must see how the behavior of electrons in atoms is described. Quantum mechanics provides us with $|\psi|^2$, a mathematical expression of the probability of finding an electron at all points in space. This probability function is the best indication available of how the electron behaves, for as a consequence of the Uncertainty Principle, the amount we can know about the electron is limited. While quantum mechanics can tell us the exact probability of finding an electron at any two particular points, it does not tell us how the electron moves from one of these points to the other. Thus the idea of an electron orbit is lost; it is replaced with a description of where the electron is most likely to be found. This total picture of the probability of finding an electron at various points in space is called an *orbital*.

There are various types of orbitals possible, each corresponding to one of the possible combinations of quantum numbers. These orbitals are classified according to the values of n and l associated with them. In order to avoid confusion over the use of two numbers, the numerical values of l are replaced by letters; electrons in orbitals with $l = 0$ are called *s*-electrons, those occupying orbitals for which $l = 1$ are *p*-electrons, and those for which $l = 2$ are called *d*-electrons. The numerical and alphabetical correspondences are summarized in Table 10.3. Using the alphabetical notation for l, we would say that in the ground state of the hydrogen atom ($n = 1, l = 0$) we have a 1*s*-electron, or that the electron moves in a 1*s*-orbital.

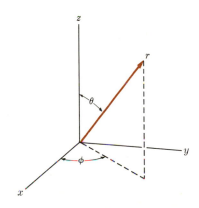

The relation of the spherical polar coordinates r, θ, and ϕ to cartesian coordinates x, y, and z.

FIG. 10.13

To make the concept of an orbital more meaningful, it is helpful to examine the actual solutions of the wave function for the one-electron atom. Because of the spherical symmetry of the atom, the wave functions are most simply expressed in terms of a spherical polar-coordinate system, shown in Fig. 10.13, which has its origin at the nucleus. It is found that the wave functions can be expressed as the product of two functions, one of which (the "angular part" χ) depends only the angles θ and ϕ, the other of which (the "radial part" R) depends only on the distance from the nucleus. Thus we have

$$\psi(r, \theta, \phi) = R(r)\chi(\theta, \phi).$$

Table 10.4 Angular and radial parts of hydrogen-atom wave functions.

Angular part $X(\theta, \phi)$	Radial part $R_{n,l}(r)$
$X(s) = \left(\dfrac{1}{4\pi}\right)^{1/2}$	$R(1s) = 2\left(\dfrac{Z}{a_0}\right)^{3/2} e^{-\sigma/2}$
$X(p_x) = \left(\dfrac{3}{4\pi}\right)^{1/2} \sin\theta\cos\phi$	$R(2s) = \dfrac{1}{2\sqrt{2}}\left(\dfrac{Z}{a_0}\right)^{3/2}(2-\sigma)e^{-\sigma/2}$
$X(p_y) = \left(\dfrac{3}{4\pi}\right)^{1/2} \sin\theta\sin\phi$	$R(2p) = \dfrac{1}{2\sqrt{6}}\left(\dfrac{Z}{a_0}\right)^{3/2}\sigma e^{-\sigma/2}$
$X(p_z) = \left(\dfrac{3}{4\pi}\right)^{1/2} \cos\theta$	
$X(d_{z^2}) = \left(\dfrac{5}{16\pi}\right)^{1/2}(3\cos^2\theta - 1)$	
$X(d_{xz}) = \left(\dfrac{15}{4\pi}\right)^{1/2} \sin\theta\cos\theta\cos\phi$	$R(3s) = \dfrac{1}{9\sqrt{3}}\left(\dfrac{Z}{a_0}\right)^{3/2}(6 - 6\sigma + \sigma^2)e^{-\sigma/2}$
$X(d_{yz}) = \left(\dfrac{15}{4\pi}\right)^{1/2} \sin\theta\cos\theta\sin\phi$	$R(3p) = \dfrac{1}{9\sqrt{6}}\left(\dfrac{Z}{a_0}\right)^{3/2}(4 - \sigma)\sigma e^{-\sigma/2}$
$X(d_{x^2-y^2}) = \left(\dfrac{15}{4\pi}\right)^{1/2} \sin^2\theta\cos 2\phi$	$R(3d) = \dfrac{1}{9\sqrt{30}}\left(\dfrac{Z}{a_0}\right)^{3/2}\sigma^2 e^{-\sigma/2}$
$X(d_{xy}) = \left(\dfrac{15}{4\pi}\right)^{1/2} \sin^2\theta\sin 2\phi$	

$$\sigma = \frac{2Zr}{na_0}; \qquad a_0 = \frac{h^2}{4\pi^2 me^2}$$

This factorization helps us to visualize the wave function, since it allows us to consider the angular and radial dependences separately.

Table 10.4 contains the expressions for the angular and radial parts of the one-electron-atom wave functions. Note that the angular part of the wave function for an s-orbital is always the same, $(1/4\pi)^{1/2}$, regardless of principal quantum number. It is also true that the angular dependence of the p-orbitals and of the d-orbitals is independent of principal quantum number. Thus all orbitals of a given type (s, p, or d) have the same angular behavior. The table shows, however, that the radial part of the wave function depends both on the principal quantum number n and on the angular momentum quantum number l.

To find the wave function for a particular state, we simply multiply the appropriate angular and radial parts together. Thus the wave function for a

1s-orbital is

$$\psi(1s) = \frac{1}{\pi^{1/2}} \left(\frac{Z}{a_0}\right)^{3/2} e^{-Zr/a_0},$$

where a_0 is the Bohr radius, 0.529×10^{-8} cm. By squaring this function, we obtain an expression which gives the probability of finding the electron in a unit volume as a function of r, the distance from the nucleus:

$$\psi^2(1s) = \frac{1}{\pi} \left(\frac{Z}{a_0}\right)^3 e^{-2Zr/a_0}.$$

From this expression we can see that the probability of finding an electron in a 1s-orbital is independent of the angular coordinates θ and ϕ, and decreases monotonically as r increases.

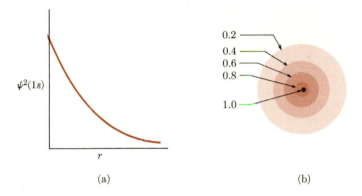

(a)

(b)

Representation of the hydrogen 1s-orbital: (a) ψ^2 as a function of r, and (b) contours of constant ψ^2 measured relative to ψ^2 at the origin.

FIG. 10.14

For purposes of qualitative discussion, it is often useful to have a graphical representation of an orbital. One possible way to show what an orbital looks like is to plot a "cross section" of the probability of finding the electron. That is, we imagine starting at the nucleus and proceeding outward along a radius, plotting the probability of finding the electron as a function of the distance from the nucleus. This type of graph is shown in Fig. 10.14(a) for a hydrogen atom in the $n = 1$, $l = 0$ state. We see that there is a finite probability of finding the electron at any value of r between zero and infinity. This contrasts sharply with Bohr's theory, which pictured the electron as fixed at one radius.

The "cross-section" representation of the orbital does not tell us how the probability of finding the electron depends on the angular coordinates, which, along with r, specify the location of a point in space. One way to represent the angular properties of an orbital is to plot contour maps of the probability of finding an electron, as shown in Fig. 10.14(b). The fact that regions of constant probability are concentric shells shows that for the ground state of the hydrogen atom, the orbital has a spherical shape. A somewhat simpler way to represent the shape of the orbital is to draw a single surface along which the probability of finding the electron is a constant; for a hydrogen atom in its ground state this surface of constant $|\psi|^2$ is a sphere.

Now let us examine the radial parts of the 2s- and 3s-orbitals. Apart from a constant multiplier, the 2s wave function behaves as

$$\psi(2s) \propto \left(2 - \frac{Zr}{a_0}\right) e^{-Zr/2a_0}.$$

The fact that $Zr/2a_0$ appears in the exponential shows that as r increases, the 2s-function decreases in amplitude more *slowly* than does the 1s-function, which has Zr/a_0 in its exponential factor. This is one of the reasons that the 2s-electron tends to stay farther from the nucleus, and has higher energy than does the 1s-electron.

The factor $(2 - Zr/a_0)$ in the 2s wave function controls the sign of the function. For small values of r, Zr/a_0 is smaller than 2, and the wave function is positive, but for large values of r, Zr/a_0 is greater than 2, and the function is negative. At $r = 2a_0/Z$, the pre-exponential factor is zero. Since the radial function vanishes on the circle of radius $r = 2a_0/Z$, this is said to be the locus of a *radial node*.

A similar analysis can be applied to the 3s-function. The exponential factor is now $e^{-Zr/(3a_0)}$, which decreases even more slowly with increasing r than the exponential factors for $\psi(1s)$ and $\psi(2s)$. Therefore, the 3s-electron is, on the average, farther from the nucleus than a 1s- or 2s-electron. Again, the radial nodes for $\psi(3s)$ are found at radii for which the pre-exponential factor vanishes. Thus solving the equation (see $\psi(3s)$ in Table 10.4)

$$6 - \frac{4Zr}{a_0} + \frac{4}{9} \frac{Z^2 r^2}{a_0^2} = 0$$

will give the positions of the radial nodes. Since this is a quadratic equation in r, we expect two solutions, and thus two radial nodes. In general, for an ns-orbital, there are $n - 1$ radial nodes. Notice that the number of nodes increases with increasing energy, just as was true for the functions for the particle in a box.

Let us now examine the 2p wave functions in detail. Table 10.4 shows that the radial part of $\psi(2p)$ is

$$R(2p) = \frac{1}{2\sqrt{6}} (Z/a_0)^{3/2} (Zr/a_0) e^{-Zr/(2a_0)}.$$

Thus the 2p wave function has no nodes at finite values of r. In contrast to the s-functions, which were *nonzero* at $r = 0$, the p-functions vanish at $r = 0$.

This difference is important, and is sometimes described by saying that the s-electron has a greater ability to *penetrate* to the nucleus than does the p-electron. We shall find that a d-electron has even less ability to penetrate to the nucleus. This difference between the s, p, and d-electrons is used to explain one of the important features of the energy levels of many-electron atoms, as we shall see subsequently.

In contrast to s-orbitals, the p-orbitals are not spherically symmetric. This is most simply seen by examining the angular part of the $2p_z$-function. We see from Table 10.4 that $\psi(2p_z)$ is proportional to $\cos\theta$. Thus it has an angular maximum along the positive z-axis, for there $\theta = 0$, and $\cos(0) = +1$, which is the maximum value of the cosine. Similarly, along the negative z-axis, the p_z function has its most negative value, for there $\theta = \pi$, and $\cos(\pi) = -1$, the most negative value of the cosine. The fact that the angular part has its maximum magnitude along the z-axis is responsible for the designation p_z for the function. Everywhere in the xy-plane $\theta = \pi/2$, and $\cos\theta = 0$. Thus the xy-plane is the locus of an angular node of the p_z function.

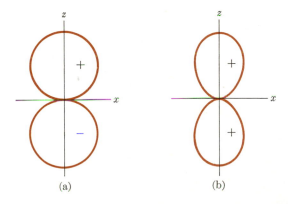

(a)

(b)

The angular part of the $2p_z$-orbital. (a) A plot of $\cos\theta$ in the zx-plane, which represents the angular part of the $2p_z$ wave function. Note the difference in the sign of the function in the two lobes. (b) A plot of $\cos^2\theta$ in the zx-plane, which represents the square of the wave function, and hence the probability density for finding an electron.

FIG. 10.15

A similar analysis is possible for the other 2p-functions. The p_x function has the yz-plane as an angular node, since the function is proportional to $\sin\theta\cos\phi$, and $\cos\phi = 0$ everywhere in the yz-plane. The maximum values of 1 for $\sin\theta$ and $\cos\phi$ occur along the positive x-axis. The p_y function, proportional to the $\sin\theta\sin\phi$, vanishes in the xz-plane, where $\sin\phi = 0$, and has a maximum along the positive y-axis, where both $\sin\theta$ and $\sin\phi$ are unity.

Two of the ways of representing the angular part of the p_z function are shown in Fig. 10.15. In the first instance, $\cos\theta$ is plotted as a function of θ, and the result is two tangent circles. The node in the xy-plane (perpendicular to the page) is clear, as is the maximum magnitude along the z-axis. The cosine function is positive for positive z and negative for negative z, as is indicated. If the square of the angular part, $\cos^2 \theta$, is plotted as in Fig. 10.15b, a double teardrop appearance results. The node and the location of the maxima in the function are perhaps clearer than in the $\cos \theta$ plot. Both these representations of the angular part of the p-functions are encountered frequently.

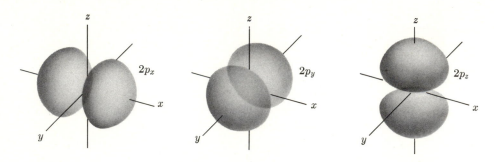

FIG. 10.16   The 2p-orbitals of the hydrogen atom. (Adapted from K. B. Harvey and G. B. Porter, *An Introduction to Physical Inorganic Chemistry.* Reading, Mass.: Addison-Wesley, 1963.)

The simultaneous representation of the radial and angular parts of $|\psi|^2$ for the p-orbitals is more difficult, but is shown in Fig. 10.16. The surfaces of constant $|\psi|^2$ are two spheroidal lobes, with the nucleus located between them in the nodal plane. The three p-orbitals are identical except for the direction of their symmetry axes, which, as we have indicated above, lie along the axes of a cartesian coordinate system. Accordingly, it is often convenient to distinguish between the orbitals by labeling them p_x, p_y, and p_z.

For an electron with $n = 3$, l may be 0, 1, or 2. Thus we might have a 3s-, 3p-, or 3d-electron. Corresponding to the fact that when $l = 2$, m_l can assume one of five values, there are five d-orbitals. The approximate shapes of these orbitals are shown in Fig. 10.17. The significant feature is that two of these orbitals point along the coordinate axes, while the symmetry axes of the other three are in the coordinate planes, but pointing between the cartesian axes. The labeling of the d-orbitals as given in Fig. 10.17 is derived from the directions or planes in which the orbitals have their maximum density.

We have now seen instances where the wave function depends only on the radial distance r (s-orbitals) and on both r and the angles ϕ and θ (p- and d-orbitals). One way of remembering the general behavior of the wave functions is to systematize their nodal properties. For a hydrogen atom wave function of principal quantum number n, there is a total of $n - 1$ nodes which occur at *finite* values of the radial distance r. Of these $n - 1$ nodes, some are encountered as we proceed radially out from the nucleus at any fixed angle. These are called *radial nodes*. Others are encountered as we proceed around the atom at a fixed distance from the nucleus. These are called *angular nodes*. Examination of the wave functions shows that the number of angular nodes is just equal to l, the angular momentum quantum number. Thus we have

$$\text{total nodes} = n - 1,$$
$$\text{angular nodes} = l,$$
$$\text{radial nodes} = n - l - 1.$$

With these relations in mind, it is easier to interpret the various qualitative pictures of orbitals that are encountered. Also, the nodal properties of the wave functions prove to be very important in the theory of chemical bonding, so it is advisable to analyze and understand these properties thoroughly. Finally, we should note that it is sometimes stated that the total number of nodes in an atomic wave function is n, rather than $n - 1$. In this case, the node which always occurs at $r = \infty$ is being included in the count.

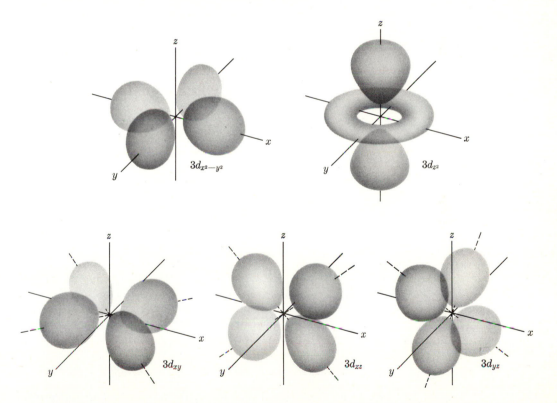

The 3d-orbitals of the hydrogen atom. Note the relation between the labeling of the d-orbitals and their orientations in space. (Adapted from K. B. Harvey and G. B. Porter, *An Introduction to Physical Inorganic Chemistry.* Reading, Mass.: Addison-Wesley, 1963.)   **FIG. 10.17**

Now that we have the general shape or angular properties of the orbitals in mind, we can examine what is called the *radial probability distribution* of the electron: the probability of finding the electron anywhere in a spherical shell of radius r and thickness dr. This radial probability differs from the probability we used earlier to plot the "cross section" of the atom. Previously we asked only for the probability of finding the electron at one particular point a distance r from the nucleus; for the radial probability we ask what is the chance of finding the electron at any of all the points which are a distance between r and $r + dr$ from the nucleus. Thus the radial probability function is $|R|^2$, the radial part of the wave function squared and multiplied by the volume of a spherical shell, $4\pi r^2\, dr$.

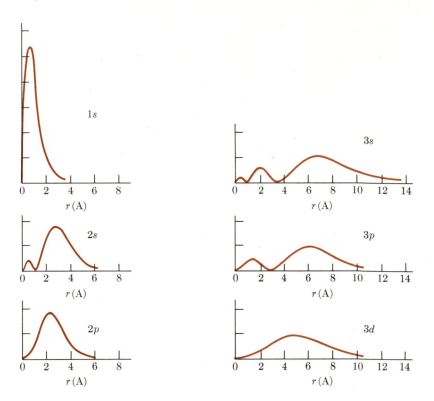

FIG. 10.18   Radial probability density for some orbitals of the hydrogen atom. Ordinate is proportional to $4\pi r^2 R^2$, and all distributions are to the same scale.

Figure 10.18 shows how this radial probability depends on the distance from the nucleus for various orbitals. This probability of finding an electron very near the nucleus is small, for in this region $4\pi r^2$ is small. The maxima in the radial-probability curves occur at radii where the electron is most likely to be found. For the 1s-electron of the hydrogen atom, this radius of maximum probability is 0.529 Å. The curves show that on the average the 2s-electron spends its time a greater distance from the nucleus than does a 1s-electron. This is qualitatively consistent with the relative energies of the 1s- and 2s-states, for the electron which on the average is close to the nucleus is bound to have the lower energy. Comparison of the radial probability curves of electrons with the same n but different l shows that their average distance from the nucleus is approximately the same. However, an s-electron has a greater chance of being very close to the nucleus than does a p-electron, which in turn is more likely to be near the nucleus than is a d-electron. The different abilities of s-, p-, and d-electrons to penetrate to the nucleus should be noted carefully since they persist in atoms with many electrons, and are responsible for many of the details of the structure of the periodic table.

To characterize the orbitals of the one-electron atom further, we draw upon two equations that result from the quantum-mechanical treatment. For a single electron moving around a central nuclear charge Ze in an orbital whose quantum numbers are n, l, the average value of the reciprocal of the electron-nucleus separation is

$$\frac{\bar{1}}{r} = \frac{4\pi^2 me^2 Z}{n^2 h^2}. \tag{10.21}$$

Note that this expression does not contain the quantum number l. This begins to reveal the reason that the orbitals with the same n but different l have the same energy in the one-electron atom. The average potential energy of the electron-nucleus system can be obtained by multiplying Eq. (10.21) by $-Ze^2$ to give

$$\overline{V} = -\frac{\overline{Ze^2}}{r} = -\frac{4\pi^2 me^4 Z^2}{n^2 h^2}.$$

If, as was true in the Bohr treatment of the atom, the total energy is one-half the potential energy, we might write

$$E = -\frac{1}{2}\frac{\overline{Ze^2}}{r}$$
$$= -\frac{2\pi^2 me^4 Z^2}{n^2 h^2},$$

which is in fact Eq. (10.20), the correct expression for the allowed energies. Thus while we can make the general observation that the energy of an orbital increases as its size increases, the important factor which determines the energy is the average of $1/r$.

This conclusion is further supported by the following expression for the average value of r, again derived from the quantum-mechanical treatment of the one-electron atom:

$$\bar{r} = \frac{n^2}{Z} \left\{ 1 + \frac{1}{2} \left[1 - \frac{l(l+1)}{n^2} \right] \right\} \frac{h^2}{4\pi^2 me^2} ; \qquad (10.22)$$

while $\bar{r}$ increases as n increases, it decreases as l increases. The form of the orbital at distances far from the nucleus is important in determining the value of $\bar{r}$. Since $\bar{r}$ is less for a p-orbital than for an s-orbital, it must be that the electron density in a p-orbital decreases faster at great distances than does the density of the corresponding s-orbital. Reference to Fig. 10.18 shows that this is true.

10.6 MULTI-ELECTRON ATOMS

The application of quantum mechanics to atoms with many electrons is a difficult mathematical procedure, but theoretical results which are in extremely good agreement with experiment have been obtained. Consequently, we are confident that quantum mechanics provides a completely satisfactory description of even the most complicated atoms. The simplest procedures used for the approximate qualitative description of multi-electron atoms are natural extensions of those used to describe the hydrogen atom. Electrons are associated with atomic orbitals which are qualitatively similar to the orbitals of the hydrogen atom. Each orbital is labeled with a set of quantum numbers which are just the same as those used for the hydrogen atom. As before, the principal quantum number n is most important in determining the energy of the orbital, and once again the value of l determines its shape or angular properties. However, the value of l for an electron in a multi-electron atom also affects the energy. Thus in a multi-electron atom a $2p$-electron has a higher energy than a $2s$-electron, and the energy of a $3d$-electron is greater than that of a $3p$-electron, which is in turn greater than the energy of a $3s$-electron. This is illustrated in Fig. 10.19, which compares the allowed energy states of the lithium and sodium atoms with those of the hydrogen atom. While the energy-level patterns of the various atoms differ in their quantitative details, all are at least qualitatively similar to those of lithium and sodium.

Another important qualitative feature of multi-electron atoms is that each electron in the atom has a *unique* set of quantum numbers. That is, each electron has a combination of n, l, m_l, and m_s which is in some way different from those of all other electrons in the atom. This important universal observation is called the **Pauli Exclusion Principle.** It is an experimental fact which really has no fundamental explanation, just as there is no explanation of why two like charges repel each other with a force given by Coulomb's law. Another way of stating the Pauli Principle is that no more than two electrons can occupy the same atomic orbital. Two electrons in the same orbital have the same values of n, l, and m_l; thus in order for each electron to have a unique set of quantum numbers, one must have spin $+\frac{1}{2}$, the other spin $-\frac{1}{2}$. In short, two electrons can occupy the same orbital *if and only if their spins are different.*

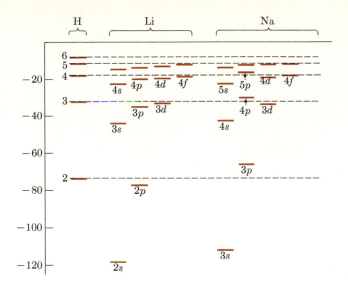

Comparison of the energy levels of the hydrogen atom, lithium atom, and sodium atom.   **FIG. 10.19**

The energy-level patterns of the lithium and sodium atoms reflect some important features of the behavior of electrons in multi-electron atoms. For example, why is it that the valence electron of sodium, normally in the 3s-orbital, has higher energy when it is excited to the 3p-orbital, and still higher energy when it is in the 3d-state?

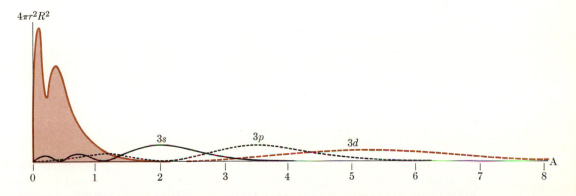

The radial distribution of electron density in the sodium atom in its ground and excited   **FIG. 10.20**
states. The shaded area represents the core electrons. The distribution of the valence electron when it is in the 3s-, 3p-, or 3d-orbitals is also shown.

The answer can be deduced from Fig. 10.20, which shows the radial distribution of the ten core electrons in the sodium atom, and for the valence electron in the 3s-, 3p-, or 3d-orbitals. We see that when the electron is in the 3d-orbital, it spends virtually all its time relatively far from the nucleus, well outside the regions where the ten core electrons spend most of their time. As a result, the 3d-electron is *shielded* or *screened* from the full nuclear charge of +11 by the ten core electrons. To a fairly good approximation, the 3d-electron moves under the influence of an "effective" nuclear charge of approximately $11 - 10 = 1$. Consequently, the 3d-electron in the sodium atom has nearly the same energy as a 3d-electron in a hydrogen atom, as Fig. 10.19 shows.

Figure 10.20 also shows that an electron in the 3p- or 3s-orbital of sodium also spends most of its time outside the inner core of ten electrons. However, in contrast to the 3d-orbital, the 3p-orbital has a subsidiary maximum at a fairly small distance from the nucleus, and has a noticeable density in regions even closer to the nucleus, well within the distribution of core electrons. When the valence electron penetrates the inner-core electrons, it is no longer shielded or screened from the nucleus, and feels an increasing amount of the full +11 charge, the nearer it gets to the nucleus. This causes a lowering of the energy of the electron. The effect is even more extreme for the 3s-electron, which has two subsidiary maxima at small radii, and which penetrates to the nucleus most effectively. Consequently, the energy of the 3s-orbital is lower than that of the 3p-, which is in turn lower than that of the 3d-orbital. The type of orbital-energy splitting which occurs in sodium prevails in every multi-electron atom, and for the same reason.

Why, as Figure 10.19 shows, do the energies of the s-, p-, d-, f-, etc., orbitals of a very high principal quantum number in sodium and lithium lie very close to each other, and very close to the hydrogen atom level of the same principal quantum number? The answer again involves the penetration effect. An electron excited to the 6s-state of Na spends most of its time very far from the nucleus, and thus is very well screened from the full nuclear charge by the inner core. Because it is so far from the nucleus, its penetration of the core is extremely rare, and so virtually all the time this electron is moving under an effective nuclear charge of +1. Consequently its energy is very close to that of an electron in the hydrogen atom with $n = 6$. The same argument applies to the electron in the p-, d-, f-, etc., orbitals of the same principal quantum number. Electrons in any of these orbitals are far from the nucleus, do not penetrate the core, and have approximately the same energy.

Electron Configurations

A number of aspects of the properties and behavior of atoms can be understood in terms of the number of electrons they have, and the relative energies of the orbitals they occupy. To see how this comes about, we need the electron configurations of the gaseous atoms of the elements, which tell us the atomic orbitals that are occupied by electrons. To obtain these electronic configurations, we first imagine an atomic nucleus of appropriate charge surrounded by empty atomic orbitals. Then we feed electrons into these orbitals, filling them in order of increasing energy, always remembering that according to the Exclusion Principle, each orbital can accommodate only two electrons whose spins must be opposite. The order of increasing energy for the atomic orbitals is shown in Fig. 10.21. This diagram is *qualitatively* correct for almost every neutral atom, and can be used to find the electron configuration of all but a very few elements.

A word of caution concerning the interpretation of Fig. 10.21 and the idea of "feeding" electrons into orbitals is in order. While it is useful to describe atoms qualitatively by saying that there are electrons "in" certain orbitals, and while it is sometimes helpful to think of atoms as being built up by "placing" electrons into a set of vacant orbitals, this *language* must not be taken too literally. The orbitals of an atom are not a permanent set of "boxes" rigidly placed on an energy scale as Fig. 10.21 might seem to suggest. When we say an electron is "in an orbital" we are saying only that an electron is behaving in a certain manner, and in this sense an orbital exists physically only if an electron is "in" it. Moreover, each atom and ion has a unique set of energy levels determined by its nuclear charge and number of electrons. Consequently, the energy associated with a given orbital depends on what other orbitals are occupied, and is not the same for all atoms. Thus the pattern of orbital energies shown in Fig. 10.21, while useful, has only qualitative significance.

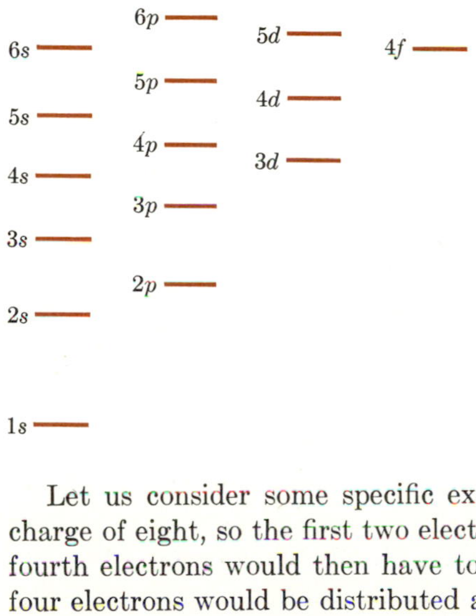

Schematic valence-orbital energy diagram for neutral atoms.

FIG. 10.21

Let us consider some specific examples. The oxygen atom has a nuclear charge of eight, so the first two electrons would fill the $1s$-orbital, the third and fourth electrons would then have to go into the $2s$-orbital, and the remaining four electrons would be distributed among the three $2p$-orbitals. The resulting configuration is described by writing $1s^2 2s^2 2p^4$, where $1s$, $2s$, $2p$ denote the type of orbital, and the exponents give the number of electrons occupying these

Table 10.5 Electron configuration of gaseous atoms

Atomic number	Element	Electronic configuration	Atomic number	Element	Electronic configuration
1	H	$1s$	27	Co	$-3d^7 4s^2$
2	He	$1s^2$	28	Ni	$-3d^8 4s^2$
3	Li	[He] $2s$	29	Cu	$-3d^{10} 4s$
4	Be	$-2s^2$	30	Zn	$-3d^{10} 4s^2$
5	B	$-2s^2 2p$	31	Ga	$-3d^{10} 4s^2 4p$
6	C	$-2s^2 2p^2$	32	Ge	$-3d^{10} 4s^2 4p^2$
7	N	$-2s^2 2p^3$	33	As	$-3d^{10} 4s^2 4p^3$
8	O	$-2s^2 2p^4$	34	Se	$-3d^{10} 4s^2 4p^4$
9	F	$-2s^2 2p^5$	35	Br	$-3d^{10} 4s^2 4p^5$
10	Ne	$-2s^2 2p^6$	36	Kr	$-3d^{10} 4s^2 4p^6$
11	Na	[Ne] $3s$	37	Rb	[Kr] $5s$
12	Mg	$-3s^2$	38	Sr	$-5s^2$
13	Al	$-3s^2 3p$	39	Y	$-4d 5s^2$
14	Si	$-3s^2 3p^2$	40	Zr	$-4d^2 5s^2$
15	P	$-3s^2 3p^3$	41	Nb	$-4d^4 5s$
16	S	$-3s^2 3p^4$	42	Mo	$-4d^5 5s$
17	Cl	$-3s^2 3p^5$	43	Tc	$-4d^5 5s^2$
18	Ar	$-3s^2 3p^6$	44	Ru	$-4d^7 5s$
19	K	[Ar] $4s$	45	Rh	$-4d^8 5s$
20	Ca	$-4s^2$	46	Pd	$-4d^{10}$
21	Sc	$-3d 4s^2$	47	Ag	$-4d^{10} 5s$
22	Ti	$-3d^2 4s^2$	48	Cd	$-4d^{10} 5s^2$
23	V	$-3d^3 4s^2$	49	In	$-4d^{10} 5s^2 5p$
24	Cr	$-3d^5 4s$	50	Sn	$-4d^{10} 5s^2 5p^2$
25	Mn	$-3d^5 4s^2$	51	Sb	$-4d^{10} 5s^2 5p^3$
26	Fe	$-3d^6 4s^2$	52	Te	$-4d^{10} 5s^2 5p^4$

orbitals. In the same manner, we find that the electron configuration of the sodium atom is $1s^2 2s^2 2p^6 3s$. In discussing an atom with many electrons, it is often convenient to omit writing the assignments of all but the valence electrons. Thus for iron we could write the important part of its electron configuration as $4s^2 3d^6$, and assume that it is known that the $1s$-, $2s$-, $2p$-, $3s$-, and $3p$-orbitals are filled. Table 10.5 gives all the known configurations of the lowest energy states of the free gaseous atoms of the elements.

The Periodic Table

Table 10.6 is intended to show how the structure of the periodic table is related to the electron configurations of the atoms. Each of the periods starts with an element which has one valence electron in an s-orbital. The first period is only two elements long, since the $1s$-orbital can accommodate only two electrons. The third electron in lithium must enter the $2s$-orbital, and the second period begins. Since there are one $2s$-orbital and three $2p$-orbitals, each capable of

Table 10.5 (Continued)

Atomic number	Element	Electronic configuration	Atomic number	Element	Electronic configuration
53	I	$-4d^{10}5s^25p^5$	78	Pt	$-4f^{14}5d^96s$
54	Xe	$-4d^{10}5s^25p^6$	79	Au	[ ] $6s$
55	Cs	[Xe] $6s$	80	Hg	$-6s^2$
56	Ba	$-6s^2$	81	Tl	$-6s^26p$
57	La	$-5d6s^2$	82	Pb	$-6s^26p^2$
58	Ce	$-4f^26s^2$	83	Bi	$-6s^26p^3$
59	Pr	$-4f^36s^2$	84	Po	$-6s^26p^4$
60	Nd	$-4f^46s^2$	85	At	$-6s^26p^5$
61	Pm	$-4f^56s^2$	86	Rn	$-6s^26p^6$
62	Sm	$-4f^66s^2$	87	Fr	[Rn] $7s$
63	Eu	$-4f^76s^2$	88	Ra	$-7s^2$
64	Gd	$-4f^75d6s^2$	89	Ac	$-6d7s^2$
65	Tb	$-4f^96s^2$	90	Th	$-6d^27s^2$
66	Dy	$-4f^{10}6s^2$	91	Pa	$-5f^26d7s^2$
67	Ho	$-4f^{11}6s^2$	92	U	$-5f^36d7s^2$
68	Er	$-4f^{12}6s^2$	93	Np	$-5f^46d7s^2$
69	Tm	$-4f^{13}6s^2$	94	Pu	$-3f^67s^2$
70	Yb	$-4f^{14}6s^2$	95	Am	$-5f^77s^2$
71	Lu	$-4f^{14}5d6s^2$	96	Cm	$-5f^76d7s^2$
72	Hf	$-4f^{14}5d^26s^2$	97	Bk	$-5f^97s^2$
73	Ta	$-4f^{14}5d^36s^2$	98	Cf	$-5f^{10}7s^2$
74	W	$-4f^{14}5d^46s^2$	99	Es	$-5f^{11}7s^2$
75	Re	$-4f^{14}5d^56s^2$	100	Fm	$-5f^{12}7s^2$
76	Os	$-4f^{14}5d^66s^2$	101	Md	$-5f^{13}7s^2$
77	Ir	$-4f^{14}5d^76s^2$	102	No	$-5f^{14}7s^2$
			103	Lr	$-5f^{14}6d7s^2$

accepting two electrons, $2 \times (1 + 3) = 8$ elements enter the table before the $2s$- and $2p$-orbitals are filled in the element neon. The third period is also eight elements long and ends when the $3s$- and $3p$-orbitals are filled in argon.

Since the $4s$-orbital is lower in energy than the $3d$-orbitals, a new period starts with potassium before any electrons enter the $3d$-orbitals. After the $4s$-orbital is filled in calcium, the five $3d$-orbitals are the next available in order of increasing energy. These five orbitals accommodate ten electrons, and therefore there are 10 transition-metal elements which enter the table at this point. Once these 10 elements have entered, the fourth period is completed by filling the $4p$-orbitals. In the fifth period the $5s$-, $4d$-, and $5p$-orbitals are filled in succession. The sixth period is different in that after the $6s$-orbital is filled, and one $5d$-electron enters, the $4f$-orbitals are the next available in order of increasing energy. Since an f-orbital corresponds to $l = 3$, the quantum number m can assume integral values from -3 to $+3$, for a total of seven different choices. Thus there are seven $4f$-orbitals, and we expect $7 \times 2 = 14$ elements to appear before any more $5d$-orbitals are filled, as is observed.

Table 10.6 The periodic table, showing the separation into the *s*-, *p*-, *d*-, and *f*-blocks

1s	1 H									
2s	3 Li	4 Be								
3s	11 Na	12 Mg								
4s	19 K	20 Ca								
5s	37 Rb	38 Sr								
6s	55 Cs	56 Ba								
7s	87 Fr	88 Ra								

d-block										
3d	21 Sc	22 Ti	23 V	24 Cr	25 Mn	26 Fe	27 Co	28 Ni	29 Cu	30 Zn
4d	39 Y	40 Zr	41 Nb	42 Mo	43 Tc	44 Ru	45 Rh	46 Pd	47 Ag	48 Cd
5d	57– La	72 Hf	73 Ta	74 W	75 Re	76 Os	77 Ir	78 Pt	79 Au	80 Hg
6d	89– Ac	103								

p-block						
						2 He
2p	5 B	6 C	7 N	8 O	9 F	10 Ne
3p	13 Al	14 Si	15 P	16 S	17 Cl	18 Ar
4p	31 Ga	32 Ge	33 As	34 Se	35 Br	36 Kr
5p	49 In	50 Sn	51 Sb	52 Te	53 I	54 Xe
6p	81 Tl	82 Pb	83 Bi	84 Po	85 At	86 Rn
7p						

f-block														
4f	58 Ce	59 Pr	60 Nd	61 Pm	62 Sm	63 Eu	64 Gd	65 Tb	66 Dy	67 Ho	68 Er	69 Tm	70 Yb	71 Lu
5f	90 Th	91 Pa	92 U	93 Np	94 Pu	95 Am	96 Cm	97 Bk	98 Cf	99 Es	100 Fm	101 Md	102	103 Lr

After the 14 rare-earth elements have entered the table, the last transition metals appear as the 5d-orbitals are occupied. These in turn are followed by the six elements required to fill the three 6p-orbitals, and the sixth period ends with radon. The seventh period starts by filling the 7s-orbital and after one 6d-electron appears, subsequent electrons enter the 5f-orbitals. Thus the periodic table ends with the actinide series, a group of 14 elements analogous in properties and electronic structure to the rare earths.

The gaseous atoms of elements in the same column of the periodic table have, for the most part, the same configuration for their valence electrons, and as is well known, the elements in the same column resemble one another chemically. Furthermore, whenever "horizontal" chemical similarity exists, such as among the rare earths or transition metals, the elements which are chemically similar differ only by the number of electrons in a particular type of orbital, such as 4f, or 3d. In addition to these general relations between electron configuration and chemical properties, there are many more detailed correlations which we shall examine in our later discussions of the chemical properties of the elements.

The structure of the periodic table raises some interesting questions about electronic behavior. Why is it that the third short period stops with argon in the valence electron configuration $3s^2 3p^6$, and the 4s-electrons are added before the 3d-orbitals start to fill? The answer lies in the penetration effect discussed earlier. The 3d-orbital in the potassium atom is concentrated outside the inner core of 18 electrons, and an electron in this orbital is very well screened from the nucleus. Since the 4s-orbital penetrates the core, an electron in it can feel nearly the full nuclear charge some of the time, and thereby lower its energy. This effect is so pronounced that despite its higher principal quantum number, the energy of the 4s-orbital is lower than that of the 3d-orbital. Therefore, the element 19, potassium, has the configuration [Ar]4s, and displays the general chemical properties of an alkali metal.

Why do the 3d-orbitals suddenly become lower in energy and begin to fill immediately after the 4s-shell is completed? The qualitative explanation is based on the fact that even though the 4s-electrons penetrate the core and the 3d-electrons do not, the *major* parts of these orbitals occupy approximately the same region of space. Because the two 4s-electrons are not any closer to the nucleus, they do not screen the 3d-electrons from the nuclear charge. As a result, once the nuclear charge is increased to accommodate the two 4s-electrons, the effective nuclear charge ready to act on the 3d-electrons increases noticeably. In short, the increase in the nuclear charge which occurs in the sequence Ar, K, Ca, is not screened from the 3d-orbital by the added 4s-electrons because these electrons are not part of the inner core. Consequently, electrons added to the 3d-orbitals after the 4s-orbital is filled feel an increased nuclear charge, their energy is lowered, and the first transition series begins. We can make use of a similar argument to explain the occurrence of the second and third transition series.

Because the 4s- and 3d-electrons have somewhat similar energies in the first half of the transition series, these elements generally show a number of different oxidation states in their compounds. However, the lowering of the energy of the 3*d*-electrons relative to the 4*s*-electrons continues through the transition series. As the nuclear charge increases, the energy of the 3*d*-electrons falls well below the energy of the 4*s*-electrons, and becomes low enough so that in the latter half of the transition series, the typical oxidation state displayed is +2, which corresponds to removal of the two 4*s*-electrons only. Higher oxidation states of these elements (Fe–Cu) are produced only with difficulty. When we reach the element zinc, the energy of the 3*d*-electrons becomes so low that they no longer are directly involved in the chemistry of this and subsequent elements.

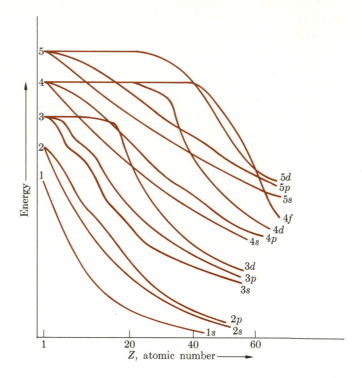

FIG. 10.22   Schematic diagram of the variation of orbital energies with atomic number.

We see that the variation of the relative energies of the orbitals is responsible for the detailed structure of the periodic table. These energy variations can be summarized conveniently in a diagram like that in Fig. 10.22. We see that as atomic number increases, the energy of all orbitals tends to fall. The differences in penetrating power cause a splitting of the energies of the *s*-, *p*-, *d*-, and *f*-orbitals of a given principal quantum number. In the valence shell, the *d*-orbitals sometimes are of higher energy than the *s*- and *p*-orbitals of the next higher principal quantum number. The 4*f*-orbitals for a time lie higher in energy than the 6s-orbitals, and at a point fall rapidly to begin the rare earth series. As orbitals become part of the core, the differentiation in their energies caused by penetration decreases, and although the order of energy is still $s < p < d < f$, orbitals of a given value of n are lower in energy than all of those orbitals with principal quantum number equal to $n + 1$ or greater.

Ionization Energies

We have made use of Fig. 10.21, a *qualitative* indication of the relative energies of the various orbitals in a multi-electron atom. To understand the finer details of the periodic table and chemical behavior we must have a more quantitative indication of the energy with which an atom binds its electrons. This we obtain from measurements of the ionization energy: the minimum energy required to remove an electron from a gaseous atom to form a gaseous ion. Since in the gas phase the atom and the ion are isolated from all external influences, the energy necessary to effect the ionization is exactly the energy with which the atom binds its electron. Thus the magnitude of the ionization energy gives a quantitative measure of the stability of the electronic structure of the isolated atom. The ionization energies of the gaseous atoms of the elements are given in the first column of Table 10.7, and these energies are plotted as a function of atomic number in Fig. 10.23.

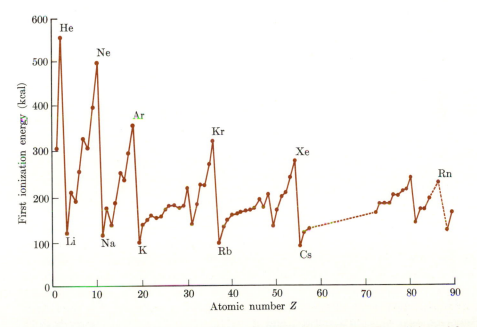

First ionization energy of the elements as a function of atomic number. (Adapted from K. B. Harvey and G. B. Porter, *Physical Inorganic Chemistry*. Reading, Mass.: Addison-Wesley, 1963.)

FIG. 10.23

Table 10.7 Ionization energies of gaseous atoms (kcal/mole)

Atomic number	Element	I_1	I_2	I_3	I_4
1	H	313.5			
2	He	566.9	1254		
3	Li	124.3	1744	2823	
4	Be	214.9	419.9	3548	5020
5	B	191.3	580.0	874.5	5980
6	C	259.6	562.2	1104	1487
7	N	335.1	682.8	1094	1786
8	O	314.0	810.6	1267	1785
9	F	401.8	806.7	1445	2012
10	Ne	497.2	947.2	1500	2241
11	Na	118.5	1091	1652	2280
12	Mg	176.3	346.6	1848	2521
13	Al	138.0	434.1	655.9	2767
14	Si	187.9	376.8	771.7	1041
15	P	254	453.2	695.5	1184
16	S	238.9	540	807	1091
17	Cl	300.0	548.9	920.2	1230
18	Ar	363.4	637.0	943.3	1379
19	K	100.1	733.6	1100	1405
20	Ca	140.9	273.8	1181	1550
21	Sc	151.3	297.3	570.8	1700
22	Ti	158	314.3	649.0	997.2
23	V	155	328	685	1100
24	Cr	156.0	380.3	713.8	1140
25	Mn	171.4	360.7	777.0	
26	Fe	182	373.2	706.7	
27	Co	181	393.2	772.4	
28	Ni	176.0	418.6	810.9	
29	Cu	178.1	467.9	849.4	
30	Zn	216.6	414.2	915.6	
31	Ga	138	473.0	708.0	1480
32	Ge	182	367.4	789.0	1050
33	As	226	466	653	1160
34	Se	225	496	738	989
35	Br	273.0	498	828	
36	Kr	322.8	566.4	851	
37	Rb	96.31	634	920	
38	Sr	131.3	254.3		1300

By examining Fig. 10.23, we find that there is a periodicity in the value of the ionization energy that parallels the periodicity in the chemical properties of the elements. Starting with one of the alkali metals, there is a general tendency for the ionization energy to increase until a maximum is reached at the subsequent rare gas; this is repeated along each row of the periodic table. Superimposed on this general trend is a "fine structure," subsidiary maxima and minima, which we shall explain in terms of electron configurations.

Table 10.7 (Continued)

Atomic number	Element	I_1	I_2	I_3	I_4
39	Y	147	282.1	473	
40	Zr	158	302.8	530.0	791.8
41	Nb	158.7	330.3	579.8	883
42	Mo	164	372.5	625.7	1070
43	Tc	168	351.9		
44	Ru	169.8	386.5	656.4	
45	Rh	172	416.7	716.1	
46	Pd	192	447.9	759.2	
47	Ag	174.7	495.4	803.1	
48	Cd	207.4	389.9	864.2	
49	In	133.4	435.0	646.5	1250
50	Sn	169.3	337.4	703.2	939.1
51	Sb	199.2	380	583	1020
52	Te	208	429	720	880
53	I	241.1	440.3		
54	Xe	279.7	489	740	
55	Cs	89.78	579		
56	Ba	120.2	230.7		
57	La	129	263.6	442.1	
72	Hf	160	344		
73	Ta	182	374		
74	W	184	408		
75	Re	182	383		
76	Os	200	390		
77	Ir	200			
78	Pt	210	4280		
79	Au	213	473		
80	Hg	240.5	432.5	789	
81	Tl	140.8	470.9	687	1170
82	Pb	171.0	346.6	736.4	975.9
83	Bi	168.1	384.7	589.5	1040
84	Po	194			
85	At				
86	Rn	247.8			
87	Fr				
88	Ra	121.7	234.0		
89	Ac	160	279		

It is easy to understand why the ionization energy of helium is greater than that of hydrogen, particularly if we refer to Eq. (10.20), the expression for the binding energy of a one-electron atom. It is clear that the binding energy is sensitive to the nuclear charge if n is a constant. While this expression does not apply quantitatively to atoms with more than one electron, we can use it as a qualitative indication that as we go from hydrogen to helium, we expect the increase in nuclear charge to increase the binding energy or ionization energy of the 1s-electron. If Eq. (10.20) were correct, changing Z from 1 to 2 should increase the ionization energy from 314 to 1254 kcal. The fact that the observed ionization energy of helium is only 567 is a result of the repulsion of the two electrons, which makes the He atom less stable than might be expected from Eq. (10.20). Thus in thinking about ionization energies we must keep in mind the effects both of increasing nuclear charge and of repulsion between electrons.

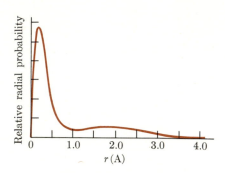

FIG. 10.24 The electron distribution in the lithium atom. The first maximum corresponds to the 1s-electrons; the second, to the 2s-electron.

Why is the ionization energy of lithium, with a nuclear charge of three, so much smaller than that of helium? The electron configuration of lithium is $1s^2 2s$, so in order to form the ion a 2s-electron must be removed. Equation (10.20) shows that if all else is held constant, the binding energy of an electron decreases as n increases, and this factor alone would tend to lower the ionization energy of lithium relative to helium. But why doesn't the increased nuclear charge of lithium offset the change in principal quantum number? The reason can be found in Fig. 10.24, which gives the radial distribution of the electrons in the lithium atom. It is clear that the 1s-electrons spend most of their time very close to the nucleus, while the 2s-electron is for the most part found at much greater radii. This effect is so extreme that it is reasonable to say that the 1s-electrons "screen" the 2s-electron from the nucleus. That is, most of the time the 2s-electron feels not a charge of +3, but a *net* positive charge of approximately $3 - 2 = 1$. Only rarely when the 2s-electron moves very close to the nucleus does it "see" the full +3 nuclear charge. This screening effect by the inner electrons, together with the increase in the principal quantum number, provides a satisfactory explanation of the relative ionization energies of helium and lithium.

Let us continue to analyze the trends in ionization energies shown in Fig. 10.23. We find that the ionization energy of beryllium is somewhat greater than that of lithium, a fact which we can now attribute to the increased nuclear charge. Yet we find that increasing the nuclear charge one more unit and adding one more electron to form the boron atom produces a slight *decrease* in ionization energy. The electron configuration of boron is $1s^2 2s^2 2p$, and the surprisingly low ionization energy is an indication that p-electrons tend to be slightly higher in energy than s-electrons of the same principal quantum number, and thus require less energy for their removal. This effect, too, can be understood in terms of the screening effects of the $1s$-electrons. As we noted earlier in examining the form of the hydrogen orbitals shown in Fig. 10.18, a $2s$-electron has a greater probability of being very close to the nucleus than does a $2p$-electron. This means that a $2s$-electron is better able to penetrate the $1s$-screen than is a $2p$-electron. Thus $2s$-electrons feel the full nuclear charge more often than do $2p$-electrons, and consequently the energy of the $2s$-electrons is always lower than the energy of the $2p$-electrons.

The addition of the second and third $2p$-electrons in carbon and nitrogen is accompanied by increases in ionization energy which we once again attribute to the increasing nuclear charge. To understand the slight drop in ionization energy which occurs at the oxygen atom, we must investigate the filling of the $2p$-orbitals more carefully. Because of the repulsion between like charges, electrons try to avoid each other as much as possible. This can be most effectively accomplished if each of the first three p-electrons is placed in a different p-orbital. Thus the outer electron configuration of carbon is $2s^2 2p_x^1 2p_y^1$, and that of nitrogen is $2s^2 2p_x^1 2p_y^1 2p_z^1$. When the fourth p-electron enters in oxygen, it must be placed in a p-orbital which already has an electron in it. Apparently the extra repulsion which results from two electrons occupying the same orbital offsets the increased nuclear charge, and the ionization energy of oxygen is slightly less than that of nitrogen. As the fifth and sixth p-electrons are added, the effect of increasing nuclear charge overcomes electron repulsion, and the ionization potential rises to a maximum at neon.

Figure 10.23 shows that the third period repeats the behavior found in the second period. In the fourth period, a new feature is introduced. After the two $4s$-electrons have appeared in potassium and calcium, the ionization energy rises very slowly as electrons are added in the transition-metal series. It would appear from the order in which the $4s$- and $3d$-orbitals are filled that the $3d$-orbitals are the higher in energy of the two. Yet, when one of the transition metals is ionized it is a $4s$-*electron* which is removed. This indicates that the energies of the $4s$- and $3d$-orbitals are very close, and that a slight change in the structure of the atom can change their relative energies. Note that while the electron configurations of most of the transition metals are of the type $3d^n 4s^2$, chromium has the configuration $3d^5 4s^1$, and copper $3d^{10} 4s^1$. The fact that the energy of the $3d$-orbitals is nearly the same as that of the $4s$-orbitals is in large measure responsible for the large number of oxidation states displayed by the transition metals.

The features found in the fourth period are repeated in the fifth. In the sixth period we find that the ionization energies of the transition metals are higher than the corresponding elements in the fourth and fifth periods. The cause of this is the appearance of the fourteen rare-earth elements immediately before the $6d$-orbitals are filled. Thus the ionization energies of the transition metals of the sixth period reflect an "extra" amount of nuclear charge introduced with the rare-earth elements.

So far we have been exclusively concerned with the energy necessary to remove the most weakly bound electron from the atom. This is called the first ionization energy of the element. That energy required to remove the second electron, as in

$$Li^+(g) \rightarrow Li^{++}(g) + e,$$

is called the second ionization energy, and values for the elements are given in the second column of Table 10.7. The magnitudes of the second and higher ionization energies can also be understood in terms of electronic configurations and nuclear charge. Consider, for example, that He, Li^+, and Be^{++} all have the electron configuration $1s^2$; we say that they are *isoelectronic*. Comparison of their ionization energies should give us a good indication of the effect of nuclear charge on the binding energy of an electron. The required data from Table 10.7 are the first ionization energy of He, 567 kcal; the second ionization energy of Li, 1743 kcal; and the third ionization energy of Be, 3547 kcal. These are the energies required to remove one of the two $1s$-electrons. The effect of nuclear charge on binding energy of an electron is quite clear.

To see how changing the principal quantum number affects the binding energy of the electron, we need only compare the first and second ionization energies of any of the alkali metals. For lithium we have 124 and 1743 kcal, for sodium 118 and 1090 kcal, and so on. There is an enormous difference between the energy required to remove the outermost s-electron ($\sim$100 kcal) and the 1000 kcal needed to eject an electron of next lower principal quantum number. Since such a huge energy is required to remove an inner electron it is not surprising that the highest positive oxidation states of the metals are never greater than the number of valence electrons.

Ionization of the transition elements displays what may seem at first to be a surprising feature. In the sequence K, Ca, Sc, the $4s$-orbital fills before the $3d$-orbitals. However, when scandium is ionized sequentially, the following electron configurations occur

$$Sc(4s^2 3d) \rightarrow Sc^+(4s3d) + e^-$$
$$Sc^+(4s3d) \rightarrow Sc^{++}(3d) + e^-$$
$$Sc^{++}(3d) \rightarrow Sc^{+3} + e^-$$

That is, when scandium is ionized sequentially the s-orbital is emptied before the d-electron is removed. This merely indicates that in Sc^{++}, the $3d$-orbital is lower in energy than the $4s$-orbital. This same phenomenon occurs throughout the transition series. Thus, vanadium, whose neutral atom has the configuration $4s^2 3d^3$, gives V^+ with a configuration $3d^4$, and when $Co(4s^2 3d^7)$ is ionized, $Co^+(3d^8)$ is formed.

To understand the explanation of the increasing stability of the $3d$ orbitals with increasing ionic charge, it helps to refer to Fig. 10.25, which shows the relative energies of the $4s$-, $4p$-, and $3d$-orbitals in the isoelectronic sequence K, Ca^+, Sc^{++}, and Ti^{+3}. In the potassium atom, the $4s$- and $4p$-orbitals lie lower in energy than the $3d$-orbital because of the penetration effect discussed earlier. The ion Ca^+ has the same number of electrons as K, but has an increased nuclear charge. As a result, the electron core of Ca^+ is somewhat contracted. As the core contracts, the $4s$- and $4p$-orbitals lie more and more outside the region occupied by the core electrons, and consequently their ability to lower their energy by penetrating the more compact core decreases. In Ca^+ the penetration of the $4p$-orbital is no longer sufficient to lower its energy below that of the $3d$-orbital, and we find the $3d$-orbital above the $4s$-, but below the $4p$-orbital. Further contraction of the core occurs upon passing to Sc^{++}, and in this case the penetration of even the $4s$-orbital is small. Consequently the $3d$-orbital lies lower in energy than $4s$- and $4p$-, because of its smaller principal quantum number. These effects occur again in Ti^{+3}.

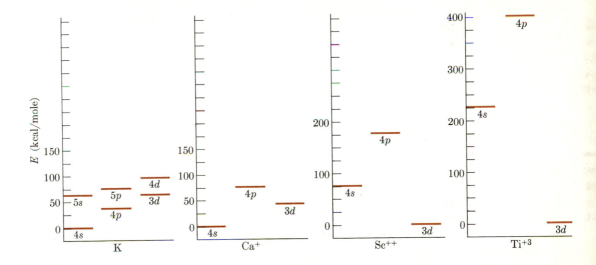

The energies of the $4s$-, $4p$-, and $3d$-orbitals in the isoelectronic sequence K, Ca^+, Sc^{++}, and Ti^{+3}.

FIG. 10.25

In essence, the reason that the d-orbitals become more stable with increasing ionic charge is the same as the reason why d-orbitals become increasingly stable as one proceeds sequentially through the transition elements: the contraction of the core diminishes the importance of penetration, and increases the importance of the principal quantum number in determining the energy.

There are several useful generalizations we can draw from our study of ionization energy that will help us understand the behavior of electrons in atoms and molecules. First, it is apparent that as electrons are added to orbitals of the same principal quantum number in successive elements, the ionization energy increases due to the increase in nuclear charge. This explains the general trend in ionization energy along any row of the periodic table. Second, electrons of the highest principal quantum number are shielded from the nucleus by the inner or core electrons. This is one of the reasons that the ionization energies of the alkali metals are so low. Third, when several p- or d-orbitals are available, one electron enters each orbital until all are half-filled. It is found that the electrons in these half-filled orbitals all have the same spin. This half-filled set of orbitals with all spins the same seems to be particularly stable, for addition of another electron often results in a decrease in the ionization energy. Fourth, among elements in the same periodic column or group, with the exception of the transition metals, there is a tendency for the ionization energy to decrease as the atomic number increases. Thus among elements of the same periodic family, the ones with higher atomic number tend to be oxidized more easily.

Electron Affinities

The electron affinity is the amount of energy required to remove an electron from a gaseous negative ion, as in

$$Cl^-(g) \rightarrow Cl(g) + e^-(g).$$

Table 10.8 Electron affinities of gaseous atoms (kcal/mole)

Atomic number	Element	Affinity
1	H	17.4
3	Li	(14)*
5	B	(7)
6	C	29
7	N	(0.9)
8	O	34
9	F	79.5
11	Na	12.5
13	Al	12
14	Si	(32)
15	P	(18)
16	S	48
17	Cl	83.4
19	K	11.5
29	Cu	28.3
34	Se	46.6
35	Br	77.3
47	Ag	30.1
53	I	70.5
78	Pt	49.1
79	Au	53.1

*Values in parentheses are estimated by quantum mechanical calculation and have not been verified experimentally.

As its name implies, the energy necessary to effect this electron detachment is a measure of the affinity, or attraction, of the atom for its extra electron. A positive electron affinity means energy is required to remove the electron from the ion, and a negative electron affinity means the isolated negative ion is unstable. Table 10.8 shows that the electron affinities of the halogen atoms are greater than those of the other elements. In the halogens, there is one vacancy in the valence p-orbitals. As the ionization energies of these elements show, the large nuclear charge binds the p-electrons strongly, so it is not surprising that there is a large residual affinity for an extra electron. This argument also accounts for the electron affinity of the oxygen, sulfur, and hydrogen atoms. In contrast, the rare gases have no vacancies in their valence orbitals, and any electron added to them would have to be placed in an orbital of next higher quantum number. Because of the screening of the inner electrons, this added electron would feel very little net attraction of the atom, and consequently the electron affinities of the rare gases are essentially zero. This argument helps to show why the rare gases tend to be so inert. Since their electron affinities are so small, they never enter compounds as electron acceptors. On the other hand, their ionization energies are so high that they are oxidized with great difficulty, and consequently they form only a limited number of compounds.

10.7 CONCLUSION

In Chapter 1 we learned that the general acceptance and development of the atomic theory of Dalton was the result of the performance and critical analysis of a large number of experiments. Here in Chapter 10 we have found that the same can be said of our theories of the electronic structure of atoms. It took many years and many experiments to ascertain the particulate nature of electricity, the qualitative arrangement of electrical particles in the atom, and the quantitative laws of behavior of atomic systems. Today we can regard the problems associated with the gross features of atomic electronic structure as having been solved. However, chemists maintain an active interest in this subject, because of the obvious, but incompletely understood, relation between the chemical and physical properties of matter and the electronic structure of atoms. Our hope is to be able to understand or to explain quantitatively much of the chemistry of the elements in terms of the electronic properties of their atoms. The subsequent chapters of this book will discuss some of the progress that has been made in solving this fascinating problem.

SUGGESTIONS FOR FURTHER READING

Coulson, C. A., *Valence*. New York: Oxford, 1961.

Eisberg, R. M., *Fundamentals of Modern Physics*. New York: Wiley, 1961.

Gerhold, G. A., L. McMurchie, and T. Tye, *American Journal of Physics* **40**, 988 (1972).

Harvey, K. B., and G. B. Porter, *Introduction to Physical Inorganic Chemistry.* Reading, Mass.: Addison-Wesley, 1963.

Hochstrasser, R. M., *Behavior of Electrons in Atoms.* New York: Benjamin, 1964.

Perlmutter-Hayman, B., *Journal of Chemical Education* **46**, 429 (1969).

Shamos, M. H., *Great Experiments in Physics.* New York: Holt, 1959.

PROBLEMS

10.1 Energy of 118.5 kcal/mole is required to ionize sodium atoms. Calculate the lowest possible frequency of light that can ionize a sodium atom and the corresponding wavelength. One kcal/mole corresponds to 6.95×10^{-14} ergs/atom, and Planck's constant h is equal to 6.62×10^{-27} erg-sec.

10.2 In the classical wave theory of light, intensity was associated with the squares of the maximum amplitudes of the electric and magnetic fields. In the particle theory of light, what property of the model is associated with the intensity of light?

10.3 When light of 4500-A wavelength impinges on a clean surface of metallic sodium, electrons whose maximum energy is 0.4 ev, or 0.64×10^{-12} erg, are ejected. What is the maximum wavelength of light which will eject electrons from metallic sodium? What is the binding energy of an electron to a sodium crystal?

10.4 Plot the angular part of $f(\theta)$ for the scattering of α-particles as given by Eq. (10.6), by calculating the value of $\sin\theta/\sin^4(\theta/2)$ at a few angles. Is most of the scattering in the forward or backward direction with respect to the original beam? When all other factors are constant, are more particles scattered from a high-velocity beam or from a low-velocity beam of α-particles? Does the scattering increase or decrease as the nuclear charge of the target atoms is increased?

10.5 Write the electronic configurations of the following species and indicate those that are isoelectronic, or have the same number of electrons: Ne, Al, $O^=$, Cl^-, K^+, Ti, Ar.

10.6 Without consulting the periodic table, deduce the atomic numbers of all the inert gases from the fact that except for helium, all have a valence-electron configuration ns^2np^6.

10.7 Without referring to the periodic table, write the electron configurations and give the group of the periodic table to which the elements with the following atomic numbers belong: 3, 14, 8, 17, 37, 56.

10.8 Plot a graph of the square roots of the ionization energies versus the nuclear charges for the series Li, Be^+, B^{++}, C^{+3}, and Na, Mg^+, Al^{++}, Si^{+3}. Explain the observed relationship with the help of Bohr's expression for the binding energy of an electron in a one-electron atom.

10.9 By using the mathematical expression for a $2p_z$-wave function, show that the probability of finding a $2p_z$-electron anywhere in the xy-plane is zero.

10.10 Use the expressions in Table 10.4 to show that when one electron occupies the $2p_x$-orbital, and another the $2p_y$-orbital, the resulting electron distribution

$[\psi^2(p_x) + \psi^2(p_y)]$ is cylindrically symmetric about the z-axis. Show also that if there is an electron in each of the $2p_x$-, $2p_y$- and $2p_z$-orbitals, the atom is spherically symmetric.

10.11 In the previous problem it was shown that if the p-orbitals are each equally occupied, the charge distribution in the atom is spherically symmetric. A similar conclusion can be reached concerning d-orbitals: a filled or half-filled set of d-orbitals is spherically symmetric. Which of the following species has spherical symmetry: Na, Na$^+$, Al, Zn, N, F, O$^=$, Cr?

10.12 The $2s$-orbital has a node, or a region where the probability of finding the electron is zero. From the expression for $\psi(2s)$ given in Table 10.4, find the value of r in terms of a_0 at which this node occurs.

10.13 What is the difference in energy between the $1s$- and $2p$-orbitals in the hydrogen atom? In the x-ray spectrum of copper, radiation of 1.54-A wavelength is emitted when an electron changes from the $2p$- to the $1s$-orbital. What is the energy difference between these orbitals in copper?

10.14 From the table of second and third ionization energies, cite some examples showing that a half-filled set of p- or d-orbitals has a noticeable extra stability.

10.15 The following are a few elements and their characteristic x-ray wavelengths:

Mg	9.87 A	Cr	2.29 A
S	5.36 A	Zn	1.43 A
Ca	3.35 A	Rb	0.93 A

Convert these wavelengths to frequencies, and then plot the square root of the frequency as a function of the position of the element in the periodic table. Determine the constants c and b which occur in Moseley's relation between ν and Z, the atomic number. Compare your value of c with the evaluated factor $2\pi^2 me^4/h^3$, which is taken from the Bohr expression for the frequencies emitted by a one-electron atom. Here m is the electron mass in grams, and e is the electron charge in electrostatic units.

CHAPTER 11

THE CHEMICAL BOND

The existence of stable polyatomic species, whether elemental or compound, implies that atoms can act upon each other to form aggregates which have lower energy than separated fragments. When this energy lowering exceeds approximately 10 kcal/mole of atoms, we say that chemical bonds exist, since stabilization energies of this magnitude produce species which have distinct and characteristic chemical properties. The existence of structural isomers like ethyl alcohol and dimethyl ether, which both have the same molecular formula (C_2H_6O) but very different chemical and physical properties, emphasizes that the properties of a compound are dictated not only by its empirical composition, but by the way its atoms are bonded. A chemical reaction is really just a process which exchanges one bonding arrangement for another. Consequently an understanding of chemical bonding is necessary if we are to understand the chemical and physical properties of elements and compounds.

In science "understanding" means being able to predict or rationalize a variety of facts in terms of a few general principles. Before we start developing principles of chemical bonding, we should specify what facts we must explain. Perhaps the minimum requirement of a chemical bonding theory is that it show why compounds have their particular formulas. If we approach this problem historically, we encounter one of the most primitive concepts associated with chemical bonding: valence. According to the definition introduced in 1850, valence is the combining capacity of an element; the number of atoms of hydrogen or chlorine with which one atom of the element combines. By using this definition, it was possible to characterize *some* elements with a valence which aided in predicting the formulas of *some* of their compounds. But to say that sodium has a valence of one explains nothing; it is just a restatement of the fact that sodium and chlorine form a compound whose formula is NaCl. We want to know *why* this formula is NaCl and not something else. We shall find that our theory of chemical bonding can relate molecular formulas to the electronic structures of the constituent atoms. Thus the concept of valence as just defined is not really needed, and it has been largely abandoned and replaced by more specific and informative terms. The modern use of the word "valence" is not as a noun, but as an adjective meaning "associated with chemical bonding." Thus we speak of valence electrons, meaning the electrons most weakly bound to the atom which may be involved in the formation of chemical bonds.

As a second requirement, a satisfactory theory should tell us why chemical bonds form. We have already given the general answer to this question. Chemical bonds are formed because in so doing atoms can follow the universal tendency of all mechanical systems to reach the state of lowest energy. Since by forming a bond, a pair of atoms release a certain amount of energy to their surroundings, this same amount, called the bond dissociation energy, must be delivered to the molecule in order to break the bond. We expect to find in our theory of chemical bonding an explanation of how and why bond formation lowers the energy of a system of atoms. We should even hope to be able to calculate this bond dissociation energy, or at least to be able to understand its magnitude qualitatively. The calculation of bond energies has in fact been accomplished, but it is a difficult process. A qualitative rationalization of bond energies is more nearly within our grasp.

Another feature that a theory of chemical bonding should explain is the geometry of molecules. Why do carbon dioxide and water have the structures

$$O = C = O \qquad \overset{\displaystyle O}{\underset{\displaystyle H \qquad H}{\diagup \diagdown}} \; ?$$

What is it that makes one molecule linear and the other bent? We shall find that there are qualitative answers to this question.

11.1 PARAMETERS OF MOLECULAR STRUCTURE

Although the complete quantitative theory of the chemical bond involves the rigorous application of quantum mechanics, the ideas about chemical bonding which most chemists work with are primarily qualitative in nature and have been developed by trying to use the qualitative concepts of quantum mechanics to understand experimental facts. This is what we shall attempt to accomplish here. In the process of developing a simple theory of chemical bonding, the experimental facts about molecular structure have been extremely important guides. Therefore, before we approach the theory of chemical bonding, let us examine some experimentally determined values of bond energies, bond lengths, and bond angles to see whether there are any obvious regularities that will aid in constructing and understanding our theory.

Bond Energies

For a diatomic molecule, the bond dissociation energy D is the enthalpy change of the reaction in which the gaseous molecule is separated into gaseous atoms. For example,

$$H_2(g) = 2H(g), \qquad D(H\!-\!H) = \Delta H = 104 \text{ kcal/mole.}$$

Usually bond dissociation energies are given in units of kilocalories per mole of bonds broken. Table 11.1 contains a list of the bond dissociation energies for some common diatomic molecules.

Table 11.1 Dissociation energies of diatomic molecules (kcal/mole)

Li_2	25	LiH	58
Na_2	17	NaH	47
K_2	12	KH	43
Rb_2	11	RbH	39
Cs_2	10.4	CsH	42
F_2	37	HF	135
Cl_2	59	HCl	103
Br_2	46.1	HBr	87.4
I_2	36.1	HI	71.4
N_2	226	NO	150
O_2	119	CO	256
H_2	104		

Some groups of molecules have similar dissociation energies which display an obvious trend among successive members of the group. For example, consider the diatomic molecules of the alkali metals. The bond energy of each member of the series is comparatively small and decreases as the atomic number of the alkali atom increases. Among the hydrogen halides, the bond energies are comparatively large and again decrease as the atomic number of the halogen increases. In contrast to these smooth trends among chemically related molecules, there can be notable differences between molecules of atoms that are near to each other in the periodic table. The dissociation energy of O_2 is only slightly over half that of its neighbor N_2, but is more than three times as great as the dissociation energy of F_2. We shall find that there are remarkably simple explanations for some of the relations between bond energies, while others are not understood and offer us a fine chance to use our imagination.

It is also possible to define bond dissociation energy for the bonds in poly-atomic molecules. The dissociation may involve fragmenting the molecule into an atom and a group of atoms, called a radical, as in

$$H_2O(g) = H(g) + OH(g), \qquad D(H\text{—}OH) = 119.7 \text{ kcal.}$$

In other cases, the dissociation may be into two radicals:

$$HO\text{—}OH(g) = 2OH(g), \qquad D(HO\text{—}OH) = 48 \text{ kcal.}$$

Now a bond between two particular atoms such as O and H may occur in a variety of compounds, and it is interesting to see what effect these different environments have on the bond dissociation energy. We already know that $D(H\text{—}OH) = 119.7$ kcal, and other experiments show that

$$OH(g) = O(g) + H(g), \qquad D(O\text{—}H) = 101.5 \text{ kcal;}$$
$$HOOH(g) = HOO(g) + H(g), \qquad D(HOO\text{—}H) = 103 \text{ kcal.}$$

It is clear that the dissociation energy of the O—H bond is sensitive to its environment, but still the *fractional* variation in the dissociation energy usually is not very large. Variations of a similar magnitude occur in a series of C—H bonds:

$$CH_4(g) = CH_3(g) + H(g), \qquad D(H\text{—}CH_3) = 103 \text{ kcal;}$$
$$CH_3CH_3(g) = CH_3CH_2(g) + H(g), \qquad D(H\text{—}CH_2CH_3) = 96 \text{ kcal;}$$
$$(CH_3)_3CH = (CH_3)_3C(g) + H(g), \qquad D(H\text{—}C(CH_3)_3) = 90 \text{ kcal.}$$

Other C—H dissociation energies lie near or in the range between 90 and 103 kcal.

The approximate constancy of bond dissociation energies is very significant, for it suggests that the principal factors that determine the energy of a particular bond are the intrinsic properties of the two bonded atoms and are to only a lesser extent properties of the environment provided by the rest of the atoms in the molecule. Consequently we can hope to build a theory that explains most of the features of chemical bonding in terms of the properties of the bonded atoms.

Use of Bond Energies

The near constancy of the dissociation energy of a particular type of bond has an important practical consequence. It is possible to characterize the C—H bond, or any other chemical bond, by an **average bond energy** ϵ which is the *approximate* energy needed to break that bond in any compound in which it occurs. This *average* bond energy ϵ is different from the bond dissociation energy D which refers to the energy needed to break a *particular* bond in a *particular* molecule. Table 11.2 is a short list of average bond energies.

Table 11.2 Average bond energies (kcal/mole)

C—H	98.7	C—C	82.6
C—F	~110	C=C	145.8
C—Cl	80	C≡C	199.6
C—Br	69	C—O	85
C—I	55	C=O	178
C—N	80	O—H	110.6

By using average bond energies, it is possible to estimate the energy released when a gaseous molecule is formed from its gaseous atoms. For example, ΔH, the energy released at constant pressure for the reaction

$$3H(g) + C(g) + Cl(g) = CH_3Cl(g),$$

is the sum of the energies of three C—H bonds and one C—Cl bond, all taken with a negative sign because energy is released. That is,

$$\Delta H = -3\epsilon(C—H) - \epsilon(C—Cl)$$
$$= -296 - 80$$
$$= -376 \text{ kcal/mole.}$$

To find the ΔH for the formation of CH_3Cl from the elements hydrogen, chlorine, and carbon in their more usual forms, we must write

$$\tfrac{3}{2}H_2(g) + \tfrac{1}{2}Cl_2(g) + C(\text{graphite}) = CH_3Cl(g).$$

This reaction is the sum of two processes:

$$\tfrac{3}{2}H_2 + \tfrac{1}{2}Cl_2 + C(\text{graphite}) = 3H(g) + Cl(g) + C(g),$$
$$3H(g) + Cl(g) + C(g) = CH_3Cl(g).$$

We have already computed the ΔH for the second of these processes. For the first, ΔH can be expressed in terms of the H_2 and Cl_2 bond dissociation energies and the heat of vaporization of graphite to carbon atoms. That is

$$\Delta H = \tfrac{3}{2}D(H—H) + \tfrac{1}{2}D(Cl—Cl) + \Delta H_v(C)$$
$$= \tfrac{3}{2}(104) + \tfrac{1}{2}(57.9) + 170.9$$
$$= 356 \text{ kcal.}$$

Finally, for the overall reaction,

$$\tfrac{3}{2}H_2(g) + \tfrac{1}{2}Cl_2(g) + C(\text{graphite}) = CH_3Cl(g),$$
$$\Delta H = -376 + 356$$
$$= -20 \text{ kcal.}$$

The value of ΔH found by more direct calorimetry is -19.6 kcal, which differs only slightly from the result we have obtained. In some cases, there can be discrepancies of a few kilocalories between values of ΔH calculated from bond energy values and those measured calorimetrically, since the *average* bond energy ϵ is only an approximation to the true dissociation energy of a bond in a *particular* molecule. Nevertheless, bond energies do provide a very useful indication of the strengths of chemical bonds, and can be used to estimate the energetics of chemical reactions when direct calorimetric data are not available.

Bond Lengths

In molecules, atoms are always vibrating with respect to each other, so there is no single fixed distance between any pair of atoms. However, there is a well-defined *average distance* between the nuclei of two bonded atoms, and this is called the bond length or bond distance. If a substance can be obtained in crystalline form, it is possible to measure the distances between its atoms by x-ray diffraction, and many of the bond distances we shall discuss have been obtained from x-ray data.

Table 11.3 Bond lengths for some diatomic molecules (angstroms)

F_2	1.42	HF	0.92
Cl_2	1.99	HCl	1.27
Br_2	2.28	HBr	1.41
I_2	2.67	HI	1.61
ClF	1.63	H_2	0.74
BrCl	2.14	N_2	1.094
BrF	1.76	O_2	1.207
ICl	2.32	NO	1.151
		CO	1.128

If a substance does not crystallize conveniently, there are other techniques available to measure its bond distances. The most important of these is molecular spectroscopy. Just as the spectrum of the hydrogen atom is determined by the mechanics of the electron-nucleus system, the spectrum of a molecule is determined by the mechanics of its several nuclei and electrons. By analyzing molecular spectra, it is possible to locate very accurately all the nuclei in a molecule relative to one another and thus to obtain a very detailed picture of the nature of the structure of a molecule. Subsequently we shall indicate some of the details of how molecular spectra provide structural information. For the present, however, we shall concentrate on the results of such measurements.

Table 11.3 lists bond distances for several common diatomic molecules. Note that in a related series of molecules such as the halogens or the hydrogen halides, the bond distance increases with increasing atomic number. Such a trend has a reasonable qualitative explanation if we recognize that the bond length is the position of greatest stability, or of minimum energy for a pair of

Table 11.4 Variation of O—H, C—C, and C—H bond lengths (angstroms)

Bond	Molecule	Bond length
O—H	Water, H_2O	0.96
O—H	Hydrogen peroxide, H_2O_2	0.97
O—H	Methanol, CH_3OH	0.96
O—H	Formic acid, CHOOH	0.96
O—H	Hydroxyl radical, OH	0.97
C—C	Diamond	1.54
C—C	Ethane, C_2H_6	1.54
C—C	Propane, C_3H_8	1.54
C—C	Ethanol, C_2H_5OH	1.55
C—C	Neopentane, $(CH_3)_3CH$	1.54
C—H	Methane, CH_4	1.095
C—H	Ethane, C_2H_6	1.095
C—H	Ethylene, C_2H_4	1.087

atoms. The energy lowering associated with bond formation has its origin in the way the valence electrons of the bonded atoms behave, and is opposed by electrostatic repulsions between the two nuclei and between the inner electron shells of the two atoms. The strength of both these sources of repulsion increases as the atoms are brought together. Now as atomic number increases in one column of the periodic table, the valence electrons lie at successively greater distances from the nuclei. Also, the repulsion between two nuclei must increase as their charges increase. Consequently the distance at which the energy of bonded atoms is a minimum, or the distance at which the bond is strongest, tends to increase as atomic number increases.

In discussing bond energies we found that the dissociation energy of a particular type of bond was *largely* independent of the molecule in which the bond occurred. Let us see whether the same is true for bond distances. Table 11.4 gives a comparison of the O—H, C—C, and C—H bond lengths in various compounds. The constancy of each of these bond lengths is remarkable, and this supports our earlier hypothesis that the properties of a bond are largely determined by the nature of the bonded atoms.

We must now admit that it is possible to find bonds between the same pair of atoms that have quite different lengths and energies in various compounds. Consider the data in Table 11.5. In the compounds ethane, ethylene, and acetylene there is considerable variation in the length and energy of the carbon-carbon bond. Rather than look upon this as a violation of the idea that bond properties are independent of molecular environment, it is profitable to take the bonds in these compounds as representative of *three different types* of carbon-carbon bond. In the first type, the carbon atoms are each bonded to a total of four atoms, in the second, each carbon atom is bonded to a total of three atoms, and in the third, each is bonded to two atoms. When more data are

Table 11.5 Variations in bond lengths and energies

Bond	Molecule	Bond length, (A)	Bond energy (kcal/mole)
C—C	Ethane, H_3CCH_3	1.54	83
	Ethylene, H_2CCH_2	1.34	146
	Acetylene, HCCH	1.20	200
O—O	Hydrogen peroxide, H_2O_2	1.48	48
	O_2^- in BaO_2	1.49	
	O_2^- in KO_2	1.28	
	O_2	1.21	118
	O_2^+	1.12	150

examined, it is found that the length and energy of each particular *type* of carbon-carbon bond are approximately constant in a variety of compounds. Thus our rule of the invariance of bond properties is preserved, and in fact it has led us to a discovery that two atoms may be bonded together in more than one way. We shall see later in the chapter how the existence of different bond types can be explained.

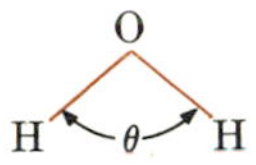

The bond angle θ is the internal angle between lines drawn through the nuclei of the bonded atoms.

FIG. 11.1

Bond Angles

Figure 11.1 shows that a bond angle is the internal angle of intersection between two lines drawn through the nucleus of a central atom from the nuclei of two atoms bonded to it. Because atoms are in constant vibration, there is no definite fixed value for a bond angle just as there is no fixed bond length. However, the average angle about which the three atoms vibrate is well defined, and it is this to which the term bond angle refers. Bond angles, like bond lengths, are determined principally from x-ray diffraction measurements and molecular spectroscopy.

Table 11.6 The bond angle about the oxygen atom

	Molecule	∠ X—O—Y (degrees)
H_2O	Water	104.5
F_2O	Oxygen difluoride	103.2
Cl_2O	Oxygen dichloride	111
$(CH_3)_2O$	Dimethyl ether	111
CH_3OH	Methanol	109

Table 11.6 gives the bond angle about the oxygen atom in a number of its compounds. While there is some variation, most of the angles lie in the range from 104 to 111 degrees. Similarly, Table 11.7 shows that the H—C—H bond angle is near 110° in several simple carbon compounds. As we continue the study of descriptive chemistry, we shall find that such regularities in the geometries of molecules of a given element are very common, and consequently it is of great interest to explain why these regularities occur.

Table 11.7 Variation of the
H—C—H bond angle

Molecule	∠ H—C—H (degrees)
CH_4	109.5
CH_3Cl	110.5
CH_2Cl_2	112.0
CH_3Br	111.2
CH_3I	111.4
CH_3OH	109.3
C_2H_6	109.3

When we examine the bond angles in similar compounds of successive members of a group in the periodic table, some more striking resemblances appear. Table 11.8 gives the bond angles of the hydrides of Groups IV, V, and VI. We find that all the hydrides in Group IV have bond angles of 109.5°, which is called the tetrahedral angle because it corresponds to having the atoms located at the apices of a regular tetrahedron. In Group V, all the hydrides have the structure of a regular trigonal pyramid. The bond angle about the central atom is 107° for NH_3 and decreases to 91° for SbH_3. In Group VI, there is a similar trend as the angle decreases from 104° in water to 89° in H_2Te.

The bond-angle data we have considered point to the fact that the bond angles about a central atom are determined largely by the properties of that atom alone. This idea is consistent with the *approximate* constancy of the bond

Table 11.8 Bond angles for several hydrides

	∠ H—X—H (degrees)		∠ H—X—H (degrees)		∠ H—X—H (degrees)
CH_4	109.5	NH_3	107.3	H_2O	104.5
SiH_4	109.5	PH_3	93.3	H_2S	92.2
GeH_4	109.5	AsH_3	91.8	H_2Se	91.0
SnH_4	109.5	SbH_3	91.3	H_2Te	89.5

angle about a given atom when different groups are bonded to it. It is also consistent with the fact that similar compounds of atoms in a given group of the periodic table have very similar geometries. In particular, this latter observation suggests that the bond angles about an atom are largely determined by its *number* of valence electrons, for number of valence electrons is the most obvious property that members of the same periodic group have. This is a simple idea that can be expanded to a satisfactory explanation of the relations between bond angles.

Molecular Spectroscopy

Molecules, like atoms, have quantized electronic-energy levels. In addition, however, molecules have energy levels that are associated with the motions of their atoms *relative* to each other. These internal or relative modes of motion usually can be divided into two groups: *vibrational* motions associated with the stretching and bending of bonds, and *rotational* motions, in which the molecule tumbles in space. To a good first approximation, the total energy of a molecule (apart from its translational energy) can be written as

$$E = E_{\text{electronic}} + E_{\text{vibrational}} + E_{\text{rotational}}.$$

Each of these modes of motion is quantized, and thus the total energy of a molecule assumes only certain definite values.

The spacing of the electronic-energy levels is determined by the nature of the molecular electronic orbitals that are filled and unfilled, while the stiffness or vibration frequency of the bonds determines the spacing of the vibrational-energy levels. The masses of the atoms and the distances between them determine the difference in energy between rotational levels. An experimental determination of the spacing of the various energy levels allows one to calculate what the bond distances, bond angles, and vibration frequencies of the molecule are, and what its molecular orbital energy pattern is. Molecular spectroscopy is the study of these energy levels through the interaction of molecules with light.

In subsequent sections we shall discuss the molecular orbitals of diatomic and polyatomic molecules and the energy-level patterns in which they fall. For the present, let us concentrate on the vibrational and rotational levels of molecules which are in their most stable electronic state. These internal modes of motion for a diatomic molecule are illustrated in Fig. 11.2. We see that there is only one vibrational mode of motion, the stretching of the chemical bond. There are two rotational motions which are identical, except that the planes in which the two rotations occur are mutually perpendicular. The expressions for the rotational- and vibrational-energy levels are in this case quite simple, as we shall now see.

Just as the translational kinetic energy of a molecule can be expressed in classical mechanics as the square of the linear momentum divided by two times its mass,

$$E_{\text{trans}} = \frac{p_{\text{linear}}^2}{2m},$$

the classical rotational energy can be written as the square of its rotational or *angular* momentum divided by two times its moment of inertia I:

$$E_{\text{rot}} = \frac{p_{\text{angular}}^2}{2I}.$$

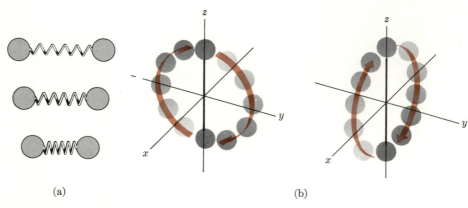

(a)   (b)

FIG. 11.2   The internal modes of motion of a diatomic molecule. (a) Vibration of the atoms along their line of centers. (b) Rotation about their center of mass.

The moment of inertia of a diatomic molecule consisting of atoms of masses m_1 and m_2 and bond distance r_0 is given by

$$I = \frac{m_1 m_2}{m_1 + m_2} \, r_0^2 \equiv \widetilde{m} r_0^2,$$

where $\widetilde{m}$ is the so-called *reduced mass* $(m_1 m_2)/(m_1 + m_2)$. This classical formula can be turned into the correct quantum expression if we replace the classical angular momentum by the values allowed by quantum mechanics:

$$p_{\text{angular}} = \sqrt{J(J+1)}\,\hbar, \qquad J = 0, 1, 2, 3, \ldots$$

We then get

$$E_{\text{rot}} = \frac{J(J+1)\hbar^2}{2I}, \qquad J = 0, 1, 2, 3, \ldots$$

for the allowed rotational energy levels. The absorption of a photon of appropriate energy $h\nu$ can cause the rotational quantum number of a molecule to increase by one. The energy change of the molecule is

$$\Delta E_{\text{rot}} = h\nu = \frac{\hbar^2}{2I} [J(J+1) - J'(J'+1)].$$

If $J' = J - 1$, we have

$$\Delta E_{\text{rot}} = h\nu = \frac{\hbar^2}{2I} [J(J+1) - (J-1)J]$$

$$= \frac{\hbar^2}{2I} [2J] = \frac{\hbar^2}{I} J.$$

Therefore, if the energy of the photon absorbed is known, and the initial quantum number J is known, the moment of inertia I and bond distance r_0 can be calculated.

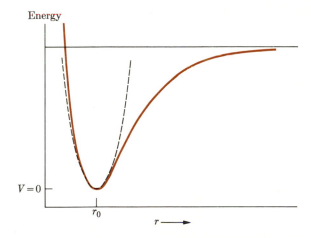

The potential energy curve for the nuclear motion in a diatomic molecule. Near the minimum, the parabola $V = \frac{1}{2}k(r - r_0)^2$ closely approximates the true curve.

FIG. 11.3

Let us now consider the vibrational energy-level pattern. Figure 11.3 shows how the potential energy of a diatomic molecule varies as the bond is stretched or compressed from r_0, the position of lowest energy. Near the minimum, the true potential-energy curve can be fairly closely approximated by a parabola whose algebraic expression is

$$V = \frac{1}{2}k(r - r_0)^2 = \frac{1}{2}k(\Delta r)^2.$$

The quantity k is called the *force constant* of the bond. The origin of this term becomes clear if we find, by differentiation, the force exerted by the atoms. Since, in general, force is the negative derivative of a potential energy,

$$F = -\frac{dV}{d(\Delta r)} = -\frac{d}{d(\Delta r)}\,[\tfrac{1}{2}k(\Delta r)^2]$$
$$= -k(\Delta r).$$

Thus, k is the proportionality constant between force and displacement. The minus sign indicates that the force exerted by the atoms is in the opposite direction from their displacement, and tends to restore them to their most stable condition.

The force constant k has another interpretation. The first derivative of the potential energy V is the slope of the V–Δr curve:

$$\frac{dV}{d(\Delta r)} = \frac{d}{d(\Delta r)}\,[\tfrac{1}{2}k(\Delta r)^2] = k(\Delta r).$$

The second derivative $d^2V/d(\Delta r)^2$ is the curvature of V:

$$\frac{d^2V}{d(\Delta r)^2} = \frac{d}{d(\Delta r)}\,k(\Delta r) = k.$$

Thus k is the curvature of the potential-energy curve near its minimum, where the parabola is a good approximation to the true curve.

According to classical mechanics, a system with the potential energy $\tfrac{1}{2}k(\Delta r)^2$ will oscillate sinusoidally or harmonically with a frequency ν given by

$$\nu = \frac{1}{2\pi}\,(k/\widetilde{m})^{1/2},$$

where k is the force constant, and $\widetilde{m}$ is the reduced mass $m_1 m_2/(m_1 + m_2)$. According to quantum mechanics, such an oscillator has quantized energy levels given by

$$E_{\text{vib}} = (v + \tfrac{1}{2})h\nu, \qquad v = 0, 1, 2, 3, \ldots,$$

where v is the (integral) vibrational quantum number. Thus the larger k is, the stiffer is the bond, and the larger is the spacing between vibrational levels. It is clear that determination of the vibrational frequency by molecular spectroscopy allows us to learn the general shape of the molecular potential-energy curve near the most stable position.

We see now that the spacing of the vibrational- and rotational-energy levels is determined by the stiffness and length of the bond, as well as the atomic masses. The bond distance and force constant are themselves determined by the electronic structure of the molecule. If the molecule becomes electronically excited, it is very likely that the values of k and r_0 in the excited electronic state will be significantly different from their values in the ground electronic state. By careful analysis of spectra in which electronic excitation occurs, it is possible to learn the values of k and r_0 for both electronic states. This is how the geometric and bonding properties of excited electronic states are determined.

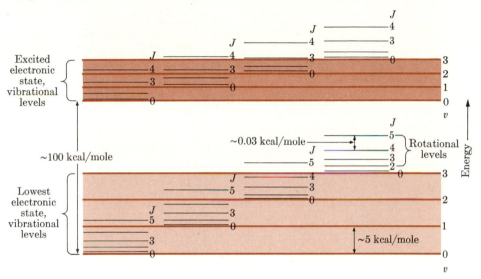

Energy levels of a diatomic molecule. Two electronic states are shown, each with a few of its vibrational and rotational levels. The level spacings are not drawn to scale, but typical energy differences are indicated.

FIG. 11.4

Figure 11.4 is a graphical summary of our discussion. The electronic energy levels of diatomic molecules are widely separated, often by as much as 100 kcal/mole. The vibrational-energy spacings for each electronic state are frequently between 1 and 7 kcal/mole. Rotational-energy levels are much more closely spaced, differing in energy by less than approximately 0.03 kcal.

Transitions between rotational-, vibrational-, and electronic-energy levels involve quite different amounts of energy, and therefore, are induced by photons of quite different frequencies. Figure 11.5 shows the relationship between the frequency and wavelength of photons and the molecular energy changes they induce. Because rotational-energy levels are closely spaced, transitions between them involve low-energy photons that are produced in the radio-frequency region of the spectrum, and have associated wavelengths of about 1 cm. Vibrational changes are induced by photons from the infrared region, whose associated wavelengths are 10^{-4} to 10^{-3} cm. Photons that produce electronic transitions have wavelengths of approximately 5000 A (5×10^{-5} cm) or shorter, and lie, therefore, in the visible or ultraviolet regions of the spectrum.

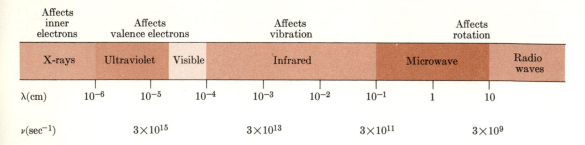

FIG. 11.5 The regions of the electromagnetic spectrum and the molecular energy levels they affect.

The spectra of polyatomic molecules are somewhat similar to, but much more complicated than, diatomic molecule spectra. A polyatomic molecule may have three different moments of inertia, and the expressions for its rotational-energy levels can be quite complex. Nevertheless, bond distances and bond angles can be derived from polyatomic molecular spectra. The vibrational spectra of polyatomic molecules are simplified somewhat by the fact that certain molecular groups, like C—H, C=O, C=C, —NO$_2$, etc., have very characteristic vibrational frequencies. The presence of such groups in a molecule of unknown structure can be inferred from the vibrational spectrum; and consequently the infrared spectrum of a molecule is an extremely valuable tool for the determination of molecular structure and for chemical analysis.

We have not mentioned the molecular energy levels which arise from the interaction of the nuclear spin or electron spin with external magnetic fields, and the interaction of the nuclear asymmetry (quadrupole) with the electric field produced by the molecular electrons. Study of these levels reveals some rather subtle features of the distribution of electrons in a molecule. From these, and the foregoing considerations, it is easy to see why the general area of molecular spectroscopy is such an active field of research for chemists.

11.2 IONIC BONDS

The variety of chemical formulas, bond energies, and molecular geometries suggests that a detailed theory of chemical bonding should be very complex. This is certainly true. Consequently, in discussing chemical bonds it is common to use "models," or conceptual pictures which sacrifice some accuracy to gain considerable simplicity. Accordingly, our discussions will involve two different bond types: the ionic bond and the covalent bond. There are a few situations in which either of these extremes is found, but the true value of these two models is that most chemical bonds have properties which are intermediate but *close* to one or the other. Consequently we will be able to construct explanations of most bonding phenomena in terms of these two extreme bond types.

In the ionic-bond model we imagine the particles that are bonded to be spherical entities possessing a net positive or negative charge. Now it is a fundamental result of electrostatic theory that a spherical distribution of charge behaves as though the net charge were concentrated at the center of the sphere. Consequently the major simplification of the ionic-bond model is that we can calculate the electrostatic forces acting between ions by using Coulomb's law, just as though the ions themselves were point charges. This approach is not exact, and must be refined slightly, but is a good example of how a slight simplification can lead to some very useful results.

Ionic bonding is found in the compounds of very electropositive elements, such as the alkali metals, and with very electronegative elements, such as the halogens. How do we know that a compound such as solid sodium chloride consists of a lattice of positive and negative ions? The fact that the fused salt and its aqueous solutions conduct electricity surely is not proof that ions exist in the crystalline compound. The best independent evidence for the presence of ions comes from spectroscopic investigations which show that the chlorine nucleus is surrounded by a complete octet of valence electrons; this is surely consistent with the existence of a chloride ion. Supporting evidence comes from careful x-ray studies which measure the density of electrons at all points in the crystal. It is found that sodium chloride is made up of spherical groups of 10 and 18 electrons which correspond to the sodium and chloride ions respectively. Thus there is no question that ions exist, and that the cohesive forces in the crystal are due to the mutual attraction of the oppositely charged species.

Ionic Lattice Energies

Let us think about the energetics of the formation of crystalline sodium chloride from the elements. The reaction

$$Na(s) + \tfrac{1}{2}Cl_2(g) = NaCl(s)$$

is really a very complicated process, since it involves the destruction of the bonds both in the chlorine molecules and in metallic sodium, as well as the formation of the sodium-chloride crystal lattice. Thus the energy released when this reaction occurs, a quantity that measures the stability of NaCl relative to metallic sodium and gaseous chlorine, is determined not only by the properties of sodium chloride itself, but by the strength of the bonding in metallic sodium and molecular chlorine as well.

We can see this even more clearly by referring to Fig. 11.6. There we find two processes for the formation of sodium chloride from its elements. One is the direct conversion of elements to compound, the other is a hypothetical three-step sequence that accomplishes the same thing. If energy is conserved, the energy released in the direct one-step conversion of elements to the compound must be the same as is released by following the alternative three-step process.

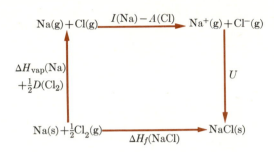

FIG. 11.6   Alternative paths for the formation of solid sodium chloride from its elements.

The first step in this latter path consists of vaporizing the metallic sodium and dissociating the chlorine to atoms. To carry out this step we would have to put energy into the system, and the amount required per mole of each kind of atom is equal to the sum of the enthalpy of vaporization of sodium, ΔH_{vap}, and one-half the bond dissociation energy of chlorine, $\frac{1}{2}D(\text{Cl}-\text{Cl})$. Thus the energetics of this first step show how the stability of metallic sodium and chlorine molecules affects the energy released by the overall reaction.

In the second step the gaseous atoms are converted to gaseous ions. To do this requires that we put into the system an amount of energy equal to the ionization energy of sodium, $I(\text{Na})$, but we get back an amount equal to the electron affinity of chlorine. The *net* energy required is then $I(\text{Na}) - A(\text{Cl})$.

Finally in the third step the gaseous ions condense to the sodium-chloride crystal lattice. The energy released by the system in this step is called the **ionic crystal lattice energy** and is a direct measure of the stability of the ionic crystal.

Now let us examine the formation of crystalline sodium chloride in more detail. As our starting or reference point we choose the gaseous sodium and chlorine *atoms*. By making this choice, we will be able to concentrate entirely on the energetic factors that influence the strength of the ionic bond and we will not have to worry about the bonding in metallic sodium or molecular chlorine. In order to expose the factors which control the strength of ionic bonds, we shall imagine that the formation of the sodium chloride crystal is carried out in three steps: the first is the formation of the isolated gaseous ions from the atoms, the second is the formation of the gaseous sodium-chloride diatomic molecule from the ions, and the third is the formation of the ionic crystal.

As we have remarked, to form a gaseous sodium ion and an electron from a gaseous atom we must supply an amount of energy equal to the ionization energy of sodium, 118.4 kcal/mole. However, by transferring this electron to a chlorine atom we *obtain* an amount of energy equal to the electron affinity of chlorine, 83.4 kcal/mole. Consequently, the net energy required to form the ions from the gaseous atoms is given by

$$
\begin{array}{lll}
Na = Na^+ + e^- & \Delta E = I(Na) & = 118.4 \\
\underline{e^- + Cl = Cl^-} & \underline{\Delta E = -A(Cl)} & = -83.4 \\
Na + Cl = Na^+(g) + Cl^-(g) & \Delta E = & 35.0
\end{array}
$$

To form the ions from the atoms *requires* 35 kcal/mole of ion pairs. This result by itself is very significant. One "explanation" of chemical bonding which has been offered is that bonds are formed because atoms have a "desire" to form octets of valence electrons. Our calculation shows that 35 kcal are *required* if sodium and chlorine are to do nothing more than form completed valence octets. Thus the atoms have no mutual "urge" simply to reach the octet structure, and we must look further for a more concrete reason for bond formation.

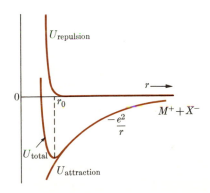

Variation with distance of the potential energy of oppositely charged ions.

FIG. 11.7

We will consider the formation of a gaseous sodium-chloride molecule from the ions. Coulomb's law of electrostatics shows that bringing two opposite charges of magnitude e from infinity to a distance r of each other lowers their potential energy by an amount $-e^2/r$. This "attractive" potential energy, plotted as a function of distance, is shown in Fig. 11.7. When the two ions are quite close together, their outermost electrons start to occupy the same space, and a strong repulsive force develops. Correspondingly, a "repulsive" potential energy rises abruptly. The sum of the attractive and repulsive contributions gives the net potential energy of the two ions, as is also shown in Fig. 11.7. The distance r_0 at which the potential energy is a minimum is the distance at which the ions would come to rest if all their kinetic energy were removed, and is the equilibrium bond distance.

From the shape of the net potential-energy curve, we can see that the potential energy of the ions at their equilibrium separation is closely approximated by the attractive contribution, $-e^2/r_0$, alone. The gaseous diatomic sodium-chloride molecule, produced by evaporation of solid sodium chloride, has been studied and its internuclear separation is found to be 2.38 A. Thus the Coulomb potential energy is

$$-\frac{e^2}{r_0} = \frac{(4.8 \times 10^{-10}\, \text{esu})^2}{2.38 \times 10^{-8}\, \text{cm}} = -9.68 \times 10^{-12}\, \text{ergs}$$

$$= -139.3\, \text{kcal/mole}.$$

We see that to this approximation the energy of a mole of diatomic molecules is 139.3 kcal lower than that of the separated ions. To find the energy released when the molecule is formed from the gaseous *atoms*, we need only combine the three steps:

$$
\begin{array}{ll}
\text{Na(g)} = \text{Na}^+\text{(g)} + \text{e}^- & \Delta E = 118.4\ \text{kcal/mole} \\
\text{e}^- + \text{Cl(g)} = \text{Cl}^-\text{(g)} & \Delta E = -83.4 \\
\underline{\text{Na}^+\text{(g)} + \text{Cl}^-\text{(g)} = \text{NaCl(g)}} & \underline{\Delta E = -139.3} \\
\text{Na(g)} + \text{Cl(g)} = \text{NaCl(g)} & \Delta E = -104.3\ \text{kcal}
\end{array}
$$

The fact that the energy change is negative shows that the energy of the ionically bonded sodium and chlorine is *lower* than that of the separated atoms. The reason for the stability of the molecule is clear. Even though some investment of energy is required to form the ions, this is more than compensated by the energy due to their mutual Coulomb attraction.

FIG. 11.8   Part of a one-dimensional sodium chloride "crystal."

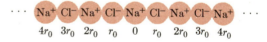

While it is possible to obtain and study the gaseous diatomic molecules of the alkali halides at high temperature, these ionic compounds are solids with extremely low vapor pressures at room temperature. It is not difficult to understand this if we examine the simplified model of the solid state shown in Fig. 11.8. This is a hypothetical one-dimensional "crystal" consisting of alternate sodium and chloride ions. To find the energy of formation evolved when such a crystal is formed from the gaseous ions, we must calculate the Coulomb potential energy of one sodium ion as it is acted upon by all other ions in the crystal.

To begin, the two neighboring chloride ions located at a distance r_0 contribute $-2e^2/r_0$ to the potential energy, while the two nearest sodium ions give $+2e^2/2r_0$. The positive sign arises because of the repulsion between like charges on the sodium ions. Continuing this procedure for all ions in the crystal gives the potential energy as a sum of an infinite number of terms, which we write as

$$U = -\frac{2e^2}{r_0} + \frac{2e^2}{2r_0} - \frac{2e^2}{3r_0} + \cdots$$

$$= -\frac{2e^2}{r_0}\left[\left(1 - \frac{1}{2}\right) + \left(\frac{1}{3} - \frac{1}{4}\right) + \left(\frac{1}{5} - \frac{1}{6}\right) + \cdots\right].$$

Since each term in the brackets is a positive number, the value of the bracket must be greater than $\frac{1}{2}$, the value of the first term. Therefore, U, the potential energy, must be more negative than $-e^2/r_0$. Consequently, the potential energy of a sodium ion in this one-dimensional crystal is lower than it is in the diatomic molecule.

The calculation we have outlined can be extended to real three-dimensional crystals, and the sum of the infinite series evaluated. The attractive Coulomb energy of any ionic lattice can be expressed as $-Me^2/r_0$, where M, called the Madelung constant, depends on the geometric arrangement of the ions. For the sodium-chloride crystal lattice, M is 1.75; thus if all other things are constant, the ionic solid has a 75% lower energy than the gaseous diatomic molecule. This extra energy lowering occurs because a sodium ion in the solid is bonded through Coulomb forces to *all chloride ions in the crystal*. The separation of the ions in the sodium-chloride crystal is 2.80 A, somewhat larger than that found in the diatomic molecule. Consequently the numerical value of the Coulomb energy for the sodium-chloride lattice is

$$U = -1.75\,\frac{e^2}{r_0} = -1.75\,\frac{(4.80 \times 10^{-10})^2}{2.80 \times 10^{-8}}$$

$$= -1.44 \times 10^{-11} \text{ ergs}$$

$$= -207 \text{ kcal/mole.}$$

Strictly, the lattice energy U is the energy evolved when the solid is formed from its gaseous ions. We have evaluated only the contribution due to Coulomb forces between the ions considered as point charges. Because of the finite size of the electron cloud around each ion, there exist repulsive forces between neighboring ions that we have not included in our calculation. The effect of these repulsions is to make the actual lattice energy about 10% less negative than the value given by the Coulomb forces alone. Thus the true lattice energy of sodium chloride is -183 kcal/mole. Comparing this number with -139.5 kcal/mole, the Coulomb energy calculated for the diatomic molecule, leaves no question that the crystal is more stable. This is the reason that sodium chloride is a solid with a very low vapor pressure at room temperature.

Now that we have a value for the energy evolved when the crystal lattice is formed from the gaseous *ions*, we can find how much energy is released when solid sodium chloride is formed from gaseous sodium and chloride atoms. We have

$$\begin{array}{lll} Na(g) + Cl(g) \rightarrow Na^+(g) + Cl^-(g) & \Delta E = & 35 \text{ kcal} \\ \underline{Na^+(g) + Cl^-(g) \rightarrow NaCl(s)} & \underline{\Delta E = -183 \text{ kcal}} \\ Na(g) + Cl(g) \rightarrow NaCl(s) & \Delta E = -148 \text{ kcal} \end{array}$$

Since ΔE for the overall process is negative, energy is evolved when solid sodium chloride is formed from its gaseous atoms. Consequently we can say that crystalline sodium chloride is more stable than its gaseous atoms, and the source of this stability is Coulomb attraction of the ions in the crystal lattice.

Now that we have completed our analysis of the energetics of formation of sodium chloride from its atoms, it is interesting to see how the ionic lattice energy is calculated when the repulsion between neighboring ions is taken into account. As we have remarked, this repulsion is a result of the finite size of the ions and is of the same nature as the van der Waals repulsions between neutral atoms. The refined expression for the potential energy of the crystal is now

$$U = -\frac{Me^2}{r} + \frac{B}{r^n}.$$

The term B/r^n represents the repulsions between neighboring atoms, and the value of n ranges from 9 to 12, depending on the types of ions in the crystal. Often n can be evaluated by studying the compressibility of the crystal. The coefficient B measures the strength of the repulsive forces and differs for different ions. However, B can be eliminated from the expression for the lattice energy by recognizing that at the value of r for which the crystal is most stable, U is a minimum and $dU/dr = 0$. Therefore we carry out the differentiation of U:

$$\frac{dU}{dr} = \frac{Me^2}{r^2} - \frac{nB}{r^{n+1}}.$$

Setting this expression equal to zero we get

$$B = \frac{Me^2}{n} r_0^{n-1},$$

where r_0 is the distance at which dU/dr vanishes. Substituting the expression for B back into the equation for U gives us

$$U = -\frac{Me^2}{r_0} + \frac{M}{n} \frac{e^2}{r_0}$$
$$= -\frac{Me^2}{r_0} \left(1 - \frac{1}{n}\right).$$

This is the potential energy evaluated at the most stable ionic distance, r_0. Since n is approximately 10, we see that the true lattice energy is only about 10% different from the energy of Coulomb interaction alone.

In general there is very satisfactory agreement between the calculated and measured values of the lattice energies of the alkali-metal halides. Having accounted for the bond energy of these substances, we can now try to explain their formulas. We shall suppose that the formulas found have the lowest energy of all those that are possible. Possible ionic compounds are those that are stable with respect to their gaseous *atoms*; this requires that the Coulomb potential energy *obtained* upon formation of the crystal lattice be larger than the energy which must be *expended* to form the gaseous ions from the atoms.

To see how these two energies depend on the ionic charge, let us examine the formation of solid calcium oxide from its gaseous atoms. The reactions and associated energies necessary to form the gaseous ions are

$$
\begin{aligned}
Ca(g) &= Ca^+(g) + e^- & \Delta E &= 141 \text{ kcal/mole} \\
Ca^+(g) &= Ca^{++}(g) + e^- & \Delta E &= 273 \\
O(g) + e^- &= O^-(g) & \Delta E &= -34 \\
\underline{O^-(g) + e^- = O^=(g)} & & \underline{\Delta E} &= \underline{210} \\
Ca(g) + O(g) &= Ca^{++}(g) + O^=(g) & \Delta E &= 590 \text{ kcal/mole}
\end{aligned}
$$

Compared with the analogous process for sodium chloride, an enormous energy investment is required. Considerable energy is needed to remove both electrons from the calcium atom. Moreover, note that a net of 176 kcal/mole is *required* to attach both electrons to the oxygen atom. Unlike the chloride ion, gaseous $O^=$ is unstable with respect to the loss of an electron.

Now let us see whether the lattice energy of calcium oxide can compensate for our expenditure. The Madelung constant for CaO is 1.75, and the interionic distance is 2.4 A. Since there is a charge of magnitude $2e$ on each ion, the lattice energy is approximately

$$
U = -1.75 \frac{(2e)^2}{r_0}
$$

$$
= -1.75 \times \frac{4 \times 23 \times 10^{-20}}{2.4 \times 10^{-8}} = -6.7 \times 10^{-11} \text{ ergs}
$$

$$
= -967 \text{ kcal/mole.}
$$

Thus when one mole of solid calcium oxide is formed from the atoms, the energy change is $590 - 967 = -377$ kcal. This is a considerable overestimate, since we have neglected the repulsion of neighboring ions. The correct value of U is -862 kcal/mole. Although our calculation is crude, it shows that the lattice energy increases as the product of the charges on the two ions. This suggests that in ionic compounds, each atom would try to assume as great a charge as possible. This tendency toward a high charge is limited by the increasing values of the successive ionization energies of the atoms. For instance, reference to Table 10.6 shows that to form gaseous Na^{++} requires an amount of energy equal to the sum of the first and second ionization energies of sodium, or $118 + 1091 = 1209$ kcal/mole. To form $O^=$ requires 176 kcal more. Our calculations for calcium oxide suggest that a lattice of Na^{++} and $O^=$ ions could not supply more than 1000 kcal/mole lattice energy. Hence NaO cannot be formed: the energy required to make the gaseous ions is greater than the expected lattice energy of the crystal. The same conclusion can be reached for all the alkali metals. The second ionization energy of these elements is so large that the characteristic charge of the alkali-metal ions is never more than plus one. The alkali-earth metals have two valence electrons which can be removed without excessive energy expenditure; therefore they can assume a charge of $+2$ and form extremely stable ionic compounds.

Just as the maximum positive charge which can be assumed by an element is limited by its number of valence electrons, the maximum negative charge of an ion is set by the number of vacancies in low-energy valence orbitals. For example, all the halogen atoms can accept one electron to fill their valence p-orbitals. There are then no more low-energy orbitals available to accept additional electrons. The oxygen and sulfur atoms have two vacancies in their valence p-orbitals. Both these atoms readily accept one additional electron, but as we have noted above, $O^=$ and $S^=$ are unstable with respect to loss of an electron when they are isolated. However, the doubly negative ions exist in ionic crystals where the extra lattice energy from the double charge compensates for the instability of the isolated ion.

A possibility for formation of a triply negative ion exists for nitrogen and phosphorus, which have three half-filled valence p-orbitals. In some compounds, such as Mg_3N_2 and Mg_3P_2, the nitride and phosphide ions, N^{-3} and P^{-3}, are believed to exist. Judging from the relative instability of $O^=$ and $S^=$, it is not surprising that examples of triply negative ions are rare. In general, then, the number of valence orbital vacancies appears to limit the number of electrons which an atom can acquire, and the more highly charged negative ions tend to be unstable.

Crystal Lattice Geometry

We have called attention to the abrupt rise in the potential energy that occurs when two ions are brought closer than their equilibrium separation r_0, and have attributed this rise to the overlapping and repulsion of the outermost electrons of the ions. We would expect then that the distance at which the repulsion becomes important should be determined in large measure by the extension of the electron "cloud" that surrounds each ion. Of course, this charge cloud extends to infinity, but the electron density at any distance from a particular isolated ion is determined by the electronic structure of that ion. This in turn suggests that we might try to characterize each ion by a radius, and hope that the internuclear separation in any ionic compound could be calculated to a good approximation by adding the radii of the positive and negative ions. This scheme is in fact workable.

Table 11.9 Ionic radii (angstroms)

		Li$^+$ 0.68	Be^{++} 0.30		
O$^=$ 1.45	F$^-$ 1.33	Na$^+$ 0.98	Mg^{++} 0.65		Al^{+3} 0.45
S$^=$ 1.90	Cl$^-$ 1.81	K$^+$ 1.33	Ca^{++} 0.94	Sc^{+3} 0.81	Ga^{+3} 0.60
Se$^=$ 2.02	Br$^-$ 1.96	Rb$^+$ 1.48	Sr^{++} 1.10	Y^{+3} 0.90	In^{+3} 0.81
Te$^=$ 2.22	I$^-$ 2.19	Cs$^+$ 1.67	Ba^{++} 1.29	La^{+3} 1.06	Tl^{+3} 0.91

From x-ray measurements on a crystal it is possible to determine the separation of the nuclei of adjacent ions. In order to assign ionic radii, one must find some basis for dividing the observed internuclear separation for sodium chloride, for example, into a contribution from Na$^+$ and a contribution from Cl$^-$. We will not explore the detailed basis for this division, but merely say that from a few observed internuclear distances it has been possible to generate the set of ionic radii shown in Table 11.9. These ionic radii should not be taken to mean that the electronic cloud does not extend beyond a certain distance; the only quantitative significance of ionic radii is that when they are added together they give a good approximation to the observed interionic spacing. As the comparison in Table 11.10 shows, the agreement between calculated and observed internuclear separation is good, but not exact.

Table 11.10 Interionic distances in some alkali-halide crystals (angstroms)

	Li$^+$	Na$^+$	K$^+$	Rb$^+$	Cs$^+$
Radius sum Cl$^-$	2.49	2.79	3.14	3.29	3.48
Observed distance	2.57	2.81	3.14	3.29	3.47
Radius sum I$^-$	2.87	3.17	3.52	3.67	3.86
Observed distance	3.02	3.23	3.53	3.66	3.83

Examination of Table 11.9 shows that the trends in ionic size are what we might expect on the basis of the electronic structures of the ions. In each column of the periodic table the ionic radius increases with atomic number. This is consistent with the fact that electrons of successively higher principal quantum number are found, on the average, at successively greater distances from the nucleus. Furthermore, comparison of positive and negative ions which have the same number of electrons, for example, Na^+ and F^-, K^+ and Cl^-, shows that the negative ion is always the larger. The reason for this is that the negative ion has a smaller nuclear charge and consequently a more expanded charge cloud than the positive ion. The effect of increasing nuclear charge can also be seen by comparing the sizes of isoelectronic positive ions. For example, Al^{+3} is smaller than Mg^{++}, which in turn is smaller than Na^+; all have the $(1s)^2(2s)^2(2p)^6$ configuration.

Table 11.11 Crystal-lattice energies of the alkali halides* (kcal/mole)

LiF	246.7	NaF	219.0	KF	194.3
LiCl	202.3	NaCl	186.0	KCl	169.3
LiBr	193.0	NaBr	177.0	KBr	161.8
LiI	180.0	NaI	165.7	KI	153.3
RbF	185.4	CsF	173.6		
RbCl	162.3	CsCl	155.5		
RbBr	155.9	CsBr	149.3		
RbI	147.3	CsI	141.4		

* The positive values given here are the energies required to vaporize the solid to the separated ions.

The ionic radii can be used to understand the variations in the lattice energies of the alkali halides. Table 11.11 shows that, for any given positive ion, the lattice energy becomes smaller as the negative ion becomes larger. The same trend is found if we examine the compounds of any one negative ion: the lattice energy decreases as the size of the positive ion increases. An even more detailed interpretation is possible. Since the magnitude of the lattice energy depends on $1/(r_+ + r_-)$, the lattice energy will be relatively insensitive to the size of the positive ion, if $r_- \gg r_+$. A comparison of the difference in the lattice energies of LiI and NaI with the corresponding number for LiF and NaF illustrates this effect.

Besides determining internuclear separation, ionic radii influence the co-ordination number, or the number of immediate neighbors which can be grouped around a central ion. Figure 11.9 shows the sodium-chloride crystal lattice, in which each ion has a coordination number of six. A sodium ion is surrounded by six chloride ions, and each chloride ion is surrounded by six sodium ions. We might expect that increasing the coordination number would increase the stability of the ionic lattice, since each ion would then have more near neighbors of the opposite charge. In the cesium-chloride lattice shown in Fig. 11.10 the coordination number is eight. Accordingly, the Madelung constant for this lattice is 1.763, while that of the sodium-chloride lattice is 1.748. Thus, for equal internuclear distances, the cesium-chloride type of lattice is about 1% more stable than the sodium-chloride lattice.

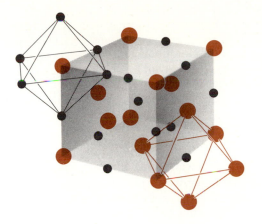

The sodium-chloride lattice, showing the sixfold coordination of the ions.   **FIG. 11.9**

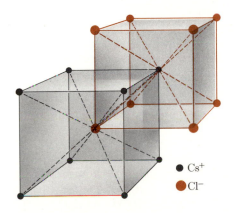

Part of the cesium-chloride lattice, showing the eightfold coordination of the ions.   **FIG. 11.10**

● Cs$^+$
● Cl$^-$

Despite this apparent extra stability of the cesium-chloride lattice structure, most of the alkali halides crystallize in the sixfold coordinated sodium-chloride structure. The reason for this preference for the seemingly less stable lattice is not difficult to understand. When one tries to pack eight large negative ions about a small central positive ion, it is found that the distance of closest approach of the positive and negative ions is not determined by $r_+ + r_-$, but by the radius of the negative ion alone. This is illustrated schematically in Fig. 11.11. Because of their large size, eight negative ions will "touch" each other before any of them come very close to the positive ion. In this situation the lattice energy is not so large as possible, because the distance between positive and negative ions is not so small as $r_+ + r_-$, the sum of the crystal radii. The difficulty is somewhat relieved if fewer negative ions are placed around the positive ion; this allows ions of opposite charge to approach each other more closely.

FIG. 11.11   Schematic representation of how the size of the negative ion alone may determine the distance between ions of opposite charge.

Thus a coordination number of six is preferred when the ionic sizes are quite different, which is the general situation for most of the alkali halides. When the ions are of comparable sizes, as in CsCl, CsBr, and CsI, the ions can enjoy the slightly greater stability of eightfold coordination. Thus by using the geometric properties of the lattices, it is possible to predict which lattice will be preferred for a given ratio of ionic radii. In the cesium-chloride lattice all the positive and negative ions touch simultaneously if $r_+/r_- = 0.73$. If the cation becomes any smaller relative to the anions, the sodium chloride lattice is preferred. More details of the correlation between ionic size and lattice geometry have been discussed in Section 3.5.

11.3   THE SIMPLEST COVALENT BONDS

The essential feature of the ionic bond is electrical asymmetry. Transfer of electrons from atoms of low ionization energy to atoms of high electron affinity produces oppositely charged ions whose mutual Coulomb attraction results in a stable crystal. However, such a simple picture cannot explain the strong bonds in such homonuclear diatomic molecules as H_2, N_2, O_2, and Cl_2. In these instances *both* bonding partners have the *same* ionization energy and the same electron affinity. Consequently there seems to be no reason to expect a permanent transfer of charge from one atom to the other, and indeed the measured properties of these molecules show that electrons are symmetrically divided between the two nuclei. The formation and stability of these symmetrical molecules are associated with an equal sharing of valence electrons; hence they are said to be examples of *covalent* bonding. In this section we shall examine the nature of this electron sharing and see why it produces chemical bonding. We must recognize, however, that the covalent bond has many subtle features, not all of which can be explained in simple language. Therefore we shall try to expose only the aspects of covalent bonding which have the most general applicability, and shall reserve detailed discussion of some special situations for later chapters.

The simplest of all molecules is the hydrogen-molecule ion, H_2^+, which occurs in electrical discharges through hydrogen gas. To those accustomed to thinking of covalent bonding only in terms of pairs of electrons, the existence of H_2^+ may seem surprising. Nevertheless, the bond in H_2^+ is quite strong; 64 kcal/mole are required to dissociate the molecule into its constituent proton and hydrogen atom. The equilibrium separation of the two nuclei in H_2^+ is 1.07 A, a distance which is comparable to other covalent-bond distances. These "normal" values of bond energy and bond length suggest that in the electronic structure of H_2^+, we will find at least some of the important properties which are responsible for covalent bonding in all molecules.

Because of its simplicity, H_2^+ can be treated in exact detail by the methods of quantum mechanics. From these theoretical calculations, values of the bond energy and bond length can be derived which are in exact agreement with those found experimentally. This success very strongly suggests that quantum mechanics provides a completely adequate theoretical framework for understanding the covalent chemical bond.

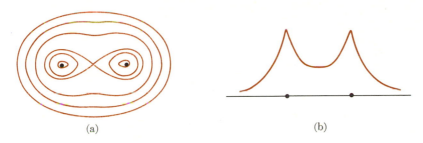

(a) (b)

Representations of the electron density in H_2^+: (a) contours of constant density and (b) variation of density along internuclear axis. **FIG. 11.12**

In order to extract the qualitative reason for covalent-bond formation from the mathematics of quantum mechanics, we can examine the probability of finding the electron at all points in the H_2^+ molecule. The graphical representation of the distribution found from the quantum-mechanical treatment is shown in Fig. 11.12. Figure 11.12(a) shows the lines of constant electron density which lie in a plane that contains both nuclei. The other graph (Fig. 11.12b) shows how the chance of finding the electron varies as we proceed along a line that runs through the two nuclei. Both representations show that the electron distributes itself symmetrically about both nuclei in the molecule. Consequently we say that the electron moves in a **molecular orbital** and belongs to the molecule as a whole, rather than to either nucleus.

The exact electron distribution in this simplest of covalent bonds is consistent with the qualitative concept that in covalent bonds an electron is shared by two nuclei. To get a better idea of what the word "sharing" really implies, it is profitable to compare the electron distribution in H_2^+ with the electron density around two hydrogen atoms which are *not bonded*. This is shown in Fig. 11.13. The electron density around each nonbonded atom has been drawn to half scale, so that we are really comparing one electron distributed between two nonbonded atoms with the one-electron distribution in H_2^+. The difference in density curves then reveals the difference between the behavior of an electron in a bond, and an electron which spends half its time near each of two non-bonded nuclei. The figure shows that forming the bond moves some of the electron density from the regions outside both nuclei to the regions near to, and in between, the nuclei.

FIG. 11.13   Electron sharing in the H_2^+ molecule. The dashed lines represent one electron distributed between two nonbonded atoms, while the solid line represents the actual variation of electron density in H_2^+.

The potential energy of an electron located at a distance r_A from nucleus A and r_B from nucleus B is

$$-e^2 \left(\frac{1}{r_A} + \frac{1}{r_B} \right).$$

Consequently the potential energy of the system is lowest (most negative) when the electron is very close to either nucleus, or when it is in a region relatively near *both* nuclei at the same time. It would appear then that bond formation, or sharing of the electron by two nuclei, permits the electron to spend more time in regions of space where its Coulomb potential energy is low, thereby lowering the total energy of the molecule. Detailed quantum-mechanical calculations substantiate this qualitative assessment of the origin of bond energy.

We can make a slightly more detailed analysis of the energy changes which accompany bond formation if we make use of a general principle called the *virial theorem*. As is shown in Section 2.2, the general form of the virial theorem relates the average value of the kinetic energy to the virial, which is the sum of $\overline{rF}$ over all particles and coordinates r:

$$\overline{KE} = \tfrac{1}{2} \sum \overline{rF}.$$

If the force F is given by Coulomb's law

$$F = q_1 q_2 / r^2,$$

where q_1 and q_2 are the charges on the particles, we get

$$\overline{\text{KE}} = \tfrac{1}{2} \sum \overline{q_1 q_2/r}.$$

The quantity $q_1 q_2/r$ is just the potential energy of interaction of two charged particles, and the sum of this quantity over all particles gives the total Coulomb potential energy of the system. Therefore, for a system which obeys Coulomb's law, the *average* potential energy and the *average* kinetic energy are related by

$$\overline{\text{KE}} = -\tfrac{1}{2}\overline{\text{PE}}.$$

The total energy of the system is $E = \overline{\text{KE}} + \overline{\text{PE}}$, so using the virial theorem, we can write

$$E = \overline{\text{KE}} + \overline{\text{PE}} = -\overline{\text{KE}} = \tfrac{1}{2}\overline{\text{PE}}. \qquad (11.1)$$

Now let us ask what happens to the total energy when we bring together two atoms to form a bond. If the bond is stable, E must decrease, or ΔE, the change in energy, must be negative. According to Eq. (11.1), ΔE is related to the change in the average potential energy by

$$\Delta E = \tfrac{1}{2} \, \Delta \overline{\text{PE}}.$$

If ΔE is negative, $\Delta \overline{\text{PE}}$, the change in the average potential energy, must also be negative. Thus formation of the covalent bond is accompanied by a decrease in the potential energy, and this, as we noted, is a result of the shared electron spending more time close to, and in between, the nuclei.

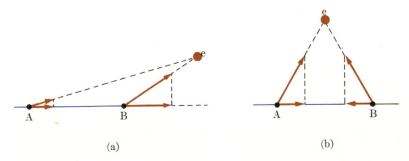

(a) (b)

Forces exerted by an electron on the nuclei A and B of a diatomic molecule. In (a) the electron exerts a force which separates the nuclei, while in (b) the electron tends to bind the nuclei together.

FIG. 11.14

There is another way to show why electron sharing tends to hold atoms together. Instead of thinking about the *energy* of the molecule, we consider the *forces* that an electron exerts on the nuclei. If, as shown in Fig. 11.14(a), an electron is in the region "outside" of both nuclei, the force, e^2/r_n^2, which it exerts on the nearer nucleus is greater than the force, e^2/r_f^2, it exerts on the farther nucleus. If we resolve those forces into components perpendicular and parallel to the internuclear axis, as shown in Fig. 11.14(a), we find that the electron tends to draw both nuclei in the direction of the internuclear axis, but with different forces. The difference between these two forces is a net force which tends to separate the nuclei. Therefore, whenever an electron is in the region "outside" both nuclei, it exerts forces which tend to *oppose* bond formation. However, when an electron is between the nuclei, the forces it exerts tend to draw the nuclei together, as shown in Fig. 11.14(b). The hyperbolic surfaces which separate the regions where the electron tends to bind or separate the nuclei are shown in Fig. 11.15. If we compare these boundary surfaces with the electron distribution in H_2^+, we see that bonding is accomplished by allowing the electron to spend time in regions between the nuclei, where the forces it exerts draw the nuclei together.

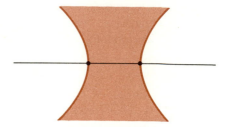

FIG. 11.15 Boundary surfaces for electron binding in a homonuclear diatomic molecule AB. Any electrons in the shaded region serve to bind the nuclei together.

A little reflection shows that covalent and ionic bonding are alike in that both result from a redistribution of electron density which causes the total energy of the system to decrease. The difference between the two types is that we can easily characterize the redistribution associated with ionic bonding by saying that an electron is transferred from one atom to another, while the redistribution associated with covalent bonding is much more subtle and difficult to describe. While we can distinguish between the complete electron transfer of ionic bonding and the equal electron sharing which occurs in covalent molecules like H_2^+, we should realize that these are not two rigid classifications into which all bonds can be forced. We will find that in most chemical bonds there is an *unequal* sharing, or *partial* transfer of electrons; distributions result which are intermediate between the two extreme models. The existence of a *continuous* range of bond properties from extreme ionic to extreme covalent is not hard to understand if we recognize and remember the similarities, as well as the differences, between ionic and covalent bonding.

The molecular orbital that we have examined is only one of many possible orbitals of H_2^+ which the electron may occupy. It is, however, the orbital of lowest energy, and since its occupancy leads to a stable bond it is called a **bonding molecular orbital.** The lowest-energy *excited* electronic state of H_2^+ corresponds to the electron occupying the molecular orbital shown in Fig. 11.16. Note that once again the electron divides its time equally between the two ends of the molecule. However, in the important regions between the nuclei there is a *deficiency* of electron density. The electron spends most of its time in the peripheral regions, relatively far from both nuclei. All of these features contrast with the properties of the low-energy bonding orbital discussed earlier. Consequently it should not be surprising to find that H_2^+ in its lowest excited state is *unstable* with respect to dissociation to a proton and a hydrogen atom. Not only does no bond exist in this state, but there is a strong repulsive force between the two fragments. This excited electron distribution is therefore called an **antibonding orbital.** Its properties emphasize that the mere sharing of an electron by two nuclei *does not automatically lead to bond formation.* The important factor in bond formation is that the electron is shared in such a way that the total energy of the system decreases. This occurs when an electron occupies a bonding molecular orbital, but does not when the electron enters an antibonding orbital.

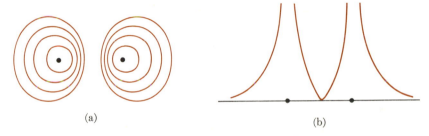

(a)

(b)

The antibonding orbital of H_2^+: (a) contours of constant electron density and (b) variation of electron density along internuclear axis.   **FIG. 11.16**

The consequences of putting the electron in the bonding or antibonding molecular orbital of H_2^+ are summarized in Fig. 11.17. We see there the total energy of the system $H^+ + H$ plotted as a function of the separation of the two nuclei. At great internuclear distances the energy of the system is rather insensitive to the magnitude of the separation. When the nuclei are near each other, there are two possibilities. If the electron is in the bonding orbital, the total energy of the system is less than that of the separated particles $H^+ + H$, and consequently we say a bond has formed. The minimum in the energy corresponds to the most stable configuration, and occurs at the equilibrium bond distance. The depth of this energy "well," corrected for the zero-point vibrational energy $\frac{1}{2}h\nu$, is the bond dissociation energy $D(H_2^+)$. If the electron in H_2^+ occupies the antibonding orbital, the total energy of H_2^+ at any finite internuclear distance is greater than the energy of the separated fragments $H^+ + H$. This is represented by the upper, or "repulsive" energy curve in Fig. 11.17.

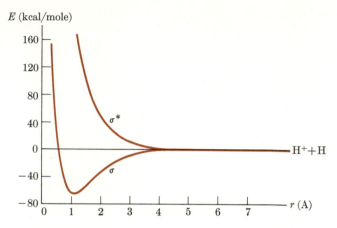

FIG. 11.17 The total energy E of the H_2^+ system as a function of internuclear distance r. The lower curve represents the situation when the electron is in the σ-bonding orbital, while the upper curve gives the behavior of the energy when the electron is in the σ^*-antibonding orbital.

The Hydrogen Molecule

Our examination of H_2^+ revealed the basic phenomenon from which the covalent bond derives its stability. However, the existence of the covalent bond has been commonly associated with the sharing of *two* electrons whose spins are opposite, or paired. Consequently the idea of "pairing two electrons to form a bond" has been an important empirical concept in valence theory.

Why is it important that the two electrons involved in a covalent bond have opposite spin? To find out, we need only examine the bonding in the hydrogen molecule from the point of view of the Pauli Exclusion Principle. For atoms, the Exclusion Principle states that no two electrons may have the same set of quantum numbers, and we saw in Chapter 10 that this means that two electrons may occupy the same atomic orbital only if they have different spin. The Exclusion Principle applies to the occupancy of molecular orbitals in the same way. Thus the consequence of the Exclusion Principle is that *only two electrons may occupy the same molecular orbital, and then only if they have opposite spin.* With this in mind, let us build the hydrogen molecule by starting with the two nuclei and feeding the electrons into the available molecular orbitals.

Just as atoms can be constructed by using orbitals resembling those of the hydrogen atom, molecules can be made up of molecular orbitals resembling those of H_2^+. At this point we are aware of two of these molecular orbitals; one is the low-energy bonding orbital, and the other is the higher-energy antibonding orbital. The first electron in H_2 will then surely occupy the low-energy bonding orbital. If the spin of the second electron were the same as that of the first electron, only one of them could occupy the bonding orbital, and to avoid violation of the Exclusion Principle, the second electron would be forced into the antibonding orbital. The result would then be an energized molecule, and experiments show that this molecule is unstable with respect to dissociation to the atoms. On the other hand, if the spins of the two electrons are opposite, or paired, both electrons can occupy the low-energy bonding orbital without violating the Pauli Exclusion Principle and thus both can contribute to the bonding of the molecule. The role of electron-spin pairing in chemical bond formation is therefore somewhat indirect. Pairing of the spins *allows* both electrons to behave in a way which strengthens the bond.

The increased effect of two bonding electrons is reflected by the bond energy and bond length of the hydrogen molecule. It takes 104 kcal/mole to dissociate hydrogen to its atoms; this is over 50% greater than the bond energy of H_2^+. Furthermore, the internuclear separation of H_2 is 0.74 A, which is less than the bond distance in H_2^+ by 0.3 A. Thus in H_2 the two electrons are better able to overcome the nuclear Coulomb repulsion and bring the two nuclei closer together than their equilibrium separation in H_2^+.

We have developed an argument that rationalizes the fact that many covalent bonds involve the sharing of a pair of electrons by two atoms. Can we use our argument to discover why two helium atoms do not form a stable diatomic molecule? Imagine building the hypothetical He_2 molecule by starting with the two nuclei and feeding the four electrons into the available molecular orbitals. The first two electrons, with opposed spins, would occupy the low-energy bonding molecular orbital which we have used in H_2^+ and H_2. The third and fourth electrons would then be accommodated in the antibonding orbital. The net result of two bonding electrons and two antibonding electrons is no bond at all. In fact, experience shows that the effect of antibonding electrons is stronger than that of bonding electrons, and consequently there is a very strong repulsive force between the two atoms when they are closer than 2.5 A.

It is of interest now to compare the energies of the species H_2^+, H_2, He_2^+, and He_2 as a function of their internuclear separation, since in these molecules there are respectively 1, 2, 3, and 4 electrons in the bonding and antibonding molecular orbitals. Figure 11.18 shows that as one proceeds from H_2^+ to H_2, the bond strength increases. In passing to He_2^+, the bond energy decreases, since He_2^+ has one antibonding electron in addition to the two bonding electrons. Finally in He_2, the presence of two antibonding electrons prevents bond formation, as discussed above.

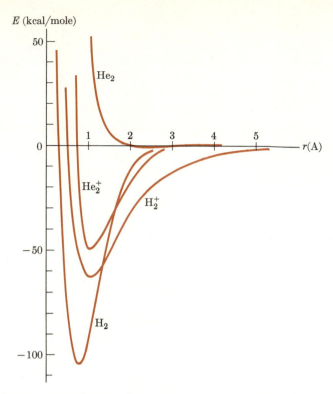

FIG. 11.18 Total energy as a function of distance for H_2^+, H_2, He_2^+, and He_2. For He_2, the bonding due to van der Waals forces is too small to be seen on this scale. Note that the $(\sigma)^2(\sigma*)^1$ configuration in He_2^+ gives a weaker bond than the one σ electron in H_2^+.

11.4 ATOMIC AND MOLECULAR ORBITALS

As we examine the electronic structures of different molecules we will encounter a variety of molecular orbitals that must be described and classified if they are to help us understand chemical properties. The first step in classifying a molecular orbital is to specify whether it is bonding or antibonding. Second, something must be said about the general shape of the electron distribution. The orbitals we discussed in Section 11.3 were cylindrically symmetric; the line which joined the two nuclei formed a natural symmetry axis for the electron distribution. Molecular orbitals that have *cylindrical symmetry* and are *bonding* are labeled as σ-orbitals; those that are cylindrically symmetric and *antibonding* are denoted by $\sigma*$.

The third aspect of the classification of molecular orbitals has to do with the behavior of the molecular orbital as the bond is broken. Imagine pulling the hydrogen molecule apart into its constituent atoms. As the atoms separate, the molecular orbital occupied by the two electrons changes shape and eventually separates into two atomic 1s-orbitals. This suggests that we might describe any molecular orbital by giving the atomic orbitals into which it separates when the atoms are removed from each other. Thus the molecular orbital in the hydrogen molecule is described as $\sigma 1s$, which indicates that it is cylindrically symmetric, bonding, and separates to give $1s$ atomic orbitals. In a similar manner, the bonding orbital in the fluorine molecule is designated as $\sigma 2p$, which shows that it is cylindrically symmetric, bonding, and separates into two fluorine $2p$ atomic orbitals.

The idea of classifying molecular orbitals by the atomic orbitals into which they separate has a useful dividend. Note in Fig. 11.13 that the $\sigma 1s$-orbital of molecular hydrogen resembles two $1s$ atomic orbitals which have been partially superimposed. Consequently we might picture the molecular orbital as being "built" of *two overlapping atomic orbitals*. This description of a molecular orbital is not exact, but it does have some useful consequences which offset its lack of rigor. Picturing a molecular orbital as built of two atomic orbitals gives us a scheme by which we can describe, at least approximately, something which is complicated (the molecular orbital) in terms of something reasonably familiar (the atomic orbital). Moreover, the idea suggests a useful empirical rule for predicting the number of covalent bonds which an atom can form. Since our discussion suggests that formation of a fully occupied bonding orbital can arise from a half-filled atomic orbital of each of two atoms, *the number of covalent bonds formed by an atom should be equal to the number of half-filled valence orbitals it can have*. For example, the fluorine and hydrogen atoms each have only one half-filled atomic valence orbital, and each almost invariably forms only one covalent bond. The oxygen atom has two half-filled orbitals and characteristically forms two covalent bonds, as in H_2O and F_2O. Similarly, the nitrogen atom has three half-filled p-orbitals and forms NH_3 and NF_3.

The carbon atoms provides an interesting test of the rule since in its normal electron configuration carbon has only two half-filled orbitals: $1s^2 2s^2 2p_x^1 2p_y^1$. However, the maximum number of half-filled orbitals it can have is four: $1s^2 2s^1 2p_x^1 2p_y^1 2p_z^1$. Indeed, carbon characteristically forms four covalent bonds, as in CH_4 and CF_4. Boron provides a similar example. Even though in its normal electron configuration, $1s^2 2s^2 2p^1$, it has only one half-filled orbital, the maximum possible number of partially filled orbitals is three. Correspondingly, boron does form three covalent bonds in BF_3, $B(OH)_3$, and other of its compounds. Both these examples show that the number of covalent bonds which an atom will form may be given by the *maximum* number of half-filled orbitals which the atom can have, rather than by the actual number in the ordinary electronic configuration.

The ion NH_4^+ might appear at first to violate our simple rule, since it suggests that the nitrogen atom, which has a maximum of three half-filled orbitals, can form four covalent bonds. This difficulty is removed if we recognize that NH_4^+ can be thought of as a compound of N^+ and four hydrogen atoms. Like the carbon atom, N^+ can have a maximum of four half-filled valence orbitals, and this is consistent with the formation of NH_4^+. Similarly, the ions BH_4^- and

BF_4^- present no problem if they are thought of as compounds of B^- and four hydrogen or fluorine atoms.

Electron-Dot Structures

A primitive but often convenient way to represent the disposal of electrons in molecules is to use electron-dot structures. Thus the equation

$$H\cdot + H\cdot \rightarrow H:H$$

symbolizes the formation of an occupied bonding orbital by the overlap of two half-filled atomic orbitals. Similarly, the structure

$$H:\overset{\cdot\cdot}{\underset{\cdot\cdot}{F}}:$$

is a simple way of saying that in HF there is an electron-pair bond and that six of the valence electrons of fluorine are "nonbonding"; that is, they remain localized near the fluorine atom and do not participate in the bonding. Electron-dot formulas allow us to decide rapidly how many bonds an atom may form. If we write the nitrogen atom as

$$:\overset{\cdot}{\underset{\cdot}{N}}\cdot$$

then it is immediately clear that three covalent bonds can be formed:

$$:\overset{\cdot}{\underset{\cdot}{N}}\cdot + 3H\cdot \rightarrow \quad \overset{\textstyle H}{:\overset{\cdot\cdot}{\underset{\cdot\cdot}{N}}:H}$$
$$\underset{\textstyle H}{}$$

While electron-dot structures can be helpful, they must be used and interpreted carefully. The true geometry of molecules is often ignored in electron-dot structures for typographical reasons. Molecules like NH_3, which have a three-dimensional structure, are often represented as planar, as we have done. We will encounter other limitations of electron-dot structures in subsequent sections, where we will find that some molecules cannot be represented by only one electron-dot structure.

The Octet Rule

Another simple device that can sometimes be used to decide how to represent the electronic structure of a molecule is the octet rule, which states that an atom other than hydrogen tends to form bonds until it is surrounded by eight electrons. As the formulas

$$\left[\begin{array}{c} \text{H} \\ \overset{..}{\text{H}:\text{B}:\text{H}} \\ \text{H} \end{array}\right]^{-} \qquad \overset{..}{\underset{..}{:}\text{F}\underset{..}{:}} \\ \overset{..}{:}\text{F}\overset{..}{:}\text{C}\overset{..}{:}\text{F}\overset{..}{:} \qquad \overset{..}{:}\text{N}:\text{H} \qquad \overset{..}{:}\text{O}:\text{H} \qquad \text{H}:\overset{..}{\underset{..}{\text{F}}}: \qquad \overset{..}{:}\text{Ne}\overset{..}{:}$$

show, the rule is consistent with the behavior of atoms in the second row of the periodic table. For these atoms the octet rule is equivalent to our statement that the number of covalent bonds equals the maximum number of half-filled atomic orbitals. Atoms of the second periodic row have only four valence orbitals and thus cannot form more than four covalent bonds. In forming these bonds they can surround themselves by no more than $4 \times 2 = 8$ electrons, exactly as suggested by the octet rule.

The octet rule is often not obeyed by atoms outside the second row of the periodic table, and thus it is of somewhat limited value. For example, phosphorus forms two chlorides, PCl_3 and PCl_5. The first of these is analogous to NH_3 or NCl_3 and obeys the octet rule. However, in PCl_5 the phosphorus atom is surrounded by ten electrons. While this compound violates the octet rule, its formation can be understood in terms of the maximum possible number of half-filled valence orbitals available in the phosphorus atom. Usually the configuration of the valence electrons in phosphorus is $3s^2 3p_x^1 3p_y^1 3p_z^1$. However, the $3d$-orbitals of the atom are not much higher in energy than the $3p$-orbitals, and consequently we can say that the maximum number of half-filled valence orbitals is five, corresponding to the configuration $3s^1 3p_x^1 3p_y^1 3p_z^1 3d^1$.

Other "violations" of the octet rule, such as the existence of SF_6 and SF_4, can be rationalized in a similar manner. The sulfur atom has the valence-electron configuration $3s^2 3p_x^2 3p_y^1 3p_z^1$, and does form compounds like H_2S, whose formulas seem directly related to this electron configuration. However, the sulfur atom can also have four half-filled valence orbitals as in the configuration $3s^2 3p_x^1 3p_y^1 3p_z^1 3d^1$, or six half-filled orbitals as in $3s^1 3p_x^1 3p_y^1 3p_z^1 3d^1 3d^1$. The existence of these possibilities provides a simple way of explaining the formulas of SF_4 and SF_6, respectively. Consequently the rule that the number of covalent bonds an atom forms is related to the possible number of unpaired valence electrons it can have is consistent with the formulas of a much larger number of compounds than is the octet rule. Nevertheless, the octet rule does provide a useful guide to the electronic structures of many compounds, particularly those containing elements from the second row of the periodic table.

It is appropriate to remark that the trick of rationalizing apparent violations of the octet rule by invoking the participation of d-orbitals in the bonding scheme is a matter of some controversy. Some scientists feel that the d-orbitals lie so high in energy that they should not be treated as valence orbitals, but as excited orbitals which cannot confer any appreciable stability to chemical bonds.

Others feel that while *d*-orbitals are of high energy in free atoms, their energy decreases as other atoms approach to make bonds. This author has not been convinced of the absolute validity of either side of this argument, and interprets the controversy merely as more evidence that the description of chemical bonds in terms of a few atomic orbitals is a highly approximate procedure. The most convincing reason for invoking *d*-orbitals in chemical bonding is that it is a simple idea which works, if it is used carefully and not overinterpreted.

At this point it is well to reconsider and review the idea that molecular bonding orbitals are built of overlapping atomic orbitals. Originally we introduced this point of view so that we could describe the electron-charge cloud in a molecular orbital in terms of the simpler atomic orbitals with which we were familiar. We noted that this description is not perfect. Then we realized that the formulas of compounds like H_2O, NH_3, and HF could be immediately explained if the number of bonds that an atom formed was equal to the number of its atomic orbitals that were half-filled with electrons. The idea behind this statement was that a bonding orbital could be "built" from one atomic orbital of each of two atoms, and then occupied by the two electrons originally associated with the atoms. Finally, to explain the formulas of CH_4, BF_3, and the halides of phosphorus and sulfur, we had to say that the number of covalent bonds an atom forms is not exclusively related to the number of half-filled orbitals it has in the *lowest* electronic configuration, but is related to the *various* numbers of half-filled orbitals it can possibly have. The rule in this form is useful, for it is consistent with the formulas of a large number of compounds.

The idea that molecular orbitals are built from atomic orbitals has more than qualitative significance. In the quantum-mechanical treatment of the chemical bond, this idea is the basis for one method of finding the mathematical functions that describe molecular orbitals. The results of these mathematical treatments of chemical bonding show that picturing molecular orbitals as being built from overlapping atomic orbitals is a severe approximation. Therefore, although this approach gives us a simple rule for rationalizing the formulas of molecules, we must always keep in mind that the fundamental reason atoms form bonds is that in so doing they reach a state of lower energy.

Because of the complications and subtlety of electron behavior, the criterion that bonds form when half-filled atomic orbitals are possible may be too crude to lead to correct predictions in all instances. Therefore, while we can use this idea profitably, we must be careful not to use it too rigidly.

11.5 MOLECULAR GEOMETRY

Having discussed the basis for understanding bond formation and molecular formulas, we can now turn to the problem of explaining the various experimentally observed bond angles. The bond angles in a molecule represent the condition of minimum energy for the molecule, and therefore, to rationalize the observed structures, we should start by deciding why varying the bond angles might affect energy. Since we have pictured the covalent chemical bond as a pair of electrons in a σ-molecular orbital largely confined near to and in between two nuclei, we can take the attitude that in fixing the angle between bonds, we really are fixing the distance between the regions in which the electrons in the two bond orbitals spend most of their time. How can this distance affect the energy of a molecule?

Our general experience indicates that electrons try to avoid each other as much as possible. Of the two reasons for this, one is obvious: there is a strong repulsive Coulomb force between two electrons, and the energy of two electrons is lowered when the electrons are kept apart. The second reason electrons tend to stay apart is more subtle. While the Pauli Exclusion Principle can be interpreted to mean that two electrons with the same spin cannot occupy the same atomic or molecular orbital, a more detailed intepretation is that two electrons with the same spin have a vanishing probability of being in the same region of space. In other words, any two electrons with the same spin tend to avoid each other. It is important to realize that this would be true *even if the electrons were uncharged*, for this property is a consequence of the Pauli Exclusion Principle, and *not* of Coulomb's law.

We find then that all electrons tend to avoid each other because of their like charge, but those with the same spin have a particularly low probability of being close to each other. These considerations suggest that there is a *correlation* in the instantaneous locations of electrons in atoms and molecules. If one electron is localized in a particular region of space, any other electron, especially one with the same spin, is most likely to be found in some other region of space. It is this correlation between the electron locations that is used to explain the geometry of molecules.

Let us consider mercuric chloride, which when vaporized exists as discrete gaseous molecules that have a linear arrangement of atoms, Cl—Hg—Cl. In the free mercury atom there are two valence electrons in the configuration $6s^2$, and it is these electrons that are involved in a pair of σ-bonds in the $HgCl_2$ molecule. Why does this molecule have linear geometry? We can see that the linear arrangement puts the two pairs of electrons involved in the two bonding molecular orbitals as far away from each other as is possible. Moreover, it allows electrons with the same spin to have a high probability of being on opposite sides of the mercury atom, as we can see by representing the molecule by

$$Cl \, \uparrow\downarrow \, Hg \, \uparrow\downarrow \, Cl.$$

Thus, if there are two pairs of bonding valence electrons around the mercury atom, the linear configuration is likely to be the arrangement of lowest energy. We might expect to find this geometry also in other molecules that have only two pairs of valence electrons about a central atom, and indeed, the molecules

$$Cl—Zn—Cl, \qquad Cl—Be—Cl, \qquad Cl—Mg—Cl, \qquad \text{and} \qquad H_3C—Hg—CH_3$$

all have the linear geometry indicated.

Now let us examine some molecules that have three pairs of valence electrons about a central atom. For example,

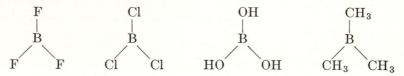

and the other halides of boron all are symmetric molecules in which the bond angles about the boron atom are 120°. Consequently the boron atom and the three atoms bonded to it lie in the same plane. A little reflection shows that this is the configuration in which the three pairs of bonding electrons can spend much of their time farthest from each other.

On the basis of the foregoing arguments and data, it is easy to understand the geometry of methane, CH_4. There are four pairs of valence electrons in bonds around the central carbon atom. Electrostatic repulsions will be minimized, and electrons with the same spin will be farthest apart when the bond orbitals are directed toward the corners of a regular tetrahedron, as illustrated in Fig. 11.19. Other molecules in which there are four pairs of valence electrons, such as NH_4^+, CCl_4, and SiF_4, also have tetrahedral geometry.

FIG. 11.19   The geometry and schematic electronic structure of methane. The hydrogen atoms lie at the apices of a regular tetrahedron, and the electrostatic repulsion between electrons is minimized.

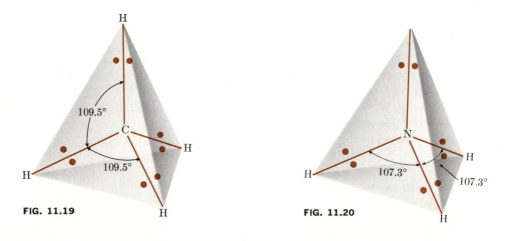

FIG. 11.19

FIG. 11.20

FIG. 11.20   The geometry and schematic electronic structure of ammonia. The hydrogen atoms lie at three of the apices of a slightly irregular tetrahedron, with a nonbonded electron pair in the region of the fourth apex.

How can we rationalize the structure of ammonia from this point of view? As noted in Section 11.2, ammonia is a pyramidal molecule in which the H—N—H bond angle is 107°, not much different from the value for the H—C—H bond angle in methane. Like methane, the ammonia molecule has four pairs of valence electrons about a central atom. However, in ammonia three pairs of valence electrons are involved in bonding while the other pair is not. It is still reasonable to expect that the central nitrogen atom will direct the three pairs of bonding electrons to three corners of a tetrahedron, and the pair of nonbonding electrons to the fourth corner, as illustrated in Fig. 11.20. The tetrahedral geometry is not expected to be regular, since the four pairs of electrons are not equivalent. Consequently the bond angles in ammonia deviate slightly from the regular tetrahedral angles of 109.5°.

The same argument can be applied to the water molecule. The four pairs of valence electrons around the oxygen atom are directed tetrahedrally, to minimize repulsions. Since the four pairs are not equivalent, the observed H—O—H bond angle (104°) does not correspond to the regular tetrahedral angle, but is only slightly smaller.

Now we have encountered enough examples to be able to formulate a more refined working hypothesis for discussing the bond angles in more complicated molecules. It seems that the pairs of valence electrons about a central atom, whether bonding or nonbonding, lie in orbitals that are directed in a manner that minimizes Coulomb repulsions between the electron pairs and keeps electrons of the same spin as far from each other as possible. The fact that the bond angles in water and ammonia are smaller than 109.5°, the regular tetrahedral angle, suggests that electron pairs in nonbonding orbitals in effect take up more space than electron pairs in bond orbitals.

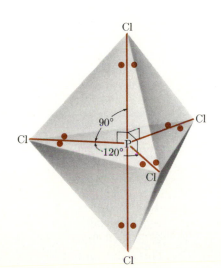

The geometry and schematic electronic structure of gaseous phosphorus pentachloride. The chlorine atoms lie at the apices of a regular trigonal bipyramid.

FIG. 11.21

As a first application of these ideas to more complicated molecules, consider phosphorus pentachloride. The PCl_5 molecule has a central phosphorus atom surrounded by five pairs of electrons that constitute the bonds to the chlorine atoms. The geometry that minimizes electron interaction is the trigonal bipyramid structure shown in Fig. 11.21. The three "equatorial" chlorine atoms lie in the same plane as the phosphorus atom and the Cl—P—Cl angle in this plane is 120°. The two "polar" chlorine atoms lie above and below this equatorial plane on the axis of the bipyramid. The angle defined by a polar chlorine atom, the phosphorus atom, and an equatorial chlorine atom is 90°. Thus the polar and equatorial chlorine atoms are not equivalent, and in fact the polar P—Cl bonds are a bit longer (2.19 A) than the equatorial bonds (2.04 A).

Other molecules with five electron pairs have structures related to that of PCl_5. The molecule SF_4 has the geometry shown in Fig. 11.22(a), where we see that one of the five electron pairs occupies an equatorial nonbonding orbital while there are two equatorial and two polar bonds to fluorine atoms. The distortion of the molecule from the regular bipyramidal geometry is in the direction expected. Figure 11.22(b) shows that the structure of ClF_3 can also be rationalized if we imagine that two of the five electron pairs occupy nonbonding orbitals in the equatorial plane.

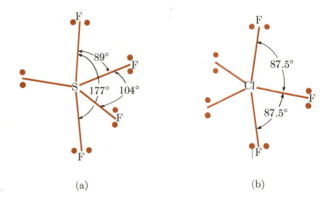

(a)   (b)

FIG. 11.22   The geometry and schematic electronic structures of (a) sulfur tetrafluoride and (b) chlorine trifluoride. In both molecules the electron pairs tend to be directed toward the apices of distorted trigonal bipyramids.

There are several compounds in which a central atom makes bonds to six other atoms, or is surrounded by six electron pairs. Sulfur hexafluoride, SF_6, has the regular octahedral structure shown in Fig. 11.23. All fluorine atoms are equivalent and all F—S—F bond angles are 90°. This octahedral geometry once again minimizes electron repulsion between pairs of electrons. Consequently the species IF_5 and ICl_4^-, which also have six electron pairs about the central atom, have structures related to that of SF_6, as Fig. 11.23 illustrates.

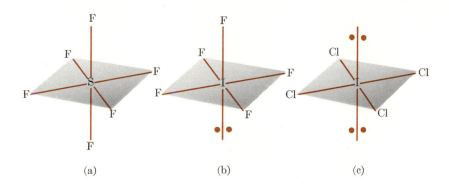

(a)   (b)   (c)

FIG. 11.23

The geometry and valence electron disposition of (a) SF_6, (b) IF_5, and (c) ICl_4^-. In each case bonding or nonbonding electron pairs lie at the apices of an octahedron.

The foregoing observations on molecular geometry can be summarized in the following way. To predict the approximate geometry of a molecule, determine how many pairs of valence electrons, bonding and nonbonding, exist around a given atom. These electron pairs will be directed in space in a manner which minimizes electron-electron repulsion. Thus 2, 3, 4, 5, and 6 electron pairs lead respectively to linear, trigonal, tetrahedral, trigonal bipyramidal, and octahedral geometries. When some electron pairs are bonding and others nonbonding, the geometry will be distorted from the regular polyhedron, with nonbonded electron pairs apparently occupying more space than bonded pairs.

While we seem to have a satisfactory scheme for explaining the bond angles found in the compounds of the nontransitional elements, we must remark that there is a good deal of conjecture in the arguments we have developed. We have assumed, for example, that we *know* that in PCl_5 all the bonds consist of electron pairs localized between two atoms so that there are five electron pairs involved in the bonding. In fact, it is not known whether this is really an accurate description of the bonding in this molecule, for not enough is known about the detailed electron distribution. The same criticism can be leveled at our assertions that the bonding in the other molecules is always of the electron-pair type; this assumption seems reasonable, but it is not known to be true. Consequently we must realize that while the scheme we have discussed allows us to see regularities and relations between bond angles in different compounds, not all the assumptions used in this application have been proved.

Hybridization

In discussing the bond angles in polyatomic molecules, we have avoided describing how the bond orbitals are built from the atomic orbitals. This is the matter to which we turn now. To analyze the bonding in polyatomic molecules, it is helpful to add to our ideas the concept of hybrid atomic orbitals. In order to understand what hybrid orbitals are, let us see why they are needed. We have remarked that the mercuric-chloride molecule is a linear, symmetric arrangement of atoms. Both Hg—Cl bonds are of equal length and strength. Can our present picture of covalent bonding explain these facts?

The mercury atom has a valence-electron configuration of $6s^2$, and thus its maximum number of half-filled atomic orbitals is two, consistent with the molecular formula $HgCl_2$. The most obvious approach to the detailed construction of the bonds would be to imagine the two electrons of Hg unpaired to the configuration $6s^1 6p^1$. Then we could picture one bond being formed from the mercury 6s-orbital and a chlorine 3p-orbital, while the other bond was made from the mercury 6p-orbital and the 3p-orbital of the other chlorine atom. This is an unsatisfactory and incorrect picture, for it implies that the two bonds in $HgCl_2$ should have different properties, since they are constructed from different atomic orbitals. This implication is inconsistent with the experimental fact that the two bonds are identical. Therefore we must modify our ideas so that they account for the equivalence of the bonds, and this is done by using the concept of hybridized atomic orbitals.

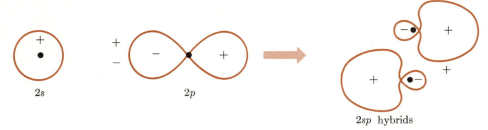

$2s$ $2p$ $2sp$ hybrids

FIG. 11.24   Schematic representation of the hybridization process. The orbitals are displaced from their common origin for clarity, and the sign of the wave function is indicated.

A hybrid atomic orbital is the result of a mathematical combination (algebraic addition) of the functions which describe two or more atomic orbitals. When the functions which represent s- and p-orbitals are added, a new hybrid function is produced; when the p-function is subtracted from the s-function, a second hybrid function results. This hybridization process and its consequences are shown in Fig. 11.24. The two ways of combining an s- and a p-function yield two *sp hybrid orbitals* which are equivalent, except that one has its greatest electron density in a direction 180° from the other. Since these orbitals effectively "point" in opposite directions, it is easy to imagine two *equivalent* bonds of a *linear* molecule being formed from them.

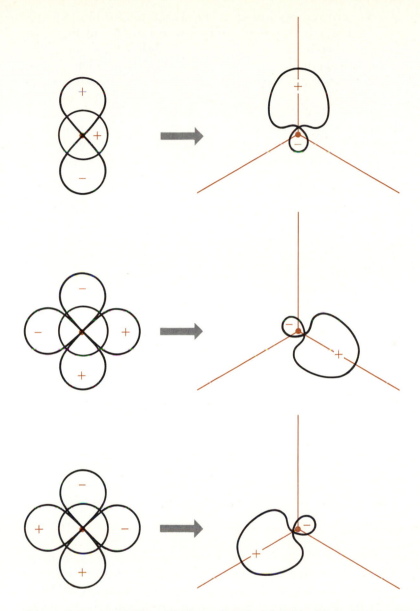

Schematic representation of the formation of sp^2-hybrid orbitals. Note how the combination of the positive s-wave function with the positive lobes of the p-wave functions produces a large positive lobe in the hybrid function. Similarly, cancellation of the positive s-function by the negative p-function lobe produces a small negative lobe in the hybrid.

FIG. 11.25

By introducing sp hybrid orbitals we can account not only for the equivalence of the bonds in $HgCl_2$, but also for the linearity of the molecule. However, it is fair to ask how "real" hybrid atomic orbitals are. Are they a good description of how electrons behave, or are they just an artificial concept used to patch up a failing theory? The answer is that they are just as real as the hydrogenlike atomic orbitals we have been using to describe many-electron atoms. The hydrogenlike orbitals are appropriate for describing the behavior of *one* electron at a time. The hybrid orbitals are a very effective way of describing the *relative* motions of two or more electrons at a time. Consequently hybrid orbitals are most useful in discussions of the bonding in polyatomic molecules.

Now let us examine hybridization in situations where there are more than two valence electrons. There are three ways of combining an *s*-orbital and two *p*-orbitals to produce three equivalent sp^2 hybrid orbitals, as shown in Fig. 11.25. The hybrid orbitals are identical in all respects, except that their directions of maximum electron density lie 120° from each other, necessarily in the same plane. The boron atom has three valence electrons, and we might expect it to display this type of hybridization. This is in fact found, for in BF_3, BCl_3, and $B(CH_3)_3$ all bonds to the boron atom are equivalent, in the same plane, and the angle between any two of them is 120°. We will find that there are other molecules that display the planar, triangular geometry about the central atom characteristic of sp^2-hybridization.

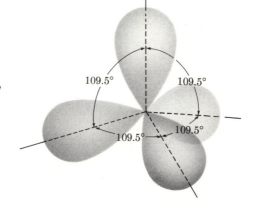

FIG. 11.26 Schematic representation of the boundary surfaces of the four sp^3 hybrid orbitals.

Hybridized atomic orbitals provide a satisfactory description of the bonding in methane and its derivatives. There are four independent ways of combining one *s*- and three *p*-orbitals so as to produce four new equivalent hybrid sp^3-orbitals. The geometric properties of the sp^3 hybrid orbitals are shown in Fig. 11.26. Like the bonds in the methane molecule, each sp^3 hybrid orbital is directed toward an apex of a regular tetrahedron. Consequently we say that the bond orbitals in methane are formed from carbon sp^3 hybrid atomic orbitals and the 1*s*-orbitals of the hydrogen atoms. This description is also appropriate for the derivatives of methane obtained by simple substitution for the hydrogen atoms, such as CH_3Cl, CF_4, etc. Furthermore, the ions NH_4^+ and BH_4^- have the same number of electrons as does methane, and the same tetrahedral geometry, so the boron and nitrogen atoms are said to display sp^3-hybridization in these compounds.

So far we have emphasized the role of hybridization in highly symmetric compounds like CH_4 and BF_3, where all valence electrons of the central atom are engaged in bonding. Let us now think of ammonia NH_3, from the point of view of hybridization. We might first imagine four sp^3 hybrid orbitals about the central nitrogen atom. Three of these orbitals would be used to make bonds to the three hydrogen atoms, and thus would account for six of the eight valence electrons in the molecule. The remaining pair of electrons would occupy the last sp^3 hybrid orbital. While an apparent consequence of this scheme is that the HNH angle should be 109.5°, the tetrahedral angle, the observed value is 107°. We have already pointed out that this deviation from tetrahedral geometry is a consequence of the nonequivalence of the four electron pairs.

The sp^3-hybridization scheme can be applied to the water molecule. The total of eight valence electrons in H_2O occupy four sp^3 hybrid orbitals centered on the oxygen atom. Two of these four pairs of electrons form the two bonds to the hydrogen nuclei, while the other two pairs are nonbonding electrons. If this were an accurate picture, the angle between the two bonds in water should be 109.5°; the observed value is 105°. Once again the deviation from the regular tetrahedral angle can be attributed to the fact that the four pairs of valence electrons are not equivalent.

To describe electron-pair bonding in molecules like PCl_5 and SF_6, d-orbitals have to be brought into the hybridization scheme. The combination of one d-, one s-, and three p-orbitals does *not* produce five *equivalent* hybrid orbitals, but rather a pair of equivalent, oppositely directed orbitals, and another group of three equivalent hybrids. The relation between these groups is shown in Fig. 11.27. The two polar hybrid orbitals make an angle of 90° with the three equatorial hybrids, which in turn lie 120° from one another. We have already remarked that the five bonds in PCl_5 are not equivalent, and the dsp^3 hybrid orbitals are consistent with this feature of the molecule.

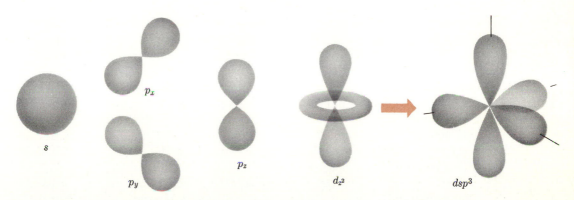

Schematic representation of the formation of dsp^3 hybrid orbitals.   **FIG. 11.27**

The situation pertaining to SF_6 and similar molecules is somewhat simpler. Combination of two d-, and one s-, and three p-orbitals does give six d^2sp^3 hybrid orbitals that are equivalent to one another. These hybrid orbitals are directed to the corners of a regular octahedron, and thus are consistent with the geometry of SF_6.

From our discussion we can conclude that it is possible to construct sets of hybrid orbitals that have directional properties consistent with the bond angles found in many molecules. As a result, we can picture an electron-pair bond in a polyatomic molecule as being formed by the overlap of the appropriate hybrid orbital of one atom with an orbital from another atom. It is best to think of hybridization as a *means* by which we can retain our picture of localized electron-pair bond formation by atomic-orbital overlap, rather than to feel that hybridization is the *reason* why molecules display a given geometry. It is not possible to see why a certain hybridization leads to a particular geometry without using mathematics, and consequently the characteristics of each hybrid set must be memorized. However, using the hybridization concept will help us uncover a number of regularities in the geometry of molecules, as we shall see in later chapters.

11.6 BOND POLARITY

Earlier we noted that one of the differences between ionic and covalent bonding is the nature of the charge distribution. In a covalent bond between two identical atoms the bonding electrons are symmetrically distributed about both nuclei, whereas in an extreme ionic bond one or more electrons are transferred from one atom to the other. We anticipated, however, that for the most part we would encounter molecules with intermediate charge distributions in which a pair of electrons was shared, *but not equally*, by two nuclei. This type of bond is called *polar covalent*, since it combines the feature of electron sharing with the existence of positive and negative regions, or electrical poles. Covalent bonds between different atoms are of this type.

Gaseous hydrogen chloride consists of individual HCl molecules which have relatively little attraction for each other; these are characteristics of a covalently bonded molecule. Yet the electronic properties of the hydrogen and chlorine atoms are different. While the ionization energy of the hydrogen atom is 313.6 kcal/mole, and that of chlorine is 299 kcal/mole, the electron affinity of hydrogen is only 17.4 kcal, much less than the 83.4 kcal/mole of the chlorine atom. This difference in electron affinities shows that the chlorine atom has a greater attraction for an additional electron than does the hydrogen atom. Consequently we expect the HCl molecule to be electrically asymmetric, or polar, with more negative charge concentrated near the chlorine atom, and an excess positive charge near the hydrogen atom.

Experiments confirm this expectation; HCl is said to have a *dipole moment*. Two equal and opposite charges of magnitude δ separated by a distance l

constitute a dipole and produce a dipole moment μ defined by $\mu = \delta \times l$. Thus the size of the dipole moment, proportional both to the magnitude and separation of the charges, is a convenient measure of charge asymmetry in a molecule. Two opposite charges of magnitude $\delta = e = 4.8 \times 10^{-10}$ esu separated by 1 A have a dipole moment of $4.8 \times 10^{-10} \times 10^{-8} = 4.8 \times 10^{-18}$ esu-cm. Consequently, 10^{-18} esu-cm is a convenient unit in which to measure the dipole moments of molecules, and one debye (abbreviated D) is defined as 10^{-18} esu-cm. Thus the dipole moment of two fundamental charges separated by one angstrom unit is 4.8 D.

The dipole moment of the HCl molecule is 1.03 D. We can better appreciate what this implies for the charge asymmetry in HCl if we *imagine* that in the molecule a net charge $+\delta$ is located at the hydrogen atom and a net charge $-\delta$ at the chlorine atom. Since the bond distance of HCl is 1.27 A, the magnitude of δ can be found by dividing the measured dipole moment by 1.27×10^{-8} cm. Thus

$$\delta = \frac{\mu}{l} = \frac{1.03 \times 10^{-18} \text{ esu-cm}}{1.27 \times 10^{-8} \text{ cm}} = 0.81 \times 10^{-10} \text{ esu.}$$

Our calculation shows that the electron distribution in HCl is equivalent to net opposite charges of 0.81×10^{-10} esu residing at each nucleus. This is the same as $0.81/4.8 = 0.17$ of a full fundamental charge, which shows that while the electron distribution is asymmetric, it is surely not accurate to say that one electron has been transferred from hydrogen to chlorine. Accordingly, we say that the bond in HCl is polar, but covalent, not ionic.

All of the hydrogen halides have dipole moments caused by the relatively great attraction that halogen atoms have for electrons. However, the atomic quantities which measure this attraction, the ionization energy and the electron affinity, show a general decrease in the sequence F, Cl, Br, I. Thus we can expect the dipole moments of the hydrogen halides to decrease as the atomic number of the halogen increases. Table 11.12 shows that this is what is found experimentally. Since the ionization energies and electron affinities of atoms close to each other in the periodic table are often very similar, we might expect the diatomic molecules of neighboring elements to have relatively small polarity. This is often true, as the examples of CO, NO, and ClF in Table 11.12 show.

Table 11.12 Dipole moments of gaseous molecules (debyes)

NH_3	1.47	H_2O	1.86	HF	1.98
PH_3	0.55	H_2S	1.1	HCl	1.03
AsH_3	0.22	H_2Se	0.4	HBr	0.79
SbH_3	0.12	H_2Te	<0.2	HI	0.38
ClF	0.88	N_2O	0.14	NO	0.16
ClBr	0.57	NO_2	0.3	CO	0.13
BrF	1.29	O_3	0.52		

In general, we can expect a dipole moment to be associated with any covalent bond between two *different* atoms. This principle can be used, together with measured dipole moments, to make conclusions about the structure of polyatomic molecules. For example, the existence of the dipole moment of carbon monoxide suggests that all carbon-oxygen bonds should be polar, yet the measured dipole moment of carbon dioxide is zero. These two facts can be reconciled if the CO_2 molecule has a symmetric structure in which the polarities of the two carbon-oxygen bonds cancel each other. Thus carbon dioxide must have a linear symmetric structure,

$$\overset{\delta -}{O}\text{————}\overset{2\delta +}{C}\text{————}\overset{\delta -}{O},$$

rather than a bent structure,

$$\left.\begin{array}{c}\overset{2\delta +}{C}\\ \overset{\delta -}{O}\diagup\;\diagdown\overset{\delta -}{O}\end{array}\right\}\begin{array}{c}+\\ \\ -\end{array}\quad\text{Net dipole,}$$

which would have a dipole moment. Similarly, the boron-fluorine covalent bond must be polar, because the ionization energies and electron affinities of boron and fluorine are different. However, the boron-trifluoride molecule has no dipole moment, because BF_3 has a planar symmetrical structure in which the effects of bond dipole moments cancel each other:

$$\begin{array}{c}\overset{\delta -}{F}\\ \mid 3\delta +\\ B\\ \overset{\delta -}{F}\diagup\;\diagdown\overset{\delta -}{F}\end{array}$$

Quite a different situation is found in PF_3, which has a pyramidal structure,

$$\left.\begin{array}{c}\overset{3\delta +}{P}\\ \underset{\delta -}{F}\diagup\mid\diagdown\underset{\delta -}{F}\\ \underset{\delta -}{F}\end{array}\right\}\begin{array}{c}+\\ \\ \\ -\end{array}\quad\text{Net dipole.}$$

Here the dipole moments of the individual bonds add constructively, and a net molecular dipole moment is produced. *If* PF_3 *were* planar, it would have no dipole moment. These examples show that the absence or presence of dipole moment in a polyatomic molecule can be a revealing clue to the structure, or atomic arrangement, of the molecule.

While we have chosen to speak of polar covalent bonds, we might ask how polar must a bond be before it is considered to be ionic. This is often a difficult question to answer. It might be supposed that the electrical conductivity of a pure substance would indicate whether or not ions were present. The equivalent conductivity of a fused salt is the current that flows when 1 volt is applied to a pair of plates 1 cm apart between which there is one equivalent of salt. Table 11.13 shows the equivalent conductivity of various halides at their melting points. It is clear that the alkali halides, our best examples of ionic compounds, are good electrical conductors and that the halides of the Group IV elements, normally thought of as covalently bonded, are good insulators. Furthermore, electrical conductivity, and hence ionic bond character of halides, tends to decrease from left to right along a row, and except for the alkali halides, increase from top to bottom in a column of the periodic table. However, there is no clear dividing line between those compounds which are conductors and those which are not. Both $BeCl_2$ and $AlCl_3$ show small, but finite, electrical conductivity, demonstrating the presence of a few ions, but suggesting that most of the fused salt consists of polar covalent molecules.

Table 11.13 Equivalent conductivities of fused chlorides (at their m.p., in ohm^{-1})

LiCl	166	$BeCl_2$	0.086	BCl_3	0	CCl_4	0
NaCl	134	$MgCl_2$	29	$AlCl_3$	1.5×10^{-5}	$SiCl_4$	0
KCl	104	$CaCl_2$	52	$ScCl_3$	15	$TiCl_4$	0

Earlier it was suggested that a compound consisting of small covalently bonded molecules would have a relatively low boiling point. In contrast, ionic substances, where the forces which hold the condensed phase together are very strong, should have high boiling points. Table 11.14 shows the boiling points of several halides. As we suggested, the boiling temperatures of the alkali halides are high, while those of the Group IV molecules are low. The trends along rows and columns roughly parallel the trends in electrical conductivity. Once again, $BeCl_2$ and $AlCl_3$ have intermediate properties, and it is not possible to classify either as an obviously covalent or ionic compound. The failure of compounds to fit into either of the two extreme classifications should neither surprise nor disappoint us. On the contrary, it is satisfying that there is an essentially continuous change in bond properties as the differences in the electrical characteristics of the bonded atoms steadily become more pronounced, and we must expect to find other examples of this situation.

Table 11.14 Boiling points of some chlorides (°C)

LiCl	1380	$BeCl_2$	490	BCl_3	12.5	CCl_4	76
NaCl	1440	$MgCl_2$	1400	$AlCl_3$	183	$SiCl_4$	57
KCl	1380	$CaCl_2$	1600	$ScCl_3$	1000	$TiCl_4$	136

In some molecules, two or even three pairs of electrons serve to bind two atoms together. For example, the bond dissociation energy of the nitrogen molecule is 226 kcal/mole, which is nearly three times as large as the dissociation energy of most covalent electron-pair bonds. Since one nitrogen atom commonly forms single bonds to three other atoms, it is not difficult to imagine that a pair of nitrogen atoms would form three electron-pair bonds with each other. The electron-dot representation of the bonding in N_2 is

$$: N ::: N :$$

which shows that the triple-bond picture is consistent with the octet rule.

To obtain a more detailed picture of the triple bond in the nitrogen molecule, we bring together 2 nitrogen atoms, each stripped of its valence electrons, and then feed these 10 valence electrons into the available molecular orbitals. To form the molecular orbitals, it is best to start by arranging on each nitrogen atom a pair of sp hybrid atomic orbitals, as in Fig. 11.28(a). Two of these sp-hybrids overlap and form a σ-molecular orbital which accommodates 2 valence electrons. The other two sp-hybrids point away from the bonding region and thus are not involved in bond formation. Each accommodates 2 electrons, and consequently we have found orbitals for 6 of the 10 valence electrons.

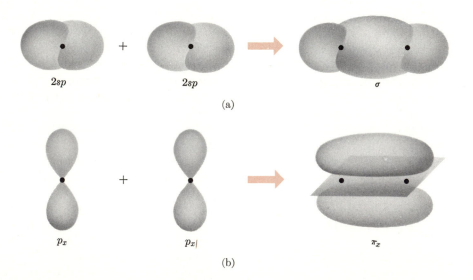

(a)

(b)

FIG. 11.28   Schematic representation of the formation of (a) the σ-bond orbital and two nonbonding orbitals, and (b) the π_x-bonding orbital of nitrogen. Another π-bonding orbital is formed from p_y atomic orbitals.

Two of the remaining valence electrons enter a bonding molecular orbital formed by the overlap of the p_x-orbitals of the two atoms. This is a new type of electron distribution which is called a π-bonding molecular orbital, and the two electrons which occupy it are said to form a π-bond. As Fig. 11.28(b) shows, the π-bond does not have the cylindrical symmetry associated with a σ-bond. A second π-bonding orbital is formed by the overlap of the atomic p_y-orbitals, and accommodates the last pair of valence electrons in the nitrogen molecule. Thus, of the six electrons which link the two atoms, one pair is in a σ-orbital, the second pair is in a π_x-orbital with greatest density in the x-direction above and below the internuclear axis, and the third pair is in a π_y-orbital with its greatest density in the y-direction. In other words, a triple bond consists of one σ-bond and two π-bonds. This is what we represent when we write

$$N\equiv N \qquad \text{or} \qquad :N:::N:$$

The bonding in other diatomic molecules which have 10 valence electrons is also properly described by this scheme. The most obvious example is carbon monoxide, which has 4 valence electrons contributed by the carbon atom and 6 by the oxygen atom. These 10 electrons are disposed among the 5 orbitals just as in the nitrogen molecule, and the large dissociation energy of CO, 256 kcal/mole, attests to the presence of a triple bond. The ion NO^+, observed in gaseous discharges and in crystalline salts, also has 10 valence electrons and also has a substantial dissociation energy of 244 kcal/mole. Finally, the bond energy of cyanide ion, CN^-, also a 10-valence electron molecule, is not known exactly but has been estimated to be greater than 200 kcal/mole. All other known properties of CN^- suggest that the carbon and nitrogen atoms are triply bonded.

There is a slightly different description of the triple bond, which may be more accurate for some molecules. Instead of regarding the σ component as made from the overlap of two sp-hybrid orbitals, we simply take it as generated from the overlap of the two pure p-orbitals. In addition, the $2s$-orbitals are pictured as nonbonding. The corresponding description of N_2 would be that, of the 10 valence electrons, four occupy nonbonding $2s$-orbitals, two occupy the σ $2p$-bonding orbital, and four occupy the two π $2p$-bonding orbitals. The true picture of nitrogen probably lies somewhere between the extremes of sp hybridization and pure p character for the σ bond.

There is a double bond between the two carbon atoms in the ethylene molecule, whose structure is shown in Fig. 11.29. To explain the geometry and electronic structure of ethylene, we can start with sp^2 hybrid atomic orbitals on each carbon atom. Two of these sp^2-hybrids overlap to form a σ-bond between the two carbon atoms. Four other σ-bonds to the 4 hydrogen atoms are then made from the remaining lobes of the sp^2-hybrids and the $1s$-orbitals of the hydrogen atoms. This system of σ-bonds accommodates 10 of the 12 valence electrons of ethylene. The last pair of electrons enters a π-bonding orbital which results from the overlap of the p_x atomic orbitals of the 2 carbon atoms. Thus the double bond in this and other molecules can be pictured as one σ-bond and one π-bond.

Besides accounting qualitatively for the large dissociation energy (145 kcal/mole) of the carbon-carbon bond in ethylene, our bond description is consistent with the geometry of the molecule. The use of sp^2 hybrid orbitals on the carbon atom was, indeed, suggested by the fact that the angles between the bonds in ethylene are near to 120°, as Fig. 11.29 shows. Moreover, the formation of the π-bond can occur only when the contributing p-orbitals are aligned parallel, which happens only when all the atoms lie in the same plane. Rotating the two CH_2 groups in opposite directions would destroy the π-bond, which is consistent with the experimental fact that ethylene is a planar molecule that resists rotation about the C=C bond.

FIG. 11.29 The geometry and schematic electronic structure of ethylene. The σ-orbitals are directed along the bond axes and the π-orbital has its electron density above and below the plane that contains all the atoms.

The rigidity of the double bond has an interesting consequence which is displayed by the N_2F_2 molecule. The two nitrogen atoms can be pictured to be sp^2-hybridized, and linked by a σ—π double bond. On each nitrogen atom one of the sp^2 hybrid orbitals is used to form a bond to fluorine, and the other is occupied by a nonbonding pair of electrons. As shown in Fig. 11.30, the fluorine atoms may be on the same or on opposite sides of the double bond leading, in this case, to two isomers which are separable and distinguishable by physical methods. Since the isomers differ only in their geometry and not in the sequence in which atoms are arranged, they are said to be *geometric isomers*. The compound which has both fluorine atoms on the same side of the bond is called the *cis isomer*; that with fluorine atoms on opposite sides is called the

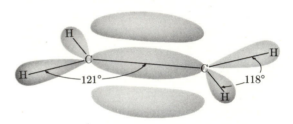

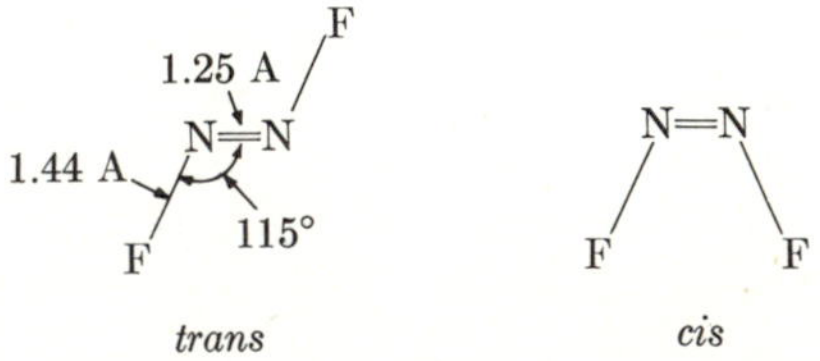

FIG. 11.30 The structures of the *trans* and *cis* isomers of N_2F_2.

trans isomer. The two are separable because they can only be interconverted by destroying the double bond, which requires considerable energy.

An analogous situation is found in compounds related to ethylene. There are *cis* and *trans* isomers of 1,2-dichloroethylene, CHClCHCl, whose structures are

Cl   Cl   H   Cl

C=C   and   C=C

H   H   Cl   H

cis dichloroethylene   *trans* dichloroethylene
$\mu = 1$ D $\mu = 0$

Note that the *trans* isomer of dichloroethylene has a symmetric structure in which the dipole moments of the bonds point in opposite directions and cancel each other, giving a molecular dipole moment of zero. In *cis* dichloroethylene the dipole moments of the two C—Cl bonds and the C—H bonds are aligned roughly in the same direction, and consequently the molecule has a dipole moment.

The lowest-energy molecular orbital of H_3^+. The nuclei lie at the apices of an equilateral triangle, and the electron density is distributed symmetrically between them.

FIG. 11.31

11.8 MULTICENTER BONDS

So far we have encountered situations in which one pair of electrons bond only *two* atoms. However, there are many instances in which one electron pair holds several atoms together. In these cases we say that multicenter, or delocalized, bonding exists. The simplest molecule which displays multicenter bonding is H_3^+, an ion found in electrical discharges through hydrogen gas. There have been no experimental determinations of the structure of H_3^+, but because of its simplicity, theoretical predictions about it are very reliable. These show that H_3^+ is an equilateral triangular molecule and that the two electrons occupy a molecular orbital which covers all three nuclei, as pictured in Fig. 11.31. Consequently, in H_3^+, two electrons bond three atoms together. The representation of the molecular orbital in Fig. 11.31 is a reasonably accurate, but not very convenient, way of showing the electronic behavior. However, our usual method of representing an electron-pair bond by a line or a pair of dots, as in

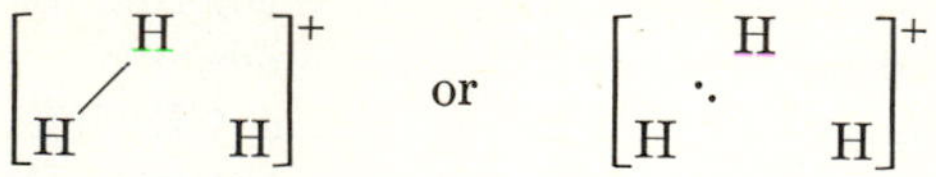

$$\begin{bmatrix} & H & \\ H & & H \end{bmatrix}^+ \quad \text{or} \quad \begin{bmatrix} & H & \\ H & & H \end{bmatrix}^+$$

completely fails to show the most important features of the molecule: both electrons visit all three nuclei, and identical forces hold all pairs of atoms together. To retain the simplicity of electron-dot or line pictures and still adequately describe the bonding in H_3^+, we adopt what is known as *resonance notation*. This is a scheme in which the *true* electron distribution is represented by a *superposition or blend* of several electron-dot or line pictures. For H_3^+ these are

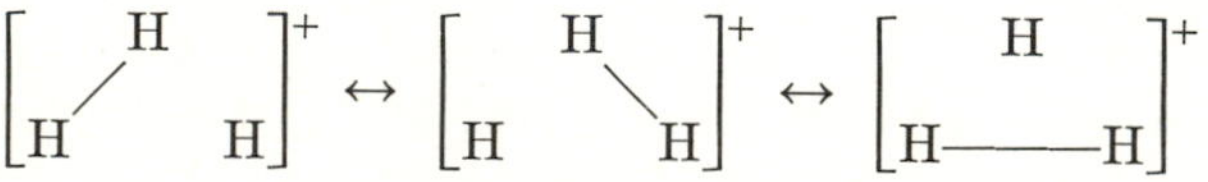

$$\begin{bmatrix} & H & \\ H & & H \end{bmatrix}^+ \leftrightarrow \begin{bmatrix} & H & \\ H & & H \end{bmatrix}^+ \leftrightarrow \begin{bmatrix} & H & \\ H\!-\!\!-\!\!-\!\!H \end{bmatrix}^+$$

The double-headed arrows are meant to indicate that no one of these structures is the true electron distribution, but that the three pictures are to be taken together to indicate that two electrons bond the three nuclei in a symmetric fashion. It is conventional to say that H_3^+ is a *resonance hybrid* of the three extreme structures shown. Note that the need for the concept of resonance arises *because we choose to represent the behavior of electrons by electron-dot or line pictures*. Just one of these pictures cannot show that electrons visit more than two nuclei, so we resort to drawing several resonance structures. Despite this rather artificial nature, the resonance concept does provide a very helpful means of discussing the electronic structure of many molecules.

The bond structure of the carbonate ion, $CO_3^=$, can be conveniently discussed in terms of resonance structures. The ion has a planar structure in which all carbon-oxygen bonds are identical:

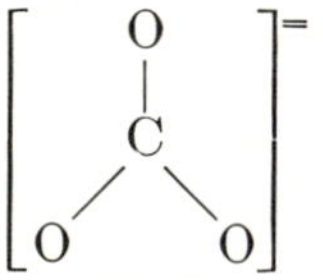

$$\begin{bmatrix} & O & \\ & | & \\ & C & \\ O & & O \end{bmatrix}^=$$

Since only elements of the second period are involved, we might expect the octet rule to be satisfied. However, a little experimentation shows that it is impossible to draw electron-dot structures which at the same time satisfy the octet rule and show that all C—O bonds are identical. We can only draw

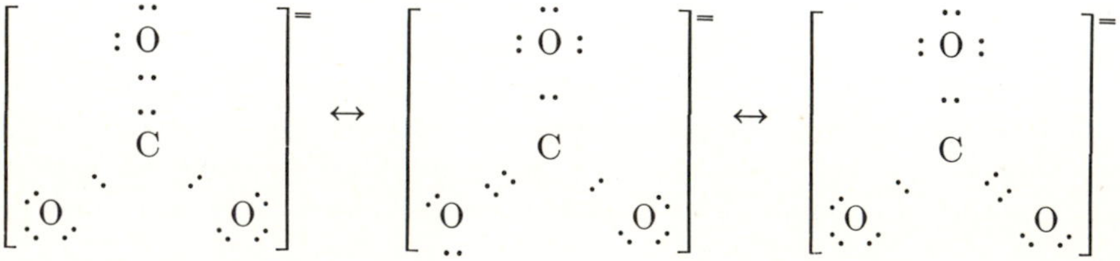

In each of these resonance structures one of the carbon-oxygen links is a double bond while the other two are single bonds. Taken together the structures mean that all carbon-oxygen bonds are *identical;* there is a σ-bond from the carbon atom to each oxygen atom, and in addition there is one pair of electrons occupying a molecular orbital which binds all three oxygen atoms to the central carbon atom. Each C—O link is intermediate between a single and double bond; in a sense, each is a $1\frac{1}{3}$ bond.

There are other four-atom chemical species which, like $CO_3^=$, have a symmetrical planar structure and 24 valence electrons. Two of the more common examples are nitrate ion, NO_3^-, and sulfur trioxide, SO_3. Since both these molecules have the same structure and number of valence electrons as $CO_3^=$, their bonding can be represented by resonance structures:

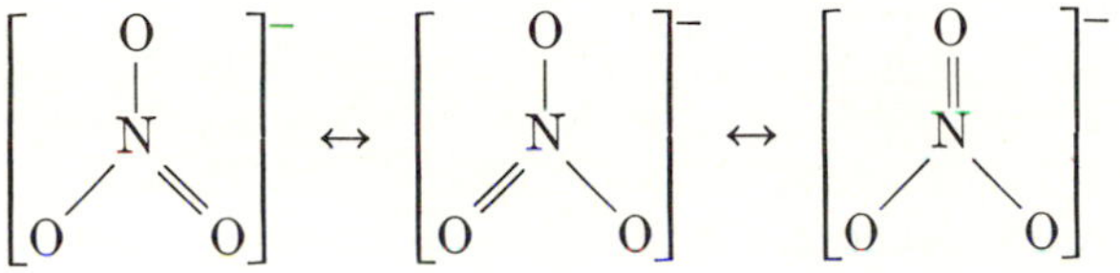

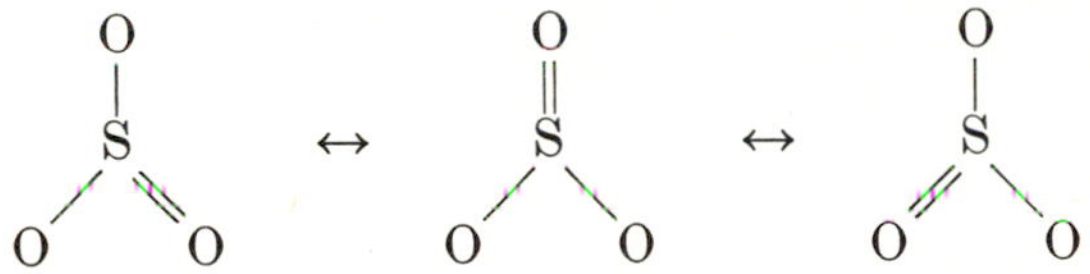

As another example of a group of isoelectronic molecules which require resonance notation, consider O_3, SO_2, and NO_2^-. All have 18 electrons, and all are symmetrical triangular molecules in which the two bonds are identical:

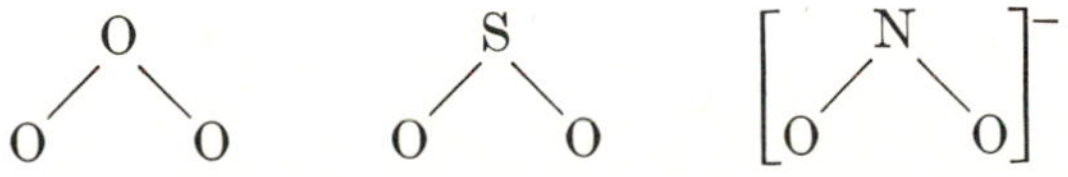

The resonance structures for ozone which satisfy the octet rule and which taken together show the equivalence of the two bonds are

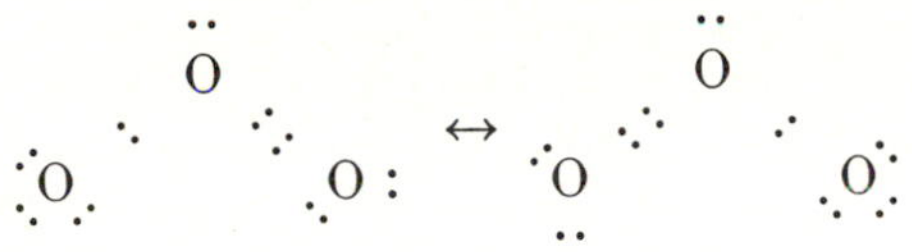

Analogous structures for SO_2 and NO_2^- are

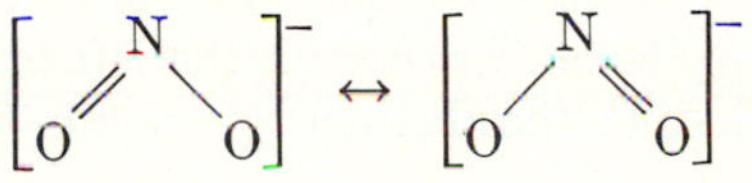

The resonance concept is particularly helpful in discussing the properties of organic molecules. To take a specific example, a variety of chemical and physical measurements show that benzene, C_6H_6, is a planar regular hexagon of carbon atoms, to each of which is attached a hydrogen atom. The detailed bond structure cannot be one of the two obvious possibilities:

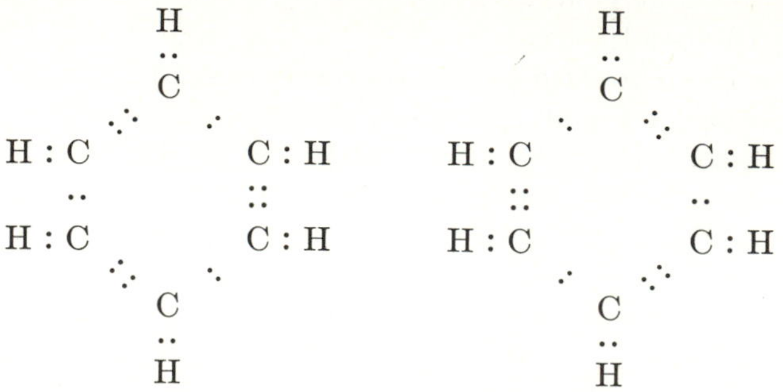

for in each of these structures there are two *different* types of carbon-carbon bonds. Consequently the molecule is pictured as a resonance hybrid of the two structures, which we can briefly indicate by

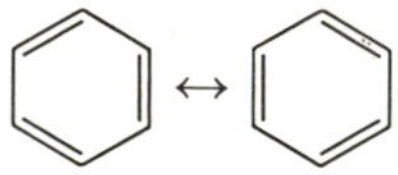

where the presence of the carbon and hydrogen atoms is understood. The idea that benzene is a resonance hybrid of two conventional structures is consistent not only with the geometric structure, but with its chemical properties, as we will see in Chapter 16.

11.9 METALLIC BONDING

Since three-quarters of the elements in the periodic table are metals, it is important to analyze the nature of the metallic bond. Once again we would like to be able to relate the nature and strength of the bonding to the properties of the individual atoms, and there are two particularly significant characteristic common to virtually all atoms of metallic elements. First, the ionization energies of the free atoms of the metallic and semimetallic elements are generally small, almost always less than 220 kcal/mole. The exception is mercury, whose ionization energy is 240 kcal/mole. In contrast, the atoms of the nonmetals usually have ionization energies greater than 220 kcal/mole. The second characteristic of a metallic atom is that the number of its valence electrons is less than the number of its valence orbitals. This observation is consistent with the fact that the metallic elements occur on the left-hand side of the periodic table, and are separated from the distinct nonmetals by the elements boron, silicon, germanium, and antimony. Let us investigate the consequences of these characteristics and see how they are related to metallic bonding.

A low ionization energy means that an atom has relatively little attraction for its valence electrons and suggests that it has very little affinity for any additional electrons. We have stressed that the stability of a covalent bond results from the potential-energy lowering that valence electrons experience when they move under the influence of more than one nucleus. If each of the two atoms that are being bound has rather slight attraction for electrons, we cannot expect the energy lowering of their valence electrons in the molecule to be at all substantial. Consequently we should not be surprised to find that the atoms of metallic elements form relatively weak electron-pair bonds with each other. Examination of Table 11.15 shows that this is true. The bond energies of the *known* diatomic molecules of the metallic elements are all very small, and there are many *possible* diatomic molecules of metallic elements that are unknown. Presumably this is true because they are not energetically stable.

Table 11.15 Dissociation energies of molecules of metallic elements (kcal/mole)

Li_2	25	Zn_2	5.7
Na_2	17	Cd_2	2.0
K_2	12	Hg_2	1.4
Rb_2	11	Pb_2	16
Cs_2	10.4	Bi_2	39
NaK	14	NaRb	13

While the interaction of a metallic atom with *one* other atom does not often lead to significant energy lowering, it is possible that greater stability can be achieved if the valence electrons of one atom move under the influence of *several* other nuclei. It is the second characteristic of metallic atoms, fewer valence electrons than valence orbitals, that makes this type of interaction possible. The fundamental limitation on the number of electrons that can be close to a given nucleus is imposed by the Pauli Exclusion Principle. For free atoms, the Exclusion Principle tells us exactly how many electrons of a given principal quantum number there can be. For aggregates of atoms, the application of the Pauli Principle is more difficult, but general observations and theoretical arguments suggest that the greatest number of valence electrons of *low energy*, shared or unshared, that can surround a given atom in any aggregate is equal to twice the number of its atomic valence orbitals. This is the basic reason for saturation of valence—the reason why NF_3 but not NF_5, PCl_3 and PCl_5 but not PCl_7, exist. The fact that the atoms of metallic elements have few valence electrons means that when in a condensed phase, each atom may share the electrons of many nearest neighbors in a manner that is energetically favorable, without violating the Pauli Exclusion Principle. Indeed, the characteristic feature of metallic crystals is that the coordination number of the atoms is high: 8 in the body-centered cubic lattice, and 12 in the hexagonal and cubic closest-packed structures.

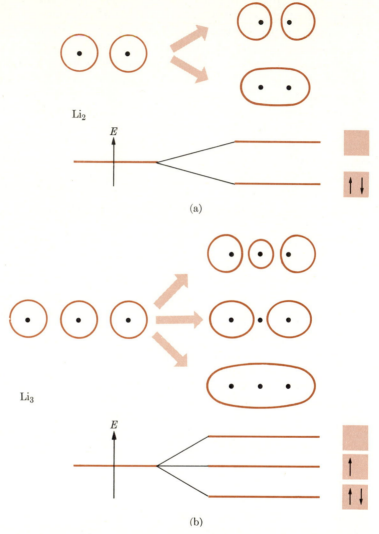

FIG. 11.32   Formation of molecular orbitals and the accompanying energy change for (a) Li_2 and (b) Li_3.

As an example of how the high coordination number in metals does not lead to a violation of the Exclusion Principle, consider the element aluminum, which has a cubic closest-packed lattice. We can take the attitude that each atom shares its 3 valence electrons with each of its 12 nearest neighbors. On the average, then, a given atom receives $\frac{3}{12}$ of an electron from each of its neighbors, or a total of 3 from all its neighbors. Consequently the total average number of electrons shared by any atom is 6, 3 of which come from itself and 3 from its neighbors. We see, therefore, that despite the high coordination number, the average number of electrons near a single atom does not exceed twice the number of valence orbitals.

So far our argument suggests that the reason metallic crystals are more stable than the separated atoms is that in the crystal, the atomic valence electrons can move in the electric field of several nuclei. This idea can be extended in a way that shows that metals are extreme examples of the delocalized or multi-center bonding we discussed in Section 11.8. To see how this comes about, let us imagine building a one-dimensional lithium crystal by bringing together two, three, and then many lithium atoms.

Figure 11.32(a) illustrates the formation of Li_2 from two lithium atoms. The figure shows both the potential energy of an electron and its total energy in both the separated atoms and the diatomic molecule. When the atoms are brought together, the energy levels of the $2s$ atomic electrons split into two new levels, labeled $\sigma 2s$ and $\sigma^* 2s$, which correspond to bonding and antibonding molecular orbitals. In Li_2, both electrons occupy the $\sigma 2s$-orbital, but any electrons in the $\sigma 2s$- or $\sigma^* 2s$-orbitals are the property of the molecule as a whole. On the other hand, the $1s$-electrons are largely localized near particular atoms, since their total energy is less than that needed to pass over the potential-energy barrier that exists between the two atoms.

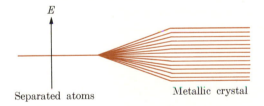

The behavior of the energies of the valence orbitals of lithium when the metallic crystal is formed from many separated atoms.

FIG. 11.33

When three atoms are brought together, the situation is as pictured in Fig. 11.32(b). There are three molecular orbitals, and once again electrons in any of these orbitals are the property of the molecule as a whole, and not of any one particular atom. The situation that obtains when many lithium atoms are brought together is shown in Fig. 11.33. For N atoms there are N molecular orbitals that arise from the overlap of the $2s$-orbitals alone, and the energies of these orbitals now form a *band* of closely spaced levels. More orbitals are contributed by the overlap of the $2p$ atomic orbitals, and the energies of these also lie in a dense band that is joined continuously to the band formed from the $2s$-orbitals. Any electron in one of these orbitals is the property of the crystal as a whole, and serves to bind many nuclei together. It is in this sense that metals represent an extreme case of multicenter bonding.

The qualitative conclusions we have drawn hold true for three-dimensional crystals. In general, all the valence orbitals of the free atoms are converted to a group of nonlocalized or multicenter orbitals in the metallic crystal, and the energies of these orbitals form a closely spaced group called the *valence band*. One way to describe the electronic situation in metals, then, is to picture the crystal as a collection of ions like Li^+ immersed in a "sea" of mobile valence electrons. This electron sea is responsible for the cohesion of metals, and for their unique mechanical, electrical, and thermal properties, as we shall see in subsequent chapters.

11.10 CONCLUSION

Chemical bonding is an interesting but complex subject. In this chapter we have exposed the most useful fundamental concepts, and have suggested a few rules which will be helpful in discussing and understanding the relation between bonding and chemical properties. However, it must be realized that bonding phenomena are so complex that there exist "violations" of almost every simple bonding rule. In particular, it was long believed that the rare gas atoms could never form any true compounds because of their "complete valence octet." The synthesis in 1962 of the xenon fluorides removed this last "safe" application of the otherwise flimsy octet rule. Consequently, in order to deepen our understanding of the valence rules, and to be sure that they do not need revision or modification, we must constantly test them by application to known chemical phenomena. This will be our purpose in subsequent chapters.

SUGGESTIONS FOR FURTHER READING

Cartmell, E., and G. W. A. Fowles, *Valency and Molecular Structure.* London: Butterworth, 1961.

Companion, A. L., *Chemical Bonding.* New York: McGraw-Hill, 1964.

Coulson, C. A., *Valence.* New York: Oxford, 1961.

Day, M. C., and J. Selbin, *Theoretical Inorganic Chemistry.* New York: Reinhold, 1962.

Gray, H. B., *Chemical Bonds.* Menlo Park, Calif.: W. A. Benjamin, 1973.

Griswold, E., *Chemical Bonding and Structure.* Lexington, Mass.: Raytheon Education Co., 1968.

Harvey, K. B., and G. B. Porter, *Introduction to Physical Inorganic Chemistry.* Reading, Mass.: Addison-Wesley, 1963.

Linnett, J. W., *Wave Mechanics and Valency.* London: Methuen, 1960.

Pauling, L., *The Nature of the Chemical Bond.* Ithaca, New York: Cornell, 1960.

Ryschkewitsch, G. E., *Chemical Bonding and the Structure of Molecules.* New York: Reinhold, 1963.

11.1 In which of the following compounds would you expect to find the smallest separation between the nuclei of neighboring ions? All have the sodium chloride lattice structure: NaI, KCl, LiF. Which should have the most stable crystal lattice?

11.2 If the alkali-halide crystals were truly ionically bonded, their crystal-lattice energies would be proportional to $1/r_0$, where r_0 is the separation between neighboring ions. Use the following data for r_0 and those in Table 11.11 to show that a plot of crystal-lattice energy as a function of $1/r_0$ is linear:

LiF	2.01 A	KF	2.66 A
LiCl	2.57 A	KBr	3.29 A
LiBr	2.75 A	KI	3.53 A
LiI	3.02 A		

11.3 All of the alkaline-earth oxides have the sodium-chloride crystal lattice structure. Calculate the contribution to the lattice energy due to the Coulomb forces between ions. The separation in angstroms of neighboring nuclei are:

MgO	2.10	SrO	2.57
CaO	2.4	BaO	2.76

11.4 By considering the energy factors that determine the stability of ionic crystal lattices, and the ease of formation of the free ions themselves, try to explain why although both TlF and TlF_3 exist, TlI exists but TlI_3 does not.

11.5 Predict the geometries of the following molecules and ions:

BF_4^- PF_6^-
$TeCl_4$ XeF_4
I_3^- $CuCl_2^-$
$PbCl_2$

Justify your predictions by citing for each case an analogous or isoelectronic species whose geometry you know.

11.6 The sulfate ion, $SO_4^=$, consists of a central sulfur atom surrounded by four oxygen atoms located at the corners of a regular tetrahedron. All the sulfur-oxygen bonds are equivalent. Draw the electron-dot structure(s) that are consistent with the equivalence of the bonds.

11.7 What feature of electron sharing is it that gives the covalent bond in H_2 its stability? Is the sharing of an electron by two atoms a *sufficient* condition for the formation of a stable bond?

11.8 Phosphorus exists in one of its allotropic forms as P_4 molecules. In a P_4 molecule, the phosphorus atoms are at the corners of a regular tetrahedron, each atom is bonded to three others, and all bonds are equivalent. Draw the P_4 molecule, and indicate its electron-dot structure. What is the P—P—P bond angle? How does this compare with the bond angles in PH_3? On this basis do you find the existence of P_4 tetrahedra surprising or something to be expected?

11.9 The rule that the number of covalent bonds an atom can form is related to the possible number of unpaired valence electrons it can have is consistent with the formulas of a large number of known compounds. Show that this rule leads to the prediction of the existence of compounds like XeF_2 and XeF_4.

11.10 Discuss how electron spin influences (a) the number of covalent bonds an atom can form, and (b) the geometries of covalent compounds.

11.11 The linear geometry of $HgCl_2$ suggests that we think of the central mercury atom as displaying *sp*-hybridization. What form of hybridization is associated with the central atom when the atoms bonded to it are located at the corners of (a) an equilateral triangle, (b) a regular tetrahedron, (c) an octahedron, (d) a trigonal bipyramid?

11.12 The antibonding σ^*1s differs from the bonding $\sigma1s$-orbital of H_2^+ in that the antibonding orbital has a deficiency of electron density in the region between the nuclei. In effect, the σ^*1s-orbital is divided by a nodal plane of zero electron density which is perpendicular to the internuclear axis. With this in mind, draw what you think an antibonding π^*2p-orbital should look like.

11.13 The molecules of nitrogen $N{\equiv}N$ and acetylene $HC{\equiv}CH$ are isoelectronic, and both incorporate a triple bond. By reviewing the electronic structure of nitrogen, predict the geometry of acetylene, and discuss its electronic structure in terms of hybridization about the carbon atom, and the type of molecular orbitals occupied by electrons.

11.14 The molecule PF_3 is polar, with a dipole moment of 1.02 D, and thus the P—F bond is polar. Judging from the proximity of silicon and phosphorus in the periodic table, we expect that the Si—F bond would also be polar, but the molecule SiF_4 has no dipole moment. Explain why this is so.

11.15 The gaseous potassium chloride molecule has a measured dipole moment of 10.0 D, which indicates that it is a very polar molecule. The separation between the nuclei in this molecule is 2.67×10^{-8} cm. What would the dipole moment of a KCl molecule be if there were opposite charges of one fundamental unit $(4.8 \times 10^{-10}$ esu) located at each nucleus? Is the picture of a completely ionic KCl molecule entirely satisfactory?

11.16 In nitryl chloride, O_2NCl, the chlorine atom and the two oxygen atoms are bonded to a central nitrogen atom, and all atoms lie in a plane. Draw the electron-dot resonance structures that satisfy the octet rule and which together are consistent with the fact that the two nitrogen-oxygen bonds are equivalent.

11.17 In the formulas of the following molecules, the element given first is a "central" atom to which the other atoms are attached. What do you expect their geometric structure to be? AsF_3, AsF_5, XeF_2, AlF_4^-, PCl_4^+, PCl_6^-.

CHAPTER 12

MOLECULAR ORBITALS

In Chapter 11 we examined a number of simple descriptions of the chemical bond. We emphasized that, regardless of the model or picture of the bond, the physical phenomenon responsible for molecular stability is the lowering of the Coulomb potential energy that occurs when valence electrons can move under the attraction of two or more nuclei. The wave functions that describe how the electrons in molecules are distributed are called molecular orbitals. These molecular orbitals can be used to calculate the geometry, energy levels, and other properties of molecules. It is possible, however, from the qualitative characteristics of molecular orbitals, to deduce or rationalize some important qualitative properties of molecules. In this chapter we shall extend and broaden our ideas about molecular orbitals so that we will be able to use them to understand the structure and stability of a variety of molecules.

12.1 ORBITALS FOR HOMONUCLEAR DIATOMIC MOLECULES

In Section 11.3 we discussed the characteristics of the two molecular orbitals of H_2^+ which have the lowest energy, $\sigma 1s$ and $\sigma^* 1s$. In preparation for our encounter with more complex systems, it is desirable to review and summarize what was said about these simplest molecular orbitals.

By solving the Schrödinger equation for the motion of one electron in the field of two fixed protons, it is possible to obtain exact, but quite complicated, descriptions of the molecular orbitals of H_2^+. A description which is mathematically much simpler, though definitely an approximation, can be obtained by regarding these molecular orbitals as Linear Combinations of Atomic Orbital Molecular Orbitals, conventionally denoted as LCAO-MO's. This means simply that to find the molecular-orbital wave function, we linearly combine (add or subtract) the atomic-orbital wave function of one atom with the atomic-orbital wave function for the other atom. We have then, for two protons a and b,

$$\sigma 1s \cong \frac{1}{\sqrt{2(1+S)}} \, [\psi_a(1s) + \psi_b(1s)],$$

$$\sigma^* 1s \cong \frac{1}{\sqrt{2(1-S)}} \, [\psi_a(1s) - \psi_b(1s)].$$

The factors of $1/\sqrt{2(1 \pm S)}$ normalize the *molecular* wave function so that the probability of finding an electron somewhere in all space is unity. The quantity S is called the overlap integral, which is explicitly

$$S = \int \psi_a \psi_b \, d\tau,$$

where $d\tau$ is the differential element of volume. The overlap integral S is a measure of how closely the two atomic wave functions coincide. Usually S is approximately equal to 0.2 or 0.3.

A pictorial representation of this LCAO procedure appears in Fig. 12.1. When the two $1s$ functions are added they reenforce each other everywhere, and most notably in the region between the two nuclei. This buildup of electron density between the nuclei helps to lower the Coulomb potential energy. As a result, this orbital has bonding characteristics and is denoted $\sigma 1s$. When one atomic orbital is subtracted from the other, they exactly cancel each other in a plane midway between the nuclei, and thereby produce a nodal plane. The molecular wave function is of opposite sign on either side of this nodal plane. When the wave function is squared, the resulting probability density is of course everywhere positive, except on the nodal plane, where it is zero. This deficiency of electron density in the internuclear region helps to raise the Coulomb potential energy of the system, and the node in the wave function produces an increase in the electron kinetic energy. Consequently the total energy is high, the molecule is not bound, and the orbital is described as antibonding.

The phenomena associated with the formation and description of these two molecular orbitals should be noted carefully, since they will aid us in the construction of other molecular orbitals. Briefly, we can expect the following: Linear combination of *two* atomic orbitals will produce *two* molecular orbitals, one of higher and one of lower energy than the atomic orbitals. If the molecular orbital has a node *between* the nuclei, it will tend to be antibonding. If there is a buildup of electron density between the nuclei, the molecular orbital will tend to be bonding.

We can now examine the molecular orbitals needed to describe the homonuclear diatomic molecules of the second row of the periodic table, Li_2, N_2, O_2, etc. To begin, we note that the $1s$ atomic orbitals of these atoms are held quite close to the nuclei, and are very little affected by whether the atom is free or bonded chemically. We can, therefore, regard these inner-shell electrons as nonbonding, and concentrate only on molecular orbitals which we can generate from the valence atomic orbitals.

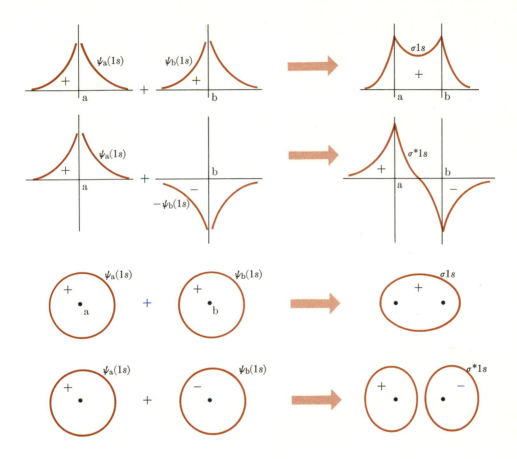

Two ways of schematically representing the formation of bonding and antibonding molecular orbitals by addition and subtraction of atomic orbitals.

FIG. 12.1

By linearly combining a $2s$-orbital on atom a with a $2s$-orbital on atom b we get approximations to the bonding and antibonding $\sigma 2s$ molecular orbitals:

$$\sigma 2s \cong N[\psi_a(2s) + \psi_b(2s)],$$
$$\sigma {*}2s \cong N{*}[\psi_a(2s) - \psi_b(2s)].$$

The procedure is completely analogous to that used for $\sigma 1s$ and $\sigma^* 1s$ of H_2^+. The quantities N and N^* are the normalization factors. The $\sigma^* 2s$-orbital has a nodal plane between the two nuclei. Consequently it is antibonding and is of higher energy than $\sigma 2s$, which does not have this nodal plane and is bonding. Formation of the $\sigma 2s$- and $\sigma^* 2s$-orbitals is illustrated in Fig. 12.2. Note that there is a nodal surface that surrounds the nuclei in both the $\sigma 2s$- and $\sigma^* 2s$-orbitals, which distinguishes them from the σ- and σ^*-orbitals generated from $1s$ atomic functions.

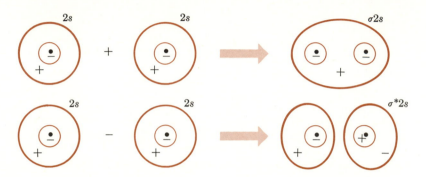

FIG. 12.2   Formation of the $\sigma 2s$ bonding and $\sigma^* 2s$ antibonding orbitals by addition and subtraction of $2s$ atomic orbitals. The plus and minus signs refer to the sign of the wave functions, and not to nuclear or electronic charges.

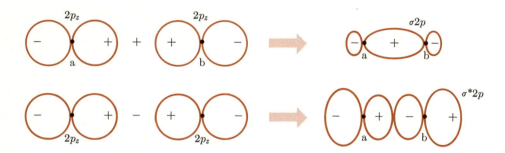

FIG. 12.3   Schematic representation of the formation of the $\sigma 2p$ bonding and $\sigma^* 2p$ antibonding orbitals by linear combination of $2p_z$ atomic orbitals.

Proper combination of the $2p$-orbitals associated with the two nuclei produces another pair of σ molecular orbitals, $\sigma 2p$ and $\sigma^* 2p$. If we take the internuclear line to be the z-axis, and then recognize that the $2p_z$-orbital of each nuclear center has cylindrical symmetry about this axis, we see that combining such atomic orbitals will indeed produce a cylindrically symmetric, or σ, molecular orbital.

In forming the linear combinations of $2p_z$-orbitals, we must be careful to take proper account of the fact that the sign of $\psi(2p_z)$ is different in the two lobes of the wave function. To avoid confusion we first set up the atomic orbitals as shown in Fig. 12.3, with the positive lobe of each function pointed into the internuclear region. Then, since the overlapping lobes of the two orbitals are both positive, adding the two functions together will increase the electron density in the internuclear region, and produce a bonding $\sigma 2p$-orbital. Subtraction of one function from the other produces a nodal plane midway between the nuclei, and the resulting molecular orbital, designated $\sigma*2p$, is antibonding. Thus we have

$$\sigma 2p \cong N[\psi_a(2p_z) + \psi_b(2p_z)]$$

and

$$\sigma*2p \cong N*[\psi_a(2p_z) - \psi_b(2p_z)]$$

as the LCAO approximations to these two molecular orbitals. Figure 12.3 shows a schematic representation of these orbitals.

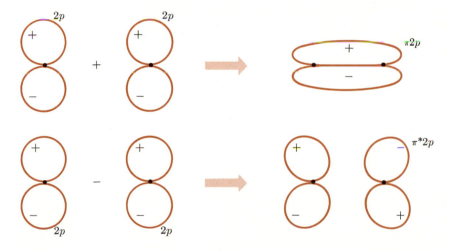

Formation of the bonding π and antibonding $\pi*$ molecular orbitals by linear combination of atomic orbitals.

FIG. 12.4

Formation of π molecular orbitals from p atomic orbitals has been discussed in Section 11.7. As indicated there, adding the p_x-orbital on nucleus a to the p_x-orbital on nucleus b in such a way that the positive and negative lobes of one orbital overlap with respectively the positive and negative lobes of the other orbital produces a $\pi 2p_x$-bonding orbital. The LCAO approximation to this orbital is

$$\pi 2p_x \cong N[\psi_a(2p_x) + \psi_b(2p_x)],$$

and its pictorial representation is given in Fig. 12.4. While the π-orbital has a nodal (yz) plane which *contains* the nuclei, there is an increase in electron density *between* the nuclei, and the orbital is bonding.

Subtraction of one $2p_x$-function from the other produces an approximation to a π^* antibonding orbital

$$\pi^*2p_x \cong N^*[\psi_a(2p_x) - \psi_b(2p_x)],$$

which is shown in Fig. 12.4. Now in addition to the nodal plane which contains the nuclei, we have a node between the nuclei, and concomitant antibonding character.

An exactly analogous argument can be applied to the combination of atomic $2p_y$-orbitals. A bonding $\pi2p_y$-orbital and an antibonding π^*2p_y-orbital can be generated which are oriented perpendicular to the $\pi2p_x$- and π^*2p_x-orbitals. Thus the total of four atomic p-orbitals which are perpendicular to the internuclear axis generates a total of four π molecular orbitals—two of which are bonding, and two of which are antibonding.

The eight molecular orbitals which we have discussed are all that can be generated for a diatomic molecule from the total of eight atomic orbitals of principal quantum number $n = 2$. Molecular orbitals of higher energy can be formed from linear combinations of $3s$- and $3p$-orbitals, but no new ideas are involved, and it is not necessary to discuss these higher-energy orbitals explicitly.

We turn now to the problem of determining the order of increasing energy of the molecular orbitals that we have discussed. Three general rules are helpful. (1) The energy of the molecular orbitals is strongly influenced by the energy of the atomic orbitals to which they are related. (2) If two atomic orbitals are largely confined to regions near their respective atomic nuclei and, therefore, do not overlap extensively, the molecular orbitals formed from them will be neither very strongly bonding nor strongly antibonding. (3) If the atomic orbitals do overlap extensively, the bonding orbital will have an energy quite a bit lower than that of the atomic orbitals, and its antibonding partner will have a correspondingly higher energy. The quantitative description of molecular-orbital energies can come only from experiment or, in favorable cases, from extensive quantum-mechanical calculation; and each molecule and ion has its own unique energy-level pattern. Just as is true for atomic energy levels, however, certain useful qualitative generalizations can be made.

Figure 12.5a shows the molecular-orbital energy pattern which applies to the homonuclear diatomic molecules O_2, F_2, and their positive and negative ions. The valence orbitals of lowest energy are the σ-σ^* bonding-antibonding pair generated from the $2s$ atomic orbitals. These lie lowest in energy, principally because the $2s$ atomic orbitals from which they are formed lie well below the $2p$-orbitals in the free atoms. The $2s$-orbitals, particularly in fluorine, are so low in energy that they do not overlap and interact extensively. As a result the $\sigma 2s$-orbital is not strongly bonding, nor is the $\sigma^* 2s$-orbital very strongly antibonding in oxygen and fluorine.

Since the atomic $2p$-orbitals all have the same energy, the molecular orbitals generated from them have somewhat similar energies. The overlap of the $2p_z$-orbitals along the internuclear line is relatively large, and consequently, the $\sigma 2p_z$ bonding orbital has lower energy, and its antibonding partner $\sigma^* 2p_z$ has higher energy, than the other molecular orbitals from atomic $2p$-orbitals. The π_x and π_y bonding orbitals have the same energy, since they are equivalent except for their orientation in space. They lie somewhat below, and their antibonding partners π_x^* and π_y^* lie somewhat above, the energy of the atomic orbitals used to generate them.

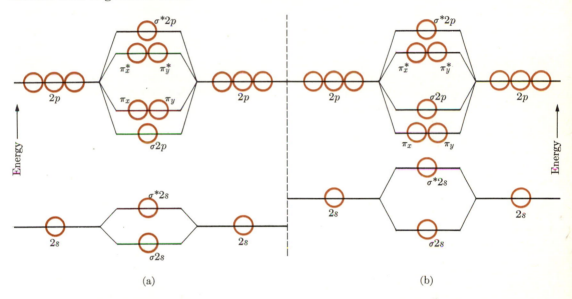

Molecular-orbital energy patterns for homonuclear diatomic molecules. (a) Diagram for molecules with low-lying 2s-orbitals. (b) Diagram for N₂ and lighter homonuclear diatomics.

FIG. 12.5

For the diatomics Li_2, Be_2, B_2, C_2, and N_2, the molecular orbitals fall in a very slightly different pattern, shown in Fig. 12.5(b). In these molecules, the $\sigma 2p$-orbital lies a bit higher in energy than the two bonding $\pi 2p$-orbitals. This feature is a consequence of the repulsion between electrons that occupy the $\sigma 2s$- and $\sigma^* 2s$-orbitals and any electrons in the $\sigma 2p$-orbital. These repulsions, and the resultant elevation of the energy of the $\sigma 2p$-orbital, occur because both the $\sigma 2p$ and $\sigma 2s$ electrons tend to occupy the same region of space in these lighter diatomic molecules. This effect is diminished in O_2 and F_2 because in the atoms O and F, the $2s$ atomic orbitals are quite low in energy, and are largely confined to regions close to the nucleus. This characteristic is maintained in the $\sigma 2s$ bonding and antibonding orbitals of O_2 and F_2. Consequently, the $\sigma 2s$

electrons do not seriously interfere with the $\sigma 2p$ electrons, and in the heavier diatomics, the energy of $\sigma 2p$ is below the level of the $\pi 2p$-orbitals.

Now we can examine the electronic structure and bonding of the homonuclear diatomic molecules of the second-row elements by feeding the appropriate number of electrons into the orbital energy-level patterns of Fig. 12.5. We begin with Li_2, which has a total of six electrons. Four of these are $1s$ electrons, two on each atom. The energy of these "core" electrons is very low, and since they are largely confined to regions near the nuclei, they do not contribute to chemical bonding. We shall therefore ignore them in Li_2 and in the other diatomics of higher atomic number. We are left then with two electrons which can be accommodated with paired spins in the $\sigma 2s$-orbital. Since Li_2 has two electrons in a bonding orbital, we say it has a single electron-pair bond.

For Be_2, we have in addition to the core $1s$ electrons, four valence electrons to accommodate. Two of them enter the $\sigma 2s$ bonding orbital. The other two must enter the next available orbital of lowest energy, which is the $\sigma^* 2s$ antibonding orbital. Since Be_2 has two bonding and two antibonding electrons, its situation is much like that of He_2. We expect, therefore, that Be_2 will not be a bound molecule, and in fact no stable Be_2 is known.

The molecule B_2 is known to exist as a gaseous species, and to have two unpaired electrons. Let us see if this is consistent with our molecular-orbital energy-level diagram. Of the six valence electrons, a total of four enter the $\sigma 2s$- and $\sigma^* 2s$-orbitals. The remaining two electrons would pair and enter the $\sigma 2p$-orbital, if that were the orbital of lowest energy available. In B_2, however, the $\pi 2p_x$- and $\pi 2p_y$-orbitals lie lower in energy than $\sigma 2p$, as we have discussed earlier. Since the π-orbitals have the same energy, the most favorable situation is for each to be occupied by one electron. The spins of these two electrons are parallel, just as are the spins of the electrons in the half-filled $2p$ atomic orbitals in C, N, and O. In B_2 then, we find four bonding valence electrons, and two antibonding electrons, for a net of $4 - 2 = 2$ bonding electrons. Thus we can say that there is a single bond in B_2. However, a more accurate description would be that there are two half bonds, since the last two bonding electrons are in different orbitals.

The molecule C_2 occurs in flames and electrical discharges through carbon-containing gases. It has a bond energy of 150 kcal/mole, which suggests that the atoms are linked by a double bond. The molecule has eight valence electrons, two more than B_2. The most stable valence-electron configuration of C_2 has been found to be $(\sigma 2s)^2(\sigma^* 2s)^2(\pi 2p_x)^2(\pi 2p_x)^2$. The net number of bonding electrons is four, and thus the molecule has a double bond. The bonding is rather unusual in that the double bond consists of two π bonds, rather than the more usual σ-π combination. Another unusual feature of C_2 is that the electron configuration $(\sigma 2s)^2(\sigma^* 2s)^2(\pi 2p_x)^2(\pi 2p_y)^1(\sigma 2p)^1$ is only 2.3 kcal/mole higher in energy than the ground state. This shows that the $\pi 2p$- and $\sigma 2p$-orbitals are quite close in energy.

The next molecule in the sequence is N_2. As indicated in Section 11.7, the electron configuration of N_2 is $(\sigma 2s)^2(\sigma^* 2s)^2(\pi 2p_x)^2(\pi 2p_y)^2(\sigma 2p)^2$, which corresponds to a triple bond. The $\sigma 2s$-σ^*2s bonding-antibonding pair can almost as well be regarded as two nonbonding $2s$ atomic orbitals, only slightly distorted by interaction with each other. Thus the triple bond consists of two π bonds and a σ bond.

In N_2, all bonding orbitals are filled, and any additional electrons in subsequent molecules must enter antibonding orbitals. The O_2 molecule, for example, has 12 valence electrons, two more than N_2. The first ten of these can be accommodated as were the electrons in N_2, but the last two must enter the π^* antibonding orbitals. The configuration of lowest energy has one of these last two electrons in the π_x^*-orbital, and the other in the π_y^*-orbital. The configuration of O_2 is therefore

$$(\sigma 2s)^2(\sigma^*2s)^2(\sigma 2p)^2(\pi_x 2p)^2(\pi_y 2p)^2(\pi_x^*2p)^1(\pi_y^*2p)^1,$$

and there is a net of $8 - 4 = 4$ bonding electrons. We find what is in effect a double bond in O_2, which is consistent with its fairly large bond energy of 118 kcal/mole. The double bond is somewhat unique, however, in that it seems to consist of a triple bond opposed by two half antibonds. The molecular-orbital description of O_2 provides a simple explanation for the paramagnetic properties of the molecule, since the two electrons which occupy the two separate antibonding orbitals have parallel spins.

The fluorine molecule F_2 has two more electrons than O_2, and consequently has the valence electron configuration $(\sigma 2s)^2(\sigma^*2s)^2(\sigma 2p)^2(\pi 2p)^4(\pi^*2p)^4$. The net of $8 - 6 = 2$ bonding electrons corresponds to a single electron-pair bond. This bond has a dissociation energy of only 35 kcal/mole, which is comparatively small. One possible explanation of the small bond energy is that the four electrons in the π^* antibonding orbitals exert a greater antibonding effect than the bonding effect of the four electrons in the π bonding orbitals.

In contrast to the molecular-orbital approach, the valence-bond picture of F_2 treats the $2s$-orbitals and two of the $2p$-orbitals on each atom as nonbonding or atomic in character. These nonbonding orbitals accommodate six electrons on each atom, and the single electron-pair bond is formed by the overlap of the two remaining p-orbitals. Thus the treatment of the bond is essentially the same in the two pictures, but the electrons which are treated in the molecular-orbital method as bonding and antibonding pairs are described simply as nonbonding or atomic in the electron-pair valence-bond method.

Question. In the following pairs of molecules, which would have the greater dissociation energy? O_2^+, O_2^-; Be_2, Be_2^+; B_2, B_2^+; C_2, C_2^-.

12.2 HETERONUCLEAR DIATOMIC MOLECULES

In forming the LCAO approximations to the molecular orbitals for homonuclear diatomic molecules we combined with each other *atomic orbitals of the same type* from each atom. We did this because from the symmetry of homonuclear molecules we expect that electrons in a given molecular orbital are shared equally between the two identical nuclear centers. The molecular orbitals of heteronuclear molecules do not have this symmetric character. However, if the diatomic molecule is composed of atoms of rather similar atomic numbers, as is true for CO, NO, and CN, the asymmetry is not pronounced, and the electronic structure can be described satisfactorily in terms of the molecular orbitals which we used for homonuclear diatomics. Therefore, the CO molecule, which has ten valence electrons, has the valence-electron configuration

$$(\sigma 2s)^2(\sigma^* 2s)^2(\pi 2p)^4(\sigma 2p)^2,$$

just as does N_2. The qualitative difference is that because of the greater charge on the oxygen atom, the bonding molecular orbitals put more electron density near the oxygen atom. For the antibonding orbitals, the opposite is true, as we shall see.

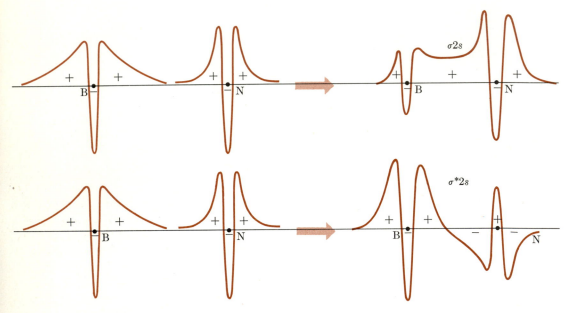

FIG. 12.6   Schematic representation of the formation of $\sigma 2s$ and $\sigma^* 2s$ in BN. The bonding σ orbital is more concentrated near the more electronegative nitrogen, and the antibonding orbital has greatest amplitude near boron.

Let us consider forming the σ and σ^* pair of molecular orbitals from two atomic orbitals of different energy, such as the 2s-orbital of B and the 2s-orbital of N. Because of the greater nuclear charge of nitrogen, the 2s atomic orbital on this nucleus lies lower in energy than does the 2s-orbital on the boron nucleus. We expect, consequently, that the lowest-energy or bonding σ orbital formed by this combination will be concentrated largely at the nitrogen atom, since this is the region of low potential energy. In the mathematical description of this orbital, this asymmetry can be produced by adding the two atomic orbitals together with coefficients chosen to ensure that N $2s$ is more important than B $2s$. Thus the simplest LCAO approximation for the σ-bonding molecular orbital is

$$\sigma = C_B\psi_B(2s) + C_N\psi_N(2s),$$

where $C_N > C_B > 0$, and ψ_B and ψ_N are atomic wave functions centered on boron and nitrogen, respectively.

The corresponding antibonding orbital will have a node between the nuclei, and will be of higher energy. Because of the smaller nuclear charge on boron, the region around this nucleus is of higher potential energy than the region around the nitrogen nucleus. Consequently we expect that the antibonding orbital will be concentrated largely near the boron nucleus. The mathematical description is

$$\sigma^* = C'_B\psi_B(2s) - C'_N\psi_N(2s),$$

with $C'_B > C'_N$. A pictorial description of the formation of σ and σ^* for BN is given in Fig. 12.6.

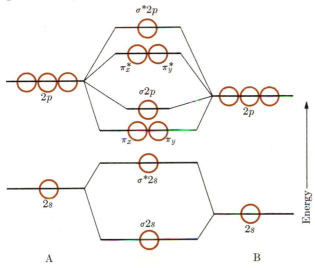

Molecular-orbital energy-level diagram for a heteronuclear diatomic molecule AB in which B is more electronegative than A.

FIG. 12.7

Formation of the other bonding-antibonding molecular-orbital pairs for diatomic molecules whose nuclei do not differ greatly in atomic number follows the pattern we have just discussed. The bonding orbitals are more concentrated around the nucleus of higher atomic number, and antibonding orbitals have greater density near the nucleus of lower charge. An orbital energy-level diagram is given in Fig. 12.7. From it we can deduce that the valence-electron configuration of BO is $(\sigma 2s)^2(\sigma^*2s)^2(\pi_x 2p)^2(\pi_y 2p)^2(\sigma 2p)^1$. The configuration of BN, which is not obvious from the diagram, is

$$(\sigma 2s)^2(\sigma^*2s)^2(\pi_x 2p)^2(\pi_y 2p)^1(\sigma 2p)^1.$$

When the difference in the atomic numbers of the combining atoms is large, one must take special care in describing the molecular orbitals. In these cases, molecular orbitals are formed not by combining atomic orbitals of similar designation (such as $2s$ with $2s$, etc.) but by combining orbitals of *similar energy*. The molecule HF provides a good example. The hydrogen $1s$ atomic orbital does not combine with the fluorine $1s$-orbital to form a molecular orbital because the $1s$ electrons of fluorine are *very* strongly bound, and are confined to regions quite near the fluorine nucleus. The same is true for the $2s$ electrons of fluorine. Because the binding energy associated with the $1s$-orbital of hydrogen and the $2p$-orbitals of fluorine are somewhat similar, overlap and interaction of $H(1s)$ and $F(2p_z)$ does occur and produces a bonding-antibonding pair of molecular orbitals. The $2p_x$- and $2p_y$-orbitals of fluorine remain as nonbonding atomic orbitals in HF. Since the ionization energy of H is 313.6 kcal/mole, and that of F is 402 kcal/mole, we expect that the electron pair in the σ bonding orbital of HF will spend more time near the fluorine nucleus. Consequently, HF should be a polar molecule, as is indeed observed.

The gaseous molecule LiF provides an example of extremely unequal sharing of electrons in a molecular orbital. A σ-σ^* pair is generated by the interaction of the lithium $2s$-orbital and a fluorine $2p$-orbital. Because the ionization energy of lithium is so much smaller than that of fluorine, the electron pair in the bonding orbital spends almost all of its time in the vicinity of the fluorine atom. Therefore, LiF is a very polar molecule, so much so that we say that the bonding is nearly purely ionic. Thus, by proper choice of the coefficients of the contributing atomic orbitals, the molecular-orbital concept can be used to represent the pure covalent bond of homonuclear diatomics, and as well, the ionic bond of the alkali halides.

Question. The energy level diagram in Fig. 12.7 is appropriate for molecules composed of the lower atomic number elements Be, B, C, and N. What modification should be made to make it satisfactory for a molecule like OF?

12.3 TRIATOMIC MOLECULES

We begin with the simplest triatomic neutral molecule which we can imagine, H_3. Let us assume that this molecule has a linear symmetric geometry, and later justify this assumption. For simplicity, we deal only with molecular orbitals built from hydrogen $1s$-orbitals. Since there are three atomic orbitals, we expect to be able to generate three such molecular orbitals. The molecular orbital of lowest energy, shown in Fig. 12.8, consists of a linear combination of the three $1s$-orbitals, all taken with the same sign. Because it is cylindrically symmetric with no nodes between the nuclei, it is a σ-bonding orbital. Even though it covers and binds together three nuclei, it accepts only two electrons with paired spins.

The orbital of next higher energy in Fig. 12.8 has a node at the central hydrogen atom. As a result, the orbital does not produce bonding between the outer and center atoms, and, in fact, the orbital is weakly antibonding between the end atoms. Thus an electron in this orbital tends to force the molecule apart.

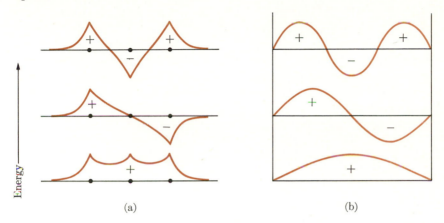

(a) Molecular orbitals of linear H_3 and (b) wave functions for a particle in a box.   **FIG. 12.8**

The third orbital of H_3 is obtained by combining the hydrogen $1s$-orbitals with alternate signs:

$$\sigma^* \propto +\psi_a(1s) - \psi_b(1s) + \psi_c(1s).$$

This produces an orbital with nodes between adjacent nuclei, as Fig. 12.8 shows. Consequently this orbital has very strong antibonding character.

Figure 12.8 also shows a comparison between the molecular orbitals which we have generated for H_3 and the wave functions for a particle in a box. The resemblance between the two sets of functions is not accidental, since a collection of three protons does produce a Coulomb field which is somewhat like a box for electrons. The major difference is that the electron potential energy in H_3 is not constant along the internuclear axis, but becomes markedly more negative near each nucleus. This is what is responsible for the sharp peaks in the H_3 wave functions which do not appear in the particle-in-a-box function. We see, on the other hand, that the energy of the σ orbitals of H_3 increases as the number of nodes increases, just as is found for the particle-in-the-box functions.

In H_3, two electrons enter the lowest σ-bonding orbital. The third electron enters the next higher σ orbital, which is relatively weakly antibonding between the end atoms and nonbonding between end and center atoms. The net effect of this orbital occupation is that H_3 is stable with respect to dissociation to three atoms by approximately 94 kcal/mole:

$$H_3 = 3H, \qquad \Delta H = 94 \text{ kcal.}$$

This is approximately what we should expect for a molecule which has two bonding electrons and one electron which is weakly antibonding. We must not conclude, however, that H_3 is a stable molecule. The antibonding effect between the end atoms is great enough so that H_3 is slightly unstable with respect to H_2 and H:

$$H_3 = H_2 + H, \qquad \Delta H = -8 \text{ kcal.}$$

Consequently, H_3 is a transient species which cannot be isolated in quantity. The energy and structure of H_3 are important because it does occur as a transient intermediate in the reaction between a hydrogen or deuterium atom and hydrogen molecules:

$$D + H_2 \rightarrow [DHH] \rightarrow DH + H.$$

From this we see that the activation energy of the hydrogen atom–hydrogen molecule reaction is just the energy needed to form H_3 from H and H_2, or +8 kcal.

The most stable geometry of H_3 is linear, rather than bent, since the last electron is in an orbital which is antibonding between the two end atoms. Any bending of the molecule brings the two end atoms together, and this is opposed by the antibonding electron.

The situation in H_3^+ is different, because this molecule-ion has no electron in the antibonding orbitals. There are two major consequences of this. First, H_3^+ is stable with respect to dissociation:

$$H_3^+ = H^+ + H + H, \qquad \Delta H = 204 \text{ kcal,}$$
$$H_3^+ = H^+ + H_2, \qquad \Delta H = 105 \text{ kcal.}$$

Second, H_3^+ is not linear, but has the geometry of an equilateral triangle. The molecular orbital of lowest energy, occupied by two electrons, is formed by the mutual overlap of three $1s$ atomic orbitals. Another way of describing the bonding in H_3^+ was discussed in Section 11.8.

In describing H_3 we have made the first use of molecular orbitals which are delocalized (i.e., which cover more than two atoms). This technique is a necessary extension of the simpler idea that molecules are bonded by electron pairs *localized* between pairs of nuclei. In Section 11.8 we saw how multicenter bonding can be described in terms of resonance structures. The delocalized molecular-orbital method is the more useful and natural technique for mathematically describing multicenter bonding; but the localized electron-pair idea with resonance structures is often more convenient for generalizing a qualitative picture of molecular binding. In subsequent discussions, we shall try to draw attention to the relation between the localized and delocalized pictures.

In comparing H_3 with H_3^+ we have encountered a very important general phenomenon. The geometry of molecules can be profoundly influenced by a relatively small change in the number of electrons or in the molecular orbitals which a given number of electrons occupy. Our next example again illustrates this idea.

We shall now examine the electronic structure of methylene, CH_2. Methylene is a stable but very reactive molecule, and consequently exists only as a transient species in certain chemical reactions. Nevertheless, high-speed molecular spectroscopy has provided a fairly clear picture of its molecular structure.

A triatomic molecule like CH_2 may have either a linear or bent structure, and it is not at all immediately clear which is the more stable. To begin, we shall assume a *linear* geometry, construct the molecular orbitals, and then repeat the process for a *bent* CH_2. Then we shall compare the electronic structures of the two forms, and draw conclusions.

For the linear H—C—H we can construct a sigma bonding-antibonding pair of molecular orbitals from the carbon $2s$-orbitals and the hydrogen $1s$-orbitals. The forms of these orbitals would be

$$\sigma_s = C_1 1s_a + C_2 2s_c + C_1 1s_b,$$

and

$$\sigma_s^* = C_3 1s_a - C_4 2s_c + C_3 1s_b,$$

where $2s_c$ stands for the $2s$ wave function of carbon, and $1s_a$, $1s_b$ stand for the $1s$-orbitals of hydrogen atoms a and b. The coefficients of the two hydrogen-atom functions in a given molecular orbital are the same, since the methylene molecule is symmetric. Notice that these orbitals are very much like the lowest- and highest-energy orbitals of H_3, except that the carbon $2s$-orbital substitutes for the central hydrogen orbital.

A second σ bonding-antibonding pair can be constructed for CH_2, this time using the $2p_z$-orbital of carbon. The mathematical forms of these orbitals are

$$\sigma_p = C_5 1s_a + C_6 2p_c - C_5 1s_b,$$
$$\sigma_p^* = C_7 1s_a - C_8 2p_c - C_7 1s_b.$$

A pictorial representation of these orbitals is given in Fig. 12.9. Note that since the carbon $2p_z$-orbital has two lobes of different sign, the hydrogen-atom orbitals must be added in with opposite signs to give a bonding orbital with no internuclear nodes. The antibonding orbital is obtained by using the same combination of hydrogen orbitals, but with the sign of the carbon $2p_z$-orbital reversed. Note also that so far as its nodal properties are concerned, the σ_p-orbital is somewhat analogous to the σ-orbital of intermediate energy in H_3. The methylene orbital is different in that it is bonding between the end and center atoms, because it includes a contribution from the carbon $2p_z$-orbital. As a result it is bonding between end and center atoms, but antibonding between end atoms.

From the four atomic orbitals $2s$, $2p_z$ of carbon, and $1s$ from the hydrogen atoms, we have generated four molecular orbitals. Hydrogen has no low-energy p-orbitals to interact with the $2p_x$- and $2p_y$-orbitals of carbon, and so these remain *as* localized nonbonding orbitals in linear methylene.

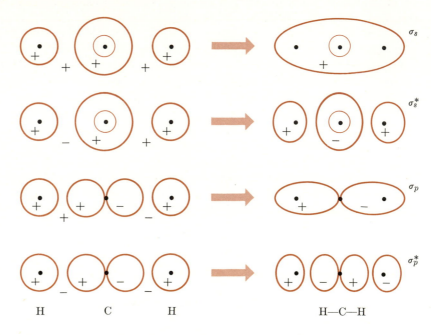

FIG. 12.9   Schematic representation of the formation of the σ and $\sigma*$ molecular orbitals of linear methylene.

The orbital energy-level diagram for linear methylene is shown in Fig. 12.10. The total of six valence electrons fill the two σ-bonding orbitals, and half-fill each of the two nonbonding carbon $2p_x$- and $2p_y$-orbitals, with the latter two electrons having parallel spin. Thus we can describe methylene as held together by two three-center σ bonds. We can also conclude that linear methylene will have unpaired electron spins, and this has been experimentally verified.

It is edifying to construct the localized valence-bond picture of methylene, and compare it to the molecular-orbital picture. We begin by constructing sp hybrid atomic orbitals on the carbon atom. We then construct two localized electron-pair bonds to the hydrogen atoms by combining each sp hybrid with the appropriate H $1s$ function. Each of these valence-bond orbitals accommodates two electrons with paired spins, and the last two valence electrons remain in nonbonding carbon $2p$-orbitals. Thus the delocalized molecular-orbital picture and the valence-bond sp-hybrid picture are fairly closely related, in that in both models the bonding electrons occupy σ orbitals generated from carbon $2s$- and $2p_z$-orbitals in combination with atomic hydrogen $1s$-orbitals.

We can generate the orbital-energy diagram for bent CH_2 simply by examining what happens to the energy of the individual molecular orbitals as linear CH_2 is bent. The lowest-energy σ orbital constructed of the nondirectional carbon $2s$- and hydrogen $1s$-orbitals is relatively unchanged in energy by bending the molecule. However, the next higher σ orbital becomes weaker bonding and higher in energy as the molecule bends, because the overlap between the carbon $2p_z$- and the hydrogen $1s$-orbitals decreases as the hydrogen atoms move off the z-axis, and because this orbital, being antibonding between end atoms, increases in energy as the hydrogen atoms approach each other.

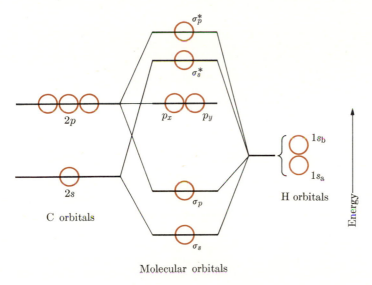

Molecular-orbital energy-level diagram for linear methylene.   **FIG. 12.10**

If we imagine that the bending of the molecule takes place in the xz-plane, we can conclude that the energy of the carbon $2p_y$-orbital is left unchanged. This orbital has a node in the xz-plane, and since it is cylindrically symmetric about the y-axis, it is insensitive to the angular positions of the hydrogen atoms. The situation is quite different for the carbon $2p_x$-orbital, however. Figure 12.11 shows that, as the molecule bends, the carbon $2p_x$-orbital begins to overlap with the $1s$-orbitals of the hydrogen atoms. This overlap lowers the energy of the $2p_x$-orbital, for instead of being nonbonding, the $2p_x$-orbital acquires bonding character in the bent methylene molecule.

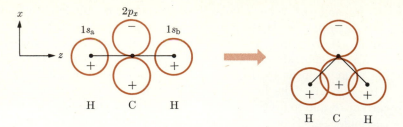

FIG. 12.11 Schematic demonstration of the increase in overlap and bonding character that occurs between hydrogen 1s-orbitals and the carbon $2p_x$-orbital when methylene is bent in the xz-plane.

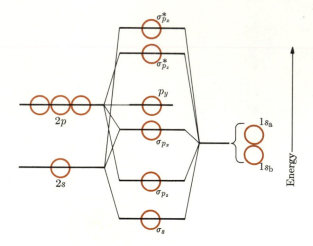

FIG. 12.12 The molecular-orbital energy-level diagram for bent methylene and other bent XH_2 molecules. The σ designation of the orbitals is not fully accurate since cylindrical symmetry is lost in a bent molecule. These σ orbitals are further described by giving the *principal* contributing atomic orbital on the central atom.

Figure 12.12 gives the resulting orbital energy-level diagram for the bent methylene molecule, which also applies to other bent molecules of the XH_2 type. We see that the energy of σp_z has been increased and that of p_x has been lowered relative to their energies in linear CH_2, but otherwise the pattern differs relatively little from that shown in Fig. 12.10. One important detail is that in the bent molecule the bonds no longer have cylindrical symmetry, so the notation σ and σ^* is strictly no longer applicable. For the present, however, it is better to retain the improper σ-σ^* notation in order to emphasize the relationship between the orbitals in the bent and linear molecules.

The six valence electrons of bent CH_2 can enter the three low-energy orbitals with paired spins, and leave the nonbonding $2p_y$-orbital empty. Thus, in contrast to the linear molecule, bent methylene should have no unpaired electron spins. This has been found experimentally. The question of whether the linear or bent molecule is the more stable cannot be answered merely by examining the orbital-energy diagrams. Bent CH_2 achieves some stability by having, in addition to its two pairs of bonding electrons, a third pair of electrons in the σ_{p_x}-orbital which has some bonding character. On the other hand, while linear methylene has only two pairs of bonding electrons, it avoids some electron repulsion by having only one electron in the $2p_x$- and $2p_y$-orbitals. The actual condition of the CH_2 molecule found experimentally is a compromise between these two effects. In its lowest electronic state, the molecule has one electron each in the $2p_x$- and $2p_y$-orbitals, *and* is slightly bent so that the $2p_x$-orbital energy is lowered somewhat.

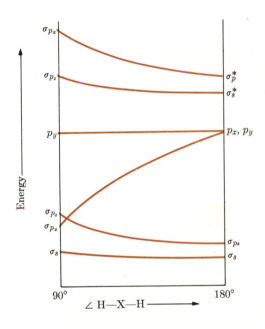

The qualitative variation of the orbital energies of the XH_2 molecules as a function of the XH_2 bond angle.

FIG. 12.13

A localized valence-bond model for bent methylene would involve three sp^2 hybrid orbitals on the carbon atom. Two of these would be used to form electron-pair bonds to each hydrogen atom. The third sp^2 hybrid would be occupied by a pair of nonbonding electrons, but would be of fairly low energy because of the partial s-orbital character. The carbon $2p$-orbital perpendicular to the plane of the molecule would be unoccupied. Thus again the localized valence-bond model is fairly closely related to the delocalized molecular-orbital picture.

By plotting the qualitative dependence of orbital energy as a function of bond angle, we can generate the diagram shown in Fig. 12.13 which can be used to predict or rationalize the geometries of triatomic hydrides. For example, the transient molecule BeH_2 has four valence electrons, just enough to occupy the two low-energy σ bonding orbitals. As Fig. 12.13 shows, these orbitals have lowest energy in the linear configuration, and hence we expect BeH_2 to be a linear molecule.

Passing to BH_2, we have five valence electrons, four of which enter the σ bonding orbitals. The fifth electron enters what is an atomic $2p$-orbital of boron in the linear configuration. Figure 12.13 shows, however, that the energy of this orbital is lower if the molecule is bent. In BH_2 this energy-lowering upon bending is enough to overcome the rise in energy of the σp_z orbital, and BH_2 is a bent molecule. In electronically excited BH_2, however, the last electron occupies the p_y-orbital, and from Fig. 12.13 we expect this excited BH_2 will be a linear molecule. This is found experimentally.

The next molecule in the series is CH_2, which we have discussed in detail, and noted that the linear configuration with the lower electron repulsion is favored. In NH_2, however, there is an additional electron, and double occupancy of orbitals cannot be avoided. Thus the bent geometry of NH_2 is favored, since after four electrons have occupied the two lowest bonding orbitals, *two* electrons must occupy the σ_x-orbital whose energy decreases as the molecule bends. The last electron enters the nitrogen p_y-orbital.

In the water molecule, the eight valence electrons occupy the two strong σ bonding orbitals, the in-plane nonbonding σ-orbital which is of low energy in the bent form, and the nonbonding p_y-orbital of the oxygen atom. Thus water should be, and is found to be, a bent molecule.

Question. Would you expect the following triatomic hydrides to be linear or bent molecules? CH_2^+, NH_2^-, BH_2^+, BH_2^-.

Triatomic Nonhydrides

We turn now to the molecular-orbital description of the linear symmetric molecule CO_2. The orbitals which we generate for it can be generalized and applied to a discussion of the structure of other triatomic molecules which do not contain hydrogen.

For simplicity we take the $2s$-orbitals of the two oxygen atoms to be nonbonding atomic orbitals even in the molecule. Then one σ-σ^* pair of molecular orbitals can be generated from overlap of the carbon $2s$-orbital with the $2p_x$-orbitals of oxygen. Another σ-σ^* pair can be made from combination of the carbon $2p_z$-orbital with the oxygen $2p_z$-orbitals. The simplest LCAO approximations to these four molecular orbitals are

$$\sigma_{2s} = C_1 2p_a(O) + C_2 2s(C) + C_1 2p_b(O),$$
$$\sigma_{2s}^* = C_3 2p_a(O) - C_4 2s(C) + C_3 2p_b(O),$$

and

$$\sigma_{2p} = C_5 2p_a(O) + C_6 2p(C) - C_5 2p_b(O),$$
$$\sigma_{2p}^* = C_7 2p_a(O) - C_8 2p(C) - C_7 2p_b(O).$$

The signs have been chosen so that there are no internuclear nodes for the bonding orbitals, but there are nodes between the nuclei in the antibonding orbitals. These orbitals have their lowest energy when the molecule is linear. The pictorial representation of these orbitals appears in Fig. 12.14.

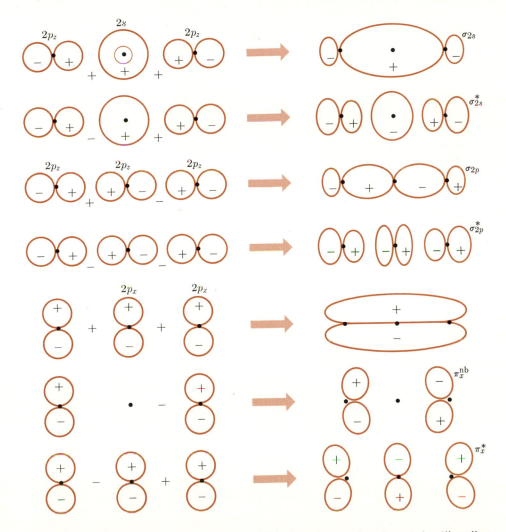

A schematic representation of the formation of the bonding, nonbonding, and antibonding orbitals of linear CO_2.

FIG. 12.14

The π molecular orbitals are generated from the overlap of the atomic
p-orbitals which are perpendicular to the internuclear axis of the molecule.
There are six such atomic orbitals, so we expect six π molecular orbitals. Three
of these will be π_x-orbitals, and three will be equivalent π_y-orbitals. The
π-orbital of lowest energy has the form

$$\pi_x = C_9 2p_a(O) + C_{10} 2p(C) + C_9 2p_b(O),$$

where p_x atomic orbitals are used. There is a π_y-orbital of the same form and
same energy. Both are strongly bonding.

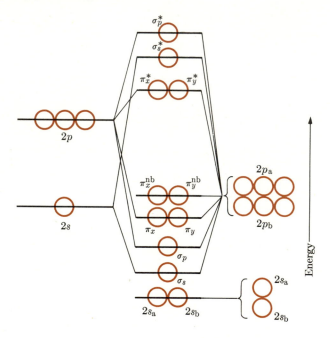

FIG. 12.15 The molecular-orbital energy-level diagram of linear symmetric CO_2 and other linear
triatomic molecules.

The π-orbital of next higher energy involves only the oxygen atoms. It is
nonbonding to the carbon atom, and weakly antibonding between the oxygen
atoms. Its form is

$$\pi_x^{nb} \propto 2p_a(O) - 2p_b(O),$$

and there is a π_y^{nb}-orbital of the same form and energy. The notation π^{nb} draws
attention to the nonbonding character of the orbital, but it must be remembered
that there is also a weak antibonding interaction between the oxygen atoms.
The third and highest-energy π-orbital is

$$\pi_x^* = C_{11} 2p_a(O) - C_{12} 2p(C) + C_{11} 2p_b(O),$$

which is antibonding between carbon and oxygen. The π_y^*-orbital has the same form and energy.

We see that the combinations of the p-orbitals produces a pair of strongly bonding π-orbitals, a pair of essentially nonbonding π^{nb}-orbitals, and a pair of strongly antibonding π^*-orbitals. The form of these π-orbitals is shown in Fig. 12.14. Comparison of this set of π orbitals with the σ-orbitals of H_3 shows that they have the same bonding, nonbonding, and antibonding pattern, and the same *internuclear* nodal properties.

Figure 12.15 shows the energy-level pattern of the molecular orbitals of linear symmetric CO_2. The 16 valence electrons fill the two oxygen $2s$ atomic orbitals, the two σ bonding orbitals, the two π bonding and the two π nonbonding orbitals. Since there is a total of eight bonding electrons, the two C—O links can be regarded as double bonds, just as is done in the electron-dot or valence-bond representations. We also notice that the orbitals which are occupied are all more stable when the molecule is linear, which is, in fact, found to be the geometry of CO_2.

The molecular orbitals which we have just discussed can be used to describe other triatomic molecules which have 16 or fewer valence electrons. Other 16-electron molecules which have the expected linear geometry and which are strongly bound are N_2O, N_3^-, CS_2, OCS, OCN^-, and NO_2^+. In molecules like CS_2, where the valence atomic orbitals of sulfur have principal quantum number $n = 3$, the form of the molecular orbitals is quite analogous to that which occurs in CO_2.

The transient reactive molecules NCO, NCN, CCN, and C_3 have respectively 15, 14, 13, and 12 valence electrons. They all have eight electrons in the two σ and two π bonding orbitals, and between three and zero π nonbonding electrons. Like CO_2, they are all linear molecules.

If more than 16 electrons must be accommodated in a linear molecule, some would have to enter the π^* antibonding orbitals, as Fig. 12.15 shows. This unfavorable situation can be relieved somewhat if the molecule departs from linearity. If the molecule is bent in the xz-plane, the π_y bonding, nonbonding, and antibonding orbitals remain largely unchanged in character and energy. The π_x-, π_x^{nb}-, and π_x^*-orbitals change considerably, however. They revert to $2p_x$ atomic nonbonding orbitals on the two end atoms, and a p_x-orbital on the central atom. This orbital is also largely nonbonding, but as we found in the discussion of CH_2, the orbital energy lowers as the molecule bends by overlapping and combining with the σ bonding system.

Figure 12.16 summarizes the behavior of the molecular orbitals of a triatomic molecule as it is bent, and Fig. 12.17 shows the orbital energy-level pattern that can be applied to most bent triatomics. If we enter the 17 electrons of NO_2 into these orbitals, for example, we find six bonding electrons, 11 nonbonding electrons, and no antibonding electrons. Of the three bonding orbitals, two are σ and one π, and all three cover all nuclei. Thus there are on the average $6/2 = 3$ electrons per chemical bond; this corresponds to two $1\frac{1}{2}$ bonds in NO_2. This is

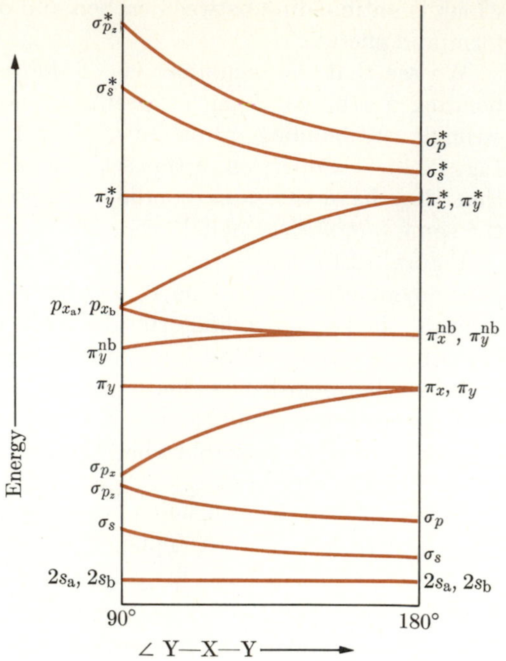

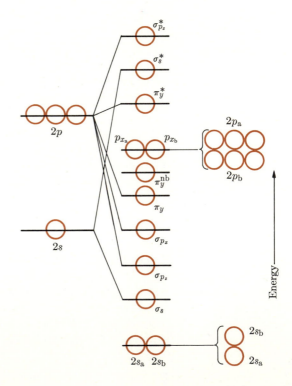

FIG. 12.16 The qualitative variation of the orbital energies of the XY$_2$ molecules as a function of the XY$_2$ bond angle.

FIG. 12.17 Molecular-orbital energy-level diagram for bent triatomic molecules.

consistent with the conclusion drawn from the electron-dot resonance structures

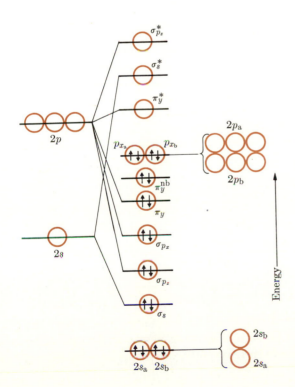

We expect that the bond energy of NO_2 is less than that of CO_2, which has two double bonds. This is found experimentally:

$$NO_2 = NO + O, \quad \Delta H = +72 \text{ kcal},$$
$$CO_2 = CO + O, \quad \Delta H = +127 \text{ kcal}.$$

Even though bending NO_2 reduces the number of π bonds, this is evidently energetically more favorable than retaining the linear geometry, and forcing the last electron to enter a π^* molecular orbital which is strongly antibonding.

In the nitrite ion, NO_2^- and ozone, O_3, there are 18 valence electrons. On the basis of the arguments just given, we expect these molecules to be bent, and to have the electron configuration given in Fig. 12.18. The bond angle of NO_2^- is 115°, smaller than that of NO_2 (135°), as expected. The bond angle in O_3 is 117°, very similar to that of NO_2^-.

The molecular-orbital occupancy for O_3, NO_2^-, and other 18-electron triatomic molecules.

FIG. 12.18

Passing to molecules like NF_2 and OF_2 with 19 and 20 valence electrons respectively, we see from Figs. 12.16 and 12.18 that electrons 19 and 20 enter the π_y^* antibonding orbital whose energy is not sensitive to bond angle. Consequently these molecules are bent, just as are molecules with 17 and 18 electrons. For OF_2, Fig. 12.18 shows that there are three pairs of bonding electrons, eight pairs of nonbonding electrons, and one antibonding pair, for a net of four electrons binding three nuclei. We could picture the binding in OF_2 as due to a pair of single bonds. The small bond energy of 45 kcal/mole suggests that the effect of the π^* antibonding electrons is quite pronounced.

The molecule OF_2 is a case in which the delocalized molecular-orbital approach provides a better qualitative description of the molecule than does the simplest valence-bond picture. In the latter, there are electron-pair bonds between the oxygen and the two fluorine atoms, and the rest of the electrons are considered nonbonding. To account for the low bond energy, repulsions between the nonbonded electron pairs of the oxygen and fluorine are postulated. In the delocalized molecular-orbital picture, these "repulsions" are described naturally by the effect of the antibonding electrons.

As examples of molecules with 22 valence electrons we have I_3^- and the other trihalide ions, as well as KrF_2 and XeF_2. From Fig. 12.16 we see that electrons 21 and 22 in these molecules must be in the strongly antibonding σ_z^*-orbital. The energy of this orbital is lowest in the linear configuration. This effect is so strong that the most stable geometry of molecules with 21 or 22 valence electrons is linear. From Fig. 12.15, we deduce that such molecules as KrF_2, XeF_2, and I_3^- have four pairs of bonding electrons, four pairs of nonbonding electrons, and three pairs of antibonding electrons. This leaves a net of two bonding electrons to hold together three atoms, or effectively one electron per internuclear linkage. As a result, these 22-electron molecules are not very stable with respect to dissociation.

We should note that this delocalized molecular-orbital description of XeF_2 is quite different from the simple valence-bond picture. In the latter, two $5d$-orbitals of Xe would be combined with the $5s$- and $5p$-orbitals to give a d^2sp^3 hybrid set. Two of these hybrids would be used to form electron-pair bonds to the two fluorine atoms, while the three other hybrids would be filled with six nonbonding electrons. In this manner, the ten valence electrons which surround xenon in XeF_2 can be accommodated. It is not yet clear whether this valence-bond description or the delocalized molecular-orbital picture is a better approximation to the actual electronic structure.

12.4 TRIGONAL PLANAR MOLECULES

We select for our discussion the nitrate ion, NO_3^-, which, as noted in Section 11.8, is a planar symmetric molecule with the oxygen atoms at the apices of an equilateral triangle. The orbital pattern we will generate applies to other molecules of this geometry, such as SO_3, BF_3, and $CO_3^=$.

The detailed mathematical description of the molecular orbitals for trigonal planar molecules is noticeably more complicated than what we have encountered for diatomic and triatomic molecules. Rather than become involved in this detail, we shall proceed intuitively, using our previous experience as a guide. We begin by dividing the molecular orbitals into two groups: the σ-orbitals which have greatest density in the plane of the molecule, and the π-orbitals, which have greatest density above and below the nuclear (or xy) plane. Again we regard the three oxygen $2s$-orbitals as atomic nonbonding orbitals.

Let us concentrate first on the σ molecular orbitals. To construct them, we have available the $2p_x$-, $2p_y$-, and $2s$-orbitals of the central nitrogen atom, as well as the one p-orbital from each oxygen atom which points along the N—O axis. This is a total of six atomic orbitals, and we expect six σ molecular orbitals, three bonding and three antibonding.

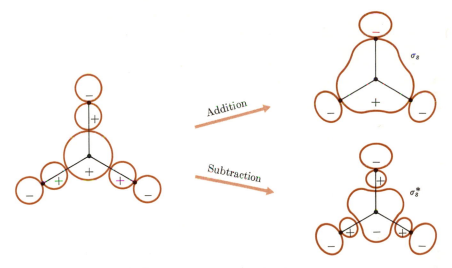

Formation of σ_s-orbitals for a planar trigonal molecule.

FIG. 12.19

One σ-σ^* pair is easy to generate, as Fig. 12.19 shows. These orbitals arise from the interaction of the $2p$-orbitals of oxygen with the $2s$-orbital of nitrogen. The strongly bonding component is designated σ_s, and its strongly antibonding partner σ_s^*.

The atomic $2p_x$- and $2p_y$-orbitals of nitrogen are equivalent, and thus we might expect to generate from them and the p-orbitals of oxygen two equivalent bonding σ_p-orbitals, and their equivalent antibonding components. The two equivalent bonding orbitals are shown in Fig. 12.20, and their antibonding partners can be generated by choosing the signs of the oxygen p-orbitals so that internuclear nodes appear. The two σ_p bonding orbitals have the same energy, although this fact is not obvious from Fig. 12.20. Likewise, the two σ_p^*-orbitals

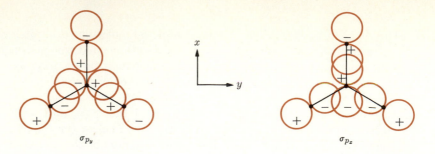

FIG. 12.20   Schematic representation of the formation of the two equivalent σ_p bonding orbitals for a planar trigonal molecule.   The two antibonding partners can be obtained by reversing the signs on the central p-orbital.

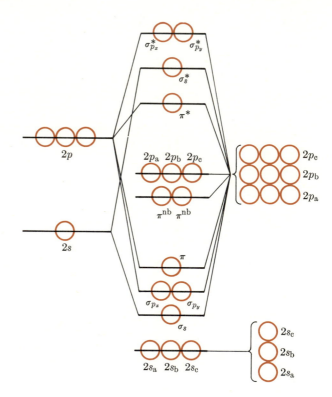

FIG. 12.21   The molecular-orbital energy levels for NO_3^- and other planar trigonal molecules.

have the same energy.   The two σ_p bonding orbitals, two σ_p^* antibonding orbitals, and the σ_s-σ_s^* bonding-antibonding pair give us the six σ molecular orbitals which we expected to generate from the six available atomic orbitals. These six molecular orbitals appear in the energy-level pattern for NO_3^- as indicated in Fig. 12.21.

Each oxygen atom has a $2p$-orbital which lies in the xy or molecular plane, but which is perpendicular to its N—O bond axis. These three oxygen orbitals are nonbonding in the molecule, and thus their energy is about the same as that of a $2p$-orbital in a free atom. This is indicated in Fig. 12.21.

We now consider the π molecular orbitals of NO_3^-. To generate them we have one p-orbital on each oxygen atom and one p-orbital on the central nitrogen atom, all with their lobes above and below the molecular plane. From these four atomic orbitals, we expect four molecular orbitals. As Fig. 12.22 shows, it is easy to generate a π-π^* bonding-antibonding pair of molecular orbitals. The remaining pair of π-orbitals is nonbonding and equal in energy. Unfortunately these are difficult to represent graphically in a way which convincingly demonstrates their equivalence. We will have to accept without proof the results of the mathematical analysis which reveals these properties.

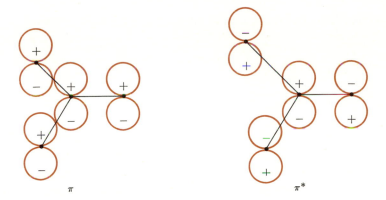

Formation of the π and π^* bonding and antibonding molecular orbitals in NO_3^- and other planar trigonal molecules.   **FIG. 12.22**

Having generated 16 molecular orbitals from the 16 atomic orbitals available, we have a complete set of orbitals for NO_3^- and other molecules of similar geometry. The energy-level pattern of these orbitals is shown in Fig. 12.21. The 24 valence electrons of NO_3^- fill four bonding orbitals and eight nonbonding orbitals. Since there are three N—O linkages, each can be called a $1\frac{1}{3}$ bond. This description is consistent with the resonance structures which occur in the valence-bond picture of NO_3^-:

$$\left[\begin{array}{c} O \\ | \\ N \\ \diagup \quad \diagdown\!\!\!\!= \\ O \qquad O \end{array} \right]^- \leftrightarrow \left[\begin{array}{c} O \\ \| \\ N \\ \diagup \quad \diagdown \\ O \qquad O \end{array} \right]^- \leftrightarrow \left[\begin{array}{c} O \\ | \\ N \\ \diagup \quad \diagdown \\ O \qquad O \end{array} \right]^-$$

The molecules SO_3, BF_3, and $CO_3^=$ also have 24 valence electrons and have electronic structures which are the same as that of NO_3^-.

In treating hydrocarbon molecules it is usually easiest to proceed as we did in Section 11.7 for ethylene. A system of localized σ bonds is generated by forming hybrid orbitals on carbon which are appropriate to the molecular geometry, and then overlapping these hybrids to form σ bonds to hydrogen atoms or carbon atoms. The second component of double bonds is then formed by the overlap of carbon p-orbitals which are perpendicular to the internuclear axis. This procedure is sufficient to describe "isolated" double bonds which occur in 1-butene or 1,4-pentadiene:

$$\overset{1}{C}H_2\!\!=\!\!\overset{2}{C}H\overset{3}{C}H_2\overset{4}{C}H_3 \qquad\qquad \overset{1}{C}H_2\!\!=\!\!\overset{2}{C}H\overset{3}{C}H_2\overset{4}{C}H\!\!=\!\!\overset{5}{C}H_2$$

1-butene   1,4-pentadiene

Special effects arise, however, when the double bonds are conjugated, or separated by only one C—C linkage, as in 1,3-butadiene:

$$\overset{1}{C}H_2\!\!=\!\!\overset{2}{C}H\overset{3}{C}H\!\!=\!\!\overset{4}{C}H_2$$

1,3-butadiene

Let us treat the π bonding system of 1,3-butadiene by the delocalized molecular-orbital method.

We begin by assuming that the σ orbital system of butadiene exists, and that we have to deal with only one p atomic orbital on each carbon atom. From these four atomic orbitals we can expect to generate four molecular orbitals. Figure 12.23 indicates how this is done.

We see that the orbital of lowest energy has no internuclear nodes, and thus is bonding between all nuclei. The orbital of next higher energy has one node between the inner carbon atoms. Consequently it is bonding between atoms a and b, and between atoms c and d, but antibonding between atoms b and c. From the nodal pattern of the third orbital, we see that it is b-c bonding, but a-b, and c-d antibonding. Finally the orbital of highest energy has nodes between all nuclei, and is totally antibonding. Notice that the internuclear nodal pattern of these π-orbitals is very much like the nodal pattern of the particle-in-a-box wave functions. This analogy can be pursued quantitatively, and for long-chain conjugated molecules, particle-in-a-box wave functions are often used in place of LCAO molecular orbitals to describe the π electrons.

In butadiene there are four electrons to be accommodated. Thus the two π orbitals of lowest energy in Fig. 12.23 each have two electrons of paired spin in them. We have then two occupied orbitals which are bonding between atoms a and b, and atoms c and d. Four π bonding electrons spread over two linkages give in effect one π bond at each linkage; this is consistent with the simple valence-bond description of butadiene. Of the two occupied orbitals, one is bonding between atoms b and c, the other antibonding; thus there is approximately no π bonding between these centers. This, too, is consistent with the simplest valence-bond picture. However, quantitative calculations show that this cancellation of π bonding between the inner atoms in butadiene is not exact, and that there is some double-bond character between them. In valence-bond terms, this is described by the resonance structures

$$\overset{+}{CH_2}-CH=CH-\overset{..}{\overset{-}{CH_2}} \leftrightarrow CH_2=CH-CH=CH_2 \leftrightarrow \overset{..}{\overset{-}{CH_2}}-CH=C-\overset{+}{CH_2}.$$

Again, the delocalized molecular-orbital method appears to be a more satisfactory method of describing these effects.

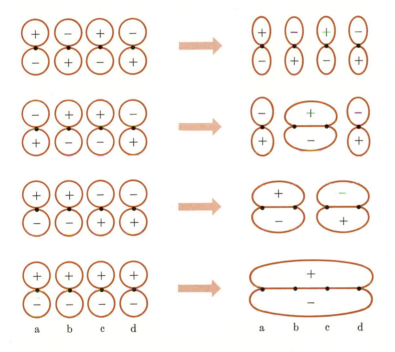

Schematic representation of the formation of the π molecular orbitals of 1,3-butadiene from atomic p-orbitals.

FIG. 12.23

In Section 11.8 we noted the special stability of the benzene molecule C_6H_6. The fact that this molecule has the geometry of a regular hexagon with six equivalent C—C bonds was rationalized by using the resonance structures in which the presence of carbon and hydrogen atoms is understood. How is this molecule described in terms of delocalized molecular orbitals?

We begin by assuming the existence of the σ bonding system, and six unused p-orbitals perpendicular to the molecular plane. From these six atomic orbitals we expect to find six π molecular orbitals. As Fig. 12.24 shows, one bonding-antibonding pair is easy to find. The strongly π bonding component consists of the overlap of all p-orbitals, with no internuclear nodes, whereas the π^* antibonding orbital has nodes between all nuclei. These are respectively the lowest- and highest-energy π-orbitals of benzene.

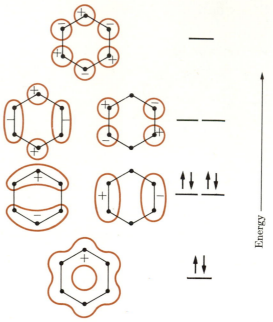

FIG. 12.24 Schematic representation of the π molecular orbitals of benzene and their relative energy levels. The top view of the orbitals is given; the signs are opposite in the bottom lobes of the orbitals which lie below the molecular plane.

There are four molecular orbitals yet to be found. Generating these properly from the atomic orbitals requires techniques which we have not developed, so we shall have to quote the results. There is a pair of molecular orbitals shown in Fig. 12.24 which have only one node. Although these orbitals do not appear equivalent, they have the same energy and are predominantly of bonding character. The members of another pair of molecular orbitals, also shown in Fig. 12.24, have three nodal surfaces. These orbitals are of equal energy, even though they do not resemble each other. They are both of predominantly antibonding character.

The orbital energy-level pattern for benzene is given in Fig. 12.24. There are six electrons to occupy the lowest three π orbitals with paired spins. Occupation of these three orbitals gives six π bonding electrons which cover the whole molecule.

12.6 CONCLUSION

In this chapter we have described the electronic structures of several types of molecules by one of the simplest approximate methods available—delocalized molecular orbitals made by linearly combining atomic orbitals. These LCAO-MO's give a useful qualitative picture of the electronic properties of molecules, but in their simplest form do not allow accurate quantum-mechanical calculation of molecular properties. Nevertheless, LCAO-MO's allow us to see more clearly the sometimes very strong relationships between the structure and bond properties of superficially different molecules like BF_3 and NO_3^-, CO_2 and NO_2^+, $O_2^=$ and F_2, etc. Consequently, LCAO-MO's, used critically and cautiously, can be an important means for qualitatively understanding and correlating chemical behavior.

SUGGESTIONS FOR FURTHER READING

Coulson, C. A., *Valence*. Oxford: Oxford University Press, 1961.

Gray, H. B., *Chemical Bonds*. Menlo Park, Calif.: W. A. Benjamin, 1973.

Gray, H. B., *Electrons and Chemical Bonding*. Menlo Park, Calif.: W. A. Benjamin, 1964.

Linnett, J. W., *Wave Mechanics and Valency*. New York: Wiley, 1960.

Royer, D. J., *Bonding Theory*. New York: McGraw-Hill, 1968.

Sebera, D. K., *Electronic Structure and Chemical Bonding*. Waltham, Mass.: Blaisdell, 1964.

PROBLEMS

12.1 Which of the following pairs of molecules would you expect to have the higher bond energy?

a) F_2, F_2^+;　　　　b) NO, NO^-;　　　　c) BN, BO;
d) NF, NO;　　　　e) Be_2, Be_2^+.

12.2 Construct a molecular-orbital energy-level diagram which would be appropriate for ionic molecules like LiF.

12.3 Construct a molecular-orbital energy-level diagram for HF.

12.4 The ion CO_2^- can be produced in radiation-damaged crystals of substances that contain the —COOH group. What would you expect the geometry of CO_2^- to be?

12.5 Predict the geometries of the following triatomic molecules.

a) CCn,　　　　b) CCO,　　　　c) FCO,　　　　d) FOO,
e) FNO,　　　　f) FCN,　　　　g) NCO.

12.6 Given that N_3^- exists as a weakly bound ion, would you expect N_3 to be a bound molecule? What about N_3^+?

12.7 Predict the geometries of the following symmetric triatomic molecules:

a) OBO, b) CNC, c) Li_3^+ d) CO_2^+,

e) O_3^+, f) F_3^+, g) O_3^-.

12.8 Discuss the relation between the bonding in the molecules BF_3, F_2CO, and FNO_2. Do you think a bound planar trigonal O_4 might be found some time in the future?

12.9 Discuss the π molecular orbitals in cyanogen, which has the valence-bond structure N≡C—C≡N.

12.10 What changes in the π molecular orbitals and energy levels occur when one of the CH groups in benzene is replaced by an N atom to give pyridine C_5H_5N? Note that nitrogen is more electronegative than CH, so orbitals with electron density at the nitrogen will be lowered.

12.11 Use the molecular orbital-energy level diagram of problem 12.3 to consider the molecules OH and OH^+. Would you expect their dissociation energies to $O + H$ and $O^+ + H$, respectively, to be similar, or very different? Why?

CHAPTER 13

PERIODIC PROPERTIES

In this and subsequent chapters we shall be dealing with the descriptive chemistry of the elements. To say that this is a vast subject is a truism, for descriptive chemistry in the broadest sense includes all we know about all matter. Even a first approach to this subject would be hopelessly confusing were it not for a most useful generalization: *the properties of the elements are periodic functions of their atomic numbers.* With the help of this *periodic law,* it is possible to organize and to systematize the chemistry of the elements into a manageable subject. Learning descriptive chemistry then becomes a process of discovery and assessment of facts, prediction and verification of chemical behavior, and evaluation of correlations and explanations. All of this leads to an understanding of why elements have the properties they do. By no means are there satisfactory explanations for all chemical behavior and there is considerable opportunity to generate new ideas about why matter behaves as it does. In this chapter we shall discuss some of the most useful systematic relations that exist in descriptive chemistry. This background will help us to organize the more detailed information in subsequent chapters and will provide us with a general view of chemical behavior.

13.1 THE PERIODIC TABLE

Since the first publications of the periodic law by Mendeleev and Meyer in the 1870's, there has been a very large number of forms proposed for the periodic table. The version that is easiest to use and which is most closely related to the

Table 13.1 Long form of the periodic table of elements

IA	IIA	IIIB	IVB	VB	VIB	VIIB	VIII			IB	IIB	IIIA	IVA	VA	VIA	VIIA	
H 1																	He 2
Li 3	Be 4											B 5	C 6	N 7	O 8	F 9	Ne 10
Na 11	Mg 12											Al 13	Si 14	P 15	S 16	Cl 17	Ar 18
K 19	Ca 20	Sc 21	Ti 22	V 23	Cr 24	Mn 25	Fe 26	Co 27	Ni 28	Cu 29	Zn 30	Ga 31	Ge 32	As 33	Se 34	Br 35	Kr 36
Rb 37	Sr 38	Y 39	Zr 40	Nb 41	Mo 42	Tc 43	Ru 44	Rh 45	Pd 46	Ag 47	Cd 48	In 49	Sn 50	Sb 51	Te 52	I 53	Xe 54
Cs 55	Ba 56	* 57–71	Hf 72	Ta 73	W 74	Re 75	Os 76	Ir 77	Pt 78	Au 79	Hg 80	Tl 81	Pb 82	Bi 83	Po 84	At 85	Rn 86
Fr 87	Ra 88	† 89															

*	La 57	Ce 58	Pr 59	Nd 60	Pm 61	Sm 62	Eu 63	Gd 64	Tb 65	Dy 66	Ho 67	Er 68	Tm 69	Yb 70	Lu 71
†	Ac 89	Th 90	Pa 91	U 92	Np 93	Pu 94	Am 95	Cm 96	Bk 97	Cf 98	Es 99	Fm 100	Md 101	No 102	Lw 103

electronic structures of the atoms is the so-called *long form* shown in Table 13.1. The elements fall into 18 vertical columns which define the chemical families or groups. The members of each group most often have valence-electron configurations that are the same, except for principal quantum numbers. While chemical similarities are most often strongest among elements in the same column, there is some resemblance between elements that are not in the same column but which do have the same *number* of valence electrons. For example, members of the scandium group have the configurations $(n - 1)d^1ns^2$ and are in some respects similar to the elements below boron, which have the configurations ns^2np^1. Consequently, the elements under scandium in the third column are said to be members of group III, subgroup B, or simply of group IIIB, while the boron family is labeled as group IIIA. Other groups in the periodic table are related and labeled in a similar manner. The elements in the three columns designated as group VIII resemble each other in many respects and separate the A subgroups from the B subgroups in the periodic table.

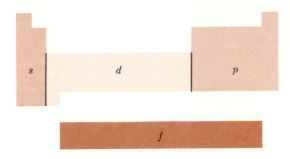

The separation of the periodic table into blocks of elements according to the filling of valence orbitals.

FIG. 13.1

To keep the periodic chart from being excessively long, the 14 elements which follow lanthanum and the 14 elements that fall after actinium are placed in separate rows at the bottom of the table. This procedure also emphasizes that the periodic table can be broken into blocks of elements on the basis of the electron configurations of the atoms. Figure 13.1 shows that the elements in which the *s*-, *p*-, *d*-, and *f*-orbitals are being filled are grouped naturally in the long form of the periodic table. The eight families of the *s*- and *p*-blocks are often called the *representative elements*, those of the *d*-block are called *transition elements*, while members of the *f*-block are known as the *inner transition elements*.

While the structure of the periodic table is designed to emphasize the existence of *vertical* relationships between members of the same group, a number of properties show regular trends along each row of the table. We have already come upon one such trend in Chapter 10: the general tendency for ionization energy to increase along any row of the periodic table. Our discussion in this and subsequent chapters will reveal other similar *horizontal* trends in chemical and physical properties. In addition, important *diagonal* relationships appear: There are often similarities between an element and its diagonal neighbor in the succeeding *column* and *row* of the periodic table. To make the existence of such relationships clear, and to emphasize the usefulness of the periodic table, in the remainder of this chapter we shall discuss some of the clearer trends in the properties of the elements and of some of their common compounds.

13.2 PERIODIC PROPERTIES

A very large number of chemical and physical properties of the elements vary periodically with atomic number. Some of these properties are related to the electron configurations of the atoms in quite obscure and complicated ways, while others are more susceptible to interpretation and explanation. These latter properties, such as electrical conductivity, crystal structure, ionization energy, electron affinity, possible oxidation states, and atomic size, are related to each other and to the general chemical behavior of the elements. Thus an appreciation of the importance of these particular properties, and of how they vary throughout the periodic table, will help us to correlate, remember, and predict the detailed chemistry of the elements.

Electrical and Structural Properties

The chemical elements can be classified as *metals*, *nonmetals*, and *semimetals* on the basis of their electrical properties alone. Metals are good conductors of electricity, and their electrical conductivity *decreases* slowly as temperature is increased. The nonmetals are electrical insulators: Their ability to conduct electricity is either extremely small or undetectable. The electrical conductivities of semimetals or semiconductors are small but measurable, and tend to *increase* as temperature increases. Electrical conductivities are usually measured in units of $ohm^{-1} \cdot cm^{-1}$, and a conductivity of $1 \, ohm^{-1} \cdot cm^{-1}$ means that if a potential difference of 1 volt is applied to opposite faces of a 1-cm cube of material, a current of 1 amp will flow. The electrical conductivities of metals are, in general, greater than approximately $1 \times 10^4 \, ohm^{-1} \cdot cm^{-1}$, as Table 13.2 shows. The shaded group of semimetals have small conductivities (in the range from 10 to $10^{-5} \, ohm^{-1} \cdot cm^{-1}$) that are sensitive to impurities, and nonmetals have even smaller conductivities (i.e., are insulators).

Table 13.2 shows that the metallic elements appear in the left-hand part of the periodic table, and are separated from the nonmetals by a *diagonal* band of semimetals that runs from boron to tellurium. The classification of elements close to this group of semimetals is not always straightforward, for several of the elements of groups IVA, VA, and VIA occur in different allotropic forms,

Li 11.8	Be 18	B	C	N	O	F
Na 23	Mg 25	Al 40	Si	P	S	Cl
K 15.9	Ca 23	Ga 2.4	Ge	As	Se	Br
Rb 8.6	Sr 3.3	In 12	Sn 10	Sb 2.8	Te	I
Cs 5.6	Ba 1.7	Tl 7.1	Pb 5.2	Bi 1.0	Po	At

Sc —	Ti 1.2	V 0.6	Cr 6.5	Mn 20	Fe 11.2	Co 16	Ni 16	Cu 65	Zn 18
Y —	Zr 2.4	Nb —	Mo 23	Tc —	Ru 8.5	Rh 22	Pd 1	Ag 66	Cd 15
La 1.7	Hf 3.4	Ta 7.2	W 20	Re —	Os 11	Ir 20	Pt 10	Au 49	Hg 4.4

each of which has different electrical properties. For example, the α-phase of tin, sometimes called grey tin, has the diamond type of crystal lattice found in silicon and germanium, and like these elements, grey tin has the electrical properties of a semimetal. On the other hand, white tin, the β-phase that is stable above 13°C, is a metallic conductor. As another example, white phosphorus, a molecular solid of P_4 units, and red phosphorus, which has a complex chain structure, are both electrical insulators and thus are of nonmetallic character. In contrast, the allotrope black phosphorus has a crystal structure made up of corrugated sheets, as shown in Fig. 13.2, and in this form phosphorus behaves like a semimetal. Similar phenomena are found for selenium.

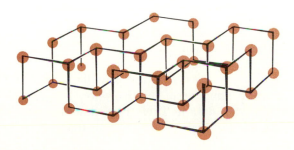

The crystal structure of the black phosphorus allotrope. **FIG. 13.2**

One allotrope is a molecular solid which consists of rings with the formula Se_8, and in this form selenium is a nonmetal. Another allotrope is made up of long covalently bonded chains of selenium atoms and has the electrical properties of a semimetal. Thus not all elements can be classified uniquely as metals, semimetals, or nonmetals without reference to the occurrence of allotropic forms.

Despite the classification difficulties imposed by allotropy, a few generalizations are clear. Metallic behavior is found among the transition elements, the members of groups I and II, and the heavier elements of groups IIIA, IVA, and VA. As noted in Chapter 3, the metallic elements have crystal structures of high coordination number, either 12 in the closest-packed lattices, or 8 in the body-centered cubic lattice. In contrast, the distinctly nonmetallic elements are the lighter members of groups IVA, VA, VIA, and VIIA. These elements usually occur as small covalently bonded molecules like N_2, S_8, and Cl_2, which form volatile molecular solids. The semimetals have complex crystal structures which may involve three-dimensional networks, infinite layer lattices, or long chain molecules. The coordination number in these crystals is small, in contrast to metallic crystals. On the other hand, semimetallic solids do not incorporate the small discrete molecular units found among the nonmetals. These correlations between electrical and structural properties are particularly evident among the elements which exist in several allotropic forms.

Ionization Energy, Electron Affinity, and Electronegativity

The very striking periodic variation of the ionization energies of the elements and its relation to the electron configurations has been discussed in Section 10.6. Here we need only recall the gross features of this variation and note their relation to the general properties of the elements. As Fig. 10.23 shows, among the elements of any row of the periodic table the ionization energy tends to increase as atomic numbers increase. As we noted in Section 11.9, metallic behavior is associated with elements of low ionization energy, and thus the increase of ionization energy along a period is related to the disappearance of metallic character that occurs eventually in any row of the periodic table. In a given family or column of the table, the ionization energy tends to decrease as the atomic numbers increase. This behavior is clearest among the representative elements, and is related to the appearance of metallic properties which occurs as the atomic numbers increase in groups IIIA, IVA, VA, VIA, and VIIA. For example, boron, which has an ionization energy of 191 kcal, is a semimetal, but the other members of group IIIA have ionization energies of 140 kcal or less, and are metals. Similar vertical trends in ionization energy and metallic properties occur in groups IVA, VA, VIA, and VIIA.

It is difficult to make any generalizations concerning the periodic behavior of electron affinities because the electron affinities of relatively few elements are known with certainty. Nevertheless, the data in Table 13.3 show that the electron affinities of the nonmetals are usually higher than those of the metals,

Table 13.3 Electron affinities of gaseous atoms (kcal/mole)

H	17.3													
Li	14	Be	$\sim$0	B	7	C	29	N	0	O	34	F	79.5
Na	19	Mg	$\sim$0	Al	12	Si	32	P	17	S	47	Cl	83.4
K	21											Br	77.3
												I	70.5

and in particular, the electron affinities of the halogen atoms are strikingly large. The variation of the electron affinity and ionization energy with atomic number makes it clear that the nonmetals have greater tendency to acquire and less tendency to release electrons than do the semimetals and metals.

An element that tends to acquire rather than lose electrons in its chemical interactions is said to be *electronegative*. Various attempts have been made to create a quantitative scale of electronegativity. Perhaps the simplest procedure is that of R. S. Mulliken, who suggested that electronegativity is proportional to the average of the ionization energy and the electron affinity. Another scale of electronegativity, proposed by Pauling, is based on the difference in the bond energies of diatomic molecules. Pauling suggested that the difference in the electronegativities X_A and X_B of two atoms A and B is given by

$$|X_A - X_B| = 0.208 \, [D_{AB} - (D_{AA}D_{BB})^{1/2}]^{1/2}, \qquad (13.1)$$

where D_{AB} is the bond energy of the diatomic molecule AB expressed in kcal/ mole, and D_{AA} and D_{BB} are the corresponding quantities for the molecules A_2 and B_2. The factor 0.208 arises from the conversion of electron-volts to kilocalories. The form of Eq. (13.1) is empirical; it is based on the observation that the bonds between atoms of qualitatively different electronegativity tend to be stronger than bonds in homonuclear molecules. Thus the bond energy of any of the hydrogen halides HX is greater than the geometric mean $(D_{HH}D_{XX})^{1/2}$ of the bond energies of the halogen and hydrogen. This extra bond energy of the polar molecule is taken by Pauling to be a measure of the electronegativity difference of the atoms, as Eq. (13.1) states.

Other electronegativity scales, based on different, less well-defined properties of atoms, have been proposed. Even though the bases of these scales may seem quite unrelated, it is often found that one scale differs from another only by a constant multiplying factor, and thus the scales may be in large measure equivalent. By use of the Pauling, Mulliken, and other definitions, it is possible to assign numerical values of electronegativity to almost every element. Table 13.4 is a partial list of these electronegativity values. The value of such a table is twofold. It provides a clear expression of the qualitative generalization that

Table 13.4
Electronegativities of the representative elements

H 2.1						
Li 0.97	Be 1.5	B 2.0	C 2.5	N 3.1	O 3.5	F 4.1
Na 1.0	Mg 1.2	Al 1.5	Si 1.7	P 2.1	S 2.4	Cl 2.8
K 0.90	Ca 1.0	Ga 1.8	Ge 2.0	As 2.2	Se 2.5	Br 2.7
Rb 0.89	Sr 1.0	In 1.5	Sn 1.72	Sb 1.82	Te 2.0	I 2.2
Cs 0.86	Ba 0.97	Tl 1.4	Pb 1.5	Bi 1.7	Po 1.8	At 1.9

the ability of elements to attract and hold electrons increases from left to right along any row, and from bottom to top in any column of the periodic table. This observation is important in the understanding of the chemical behavior of the elements. The second use of the numerical electronegativity scale is that electronegativity differences often can be related semiquantitatively, although empirically, to properties of bonds such as dipole moments and bond energies.

Oxidation States

The important oxidation states of the elements are represented graphically in Fig. 13.3. There are some very clear periodic regularities, and appreciation of these can simplify the problem of remembering the important chemistry of the elements.

The oxidation states of the representative elements bear a simple relationship to the electron configurations of the atoms. Many of the oxidation states correspond to the atom's losing or gaining enough electrons to acquire, at least formally, a "closed shell" electron configuration of the type ns^2np^6 or nd^{10}. This tendency is particularly clear in groups IA and IIA, and among the lighter members of group IIIA. In group IIIA, the valence-electron configuration of the atoms is ns^2np^1, and loss of three electrons to form the $+3$ oxidation state results in ions which have the $(n-1)s^2(n-1)p^6$ or $(n-1)d^{10}$ configurations. For the elements indium and thallium, however, the $+1$ oxidation state also occurs, and this corresponds to loss of only the p-electron and results in an ns^2-configuration for In^+ and Tl^+. Oxidation states that correspond to loss of np-electrons and retention of ns^2-electrons also occur among the heavier ele-

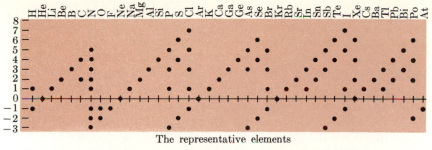

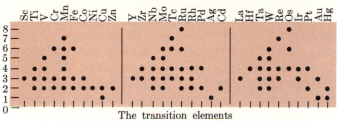

The common oxidation states of the representative elements and transition metals plotted as a function of atomic number.

FIG. 13.3

ments of groups IVA, VA, VIA, and VIIA. Thus tin and lead, which have ns^2np^2-configurations, display both +2 and +4 oxidation states; phosphorus, arsenic, antimony, and bismuth with the ns^2np^3-configuration have both +3 and +5 oxidation states, and so on, as Fig. 13.3 shows.

In any of the groups IIIA, IVA, VA, VIA, and VIIA, where two or more positive oxidation states are found, the lower oxidation states tend to become more important as one goes down a column in the table. That is, the chemistries of carbon, silicon, and germanium involve the +4 oxidation state almost exclusively, while for tin and particularly lead the +2 state is more important than the +4 state. Likewise, while the +5 state is very important in the chemistry of nitrogen, phosphorus, and arsenic, it is less so for antimony, and for bismuth the +3 state is dominant and the +5 state occurs rarely.

While positive oxidation states are of exclusive importance for the metals, and almost as important for the semimetals, negative oxidation states appear in group VA and are very common among the nonmetals. Thus nitrogen and phosphorus form nitrides and phosphides which contain the N^{-3} and P^{-3} ions respectively, but the −3 state is much less important in the chemistries of arsenic and antimony, and virtually nonexistent in bismuth chemistry. In group VIA the −2 state is important for all the elements, but it is relatively more important for the lighter than for the heavier members of the group. The same can be said for the −1 oxidation state displayed by elements in group VIIA. The importance of negative oxidation states among the lighter nonmetals is consistent with the relatively high electronegativity of these elements.

The transition elements display a large number of oxidation states, but there are still some regularities and trends to be noted. The maximum oxidation states found for the scandium, titanium, chromium, and manganese families correspond to the loss or the participation in bonding of all electrons in excess of an inert gas configuration. That is, members of the scandium family have the configuration [inert gas] $(n-1)d^1ns^2$ and display only the $+3$ oxidation state. Likewise, members of the manganese family have the configuration [inert gas] $(n-1)d^5ns^2$ and have a maximum oxidation state of $+7$. These remarks, together with the information in Fig. 13.3, show that for transition elements with a d-shell no more than half-filled, the maximum oxidation state is equal to the number of the group. For elements of the first transition series which have more than five $3d$-electrons, however, oxidation states higher than $+2$ or $+3$ are rare. Thus the chemistry of iron, which has the valence-electron configuration $3d^64s^2$, is largely confined to the $+2$ and $+3$ oxidation states, while the $+6$ state is rare and the possible $+8$ state is unknown. The $+8$ state is of importance in the chemistry of the other members of the iron family, ruthenium and osmium. In the cobalt, nickel, copper, and zinc families, the important oxidation states are all less than what would correspond to removal of all s- and d-electrons.

Another useful generalization about the transition elements is that among the members of any family, the higher oxidation states become relatively more important as the atomic number increases. For example, titanium chemistry involves the $+2$, $+3$, and $+4$ states, but the chemistry of zirconium and hafnium is almost entirely that of the $+4$ state. Similarly, the $+2$, $+3$, and $+6$ states of chromium are all important, but the chemistries of molybdenum and tungsten involve the $+6$ state primarily. In general, the members of the $3d$ transition series all have important lower oxidation states, either $+2$, $+3$, or both. For the elements of the $4d$- and $5d$-series, these states are often not important, if they exist at all. Note carefully that the increased importance of higher oxidation states that occurs going down a column of the transition metals is *opposite* to the trend observed among the representative elements.

Size Relationships

The periodic variation of the size of the atoms was first noted by Lothar Meyer in 1870. Meyer computed the "atomic volume" by dividing the atomic weight of an element by its density. When this quantity is plotted as a function of atomic number, the sawtooth curve shown in Fig. 13.4 results. The atomic volume calculated in this manner is at best only a qualitative indication of atomic size, for the density of an element depends on its temperature and its crystal structure. Elements that exist in a number of different allotropic crystalline forms would apparently have more than one atomic volume. Nevertheless, the periodic variation of atomic volume is striking, and atomic size is a very useful concept which can help us to understand the chemistry of the elements.

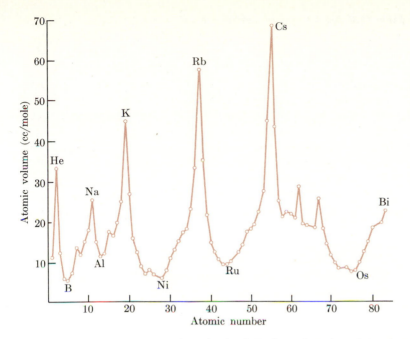

The atomic volume expressed in cc/mole plotted as a function of atomic number.

FIG. 13.4

Because the electron cloud of any atom has no definite limit, the size of an atom cannot be defined simply and uniquely. However, one valid measure of atomic size is the value of the Lennard-Jones σ-parameter, which represents the distance of closest approach of the nuclei of two free gaseous atoms. If the atoms are pictured as spheres, σ is equal to their diameter, and thus $\sigma/2$ represents an atomic radius. The values of $\sigma/2$ for the inert gases are given in Table 13.5. It is clear that in this single family, the atomic radius increases as atomic number increases.

Table 13.5 Lennard-Jones radii
of the noble gas atoms, $\sigma/2$ (A)

He	Ne	Ar	Kr	Xe
1.31	1.39	1.70	1.80	2.0

To assess the sizes of the atoms of metallic elements, the internuclear distance in the metallic crystal is determined by x-ray diffraction and divided by two to give an atomic radius. The apparent radius of an atom determined in this manner depends to a certain degree on the crystal structure of the metal. For example, the apparent radius of the titanium atom in the body-centered cubic lattice, 1.43 A, is different from the radius of 1.49 A found in the hexagonal

Table 13.6 Atomic radii of metallic elements (angstroms)

Li 1.55	Be 0.89											B 0.80
Na 1.90	Mg 1.36											Al 1.25
K 2.35	Ca 1.74	Sc 1.44	Ti 1.32	V 1.22	Cr 1.17	Mn 1.17	Fe 1.16	Co 1.16	Ni 1.15	Cu 1.17	Zn 1.25	Ga 1.25
Rb 2.48	Sr 1.91	Y —	Zr 1.45	Nb 1.34	Mo 1.29	Tc —	Ru 1.24	Rh 1.25	Pd 1.28	Ag 1.34	Cd 1.41	In 1.50
Cs 2.67	Ba 1.98	La 1.69	Hf 1.44	Ta 1.34	W 1.30	Re 1.28	Os 1.26	Ir 1.26	Pt 1.29	Au 1.34	Hg 1.44	Tl 1.55

closest-packed lattice. The difference is not usually of serious magnitude, however, and a meaningful set of atomic radii can be tabulated. Table 13.6 gives a few such data. It is clear that in a given family, size increases as atomic number increases. Among elements in a given row of the periodic table, however, size decreases as the atomic number increases. Both these trends are to be expected on the basis of the accompanying changes in electronic structure. As atomic number increases in a given family, the principal quantum number of the valence electrons increases, and consequently these electrons lie at greater and greater distances from the nucleus. Along a given row of the periodic table, the principal quantum number of the valence electrons is constant, but as the nuclear charge increases, the valence electrons tend to be drawn closer to the nucleus, and the atoms tend to become smaller.

Another quantitative expression of size which is more useful in understanding chemical properties is the ionic radius. In Section 11.2 we gave the values of several ionic radii and suggested how they were related to crystal geometry. Here we need only emphasize the regular trends in ionic size which occur in the sequences of the periodic table. In Fig. 13.5 the ionic radii of several ions are plotted as a function of atomic number. It is clear that for any isoelectronic sequence, that is, for any series of ions which have the same number of electrons, the ionic radius decreases as the atomic number increases. This is certainly to be expected, for as the nuclear charge increases, the electron cloud is bound to contract. The data also show that ionic size increases as atomic number increases in a given family. A particularly interesting feature of this trend is the noticeable discontinuity in the slope of the dashed line in Fig. 13.5 that occurs at the element potassium. The ionic radii of the members of a family do not increase as rapidly after potassium as before. One suggested explanation is that in the interval between potassium and rubidium the first transition series occurs, and as these "extra" elements enter the periodic table, the increas-

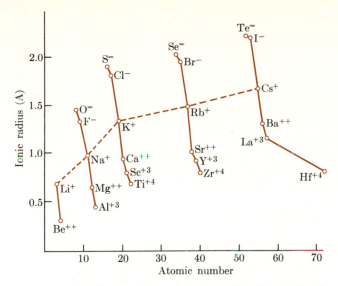

The ionic radii of several ions plotted as a function of atomic number. The solid lines link ions that are isoelectric.

FIG. 13.5

ing nuclear charge tends to cause the atoms and their ions to contract. Thus the ions that follow any of the transition series are smaller than if only eight elements had separated them from the lighter members of their family.

A very clear demonstration of how ionic size decreases along a transition series is provided by the lanthanide elements. In the 14 elements that follow lanthanum, 4f-electrons are being added to give electron configurations of the type $5s^2 5p^6 4f^n 6s^2$. All the lanthanides form +3 ions in which the two 6s-electrons and one of the 4f-electrons have been lost. The size of these ions becomes progressively smaller as the atomic number increases, as Table 13.7 shows. This decrease is known as the lanthanide contraction, and its occurrence is rather directly responsible for a number of features of the chemistry of the transition elements which follow the lanthanides in the periodic table.

Table 13.7 Ionic radii of the lanthanide elements (angstroms)

La^{+3}	1.061	Tb^{+3}	0.923
Ce^{+3}	1.034	Dy^{+3}	0.908
Pr^{+3}	1.013	Ho^{+3}	0.894
Nd^{+3}	0.995	Er^{+3}	0.881
Pm^{+3}	0.979	Tm^{+3}	0.869
Sm^{+3}	0.964	Yb^{+3}	0.858
Eu^{+3}	0.950	Lu^{+3}	0.848
Gd^{+3}	0.938		

Oxygen forms binary compounds with all the chemical elements except some of the inert gases. A comparison of the properties of the oxides reveals some of the characteristics of the elements and helps to systematize the chemistry of the more complex compounds. In the following we shall be concerned exclusively with the normal oxides—those in which oxygen displays an oxidation state of −2. The *peroxides* and the *superoxides*, which contain the $O_2^=$ and O_2^- ions respectively, will be discussed in Chapter 15.

Table 13.8 contains the standard free energies of formation of a few of the oxides. The most striking feature of these data is that virtually all the elements form at least one oxide which has a negative free energy of formation. The exceptions include some of the halogens, inert gases, and nitrogen. Thus the oxides as a group are very stable compounds, and are exceeded in this respect only by the fluorides.

Table 13.8 Standard free energy of formation of some oxides, ΔG_f^0 (kcal/mole)

Li_2O	BeO	B_2O_3	CO_2	N_2O_5	O_2	F_2O
−133.9	−136.1	−283	− 94	27.9		9.7
Na_2O	MgO	Al_2O_3	SiO_2	P_4O_{10}	SO_2	Cl_2O
− 90.4	−135.3	−377	−191	—	−71.8	22.4
K_2O	CaO	Ga_2O_3	GeO_2	As_4O_6	SeO_2	Br_2O
− 86.4	−144.4	−237	−127	−275	−41.5	
Rb_2O	SrO	In_2O_3	SnO_2	Sb_4O_6	TeO_2	I_2O_5
− 69.5	−138.8	−200	−124	−298.0	−64.6	—
Cs_2O	BaO	Tl_2O_3	PbO_2	Bi_2O_3	PoO_2	
− 65.6	−126.0	—	− 52	−118	−46	

The standard enthalpies of formation of the oxides, like their free energies of formation, range in value from very negative to positive. Nevertheless, it is possible to detect some regularities in the thermochemistry of the oxides. To do this we must compare the enthalpies of formation per gram-equivalent of oxygen. That is, for an oxide of general formula M_xO_y, we divide ΔH_f by the integer $2y$ to get a number which represents the stability of the bonds to an oxygen atom more faithfully than does the molar enthalpy of formation itself. We see from Fig. 13.6 that the enthalpy of formation per gram-equivalent of oxygen tends to become more negative as atomic number decreases along a given period. This trend indicates that, in general, oxygen forms compounds of greatest stability with elements which are well removed from it in the periodic table. We shall find that this is true for some other elements; the most stable compounds of the elements of groups VIA and VIIA are those with the elements of groups IA, IIA, IIIA, and the transition metals.

One very useful way to classify oxides is in terms of acid-base properties. In general, any compound that dissolves in, or reacts with, water to produce an excess of hydrogen ions can be called an acid, and any compound that produces a deficiency of hydrogen ions is a base. Oxides like Na_2O and BaO are clearly basic, for they dissolve in water according to the reactions

$$Na_2O(s) + H_2O = 2Na^+(aq) + 2OH^-(aq),$$
$$BaO(s) + H_2O = Ba^{++}(aq) + 2OH^-(aq).$$

There are a number of oxides that are insoluble in water, but dissolve in solutions of acids. For example, we have

$$MnO(s) + 2H^+(aq) = Mn^{++}(aq) + H_2O,$$
$$NiO(s) + 2H^+(aq) = Ni^{++}(aq) + H_2O.$$

These oxides are also considered to be basic, for they react with a known acid.

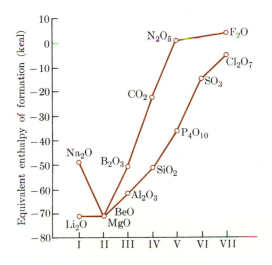

The enthalpy of formation per equivalent of oxygen $[(1/2y)\Delta H^0(M_xO_y)]$ plotted as a function of position in the periodic table.

FIG. 13.6

In contrast, there are oxides that are clearly acidic. For example, both SO_3 and P_4O_{10} react with water to produce hydrogen ions:

$$SO_3(s) + H_2O = H^+(aq) + HSO_4^-(aq),$$
$$P_4O_{10}(s) + 6H_2O = 4H^+(aq) + 4H_2PO_4^-(aq).$$

Certain other oxides like SiO_2 are insoluble in water, but react with strong bases to form soluble salts:

$$SiO_2(s) + Na_2O(s) = Na_2SiO_3(s),$$
$$Na_2SiO_3(s) + H_2O = 2Na^+(aq) + SiO_3^=(aq).$$

Such oxides are also acidic, but less so than SO_3 or P_4O_{10}.

There are also some oxides that have both acidic and basic properties. For instance, Al_2O_3 and ZnO are rather insoluble in water, but dissolve in either strong acids or strong bases:

$$Al_2O_3(s) + 6H^+(aq) = 2Al^{+3}(aq) + 3H_2O,$$
$$Al_2O_3(s) + 2OH^-(aq) + 3H_2O = 2Al(OH)_4^-(aq),$$
$$ZnO(s) + 2H^+(aq) = Zn^{++}(aq) + H_2O,$$
$$ZnO(s) + H_2O + 2OH^-(aq) = Zn(OH)_4^=(aq).$$

Oxides that react with acids and bases are said to be *amphoteric*.

Table 13.9 compares the acid-base properties of some of the oxides. It is clear that the elements in the lower left-hand region of the periodic table form basic oxides, while the acidic oxides are associated with the nonmetallic elements of the upper right-hand region. Dividing these two groups are the amphoteric oxides of Be, Al, Ga, Sn, and Pb; these lie in a diagonal band enclosed in heavy lines in Table 13.9. Along any row of the table, oxide acidity increases as atomic number increases, but in any family oxide acidity decreases as atomic number increases. In summary, we can say that among the representative elements,

Table 13.9 Acid-base properties of some oxides of the representative elements

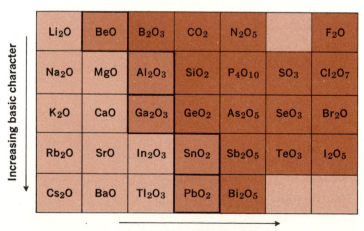

Increasing basic character

Increasing acidic character

Li_2O	BeO	B_2O_3	CO_2	N_2O_5		F_2O
Na_2O	MgO	Al_2O_3	SiO_2	P_4O_{10}	SO_3	Cl_2O_7
K_2O	CaO	Ga_2O_3	GeO_2	As_2O_5	SeO_3	Br_2O
Rb_2O	SrO	In_2O_3	SnO_2	Sb_2O_5	TeO_3	I_2O_5
Cs_2O	BaO	Tl_2O_3	PbO_2	Bi_2O_5		

the oxides of the metals usually are basic or amphoteric, those of the nonmetals are acidic, and those of the semimetals are weakly acidic.

A number of elements both in the transition series and among the representative groups form several oxides. The general observation is that in these cases the acidity of the oxides increases as the oxidation number increases. For example, we can cite the following.

VO	basic	CrO	basic	As_2O_3	weakly acidic
V_2O_3	basic	Cr_2O_3	amphoteric	As_2O_5	acidic
VO_2	amphoteric	CrO_3	acidic		
V_2O_5	acidic				

We shall encounter other illustrations of this rule in subsequent chapters.

Another, somewhat less straightforward, method of classifying the binary oxides is by bond type. The solid oxides of the alkali metals have the antifluorite lattice in which each metal atom is surrounded by four oxygen atoms, and each oxygen atom by eight metal atoms. There is no suggestion of discrete M_2O molecules, and thus these compounds can be considered as ionic oxides. Likewise, the alkaline-earth oxides and many of the transition-metal oxides of the type MO exist in the typically ionic rock-salt lattice structure. The mechanical, electrical, and thermal properties of these compounds are those associated with ionic lattices. Thus the oxides of groups IA, IIA, and of the transition metals in their lower oxidation states are ionic compounds.

Many of the oxides of the nonmetals exist as discrete molecules under all circumstances, and this indicates that the bonds in these compounds are predominantly covalent. The covalent oxides, like NO, F_2O, ClO_2, and SO_2, are formed principally by elements which, being close to oxygen in the periodic table, have electronegativities rather close to that of oxygen.

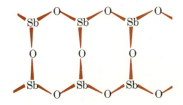

A segment of the infinite double chain of the Sb_4O_6 crystal. Each antimony atom lies at the apex of a pyramid whose base is formed by three oxygen atoms.

FIG. 13.7

The oxides of the heavier nonmetals and semimetals tend to be solids of moderately complex crystal structure. Thus, while CO_2 exists as discrete molecules, SiO_2 is an infinite three-dimensional network consisting of alternate silicon and oxygen atoms covalently bonded, and both GeO_2 and SnO_2 have complex three-dimensional structures in which oxygen is bonded covalently. Among the oxides of group VA, the transition from "molecular" to covalent lattice compounds occurs at antimony. Thus, besides the molecular oxides of nitrogen, the compounds P_4O_6, P_4O_{10}, and As_4O_6 exist as discrete molecules both in the gas and condensed phases, but Sb_4O_6 exists both as discrete molecules or as a solid of infinite covalent chains of the type shown in Fig. 13.7. In group VIA, the oxides of sulfur are discrete small molecules, but SeO_2 exists in the solid as infinite covalent chains, and TeO_2 also shows no discrete small molecules in the solid phase.

To summarize the bonding properties of the oxides, we can say that the distinctly metallic elements form ionic oxides, while the oxides of the very electronegative nonmetals are, in general, small discrete covalently bonded molecules both in the gas and condensed phases. The oxides of the heavier nonmetals and the semimetals very often exist in the intermediate condition of an infinite lattice with largely covalent linkages. There is then, a fairly close correlation between the acid-base properties of oxides, and their ionic-covalent character. The ionic oxides tend to be basic or amphoteric, the oxides of the infinite covalent lattice structures have weakly acidic, weakly basic, or amphoteric properties, and the molecular oxides tend to be acidic.

13.4 THE PROPERTIES OF HYDRIDES

The formulas of some of the binary hydrogen compounds of the elements are given in Table 13.10, together with their standard enthalpies of formation. The enthalpies of formation show that hydrogen forms energetically stable compounds with the very electropositive elements of groups IA and IIA, and with the very electronegative elements of groups VIA and VIIA. The hydrides of the heavier elements of groups IIIA, IVA, and VA tend to be unstable, and in some instances are so difficult to prepare that little is known about their properties. Several of the elements, most notably B, C, Si, and Ge each form more than one hydride, and only the simplest of these appear in Table 13.10.

On the basis of chemical and physical properties it is possible to classify the hydrides as of the *ionic, covalent,* or *interstitial* type. The compounds of hydrogen

Table 13.10 Standard enthalpies of formation for some hydrides, ΔH_f^0 (kcal/mole)

LiH	BeH$_2$	B$_2$H$_6$	CH$_4$	NH$_3$	H$_2$O	HF
-21.6	—	7.5	-17.9	-11.0	-57.8	-64.2
NaH	MgH$_2$	AlH$_3$	SiH$_4$	PH$_3$	H$_2$S	HCl
-13.7	—	—	-14.8	2.21	-4.8	-22.1
KH	CaH$_2$	Ga$_2$H$_6$	GeH$_4$	AsH$_3$	H$_2$Se	HBr
-13.6	-45.1	—	—	41.0	20.5	-8.7
RbH	SrH$_2$	InH$_3$	SnH$_4$	SbH$_4$	H$_2$Te	HI
-12	-42.3	—	—	34	18.5	6.2
CsH	BaH$_2$	TlH$_3$	PbH$_4$	BiH$_3$		
-20	-40.9	—	—	—		

with the alkali and alkaline-earth metals can be prepared by direct action of the elements at elevated temperatures, and are white crystalline solids. When any of these hydrides is dissolved in molten salts such as LiCl and KCl and electrolyzed, hydrogen gas is evolved at the *anode*. This indicates that the ion H^- is present in the mixture. Quite elaborate spectroscopic experiments have shown that the *gaseous* lithium hydride molecule is very polar, with the hydrogen atom bearing the negative charge. Available experimental data indicate then, that the hydrides of groups IA and IIA contain the hydride ion H^-. For this reason, these ionic compounds are often called the saline, or saltlike hydrides.

It is interesting to compare the energetics of formation of the hydride ion with the corresponding values for the fluoride ion. We have

$$\frac{1}{2}H_2(g) \qquad = H(g) \qquad \Delta H = \qquad 52 \text{ kcal}$$
$$\underline{H(g) + e^- = H^-(g) \qquad \Delta H = -17 \text{ kcal}}$$
$$\frac{1}{2}H_2(g) + e^- = H^-(g) \qquad \Delta H = +35 \text{ kcal}$$

$$\frac{1}{2}F_2(g) \qquad = F(g) \qquad \Delta H = \qquad 18.5 \text{ kcal}$$
$$\underline{F(g) + e^- = F^-(g) \qquad \Delta H = -80.5 \text{ kcal}}$$
$$\frac{1}{2}F_2(g) + e^- = F^-(g) \qquad \Delta H = -62.0 \text{ kcal}$$

Thus the formation of the gaseous hydride ion is an endothermic process, whereas formation of the fluoride ion, or any of the other gaseous halide ions, is an exothermic process. Because the formation of the hydride ion is so energetically unfavorable, it is not surprising that ionic or saltlike hydrides are formed only by the very electropositive elements of groups IA and IIA. The ionization energies of the other elements are high enough to prevent the transfer of their electrons to the hydrogen atoms.

The relative instability of the hydride ion suggests that saline hydrides should be good reducing agents. In fact, reactions such as

$$NaH + CO_2 = HCOONa$$
$$4NaH + Na_2SO_4 = Na_2S + 4NaOH$$

take place at elevated temperatures and demonstrate the reducing power of the hydride ion. In addition, the hydride ion reacts rapidly and completely to reduce water or any other proton donor to hydrogen, as in

$$CaH_2 + 2H_2O = Ca(OH)_2 + 2H_2.$$

Thus we can regard hydride ion as a powerful reductant and very strong base.

The representative elements of groups IVA, VA, VIA, VIIA, and boron of group IIIA form volatile molecular hydrides in which the bonds to hydrogen are largely of covalent character. Among these covalent hydrides there are a number of clear trends in thermal and chemical properties. Figure 13.8 shows that the H—X bond energies of the binary hydrides increase along any row of the periodic table. Futhermore, in any given column, the H—X bond energies decrease as the atomic numbers increase. These trends in bond energies account for the instability of the hydrides of the heavier elements of groups IVA and VA. The compounds PbH_4 and BiH_3 are so unstable that only trace amounts have ever been detected.

FIG. 13.8 The average energies of bonds between hydrogen and the atoms of nonmetallic elements.

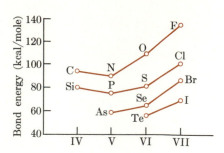

The ionic hydrides are strong bases, and along any given row of the periodic table, the covalent hydrides become increasingly acidic as the atomic numbers increase. Thus methane, CH_4, has virtually negligible acidic properties, but NH_3 donates a proton to very strong bases to form NH_2^-, H_2O loses a proton even more readily, and HF is a moderately strong acid. This trend in acidity appears again in the succeeding rows of the periodic table. In addition, the acidities of the hydrides of groups VIA and VIIA increase as one proceeds down these columns.

The boiling temperatures and the enthalpies of vaporization of some covalent hydrides are plotted as a function of position in the periodic table in Figs. 13.9 and 13.10. In the sequence H_2S, H_2Se, H_2Te there is an increase in both boiling temperature and ΔH_{vap}. This trend is to be expected, for in general, attractive intermolecular forces increase in strength as the number of electrons in the molecules increases. It is clear, however, that the enthalpy of vaporization of water and its boiling temperature are both much higher than what we might expect from the trend established by H_2S, H_2Se, and H_2Te. The same phenomenon occurs among the hydrides of groups VA and VIIA: The cohesive forces between the molecules of the hydride of the lightest member of each family are markedly larger than would be expected from the ordinary variation of intermolecular forces with number of electrons. These data, and considerably more information drawn from molecular spectroscopy, have led to the conclusion that if hydrogen is bonded to a very electronegative atom, it is capable of forming another weak bond with another electronegative atom which has a pair of nonbonded electrons. This interaction, called the *hydrogen bond*, is weaker than most chemical bonds because the dissociation energy of a hydrogen bond is only about 7 kcal/mole. On the other hand, the hydrogen bond is much stronger than the ordinary van der Waals "bonds" between molecules.

It is difficult to overemphasize the importance of the hydrogen bond. Hydrogen bonding is responsible for the high boiling point of water, and for that reason alone it exerts an enormous influence on physiological and geological processes. Moreover, hydrogen bonding helps determine the configurations of greatest stability for virtually every large biologically important molecule.

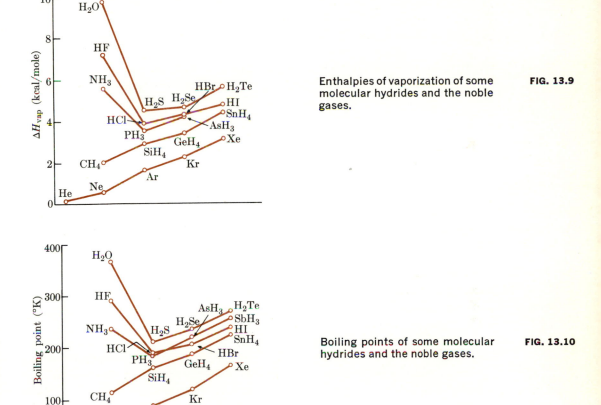

Enthalpies of vaporization of some molecular hydrides and the noble gases.

FIG. 13.9

Boiling points of some molecular hydrides and the noble gases.

FIG. 13.10

Thus even though it is an interaction that occurs only when hydrogen is bonded to a very restricted group of atoms, hydrogen bonding is responsible for the nature of life as we know it.

Despite the importance of hydrogen bonding and the great amount of work devoted to understanding it, no completely satisfactory theoretical explanation of the hydrogen bond has been found. Like other bonds, hydrogen bonds form because by their doing so, there occurs an energetically favorable redistribution of electron density among the atoms. It is the exact nature of this redistribution that is not clear at this time.

While the hydrides of the nonmetals are well-characterized molecular compounds, less is known about the transition-metal hydrides. Many of these compounds have nonstoichiometric composition. Metallic titanium and vanadium absorb hydrogen with evolution of heat, but the arrangement of the metallic atoms in the lattice remains substantially constant with only a slight increase in the distance between nearest neighbors. Thus the hydrogen seems to occupy interstitial positions in the metallic lattice, and these "compounds" are sometimes called interstitial hydrides. It is not clear in many instances whether the interstitial hydrides should be regarded as true compounds or merely as solutions of hydrogen in the metal. Despite the general lack of knowledge about the nature of the transition metal–hydrogen systems, some of them are of considerable importance in the laboratory. Nickel, platinum, and palladium absorb varying quantities of hydrogen gas and in this condition act as catalysts for reactions such as

$$H_2 + C_2H_4 \xrightarrow[\text{or Ni}]{\text{Pt, Pd}} C_2H_6,$$

where hydrogen is added to other molecules.

The hydrides of Be, Mg, Al, and the heavier members of group IIIA are nonvolatile solids which are poorly characterized, but which appear to have properties intermediate between the ionic hydrides and the covalent molecular hydrides.

13.5 CONCLUSION

In this chapter we have discussed a few of the general properties of the chemical elements and their simple compounds. In studying inorganic chemistry, it is helpful to recognize how the variations in such properties of an element as electron configuration, ionization energy, atomic size, oxidation states, and the nature of oxides and hydrides are correlated, and how these variations reflect the general nature of the elements. Therefore, let us summarize the properties which we have associated with the metallic, semimetallic, and nonmetallic elements.

Metallic elements are found in the lower and left-hand regions of the periodic table. In general, they have low ionization energies and fairly large atomic radii which tend to decrease from left to right along a row of the periodic table. In metallic crystal lattices, the coordination number is high—usually 8 or 12. In their compounds, metals display positive oxidation states virtually exclusively. The representative metals most often have only one or sometimes two oxidation states. When there are two known oxidation states, the lower tends to be the more important particularly for the heavy representative metals. The transition metals, in general, display several oxidation states. While the lower states such as +2 and +3 are important in the chemistry of the elements of the 3d transition series, these states tend to be less important than higher states in the 4d and 5d transition elements.

The oxides of the metals in their lower oxidation states are basic or amphoteric in behavior and are adequately described as having ionic lattice structures. The oxides of metals in the higher oxidation states such as +4 and above are either amphoteric or acidic in nature and sometimes exist as covalent molecular compounds rather than as infinite ionic lattices. The binary hydrides of the metals are with few exceptions nonvolatile compounds that are basic reducing agents.

The semimetals lie on a diagonal band that runs through the periodic table from boron to tellurium. The ionization energies of these elements are slightly higher than those of the metals. The crystal structures of the semimetals are in many cases complex and involve infinite chains, layer structures, and infinite three-dimensional networks of atoms. The coordination numbers of the atoms in these lattices are smaller than those displayed by metallic elements. Many of the semimetals display both positive and negative oxidation states, but the former are usually more important. The oxides of these elements are most often acidic and in some cases amphoteric. The hydrides of the semimetals are generally volatile compounds consisting of small covalently bonded molecules.

The nonmetallic elements lie in the upper right-hand area of the periodic table. The atoms of these elements have high ionization energies and large electron affinities. The elements exist most frequently as relatively small molecules in all phases. In their compounds, the nonmetals, except fluorine, display both positive and negative oxidation states, but the latter tend to be more stable in most circumstances. The oxides of the nonmetals are frequently volatile compounds that consist of small discrete molecules. Almost without exception, these oxides display acidic properties. Because the nonmetals have electronegativities comparable to that of oxygen, the nonmetallic oxides are best thought of as covalently bonded molecules. The hydrides of the nonmetals also are volatile compounds that consist of small covalently bonded molecules. Some of these have amphoteric properties and are able to accept as well as donate protons. The acidity of the hydrides does tend to increase, however, as one passes from left to right in any row of the periodic table and from top to bottom in any column of the nonmetallic elements.

While the generalizations we have mentioned are useful, they must be accepted and used cautiously. It is possible to find exceptions to, or perhaps subtle deviations from, almost every one of these general principles. Consequently they cannot be taken to be a completely reliable basis for abolute predictions of chemical behavior. However, they do form a general reference to which the detailed properties of the individual elements can be compared, and the exceptions that are noted in this process can easily be remembered.

SUGGESTIONS FOR FURTHER READING

Day, M. C., and J. Selbin, *Theoretical Inorganic Chemistry*. New York: Reinhold, 1962.

Douglas, B. E., and D. H. McDaniel, *Concepts and Models of Inorganic Chemistry*. New York: Blaisdell, 1965.

Gould, E. S., *Inorganic Reactions and Structure*. New York: Holt, Rinehart and Winston, 1962.

Moeller, T., *Inorganic Chemistry*. New York: Wiley, 1952.

Phillips, C. S. G., and R. J. P. Williams, *Inorganic Chemistry*, Vols. 1 and 2. London: Oxford Univ. Press, 1965.

Rich, R. L., *Periodic Correlations*. Menlo Park, Calif.: W. A. Benjamin, 1965.

Sanderson, R. T., *Chemical Periodicity*. New York: Reinhold, 1960.

Sisler, H. H., *Electronic Structure, Properties, and the Periodic Law*. New York: Reinhold, 1963.

PROBLEMS

13.1 Without consulting a periodic table, deduce the atomic number and electronic structure for the following atoms: (a) the third alkali metal, (b) the second transition metal, (c) the third halogen, (d) the third noble gas.

13.2 What trend in atomic size is to be expected in a given family like the alkali metals? Support your answer by using the principles governing the electronic structures of atoms.

13.3 Explain why the electron affinities of the atoms increase from left to right along a row in the periodic table. Why is the electron affinity of the nitrogen atom equal to zero, while carbon and oxygen have substantial electron affinities?

13.4 Zirconium and hafnium have virtually the same atomic and ionic radii. Why isn't hafnium a larger atom than zirconium?

13.5 In each of the following pairs, which would be the larger ion? Ti^{++}, Fe^{++}; Mn^{++}, Zn^{++}; $O^=$, F^-; S, $Se^=$; Tl^+, Tl^{+3}.

13.6 In each of the following pairs of hydrides, decide which is the more stable thermodynamically with respect to its elements: HCl, HI; PH_3, SbH_3; NH_3, H_2O.

13.7 In virtually all transition-metal families, and in many of the groups of representative elements, the elements display two or more positive oxidation states. How does the relative stability of the higher and lower oxidation states vary with increasing atomic number in

a) the transition-metal families?   b) a family of representative elements?

13.8 In the following, choose one of each pair of alternatives. The oxides of the nonmetallic elements are typically: (a) acidic or basic, composed of (b) small molecules or infinite network solids, bonded (c) covalently or ionically.

13.9 The formation of a typical metal oxide MO from its elements is exothermic:

$$M(s) + \tfrac{1}{2}O_2(g) = MO(s), \qquad \Delta H_f^0 < 0.$$

Show that this reaction can be analyzed in terms of a series of steps in which the metal is vaporized, the oxygen dissociated, the gaseous atoms converted to ions, and the ions converted to a solid. Discuss how the ΔH_f^0 of the oxide is affected by (a) the strength of the bonding in the metallic crystal, (b) the ionization energy of the metal atom, and (c) the size of the metallic ion.

CHAPTER 14

THE REPRESENTATIVE ELEMENTS: GROUPS I-IV

Having discussed many of the principles of chemistry, we are in a position to examine the detailed chemistry of the elements. Our treatment cannot be encyclopedic, yet we shall present enough information to make clear the relationship between the properties of the elements and their position in the periodic table. To this end, we shall emphasize the presence or absence of resemblances between members of a given periodic group and the existence of trends in properties that occur among elements of neighboring groups. In this chapter we shall find that there are very marked vertical resemblances among the metallic elements of groups IA and IIA; as we proceed across the periodic table from left to right the metallic properties are modified and then disappear, and the resemblance between elements of the same group becomes less obvious in groups IIIA and IVA.

14.1 THE ALKALI METALS

These metals (Li, Na, K, Rb, Cs, Fr) are never found in the elemental state in nature, for they react rapidly and completely with virtually all nonmetals. While sodium and potassium are rather abundant in nature, the others are much less common. In particular, francium occurs naturally in only trace amounts, and all its isotopes are radioactive.

Because the alkali metals are strong reducing agents, electrolysis is the only convenient way of recovering them in quantity from their compounds. On a smaller scale, laboratory preparation of the metals usually involves a reaction such as

$$Ca(s) + 2CsCl(s) = CaCl_2(s) + 2Cs(g).$$

The alkali metals are volatile, and can be distilled out of the reaction mixtures and obtained as pure products.

The freshly prepared surfaces of the alkali elements show the bright silvery luster characteristic of metals, and the substances are good conductors of electricity and heat. As a group they are the softest metals and possess some of the lowest melting temperatures. Table 14.1 shows that the melting and boiling points of the metals decrease regularly as the atomic number increases. Parallel to this trend, there is a decrease in hardness as atomic number increases; lithium can be cut with a knife with difficulty, but the succeeding metals can be cut with increasing ease. Due to their softness and reactivity the metals are never used in structural applications. Because of its high specific heat and thermal conductivity, sodium is used as a coolant in the valves of internal combustion engines and in nuclear reactors. All the metals find some use as reductants in laboratory and industrial processes.

Comparison of the first and second ionization energies of the alkali-metal atoms shows why the chemistry of all these elements involves the $+1$ oxidation state exclusively. The outermost s-electron can be removed with an ease that increases with atomic number, but the second ionization energy is so large that the $+2$ oxidation state is unstable and is never observed.

The small first ionization energies of the alkali atoms are reflected in the bond dissociation energies of the gaseous alkali diatomic molecules, which, as Table 14.1 shows, range from 25 kcal for Li_2 to 10.4 kcal for Cs_2. The energy lowering associated with covalent-bond formation comes from the extra attrac-

Table 14.1 Properties of the group IA elements

	Li	Na	K	Rb	Cs
Atomic number	3	11	19	37	55
Configuration	$2s^1$	$3s^1$	$4s^1$	$5s^1$	$6s^1$
Ionization energy, kcal $\{I_1$	124	118	100	96	90
$\quad\quad\quad\quad\quad\quad\quad\ I_2$	1744	1091	734	634	579
Atomic radius, A	1.33	1.57	2.03	2.16	2.35
Melting point, °K	454	371	336	312	302
Boiling point, °K	1640	1163	1140	970	958
ΔH_{sub}, kcal	38.4	25.9	21.5	19.5	18.7
Ionic radius, M^+, A	0.68	0.98	1.33	1.48	1.67
ΔH_{hyd}, kcal	121	95	76	69	62
$\mathcal{E}^0(M^+, M)$, volts	−3.02	−2.71	−2.92	−2.99	−2.99
$D(M_2)$, kcal	25	17	12	11	10.4

tion an electron feels when it moves in the electric field of more than one nucleus. The first ionization energies show that the alkali-metal atoms have little attraction for their own valence electrons and thus have even less attraction for extra electrons. Therefore, when two alkali atoms form a covalent bond there is only a small decrease in energy. This argument can also be used to rationalize the small sublimation energies and softness of the metals themselves. Even when the valence electrons move in the fields of several nuclei, as they do in the metals, their energy lowering is relatively slight and the metallic bonding weak.

The atomic radii in Table 14.1 are obtained by dividing the observed internuclear separation in the metals by two, and thus the interpretation that these numbers represent the "sizes" of the alkali atoms must be used cautiously. We can see, however, that the radii of the alkali atoms vary as we might expect: The radii increase as the atomic numbers increase. A parallel trend occurs among the ionic crystal radii, and we shall see that the increase in ionic size with atomic number among elements of a given group is a general occurrence that has important consequences. The first of these involves the hydration energies of the ions. Strictly, the hydration energy of an ion is the enthalpy change of the process

$$M^{+n}(aq) = M^{+n}(g) + H_2O(l).$$

The enthalpy of hydration cannot be measured directly and thus is only an estimated quantity. In general, the smaller an ion is, and the larger its charge, the stronger is the electric field it exerts on its surrounding water molecules, and the larger is its hydration energy. The alkali-metal ions as a group are the largest positive ions and have the minimum ionic charge of $+1$. Thus the hydration energies of the alkali ions are usually smaller than those of other ions. Table 14.1 shows that as the atomic number of the alkali ions increases, their hydration energies decrease. This trend is consistent with the increase in size of the alkali ions which accompanies the increase in atomic number.

The alkali metals are reducing agents, and one measure of their strength as reductants is their standard reduction potentials. Table 14.1 shows that these have relatively large negative values, as we might expect from the fact that all the alkali metals reduce water spontaneously to hydrogen by the reaction

$$M + H_2O = M^+(aq) + OH^-(aq) + \tfrac{1}{2}H_2.$$

It is informative to try to analyze the action of the alkali metals as reductants in some detail. To do this we shall examine the ΔH of the half-reaction

$$M(s) = M^+(aq) + e^-$$

by recognizing that there is another path from reactants to products, as shown in Fig. 14.1. We see that the ΔH^0 of the oxidation half-reaction in which the electron is left in the gaseous state is equal to the sum of the enthalpies of sublimation and ionization, and the negative of the enthalpy of hydration. If it is the ΔH^0 of this half-reaction that does determine the performance of the metal as a reducing agent, then ΔH^0 should be smallest for the best reducing agent. There are two points, however, that should be noted carefully. The first is that it is the Gibbs free energy change ΔG^0, not ΔH^0, which is directly related to the performance of a metal as a reductant. Nevertheless, the value of ΔG^0 for the half-reaction is largely determined by the value of ΔH^0 in the cases with which we are dealing, and for the purposes of comparison of the metals it is legitimate to ascribe variations in reducing power to variations in ΔH^0. The second point is that we are computing the ΔH^0 of a *hypothetical half-reaction*, not of an overall reaction, and thus only the relative values of ΔH^0 for the different reactions, not the absolute values, have any significance.

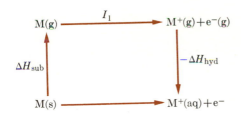

Alternative paths for the alkali-metal half-cell reaction.

FIG. 14.1

With these points in mind, we can examine the data in Table 14.2. The enthalpy change for the lithium half-reaction is the smallest, and according to the half-reaction potentials, lithium is the best reductant. The values of ΔH^0 are also consistent with the fact that sodium is the poorest reductant of the alkali metals. Since we have ignored entropy effects, however, the significance of the small variations in ΔH^0 for the potassium, rubidium, and cesium half-reactions is limited. In order to understand the small differences in the electrode potentials of these metals, we would have to consider the entropy changes as well as the enthalpy changes associated with their electrode reactions. Nevertheless, we can deduce from Table 14.2 that the reason that lithium is such a good reducing agent is that a relatively large amount of energy is evolved when the gaseous lithium ion is hydrated.

Table 14.2 Enthalpy changes for electrode half-reactions (kcal)

	Li	Na	K	Rb	Cs
ΔH_{sub}	38.4	25.9	21.5	19.5	18.7
I_1	124	118	100	96	90
$-\Delta H_{hyd}$	−121	−95	−76	−69	−62
$\Delta H[M(s) = M^+(aq) + e^-]$	41	49	46	46	47

There is an important general lesson to be learned from this analysis. Any property such as the standard reduction potential of an element may be determined by *several* more fundamental properties which, while fairly simple in themselves, work together in a complex way. Thus, while sublimation energy, ionization energy, and hydration energy may each vary in a simple manner in a given sequence of elements, their net effect can be an irregular set of standard reduction potentials. It is always tempting in studying descriptive chemistry to attribute an observed trend in chemical behavior to the smooth variation of a *single* fundamental property of atoms, but this should be done cautiously, for it is more than likely that any observed chemical property is related to *several* more fundamental factors.

The Alkali-Metal Oxides

Of the alkali metals, only lithium reacts directly with oxygen to give the simple monoxide Li_2O. The direct reaction between sodium and oxygen gives Na_2O_2, sodium peroxide. The other alkali metals react with oxygen to form *superoxides* of the general formula MO_2, which contain the superoxide ion O_2^-. The simple monoxides of all the alkali metals can be prepared, however, by reduction of their nitrates. As a typical reaction we have

$$KNO_3(s) + 5K(s) = 3K_2O(s) + \tfrac{1}{2}N_2(g).$$

The alkali-metal oxides are ionic compounds which have the antifluorite lattice discussed in Section 3.4. All the oxides are strong bases and dissolve readily in water by reactions of the type

$$K_2O(s) + H_2O(l) = 2K^+(aq) + 2OH^-(aq).$$

The crystalline alkali-metal hydroxides like NaOH also are ionic compounds which have the sodium-chloride crystal structure, are quite soluble in water, and of course are strong bases.

The Alkali Halides

These compounds are extremely stable crystalline substances of high melting and boiling temperature. As an indication of how stable these substances are, we can cite the value of the Gibbs free energy of formation of sodium chloride, which is -91.8 kcal/mole. Accordingly, the equilibrium constant for the reaction

$$Na(s) + \tfrac{1}{2}Cl_2(g) = NaCl(s)$$

at 25°C is given by

$$K = \frac{1}{(P_{Cl_2})^{1/2}} = e^{-\Delta G^0/RT} = 10^{(91,800/1360)}$$

$$= 1.6 \times 10^{67}.$$

An equilibrium constant of this magnitude means that the pressure of chlorine in equilibrium with sodium and sodium chloride at 25°C is approximately 10^{-134} atm. The other alkali halides are of comparable stability.

As we discussed in Section 11.2, the alkali halides are the outstanding examples of ionically bonded compounds, and it is interesting to examine their properties from this point of view. In Section 11.2 we noted that the ionic lattice energy, the energy required to separate a mole of solid ionic compound into its gaseous ions, is a quantitative measure of ionic bond strength. Figure 14.2 summarizes the lattice-energy data previously presented in Table 11.11. For a given alkali ion, the lattice energies of the crystals decrease as the atomic numbers of the halides increase. Also, for a particular halogen, the lattice energies decrease as the atomic numbers of the alkali atoms increase. Both these trends are consequences of the increase in ionic radius that occurs as atomic number increases.

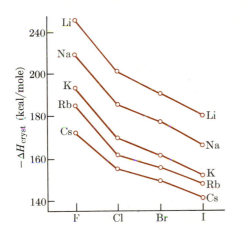

The crystal lattice energies of the alkali halides.

FIG. 14.2

Spectroscopic studies of the alkali halide vapors have provided information about the structures of the gaseous alkali halide molecules MX. Table 14.3 gives the dipole moments and bond distances of some of the gaseous MX molecules. It is interesting to note that the M—X bond distance in the gaseous alkali halides is less than the internuclear separation in the crystalline solids. This observation shows that the ionic crystal radii have limited significance— they can be used to predict the spacings in ionic crystals, but they do not truly represent the "sizes" of ions in other bonding situations.

The second point of interest is the comparison between the observed dipole moments of the gaseous alkali halide molecules and the dipole moment calculated by assuming that there is a net charge of $\pm e$ (4.8×10^{-10} esu) located at each nucleus. This assumption would be appropriate for a truly ionic bond between spherical ions. The dipole moment of such a molecule would be equal

Table 14.3 Dipole moments of some gaseous alkali halides

	Bond distance r_0 (A)	$e \times r_0$ (debyes)	μ_{obs} (debyes)
LiBr	2.17	10.4	6.19
LiI	2.39	11.5	6.25
NaCl	2.36	11.3	8.5
KF	2.17	10.4	8.6
KCl	2.67	12.8	10.0
KBr	2.82	13.5	10.6
KI	3.05	14.6	11.1

to $e \times r_0 \times 10^{18}$, where e is the value of the fundamental charge in electrostatic units, and r_0 is the *experimental* value of the internuclear separation in centimeters. The data in Table 14.3 show that the observed dipole moments are all noticeably smaller than the values calculated assuming a net fundamental charge located at each nucleus. Thus the gaseous alkali halides do not consist of perfectly spherical ions. Instead, the negative ion tends to be distorted or polarized by the positive ion as is indicated in Fig. 14.3. Some of the negative charge is drawn from the halide ion toward the alkali ion, and this distortion results in a decrease of the dipole moment of the molecule. This polarization is most extreme when a small positive ion acts on a large negative ion, and we can see from Table 14.3 that the discrepancy between the observed dipole moment and that calculated for spherical ions is most serious in such compounds as LiBr and LiI. The polarization effect represents the start of conversion from ionic to covalent bonding. The fact that it is present even in the alkali halides means that we should expect to find it important in other "ionic" compounds.

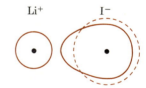

FIG. 14.3 The distortion or polarization of the iodide ion by the lithium ion in the lithium-iodide diatomic molecule.

With the exception of lithium fluoride, the alkali halides are all quite soluble in water. Table 14.4 lists the solubilities of some alkali halides, together with ΔH_{sol}, the enthalpy change that accompanies the dissolution reaction

$$MX(s) + H_2O = M^+(aq) + X^-(aq).$$

Also listed are $-\Delta H^0_{cryst}$ and $-\Delta H^0_{hyd}$, respectively, the enthalpy changes that accompany the vaporization of the halide to gaseous ions, and the hydration of these ions. The sum of $-\Delta H^0_{cryst}$ and $-\Delta H^0_{hyd}$ must equal ΔH^0_{sol}. We see from

Table 14.4 Some enthalpies of hydration and crystallization (kcal)

	$-\Delta H^0_{\text{cryst}}$	$-\Delta H^0_{\text{hyd}}$	ΔH^0_{sol}	solubility (moles/liter)
LiF	246.7	−240.1	6.6	0.11
NaCl	186.6	−185.3	0.9	6.1
NaI	165.7	−167.5	−1.8	11.0
KCl	169.3	−165.2	4.1	4.6
RbCl	162.3	−159.3	3.0	7.5

the data in the table that the enthalpy change that accompanies the solution process is really a small difference between two large quantities. This is true generally for other salts. Consequently, although the lattice enthalpies and hydration enthalpies may vary smoothly for a particular sequence of compounds, the enthalpy of solution may fluctuate in an unpredictable fashion. Since this quantity, along with the entropy change that accompanies dissolution, determines the salt solubility, we can expect that the relative solubilities of a series of salts might be difficult to predict without careful consideration of thermodynamic data. Since this is true, it is often most efficient just to remember general trends in solubilities, rather than try to relate them to more fundamental quantities like lattice enthalpies and hydration enthalpies.

The circulatory and intracellular fluids of living matter are aqueous solutions which contain, among many other things, significant amounts of Na^+ and K^+. Living cells are separated from one another and from the circulatory system by their wall membranes. These membranes are combinations of protein and lipid macromolecules, neither of which is a good solvent for ionic substances. However, Na^+ and K^+, as well as other charged species, pass through cell walls quite rapidly. One of the most challenging problems in physiological chemistry is to determine how Na^+ and K^+ are transported through the cell walls which, on the basis of their structure, would be expected to be rather impermeable to ions.

Recently, it has been discovered that the salts of alkali metals can be made quite soluble in organic solvents if they are treated with organic chemicals called cyclic polyethers. An example of how cyclic polyethers can interact with alkali metal salts is shown in Fig. 14.4. At each apex in the drawing there is a carbon atom to which is attached either one or two hydrogen atoms so that there is a total of four bonds to each carbon atom. The geometric structure of the organic molecule is such that the six oxygen atoms (the ether functional groups) form a cavity into which the potassium ion can fit. The result is a solvation or complexing of the cation by the polyether molecule. The negative ion remains in the vicinity of the complexed cation. The aggregate is soluble in organic solvents, since most of the polyether molecule is made up of its hydrocarbon skeleton, which is compatible with other organic molecules, and which shields the ion pair from interaction with the solvent. By varying the molecular structure of the polyether, different sized oxygen cavities can be constructed. The most stable ion-ether complexes are formed when there is a match between the size of the cavity and the cation. As a result, it is possible to use cyclic polyethers to dissolve certain alkali metal ions preferentially. Complexing of ions by cyclic polyethers may be responsible for the ease with which these species pass through cell wall membranes.

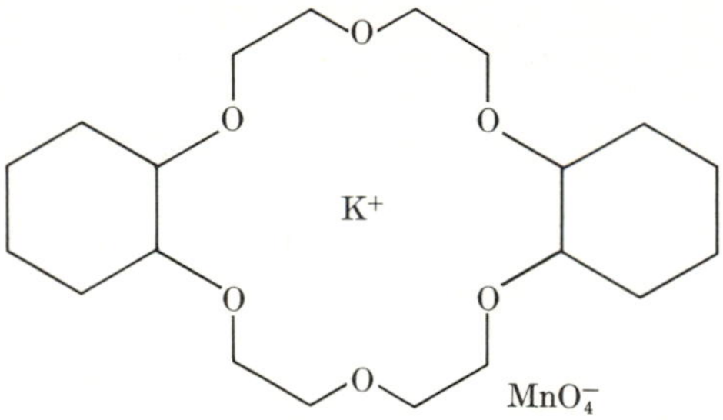

FIG. 14.4   The interaction of a cyclic polyether with potassium permanganate.

14.2 THE ALKALINE-EARTH METALS

These elements of group IIA (Be, Mg, Ca, Sr, Ba, Ra) are never found in the metallic form in nature, because, like the alkali metals, they are active reductants and react readily with a variety of nonmetals. Magnesium is the second most abundant metallic element in the sea and also occurs in a variety of silicate minerals. Calcium is found abundantly as $CaCO_3$ in marble, limestone, and chalk. The most common source of beryllium is the mineral beryl, $Be_2Al_2(SiO_3)_6$, while barium and strontium are found most frequently as the sulfates $BaSO_4$ and $SrSO_4$. All isotopes of radium are radioactive; the element itself is formed by a radioactive decay chain that starts with U^{238}, and consequently all uranium minerals contain very small amounts of radium.

All the alkaline-earth metals can be prepared by electrolysis of their fused halides. Most of the magnesium prepared commercially is obtained this way, although some use is made of the direct reduction of magnesium oxide by carbon:

$$MgO + C = Mg + CO.$$

In general, the most convenient way to prepare small amounts of the other alkaline-earth metals is reduction of their oxides by more readily available reducing metals, as in the reaction

$$3BaO + 2Al = 3Ba + Al_2O_3.$$

The group IIA metals are all considerably harder than the alkali metals, but the trend of increasing softness with increasing atomic number occurs in group IIA as it did in group IA. Despite a tendency toward brittleness, the alkaline-earth metals can be hammered and rolled without fracture. As a structural material, however, only magnesium has important applications. Alloyed with aluminum, zinc, and manganese, it forms lightweight and moderately strong materials that are used principally in aircraft construction. Calcium and barium react readily with oxygen and nitrogen at elevated temperatures, and consequently are used as "getters" to remove the last traces of air from vacuum tubes.

Table 14.5 Properties of the group IIA elements

	Be	Mg	Ca	Sr	Ba
Atomic number	4	12	20	38	56
Configuration	$2s^2$	$3s^2$	$4s^2$	$5s^2$	$6s^2$
Ionization energy, kcal $\{I_1$	214	175	140	132	120
$\quad\quad\quad\quad\quad\quad I_2$	429	345	274	253	230
Atomic radius, A	0.89	1.36	1.74	1.91	1.98
Melting point, °K	1556	929	1123	1043	983
Boiling point, °K	2750	1400	1750	1640	1950
ΔH_{sub}, kcal	77.9	35.6	42.2	39.1	42.5
Ionic radius, M^{++}, A	0.30	0.65	0.94	1.10	1.29
ΔH_{hyd}, kcal	570	460	395	355	305
$\mathcal{E}^0(M^{++}, M)$, volts	−1.70	−2.34	−2.87	−2.89	−2.90

The comparative hardness of the alkaline-earth metals suggests that the metallic bonding in group IIA elements is stronger than in the group IA elements; this is confirmed by some of the data in Table 14.5. The melting and boiling temperatures and the enthalpies of vaporization of the group IIA metals are much higher than those of the alkali metals. Although the values of the enthalpies of vaporization fluctuate somewhat, the same gross trend is found in group IIA as in group IA: beryllium, the lightest member of the alkaline-earth family, has a larger enthalpy of vaporization than does barium.

Table 14.5 also shows that both the atomic radii and the ionic radii of the group IIA elements increase as atomic number increases. The atomic radii and particularly the ionic radii are smaller than the corresponding quantities for the immediate neighbor elements in group IA. The contraction in ionic size in going from group IA to group IIA has an obvious explanation. The ions Li^+ and Be^{++}, Na^+ and Mg^{++}, K^+ and Ca^{++}, etc., are isoelectronic; each pair has the electronic structure of the preceding inert gas. The alkaline-earth ion, however, has a higher nuclear charge than the corresponding alkali-metal ion, and this factor causes the decreased size of the alkaline-earth ion. The nuclear-charge effect must also be responsible for the comparative atomic sizes of the group IA and group IIA elements.

The enthalpies of hydration for the group IIA ions have two notable features. First, the hydration enthalpy becomes smaller as the size of the ion increases. More striking, however, are the magnitudes of the hydration enthalpies. Far more heat is evolved upon hydration of the ions of group IIA than with the corresponding alkali-metal ions. This difference can be attributed to the increased charge on the alkaline-earth ions: Even though Na^+ and Ca^{++} have nearly the same radii, the enthalpy of hydration of Ca^{++} is nearly four times that of Na^+. This suggests that in general the hydration enthalpy is proportional to the square of the charge on the ion, and this relation is obeyed in fact to a fair approximation.

It is the increase in hydration enthalpy with ionic charge that is responsible for the existence of the group IIA ions exclusively in the $+2$ oxidation state in aqueous solutions. The first ionization energies of these elements are not particularly large, but the second ionization energies are substantial. This observation alone might lead us to expect the $+1$ oxidation state to be important in the chemistry of group IIA. The following energetic relationships show that *gaseous* Ca^+ is surely stable with respect to gaseous Ca and Ca^{++}.

$$
\begin{array}{ll}
Ca(g) = Ca^+(g) + e^- & \Delta H = 140 \text{ kcal} \\
\underline{e^- + Ca^{++}(g) = Ca^+(g)} & \underline{\Delta H = -274 \text{ kcal}} \\
Ca(g) + Ca^{++}(g) = 2Ca^+(g) & \Delta H = -134 \text{ kcal}
\end{array}
$$

The situation is quite different, however, when we consider the energetic relations between the *aqueous* ions and the *solid* metal. If we estimate the enthalpy of hydration of the hypothetical aqueous Ca^+ to be equal to that of K^+, we can write

$$
\begin{array}{ll}
2Ca^+(g) = 2Ca^+(aq) & \Delta H \cong -152 \text{ kcal} \\
Ca(s) = Ca(g) & \Delta H = 42 \text{ kcal} \\
Ca^{++}(aq) = Ca^{++}(g) & \Delta H = 395 \text{ kcal} \\
\underline{Ca(g) + Ca^{++}(g) = 2Ca^+(g)} & \underline{\Delta H = -134 \text{ kcal}} \\
Ca(s) + Ca^{++}(aq) = 2Ca^+(aq) & \Delta H = +151 \text{ kcal}
\end{array}
$$

Now it is clear that aqueous Ca^+ is energetically unstable with respect to $Ca^{++}(aq)$ and calcium metal, and the analogous conclusion can be reached for any of the other alkaline-earth systems. Examination of the enthalpy changes involved shows clearly that it is the enthalpy of hydration of the dipositive ion that is responsible for its stability in aqueous solution. The rationalization of the occurrence of the $+2$ state in the ionic crystals of the alkaline-earth compounds is similar to the above argument, and has been given in Section 11.2. In effect, the gain in the crystal-lattice energy obtained by forming a dipositive ion more than compensates for the extra energy required to remove the second electron from an alkaline-earth atom.

The standard reduction potentials of the alkaline-earth elements given in Table 14.5 show that beryllium and magnesium are poorer reductants than the

Table 14.6 Enthalpy changes for electrode half-reactions

	Be	Mg	Ca	Sr	Ba
ΔH_{sub}	77.9	35.6	42.2	39.1	42.5
$I_1 + I_2$	643	520	414	385	350
$-\Delta H_{hyd}$	−570	−460	−395	−355	−305
$\Delta H[M(s) = M^{++}(aq) + e^-]$	151	96	61	69	87

heavier elements. The data which in part explain these differences are given in Table 14.6, where the enthalpies associated with the reaction sequence

$$M(s) = M(g) = M^{++}(g) + 2e^- = M^{++}(aq) + 2e^-$$

are given for all the alkaline-earth elements. For the electrode reaction written as an oxidation, the enthalpy change is most positive for beryllium and magnesium. This is consistent with the observation that these metals are the poorest reducing agents in group IIA. The enthalpy data show that the principal reason for the inferiority of beryllium and magnesium as reductants is their relatively large ionization energies, although the large sublimation energy of beryllium also makes an important contribution to this poor performance as a reductant. The observation that the electrode potentials of calcium, strontium, and barium are nearly the same, but the enthalpy changes associated with the reactions fluctuate, shows that the analysis of electrode potentials on the basis of enthalpy changes alone has limitations imposed by neglect of entropy changes and possible inaccuracies in the values of enthalpies of hydration.

The Oxides and Hydroxides

The oxides of magnesium and the heavier group IIA elements can be prepared by direct combination of the elements, or by thermal decomposition of the carbonates:

$$M + \tfrac{1}{2}O_2 = MO,$$
$$MCO_3 = MO + CO_2,$$

where M = Mg, Ca, Sr, Ba. As can be seen from Table 14.7, the oxides are extremely stable; their negative enthalpies and free energies of formation are consequences of the very large ionic crystal-lattice energy obtained by packing doubly charged ions in a rock-salt or sodium-chloride type of lattice.

Table 14.7 Thermodynamic properties of alkaline-earth oxides (298°K)

	BeO	MgO	CaO	SrO	BaO
ΔH_f^0, kcal/mole	−143.1	−143.8	−151.9	−141.1	−133.4
ΔG_f^0, kcal/mole	−136.1	−136.1	−144.4	−133.8	−126.3

Table 14.8 Solubility products of the group IIA hydroxides

	K_{sp}
Be(OH)$_2$	1.6×10^{-26}
Mg(OH)$_2$	8.9×10^{-12}
Ca(OH)$_2$	1.3×10^{-6}
Sr(OH)$_2$	3.2×10^{-4}
Ba(OH)$_2$	5×10^{-3}

The heavier oxides react with water to form hydroxides of the general formula M(OH)$_2$. These hydroxides are strong bases, for they react with acids and also dissolve in water as M^{++} and OH$^-$ ions. Their solubility in water is somewhat limited, but increases with the increasing atomic number of the cation, as is shown by the solubility products in Table 14.8.

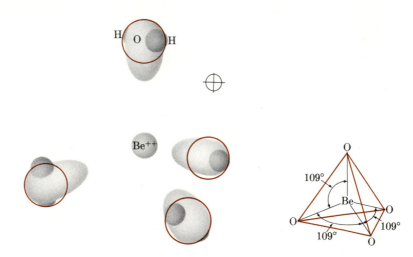

FIG. 14.5

The polarization of water molecules by a beryllium ion.

The properties of beryllium oxide distinguish it from the other alkaline-earth oxides. Beryllium oxide is harder and higher melting than the oxides of the heavier metals. More distinctive, however, is its acid-base behavior. Beryllium oxide is amphoteric; it reacts slowly with only very concentrated strong acids to give solutions of hydrated ions [Be(H$_2$O)$_4$]$^{++}$, and it also reacts with strong bases and enters solution as an anion whose formula is probably Be(OH)$_4^=$.

The amphoteric behavior of beryllium oxide can be thought of as a consequence of the small size and relatively large charge of the beryllium ion. The electric field around such a small ion is particularly intense, and consequently

Table 14.9 Properties of some group IIA chlorides

	$T_{mp}(°C)$	$T_{bp}(°C)$	Equivalent conductivity* (ohm^{-1})
$BeCl_2$	405	490	0.086
$MgCl_2$	715	1400	29
$CaCl_2$	780	1600	52

*At the melting point.

the beryllium ion polarizes surrounding molecules by drawing electrons from them. Thus the situation in $[Be(H_2O)_4]^{++}$ might be as represented in Fig. 14.5. The effect of the beryllium ion is to withdraw electronic charge from the surrounding water molecules and thereby facilitate removal of their protons and the formation of $Be(OH)_4^{=}$. Thus $[Be(H_2O)_4]^{++}$ is an acid while $Be(OH)_4^{=}$ is a base. We shall find other instances in which a small, highly charged ion exists as a hydrated positive ion in acidic solutions, and as a hydroxy-anion in basic solutions.

The Halides

All the group IIA metals combine directly with the halogens to form the metal halides of the general formula MX_2. The chlorides in particular are interesting, for their properties display trends which can be interpreted as indicating a change from covalent to ionic bonding as the atomic number of the metal increases. Table 14.9 shows that beryllium chloride has lower melting and boiling points, and much lower electrical conductivity in the fused state than do the other alkaline-earth chlorides. Moreover, beryllium chloride is soluble in some organic solvents, which suggests that the charge distribution in $BeCl_2$ tends to be rather uniform, like the distribution in the covalently bonded organic molecules. The appearance of covalent character in the beryllium halides should not be too surprising in view of our earlier remarks about how a small, highly charged ion polarizes neighboring negative ions. The properties of beryllium chloride indicate qualitatively that this polarization effect is so extreme that it is more accurate to think of the beryllium halides as covalently bonded molecules.

In the gas phase, $BeCl_2$ is a linear symmetric molecule, which is consistent with a description of the bonding in terms of sp-hybrid orbitals around the beryllium atom. In the solid phase, beryllium chloride displays the bridged structure shown in Fig. 14.6. Each beryllium atom is surrounded by four chlorine atoms, with a Cl-Be-Cl angle of 98°. This type of bridge structure, where a halogen atom is associated with two metal atoms, also occurs in the aluminum and gallium halides, and in other systems where a small, highly charged cation of low coordination number can associate with ligands which have an unshared pair of electrons.

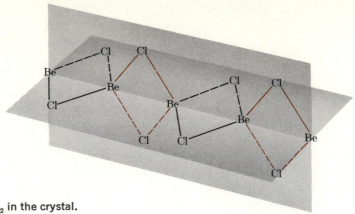

FIG. 14.6 The structure of BeCl$_2$ in the crystal.

The chlorides, bromides, and iodides of the heavier alkaline-earth metals are ionic solids which have appreciable solubility in water. The fluorides, however, are relatively insoluble, and as Table 14.10 shows, the trends in the solubilities of the fluorides are similar to the trends in solubilities of the hydroxides.

Table 14.10 Solubility products of some alkaline-earth salts

	F$^-$	SO$_4^=$	CO$_3^=$	CrO$_4^=$
Mg^{++}	8×10^{-8}	—	10^{-5}	—
Ca^{++}	1.7×10^{-10}	2.4×10^{-5}	4.7×10^{-9}	7.1×10^{-4}
Sr^{++}	8×10^{-10}	8×10^{-7}	7×10^{-10}	3.6×10^{-5}
Ba^{++}	2.4×10^{-5}	1×10^{-10}	1.6×10^{-9}	8.5×10^{-11}

Other Salts

The carbonates, sulfates, and chromates of the alkaline-earth metals show similar solubility trends, as the data in Table 14.10 demonstrate. The solubilities of all these salts decrease as the atomic number of the metal ion increases, and this behavior is opposite to that observed for the fluorides and hydroxides. Although we have remarked about the uncertainty of relating solubility trends to fundamental properties such as lattice and hydration enthalpies, we can make a meaningful analysis of the solubilities of the alkaline-earth carbonates, sulfates, and chromates in these terms. As we proceed in sequence from BeSO$_4$ to BaSO$_4$, the enthalpy of hydration of the positive ion becomes smaller because of the increase in ionic size. This tends to make the salts of the heavier metal ions less soluble than those of the lighter ions. The lattice energies of the sulfates, and of the carbonates and chromates as well, do not change greatly in sequence from beryllium to barium, for the lattice energies are determined principally by the reciprocal of the sum of the ionic radii, $1/(r_+ + r_-)$. This quantity changes rather slowly in the sequence from beryllium to barium,

Table 14.11 Thermodynamic data for $MCO_3(s) = MO(s) + CO_2(g)$

	ΔH^0 (kcal)	ΔG^0 (kcal)	$T(P_{CO_2} = 1\ atm)$
$MgCO_3 = MgO + CO_2$	28	16	540°C
$CaCO_3 = CaO + CO_2$	42	31	900°C
$SrCO_3 = SrO + CO_2$	57	45	1290°C
$BaCO_3 = BaO + CO_2$	64	52	1360°C

because r_-, the radius of the negative ion, is much greater than that of any of the positive ions, and thus $r_+ + r_-$ is insensitive to variations in r_+. Therefore the trend in the solubilities of the group IIA metal salts of *large* anions like $SO_4^=$, $CO_3^=$, and $CrO_4^=$ is dictated predominantly by the trend in the enthalpies of hydration.

The thermal stabilities of the group IIA metal carbonates provide another demonstration of how ionic size and lattice energies influence chemical behavior. At elevated temperatures all the carbonates decompose to the oxides according to the reaction

$$MCO_3(s) = MO(s) + CO_2(g).$$

The temperature at which the equilibrium pressure of CO_2 is equal to 1 atm increases as the atomic number of the metal ion increases, as the data in Table 14.11 show. Thus there is a rather smooth increase in the stabilities of the carbonates as the atomic number of the metal ion increases. The values of ΔG^0 and ΔH^0 for the decomposition reactions, also given in Table 14.11, are all positive, but tend to become larger as one proceeds toward the heavier metals. The trend in ΔG^0 is due virtually entirely to the fact that the values of ΔH^0 become larger as atomic number increases. Thus, in attempting to explain the trend in carbonate stabilities we can ignore entropy effects and concentrate on enthalpies.

The trend of increasing ΔH^0 with increasing atomic number of the cation must be a result of the variation in the lattice energies of MCO_3 and MO as the cation size increases. We have remarked that the lattice energy of a salt of a *large* anion is not particularly sensitive to the size of the cation, for the ionic separation $r_+ + r_-$ is determined principally by r_-, the radius of the negative ion. Therefore the carbonate lattice becomes only slightly less stable as the atomic number and size of the cation increase. On the other hand, the lattice energy of an oxide *is* sensitive to the size of the cation, because the radii of the oxide ion and the cations are quite comparable. Thus the oxides of the smaller cations should be much more stable than those of the larger cations, and therefore the decompositions of the carbonates of the smaller cations should be less endothermic than those of the larger cation, as is observed. It is the decreasing size of the cation that stabilizes the lighter alkaline-earth oxides and favors the decomposition of their carbonates.

There is an important industrial process which involves the thermal decomposition of calcium carbonate. The carbonates and bicarbonates of the alkali metals, especially Na_2CO_3 and $NaHCO_3$, are important industrial chemicals which are used in the manufacture of glass, paper, water softeners, detergents, and soap. While there is an ample supply of sodium chloride deposited in readily available marine evaporites, there are no large deposits of Na_2CO_3. Most of the Na_2CO_3 and $NaHCO_3$ used by the chemical industry is manufactured from $NaCl$ and $CaCO_3$ by the Solvay process.

In the Solvay process, a concentrated aqueous solution of sodium chloride is first saturated with ammonia, and then carbon dioxide is bubbled through it. The results can be represented by the reactions

$$NH_3(aq) + CO_2(aq) + H_2O = NH_4^+(aq) + HCO_3^-(aq),$$

$$NH_4^+(aq) + HCO_3^-(aq) + Na^+(aq) + Cl^-(aq)$$
$$= NaHCO_3(s) + NH_4^+(aq) + Cl^-(aq).$$

The solution is chilled to about 15°C in order to decrease the solubility of sodium bicarbonate, which separates as a fairly pure solid. If sodium carbonate is desired, it can be obtained by heating the bicarbonate salt.

The raw materials used in these reactions are sodium chloride, ammonia, and carbon dioxide. A convenient source of CO_2 is the pyrolysis of limestone, $CaCO_3$, of which ample supplies are available:

$$CaCO_3(s) = CaO(s) + CO_2(g).$$

The lime (CaO) which is obtained from this reaction can be "slaked" to give calcium hydroxide:

$$CaO(s) + H_2O(l) = Ca(OH)_2(s).$$

In turn, this calcium hydroxide is used to recover gaseous ammonia from the ammonium chloride solution produced in the first steps of the process:

$$2NH_4^+(aq) + 2Cl^-(aq) + Ca(OH)_2(s)$$
$$= 2NH_3(g) + Ca^{++}(aq) + 2Cl^-(aq) + 2H_2O.$$

Thus the overall result of the Solvay process is the conversion of $NaCl$ and $CaCO_3$ to $NaHCO_3$ and $CaCl_2$. Calcium chloride is used as a drying agent, a de-icing agent, and a soil conditioner. The Solvay process illustrates a number of desirable features of an industrial chemical conversion: readily available raw materials ($NaCl$ and $CaCO_3$) are used, the reactions are rapid, the more expensive intermediate material (NH_3) is recycled, and uses for the by-product ($CaCl_2$) exist.

The striking feature of the chemistry of the alkali and alkaline-earth metals is the close resemblance among members of the same family. There are some similar points of resemblance among the elements of group IIIA (B, Al, Ga, In, Tl), but in addition, these elements display a range of properties and some notable contrasts. In passing from boron to thallium we encounter a change from semimetallic to metallic properties, from acidic to amphoteric to basic oxides, and from halides in which the bonding is distinctly covalent to those in which it is more nearly ionic. Such contrasts in the chemical properties of the elements of a single family occur again particularly in groups IV and V, and to some degree in group VI, as we shall see.

The principal natural source of boron is the deposits of borax, $Na_2B_4O_7 \cdot 10H_2O$, and recovery of the pure element from this compound is difficult. One method that is used is the conversion of borax to the oxide B_2O_3, which is then reduced with magnesium. This process does not give a particularly pure product, since the reduction of the oxide is never quite complete. The reduction of boron trichloride with hydrogen gives a product of better quality, but this process is less well suited for production of the element in quantity.

Aluminum is the most abundant metal in the earth's crust, and is recovered pure and in quantity from its oxide by electrolytic reduction. In contrast, gallium, indium, and thallium are quite rare, and are obtained only as by-products in the recovery of other more important metals like aluminum, zinc, cadmium, and lead.

Table 14.12 gives some of the physical properties of the group III elements. The trends in hardness, boiling temperature, and ΔH_{sub} parallel those found in groups I and II. It is clear from the boiling points and the values of ΔH_{sub} that all the group III elements are bound more strongly in the condensed state than

Table 14.12 Properties of the group IIIA elements

		B	Al	Ga	In	Tl
Atomic number		5	13	31	49	81
Configuration		$2s^22p^1$	$3s^23p^1$	$4s^24p^1$	$5s^25p^1$	$6s^26p^1$
Ionization energy, kcal,	I_1	191	138	138	133	141
	I_2	580	434	473	435	471
	I_3	874	656	708	646	687
Atomic radius, A		0.80	1.25	1.24	1.50	1.55
Melting point, °K		2300	932	312	429	577
Boiling point, °K		4000	2700	2500	2300	1740
ΔH_{sub}, kcal		135	77.5	65.3	58	43
Ionic radius, M^{+3}, A		—	0.45	0.60	0.81	0.95
ΔH_{hyd}, kcal		—	1121	1124	994	984
$\mathcal{E}^0(M^{+3}, M)$, volts		—	−1.67	−0.52	−0.34	0.72

the metals of groups I and II. With the exception of boron, the group III elements do not have exceptionally high melting points; on the contrary, gallium melts at 29°C. Because of the enormous temperature interval between the melting and boiling points of gallium, it is sometimes used as a thermometer liquid.

Boron

Boron is the first of the elements we have encountered that is not a metal. Its electrical conductivity is small and *increases* as temperature rises, which is opposite to the behavior observed for metals. Although boron is formally in the +3 oxidation state in its oxide and halides, there is no chemistry associated with a free B^{+3} species. On the contrary, boron is bonded covalently to nonmetals and is an electron *acceptor* in borides like MgB_2 and AlB_2. Inspection of the ionization energies in Table 14.12 shows why the free B^{+3} species is not chemically important. The energy required to remove three electrons from the boron atom is very large, and the ion so formed would be extremely small and would exert large polarizing forces on neighboring atoms. This would result in transfer of electron density to the boron from its neighbors, and thus the energetically favored situation would be electron sharing, or covalent bonding to boron.

Table 14.13 gives some of the properties of the boron halides. As liquids they do not conduct electricity, and their boiling points are all very low compared with those of the halides of the group I and II elements. In the gas, liquid, and solid phases all the boron halides exist as discrete molecular species BX_3. All these facts are in contrast with the behavior expected from ionically bonded substances and constitute the principal justification for picturing the boron-halogen bond as covalent. Further confirmation of this picture lies in the observation that the boiling points of the boron halides increase as the atomic number of the halogen increases. This is the behavior expected of a series of compounds in which the forces of attraction between molecules are of the van der Waals type, for these forces *increase* as the number of electrons in a molecule increases. In contrast, the strength of ionic attractions *decreases* as the ions become larger.

All the boron halides act as electron acceptors, as for example, in the reactions

$$BF_3 + NH_3 = F_3BNH_3,$$
$$BF_3 + F^- = BF_4^-.$$

Table 14.13 Properties of the boron trihalides

	Melting point (°C)	Boiling point (°C)	ΔH_f^0(kcal)
BF_3	−127	−101	−270
BCl_3	−107	12	− 95.7
BBr_3	− 46	91	− 44.6
BI_3	43	210	—

In these reactions, BF_3 accepts a pair of electrons donated by NH_3 or F^-. Thus BF_3 and the other boron halides are Lewis acids, and BF_3 in particular is often used as an acid catalyst in the reactions of organic compounds.

The oxides of semimetals are, in general, acidic, and the oxide of boron is no exception. Boron trioxide, B_2O_3, when hydrated, forms boric acid, $B(OH)_3$. Despite its formula, boric acid is a *monobasic* acid and is quite weak. The acid reaction of $B(OH)_3$ is not simply a loss of a proton, but rather

$$B(OH)_3 + H_2O = B(OH)_4^- + H^+.$$

In this reaction boron again displays its tendency to accept electrons, for boric acid is evidently a Lewis acid and accepts electrons from OH^-.

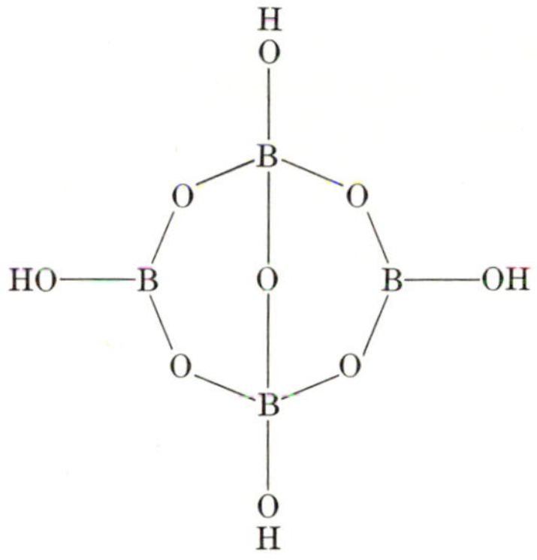

The structure of the $B_4O_5(OH)_4^=$ ion. **FIG. 14.7**

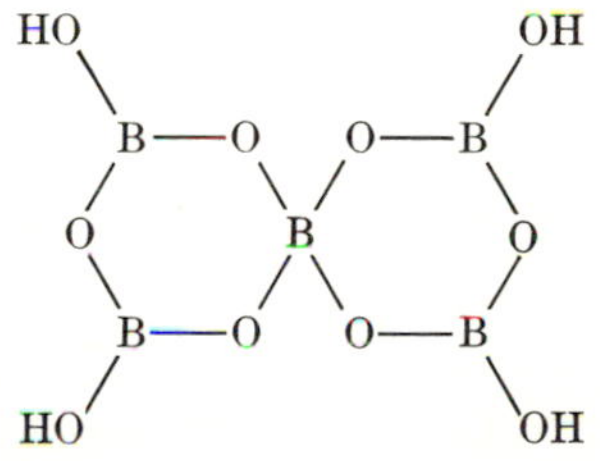

The structure of the $B_5O_6(OH)_4^-$ anion. **FIG. 14.8**

The salts of boric acid have complex structures and, at least in aqueous solution, never contain the simple anion BO_3^{-3}. When a solution of boric acid which is more concentrated than approximately $0.02\ M$ is made basic, the reaction

$$2B(OH)_3 + B(OH)_4^- = B_3O_3(OH)_4^- + 3H_2O$$

occurs to form the triborate ion, $B_3O_3(OH)_4^-$. This ion, as well as the tetraborate ion $B_4O_5(OH)_4^=$, which occurs in borax, have structures which involve boron-oxygen rings, as Fig. 14.7 shows. In both these anions there are boron atoms which display trigonal coordination, and in $B_4O_5(OH)_4^=$, two of the boron atoms are tetrahedrally coordinated by oxygen. Even more complicated structures occur. In the salt $KB_5O_8 \cdot 4H_2O$, the anion has the double-ring structure shown in Fig. 14.8, while in CaB_2O_4, the anion is an infinite chain of BO_2^- units which is pictured in Fig. 14.9. This great variety of stable boron anion structures makes it possible for boric oxide B_2O_3 to incorporate other metal oxides with consequent formation of boric oxide-metal borate glasses. This is the basis for the action of B_2O_3 and the use of borates as cleansing fluxes in soldering and brazing.

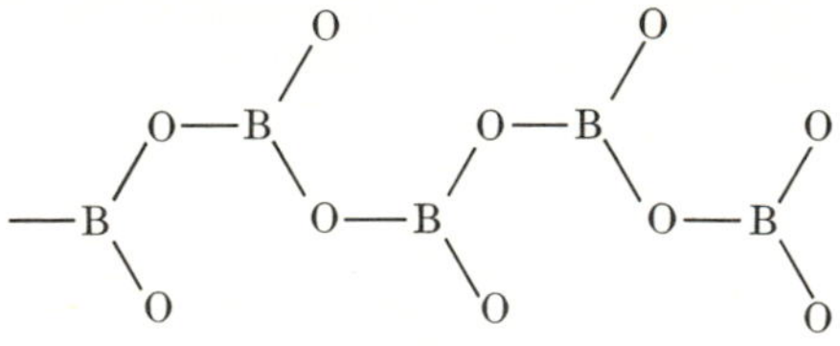

FIG. 14.9   The structure of the BO_2^- chain anion.

Boron forms a series of volatile compounds with hydrogen called the *boranes*. Diborane, B_2H_6, is the simplest of these hydrides, and can be prepared by the reaction of lithium hydride with boron trifluoride,

$$6LiH + 8BF_3 = 6LiBF_4 + B_2H_6.$$

When diborane is heated to temperatures between 100 and 250°C, it is converted to a number of other boranes:

$$B_2H_6 \xrightarrow{\text{heat}} B_4H_{10}, B_5H_9, B_5H_{11}, B_6H_{10}, B_9H_{15}, B_{10}H_{14}, B_{10}H_{16}.$$

The formulas of the boranes are of two types, B_nH_{n+4} and B_nH_{n+6}. Compounds of the first type seem to be the most stable.

Most of the boranes are spontaneously inflammable in air and are rapidly hydrolyzed by water to boric acid. The exceptions to this rule are B_9H_{15} and $B_{10}H_{14}$, which are stable in air and hydrolyze only very slowly. There has been some interest in the boranes as fuels, for the energy evolved by their reactions with oxygen is considerable:

$$B_2H_6 + 3O_2 = B_2O_3 + 3H_2O, \qquad \Delta H = -482 \text{ kcal.}$$

The molecular structure and bonding in the borane series is quite unique. Figure 14.10 shows the structure of diborane. The boron atoms and four of the six hydrogen atoms lie in the same plane with the remaining two hydrogens occupying "bridge" positions between the boron atoms. The B—H—B system is a three-center electron-pair bond. To see that two electrons do indeed bind three atoms together, we construct in Fig. 14.11 the electron-dot structure of diborane, starting with separated BH_2 fragments and the bridge hydrogen atoms. The two electrons in each bridge bond visit both boron atoms and the

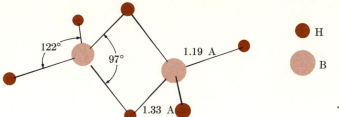

The structure of diborane. **FIG. 14.10**

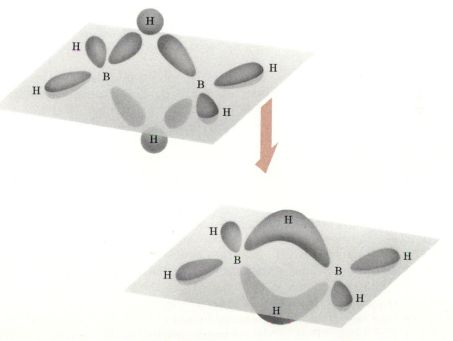

Schematic construction of diborane from its fragments.

FIG. 14.11

hydrogen nucleus. This can be seen more clearly in Fig. 14.12, where the molecular orbital which corresponds to the bridge bond, and its construction from the hydrid orbitals of boron and the 1s-orbital of hydrogen are shown. The structures of the other boranes also involve the hydrogen bridge bond. In addition, we shall find that there are compounds of other elements in which bridge bonds formed by halogen atoms are important.

Formation of three-center bonds in diborane. **FIG. 14.12**

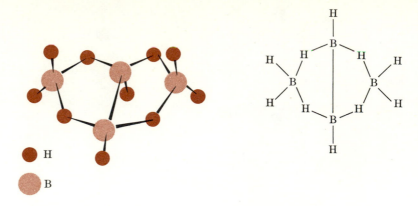

FIG. 14.13   The structure and line representation of the bonding in tetraborane, B_4H_{10}.

The structure of tetraborane, B_4H_{10}, can be understood in terms of conventional two-center electron pair bonds and the three-center, two-electron or B—H—B bridge bond. In Fig. 14.13 we show both the structure of B_4H_{10} and a line drawing representation of its bonding. We see that there are six conventional B—H electron pair bonds, and one B—B electron pair bond. In addition, there are four B—H—B bridge bonds, and together these account for the twenty-two valence electrons contributed by four boron and ten hydrogen atoms.

In order to understand the bonding in B_5H_9 and other of the boron hydrides, we must consider another type of three-center bond. This involves two electrons bonding three boron atoms arranged in an equilateral triangle. Formation of this three-center, two-electron bond is represented in Fig. 14.14. Each boron atom contributed one hybrid orbital to form a molecular orbital which has no nodes between the boron nuclei. Thus this bonding orbital is very much like the lowest molecular orbital in H_3^+. In line representations of the bonding in boron hydrides, this three-center, two-electron boron bond is indicated by the construction shown in Fig. 14.14.

The structure of pentaborane-9, B_5H_9, is given in Fig. 14.15. The boron atoms form a square-based pyramid. There are five conventional two-electron B—H bonds, and four B—H—B bridge bonds which together account for eighteen of the twenty-four valence electrons in the molecule. The remaining six electrons bond the boron atom at the apex to the four boron atoms at the base of the pyramid. Since all four sides of the pyramid are equivalent, the molecule must be pictured as a resonance hybrid of bond structures of the type shown in Fig. 14.15. That is, the electronic distribution is a superposition of

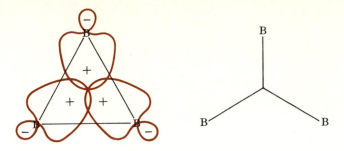

Formation of the three-center, two-electron bond, and its line representation.

FIG. 14.14

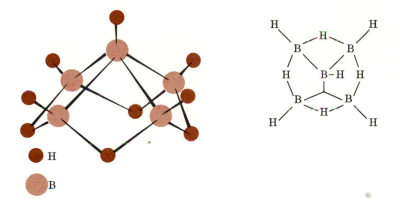

The geometrical structure of B_5H_9, and one of its four-bond resonance structures.

FIG. 14.15

four equivalent structures in which there are two conventional B—B electron pair bonds and one three-center bond between the apex and base boron atoms.

Other species in which hydrogen and boron are bonded include the borohydride ion, BH_4^-, and more complicated ions such as $B_3H_8^-$ and $B_{10}H_{13}^-$. Lithium borohydride can be made by the reaction of lithium hydride with diborane:

$$2LiH + B_2H_6 = 2LiBH_4.$$

In general, the alkali-metal borohydrides are ionic compounds consisting of M^+ and BH_4^-. On the other hand, aluminum borohydride, $Al(BH_4)_3$, is a rather volatile compound and is apparently best described as BH_2 units covalently bonded to aluminum by Al—H—B bridge bonds.

Aluminum

In contrast to boron, elemental aluminum is definitely metallic. Nevertheless, in some of its compounds aluminum, to a slight degree, displays properties that are often associated with the semimetals: It forms a markedly amphoteric oxide, and halides that are rather volatile.

In the recovery of aluminum from its ore bauxite, the amphoteric properties of Al_2O_3 are used to advantage. The ore must be freed of iron impurities before it is electrolytically reduced to aluminum. To do this, crude Al_2O_3 is treated with hot alkali solution, and the aluminum oxide dissolves as $Al(OH)_4^-$. Because Fe_2O_3 is not amphoteric, the iron impurities remain undissolved and can be removed by filtration. The hot solution of $Al(OH)_4^-$ is then cooled and agitated, and $Al_2O_3 \cdot 3H_2O$ precipitates. The purified $Al_2O_3 \cdot 3H_2O$ is heated to produce Al_2O_3, which is then dissolved in a fused mixture of cryolite, Na_3AlF_6, NaF, and CaF_2 and then electrolyzed to produce metallic aluminum.

The electrode potential of aluminum,

$$Al^{+3}(aq) + 3e^- = Al(s), \qquad \varepsilon^0 = -1.66,$$

shows that the metal is a strong reducing agent. In ordinary circumstances, however, the surface of aluminum is covered with a dense, tough, transparent coating of oxide which protects the metal from further chemical attack. The oxide coating can be destroyed by amalgamation, and in this condition aluminum displays its true reducing properties and dissolves readily in water with the evolution of hydrogen.

The enthalpy of formation of Al_2O_3 is negative, and its magnitude indicates the great stability of this compound:

$$2Al + \tfrac{3}{2}O_2 = Al_2O_3, \qquad \Delta H = -399 \text{ kcal.}$$

Aluminum oxide is so stable that metallic aluminum will reduce any metallic oxide to the metal in the *thermite* process:

$$2Al + Cr_2O_3 = Al_2O_3 + 2Cr, \qquad \Delta H = -126 \text{ kcal,}$$
$$2Al + Fe_2O_3 = Al_2O_3 + 2Fe, \qquad \Delta H = -203 \text{ kcal.}$$

The most important form of anhydrous aluminum oxide, Al_2O_3, is α-alumina. In this solid, the oxide ions form a hexagonally close-packed array, and aluminum ions occupy two thirds of the octahedral interstitial sites, in such a way that each oxygen ion is surrounded by four aluminum ions. In its pure crystalline state, α-alumina (the mineral corundum) is highly transparent, mechanically strong, thermally stable, an excellent electrical insulator, and very resistant to attack by acids or bases. When crystalline alumina contains traces of certain transition metal ions, it can become very attractively colored. The gemstone ruby is α-alumina, which contains small amounts of Cr^{+3}, while blue sapphire is α-alumina contaminated by Fe^{++}, Fe^{+3} and Ti^{+3}.

Table 14.14 Properties of the aluminum trihalides

	Melting point (°C)	Boiling point (°C)	ΔH_f^0(kcal)
AlF_3	—	1291 (sub)	−311.0
$AlCl_3$	192	180 (sub)	−166.2
$AlBr_3$	97	255	−125.8
AlI_3	180	381	− 75.2

Aqueous solutions of aluminum salts are acidic because of the hydrolysis of Al^{+3}. The intense electric field of the relatively small, highly charged ion apparently draws electrons away from the neighboring water molecules and enables them to become proton donors. The ionization constant for the reaction

$$[Al(H_2O)_6]^{+3} = [Al(H_2O)_5OH]^{+2} + H^+, \qquad K = 1.1 \times 10^{-5},$$

shows that aqueous Al^{+3} is nearly as strong an acid as acetic acid. This pronounced hydrolysis is to be expected from any species that forms an amphoteric oxide.

Aluminum fluoride is a high-melting compound of low volatility, but the other halides of aluminum melt at relatively low temperatures, as Table 14.14 shows. In the vapor phase the chloride, bromide, and iodide of aluminum exist as Al_2X_6 molecules that have the halide bridge structure shown in Fig. 14.16.

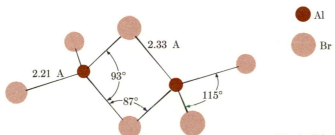

● Al

● Br

The halide bridge structure of Al_2Br_6.   **FIG. 14.16**

This bridge structure is apparently a fairly general characteristic of electron-deficient systems, where the formation of bonds by the usual electron-sharing process leads to a *nearly* completed valence-electron shell.

Gallium, Indium, and Thallium

The chemistry of gallium is very similar to that of aluminum. The electrode potential

$$Ga^{+3}(aq) + 3e^- = Ga(s), \qquad \mathcal{E}^0 = -0.52 \text{ volts},$$

indicates that gallium is a good reducing agent, but not as powerful as aluminum. Like aluminum, gallium metal is protected by an oxide coating, but will dissolve slowly with evolution of hydrogen both in acids and bases to give Ga^{+3} and $Ga(OH)_4^-$, respectively. The Ga^{+3} ion is extensively hydrolyzed in aqueous solution, as shown by the equilibrium constant for the reaction

$$[Ga(H_2O)_6]^{+3} = [Ga(H_2O)_5OH]^{++} + H^+, \qquad K = 2.5 \times 10^{-3}.$$

Indium is a rare metal which because of its softness and scarcity has no important structural uses. The metal does take a high polish and is sometimes used in the construction of special mirrors. The electrode potential

$$In^{+3}(aq) + 3e^- = In(s), \qquad \varepsilon^0 = -0.34 \text{ volts},$$

indicates that indium is a poorer reducing agent than aluminum and gallium. Although solutions of In^{+3} are extensively hydrolyzed, the oxide In_2O_3 is primarily basic and not amphoteric like Al_2O_3 and Ga_2O_3.

Indium also differs from aluminum and gallium in that it forms compounds in the +1 oxidation state. The halides InCl, InBr, and InI are known, as is In_2O.

Thallium is also a rare metal and, like indium, is very soft. Unlike the other group III metals, thallium is not protected by an oxide coating and consequently is oxidized readily by air. Another point of difference is that the +3 state of thallium has considerable oxidizing power, as is shown by the electrode potential

$$Tl^{+3}(aq) + 2e^- = Tl^+(aq), \qquad \varepsilon^0 = 1.25 \text{ volts}.$$

While compounds of thallium in the +3 oxidation state such as Tl_2O_3, the trihalides except for the iodide, and $Tl_2(SO_4)_3$ are well known, much of the chemistry of thallium involves the +1 oxidation state. Except that TlCl is insoluble in water, the behavior of thallous salts is, in general, similar to that of the alkali-metal salts.

14.4 THE ELEMENTS OF GROUP IVA

The close resemblance between elements of the same family, so obvious in groups I and II, and somewhat less evident in group III, is even less apparent in group IVA. Carbon is indisputably a nonmetal, and while the chemistry of silicon is in some respects characteristic of a nonmetal, the electrical and other physical properties of the element are those of a semimetal. Germanium is from all points of view a semimetal. Although one allotropic form of tin has the electrical properties of a semimetal, tin and particularly lead display the chemical and physical characteristics of metals. The data in Table 14.15 illustrate the considerable differences in properties like atomic size and melting point that occur in group IV.

The elements of group IV have in common the oxidation states of +2 and +4, but while the +4 state is of overwhelming importance for carbon and silicon, the +2 state becomes increasingly important for germanium and tin

Table 14.15 Properties of the group IVA elements

	C	Si	Ge	Sn	Pb
Atomic number	6	14	32	50	82
Configuration	$2s^22p^2$	$3s^23p^2$	$4s^24p^2$	$5s^25p^2$	$6s^26p^2$
Ionization energy, kcal	260	188	182	169	171
Atomic radius, A	0.77	1.17	1.22	1.41	1.54
Melting point, °K	>4000	1685	1210	505	601
Boiling point, °K	3900	3000	3100	2960	2024
ΔH_{sub}, kcal	171	108	90	72	47
Ionic radius M^{++}, A	—	—	—	1.10	1.32
$\mathcal{E}^0(M^{++}, M)$, volts	—	—	—	−0.136	−0.126

and is the most important oxidation state of lead. Thus instead of a strong resemblance between elements, one finds in group IV fairly smooth trends from one type of behavior to another.

Carbon occurs naturally in the allotropic forms diamond and graphite and much more abundantly in contaminated forms like coal. Silicon is the second most abundant element after oxygen, and is found in an enormous variety of silicate minerals. In contrast to carbon and silicon, germanium, tin, and lead are rather rare elements. Tin and lead are familiar, however, because they are easily recovered from their ores and are technologically important as pure metals and in alloys.

Ultrapure silicon and germanium are used in the electronics industry for the manufacture of transistors. The methods used to obtain each of these elements in the pure form are essentially the same. For example, silicon is obtained as the dioxide, which is then reduced with carbon,

$$SiO_2 + 2C = Si + 2CO.$$

This rather crude silicon is converted to the tetrachloride by direct reaction with chlorine,

$$Si + 2Cl_2 = SiCl_4.$$

Silicon tetrachloride is volatile and can easily be purified by distillation. Purified silicon is then recovered by reducing the chlorine with hydrogen,

$$SiCl_4 + 2H_2 = Si + 4HCl.$$

The silicon is purified further by the zone melting process. In this operation a short length of a rod of silicon is melted, and this melted zone is moved slowly along the rod. Impurities tend to collect in the melted zone as the element recrystallizes, and are transported to the end of the rod as the melted zone moves. After repetition of this process the "impure" end of the rod is sawed off, and the remaining silicon (or germanium) may have an impurity level lower than 10^{-11} mole fraction.

The common ore of lead is galena, PbS. To recover the lead, the sulfide is roasted in air and converted to PbO, which is then reduced with carbon. Tin occurs as the oxide SnO_2, which can be reduced directly with carbon. Any further purification of lead or tin is usually carried out by dissolving the metal and then depositing it electrolytically.

Carbon

Of the two allotropic forms of carbon, graphite is slightly more stable at 25°C and 1 atm. The enthalpy change for the reaction

$$C(\text{diamond}) = C(\text{graphite})$$

is only −453.2 cal/mole.

The structures of diamond and graphite are shown in Fig. 14.17. In diamond, each carbon atom is covalently bonded to four others which are located at the apices of a regular tetrahedron, and thus we can regard the carbon atom as displaying sp^3-hybridization. The C—C bond length (1.54 A) and bond energy (85 kcal) in diamond are virtually the same as they are in compounds like ethane, H_3C—CH_3. In graphite the atoms form planar layers within which they have a regular hexagonal arrangement. The distance between planes is 3.40 A, which suggests that the planes are not covalently bonded to each other, but only held by forces of the van der Waals type. The separation of neighboring atoms in the planes is 1.415 A, which is small enough to suggest that the carbon atoms are bound to each other by multiple bonds. No single set of bond structures can be drawn that satisfies the octet rule and also indicates that all bonds are equivalent. In Fig. 14.18 are three resonance structures which can be taken together to represent the bond structure in the planes of graphite. We might say that the bond in graphite is neither a single nor a double bond, but a $1\frac{1}{3}$ bond.

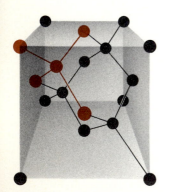

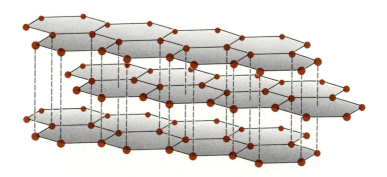

FIG. 14.17   Arrangement of atoms in (a) diamond and (b) graphite.

While diamond is a transparent insulator, graphite is a dark opaque solid with a slight metallic luster and an electrical conductivity that is fairly large in the directions parallel to the planes of atoms, but small in the direction perpendicular to the planes. Apparently the multicenter bonding within the planes is like the "free-electron sea" type of bonding found in metals, and this accounts for the luster and conductivity of graphite.

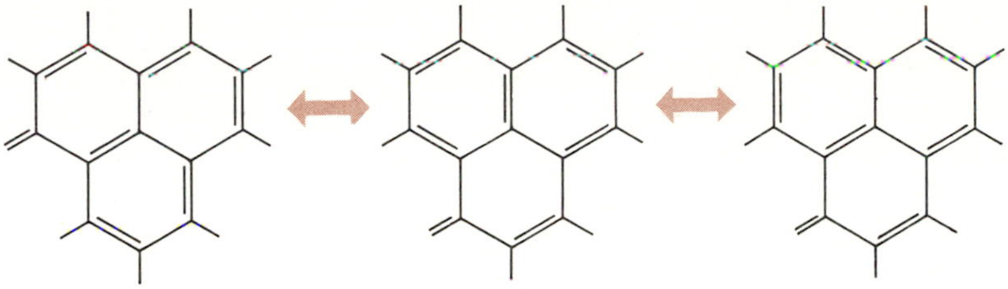

Three resonance structures for a graphite fragment.

FIG. 14.18

The outstanding chemical characteristic of carbon is that it forms a virtually unlimited number of compounds in which carbon atoms are bonded to each other. These compounds, of which the hydrocarbons C_nH_{2n+2} are an example, apparently owe their stability to the uniquely strong bonds that carbon forms with itself. Consider the following bond energies.

C—C	82 kcal	C—H	98 kcal	Si—H	75 kcal
Si—Si	53 kcal	C—F	110 kcal	Si—F	135 kcal
Ge—Ge	45 kcal	C—Cl	80 kcal	Si—Cl	91 kcal
Sn—Sn	37 kcal				

Thus the energy of the C—C bond is roughly of the same magnitude as the energies of the bonds that carbon forms to other elements. The reverse is more nearly true for silicon and particularly for germanium and tin, whose atoms form only very weak bonds to each other. The chemistry of carbon bonded to itself and to hydrogen, oxygen, nitrogen, and a few other elements is the subject of organic chemistry, which is treated in Chapter 17. In the present section we will deal only with those compounds of carbon that are usually considered to be "inorganic," principally because they do not contain hydrogen.

Carbon forms three oxides that are well characterized: carbon monoxide, carbon dioxide, and carbon suboxide, C_3O_2. Carbon monoxide has the largest bond energy of any diatomic molecule, 256 kcal. It is isoelectronic with nitrogen, and just as for nitrogen, the simple electron-dot structure that satisfies the octet rule also indicates that carbon monoxide is triply bonded.

$$: C ::: O :$$

The triple bond can be pictured as two π-bonds formed by p_x and p_y atomic orbitals and a σ-bond formed from sp hybrid orbitals, as was discussed in Section 11.7.

The nonbonded electrons on the carbon atom in carbon monoxide can in certain circumstances be donated to electron acceptors. Thus carbon monoxide reacts with the electron-deficient molecule B_2H_6 to form borine carbonyl, BH_3CO,

$$2CO + B_2H_6 = 2BH_3CO.$$

We can regard borine carbonyl as a molecule in which the electron deficiency of a BH_3 fragment has been removed by the pair of electrons donated by carbon monoxide, as shown by the following structure.

$$
\begin{array}{c}
H \\
\cdot\cdot \\
H : B : C :::O : \\
\cdot\cdot \\
H
\end{array}
$$

Carbon monoxide also forms an important series of carbonyl compounds with the transition metals. The most common example of these is nickel carbonyl, $Ni(CO)_4$, which can be formed by the direct reaction

$$Ni + 4CO = Ni(CO)_4.$$

Other compounds of this type will be discussed in Chapter 16.

Despite the stability of carbon monoxide, it reacts exothermically with oxygen to form carbon dioxide:

$$CO + \tfrac{1}{2}O_2 = CO_2, \qquad \Delta H = -67.6 \text{ kcal.}$$

Carbon dioxide is unique among the dioxides of the group IVA elements: It is a volatile molecular compound, whereas SiO_2, GeO_2, SnO_2, and PbO_2 are all nonvolatile solids with relatively complicated crystal structures.

Carbon dioxide is only moderately soluble in water, but is the anhydride of carbonic acid,

$$CO_2 + H_2O = H_2CO_3(aq).$$

The total solubility of CO_2 in water is approximately $0.034\ M$, but of this amount, 99.63% is present in the form of CO_2 molecules, and only 0.37% is actually H_2CO_3. The first ionization constant of "carbonic acid" is usually written as

$$\frac{[H^+][HCO_3^-]}{[H_2CO_3]} = 4.3 \times 10^{-7},$$

where $[H_2CO_3]$ stands for the *total* concentration of all neutral carbonic material,

H_2CO_3 *and* CO_2. Because most of this is CO_2, the equilibrium constant might better be written as

$$\frac{[H^+][HCO_3^-]}{[CO_2]} = 4.3 \times 10^{-7}.$$

Carbon unites with a number of metals to form carbides, some of which have properties of practical value. The carbides can be classified as saltlike, interstitial, and covalent. Two examples of saltlike carbides are Be_2C and Al_4C_3. These compounds both yield methane when hydrolyzed. For example,

$$Al_4C_3 + 12H_2O = 3CH_4 + 4Al(OH)_3.$$

The structure of Be_2C is the same antifluorite lattice found in Na_2O, with the beryllium and carbon atoms replacing the sodium and oxygen atoms, respectively.

The acetylides, which contain the $C_2^=$ unit, constitute another class of saltlike carbides. Calcium carbide, CaC_2, is of this type, and has a rock-salt crystal structure in which the Ca^{++} and $C_2^=$ replace Na^+ and Cl^-. Upon hydrolysis it yields acetylene, C_2H_2:

$$CaC_2 + 2H_2O = Ca(OH)_2 + C_2H_2.$$

Direct combination of carbon with some of the transition metals like titanium and tungsten yields interstitial carbides. These compounds are electrical conductors and have a metallic luster, and it is for this reason that they are pictured as metallic lattices which contain carbon atoms in the interstitial sites. The interstitial carbides are, in general, extremely hard and high-melting substances. For example, both TiC and W_2C are nearly as hard as diamond and melt at temperatures higher than 3000°K.

The compounds of carbon with elements of similar electronegativity are called covalent carbides. The most important of these is silicon carbide or carborundum, SiC. This substance is nearly as hard as diamond and has the same infinite three-dimensional lattice structure.

Silicon

Elemental silicon has a silvery metallic luster, but its electrical conductivity is substantially smaller than that of the metals. Its crystal lattice is the same as that of diamond. While silicon is rather inert at room temperature, it reacts at higher temperatures with all the halogens to form tetrahalides, with oxygen to form SiO_2, and with nitrogen to form Si_3N_4. When treated with strong base, it dissolves with evolution of hydrogen,

$$Si(s) + 2OH^- + H_2O = SiO_3^= + 2H_2.$$

Silicon dioxide or silica exists as a three-dimensional network solid of enormous stability. In one crystal form of SiO_2, the silicon atoms are arranged exactly as are the carbon atoms in diamond, except that oxygen atoms are midway between them. The crystal structure of quartz, the most familiar form of SiO_2, is a slight modification of this arrangement.

Silica melts at 1983°K, and when molten silica is cooled it most often sets to a glassy material, rather than to a crystalline solid. This *fused quartz* has a number of valuable properties. Because of the great strength of the silicon-oxygen bond (108 kcal), it is thermally stable and chemically inert to all substances except HF, F_2, and hot alkali. In addition it is an excellent electrical insulator even at high temperatures, has a very small coefficient of thermal expansion, is highly transparent to ultraviolet light, and when drawn into fibers, has excellent elastic properties.

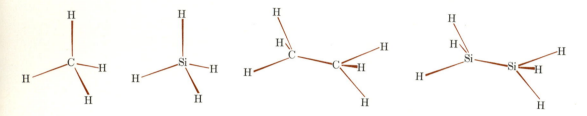

FIG. 14.19   Structural similarities between hydrocarbons and silanes.

Several silicon hydrides, or *silanes*, are known. All have the general formula Si_nH_{2n+2}, and thus are analogous to the saturated hydrocarbons C_nH_{2n+2}. Figure 14.19 shows the structural similarities between the simplest silanes and hydrocarbons.

In the hydrocarbon series, the number of carbon atoms in a chain can apparently have any value, but among the silanes, the most complicated known compound is Si_6H_{14}. Apparently the relatively weak Si—Si bond makes molecules with many silicon atoms linked together quite unstable. The silicon hydrides differ from the hydrocarbons in another important respect. Two carbon atoms may be linked by a double bond, as they are in ethylene, $H_2C=CH_2$. Thus there is a series of compounds with the general formula C_nH_{2n}. There are no analogous silicon hydrides, and apparently double and triple bonds between silicon atoms are so unstable that they are not known.

The silanes are colorless relatively volatile substances. They are all very reactive; for example, they ignite spontaneously in air,

$$Si_3H_8 + 5O_2 = 3SiO_2 + 4H_2O.$$

Silanes also react explosively with halogens, but in the presence of Al_2Cl_6, they react in a controlled manner with HCl to give chlorosilanes:

$$HCl(g) + SiH_4(g) \xrightarrow{Al_2Cl_6} SiH_3Cl(g) + H_2(g).$$

It is possible to synthesize such molecules as $(CH_3)_3SiCl$, $(CH_3)_2SiCl_2$, and CH_3SiCl_3. These and similar substituted chlorosilanes are technologically important, for upon hydrolysis they yield polymeric molecules of high molecular weight called *silicones*. Hydrolysis of $(CH_3)_2SiCl_2$ gives a chain of silicon and oxygen atoms.

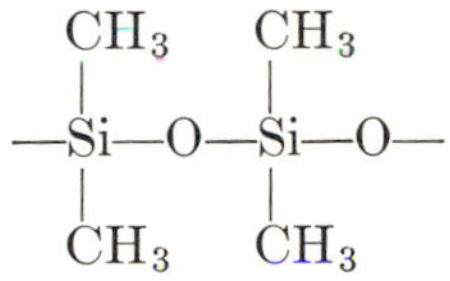

$$\begin{array}{ccc}
CH_3 & & CH_3 \\
| & & | \\
-Si-O- & & Si-O- \\
| & & | \\
CH_3 & & CH_3
\end{array}$$

Hydrolysis of CH_3SiCl_3 gives a cross-linked chain, or a two-dimensional network, as shown below.

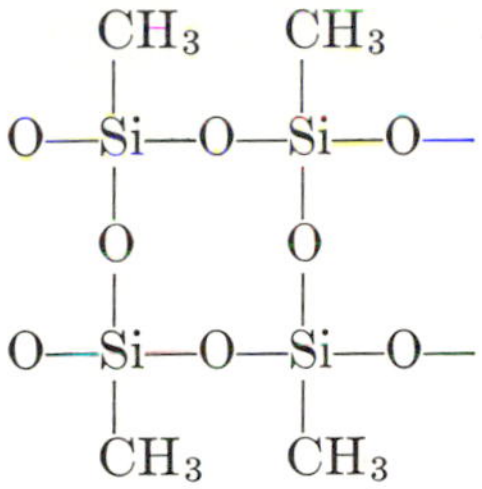

$$\begin{array}{ccc}
CH_3 & & CH_3 \\
| & & | \\
O-Si-O- & & Si-O- \\
| & & | \\
O & & O \\
| & & | \\
O-Si-O- & & Si-O- \\
| & & | \\
CH_3 & & CH_3
\end{array}$$

The amount of cross linking and the identity of the hydrocarbon substituent control the properties of the polymer. All the silicones tend to be water repellent, heat resistant, electrically insulating, and chemically inert, and these properties make them useful as lubricants, insulators, and protective coatings.

When the oxides or carbonates of the alkali metals are fused with silica, SiO_2, various alkali silicates are formed. The simplest silicate is Na_4SiO_4, in which the SiO_4^{-4} anion consists of a silicon atom surrounded by four oxygen atoms located at the corners of a regular tetrahedron. However, there are many other alkali silicates, such as $Na_2Si_2O_5$, $Na_6Si_2O_7$, and Na_2SiO_3. Despite the diverse empirical formulas of these compounds, we shall find that their structures can be understood in terms of a repetition of the fundamental SiO_4 tetrahedral unit.

Silicates occur widely in Nature. Together, silicon and oxygen make up 74 percent of the mass of the Earth's crust, and they occur mostly as the silicates of the abundant metals Al, Fe, Ca, Mg, Na, and K. It is the variety of molecular structures of the silicate anions which produces the enormous range of mechanical properties displayed by such minerals as asbestos, the micas, talc, kaolin, the feldspars, serpentine, the garnets, and the olivines. Consequently, silicate structures are of interest not only as examples of polymeric anions, but as the

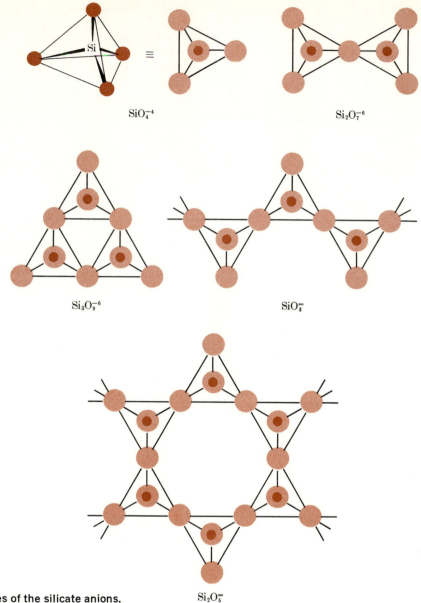

SiO_4^{-4}

$Si_2O_7^{-6}$

$Si_3O_9^{-6}$

$SiO_3^{=}$

$Si_2O_5^{=}$

FIG. 14.20   Structures of the silicate anions.

means by which the macroscopic properties of the ubiquitous Earth materials can be understood.

Figure 14.20 shows how the structures of the various silicate anions are built up from the fundamental SiO_4 tetrahedral unit. When two tetrahedra share one oxygen atom, the ion $Si_2O_7^{-6}$ results. When two oxygen atoms in each tetrahedron are shared with neighbors, either ring structures such as $Si_3O_9^{-6}$, or infinite chains of tetrahedra with the empirical formula SiO_3^{-2} can be generated. A double-chain structure can be formed when alternate tetrahedra of two single chains share oxygen atoms, and an anion of the empirical formula $Si_4O_{11}^{-6}$ results. This double chain can be extended in a similar manner to generate an infinite sheet of tetrahedra, each sharing three oxygen atoms with its neighbors. The empirical formula of this anion is $Si_2O_5^{-2}$. As we shall see, three-dimensional network anions are possible if some of the silicon atoms are replaced by Al. In the silicate minerals, positive ions occupy positions in the anion structure in such a manner that charge balance is achieved, and the anions are held together by Coulomb attraction to the cations.

In Nature, the class of minerals known as the olivines has discrete SiO_4^{-4} tetrahedral units, with no oxygen atoms shared. The chemical formula of the olivines can be written as $[Mg, Fe]_2SiO_4$. That is, Mg^{++} and Fe^{++} occur in different relative amounts, but the total number of cations is sufficient to balance the charge on the anions. Another way of describing the olivines is to imagine a hexagonal close-packed array of oxygen ions, with silicon ions occupying one-eighth of the tetrahedral interstitial sites, and Mg^{++} and Fe^{++} occupying one-half of the octahedral interstitial sites. Because Mg^{++} and Fe^{++} have similar ionic radii (0.66 Å and 0.74 Å, respectively), substitution of one ion by the other does not markedly change the structure or properties of the mineral.

Other minerals which contain discrete SiO_4^{-4} tetrahedra include willemite, Zn_2SiO_4, and zircon, $ZrSiO_4$. Garnets also contain SiO_4^{-4} tetrahedra, and have the general formula $M_3^{++} M_2^{+3}(SiO_4)_3$, where M^{++} can be Ca^{++}, Mg^{++}, or Fe^{++}, and M^{+3} is Al^{+3}, Cr^{+3}, or Fe^{+3}. Olivines and garnets are very hard and dense minerals of great thermal stability.

Cyclic silicate anions of the type $Si_3O_9^{-6}$ are found in the minerals wollastonite, $Ca_3Si_3O_9$, and benitoite, $BaTiSi_3O_9$, among others. Even larger cyclic anions occur. For example, $Si_6O_{18}^{-12}$, a ring of six tetrahedra, is found in the mineral beryl, $Al_2Be_3Si_6O_{18}$. When it is contaminated by small amounts of Cr^{+3} or V^{+3}, beryl is the gemstone emerald.

Minerals which are built around the infinite single-chain silicate anion are called pyroxenes. The best known examples are diopside, $CaMg(SiO_3)_2$, and spodumene, $LiAl(SiO_3)_2$, a major source of lithium. Double-chained silicate anions occur in the class of minerals called the amphiboles. The complete structures of these minerals can be quite complicated, since in addition to the double-chain anion, both metal and hydroxide ions are present, as in termolite, $Ca_2Mg_5(OH)_2(Si_4O_{11})_2$. Some of the amphiboles display fibrous mechanical properties which result from the great strength of the double anion chains. Some of the minerals commonly known as asbestos are amphiboles.

Combination of the silicate sheet anion $Si_2O_5^{-2}$ with various cations and hydroxide ions produces an enormous number of important and abundant minerals. Examples are talc, $Mg_3(OH)_2(Si_2O_5)_2$, and the micas, such as biotite, $K(Mg, Fe)_3(OH)_2AlSi_3O_{10}$. The close correlation between anion structure and macroscopic mechanical properties is particularly evident in the micas, which cleave very easily in directions parallel to the anion sheets.

If each SiO_4 tetrahedron shared all four oxygen atoms with its neighbors, we would have a substance with the empirical formula SiO_2 and the three-dimensional framework structure of quartz. To achieve a three-dimensional anion structure, some of the Si^{+4} ions in the tetrahedra must be replaced by Al^{+3}, and in order to maintain overall charge balance in the crystal, other cations must enter the interstices in the oxide lattice. Thus there are minerals of the general formula $M(Al, Si)_4O_8$ in which a certain fraction of the silicon atoms have been replaced by aluminum. When the ratio Si:Al is 3:1, M is an alkali metal ion, and when the ratio is 2:2, M is a doubly charged ion such as Ca^{++} or Ba^{++}. These three-dimensional alumino-silicate structures occur in the feldspars, which are the most common minerals of the Earth's crust. As might be expected from the molecular structures, feldspars are very hard and thermally stable minerals.

Germanium

This element has chemical properties that are similar to those of silicon. Solid germanium has the diamond lattice structure and displays the electrical conductivity of a semimetal. Like silicon, germanium reacts directly with halogens to form volatile tetrahalides, with oxygen to form GeO_2, and with alkalies to form germanates,

$$Ge + 2OH^- + H_2O = GeO_3^= + 2H_2.$$

Like SiO_2, GeO_2 is a weakly acidic oxide.

Germanium, like silicon, forms a series of volatile hydrides which have the general formula Ge_nH_{2n+2}. Compounds in which $n = 6, 7$, and 8 have been identified but not fully characterized. The germanium hydrides or germanes are oxidized to GeO_2 and H_2O by oxygen, but are not as flammable as the silanes.

One difference between the chemistry of germanium and silicon is that germanium exists in the $+2$ oxidation state in several of its compounds. The halides $GeCl_2$, $GeBr_2$, and GeI_2 and the sulfide GeS are known, but are rather unstable and are strong reducing agents. The synthesis of these compounds illustrates a general method often used to prepare rather unstable intermediate oxidation states. The tetrahalides or GeS_2 are first obtained by direct reaction of the elements and then treated with germanium:

$$Ge + GeCl_4 = 2GeCl_2, \qquad Ge + GeS_2 = 2GeS.$$

These elements differ from the other members of group IV in a number of respects. In the first place, lead and tin are metals, and have important $+2$ oxidation states. While the other group IV oxides are acidic, the monoxides SnO and PbO are amphoteric, as is SnO_2. Lead dioxide, PbO_2, is a powerful oxidant, and its acid-base properties are not well characterized. In contrast to the extensive series of hydrides formed by carbon, silicon, and germanium, tin and lead form only SnH_4 and PbH_4, respectively. While the tetrahalides of carbon, silicon, and germanium are stable, those of tin are less so, and PbF_4 and $PbCl_4$ are quite unstable and dissociate upon warming,

$$PbX_4 = PbX_2 + X_2.$$

Aqueous solutions of tin in the $+2$ state are formed when the metal dissolves in acid. Corresponding to the fact that SnO is an amphoteric oxide, solutions of Sn^{++} are extensively hydrolyzed:

$$Sn^{++} + H_2O = SnOH^+ + H^+, \qquad K = 10^{-2}.$$

Thus stannous ion is a moderately strong acid—as strong as HSO_4^-. Stannous ion also tends to form complexes with anions. For example, in the presence of chloride ion, $SnCl^+$, $SnCl_2$, $SnCl_3^-$, and $SnCl_4^=$ can all be formed.

The dioxide SnO_2 dissolves in base to form stannate ions, $Sn(OH)_6^=$, and in halogen acids to form complex ions such as $SnCl_6^=$. There is no evidence that Sn^{+4}, or any other monatomic ion of $+4$ charge, ever exists uncomplexed in aqueous solution.

The most common oxidation state of lead is $+2$. In aqueous solution, the plumbous ion, Pb^{++}, is hydrolyzed, but not nearly so much as the stannous ion, as the equilibrium constant for the following reaction shows:

$$Pb^{++} + H_2O = PbOH^+ + H^+, \qquad K = 10^{-8}.$$

When alkali is added to solutions of Pb^{++}, $Pb(OH)_2$ first precipitates and then redissolves in excess base as $Pb(OH)_3^-$.

Like stannous ion, Pb^{++} forms complexes with the halide ions. The equilibrium constants for

$$PbCl^+ = Pb^{++} + Cl^-, \qquad K = 0.077,$$
$$PbBr^+ = Pb^{++} + Br^-, \qquad K = 0.07,$$
$$PbI^+ = Pb^{++} + I^-, \qquad K = 0.03,$$

show that these complexes are of only moderate stability. The lead halides

PbX_2 themselves are moderately insoluble, but dissolve in excess halide ions to form such complex ions as $PbCl_3^-$ and PbI_3^-.

The sulfate, carbonate, chromate, and sulfide of lead are all quite insoluble in water, as the following equilibrium constants show:

$$
\begin{aligned}
PbSO_4 &= Pb^{++} + SO_4^=, & K &= 1.3 \times 10^{-8}, \\
PbCO_3 &= Pb^{++} + CO_3^=, & K &= 1.5 \times 10^{-13}, \\
PbCrO_4 &= Pb^{++} + CrO_3^=, & K &= 2 \times 10^{-16}, \\
PbS &= Pb^{++} + S^=, & K &= 7 \times 10^{-29}.
\end{aligned}
$$

There is very little chemistry associated with the +4 state of lead. The potential for the reaction

$$PbO_2 + 4H^+ + 2e^- = Pb^{++} + 2H_2O, \qquad \varepsilon^0 = 1.46 \text{ volts,}$$

shows that lead dioxide is a very powerful oxidizing agent. It can be produced electrolytically, as it is in the charging of a lead storage battery, or by reaction of lead (II) with hypochlorite ion in basic solution:

$$Pb(OH)_3^- + ClO^- = Cl^- + PbO_2 + OH^- + H_2O.$$

14.5 CONCLUSION

This discussion of the first four groups of representative elements has revealed some regularities that make the descriptive chemistry of these elements quite coherent. In group IA particularly, there is a strong resemblance between the elements. This marked resemblance persists in group IIA, but superimposed on it are some important trends. Beryllium is a less active reducing agent and forms a much more acidic oxide than do the other alkaline-earth elements. Moreover, its halides have properties that are easier to rationalize in terms of covalent than ionic bonding. Some of these properties appear to a lesser extent in magnesium, but are suppressed in the heavier alkaline-earth elements. In the group IIIA family, the elements display a considerable range of properties, but still retain some resemblance to each other. Boron is a semimetal and has the acidic oxide and molecular covalent halides associated with semimetals. The properties connected with metallic behavior become more obvious as the atomic weight of the elements increases. In group IVA, the trend of increasing metallic behavior with increasing atomic weight is repeated.

When elements of the same period are compared, it is clear that there is a general trend from metallic to nonmetallic behavior as one passes from left to right. In the first period, the dividing line between metals and nonmetals occurs at the element boron in group IIIA. In the second and third periods, the dividing line occurs in group IVA, while in the fourth period it is delayed to group VA.

Thus there is a diagonal region in the periodic table that separates the metals and nonmetals.

Besides the vertical relationships and horizontal trends in the periodic table, there is a resemblance between elements that are diagonal neighbors. This diagonal relationship is most noticeable in the following portion of the table.

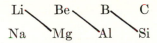

Some examples of this relationship are the semimetallic character, acidic oxides, volatile hydrides and halides of both boron and silicon, the amphoteric oxides and volatile halides of both beryllium and aluminum, and the ease with which both lithium and magnesium form nitrides. In some respects this diagonal relationship is more obvious than the vertical resemblances in a given group.

SUGGESTIONS FOR FURTHER READING

Cotton, F. A., and G. Wilkinson, *Advanced Inorganic Chemistry*, 3rd ed. New York: Interscience, 1972.

Douglas, B. E., and D. H. McDaniel, *Concepts and Models of Inorganic Chemistry*. New York: Blaisdell, 1965.

Gould, E. S., *Inorganic Reactions and Structure*. New York: Holt, Rinehart and Winston, 1962.

Heslop, R. B., and P. L. Robinson, *Inorganic Chemistry*. New York: Elsevier, 1963.

Kleinberg, J., W. J. Argersinger, Jr., and E. Griswold, *Inorganic Chemistry*. Boston: D. C. Heath, 1960.

Phillips, C. S. G., and R. J. P. Williams, *Inorganic Chemistry*, Vols. 1 and 2. London: Oxford Univ. Press, 1965.

PROBLEMS

14.1 By considering the solubilities of the alkaline-earth hydroxides, decide how the pH of a solution should affect the strength of magnesium and barium metals as reducing agents.

14.2 Using the data given in Table 14.12, analyze the energetics of the half-cell reaction $M(s) = M^{+3}(aq) + 3e^-$ for aluminum and gallium. Suggest a reason why aluminum is a better reductant than gallium.

14.3 It is generally accepted that positive ions are more extensively hydrolyzed, or act as stronger acids, the higher their charge and the smaller their radii. Cite several examples from this chapter that support this generalization.

14.4 The triiodide ion, I_3^-, is known in aqueous solution, but while the solid compound CsI_3 is stable with respect to CsI and I_2, LiI_3 is not stable with respect to LiI and I_2. To account for this difference in stabilities, reasoning similar to that used to explain the relative stabilities of the alkaline-earth carbonates can be used. Construct an argument that rationalizes the relative stabilities of CsI_3 and LiI_3.

14.5 The rock-salt crystal structure can be regarded as a cubic closest-packed anion lattice with cations in all octahedral holes. Lithium iodide has the rock-salt structure, and the distance between iodine nuclei that form an edge of a face-centered cube is 6.050 A. By assuming that the lattice dimensions are determined by anion–anion "contact," calculate the radius of the iodide ion.

14.6 How would you expect the following properties of francium to compare with those of the other alkali metals: (a) ionization energy; (b) atomic radius; (c) ionic radius? Does the occurrence of the lanthanide series have any relevance to your answers?

14.7 From the standard half-cell potentials

$$Tl^+ + e^- = Tl, \qquad \mathcal{E}^0 = -0.336 \text{ volt,}$$
$$Tl^{+3} + 2e^- = Tl^+, \qquad \mathcal{E}^0 = 1.25 \text{ volts,}$$

calculate the equilibrium constant for the reaction

$$3Tl^+ = 2Tl + Tl^{+3}.$$

Would you expect thallous ion to disproportionate in aqueous solution?

14.8 Why is Sn^{++} more extensively hydrolyzed than Pb^{++}?

14.9 With respect to decomposition by the reaction $MSO_4(s) = MO(s) + SO_3(g)$, which of the group IIA sulfates would you expect to be least stable? Which would you expect to be most stable?

14.10 Discuss, in terms of an ionic-bond model, the possible reasons why aluminum is found exclusively as Al^{+3} in solution and in its stable compounds, whereas thallium displays both $+1$ and $+3$ oxidation states. The ionic radius of Tl^+ is 1.47 A.

CHAPTER 15

THE
NONMETALLIC ELEMENTS

Groups VA, VIA, VIIA, and the inert gases include most of the nonmetallic elements. The chemical behavior of each of these elements is in general more complicated than the chemistry of the representative metals, and in some cases a nonmetallic element may show only slight resemblance to the other members of its chemical family. Nevertheless, it is possible to detect a number of regularities, particularly among the structural features of the compounds of the nonmetals, that aid us in remembering and understanding the chemistry of these elements.

15.1 THE ELEMENTS OF GROUP VA

A considerable range and variety of chemical and physical properties are encountered upon examination of the group VA elements. Some of these properties are given in Table 15.1. Nitrogen and phosphorus are nonmetals, arsenic and antimony are semimetals, and bismuth is a metal with a rather small electrical conductivity. While the characteristic oxidation states of these elements might be taken as −3, +3, and +5, there are individual peculiarities. Nitrogen appears in every integral oxidation state from −3 to +5, and as one proceeds down the family of elements, one finds that the +5 state and particularly the −3 state become increasingly unstable and rare. A similar trend was encountered in groups III and IV.

Table 15.1 Properties of the group VA elements

	N	P	As	Sb	Bi
Atomic number	7	15	33	51	83
Configuration	$2s^2 2p^3$	$3s^2 3p^3$	$4s^2 4p^3$	$5s^2 5p^3$	$6s^2 6p^3$
Ionization energy, kcal	335	254	226	199	168
Atomic radius, A	0.74	1.10	1.21	1.41	1.52
Melting point, °K	63	317	sublimes	903	545
Boiling point, °K	77	553	886	1850	1900
ΔH_f(atom), kcal	113.7	79.8	69	62	49.5

Analogous compounds of the group V elements display a range of properties. For example, ammonia, NH_3, is basic and thermodynamically stable, while PH_3 is much less basic and is thermodynamically unstable with respect to its elements. In the compounds AsH_3, SbH_3, and BiH_3, basic properties disappear entirely, and BiH_3 in particular is so unstable that only trace quantities have been identified. As a second example, consider that the +3 oxides of nitrogen and phosphorus are acidic, but the acidic properties of the corresponding oxides of arsenic and antimony are less pronounced, and the +3 oxide of bismuth, Bi_2O_3, is basic. Other trends and variations in group V will become clear from the discussion of the properties of the individual elements.

Nitrogen

Elemental nitrogen exists as a diatomic molecule whose bond dissociation energy, 225 kcal, is second in magnitude only to that of carbon monoxide. As was discussed in Section 11.7, nitrogen and carbon monoxide are isoelectronic and both possess a triple bond. Although carbon monoxide is a moderately reactive molecule, nitrogen is quite inert, particularly at temperatures near 300°K. In some cases, this inert nature is due to thermodynamic considerations. For example, all the nitrogen oxides have positive standard free energies of formation, and thus are intrinsically unstable with respect to nitrogen and oxygen. In other instances, however, the inert character of nitrogen is a result of kinetic factors, for its reactions with many reagents are quite slow. The formation of ammonia from hydrogen and nitrogen is such an example, and it is worth considering this reaction in some detail.

The standard free energy of formation of ammonia is -3.98 kcal/mole at 25°C, and thus the equilibrium constant of the reaction

$$\tfrac{1}{2}N_2(g) + \tfrac{3}{2}H_2(g) = NH_3(g)$$

is given at 298°K by

$$K = e^{-\Delta G^0 / RT} = 8.3 \times 10^2,$$

where the concentration units are atmospheres. This equilibrium constant is quite favorable to the synthesis, and ammonia should be formed in good yield at room temperature if equilibrium could be reached. The reaction is immeasurably slow, however, and no ammonia is formed when nitrogen and hydrogen are mixed at 25°C.

At elevated temperatures and in the presence of iron catalysts, the rate of reaction between nitrogen and hydrogen is large enough to make the ammonia synthesis practical. There is still a difficulty, for the standard enthalpy change for the reaction is −11.04 kcal at 298°K and is −13.3 kcal at 450°C, the temperature at which the reaction rate is conveniently large. Since the reaction is exothermic, the equilibrium constant at the elevated temperature is smaller than at 298°K, and it is important to know whether this seriously affects the maximum yield of ammonia. The value of the equilibrium constant at 723°K might be calculated from the expression

$$\ln \frac{K_2}{K_1} = -\frac{\Delta H^0}{R}\left(\frac{1}{T_2} - \frac{1}{T_1}\right), \tag{15.1}$$

but a certain amount of care is necessary. Equation (15.1) is based on the assumption that ΔH^0 is independent of temperature, and this is not true for the ammonia synthesis. We can calculate a *fair* approximation to the equilibrium constant at 723°K, however, if we use for ΔH^0 the average of the values at 298°K and 723°K, or −12.2 kcal/mole. Then we have

$$2.3 \log \frac{K_{723}}{8.3 \times 10^2} = \frac{12,200}{1.98}\left(\frac{1}{723} - \frac{1}{298}\right),$$

$$\log K_{723} = -2.35,$$

$$K_{723} = 4.5 \times 10^{-3}.$$

Because of the approximate treatment of the temperature variation of ΔH^0, this answer is not particularly accurate, and the experimentally measured value of K_{723} is 6.5×10^{-3}. Nevertheless, the approximate value would be good enough to tell us that the equilibrium yield of ammonia at 723°K is much smaller than at 298°K. This unfavorable equilibrium constant is counteracted somewhat by carrying out the reaction at a high total pressure. While this does not change the equilibrium constant, it does result in a higher percentage conversion of nitrogen and hydrogen to ammonia and makes this direct synthesis an important industrial process.

Ammonia is a colorless gas with an exceedingly pungent odor. It condenses to a liquid which has a normal boiling temperature of −33°C. As discussed in Section 13.4, both the boiling point and enthalpy of vaporization (5.64 kcal/mole) are abnormally high for a substance of this molecular weight, because of hydrogen bonding.

Liquid ammonia is in some ways similar to liquid water. Salts dissolve in ammonia to form conducting solutions, but solubilities are usually smaller in ammonia than in water. Exceptions to this generalization are the salts which contain cations which form stable ammonia complexes. Thus the silver halides, which are very sparingly soluble in water, are quite soluble in ammonia, because of the formation of the very stable $Ag(NH_3)_2^+$ complex ion. Liquid ammonia undergoes autoionization, as does water, but to a much smaller degree:

$$2NH_3(l) = NH_4^+ + NH_2^-, \qquad K_{240} \simeq 10^{-30},$$

$$2H_2O(l) = H_3O^+ + OH^-, \qquad K_{298} = 10^{-14}.$$

We see that in the liquid ammonia solvent system, NH_4^+ is the acid analogous to H_3O^+, and NH_2^- is the base analogous to OH^-. The analogy extends to amphoteric behavior as well. Thus, just as zinc hydroxide dissolves in either strong aqueous acid or strong base,

$$Zn(OH)_2(s) + 2H^+(aq) \ \ = Zn^{++}(aq) + 2H_2O,$$

$$Zn(OH)_2(s) + 2OH^-(aq) = Zn(OH)_4(aq),$$

zinc amide, $Zn(NH_2)_2$, reacts with excess NH_4^+ or NH_2^- to dissolve in liquid ammonia:

$$Zn(NH_2)_2(s) + 2NH_4^+(am) = Zn^{++}(am) + 4NH_3,$$

$$Zn(NH_2)_2(s) + 2NH_2^-(am) = Zn(NH_2)_2^-(am).$$

Perhaps the most remarkable characteristic of liquid ammonia is its ability to dissolve the alkali metals to form deep blue solutions of high electrical conductivity. The solubilities of the alkali metals range from 10 to 20 molal, depending on the temperature and the metal. In dilute (~ 0.01 M) solutions, the principal dissolved species are believed to be the alkali metal ions and independent electrons which are loosely trapped in solvent cages by electrostatic interaction with the dipole moment of NH_3. The blue color of these solutions is attributed to transitions of the quasi-free electrons between energy levels defined by their interaction with the solvent cage. As might be expected, these solutions of alkali metals in liquid ammonia are excellent reducing agents, and are frequently used for this purpose in preparative organic and inorganic chemistry.

The ammonia synthesis is the first step in the commercial "fixation" of nitrogen. Combustion of ammonia under catalytic conditions produces nitric oxide, NO, which is eventually converted to nitric acid, HNO_3. The combustion of ammonia may take either of two courses

$$4NH_3(g) + 3O_2(g) = 2N_2(g) + 6H_2O(g), \qquad K_{298} = 10^{228},$$

$$4NH_3(g) + 5O_2(g) = 4NO(g) + 6H_2O(g), \qquad K_{298} = 10^{168}.$$

Although the equilibrium constant of the first reaction is much larger than that of the second, the latter is selectively catalyzed by platinum metal, and NO is produced in quantity on the platinum surface at a temperature of about 1000°K. To complete the synthesis of nitric acid and the "fixation" of nitrogen, nitric oxide is treated with oxygen and water:

$$2NO + O_2 = NO_2,$$
$$3NO_2 + H_2O = 2HNO_3 + NO.$$

Another reaction by which elemental nitrogen can be converted to a combined form is the nitric oxide synthesis:

$$\tfrac{1}{2}N_2 + \tfrac{1}{2}O_2 = NO, \qquad \Delta H^0 = 21.60 \text{ kcal.}$$

Although ΔG^0_{298} for this reaction is 20.72 kcal and the equilibrium constant at 298°K is only 1.6×10^{-15}, the fact that ΔH^0 is positive means that at elevated temperatures the equilibrium constant will be more favorable for the synthesis. The experimental values of the equilibrium constant are, in fact, 2×10^{-2} at 2000°K, and 6×10^{-2} at 2500°K. One method of synthesizing nitric oxide is to pass nitrogen and oxygen through an electric arc discharge, which creates a high, if ill-defined, temperature.

In internal combustion engines operating on fuel-air mixtures, the combustion temperature is high enough so that small but significant amounts of nitric oxide are formed. When nitric oxide enters the atmosphere as engine exhaust, it is converted to nitrogen dioxide, NO_2, by reaction with oxygen. Photodissociation of NO_2 to NO and O initiates smog forming reactions when hydrocarbons are present in the atmosphere. Attempts to increase the performance of engines and lower the emission of unburned hydrocarbons by increasing compression ratios and combustion temperature lead to increased formation of NO. Therefore one presently favored method of decreasing both hydrocarbon and nitric oxide emission from engines is to decrease the cylinder combustion temperature to diminish NO formation, and eliminate unburned hydrocarbons from the exhaust by catalyzed combustion at low temperatures.

Elemental nitrogen does undergo direct reaction with some of the metallic elements. The reaction of nitrogen with lithium to give lithium nitride, Li_3N, proceeds slowly at room temperature, and rapidly at 250°C. The reaction of nitrogen with the alkaline-earth metals to form nitrides like Mg_3N_2 is rapid at temperatures above 500°C, and at an even higher temperature, nitrogen reacts with boron, aluminum, silicon, and many of the transition metals. Thus, while nitrogen is, in general, inert at temperatures near 25°C, it does react directly with a number of elements at somewhat elevated temperatures.

The Nitrides

Nitrogen forms an extensive series of nitrides which are usually classified as being ionic, covalent, or interstitial in nature. Lithium, the alkaline-earth elements, zinc, and cadmium form nitrides that apparently contain the N^{-3}

ion, for upon hydrolysis they yield ammonia:

$$Li_3N + 3H_2O = 3Li^+ + 3OH^- + NH_3,$$
$$Ca_3N_2 + 6H_2O = 3Ca^{++} + 6OH^- + 2NH_3.$$

Compounds of nitrogen with the elements of groups III, IV, and V are generally considered to be covalently bonded nitrides. These include BN, Si_3N_4, P_3N_5, and others. The compound BN is isoelectronic with carbon and exists in two forms that are analogous in structure to graphite and diamond. In the "graphite" form of BN there are planes which consist of alternate boron and nitrogen atoms 1.45 A apart in hexagonal rings. The distance between two neighboring planes of atoms is 3.34 A, which is large enough to suggest that there is only van der Waals bonding between the sheets of atoms. The other form of BN has the diamond structure, with alternate boron and nitrogen atoms in place of the carbon atoms. This form of BN is extremely hard— apparently harder than diamond itself.

Reaction between nitrogen and the finely divided transition metals produces interstitial nitrides such as W_2N, TiN, and Mo_2N. These compounds contain nitrogen atoms in the interstices of the metallic lattice. Like the interstitial carbides, interstitial nitrides are very hard, have high melting points, are electrical conductors, generally deviate from ideal stoichiometry, and are chemically inert.

The Oxides of Nitrogen

The known oxides of nitrogen and some of their properties are listed in Table 15.2. Every oxidation state of nitrogen from +1 to +5 is represented among the well-characterized oxides. In addition, there are two different oxides that have the empirical formula NO_3. Both of these are very reactive substances that have only been identified by spectroscopy as transient species.

Nitrous oxide, N_2O, can be synthesized by the thermal decomposition of ammonium nitrate,
$$NH_4NO_3 = N_2O + 2H_2O.$$

It is a colorless gas and is the least reactive and noxious of the nitrogen oxides. Nitrous oxide is relatively inert at room temperature, but at 500°C it decomposes to oxygen, nitrogen, and nitric oxide and will support the combustion of hydrogen and hydrocarbons. Nitrous oxide is isoelectronic with CO_2 and has the linear structure expected on this basis:

$$N\overline{}_{1.12\,A}N\overline{}_{1.19\,A}O.$$

Although nitrous oxide has a small dipole moment, 0.17 D, its physical properties are similar to those of the nonpolar carbon dioxide. Nitrous oxide boils at −88°C, while carbon dioxide sublimes at −78°C.

Table 15.2 Properties of the oxides of nitrogen

	N_2O	NO	N_2O_3	NO_2	N_2O_4	N_2O_5
Melting point, °C	−98.8	−163.6	−102	—	−9.3	30
Boiling point, °C	−88.5	−151.8	3.5	—	21.3	47
ΔH_f^0, kcal	19.5	21.6	20	8.09	2.3	3.1
ΔG_f^0, kcal	24.8	20.7	—	12.4	23.5	27.9

Nitric oxide, NO, is a colorless gas with a rather low condensation temperature of $-152°C$. The electronic structure of NO is interesting because the molecule has an odd number of electrons. Aside from the four 1s-electrons of nitrogen and oxygen, there are eleven valence electrons, one more than in N_2 and CO. The first ten of these electrons are accommodated in molecular orbitals that are qualitatively the same as those of N_2 and CO. A total of four electrons go in nonbonding orbitals: two on the nitrogen atom and two on the oxygen atom. Six electrons enter bonding orbitals: two in the σ-orbital, and two each in the π_x- and π_y-orbitals. The next orbitals in order of increasing energy are π_x^* and π_y^*: antibonding π-orbitals. The last electron in NO enters one of the π^*-orbitals, so nitric oxide has three pairs of bonding electrons and one antibonding electron. This situation is sometimes described as a $2\frac{1}{2}$ bond. This designation seems appropriate when we compare the bond lengths and bond energies of NO and NO^+.

$$: N :\!:\!: O : \qquad\qquad [: N :\!:\!:\!: O :]^+$$
$$1.15 \text{ A} \qquad\qquad\qquad 1.06 \text{ A}$$
$$150 \text{ kcal} \qquad\qquad\quad 244 \text{ kcal}$$

Thus NO^+, which does not have an electron in the π^* antibonding orbital, has a shorter stronger bond than does NO.

As the foregoing considerations might indicate, nitric oxide loses an electron rather easily and forms NO^+. For example, when a mixture of NO and NO_2 is dissolved in concentrated sulfuric acid, the reaction is

$$NO + NO_2 + 3H_2SO_4 = 2NO^+ + 3HSO_4^- + H_3O^+.$$

Compounds that contain NO^+ in ionic lattices with such species as HSO_4^-, ClO_4^-, and BF_4^- can be isolated.

Gaseous nitric oxide reacts directly with oxygen to form the brown gas nitrogen dioxide, NO_2:

$$2NO + O_2 = 2NO_2.$$

The rate of this reaction is proportional to the concentration of NO squared, and to the first power of the oxygen concentration. The mechanisms

$$NO + NO = (NO)_2 \quad \text{(fast equilibrium)}$$
$$(NO)_2 + O_2 \rightarrow 2NO_2 \quad \text{(slow)}$$

and

$$NO + O_2 = OONO \quad \text{(fast equilibrium)}$$
$$OONO + NO \rightarrow 2NO_2 \quad \text{(slow)}$$

are both consistent with the rate law. It is not now known which of these mechanisms is the more important.

Dinitrogen trioxide, N_2O_3, exists as a blue solid at low temperatures, but in the liquid and vapor state it is largely dissociated to NO and NO_2.

$$N_2O_3(g) = NO(g) + NO_2(g).$$

In effect, the chemistry of N_2O_3 under ordinary conditions is the chemistry of an equimolar mixture of NO and NO_2. Reaction of these mixtures with aqueous alkali yields solutions of the nitrite ion, NO_2^-,

$$NO(g) + NO_2(g) + 2OH^-(aq) = 2NO_2^-(aq) + H_2O.$$

The other chemical properties of N_2O_3 can be predicted from a knowledge of the chemistries of NO and NO_2.

Nitrogen dioxide and dinitrogen tetroxide, NO_2 and N_2O_4 respectively, exist as gases in equilibrium with each other:

$$\underset{\text{colorless}}{N_2O_4} = \underset{\text{red-brown}}{2NO_2}, \quad \Delta H^0 = 13.9 \text{ kcal.}$$

The reaction as written is endothermic, and consequently dissociation to NO_2 increases as temperature increases. When the mixture is condensed to a solid, the lattice is made up entirely of N_2O_4 units.

The structures of NO_2 and N_2O_4 are given in Fig. 15.1. The dimer N_2O_4 has a planar structure, and the nitrogen–nitrogen bond is remarkably long and weak $[D(N-N) = 13.9 \text{ kcal}]$ compared to single bonds between nitrogen atoms in other molecules. The bond angle in NO_2, 134°, is intermediate between the angles found in the related ions NO_2^+ (180°) and NO_2^- (116°). This is a specific example of the general observation that among triatomic molecules and ions, those with seventeen through twenty valence electrons (NO_2, NO_2^-, O_3, SO_2) are nonlinear, as has been discussed in Chapter 12.

FIG. 15.1   Structures of nitrogen dioxide and dinitrogen tetroxide.

When NO_2 dissolves in cold water, a mixture of nitrous and nitric acids is formed:

$$2NO_2 + H_2O = HNO_2 + H^+ + NO_3^-.$$

Thus this reaction is a disproportionation of nitrogen in the $+4$ oxidation state to the $+5$ and $+3$ states. Nitrous acid itself is unstable with respect to disproportionation in hot aqueous solutions, so when NO_2 dissolves in hot water, the net reaction can be written as

$$3NO_2 + H_2O = 2H^+ + 2NO_3^- + NO.$$

This reaction is used in the synthesis of nitric acid. Because NO_2 forms nitric acid upon contact with water and it also reacts directly with a number of metals, it is a very corrosive gas.

Dinitrogen pentoxide, N_2O_5, is made by the action of P_4O_{10} on nitric acid vapor:

$$4HNO_3(g) + P_4O_{10}(s) = 4HPO_3(s) + 2N_2O_5(g).$$

Thus N_2O_5 is the anhydride of nitric acid. Although the structure of N_2O_5 is not known in detail, the general arrangement of the atoms is given by

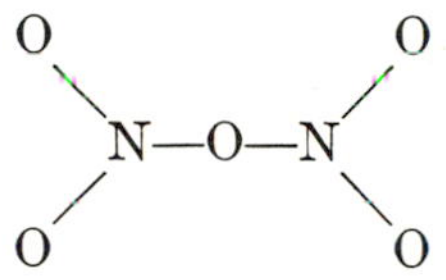

At room temperature N_2O_5 decomposes at a moderate rate according to the net reaction

$$N_2O_5(g) = 2NO_2(g) + \tfrac{1}{2}O_2(g), \qquad \Delta G^0 = -3.1 \text{ kcal.}$$

The mechanism of this decomposition involves the transient oxide of nitrogen, NO_3. In the first step of the decomposition N_2O_5 is in equilibrium with NO_2 and NO_3:

1. $\qquad N_2O_5 = NO_2 + NO_3 \qquad$ (rapid equilibrium).

Some collisions between NO_2 and NO_3 lead to formation of an oxygen molecule:

2. $\qquad NO_2 + NO_3 \rightarrow NO + O_2 + NO_2 \qquad$ (slow).

A more detailed representation of this step is

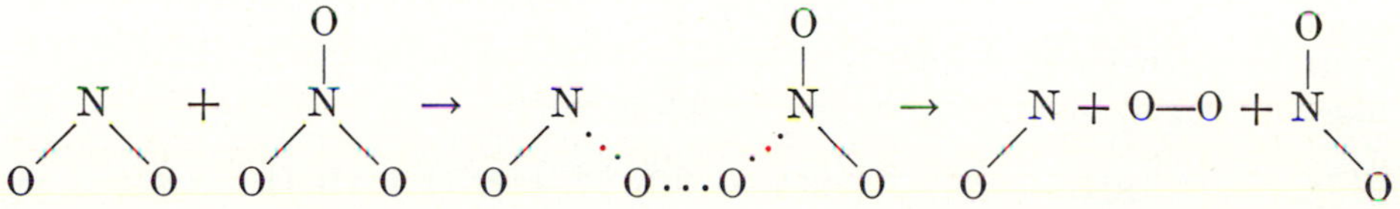

Thus the product NO_2 molecule contains a different nitrogen atom from the reactant NO_2 molecule. Finally, NO reacts very rapidly with NO_3:

$$3. \qquad NO + NO_3 \rightarrow 2NO_2.$$

Thus in the decomposition of pure N_2O_5, both NO_3 and NO are reactive intermediates, and are present only in very low concentration.

FIG. 15.2   Structure of the nitrate ion.

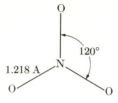

Although dinitrogen pentoxide exists as N_2O_5 molecules in the gas phase, x-ray studies show that the solid consists of the ions NO_2^+ and NO_3^-, and so it might be called nitronium nitrate. As was mentioned above, the nitronium ion, NO_2^+, is isoelectronic with CO_2 and N_2O, and has the linear structure $[O—N—O]^+$, with a nitrogen–oxygen bond distance of 1.15 A. The nitrate ion is isoelectronic with $CO_3^=$ and BF_3, and has the expected planar trigonal structure given in Fig. 15.2. As noted in Section 11.8, a single electron-dot or line structure for the nitrate ion cannot satisfy the octet rule and indicate that the three nitrogen–oxygen bonds are equivalent. Consequently, one representation of the nitrate ion is as a resonance hybrid of the structures given in Section 11.8.

FIG. 15.3   Structure of the nitric-acid molecule..

Oxyacids of Nitrogen

The most important of these is nitric acid, HNO_3. The gaseous molecule has the structure given in Fig. 15.3, which shows that the symmetry of the nitrate ion is partially destroyed when a proton is attached to it. In concentrated solutions, nitric acid is an extremely powerful oxidizing agent and reacts with metals like copper and silver with the production of NO:

$$8H^+ + 2NO_3^- + 3Cu = 3Cu^{++} + 2NO + 4H_2O.$$

Reaction of the active metals with concentrated nitric acid produces ammonium ion:

$$10H^+ + NO_3^- + 4Zn = 4Zn^{++} + NH_4^+ + 3H_2O.$$

In solutions more dilute than $2\ M$, the oxidizing power of the nitrate group is greatly diminished, and only the protons of the dissociated acid react with the active metals to evolve hydrogen. This behavior should not be too surprising, for the half-reaction

$$NO_3^- + 4H^+ + 3e^- = NO + 2H_2O, \qquad \varepsilon^0 = 0.96 \text{ volt},$$

shows that the power of NO_3^- as an oxidant should be very sensitive to the concentration of the acid, since four protons appear on the left-hand side of the equation.

We noted above that reaction of equimolar mixtures of NO and NO_2 with aqueous alkali gives solutions of nitrites. Solutions of nitrous acid can be made by acidification of solutions of nitrites. Nitrous acid is a weak acid, as the dissociation constant for

$$HNO_2(aq) = H^+(aq) + NO_2^-(aq), \qquad K = 4.5 \times 10^{-4},$$

shows. The pure liquid acid is unknown, and in the gas phase it is noticeably dissociated:

$$2HNO_2(g) = NO(g) + NO_2(g) + H_2O(g), \qquad K = 0.57 \text{ atm}.$$

As noted earlier, even aqueous solutions of nitrous acid are unstable and decompose when heated according to

$$3HNO_2(aq) = H^+(aq) + NO_3^-(aq) + H_2O(l) + 2NO(g).$$

Nitrogen appears in at least eight oxidation states in its water-soluble species, and there is a very large number of half-reactions that involve these oxidation states. In addition to the nitrate–nitric oxide half-reaction given earlier, some of the important half-reactions are the following:

$$NO_3^- + 3H^+ + 2e^- = HNO_2 + H_2O, \qquad \varepsilon^0 = 0.94 \text{ volt},$$
$$HNO_2 + H^+ + e^- = NO + H_2O, \qquad \varepsilon^0 = 1.00 \text{ volt},$$
$$2NO_3^- + 10H^+ + 8e^- = N_2O + 5H_2O, \qquad \varepsilon^0 = 1.11 \text{ volts},$$
$$N_2O + 2H^+ + 2e^- = N_2 + H_2O, \qquad \varepsilon^0 = 1.77 \text{ volts},$$
$$\tfrac{1}{2}N_2 + 4H^+ + 3e^- = NH_4^+, \qquad \varepsilon^0 = 0.27 \text{ volt}.$$

A convenient summary of this series of half-reactions is provided by the following reduction-potential diagram.

$$\overset{\displaystyle \overset{0.96}{\overline{\qquad\qquad\qquad}}}{NO_3^- \overset{0.94}{\text{——}} HNO_2 \overset{1.00}{\text{——}} NO \overset{1.59}{\text{——}} N_2O \overset{1.77}{\text{——}} N_2 \overset{0.27}{\text{——}} NH_4^+}$$
$$\underset{1.11}{\underline{\qquad\qquad\qquad\qquad}}$$

In this diagram only the nitrogen-containing species are indicated. A line represents the properly balanced half-reaction, written as a *reduction*, that interconverts the two species, and the voltage associated with the half-reaction is written over the line. Thus from this diagram it is possible to tell at a glance that nitrate ion is reduced to nitric oxide with a standard potential of $+0.96$ volt, that nitrogen gas going to ammonium ion is a spontaneous process with a standard voltage of $+0.27$ volt, and so on.

Reduction-potential diagrams make it easy to pick out species that are thermodynamically unstable with respect to disproportionation. For example, consider the following partial diagram.

$$NO_3^- \overset{0.94}{\text{——}} HNO_2 \overset{1.00}{\text{——}} NO.$$

It appears that HNO_2 is reduced to NO with a potential of $+1.00$ volt, but is *oxidized* to NO_3^- with a potential of -0.94 volt. Therefore HNO_2 should be converted to NO_3^- and NO with an associated voltage of $1.00 - 0.94 = 0.06$ volt. Writing out the relevant half-reactions in more detailed form confirms the following conclusion.

$$
\begin{array}{ll}
2 \times (HNO_2 + H^+ + e^- = NO + H_2O) & \varepsilon^0 = 1.00 \\
-1 \times (NO_3^- + 3H^+ + 2e^- = HNO_2 + H_2O) & -(\varepsilon^0 = 0.94) \\
\hline
3HNO_2 = NO_3^- + H^+ + 2NO + H_2O & \Delta\varepsilon^0 = 0.06
\end{array}
$$

In general, a species having an intermediate oxidation number will be unstable with respect to disproportionation if it appears in the reduction-potential diagram with a voltage to its right that is larger than the voltage to its left.

The reduction-potential diagrams above apply to reactions that occur in acidic solutions. It is instructive to examine the corresponding diagram for basic solutions:

$$\overset{\displaystyle \overset{0.15}{\overline{\qquad\qquad\qquad}}}{NO_3^- \overset{0.01}{\text{——}} NO_2^- \overset{-0.46}{\text{——}} NO \overset{0.76}{\text{——}} N_2O \overset{0.94}{\text{——}} N_2 \overset{-0.73}{\text{——}} NH_3}$$
$$\underset{0.10}{\underline{\qquad\qquad\qquad\qquad}}$$

The most remarkable feature of the composition of the standard potentials in acidic and basic solutions is that they show that nitrate and nitrite ions are much weaker oxidants in basic solution than they are in acidic solution. This is not too surprising, for as was noted above, the half-reactions for reduction of nitrate involve several hydrogen ions as *reactants,* as in

$$NO_3^- + 4H^+ + 3e^- = NO + 2H_2O.$$

A change from $1\,M\,H^+$ to $1\,M\,OH^-$ reduces the hydrogen-ion concentration enormously, and the oxidizing power of the nitrate ion decreases accordingly. The same generalization applies to other oxyanions: in acidic solution MnO_4^- and $Cr_2O_7^=$ are stronger oxidizing agents than they are in basic solution. An alternative way of stating this generalization is that it is easier to oxidize an element to an oxyanion of high oxidation number in basic solution than in acidic solution.

Another noteworthy feature of the reduction-potential diagram for basic solution is that it shows that nitrite ion is stable under these conditions:

$$NO_3^- \xrightarrow{\;0.01\;} NO_2^- \xrightarrow{\;-0.46\;} NO.$$

Writing out the disproportionation reaction in detail we find that

$$3NO_2^- + H_2O = NO_3^- + 2OH^- + 2NO, \qquad \Delta\varepsilon^0 = -0.47 \text{ volt,}$$

which shows that nitrites can exist without spontaneous decomposition in basic aqueous solution.

Nitrogen Halides and Oxyhalides

Nitrogen forms only four binary halides that have been obtained as pure compounds: NF_3, N_2F_2, N_2F_4, and NCl_3. Both NF_3 and N_2F_2 are obtained when NH_4F is dissolved in pure HF and electrolyzed. Dinitrogen tetrafluoride is one of the products recovered from an electrical discharge through NF_3 and mercury vapor, and is also obtained when NF_3 reacts with copper to produce CuF_2.

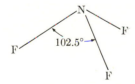

The pyramidal structure of nitrogen trifluoride.   **FIG. 15.4**

The nitrogen fluorides display some interesting structural and chemical properties. Nitrogen trifluoride is a stable, rather inert gas. It has a geometric structure very similar to that of ammonia, as is shown in Fig. 15.4. Despite the fact that NF_3 has a pair of nonbonded electrons, it is not a Lewis base, for there are no known compounds in which it donates electrons to other reagents.

Apparently this lack of basic properties is related to the electronegative nature of the fluorine atoms, which tend to draw electron density away from the nitrogen atom and make it a poor electron donor.

There are two known isomers of N_2F_2. One of them definitely has the *trans* planar FNNF structure, and the other isomer has the *cis* configuration:

trans   *cis*

The geometry of this molecule suggests that we can think of the nitrogen atoms as being sp^2-hybridized, and linked by a σ-π double bond.

Dinitrogen tetrafluoride is a gas that is partially dissociated into NF_2 fragments:

$$F_2NNF_2 = 2NF_2, \qquad \Delta H \cong 19 \text{ kcal.}$$

This situation is similar to that found for N_2O_4-NO_2, and in fact the N—N bond in N_2F_4 is nearly as weak as that in N_2O_4.

There are two series of nitrogen oxyhalides. The nitrosyl halides have the general formula XNO. Of these, the fluoride, chloride, and bromide are well known, and the iodide has been recognized as a transient species in reaction mixtures. The general formula of the nitryl halides is XNO_2, and only the fluoride and chloride are known.

None of the nitrogen oxyhalides is particularly stable, and these compounds are of interest principally because of their structural properties. The nitryl halides are isoelectronic in their valence shell with the nitrate ion, and like NO_3^-, are planar triangular molecules. In accordance with our experience with triatomic molecules which have more than 16 valence electrons, the nitrosyl halides are nonlinear and have the structural parameters given in Table 15.3. In each of these molecules the halogen is bonded to the nitrogen atom. In view of this fact, it is rather surprising that in the analogous compound NSF, the fluorine atom is attached to the sulfur atom.

Table 15.3 Properties of the nitrosyl halides

	FNO	ClNO	BrNO
Melting point, °C	−133	−65	−56
Boiling point, °C	60	− 6	0
ΔH_f^0, kcal	—	12.57	19.56
ΔG_f^0, kcal	—	15.86	19.7
X—N distance, A	1.52	1.95	2.14
N—O distance, A	1.13	1.14	1.15
X—N—O angle, deg	110	116	114

This element is the twelfth most abundant in nature, and is found most frequently in the form of phosphates like $Ca_3(PO_4)_2 \cdot H_2O$. Treatment of this phosphate with silica and coke at high temperature produces elemental phosphorus:

$$2Ca_3(PO_4)_2 + 6SiO_2 = 6CaSiO_3 + P_4O_{10}(g),$$

$$P_4O_{10}(g) + 10C = P_4(g) + 10CO.$$

The solid obtained by condensing the vapor is white phosphorus. This allotrope of phosphorus contains discrete P_4 molecules whose structure is shown in Fig. 15.5.

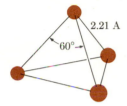

Structure of the P_4 molecule.   **FIG. 15.5**

Although white phosphorus is the allotropic form easiest to prepare, and is taken to be the standard thermodynamic state of the element, it is not the most stable allotrope. Upon heating or irradiation, white phosphorus is converted to red phosphorus, a polymeric substance whose structure is not known in detail.

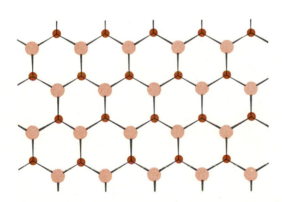

The layer structure of black phosphorus. For an alternative view, see Fig. 13.2.   **FIG. 15.6**

A third allotrope, black phosphorus, is produced by subjecting the element to high pressures or by careful recrystallization of white phosphorus. Black phosphorus is the most stable form of the element and has the structure shown in Fig. 15.6. The three allotropes of phosphorus differ in their chemical reactivity, with white phosphorus the most reactive and black the most inert.

Phosphorus forms phosphides by direct combination with some of the metals of groups I and II. Calcium phosphide also can be prepared by heating phosphorus with CaO:

$$6CaO + 2P_4 = 2Ca_3P_2 + P_4O_6.$$

When treated with water or dilute acids, Ca_3P_2 yields phosphine, PH_3:

$$Ca_3P_2 + 6H_2O = 2PH_3 + 3Ca(OH)_2.$$

This reaction is analogous to the hydrolysis of the saltlike nitrides, which yield ammonia. The compounds of phosphorus with the transition metals are grey solids with some metallic luster and conductivity, and in this respect resemble the transition-metal nitrides.

The compounds of phosphorous with the metals of group III of the periodic table (BP, AlP, GaP, InP) are interesting in that they are isoelectronic with silicon and germanium, and have the same type of crystal structure. Each atom is surrounded by four others which are located at the corners of a regular tetrahedron. All these compounds melt above 1000°C, and like silicon and germanium, are semiconductors.

Phosphorus combines with a variety of nonmetallic elements to yield covalent molecular compounds of diverse properties and structure. Of these, the oxides and halides are most important, and we will confine our discussion to them.

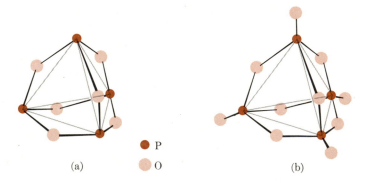

P

O

(a)   (b)

FIG. 15.7   The structures of (a) P_4O_6 and (b) P_4O_{10}.

White phosphorus reacts with atmospheric oxygen spontaneously. If the oxygen supply is somewhat limited, the principal reaction product is P_4O_6, phosphor*ous* oxide. If phosphorus burns in excess oxygen, the product is P_4O_{10}. The structures of the molecules of these two oxides are related to the P_4 structure, as Fig. 15.7 shows. For historical reasons P_4O_{10} is often written as P_2O_5 and referred to as phosphorus pentoxide. It has a large affinity for water and is often used as a drying or dehydrating agent.

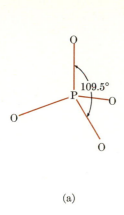

109.5°

(a)

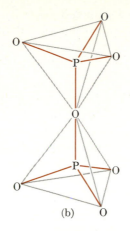

(b)

The structures of (a) PO_4^{-3}
and (b) $P_2O_7^{-4}$.

FIG. 15.8

There are several oxyacids and oxyanions of phosphorus. When a small amount of water is added to P_4O_{10}, metaphosphoric acid, $(HPO_3)_n$ is formed. This is a polymeric material in which there are anions of various chain and cyclic structures. In the trimetaphosphate ion, $(PO_3)_3^{-3}$, three PO_4 tetrahedra are linked in a cyclic structure, with each PO_4 group sharing one oxygen atom with each of its two neighboring PO_4 units. In tetrametaphosphate, $(PO_3)_4^{-4}$, there are four such tetrahedra linked in a ring. Further addition of water leads to pyrophosphoric acid, $H_4P_2O_7$, and to orthophosphoric acid, H_3PO_4. The structures of the anions of these acids are given in Fig. 15.8.

Table 15.4 Thermodynamic properties
of phosphoric acid

	ΔH_f^0(kcal/mole)	S^0(eu)
H_3PO_4(aq)	−308.2	42.1
$H_2PO_4^-$(aq)	−311.3	21.3
$HPO_4^=$(aq)	−310.4	−8.6
PO_4^{-3}(aq)	−306.9	−52
H^+(aq)	0	0

Orthophosphoric acid, or phosphoric acid as it is more commonly called, is a tribasic acid of moderate strength. A thermodynamic analysis of its successive ionization steps is instructive, for it helps reveal the factors that determine the strengths of acids in general. Table 15.4 contains the relevant data. For the first ionization we find that

$$H_3PO_4 = H^+ + H_2PO_4^-, \qquad K = 6 \times 10^{-3},$$
$$\Delta H^0 = -3.1 \text{ kcal}, \qquad \Delta S^0 = -20.8.$$

The enthalpy and entropy changes upon dissociation might seem surprising, for if the dissociation were a simple O—H bond-breaking process, we would expect ΔH^0 and ΔS^0 to be positive numbers. In aqueous solution, however, ionization of H_3PO_4 is a proton-transfer process that might better be written as

$$H_3PO_4(aq) + H_2O = H_2PO_4^-(aq) + H_3O^+.$$

Thus the reaction is a relocation of the proton, and the value of -3.1 kcal for ΔH^0 tells us that the system is in a state of lower energy when the proton has become associated with a water molecule and a pair of hydrated ions has formed.

We can also explain why the entropy change of the ionization reaction is negative. The ionization "creates" charges which tend to bind or localize water molecules. This localization of water molecules introduces order among the solvent molecules, and thereby lowers the entropy of the system.

The second ionization of phosphoric acid,

$$H_2PO_4^- = HPO_4^= + H^+, \qquad K = 7 \times 10^{-7},$$

is slightly endothermic, for $\Delta H^0 = 0.9$ kcal, but ΔS^0 is again negative, and equal to -29.9 eu. Similarly, for the third ionization, ΔH^0 is still more positive, and ΔS^0 is negative. The trend in the values of ΔH^0 for the ionization steps is not surprising, for we would expect that the energy required to remove a proton would increase as the negative charge on the parent acid increases. The value of ΔH^0 for the first ionization does show, however, that we must not think that the dissociation of an acid is *necessarily* an endothermic process. Furthermore, the values of ΔS^0 for all the ionization steps show that entropy changes can exert profound influences on acid strengths.

The simplest oxyacid of phosphorus (III) is H_3PO_3, orthophosphor*ous* acid. It can be prepared by hydrolysis of PCl_3,

$$PCl_3 + 3H_2O = H_3PO_3 + 3H^+ + 3Cl^-.$$

Despite its empirical formula, phosphorous acid is a dibasic acid, for the third hydrogen does not react with bases. The ionization reactions are

$$H_3PO_3 = H^+ + H_2PO_3^-, \qquad K_1 = 1.6 \times 10^{-2},$$
$$H_2PO_3^- = H^+ + HPO_3^=, \qquad K_2 = 7 \times 10^{-7}.$$

Similar behavior is found for *hypo*phosphorous acid, H_3PO_2, which is only a monobasic acid:

$$H_3PO_2 = H^+ + H_2PO_2^-, \qquad K = 10^{-2}.$$

The explanation of this behavior lies in the structures of $HPO_3^=$ and $H_2PO_2^-$, shown below.

hypophosphite phosphite phosphate

The hydrogen atoms attached directly to phosphorus are not acidic, whereas those bonded to oxygen are. As we remarked in Chapter 13, the oxyacids of a given element tend to become stronger as the oxidation state of the element increases. This trend is not observed in the H_3PO_2, H_3PO_3, H_3PO_4 series, and the peculiar structure of the hypophosphite and phosphite anions is the reason for this unique behavior.

Because phosphorus displays a number of oxidation states, and because its oxyanions have a tendency to polymerize, the solution chemistry of this element is moderately complex. We can get a general idea of the properties of the various oxidation states of phosphorus, however, by consulting the following reduction-potential diagrams:

$$H_3PO_4 \overset{-0.28}{\rule{1cm}{0.4pt}} H_3PO_3 \overset{-0.50}{\rule{1cm}{0.4pt}} H_3PO_2 \overset{-0.51}{\rule{1cm}{0.4pt}} P_4 \overset{-0.06}{\rule{1cm}{0.4pt}} PH_3 \quad \text{(acidic solution)},$$

with a -0.50 bridge over H_3PO_3 to H_3PO_2

$$PO_4^{\equiv} \overset{-1.12}{\rule{1cm}{0.4pt}} HPO_3^{=} \overset{-1.57}{\rule{1cm}{0.4pt}} H_2PO_2^{-} \overset{-2.05}{\rule{1cm}{0.4pt}} P_4 \overset{-0.89}{\rule{1cm}{0.4pt}} PH_3 \quad \text{(basic solution)}.$$

We can see that the oxyanions of phosphorus are very poor oxidizing agents, particularly in basic solutions. On the contrary, all but the highest oxidation state of phosphorus have strong reducing properties. In addition, elemental phosphorus is unstable with respect to disproportionation, for from the potential diagram for basic solutions we can deduce the following:

$$P_4 + 12H_2O + 12e^- = 4PH_3 + 12OH^-, \quad \varepsilon^0 = -0.89,$$
$$4H_2PO_2^- + 4e^- = P_4 + 8OH^-, \quad \varepsilon^0 = -2.05.$$

Combination of these half-reactions gives

$$P_4 + 3OH^- + 3H_2O = PH_3 + 3H_2PO_2^-, \quad \Delta\varepsilon^0 = 1.16.$$

Phosphorus Halides and Oxyhalides

The halogens react with white phosphorus to form two types of halides, PX_3 and PX_5. Iodine is an exception, since its compounds with phosphorus are PI_3 and P_2I_4. Table 15.5 lists the phosphorus halides and some of their properties. The trihalides all have pyramidal geometry similar to ammonia and PH_3, and

Table 15.5 The phosphorus halides

	PF_3	PCl_3	PBr_3	PI_3	PF_5	PCl_5	PBr_5	P_2I_4
Boiling point, °C	−95	76	173	61	−85	163(sub)	decomp	decomp
ΔH_f^0, kcal	—	−81.0	−47.5	−10.9	—	−110.7	−66.0	−19.8
ΔG_f^0, kcal	—	−68.6	—	− 8.9				
X—P—X angle, deg	104	100	101	102				

despite the differences in size of the halogen atoms, the X—P—X bond angle in all the trihalides is near 102°, as Table 15.5 shows.

Of the trihalides, PF_3 is the most stable and inert chemically. While PF_3 reacts with water only very slowly, PCl_3 is rapidly hydrolyzed to H_3PO_3:

$$PCl_3 + 3H_2O = H_3PO_3 + 3H^+ + 3Cl^-.$$

Other important reactions of PCl_3 are

$$PCl_3 + Cl_2 = PCl_5,$$
$$PCl_3 + \tfrac{1}{2}O_2 = POCl_3,$$
$$PCl_3 + 3NH_3 = P(NH_2)_3 + 3HCl.$$

Some analogous reactions may be written for the other PX_3 compounds.

In the vapor phase, the pentahalides PF_5, PCl_5, and PBr_5 are discrete molecules that have the trigonal bipyramid structure discussed in Section 11.5. In the solid phase, however, PCl_5 exists as an ionic solid consisting of PCl_4^+ and PCl_6^-, which have the structures shown in Fig. 15.9. Solid PBr_5 is made up of PBr_4^+ and Br^- ions. Apparently PBr_6^- does not form because of the difficulty of packing six large bromine atoms close to a central phosphorus atom.

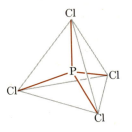 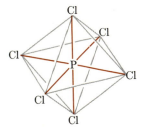

FIG. 15.9 The structures of (a) PCl_4^+ and (b) PCl_6^-.

The most important oxyhalides of phosphorus are of the type POX_3, where X may be F, Cl, or Br. As noted above, phosphoryl chloride, $POCl_3$, can be made by the reaction of oxygen and PCl_3. Phosphoryl fluoride, POF_3, is made by fluorinating $POCl_3$:

$$POCl_3 + AsF_3 = AsCl_3 + POF_3.$$

The phosphoryl halides are electron donors and form complex addition compounds with metallic halides such as Al_2Cl_6, $ZrCl_4$, $HfCl_4$, and $TiCl_4$. The geometry of the phosphoryl halides is of interest, for these compounds are isoelectronic in their valence shells with the phosphate ion, PO_4^{-3}. As expected on this basis, the phosphoryl halides have a slightly distorted tetrahedral structure, with $X\!-\!P\!-\!X$ bond angles of approximately $103°$.

Arsenic, Antimony, and Bismuth

These elements are not at all abundant and occur in nature principally as oxides and sulfides. They can be recovered by reduction of their oxides with carbon. Rapid condensation of arsenic and antimony vapors gives the yellow nonmetallic allotropes that consist of As_4 and Sb_4 tetrahedral molecules, analogous to P_4. The more stable semimetallic allotropes of arsenic and antimony exhibit metallic luster, are moderately good conductors of heat and electricity, and have crystal structures similar to that of black phosphorus.

The elements combine with oxygen and the halogens directly:

$$4As + 3O_2 = As_4O_6,$$
$$4Sb + 3O_2 = Sb_4O_6,$$
$$2Bi + \tfrac{3}{2}O_2 = Bi_2O_3,$$
$$M + 3X_2 = 2MX_3.$$

The $+3$ oxides of arsenic and antimony are discrete molecules which have structures based on the As_4 and Sb_4 tetrahedra, but in Bi_2O_3, no such discrete molecules exist. The oxides As_4O_6 and Sb_4O_6 are amphoteric, but Bi_2O_3 is basic and rather insoluble in water.

The principal oxidation states and some of the chemistry of arsenic are represented in the following reduction-potential diagrams:

$$H_3AsO_4 \xrightarrow{\;0.559\;} HAsO_2 \xrightarrow{\;0.25\;} As \xrightarrow{\;-0.60\;} AsH_3 \quad \text{(acidic solution)},$$

$$AsO_4^{\equiv} \xrightarrow{\;-0.67\;} AsO_2^- \xrightarrow{\;-0.68\;} As \xrightarrow{\;-1.43\;} AsH_3 \quad \text{(basic solution)}.$$

Once again we see that it is easier to oxidize the element to its higher oxidation states in basic solution than in acidic solution. Unlike phosphoric acid, arsenic acid, H_3AsO_4, is a moderately strong oxidizing agent in acidic solution. In basic solution, however, arsenic (V) loses its oxidizing power, and arsenic (III) becomes a good reducing agent. Basic solutions of arsenite, AsO_2^-, are often used to standardize the concentration of oxidizing solutions in quantitative analysis.

The potential diagram for antimony in acidic solution is

$$Sb_2O_5 \xrightarrow{\;0.58\;} SbO^+ \xrightarrow{\;0.21\;} Sb \xrightarrow{\;-0.51\;} SbH_3.$$

The oxide of antimony (V) is virtually insoluble in acidic solutions, but is a moderately strong oxidizing agent. In acid solution, antimony (III) exists as SbO^+ or $Sb(OH)_2^+$, which can be regarded as hydrolyzed forms of Sb^{+3}.

Bismuth appears in aqueous solution only in the +3 oxidation state. The oxide Bi_2O_3 dissolves in acids to give solutions of BiO^+ or $Bi(OH)_2^+$, which are hydrolyzed forms of Bi^{+3}. The +5 oxidation state of bismuth is obtained only by treating Bi_2O_3 with very powerful oxidizing agents like Cl_2 and Na_2O_2 in the presence of NaOH. The brown solid that results from this treatment is insoluble in water and is a very powerful oxidizing agent. It is not clear whether this solid is a true sodium bismuthate $NaBiO_3$, or a mixture of Na_2O and Bi_2O_5.

15.2 THE ELEMENTS OF GROUP VIA

Among the representative elements, it is clear that the lightest member of any periodic group has chemical properties that differ rather noticeably from those of the heavier members of the group. This behavior is particularly clear in group VIA. Oxygen, the most abundant and important member of the group, exists ordinarily as a diatomic gas, and in its chemistry displays negative oxidation states almost exclusively. Sulfur, selenium, tellurium, and polonium exist as solids with structures that are rather complex and form compounds in which they appear in a range of positive as well as negative oxidation states. Thus in most of its chemical and physical properties, oxygen is quite different from the other members of group VI, as Table 15.6 shows.

Table 15.6 Properties of the group VIA elements

	O	S	Se	Te	Po
Atomic number	8	16	34	52	84
Configuration	$2s^2 2p^4$	$3s^2 3p^4$	$4s^2 4p^4$	$5s^2 5p^4$	$6s^2 6p^4$
Ionization energy, kcal	314	239	225	208	194
Atomic radius, A	0.74	1.04	1.17	1.37	1.64
Melting point, °K	54	392	490	723	527
Boiling point, °K	90	718	960	1260	1235
ΔH_f(atom), kcal	59.1	66	49.4	46	34.5
Ionic radius, $M^=$, A	1.45	1.90	2.02	2.22	—

Even among the remaining group VI elements, there is a substantial gradation of properties. Sulfur is a nonmetal, both on the basis of its electrical properties, and the nature of its compounds. Selenium and tellurium are grey solids with some metallic luster, and because they have small electrical conductivities that increase as temperature increases, they are classified as semimetals. Polonium, a rare radioactive element, has the electrical conductivity of a metal. Thus in group VIA the vertical transition from nonmetallic to metallic behavior occurs just as it did in groups IV and V.

This element, the most abundant in nature, forms compounds with all elements except some of the inert gases. The nature of the oxides in which oxygen has an oxidation state of -2 has been discussed in Section 13.3 and throughout our treatment of descriptive chemistry. Here we need only examine the properties of the element itself and those compounds in which oxygen is found in the "abnormal" oxidation states of -1 and $-\frac{1}{2}$, respectively.

The stable allotrope of oxygen is O_2, a diatomic molecule of substantial dissociation energy (118 kcal/mole). According to magnetic measurements, each O_2 molecule has two electrons with unpaired spins. There is no simple electron-dot structure consistent with the octet rule that can account for both the high dissociation energy and the unpaired electron spins. A simple application of the molecular orbitals discussed in Section 12.1 provides a rationalization of the properties of oxygen and its diatomic molecule ions O_2^- and $O_2^=$. The O_2 molecule has, apart from the four 1s-electrons, a total of twelve valence electrons. Ten of these can be accommodated in molecular orbitals similar to those used in the nitrogen molecule: a pair of nonbonding orbitals, one on each atom, a σ bonding orbital, and two π bonding orbitals. The remaining two electrons must enter the π_x^* and π_y^* antibonding orbitals which are the next available in order of increasing energy. The energy of the oxygen molecule is lowest if the two electrons enter separate orbitals with their spins parallel, for this configuration keeps the electrons as far apart as possible and minimizes repulsion between them. Because there are three pairs of bonding electrons, and two antibonding electrons, the net number of bonding electrons in O_2 is four, and the molecule might be pictured as having a net double bond. Thus the molecular-orbital picture is consistent with the high dissociation energy and unpaired electron spins in the O_2 molecule.

Table 15.7 Bond lengths and energies

	Bond length (A)	Bond energy (kcal)
$O_2^=$ in BaO_2	1.49	—
O_2^- in KO_2	1.28	—
O_2	1.21	118
O_2^+	1.12	150

Direct combination of the heavier alkali metals with oxygen yields super-oxides of the formula MO_2. These compounds contain the superoxide ion, O_2^-, which has one more electron than O_2. This "extra" electron can be accommodated in one of the π^* antibonding orbitals, and thus the bond in O_2^- should be weaker than that in O_2. We would expect a still weaker bond in the peroxide ion, $O_2^=$, which has a total of four antibonding electrons, two each in the π_x^*- and π_y^*-orbitals. The bond energies of these ions are not known, but their bond lengths have been measured, and are given in Table 15.7. It is a fairly general and reliable principle that short bonds are strong bonds, and we see that among the diatomic oxygen species in Table 15.7, the bond length increases, and presumably the bond energy decreases, as the number of antibonding electrons increases. Thus molecular-orbital theory is consistent with the facts that are known about molecular oxygen and its diatomic ions.

The second allotrope of oxygen is ozone, O_3. Ozone is prepared by passing molecular oxygen through an electric discharge, condensing the product at 77°K, and purifying the ozone by fractional distillation and liquefaction. Ozone is dangerous, for in certain concentration ranges it is violently explosive.

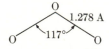

FIG. 15.10   The structure of ozone, O_3.

The structure of the ozone molecule is given in Fig. 15.10. The bond distance is almost exactly the same as that in the superoxide ion, O_2^-, and this suggests that the bonds in ozone are intermediate between single and double bonds or approximately $1\frac{1}{2}$ bonds. The resonance structures of the ozone molecule are consistent with this point of view.

Ozone is an enormously powerful oxidizing agent in aqueous solution:

$$O_3 + 2H^+ + 2e^- = O_2 + H_2O, \qquad \varepsilon^0 = 2.07 \text{ volts.}$$

In the gas phase it reacts rapidly and completely with a number of reagents:

$$NO + O_3 = NO_2 + O_2,$$
$$2ClO_2 + 2O_3 = Cl_2O_6 + 2O_2.$$

Treatment of a peroxide like BaO_2 with dilute acids yields solutions of hydrogen peroxide, H_2O_2. The commercial preparation of 30% aqueous hydrogen peroxide solution is obtained by repeated fractional distillation. Further fractionations yield even more concentrated solutions, but these are susceptible to decomposition, and must be stored and handled carefully.

Pure H_2O_2 is a viscous liquid that boils at 150°C and freezes at −0.89°C. Thus it resembles water in its physical properties, and like water it is hydrogen-bonded. It has virtually no uses as a solvent, because it is not only a powerful oxidizing agent, it is also unstable with respect to the decomposition

$$2H_2O_2 = 2H_2O + O_2.$$

Because it contains oxygen in the intermediate oxidation state of -1, hydrogen peroxide can act either as an oxidant or a reductant. The reduction-potential diagram

$$O_2 \xrightarrow{\ 0.68\ } H_2O_2 \xrightarrow{\ 1.77\ } H_2O$$

shows, however, that in acidic solutions, H_2O_2 is a much better oxidant than reductant.

Atmospheric oxygen is consumed in the oxidative metabolism of carbohydrates in animal organisms and is thereby converted to water and carbon dioxide. The photosynthetic process of plants, in which light energy from the sun is used to convert carbon dioxide and water to carbohydrate material and oxygen, restores oxygen to the atmosphere. However, in the atmosphere itself, oxygen undergoes a number of very important photochemical reactions. At altitudes of 100–150 km (1 km = 0.621 mile), oxygen is photodissociated by the very energetic short wavelength radiation ($\lambda < 2000$ Å) from the sun:

$$O_2 + h\nu \rightarrow 2O.$$

Although the pressure of oxygen is less than 10^{-6} atm at these altitudes, oxygen absorbs the short wavelength radiation so strongly that little of it penetrates below 100 km. Thus atmospheric oxygen shields the surface of the Earth from what could be very damaging radiation.

If oxygen atoms are to recombine to O_2, a third molecule M must be present to remove the energy released by chemical bond formation:

$$O + O + M \rightarrow O_2 + M.$$

This reaction is exceedingly slow at the 100 km altitude because the concentration of third molecules, O_2 or N_2, is so low. Consequently, the oxygen atoms formed by photodissociation diffuse to lower altitudes. At approximately 50 km, the density of molecular oxygen is great enough so that the reaction

$$O + O_2 + M \rightarrow O_3 + M$$

goes with appreciable speed, and starts to consume oxygen atoms. The concentration of ozone produced by this reaction reaches a maximum value of roughly 10^{12} molecules per cc at an altitude of 20 to 30 km. At lower altitudes, the ozone concentration falls, since the oxygen atoms necessary for its formation have been largely consumed at higher altitudes. At altitudes above 50 km the ozone concentration is low because the concentrations of O_2 and third body molecules M are low.

The ozone layer at 20 to 30 km performs a very important function. Molecular oxygen is transparent to light in the 2000- to 3500-Å wavelength range. The photons in this spectral region are fairly energetic, and could cause very substantial damage to plants and delicate animal tissue should they reach the surface of the Earth. Fortunately, ozone absorbs light in this spectral region, and shields the Earth from this potentially damaging radiation. Ozone is photochemically destroyed in the following manner by the reactions:

$$O_3 + h\nu \rightarrow O_2 + O,$$

$$O + O_3 \rightarrow 2O_2.$$

However, more ozone is formed by diffusion of oxygen atoms from greater altitudes, and the balance between formation and photochemical destruction maintains a stable ozone layer.

The ozone shield is susceptible to alteration by atmospheric contaminants. The reactions

$$O_3 + NO \rightarrow NO_2 + O_2,$$

$$O + NO_2 \rightarrow NO + O_2$$

are very fast, and together consume ozone and its precursor, the oxygen atom, faster than the reaction of O with O_2. If we add to these reactions the process

$$O_3 + h\nu \rightarrow O_2 + O,$$

we get a net reaction of

$$2O_3 + h\nu \rightarrow 3O_2,$$

with no net consumption of NO or NO_2. Thus these nitrogen oxides can act as catalysts for the destruction of the atmospheric ozone layer. This observation is of particular significance and concern in view of the proposals to fly supersonic transport aircraft in the upper levels of the stratosphere. The NO formed by the combustion of fuel-air mixtures leaves the stratosphere very slowly, and consequently could significantly decrease the ozone concentration through its catalytic action.

Sulfur

Sulfur occurs in nature in the elemental state and as a variety of metal sulfides. The element has several allotropic forms, and the structural properties of some of these are very complex and not well understood. In rhombic and monoclinic sulfur, its two most common forms, sulfur exists as S_8 molecules that have the puckered ring structure shown in Fig. 15.11. If sulfur is dissolved in CS_2 or organic solvents, freezing-point depression measurements show that the molecular weight of the dissolved sulfur corresponds to S_8. In another crystalline modification, sulfur exists as S_6 rings. In liquid sulfur at temperatures of about 200°C, the ring molecules open up, and long chain molecules are formed. If liquid sulfur at this temperature is poured into water, a solid plastic sulfur which contains helical chains of atoms results. Plastic sulfur is metastable and slowly reverts to the rhombic crystalline form. Sulfur vapor consists of S_8, S_4, and S_2 molecules in relative amounts that depend on temperature. Thus in its solid, liquid, and vapor phases, sulfur displays a variety of molecular structures.

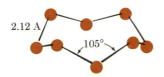

2.12 A

105°

The ring structure of S_8.   **FIG. 15.11**

Sulfur combines directly with the metallic elements to form sulfides. The sulfides of the alkali metals can be classified as ionic compounds which contain M^+ and $S^=$ ions in an antifluorite lattice. The alkaline-earth sulfides also are best pictured as ionic compounds and have, like the corresponding oxides, the rock-salt lattice. These sulfides of groups IA and IIA metals are water soluble, and the sulfide ions are extensively hydrolyzed:

$$S^=(aq) + H_2O = SH^-(aq) + OH^-(aq), \qquad K = 1.$$

Acidification of solutions of soluble sulfides leads to evolution of hydrogen sulfide, H_2S. This foul-smelling gas is very poisonous. At 25°C, a saturated solution of hydrogen sulfide has a concentration of approximately 0.1 M, and because H_2S is a weak acid, such solutions contain a small concentration of sulfide ion:

$$H_2S(aq) = H^+ + HS^-, \qquad K = 1.1 \times 10^{-7},$$
$$HS^- = H^+ + S^=, \qquad K = 10^{-14}.$$

Consequently, saturating a solution with hydrogen sulfide is an effective way to precipitate many of the very insoluble transition-metal sulfides.

The sulfides of the transition metals are not usually pictured as simple ionic compounds. The doubly and triply charged transition-metal ions are relatively small and exert large polarizing forces that tend to distort the large sulfide ions. The measured lattice energies of the transition-metal sulfides are, in general, larger than would be predicted by considering them to be ionic lattices. Consequently, in these sulfides there is a certain amount of covalent bonding between the sulfur and metal atoms. The insolubility of the transition-metal sulfides is related to the very stable lattices of these compounds.

In addition to the simple sulfides that contain the $S^=$ ion, there are poly-sulfides which contain $S_n^=$ ions, where n ranges from 2 to 6. The existence of such anions is associated with a general characteristic of sulfur chemistry: in many compounds there are chains of sulfur atoms bonded to each other. Sulfur exhibits this tendency toward *catenation*, or formation of chains of identical atoms, more than any other element except carbon. When solutions of the polysulfides are acidified, *sulfanes* of the general formula H_2S_n are formed, where n ranges from 2 to 6.

FIG. 15.12   Structure of the SO_2 molecule.

The two most important oxides of sulfur are SO_2 and SO_3. Sulfur dioxide gas (boiling point, $-10°C$) is formed by burning sulfur in air:

$$S(s) + O_2(g) = SO_2(g),$$
$$\Delta H^0 = -70.66 \text{ kcal},$$
$$\Delta G^0 = -71.99 \text{ kcal}.$$

It is clear from the thermodynamic data that sulfur dioxide is a very stable molecule. Nevertheless, the conversion of SO_2 to SO_3 is favored thermodynamically:

$$SO_2(g) + \tfrac{1}{2}O_2 = SO_3(g),$$
$$\Delta H^0 = -23.49 \text{ kcal},$$
$$\Delta G^0 = -16.73 \text{ kcal}.$$

The oxidation of sulfur dioxide is a slow reaction, but is catalyzed by vanadium pentoxide or platinum surfaces. Nearly all sulfuric acid production involves oxidizing sulfur dioxide by air in the presence of these "contact" catalysts.

Sulfur dioxide is a triatomic molecule with more than sixteen valence electrons, and consequently is nonlinear, as Fig. 15.12 shows. The two equivalent sulfur–oxygen bonds can be represented by the following resonance structures.

In the gas phase, sulfur trioxide is a planar triangular molecule with three equivalent sulfur–oxygen bonds, as Fig. 15.13 illustrates. This geometry is to be expected, for SO_3 is isoelectronic in its valence shell with BF_3, NO_3^-, and

$CO_3^=$, which are all planar symmetrical species. The bonding in SO_3 can be represented by the structures shown below.

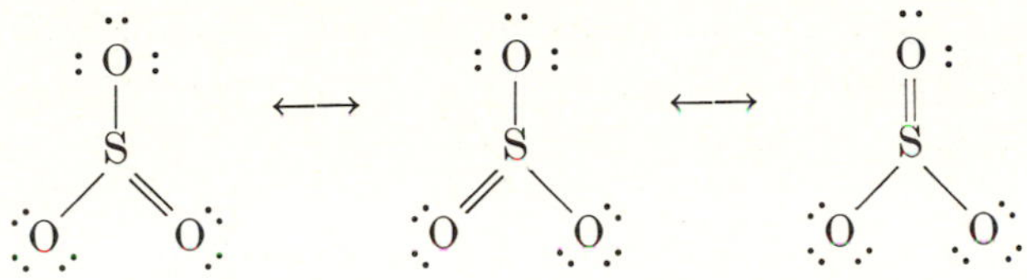

Sulfur forms a very large number of oxyacids and oxyanions. Most important of these is sulfuric acid, H_2SO_4, which is produced by hydration of SO_3 in the following two-step process:

$$SO_3(g) + H_2SO_4(l) = H_2S_2O_7(l),$$
$$H_2S_2O_7(l) + H_2O = 2H_2SO_4(l).$$

The direct reaction of SO_3 with water produces a fog which is difficult to condense, and consequently the commercial process involves dissolving SO_3 in sulfuric acid to form pyrosulfuric acid, $H_2S_2O_7$, and subsequent dilution with water to form sulfuric acid.

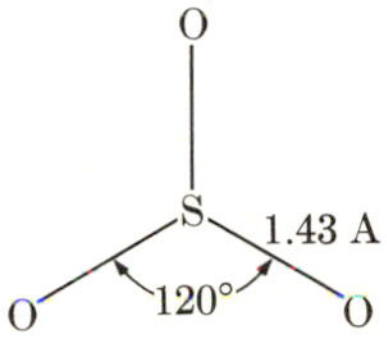

Structure of a gaseous SO_3 molecule. **FIG. 15.13**

Pure sulfuric acid is a viscous liquid that freezes at 10°C. It is a conductor of electricity because it is slightly dissociated according to

$$2H_2SO_4 = H_3SO_4^+ + HSO_4^-.$$

The acid has a great affinity for water and forms several stable hydrates. In some of its chemical reactions as well, sulfuric acid removes the elements of water from compounds:

$$HCOOH + H_2SO_4 = CO(g) + H_3O^+ + HSO_4^-,$$
$$HNO_3 + 2H_2SO_4 = H_3O^+ + 2HSO_4^- + NO_2^+.$$

Hot concentrated sulfuric acid is an oxidizing agent, and will dissolve metals like copper,

$$Cu + 5H_2SO_4 = Cu^{++} + SO_2 + 4HSO_4^- + 2H_3O^+.$$

In dilute solutions, however, the oxidizing properties associated with the sulfate group are virtually negligible.

Electrolysis of cold concentrated sulfuric acid solutions produces peroxydisulfuric acid, $H_2S_2O_8$. As its name suggests, this molecule contains an oxygen–oxygen bond,

$$H-O-\underset{\underset{O}{|}}{\overset{\overset{O}{|}}{S}}-O-O-\underset{\underset{O}{|}}{\overset{\overset{O}{|}}{S}}-O-H.$$

Peroxydisulfuric acid is an extremely powerful oxidizing agent:

$$H_2S_2O_8 + 2H^+ + 2e^- = 2H_2SO_4, \qquad \varepsilon^0 = 2.01 \text{ volts.}$$

Although direct oxidations by peroxydisulfate ions are slow, they are catalyzed by silver ions and this combination of reagents provides one of the most effective means available for converting soluble species to their highest oxidation states.

The addition of SO_2 to water produces a solution whose mild acidity is often attributed to ionization of sulfurous acid, H_2SO_3. However, sulfurous acid has never been isolated as a pure compound, and there is no evidence that the molecule H_2SO_3 exists at all. The ionization of "sulfurous acid" might better be written

$$SO_2(aq) + H_2O = HSO_3^- + H^+, \qquad K = 1.3 \times 10^{-2},$$
$$HSO_3^- = H^+ + SO_3^=, \qquad K = 5.6 \times 10^{-8}.$$

There is no doubt that both the bisulfite ion, HSO_3^-, and the sulfite ion, $SO_3^=$, exist, for salts of both are well known. Acidic solutions of SO_2 are mild reducing agents, and basic sulfite solutions are somewhat stronger reductants, for we have

$$SO_4^= + 4H^+ + 2e^- = SO_2(aq) + 2H_2O, \qquad \varepsilon^0 = \quad 0.17 \text{ volt,}$$
$$SO_4^= + H_2O + 2e^- = SO_3^= + 2OH^-, \qquad \varepsilon^0 = -0.93 \text{ volt.}$$

Solutions of the sulfite ion react with elemental sulfur to form the thiosulfate ion, $S_2O_3^=$, according to

$$S(s) + SO_3^= = S_2O_3^=.$$

The ion $S_2O_3^=$ is called the thiosulfate ion because the prefix *thio* indicates that a sulfur atom has been substituted for an oxygen atom in the parent species. The relation between the structures of the sulfate and the thiosulfate ions is shown in Fig. 15.14.

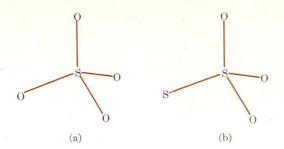

(a) (b)

The relation between the structures of (a) SO_4^- and (b) $S_2O_3^-$.

FIG. 15.14

In acidic solutions, the thiosulfate ion decomposes to sulfur and sulfite ion. Consequently, no such species as thiosulfuric acid can be isolated. In mildly acidic, neutral, or basic solutions, however, the thiosulfate ion is stable, and undergoes two important reactions. It acts as a mild reducing agent and forms the tetrathionate ion, $S_4O_6^=$:

$$2S_2O_3^= = S_4O_6^= + 2e^-, \qquad \varepsilon^0 = -0.08 \text{ volt.}$$

Thiosulfate ion will reduce iodine to iodide ion, and this reaction is used extensively in quantitative analysis. The general procedure is to allow an oxidizing agent whose concentration is to be determined to react with excess iodide ion to produce I_2. The iodine is then titrated with a solution of $S_2O_3^=$ of known concentration, and the amount of unknown oxidant calculated.

The thiosulfate ion also forms stable complexes with some metal ions. In particular, the silver–thiosulfate complex ion is very stable, as is shown by

$$Ag^+ + 2S_2O_3^= = [Ag(S_2O_3)_2]^{-3}, \qquad K = 1.6 \times 10^{13}.$$

Solutions of thiosulfate ion can dissolve the otherwise insoluble silver halides, and are used as fixing agents in the photographic process.

The following reduction-potential diagrams summarize the properties of some of the aqueous sulfur species. For acidic solutions we have

$$SO_4^= \underset{}{\overset{0.17}{\rule{1cm}{0.4pt}}} SO_2 \underset{}{\overset{0.40}{\rule{1cm}{0.4pt}}} S_2O_3^= \underset{}{\overset{0.50}{\rule{1cm}{0.4pt}}} S \underset{}{\overset{0.14}{\rule{1cm}{0.4pt}}} H_2S.$$

$$\overset{0.51}{\rule{1cm}{0.4pt}} S_4O_6^= \overset{0.08}{\rule{1cm}{0.4pt}}$$

It is clear that the sulfate ion is a poor oxidizing agent in $1\,M$ H^+. Aqueous sulfur dioxide is a moderately good oxidizing agent, but can be oxidized rather easily to sulfate ion. Thiosulfate ion is easily oxidized to tetrathionate, $S_4O_6^=$, but stronger oxidizing agents are required to convert $S_2O_3^=$ to SO_2. Thiosulfate is unstable with respect to disproportionation to sulfur and sulfur dioxide. In acidic solutions, hydrogen sulfide is a mild reducing agent.

The reduction-potential diagram for basic solutions is the following.

$$SO_4^= \xrightarrow{-0.98} SO_3^= \xrightarrow{-0.58} S_2O_3^= \xrightarrow{-0.74} S \xrightarrow{-0.51} S^=.$$
$$\underset{-0.59}{\vert\underline{\qquad\qquad\qquad}\vert}$$

It is clear that sulfate, sulfite, and thiosulfate ions are all very poor oxidants in basic solutions, and in fact $SO_3^=$ and $S_2O_3^=$ are easily oxidized in basic solutions. Sulfide ion is a reducing agent in basic solutions, and $S_2O_3^=$ is no longer unstable with respect to disproportionation to sulfur and sulfite ion.

Table 15.8 The halides of sulfur

Fluorides			Chlorides			Bromide		
	Melting point (°C)	Boiling point (°C)		Melting point (°C)	Boiling point (°C)		Melting point (°C)	Boiling point (°C)
SF_4	−121	−40	S_2Cl_2	−80	138	S_2Br_2	−46	90
SF_6	− 51	−65(sub)	SCl_2	−78	decomp			
S_2F_{10}	− 55	29	SCl_4	decomp				

Sulfur forms a number of binary compounds with fluorine, chlorine, and bromine, and these are listed in Table 15.8. The halide of most practical importance is sulfur hexafluoride, the principal product of the direct reaction between sulfur and fluorine. Sulfur hexafluoride is a thermally stable, extremely inert gas which has great resistance to electrical breakdown. It is used, therefore, as a gaseous insulator in high-voltage generators and other devices. As was discussed in Section 11.5, the six fluorine atoms in SF_6 are at the corners of a regular octahedron, with the sulfur atom at the center. Because there are six pairs of electrons around the sulfur atom, the description of the bonding in SF_6 in terms of atomic orbitals involves the $3d$- as well as the $3s$- and $3p$-orbitals of sulfur. Because of its regular octahedral geometry, with six equivalent sulfur–fluorine bonds, SF_6 is said to exhibit $sp^3\,d^2$ hybrid bonding. The sulfur–fluorine bond in SF_6 is not particularly strong: $D(SF_5{-}F) = 86 \pm 3$ kcal. Consequently, the inert nature of SF_6 is attributed not to its thermodynamic stability alone, but also to the fact that its reactions with other reagents are exceedingly slow.

The other well-characterized fluorides of sulfur are SF_4 and S_2F_{10}. Their structures are given in Fig. 15.15. Disulfur decafluoride, S_2F_{10}, is a rather unreactive compound like sulfur hexafluoride. In contrast, SF_4 is extremely reactive and is rapidly hydrolyzed by water to SO_2 and HF. It is used as a fluorinating agent for organic compounds. As was discussed in Section 11.5, SF_4 is related structurally to PCl_5, for both have a central atom surrounded by five electron pairs. The PCl_5 molecule has the trigonal bipyramid structure,

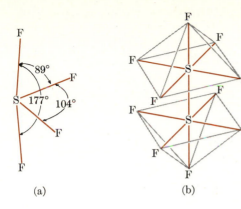

The structures of (a) SF_4 and (b) S_2F_{10}.

FIG. 15.15

and the geometry of SF_4 can be pictured as a distorted trigonal bipyramid with a pair of nonbonded electrons occupying one of the equatorial positions.

Selenium and Tellurium

These elements are quite rare and are recovered as byproducts from sulfur ores. Selenium is a poor conductor of electricity in the dark, but its conductivity increases when it is illuminated. Consequently, it is used in photoconductive cells to measure light intensity. Selenium is also used as a component of rectifiers for converting alternating to direct current. Tellurium also is used in certain electronic and light-sensitive devices.

The chemistry of selenium and tellurium resembles that of sulfur, and the differences that exist usually are associated with the metallic character that increases with atomic number. Selenides and tellurides of the metals are analogous to the sulfides, but are somewhat more covalent in nature because the ions $Se^=$ and $Te^=$ are larger and more polarizable than $S^=$. The hydrides H_2S, H_2Se, and H_2Te are all offensive, poisonous gases of moderate solubility in water. As was discussed in Section 13.4, H_2Se and H_2Te are thermodynamically unstable with respect to their elements, as is characteristic of the hydrides of the heavier elements of groups IV, V, and VI. The strength of the hydrides as acids increases in the sequence from H_2S to H_2Te, as the following equilibrium constants show:

$$H_2S = H^+ + HS^-, \qquad K = 1.1 \times 10^{-7},$$
$$H_2Se = H^+ + HSe^-, \qquad K = 2 \times 10^{-4},$$
$$H_2Te = H^+ + HTe^-, \qquad K = 2.3 \times 10^{-3}.$$

A similar trend in the acidities of the hydrides will be found in group VII.

The oxides SeO_2 and TeO_2 are both rather different physically from SO_2. Both SeO_2 and TeO_2 are solids at room temperature, and the structure of TeO_2 suggests that it is an ionic lattice. Selenium dioxide has an infinite chain structure of the type shown in Fig. 15.16. It dissolves in water to give acidic solutions, and the compound H_2SeO_3 has been prepared pure and its structure studied. In contrast, TeO_2 is quite insoluble in water and no such species as H_2TeO_3 has ever been isolated. Treatment of TeO_2 with strong bases, however, does give solutions that contain the tellurite ion, $TeO_3^=$.

FIG. 15.16
A segment of the infinite chain of SeO_2.

Oxidation of selenites gives selenates, the salts of selenic acid, H_2SeO_4. Selenic acid is rather similar in acid strength to sulfuric acid, but is a much stronger oxidizing agent, as is shown by the potential for

$$SeO_4^= + 4H^+ + 2e^- = H_2SeO_3 + H_2O, \qquad \varepsilon^0 = 1.15 \text{ volts.}$$

The $+6$ oxyacid of tellurium is quite different from those of sulfur and selenium. The formula of telluric acid is $Te(OH)_6$, and x-ray crystal structure studies show that the OH groups are at the corners of a regular octahedron with the tellurium atom at the center. Besides being structurally different from H_2SO_4 and H_2SeO_4, $Te(OH)_6$ is a weak acid, with a first ionization constant of about 10^{-7}.

15.3 THE ELEMENTS OF GROUP VIIA

The elements fluorine, chlorine, bromine, and iodine are reactive nonmetals that are always found in nature in the combined state. Although these halogens resemble each other chemically, there is a noticeable gradation of properties in the family. Fluorine is the most electronegative of the elements and displays only the oxidation state of -1. Chlorine, bromine, and iodine are also electronegative elements, but form compounds in which they are assigned positive as well as negative oxidation states. While all the halogens are oxidizing agents, their strength as oxidants decreases as the atomic number increases. Each of the halogens exists as discrete molecules in the solid, liquid, and gas phases, but the volatility of the elements markedly decreases as the atomic number increases. The variations in other properties such as ionization energy, electron affinity, and ionic size are evident from the data given in Table 15.9.

Fluorine and chlorine are the two most abundant halogens. Fluorine occurs principally as fluorspar, CaF_2, and cryolite, Na_3AlF_6. Because it is such a

Table 15.9 Properties of the group VIIA elements

	F	Cl	Br	I
Atomic number	9	17	35	53
Configuration	$2s^22p^5$	$3s^23p^5$	$4s^24p^5$	$5s^25p^5$
Ionization energy, kcal	402	300	273	241
Atomic radius, A	0.72	0.99	1.14	1.33
Melting point, °K	54	172	266	387
Boiling point, °K	85	239	331	455
ΔH_f(atom), kcal	18.6	29.01	26.71	25.48
Electron affinity, kcal	79.5	83.5	77.3	70.5
Ionic radius, X^-, A	1.33	1.81	1.96	2.19
ΔH_{hyd}, X^-, kcal	123	89	81	72

powerful oxidizing agent, fluorine is prepared commercially by electrolysis. Either fused potassium hydrogen fluoride, KHF_2, or a solution of KHF_2 in liquid hydrogen fluoride is electrolyzed to produce F_2 at the anode and H_2 at the cathode.

Chlorine also is prepared by electrolysis. The process is

$$Na^+(aq) + Cl^-(aq) + H_2O = \tfrac{1}{2}Cl_2(g) + \tfrac{1}{2}H_2(g) + Na^+(aq) + OH^-(aq)$$

and the hydrogen gas and the sodium hydroxide solution are useful byproducts of the reaction. Chlorine is a very strong inexpensive oxidizing agent, and consequently it has many industrial uses. One of these is the oxidation of bromide ion in sea water to bromine,

$$Cl_2 + 2Br^- = Br_2 + 2Cl^-.$$

While fluorine, chlorine, and bromine are found in the -1 oxidation state in nature and must be oxidized to the elemental state, iodine is obtained mainly by reduction of naturally occurring iodates. The bisulfite ion is a convenient reductant, and the reaction employed is

$$2IO_3^-(aq) + 5HSO_3^-(aq) = 3HSO_4^-(aq) + 2SO_4^=(aq) + H_2O + I_2(s).$$

The fifth member of the halogen family, astatine, is not found in nature. All its isotopes are radioactive, and the most stable, At^{210}, has a half-life of only 8.3 hours. As a result, the chemistry of astatine has been studied qualitatively, and few quantitative data on astatine are available.

The Halides

Most metallic elements react directly with the halogens to form compounds that are thermodynamically very stable. If the metal atom is relatively large and has an oxidation state of $+1$ or $+2$, the bonding in the halide is ionic, while for the higher oxidation states of the smaller metallic and semimetallic atoms the bonding in the halides tends toward a covalent nature. To see what factors determine the stability of the ionic halides, let us examine the energetics of formation of a metal halide of formula MX. The overall reaction is

$$M(s) + \tfrac{1}{2}X_2(g) = MX(s), \qquad \Delta H = \Delta H_f^0(MX).$$

This can be written as the sum of the following processes:

$$
\begin{aligned}
M(s) &= M(g), & \Delta H_{sub}, \\
M(g) &= M^+(g) + e^-, & I_1, \\
\tfrac{1}{2}X_2(g) &= X(g), & \tfrac{1}{2}D(X_2), \\
e^- + X(g) &= X^-(g), & A(X), \\
X^-(g) + M^+(g) &= MX(s), & \Delta H_{crys}.
\end{aligned}
$$

Thus the enthalpy of formation of the metal halide is determined by the enthalpies associated with sublimation and ionization of the metal, dissociation of the halogen molecule and electron attachment to the halogen atom, and formation of the ionic crystal lattice. The ionic halides as a class of compounds are very stable because the halogen molecules have relatively small bond energies and very large electron affinities. Fluorine has the smallest dissociation energy of all the halogens, and because F^- is the smallest of the halides, fluorides have the most stable crystal lattices. Consequently, the ionic fluorides are particularly stable compounds. The relative instability of the iodides must be a consequence of the large size of the iodide ion and the resulting small lattice energies of ionic iodides.

In addition to the simple monatomic halide ions, polyhalide ions are known. When iodine is added to an aqueous solution of iodide ion, the tri-iodide ion, I_3^- is formed. The ion is of moderate stability in aqueous solution as the equilibrium constant for its dissociation

$$I_3^-(aq) = I^-(aq) + I_2(aq), \qquad K = 1.3 \times 10^{-3},$$

suggests.

The corresponding dissociation constants for Br_3^- and Cl_3^- are 6×10^{-2} and 5.5, respectively, so these ions are less stable than I_3^-. Direct action of the halogens on the halides of the larger alkali metals can produce more complicated polyhalides such as KI_5CsICl_4, and $KBrF_4$.

The halogens react directly with many of the nonmetallic elements to form compounds that in general consist of small covalently bonded molecules. To see what factors determine the stability of such compounds, let us examine the energetics of formation of a halide of a nonmetal. As a specific example, consider the formation of the gaseous phosphorus trihalides, PX_3:

$$\tfrac{1}{4}P_4(s) + \tfrac{3}{2}X_2(g) = PX_3(g), \qquad \Delta H = \Delta H_f^0(PX_3).$$

Table 15.10 Bond energies of halides (kcal/mole)

B—F	154	B—Cl	109	–	–	–	–
C—F	116	C—Cl	81	C—Br	68	C—I	52
N—F	65	N—Cl	46	–	–	–	–
O—F	45	O—Cl	45	O—Br	48	–	–
Si—F	135	Si—Cl	91	Si—Br	74	Si—I	56
P—F	117	P—Cl	78	P—Br	63	P—I	44
S—F	68	S—Cl	61	S—Br	52	–	–

This overall reaction can be broken into the following steps:

$$\tfrac{1}{4}P_4(s) = P(g), \qquad \Delta H_{atom},$$
$$\tfrac{3}{2}X_2(g) = 3X(g), \qquad \tfrac{3}{2}D(X_2),$$
$$P(g) + 3X(g) = PX_3(g), \qquad -3D(P—X).$$

Thus the enthalpy of formation of the phosphorus trihalides, or any halide, is the difference between the energy required to convert the elements to atoms and the total bond energy of the compound. The relative stabilities of the gaseous halides of phosphorus depend on the strength of the P—X bonds relative to the strength of the bonds in the X_2 molecules, and an analogous conclusion holds for other nonmetals. The order of the dissociation energies of the gaseous halogen molecules is $D(F_2) \lesssim D(I_2) < D(Br_2) < D(Cl_2)$, and *if* all other factors were the same for all halogens, this would be the order of decreasing stability of the halogen compounds. The energies of the bonds formed by the different halogens vary considerably, as Table 15.10 shows. It is clear that with a given element, fluorine forms bonds that are substantially stronger than those formed by chlorine, and chlorine forms bonds that are stronger than those made by bromine and iodine. As a result of both the small dissociation energy of fluorine and the very strong bonds it forms with nonmetals, the fluorides are in general the most energetically stable of the nonmetallic halides. Even though chlorine has the largest dissociation energy of the halogen molecules, the bonds it forms with other elements are strong enough to make the nonmetallic chlorides second to the fluorides in energetic stability. Despite the small bond energy of I_2, the iodides are generally the least stable of the halides because iodine forms only very weak bonds with the nonmetals.

The Hydrogen Halides

The hydrogen halides can be prepared by the action of a nonvolatile non-oxidizing acid on a soluble halide, as in

$$NaBr + H_3PO_4 = HBr(g) + NaH_2PO_4.$$

The hydrogen halides exist as gaseous diatomic molecules under ordinary conditions, and like the other nonmetallic halides, they decrease in thermodynamic

Table 15.11 Properties of hydrogen halides

	HF	HCl	HBr	HI
Melting point, °C	−83.1	−114.8	−86.9	−50.7
Boiling point, °C	19.5	− 84.9	−66.8	−35.4
ΔH_{vap}^0, kcal	7.24	3.85	4.21	4.72
ΔH_f^0, kcal	−64.2	− 22.06	− 8.66	6.20
ΔG_f^0, kcal	−64.7	− 22.77	−12.72	0.31

stability as the atomic number of the halogen increases. Some of the properties of these compounds are listed in Table 15.11.

The electrical conductivity of pure liquid HF is rather small, but does indicate that a small amount of self-ionization occurs according to

$$2HF = H_2F^+ + F^-, \qquad F^- + HF = HF_2^-.$$

The other pure liquid hydrogen halides show very little if any self-ionization. In aqueous solutions, however, the hydrogen halides are good electrical conductors. Hydrogen fluoride is a fairly weak acid, as is shown by

$$HF + H_2O = H_3O^+ + F^-, \qquad K_1 = 7.2 \times 10^{-4},$$
$$F^- + HF = HF_2^-, \qquad K_2 = 5.1.$$

The other hydrogen halides are strong acids, and are almost totally dissociated in water. When HCl, HBr, and HI are dissolved in solvents that are poorer proton acceptors than water, they are not as extensively dissociated to ions. In fact, under these conditions it is possible to determine that the strength of the hydrogen halides as acids increases in the sequence HCl, HBr, and HI.

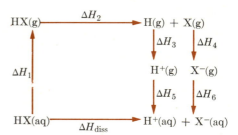

FIG. 15.17   Thermodynamic cycle for the dissociation of a halogen acid HX.

To see what factors determine the acid strength of the hydrogen halides, let us examine the thermodynamic cycle shown in Fig. 15.17. It is evident that the enthalpy of dissociation of the HX molecules is given by

$$\Delta H_{diss} = \Delta H_1 + \Delta H_2 + \Delta H_3 + \Delta H_4 + \Delta H_5 + \Delta H_6,$$

Acid	ΔH_1	$\Delta H_2(D)$	$\Delta H_3(I)$	$\Delta H_4(-A)$	$\Delta H_5 + \Delta H_6$	ΔH_{diss}
HF	11.5	134.6	315	−79.5	−381.9	− 1
HCl	4.2	103.2	315	−83.5	−348.8	−10
HBr	5.0	87.5	315	−77.3	−340.7	−11
HI	5.5	71.4	315	−70.5	−330.3	− 9

where

$$\Delta H_1 = \Delta H_{dehyd}(HX), \qquad \Delta H_2 = D(HX), \qquad \Delta H_3 = I(H),$$

$$\Delta H_4 = -A(X), \qquad \Delta H_5 + \Delta H_6 = \Delta H_{hyd}(H^+ + X^-).$$

The measured values of these quantities are given in Table 15.12, together with the computed value of $\Delta H_{diss}(HX)$. Because $\Delta H_{diss}(HF)$ is less negative than the enthalpies of dissociation of the other hydrogen halides, we are not surprised to find that HF is a weaker acid than any of the other hydrogen halides. The enthalpy of dissociation is not the only factor that determines acid strength, however. The entropy changes for the dissociation reaction have also been estimated and are given in Table 15.13. The data show that HF is the weakest of the acids not only because its dissociation is the least favored energetically, but also because the entropy change that accompanies its dissociation is the most negative. The relative importance of entropy and enthalpy effects can be assessed by comparing ΔH_{diss} with $T \Delta S_{diss}$. The data in Table 15.13 show that while the values of ΔH_{diss} set the general trend of the acidities of the hydrogen halides, entropy effects are important and in fact are what make HI a slightly stronger acid than HCl.

Table 15.13 Thermodynamics of dissociation of halogen acids

Acid	ΔH_{diss} (kcal)	ΔS_{diss} (cal/deg)	$-T \Delta S$ (kcal)	ΔG_{diss} (kcal)
HF	− 1	−21	6.3	5
HCl	−10	−13	3.9	−6
HBr	−11	− 9	2.7	−8
HI	− 9	− 3	0.9	−8

The Halogen Oxides

The known halogen oxides are listed in Table 15.14. These compounds are characteristically unstable reactive substances that exist as discrete small molecules in all phases. At room temperature they exist as gases or volatile liquids, with the exception of I_2O_5, a solid.

Table 15.14 The halogen oxides

Fluorine	Chlorine	Bromine	Iodine
F_2O	Cl_2O	Br_2O	I_2O_4
F_2O_2	ClO_2	BrO_2	I_4O_5
	Cl_2O_4	BrO_3	I_2O_5
	Cl_2O_6	$Br_2O_7(?)$	I_2O_7
	Cl_2O_7		

The compound F_2O is the only halogen oxide that is thermodynamically stable with respect to its elements. Its stability is slight, however, and it reacts readily with a variety of reducing agents. For example, when dissolved in water, it produces oxygen slowly according to the reaction

$$F_2O + H_2O = O_2 + 2HF.$$

The chlorine oxides are all small covalently bonded molecules that are rather unstable highly reactive oxidizing agents. Chlorine monoxide, Cl_2O, is prepared by the reaction

$$2Cl_2 + 2HgO = HgCl_2 \cdot HgO + Cl_2O$$

and upon heating, explodes spontaneously to give Cl_2 and O_2. Chlorine dioxide can be synthesized by the reaction

$$2ClO_3^- + SO_2 = 2ClO_2 + SO_4^=.$$

Chlorine dioxide is also spontaneously explosive, but is safe if treated carefully, and is in fact used commercially as an oxidizing agent.

Dichlorine tetroxide is not a dimer of chloride dioxide, but rather has the atomic arrangement $ClOClO_3$, which corresponds to chlorine perchlorate. It is stable for only short periods at room temperature.

Chlorine hexoxide, Cl_2O_6, is formed when ozone reacts with ClO_2,

$$2ClO_2 + 2O_3 = Cl_2O_6 + 2O_2.$$

Chlorine hexoxide is unstable and reacts explosively with organic compounds. The $+7$ oxide of chlorine, Cl_2O_7, is a volatile liquid obtained by dehydrating perchloric acid,

$$2HClO_4 \xrightarrow{P_4O_{10}} Cl_2O_7.$$

Even though it is the most stable of the chlorine oxides, it explodes when it is heated or subjected to mechanical shock.

The bromine oxides are not at all well characterized chemically or physically. Of the oxides of iodine, only I_2O_5 has been investigated extensively. It is the dehydration product of HIO_3, iodic acid, for at 200°C the reaction

$$2HIO_3 = I_2O_5 + H_2O$$

occurs. Iodine pentoxide is a stable compound and reacts in a controlled manner with a number of reducing agents. The most important of its reactions is the oxidation of carbon monoxide,

$$I_2O_5 + 5CO = I_2 + 5CO_2.$$

This reaction is quantitative, and a determination of the iodine formed allows a quantitative analysis for CO to be made.

The Halogen Oxyacids

The known oxyacids of the halogens are listed in Table 15.15. The hypohalous acids HOX, except HOF, are formed by the disproportionation of the halogens in aqueous solution,

$$X_2(aq) + H_2O = H^+ + X^- + HOX.$$

The values of the equilibrium constant of this reaction for the various halogens are: Cl_2, 4.2×10^{-4}; Br_2, 7.2×10^{-9}; and I_2, 2.0×10^{-13}. From these equilibrium constants, it can be deduced that in a saturated solution of chlorine, the concentration of HOCl is about half the concentration of chlorine, while only about 0.5% of a saturated solution of I_2 is hydrolyzed to HOI. One method of producing the hypohalous acids in greater yield is to pass the halogen into an aqueous suspension of mercuric oxide,

$$2X_2 + 2HgO + H_2O = HgO \cdot HgX_2 + 2HOX.$$

The hypohalous acids are all very weak acids, for their dissociation constants are: HOCl, 2×10^{-8}; HOBr, 2×10^{-9}; HOI, 1×10^{-11}. They are also rather unstable, and have never been isolated as pure compounds. The compound HOF is made by passing F_2 over ice and collecting the product in a cold trap. It reacts rapidly with water to produce oxygen and is thermally unstable, decomposing with a half-life of less than an hour at 25°C.

Table 15.15 Oxyacids of the halogens

Fluorine	Chlorine	Bromine	Iodine
HOF	HOCl	HOBr	HOI
	$HClO_2$	$HBrO_2(?)$	–
	$HClO_3$	$HBrO_3$	HIO_3
	$HClO_4$		HIO_4, H_5IO_6

The only known oxyacid in which a halogen appears in the +3 oxidation state is chlorous acid, $HClO_2$. Salts of chlorous acid can be made by the reaction of ClO_2 with peroxides,

$$Na_2O_2 + 2ClO_2 = 2NaClO_2 + O_2.$$

Acidification of solutions of chlorites yields $HClO_2$, which is a moderately strong acid with a dissociation constant of 10^{-2}.

The acids HXO_3 and their salts are known for all the halogens except fluorine. Solutions of chlorates, bromates, and iodates can be obtained by the disproportionation of hypohalides in basic solution, $3XO^- = 2X^- + XO_3^-$. This reaction is quantitative for all three heavier halogens. It proceeds rapidly at room temperature for iodine, and at a temperature of 75°C for chlorine and bromine.

All the halic acids are strong acids, and are essentially totally dissociated in aqueous solution. Chloric and bromic acids have never been isolated as pure compounds, but iodic acid, HIO_3, appears as white crystals when iodine is oxidized by concentrated nitric acid. The acids and their anions are strong oxidizing agents, and chloric acid in particular reacts violently with organic compounds.

The oxyacids of the halogens in the +7 oxidation state are perchloric, perbromic, and periodic acids. Perchlorates are prepared by the electrolytic oxidation of chlorates, and when a perchlorate salt is heated with concentrated H_2SO_4, perchloric acid distills from the mixture. Perchloric acid is totally dissociated to ions in aqueous solution, and is probably the strongest acid known. Thus the oxyacids of the halogens of the type $HOXO_n$ show a steadily increasing acid strength as n increases from 0 to 3. This trend of increasing acidity with increasing oxidation state of the central atom has been noted previously for other elements.

Perchloric acid is a very strong oxidizing agent and reacts explosively with organic compounds. In dilute aqueous solutions at room temperature, perchloric acid tends to be rather unreactive, for despite its very great oxidizing strength, its reactions with inorganic compounds are very slow. Because the perchlorate ion is a large ion with a small charge, it does not tend to form complexes with cations, and its salts are, in general, quite soluble in water. Consequently, perchlorate salts are often used when studies of the properties of cations in aqueous solutions are made.

Perbromates have been unknown until fairly recently. They can be synthesized by the reaction of XeF_2 or F_2 on bromates:

$$BrO_3^- + XeF_2 + H_2O = BrO_4^- + 2HF + Xe,$$

$$BrO_3^- + F_2 + 2OH^- = BrO_4^- + 2F^- + H_2O.$$

Despite their elusive nature, perbromates are stable both in aqueous solution and as solid alkali salts such as $KBrO_4$.

Periodic acid exists in several forms. In strongly acidic solutions the most important species is paraperiodic acid, H_5IO_6, a weak acid in which a central iodine atom is surrounded by five OH groups and an oxygen atom, all located at the corners of an octahedron. Paraperiodic acid is in equilibrium in aqueous solution with the anion $H_3IO_6^{--}$, and with the metaperiodate ion IO_4^-. Solutions of periodic acid are strong oxidizing agents that react smoothly and rapidly with a number of reagents. In one of the standard procedures for the analysis of manganese, periodic acid is used to oxidize manganous ion to permanganate.

Chlorine, bromine, and iodine each have extensive solution chemistry and it is helpful to summarize their properties with reduction-potential diagrams. The following diagrams apply to chlorine species in acidic and basic solutions, respectively.

$$ClO_4^- \xrightarrow{1.19} ClO_3^- \xrightarrow{1.21} HClO_2 \xrightarrow{1.64} HOCl \xrightarrow{1.63} Cl_2 \xrightarrow{1.36} Cl^-$$
$$\underset{1.47}{\rule{3cm}{0.4pt}}$$

$$ClO_4^- \xrightarrow{0.36} ClO_3^- \xrightarrow{0.33} ClO_2^- \xrightarrow{0.66} ClO^- \xrightarrow{0.40} Cl_2 \xrightarrow{1.36} Cl^-$$
$$\underset{0.50}{\rule{2cm}{0.4pt}} \qquad \underset{0.89}{\rule{2cm}{0.4pt}}$$

It is clear that in both acidic and basic solutions, all chlorine species except Cl^- are strong oxidants. Hypochlorous and chlorous acids react rather rapidly to oxidize a variety of reagents, but the reactions of chlorate and perchlorate ions with inorganic reagents are usually quite slow. There are two disproportionation reactions that are of importance in alkaline solutions:

$$Cl_2 + 2OH^- = Cl^- + ClO^- + H_2O, \qquad \Delta\varepsilon^0 = 0.96 \text{ volt},$$
$$3ClO^- = ClO_3^- + 2Cl^-, \qquad \Delta\varepsilon^0 = 0.39 \text{ volt}.$$

The first of these reactions is used to prepare hypochlorites, and the second is used to synthesize chlorates.

The reduction-potential diagrams for bromine in acidic and basic solution, respectively, are as follows.

$$\overset{1.52}{\overline{\qquad\qquad}}$$
$$BrO_3^- \xrightarrow{1.49} HOBr \xrightarrow{1.59} Br_2 \xrightarrow{1.07} Br^-$$

$$\overset{0.61}{\overline{\qquad\qquad}}$$
$$BrO_3^- \xrightarrow{0.54} BrO^- \xrightarrow{0.45} Br_2 \xrightarrow{1.07} Br^-$$
$$\underset{0.71}{\rule{3cm}{0.4pt}}$$

All species except Br^- are strong oxidizing agents. The potentials show that in basic solutions, bromine can disproportionate spontaneously to BrO^- and Br^-. Since BrO^- itself can disproportionate to Br^- and BrO_3^-, these ions are the eventual products found in alkaline solutions of bromine. In acidic solutions, however, bromine does not disproportionate, and in fact the reaction

$$BrO_3^- + 5Br^- + 6H^+ = 3Br_2 + 3H_2O, \qquad \Delta\varepsilon^0 = 0.45 \text{ volt},$$

proceeds spontaneously from left to right.

All oxidation states of iodine except the -1 state have strong or moderately strong oxidizing properties, as the following reduction-potential diagrams show.

$$H_5IO_6 \xrightarrow{1.7} IO_3^- \xrightarrow{1.14} HOI \xrightarrow{1.45} I_2 \xrightarrow{0.53} I^-$$
$$\underset{1.20}{\underline{\hspace{5cm}}}$$

$$H_3IO_6^= \xrightarrow{0.7} IO_3^- \xrightarrow{0.14} IO^- \xrightarrow{0.45} I_2 \xrightarrow{0.53} I^-$$
$$\underset{0.29}{\underline{\hspace{5cm}}}$$

Like chlorine and bromine, iodine is stable with respect to disproportionation to the $+1$ and -1 states in acidic solution, but does disproportionate in alkaline solutions. Both hypoiodous acid and hypoiodite anion are unstable with respect to self-oxidation and reduction:

$$5HOI = 2I_2 + IO_3^- + H^+ + 2H_2O, \qquad \Delta\varepsilon^0 = 0.31 \text{ volt},$$
$$3IO^- = 2I^- + IO_3^-, \qquad \Delta\varepsilon^0 = 0.35 \text{ volt}.$$

In acidic and basic solutions, the iodate ion is stable by itself, but it does react quantitatively with I^- to produce iodine in acidic solutions:

$$IO_3^- + 5I^- + 6H^+ = 3I_2 + 3H_2O, \qquad \Delta\varepsilon^0 = 0.67 \text{ volt}.$$

The Interhalogen Compounds

The known binary interhalogen compounds are listed in Table 15.16. The principal interest in these compounds lies in their molecular structure, although some practical use is made of BrF_3 as a fluorinating agent.

Table 15.16 The inter-halogen compounds

	Cl	Br	I
F	ClF	BrF	
	ClF_3	BrF_3	
	ClF_5	BrF_5	IF_5
			IF_7
Cl		BrCl	ICl
Br			IBr

The geometry of ClF_3 is shown in Fig. 15.18. As we remarked in Section 11.5, the structure of ClF_3 is related to those of PCl_5 and SF_4, for each of these molecules has a central atom that is surrounded by five pairs of valence electrons. Interelectron repulsion should be minimized if each electron pair is directed toward one of the corners of a trigonal bipyramid. In ClF_3, only the axial and one equatorial position of the bipyramid are occupied by fluorine atoms. Thus we can regard the T-shape of ClF_3 as a slightly distorted fragment of a regular trigonal bipyramid. By analogy with ClF_3 we expect BrF_3 and ICl_3 to be T-shaped molecules, and this is found experimentally.

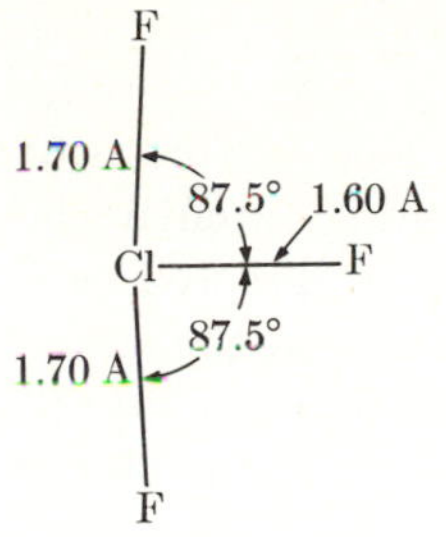

F

1.70 A

87.5°   1.60 A

Cl———F

87.5°

1.70 A

F

The structure of ClF_3.   **FIG. 15.18**

The molecules ClF_5, BrF_5 and IF_5 have respectively a central chlorine, bromine and iodine atom surrounded by six electron pairs, five of which are used to form bonds to fluorine atoms. The geometry of these molecules should be related to that of SF_6, which also has six electron pairs around a central atom. Figure 15.19 shows the structure of BrF_5. It can be rationalized by imagining the six electron pairs around the bromine atom directed toward the corners of an octahedron, with five of these corners occupied by fluorine atoms. If we assume, as we have done consistently, that the nonbonded pair of electrons occupies more space than bonding electron pairs, the departure of the geometry of BrF_5 from that of a regular octahedron is not surprising. The geometry of IF_5 is similar to that of BrF_5, for all five of the fluorine atoms lie on the same side of a plane that contains the iodine atom and which is perpendicular to the axis of symmetry of the molecule.

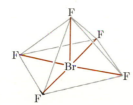

The structure of BrF_5. The bromine atom lies slightly below   **FIG. 15.19**
the plane of four of the fluorine atoms.

There are several ions of the interhalogens whose structures fit the patterns established by the interhalogens and other halides of the nonmetals. The ions I_3^-, ICl_2^-, IBr_2^-, and $BrICl^-$ are all linear and have a central iodine atom surrounded by five electron pairs. Their structures can be pictured as being related to that of PCl_5, with only two of the five electron pairs at the central atom being used to form bonds. These two bonding pairs are directed along the axis of a trigonal bipyramid, and the three nonbonding electron pairs are directed toward the equatorial corners of the bipyramid.

FIG. 15.20   The structure of ICl_4^-.

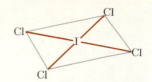

The ion ICl_4^- has the structure shown in Fig. 15.20. The iodine atom is at the center of the square formed by the four chlorine atoms, and apparently the two pairs of nonbonded electrons around the iodine atom are directed perpendicular to the plane of the ion. Thus the six electron pairs around the iodine atom can be regarded as directed toward the corner of an octahedron, and the structure of ICl_4^- is related to that of SF_6. The ion BrF_4^- also has a planar structure, as would be expected by analogy with ICl_4^-.

15.4 THE NOBLE-GAS COMPOUNDS

It has been known for a number of years that the noble-gas atoms form strong bonds to certain other atoms. For example, species such as He_2^+, Ar_2^+, ArH^+, and CH_3Xe^+ have been detected repeatedly as transient gaseous ions. The bond strength in some of these molecular ions is substantial. Dissociation of He_2^+ to He and He^+ requires 60 kcal/mole and separation of ArH^+ to Ar and H^+ requires at least 93 kcal/mole. Despite these large bond energies, such gaseous ions are only transient species, for if they acquire electrons, they dissociate immediately into atoms. The existence of these transient ions is important, however, for it shows that there is no mysterious property of the completed octet in the noble-gas atoms that absolutely prevents these atoms from being bonded to other species. Apparently strong electron acceptors, such as positive ions, can form strong bonds with the noble-gas atoms. This lesson was largely ignored by chemists before 1962.

In 1962 N. Bartlett found that molecular oxygen forms a compound with PtF_6 that can be represented as $O_2^+PtF_6^-$. Because xenon has nearly the same ionization energy as oxygen, Bartlett decided to investigate the possibility of a reaction between xenon and PtF_6. A reaction between these reagents was observed, and did demonstrate that xenon is not a totally inert gas. This observation stimulated other investigations of the chemistry of xenon and the other noble gases. Several compounds of xenon and krypton with the strong electron acceptors fluorine and oxygen are now known.

The most stable and best characterized of the noble-gas compounds are the xenon fluorides, oxyfluorides, and oxides. These are listed in Table 15.17 with some of the known thermochemical data. The xenon fluorides can be made in a variety of ways, and apparently the only requirement is that xenon be exposed to fluorine atoms. Xenon difluoride is usually formed first, and continued exposure of XeF_2 to fluorine atoms yields xenon tetrafluoride. Xenon hexafluoride is formed by the reaction of XeF_4 with a considerable excess of fluorine. These three binary fluorides of xenon have negative enthalpies of formation, and combination of these values of ΔH_f with the dissociation energy of fluorine shows that the average xenon-fluorine bond energy in these compounds is 30 ± 3 kcal/mole.

Table 15.17 Some compounds of xenon

Compound	ΔH_f (kcal/mole)	Compound	ΔH_f (kcal/mole)
XeF_2	−30	XeF_6	−96
XeF_4	−69	$XeOF_4$	–
$XeOF_2$	–	XeO_3	96

The oxygen compounds of xenon are obtained by hydrolysis of the fluorides. For example,

$$XeF_6 + H_2O = XeOF_4 + 2HF,$$
$$XeF_6 + 2H_2O = XeO_2F_2 + 4HF,$$
$$XeF_6 + 3H_2O = XeO_3 + 6HF.$$

Of these compounds, XeO_3 has been prepared in quantity and is fairly well characterized. Although it is easy to synthesize, it is violently explosive when dry. In aqueous solutions, however, it is well behaved, and as its large positive enthalpy of formation might suggest, it is a very powerful oxidizing agent. Because the only byproduct of its reduction is xenon gas, it does not introduce extra complicating chemical species into a reaction system when it is used as an oxidizing agent. Because of these features, it may be used considerably in the future as a general oxidizing agent.

When solutions containing XeO_3 are made alkaline with sodium hydroxide, disproportionation of XeO_3 occurs, xenon gas is evolved, and sodium perxenate, $Na_4XeO_6 \cdot 8H_2O$ can be recovered from the solution. In acidic solutions, the perxenate anion decomposes slowly to XeO_3 and oxygen. These acidic solutions of perxenate ion have strong oxidizing properties and can convert manganous ion to permanganate. Although standard reduction potentials are hard to determine for these systems, the following values have been obtained, which indicate clearly the powerful oxidizing properties of the xenon compounds:

$$H_4XeO_6 + 2H^+ + 2e^- = XeO_3 + 3H_2O, \qquad \varepsilon^0 = 2.36,$$
$$XeO_3 + 6H^+ + 6e^- = Xe + 3H_2O, \qquad \varepsilon^0 = 2.12,$$
$$XeF_2 + 2H^+ + 2e^- = Xe + 2HF(aq), \qquad \varepsilon^0 = 2.6V.$$

The structures of the xenon compounds fit the patterns established by other isoelectronic species, so far as is known. For example, XeF_2 is isoelectronic in the valence shell with I_3^-, ICl_2^-, and $BrICl^-$, and like these ions, is linear and symmetrical. Xenon tetrafluoride is isoelectronic with ICl_4^- and BrF_4^-, and is accordingly a symmetrical square planar molecule. Xenon trioxide is isoelectronic with the iodate ion IO_3^-, and has the same trigonal pyramid structure. The XeO_6^{-4} ion has the six oxygen atoms at the corners of a regular octahedron, as might be expected by analogy with the molecules SF_6, SeF_6, and TeF_6. Xenon tetroxide, XeO_4, is a tetrahedral molecule, as might be expected from the fact that there are four pairs of electrons around the central xenon atom, and also from analogy with the isoelectronic tetrahedral periodate ion, IO_4^-. The structure of $XeOF_4$ is related to that of the isoelectronic molecule BrF_5. In $XeOF_4$, the four fluorine atoms form the base, and the oxygen atom the apex, of a square pyramid.

The structure of xenon hexafluoride is not known. This molecule has seven pairs of electrons around the central xenon atom, but certain theoretical arguments have led to the prediction that the molecule should have the shape of a regular octahedron, while others predict that XeF_6 should have a less symmetrical structure. At the present time, the experimental evidence tends to favor the latter conclusion, but the problem has not been solved conclusively.

Xenon difluoride can act as a fluoride ion donor, and as a result, forms a number of addition compounds with fluoride acceptors such as AsF_5 and SbF_5. For example,

$$2XeF_2 + AsF_5 = (Xe_2F_3^+)(AsF_6^-).$$

The $Xe_2F_3^+$ cation has the planar structure indicated in Fig. 15.21.

FIG. 15.21 The structure of $Xe_2F_3^+$.

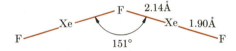

Xenon hexafluoride can act as a fluoride ion donor, and forms such compounds as $(XeF_5^+)(PtF_6^-)$. XeF_6 also acts as a fluoride acceptor, and reacts with alkali metal fluorides to give heptafluoro-or octafluoro xenates:

$$CsF + XeF_6 = CsXeF_7,$$
$$2CsXeF_7 = XeF_6 + Cs_2XeF_8.$$

Thus the chemistry of the seemingly inert element xenon is in fact quite rich.

15.5 CONCLUSION

Although the nonmetallic elements display a wide range of chemical and physical properties, certain trends and regularities in their behavior are evident. In any given family, there is a trend toward metallic behavior as the atomic numbers of the elements increase. In any row of the periodic table the elements become less metallic in nature as atomic numbers increase. All the nonmetals except the noble gases are electronegative and react with the active metals by accepting electrons to form negative ions. In addition, many of the nonmetals display a number of positive oxidation states, most commonly as oxides and oxyanions. These nonmetallic oxides are typically acidic in nature, and in general, the acidity is more marked, the higher the oxidation state of the nonmetal.

SUGGESTIONS FOR FURTHER READING

Cotton, F. A., and G. Wilkinson, *Advanced Inorganic Chemistry*, 3rd ed. New York: Interscience, 1972.

Douglas, B. E., and D. H. McDaniel, *Concepts and Models of Inorganic Chemistry*. New York: Blaisdell, 1965.

Gould, E. S., *Inorganic Reactions and Structure*. New York: Holt, Rinehart and Winston, 1962.

Heslop, R. B., and P. L. Robinson, *Inorganic Chemistry*. New York: Elsevier, 1963.

Jolly, W. L., *The Inorganic Chemistry of Nitrogen*. New York: W. A. Benjamin, 1964.

Kleinberg, J., W. J. Argersinger, Jr., and E. Griswold, *Inorganic Chemistry*. Boston: D. C. Heath, 1960.

Phillips, C. S. G., and R. J. P. Williams, *Inorganic Chemistry*, Vols. 1 and 2. London: Oxford Univ. Press, 1965.

Yost, D. M., and H. Russell, Jr., *Systematic Inorganic Chemistry*. Englewood Cliffs, N.J.: Prentice-Hall, 1946.

PROBLEMS

15.1 On the basis of van der Waals forces, and the electronic properties that determine them, explain the trend in volatility among the halogen elements.

15.2 Calculate the entropy of vaporization for each of the hydrogen halides. Do these values indicate that one of the hydrogen halides is markedly different from the others? What phenomenon could be responsible for this deviation?

15.3 Solid iodine has a dark purple color, and the crystals have a lustrous appearance. In addition, solid iodine displays a small conductivity that increases with increasing temperature. Are these observations consistent with the position of iodine in the periodic table? Explain.

15.4 Make an analysis of the energetics of the electrode reaction

$$\tfrac{1}{2}X_2(g) + e^- = X^-(aq)$$

for F_2, Cl_2, and Br_2, and suggest reasons why the strength of the halogens as oxidizing agents decreases in this sequence.

15.5 What sequence of reactions could be used to synthesize (a) $Na_2S_2O_3$ and (b) $H_2S_2O_8$ from the elements?

15.6 Nitrous oxide reacts with sodium amide to give sodium azide and water,

$$N_2O + NaNH_2 = NaN_3 + H_2O.$$

By analogy with other triatomic molecules and ions, predict the structure of the azide ion, N_3^-. Similarly, predict the geometry of the cyanate ion, OCN^-, formed by the reaction

$$(CN)_2 + 2OH^- = CN^- + OCN^- + H_2O.$$

15.7 We noted that NF_3 is not at all basic, in contrast to ammonia, NH_3. With this fact and its interpretation in mind, try to predict whether hydroxyl amine, H_2NOH, is more or less basic than ammonia.

15.8 Calculate the pH of 0.10 M Na_2SO_3.

15.9 In the compound O_2PtF_6 it has been suggested that the oxygen exists as O_2^+. By considering that bond distances in solids can be determined by x-ray methods, suggest an experiment that will help to decide whether the oxygen is present as O_2 or O_2^+.

15.10 How would you convert: (a) chlorine to $KClO_3$; (b) chlorine to $HClO_4$; (c) chlorine to ClO_2; (d) iodine to I_2O_5; (e) bromine to $NaBrO_3$?

15.11 Calculate the equilibrium constants for the following disproportionation reactions: (a) ClO^- to ClO_3^- and Cl^-; (b) Br_2 to BrO_3^- and Br^- in basic solution; and (c) I_2 to IO_3^- and I^- in basic solution.

15.12 Mercury is oxidized by NO_2 according to the reaction

$$Hg(l) + NO_2(g) = NO(g) + HgO(s).$$

In contrast, mercury is not attacked by N_2O. Is this difference in reactivity a kinetic or thermodynamic effect? Explain. $\Delta G_f^0(HgO) = -13.99$ kcal.

15.13 Complete and balance the following expressions.

$$Cu + H_2SO_4 \text{ (hot, conc.)} \rightarrow \qquad NO_2 + H_2O \text{ (cold)} \rightarrow$$
$$Zn + HNO_3 \text{ (dilute)} \rightarrow \qquad NO_2 + H_2O \text{ (hot)} \rightarrow$$
$$Zn + HNO_3 \text{ (conc.)} \rightarrow \qquad HClO_4 \xrightarrow{P_4O_{10}}$$
$$Mg + P \rightarrow \qquad Ca_3P_2 + H_2O \rightarrow$$

15.14 What sequence of reactions leads to the synthesis of N_2O, if the starting reagents are elements?

15.15 The ion SO_3^- has a pyramidal structure, and in SO_4^- the oxygen atoms are at the corners of a tetrahedron with the sulfur atom at the center. What would you expect the structures of $SOCl_2$ and SO_2Cl_2 to be? All the oxygen and chlorine atoms are bonded directly to sulfur.

15.16 Consider the following cations to be present in 0.1-M aqueous solution: Fe^{++}, Zn^{++}, Mn^{++}, Pb^{++}. Which of the ions will be precipitated as sulfides if the solution is saturated with H_2S (0.1-M) and the pH is maintained at (a) 0; (b) 2; (c) 5. Solubility products: PbS, 1×10^{-29}; ZnS, 4.5×10^{-24}; FeS, 1×10^{-19}; MnS, 7×10^{-16}.

15.17 When H_2S is bubbled into nitric acid solutions, sulfur, NO_2, NO, N_2, and NH_4^+ are formed. Write the four balanced reactions that describe this process.

CHAPTER 16

THE TRANSITION METALS

As was noted in Section 13.1, the transition elements occur between groups IIA and IIIA in the long form of the periodic table. In the first, second, and third transition series, respectively, the 3d-, 4d-, and 5d-electrons make their appearance. The third and fourth transition series also include, respectively, the lanthanides and the actinides, the "inner" transition elements; in each of these series an inner set of f-orbitals is filled. This chapter is concerned principally with the chemistry of the elements of the first transition series, but we shall also comment briefly about the properties of the other transition elements.

16.1 GENERAL PROPERTIES OF THE ELEMENTS

Before investigating the detailed chemistry of the individual elements, let us assess the general nature of the transition metals. All the transition elements are metals, and most of them have high melting points, high boiling points, and relatively large enthalpies of vaporization. The exceptional elements in this respect are those in group IIB: zinc, cadmium, and mercury. These metals have relatively low melting points and are moderately volatile. The atoms of these elements have completely filled sets of valence d-orbitals and in this respect also are different from the rest of the transition elements. This observation suggests that among the elements that have incompletely filled valence d-orbitals, the d-electrons are involved in the metallic bonding and contribute to the cohesion of the metallic crystal.

Table 16.1 Properties of the elements of the first transition series

	Sc	Ti	V	Cr	Mn	Fe	Co	Ni	Cu	Zn
Configuration	$3d^14s^2$	$3d^24s^2$	$3d^34s^2$	$3d^54s^1$	$3d^54s^2$	$3d^64s^2$	$3d^74s^2$	$3d^84s^2$	$3d^{10}4s^1$	$3d^{10}4s^2$
Ionization energy, kcal										
I_1	151	158	155	156	171	182	181	176	178	217
I_2	297	314	328	380	361	373	393	419	468	414
I_3	571	649	685	714	777	707	772	811	849	915
Atomic radius, A	1.44	1.32	1.22	1.17	1.17	1.16	1.16	1.15	1.17	1.25
Melting point, °K	1795	1950	2190	2176	1517	1812	1768	1728	1356	693
Boiling point, °K	–	3550	3650	2900	2340	3150	3150	3160	2855	1180
ΔH_f (atom)	90	113	123	95	67	99	101	103	81	31
Ionic radii, A										
M++	–	0.72	0.65	–	0.82	0.76	0.72	0.68	0.72	0.74
M+3	0.69	0.61	0.62	0.61	0.62	0.63	0.56			
ΔH_{hydr}, kcal										
M++	–	446	453	460	445	468	497	507	507	491
M+3	947	1027	1053	1105	1098	1072	1126	–	–	–
Configuration										
M++(g)	–	$3d^2$	$3d^3$	$3d^4$	$3d^5$	$3d^6$	$3d^7$	$3d^8$	$3d^9$	$3d^{10}$
M+3(g)	–	$3d^1$	$3d^2$	$3d^3$	$3d^4$	$3d^5$	$3d^6$	–	–	–
Reduction potential, volts										
M++ + 2e⁻ = M	–	–1.6	–1.2	–0.91	–1.18	–0.44	–0.28	–0.25	+0.34	–0.76
M+3 + 3e⁻ = M	–2.1	–1.20	–0.85	–0.74	–0.28	–0.04	0.4	–	–	–

Virtually all the transition metals are good conductors of heat and electricity, and as was noted in Section 13.2, the elements copper, silver, and gold of group IB are particularly outstanding in these respects. From the thermodynamic point of view, many of the transition metals, particularly those of the first transition series, are "active" metals. That is, their electrode potentials indicate that they should react spontaneously with 1-M H^+ to yield aqueous solutions of their ions. On the other hand, the rates at which many of these metals are attacked by oxidizing agents are very small, and despite their thermodynamic tendency to react, they appear to be rather inert. Moreover, some of the heavier transition metals, particularly palladium, platinum, and their close neighbors, react only with the strongest oxidizing agents. Thus while we shall find noticeable similarities between many of the transition metals, there is at the same time an enormous range of properties displayed by these elements.

The elements of the first transition series resemble each other in a number of ways. Some of the properties of these elements are summarized in Table 16.1. We see first that although there is a general decrease in the atomic radii of the elements as the atomic number increases, the radii of the elements from chromium through copper are very similar. The increase in the nuclear charge along the series tends to cause an electron cloud to contract, but the added 3d-electrons exert an opposing effect. Consequently, the general size of the atoms remains nearly constant and decreases only slowly in the transition series.

Another indication that the effects of increasing nuclear charge and addition of 3d-electrons tend to offset each other is found in the variation of the first ionization energy of the atoms. Table 16.1 shows that although the first ionization energy, in general, increases as the atomic number increases, the ionization energies of neighboring elements are very nearly the same. A similar behavior is found for the second ionization energies, which for the most part increase smoothly as the atomic number increases. The exceptions are chromium and copper; the second ionization energies of these elements are notably larger than those of their neighbors. A rationalization of this observation lies in a comparison of the electron configurations of the singly and doubly charged ions. The second ionization of chromium involves the removal of an electron from a half-filled set of 3d-orbitals, and in the second ionization of copper, an electron is removed from a filled set of 3d-orbitals. The extra stability of a filled or half-filled set of equivalent orbitals has been encountered previously, for we have noted that the atoms of the nitrogen and noble-gas families have higher ionization energies than do their neighbors. We shall see that the stability of the half-filled or filled set of d-orbitals is reflected in the chemistry of chromium and copper.

The electronic configurations of the transition-metal atoms and ions illustrate an important point concerning the orbital energy-level scheme. Because the 3d-orbitals of the neutral transition-metal atoms are filled only after the 4s-orbital is occupied, one might conclude that the 4s-orbital lies lower in energy than the $3d$-orbitals. The electron configurations of the ions show, however, that this is not always true. The configurations given in Table 16.1 indicate that the $4s$-orbitals are vacant in the gaseous transition-metal ions. In other words, the $3d$-orbitals are of lower energy than the $4s$-orbital in the ions, even though the reverse is true in the neutral atoms. This phenomenon, discussed in more detail in Chapter 10, shows that there is no *rigid* pattern of orbital energies that holds for all atoms and ions.

The ionic radii given in Table 16.1 follow the trend established by the atomic radii. For ions of a given charge, the ionic radius decreases slowly as the atomic number increases. The radii of the doubly charged ions are all somewhat smaller than that of Ca^{++}, and thus we should expect to find the oxides of the transition elements similar to, but perhaps less basic and less soluble than CaO. This is found to be true experimentally. In addition the magnitudes of the ionic radii suggest that the hydration energies of the $+2$ ions of the transition elements should be similar to but greater than that of Ca^{++} (395 kcal). Examination of the hydration energies in Table 16.1 confirms this expectation.

The radii of the $+3$ ions of the transition elements are slightly larger than that of Ga^{+3} (0.60 A). Consequently we should expect the hydration energies of the $+3$ transition-metal ions to be similar in magnitude to that of Ga^{+3} (1124 kcal). Table 16.1 shows that this expectation is realized. Similarly we might expect the $+3$ oxides of the transition metals to be similar to but slightly less acidic than Ga_2O_3. We shall see that this is in fact true.

Examination of the standard electrode potentials given in Table 16.1 shows that all metals of the first transition series, with the exception of copper, should be oxidized by 1-M H$^+$. While these transition metals are good reducing agents, they are not as strong as the metals of groups IIA and IIIA. If we recall from Section 14.1 how the enthalpies of vaporization, ionization, and hydration influence the performance of metals as reducing agents, we can understand why the transition metals are not as good reductants as the alkaline-earth metals are. The enthalpies of vaporization of all the transition metals are quite large, and it is this relatively great stability of the metallic lattices that makes the transition elements poorer reductants than magnesium or aluminum. The reason that copper is a particularly poor reductant can be found by examining the data in Table 16.1. The second ionization energy of copper is quite a bit larger than the corresponding quantity for the other transition elements, and this factor makes the aqueous Cu^{++} ion relatively less stable and copper metal a poorer reductant than the other transition metals.

Although the electrode potentials indicate that the metals of the first transition series are relatively good reductants under equilibrium conditions, the actual rate at which the metals react with oxidizing agents like hydrogen ion is sometimes immeasurably small. Several of the metals are protected from chemical attack by a thin impervious layer of inert oxide. Chromium provides the best example of this, for despite its electrode potential, it can be used as a protective nonoxidizing metal, because it is coated with a nonreactive oxide, Cr_2O_3. Thus while the transition metals can behave as active reductants under the proper circumstances, in other situations they may appear to be essentially inert because of reaction rate effects.

16.2 THE SCANDIUM FAMILY

This group includes scandium, yttrium, lanthanum, and the lanthanides, the fourteen elements that follow lanthanum in the periodic table. Although scandium is not a particularly rare element, rather little is known about its chemistry. In a number of respects scandium resembles aluminum. The metal reacts vigorously with water to liberate hydrogen as does aluminum, when it is freed of its oxide coating. The oxide Sc_2O_3 is insoluble in water as is Al_2O_3, but because of the larger size of Sc^{+3}, Sc_2O_3 is basic rather than amphoteric like Al_2O_3. Like aluminum, scandium forms stable compounds only in the $+3$ oxidation state.

Yttrium is very similar to scandium. It is an active metal, as is shown by

$$Y^{+3} + 3e^- = Y, \qquad \varepsilon^0 = -2.37 \text{ volts.}$$

The oxide of yttrium, Y_2O_3, is a white powder, insoluble in water but soluble in acids. In its compounds, yttrium displays the $+3$ oxidation state exclusively.

Lanthanum also displays only the $+3$ oxidation state and has an insoluble basic oxide and a negative standard reduction potential:

$$La^{+3} + 3e^- = La, \qquad \varepsilon^0 = -2.52 \text{ volts.}$$

The fourteen elements that follow lanthanum also display the $+3$ oxidation state and in this respect resemble the members of group IIIB. These lanthanide elements are listed in Table 16.2, together with their oxidation states and electron configurations. The configurations and the occurrence of oxidation states other than $+3$ for some of the elements suggest that there is some extra stability associated with a half-filled or completely filled set of $4f$-orbitals.

The values of the standard reduction potentials for the $+3$ ions of the lanthanides, given in Table 16.2, demonstrate the remarkable similarity of the chemistry of these elements. There is a smooth but rather slight trend toward less negative reduction potentials as the atomic numbers increase. This trend is parallel to the lanthanide contraction, the decrease in ionic radii in this series that we discussed in Section 13.2. Because of the similarity between the lanthanides, they are difficult to separate, and many of the early investigations of their chemistry were made on mixtures of the elements. Particularly pure preparations of the lanthanide elements were made in the 1940's with the aid of ion-exchange techniques, and now the properties of the individual pure elements and their compounds are well known.

Table 16.2 Some properties of the lanthanide elements

| Name | Symbol | Configurations | | | | $\mathcal{E}^0$, $M^{+3} + 3e^- = M$ |
		M	M^{++}	M^{+3}	M^{+4}	
Lanthanum	La	$5d6s^2$	–	[Xe]	–	−2.52
Cerium	Ce	$4f^26s^2$	–	$4f$	[Xe]	−2.48
Praseodymium	Pr	$4f^36s^2$	–	$4f^2$	$4f$	−2.47
Neodymium	Nd	$4f^46s^2$	$4f^4$	$4f^3$	$4f^2$	−2.44
Promethium	Pm	$4f^56s^2$	–	$4f^4$	–	−2.42
Samarium	Sm	$4f^66s^2$	$4f^6$	$4f^5$	–	−2.41
Europium	Eu	$4f^76s^2$	$4f^7$	$4f^6$	–	−2.41
Gadolinium	Gd	$4f^75d6s^2$	–	$4f^7$	–	−2.40
Terbium	Tb	$4f^96s^2$	–	$4f^8$	$4f^7$	−2.39
Dysprosium	Dy	$4f^{10}6s^2$	–	$4f^9$	$4f^8$	−2.35
Holmium	Ho	$4f^{11}6s^2$	–	$4f^{10}$	–	−2.32
Erbium	Er	$4f^{12}6s^2$	–	$4f^{11}$	–	−2.30
Thulium	Tm	$4f^{13}6s^2$	$4f^{13}$	$4f^{12}$	–	−2.28
Ytterbium	Yb	$4f^{14}6s^2$	$4f^{14}$	$4f^{13}$	–	−2.27
Lutetium	Lu	$4f^{14}5d6s^2$	–	$4f^{14}$	–	−2.25

16.3 THE TITANIUM FAMILY

The atoms of titanium, zirconium, and hafnium have valence-electron configurations of the type $(n-1)d^2ns^2$. Of the possible oxidation states, $+2$ and $+3$ are observed only in the chemistry of titanium and zirconium, while the $+4$ state is common to all the elements. The tendency of the lower oxidation states to be less important for the heavier elements in a group is typical of the transition-metal families.

The data in Table 16.3 show that all of the group IVB metals have high melting and boiling temperatures and large enthalpies of vaporization. Despite the stabilities of the metallic crystals themselves, the compounds of the elements are formed with the evolution of considerable energy. On the basis of thermodynamic properties alone, each of the metals would be considered to be very reactive. However, all three metals are protected from chemical attack by a thin transparent layer of oxide MO_2 and therefore are quite resistant to chemical attack at ordinary temperatures.

It is difficult to prepare the pure titanium, zirconium, and hafnium metals, for each element reacts readily with oxygen, nitrogen, and carbon at elevated temperatures. The commercial preparation of titanium involves conversion of the oxide TiO_2 to the tetrachloride. The latter is a volatile compound that can be purified by distillation and then reduced with magnesium metal. Thus the process can be represented by

$$TiO_2 \xrightarrow{Cl_2, C} TiCl_4 \xrightarrow{Mg} Ti.$$

Table 16.3 Properties of the group IVB elements

	Ti	Zr	Hf
Atomic number	22	40	72
Configuration	$3d^24s^2$	$4d^25s^2$	$(4f^{14})$ $5d^26s^2$
Atomic radius, A	1.32	1.45	1.44
Melting point, °K	1950	2125	2495
Boiling point, °K	3550	4700	5500
ΔH_f (atom), kcal	112.7	146	168
Ionization energy, kcal	158	158	160
Ionic radius, M^{+4}, A	0.68	0.74	0.75
ΔH_f^0 (MO_2), kcal	−218	−258	−271
ΔH_f^0 (MF_4), kcal	−370	−445	–
ΔH_f^0 (MCl_4), kcal	−179	−230	–

In the laboratory, small amounts of the very pure metals of group IVB can be prepared by the thermal decomposition of the volatile tetraiodides MI_4 on a very hot wire. The reaction for titanium is simply

$$TiI_4(g) \rightarrow Ti(s) + 2I_2(g).$$

There is no particularly strong resemblance between the metals of groups IVA and IVB. The important oxidation state of tin and particularly lead is +2, not +4, while the reverse is true in the titanium family. The tetrahalides of both groups are, however, somewhat similar. Like $SnCl_4$, $TiCl_4$, $ZrCl_4$, and $HfCl_4$ are relatively volatile compounds that exist as discrete covalently bonded molecules.

Titanium

Of the three known oxidation states of titanium, +2, +3, and +4, the +4 state is most common and most stable under many conditions. Compounds of titanium in the +2 oxidation state can be prepared by reduction of the +4 state:

$$TiO_2 + Ti = 2TiO,$$
$$TiCl_4 + Ti = 2TiCl_2.$$

The oxide TiO somewhat resembles the oxides of group IIA metals. It is basic, ionic, and has a rock-salt crystal lattice. Like many of the other transition-metal oxides, however, it is a nonstoichiometric compound and has a composition close to $TiO_{0.75}$. Titanium in the +2 state is an extremely good reducing agent, and both TiO and $TiCl_2$ will reduce water to hydrogen. Because Ti^{++} decomposes water, there is essentially no aqueous solution chemistry of this ion.

The titanous ion, Ti^{+3}, is a violet species which, although stable in aqueous solution, is a strong reducing agent. It reacts rapidly and quantitatively with oxidizing agents like Fe^{+3} and MnO_4^- and with the oxygen of the air. The oxide Ti_2O_3 can be prepared by reduction of TiO_2 with hydrogen at high temperature:

$$2TiO_2 + H_2 = Ti_2O_3 + H_2O.$$

Like many of the other $+3$ oxides of the transition metals, Ti_2O_3 is stable with respect to the elements, basic, and quite insoluble in water.

The best known of the compounds of titanium in the $+4$ oxidation state is TiO_2. This oxide is a white insoluble powder that is used as a paint pigment. Because TiO_2 has a large refractive index, crystals of the oxide have a greater brilliance than diamonds, but are rather soft and are therefore relatively unsuited for use in jewelry. Although TiO_2 is very insoluble in pure water, it does dissolve slightly in strong base to form the titanate ion, whose formula is probably $[TiO_2(OH)_2]^=$. When treated with strong acids, TiO_2 dissolves to form species such as $Ti(OH)_3^+$ and $Ti(OH)_2^{++}$. The structures of these ions are not known, and the latter is sometimes represented by the formula TiO^{++}. In any case, it is clear that TiO_2 has both acidic and basic properties, and that the simple ion Ti^{+4} does not exist in aqueous solutions.

All of the tetrahalides of titanium have been prepared. A comparison of titanium tetrachloride, $TiCl_4$, with $TiCl_2$ and $TiCl_3$ is interesting because it illustrates a useful correlation between oxidation number and physical properties. The compounds $TiCl_2$ and $TiCl_3$ are ionic crystals whose vapor pressure reaches 1 atm only at temperatures near 1000°C. On the other hand, $TiCl_4$ is a liquid at room temperature and boils at 137°C. To explain the marked increase in the volatility of the halides as the oxidation number of titanium increases from three to four, it has been suggested that the bonds in $TiCl_4$ are of a covalent nature. Certainly it is difficult to imagine that four chloride ions could surround Ti^{+4} without being so distorted as to share their electrons with the central titanium atom.

When Ti, TiF_4, or TiO_2 are treated with aqueous HF, the very stable anion $TiF_6^=$ is formed. The anion $TiCl_6^=$ is less stable: while it can be formed by reaction of $TiCl_4$ with KCl, it is rather easily hydrolyzed in aqueous solution to species which contain oxygen.

Zirconium and Hafnium

These two elements are so similar chemically and physically that for some time a mixture of them was thought to be a single element. As Table 16.3 shows, the atomic and ionic radii of zirconium and hafnium are virtually identical. The expected larger size of the heavier element is not observed, because the lanthanide elements precede hafnium in the periodic table. The contraction in size associated with these elements makes the atoms of the elements that follow them smaller than would otherwise be expected.

The best known compounds of zirconium and hafnium are the oxides ZrO_2 and HfO_2. Zirconium dioxide has an extremely high melting point (3100°K) and once it has been heated to a high temperature, it is not attacked by acids or bases, and it has very favorable mechanical properties. Consequently it is used as a refractory material in furnace linings and in the manufacture of crucibles. The oxides of zirconium and hafnium are more basic than TiO_2, and thus are less soluble in alkalis and more soluble in acids. This change to a more basic nature is to be expected from the increased size of Zr^{+4} and Hf^{+4} compared to Ti^{+4}. Because of this increase in size, the solutions of Zr(IV) and Hf(IV) are less hydrolyzed than the acid solutions of Ti(IV), and consequently the aqueous solution chemistry of the heavier elements is more extensive than that of titanium (IV).

This chemistry is quite complicated, however. Acidic solutions of Zr(IV) contain polymeric cations such as $Zr_3(OH)_4^{+8}$ and $Zr_4(OH)_8^{+8}$. In concentrated HF, the situation is simpler, and the ions $ZrF_6^{=}$ and $HfF_6^{=}$ are the principal species in solution.

Table 16.4 Properties of the group VB elements

	V	Nb	Ta
Atomic number	23	41	73
Configuration	$3d^34s^2$	$4d^45s^1$	$5d^36s^2$
Ionization energy, kcal	155	159	182
Atomic radius, A	1.22	1.34	1.34
Melting point, °K	2190	2770	3270
Boiling point, °K	3650	4900	5600
ΔH_f (atom), kcal	123	173	186
ΔH_f^0 (M_2O_5), kcal	−373	−463	−500
ΔH_f^0 (M_2O_3), kcal	−290	–	–

16.4 THE VANADIUM FAMILY

The three elements of group VB are vanadium, niobium, and tantalum. As Table 16.4 shows, these metals have high melting and boiling points and large enthalpies of vaporization. The gross features of their chemistry resemble in a general way those of the group IVB metals. All elements of the vanadium family show several oxidation states: +2, +3, +4, and +5. All of these states are important in the chemistry of vanadium, but only the +5 state and to a lesser extent the +3 state are important for niobium and tantalum. Like the metals of group IVB, the metals of the vanadium family react readily with oxygen, carbon, and nitrogen at high temperatures, and thus are difficult to prepare by conventional high-temperature reduction processes. At lower temperatures an oxide coating protects the metals from chemical attack, even though they are strong reducing agents from the thermodynamic standpoint. Another feature which is held in common with the metals of the titanium family is that the two heavier elements of the group, niobium and tantalum, have virtually the same atomic radii. Consequently niobium and tantalum, like zirconium and hafnium, resemble each other chemically and physically.

Vanadium

The chief commercial use of vanadium is as an alloying agent in steels. Its general effect is to increase the ductility and tensile strength of the alloy. Fortunately this application does not require very pure vanadium, which, as we have noted, is difficult to prepare in quantity because of its high-temperature reactivity with carbon, nitrogen, and oxygen. Small amounts of very pure vanadium can be prepared by the decomposition of VI_4 on a hot wire.

The most important compound of vanadium is the pentoxide, V_2O_5. This red solid can be prepared by the direct combination of the elements at an elevated temperature, and it is used commercially as a catalyst in the contact process for preparation of sulfuric acid. Vanadium pentoxide is amphoteric. It dissolves in acids to form the pervanadyl ion, VO_2^+.

$$V_2O_5 + 2H^+(aq) = 2VO_2^+(aq) + H_2O.$$

There is a tendency for the vanadium species to polymerize in solutions of moderate acidity according to the reaction

$$10VO_2^+ + 8H_2O = H_2V_{10}O_{28}^{-4} + 14H^+.$$

When treated with strong base, V_2O_5 dissolves as VO_4^{-3}, which also has a tendency to polymerize:

$$2VO_4^{-3} + 3H^+ = HV_2O_7^{-3} + H_2O,$$
$$HV_2O_7^{-3} + VO_4^{-3} + 3H^+ = V_3O_9^{-3} + 2H_2O.$$

Thus the aqueous solution chemistry of vanadium (V) involves some rather complex species.

If an acidic solution of vanadium (V) is treated with a reducing agent like zinc metal or ferrous ion, a blue solution of vanadium (IV) or VO^{++}, the vanadyl ion, results. The vanadyl ion occurs as a discrete unit in such salts as $VOSO_4$ and $VOCl_2$. Like titanium (IV), vanadium (IV) is amphoteric. If a solution of VO^{++} is treated with alkali, VO_2 precipitates, but further treatment of this oxide with strong base dissolves it as the ion VO_4^{-4} and its polymerization products. Another point of resemblance between titanium (IV) and vanadium

(IV) is that VCl_4, like $TiCl_4$, is a low-boiling liquid (boiling point 154°C). This again illustrates the tendency of halides to exist as discrete small molecules when the oxidation number of the metal is high or the size of the metallic atom is small.

An aqueous solution of vanadium (III), V^{+3}, can be prepared by reduction of VO^{++} with zinc. As the reduction proceeds, the color of the solution changes from the bright blue of the vanadyl ion to the green color of V^{+3}. The $+3$ state of vanadium is entirely basic in nature, and treatment of a solution of V^{+3} with alkali precipitates the insoluble V_2O_3. The salts of V^{+3} are all ionic compounds.

Exhaustive reduction of aqueous solutions of any of the higher oxidation states of vanadium yields a violet solution of V^{++}. This ion is a rapid and moderately strong reducing agent. The oxide VO is basic and insoluble and has a nonstoichiometric composition. In each of these respects, vanadium (II) resembles titanium (II).

The following reduction-potential diagram provides a concise summary of the chemistry of vanadium in acidic aqueous solution:

$$VO_2^+ \underline{\quad 1.0 \quad} VO^{++} \underline{\quad 0.36 \quad} V^{+3} \underline{\quad -0.25 \quad} V^{++} \underline{\quad -1.2 \quad} V.$$

We can see from the reduction potentials that vanadium (V) and (IV) are easily reduced, that V^{++} is a moderately good reductant, and that vanadium metal is a strong reducing agent.

The only known halide of vanadium (IV) is VF_5, a viscous liquid which boils at 48°C. Both VF_4 and VCl_4 are known and are volatile substances which can be rather easily reduced to lower halides. VCl_4 resembles $TiCl_4$ in its general physical properties and in its very rapid reaction with water to form the oxide and oxychloride $VOCl_2$. All the vanadium trihalides are known and are intensely colored solids which can be decomposed at moderate temperatures (300–500°C) to the dihalides. The latter are ionic solids of considerable stability.

Niobium and Tantalum

Niobium and tantalum are frequently found together in nature, have similar chemical and physical properties, and are difficult to separate and prepare in a very pure state. There is considerable interest in niobium metal, for it loses all electrical resistance and becomes a superconductor at low temperatures. Tantalum is a very ductile metal, is resistant to chemical attack, and has good mechanical properties at high temperatures. It forms a carbide, TaC, that is extremely hard and which is used in the fabrication of cutting tools.

The most important oxidation state of niobium and tantalum is $+5$. The oxides Nb_2O_5 and Ta_2O_5 can be dissolved by fusing them with hydroxides, but they are not attacked by hydrogen ion. Like the anions of vanadium (V), the anions of niobium (V) and tantalum (V) are polymeric. However, the only species that exist in aqueous solutions have the formula $M_6O_{19}^{-8}$, where M is Nb or Ta. In these ions, the six metal atoms occupy the corners of a regular octahedron, and are themselves surrounded by octahedra of oxygen atoms which are shared in manner consistent with the stoichiometry of the anion. This tendency to polymerize is also found in the halides: $NbCl_5$ and $TaCl_5$ are dimeric molecules in the solid, with two chlorine atoms shared by forming bridge bonds between metal atoms. In the lower halides such as $NbCl_3$ and $NbCl_2$, the basic structural unit is the $(M_6X_{12})^{+2,+3}$ cation combined with halide anions. The M_6X_{12} unit consists of an octahedron of metal atoms with a bridging halogen atom along each edge of the octahedron. Because this unit may bear either a $+2$ or $+3$ charge, both species can appear in the same crystal lattice, and the niobium and tantalum halides are typically nonstoichiometric compounds.

16.5 THE CHROMIUM FAMILY

The metals of group VIB are chromium, molybdenum, and tungsten. As Table 16.5 shows, they are high-melting, high-boiling elements and have large enthalpies of vaporization. Tungsten, in fact, is the least volatile of all the elements. All the metals are hard and corrosion resistant and have a variety of technical uses, both as pure metals and alloying agents. Again the two heavier members of the family resemble each other markedly and differ somewhat from the first element in the group. Chromium forms the ions Cr^{++}, Cr^{+3}, and $CrO_4^=$, and each oxidation state has an extensive solution chemistry.

Table 16.5 Properties of the group VIB elements

	Cr	Mo	W
Atomic number	24	42	74
Configuration	$3d^54s^1$	$4d^55s^1$	$5d^46s^2$
Ionization energy, kcal	156	164	184
Atomic radius, A	1.17	1.29	1.30
Melting point, °K	2176	2890	3650
Boiling point, °K	2900	4900	5900
ΔH_f^0 (atom), kcal	95	157	200
ΔH_f^0 (MO_3), kcal	−145	−180	−201
ΔH_f^0 (MCl_2), kcal	−94.6	− 44	− 38
ΔH_f^0 (MCl_6), kcal	-	− 90	−98.7

On the other hand, the solution chemistry of molybdenum and tungsten is largely confined to the +6 state, and in contrast to chromium (VI), molybdenum (VI) and tungsten (VI) show no oxidizing properties.

The most stable oxidation state in most circumstances is chromium (III). Chromium (II) compounds are reducing agents, and chromium (VI) compounds are strong oxidizing agents. The acid-base properties associated with these oxidation states vary in the expected manner, with the acidity increasing as the oxidation number of chromium increases. The oxide CrO and the hydroxide $Cr(OH)_2$ are basic, Cr_2O_3 is amphoteric, and CrO_3 is acidic.

The preparation of chromium from its ore illustrates some important features of its chemistry. The principal chromium ore is the mixed oxide chromite, $FeO \cdot Cr_2O_3$. Direct reduction of chromite with carbon produces an iron-chromium mixture that is used in the manufacture of steel,

$$FeO \cdot Cr_2O_3 + 4C = 2Cr + Fe + 4CO.$$

To obtain pure chromium, the chromite ore is first oxidized with air under basic conditions at high temperature:

$$FeO \cdot Cr_2O_3 \xrightarrow[\text{air}]{K_2CO_3} K_2CrO_4 + Fe_2O_3.$$

Potassium chromate, K_2CrO_4, is readily soluble in water, but Fe_2O_3 is not, so the iron and chromium can be separated. The chromate is reduced to Cr_2O_3 with carbon:

$$2K_2CrO_4 + 2C = K_2CO_3 + K_2O + CO + Cr_2O_3.$$

Finally, Cr_2O_3 is reduced with aluminum in the thermite process:

$$Cr_2O_3 + 2Al = Al_2O_3 + 2Cr.$$

From the thermodynamic standpoint, chromium metal is a good reducing agent, and in fact, when it is finely divided it reacts rapidly and completely with oxygen. In the massive state, however, chromium is protected by a thin transparent coat of Cr_2O_3, and is extremely resistant to corrosion. Consequently chromium is used as a protective and decorative coating for other metals, and when incorporated in alloys like the stainless steels, it endows them with corrosion resistance.

Aqueous solutions of chromium (III) can be obtained by dissolving Cr_2O_3 in acid or alkali:

$$Cr_2O_3 + 6H^+ = 2Cr^{+3}(aq) + 3H_2O,$$
$$Cr_2O_3 + 2OH^- + 3H_2O = 2Cr(OH)_4^-.$$

It is well established that Cr^{+3} in aqueous solution consists of a central ion surrounded by six water molecules located at the apices of a regular octahedron.

This ion is hydrolyzed, or in other words, it is a weak acid:

$$Cr(H_2O)_6^{+3} = [Cr(H_2O)_5(OH)]^{+2} + H^+, \qquad K = 1.3 \times 10^{-4}.$$

It is interesting to note the relation of Cr^{+3} to the other M^{+3} ions of the $3d$ transition series (Ti^{+3}, V^{+3}, Mn^{+3}, Fe^{+3}, and Co^{+3}). The first two of these ions are reducing agents, while the last three are oxidizing agents. Chromium (III) is intermediate in behavior, since it is neither a strong reductant nor a strong oxidant.

Perhaps the outstanding characteristic of chromium (III) is its tendency to form stable complex ions with an enormous number of electron donors. The other di- and tripositive ions of the transition metals also display this property, but the complexes of chromium (III) are particularly inert once formed, and therefore have been studied extensively. Examples include the hexaquo ion $Cr(H_2O)_6^{+3}$, the hexammine complex $Cr(NH_3)_6^{+3}$, and anion complexes such as CrF_6^{-3}. The nature of these complexes, as well as those of the other transition elements, will be considered in some detail in Section 16.11.

Alkaline solutions of chromium (III) are easily oxidized to chromium (VI), as the following standard potential shows:

$$CrO_4^- + 4H_2O + 3e^- = Cr(OH)_3 + 5OH^-, \qquad \varepsilon^0 = -0.13 \text{ volt.}$$

Solutions of the chromate ion, $CrO_4^=$, are bright yellow. When these solutions are acidified, the orange dichromate ion, $Cr_2O_7^=$, is formed:

$$2CrO_4^= + 2H^+ = Cr_2O_7^= + H_2O, \qquad K = 4.2 \times 10^{14}.$$

The dichromate ion is a very powerful oxidizing agent, as shown by

$$Cr_2O_7^= + 14H^+ + 6e^- = 2Cr^{+3} + 7H_2O, \qquad \varepsilon^0 = 1.33 \text{ volts.}$$

Comparison of this standard potential with that of

$$O_2 + 4H^+ + 4e^- = 2H_2O, \qquad \varepsilon^0 = 1.23 \text{ volts,}$$

shows that solutions of the dichromate ion are intrinsically unstable with respect to decomposition to oxygen and Cr^{+3}. This reaction is slow, however, and solutions of dichromate ion can be kept for long periods without significant decomposition.

Addition of chromate salts to concentrated sulfuric acid produces a solution of the red oxide CrO_3. The reaction can be written

$$Na_2Cr_2O_7 + 3H_2SO_4 = 2Na^+ + H_3O^+ + 3HSO_4^- + 2CrO_3.$$

These solutions have extremely powerful oxidizing properties, and are used to clean chemical glassware of grease.

The aqueous chromous ion can be obtained by reducing solutions of Cr^{+3} with zinc. The potential for the reaction

$$Cr^{+3} + e^- = Cr^{++}, \qquad \varepsilon^0 = -0.41 \text{ volt,}$$

shows that chromous ion is one of the strongest reducing agents that can exist in aqueous solution. Solutions of chromous ion react rapidly and quantitatively with oxygen, and are sometimes used to remove oxygen from a mixture of gases.

Molybdenum and Tungsten

The ores of these metals can be converted to MoO_3 and WO_3, and reduction of these oxides with hydrogen yields the pure metals as powders. Because of the extremely high melting points of these metals, the fabrication of the powders into useful objects is very difficult. Both metals are used in situations where high temperatures may develop: x-ray tubes, electron tubes, electric furnaces, and electric light filaments are some examples. When incorporated into steels, molybdenum acts as a toughening agent, and tungsten extends the temperature range in which the steel remains hard.

Both molybdenum and tungsten form halides in the $+4$, $+5$, and $+6$ oxidation states, but only in the $+6$ state do these elements have any significant solution chemistry. The oxide MoO_3 is acidic and dissolves in base to form a very complicated series of polymeric oxyanions. Such ions as $Mo_2O_7^=$ and $Mo_7O_{24}^{-6}$ occur, as well as many others. The behavior of tungsten (VI) in aqueous solution is similar.

In the lower oxidation states, the halides have structures in which the fundamental structural unit is a cluster of metal and halogen atoms. Thus in $MoCl_2$, the structural unit is $Mo_6Cl_8^{+4}$, which consists of eight Mo atoms at the corners of a regular octahedron, with one Cl atom on each triangular face of the octahedron. In WCl_3, the structural unit is the $W_6Cl_{12}^{+6}$ ion, an octahedron of tungsten atoms, with bridging chlorine atoms along each edge of the octahedron. Thus, both with regard to formation of polymeric anions and halides which contain cluster cations built on metal octahedra, Mo and W resemble Nb and Ta rather clearly.

16.6 THE MANGANESE FAMILY

Of the elements manganese, technetium, and rhenium, manganese is by far the most abundant and important. Technetium does not occur in nature, for all its isotopes are radioactive and have fairly short half-lives. Rhenium is very rare, so rare that it was not discovered until 1925. It has a few applications in high-

Table 16.6 Properties of the group VIIB elements

	Mn	Tc	Re
Atomic number	25	43	75
Configuration	$3d^5 4s^2$	$4d^6 5s^1$	$5d^5 6s^2$
Ionization energy, kcal	171	168	182
Atomic radius, A	1.17	–	1.28
Melting point, °K	1517	–	3453
Boiling point, °K	2340	–	5800
ΔH_f^0 (atom), kcal	67.2	–	187
ΔH_f^0 (MO_2), kcal	−124.2	–	−100

temperature technology and catalysis. Some properties of these elements are given in Table 16.6.

Manganese

This element is a high-melting, high-boiling metal of considerable chemical reactivity. The potential for the reaction

$$Mn^{++} + 2e^- = Mn, \qquad \mathcal{E}^0 = -1.18 \text{ volt},$$

shows that manganese should dissolve readily in dilute acids, and indeed it does. In contrast to previous members of the first transition series, the metal is not protected by an oxide coating. Manganese occurs naturally as the oxide MnO_2. Small amounts of pure manganese can be made by thermal decomposition of MnO_2 to a mixture of the oxides MnO and Mn_2O_3, followed by reduction with aluminum:

$$3MnO_2 = MnO \cdot Mn_2O_3 + 2O_2,$$
$$3MnO \cdot Mn_2O_3 + 8Al = 4Al_2O_3 + 9Mn.$$

The principal use of manganese is as an additive in steels. For this purpose the impure manganese that results when the ores are reduced directly with carbon is satisfactory. Small amounts of manganese in steels react with the oxygen and sulfur and remove them in the slag as MnO_2 and MnS. Addition of larger amounts of manganese toughens and hardens the steel.

Compounds of manganese in the +2, +3, +4, +5, +6, and +7 oxidation states are known. Dissolution of the metal in dilute acid produces Mn^{++}, an ion with a faint pink color. In contrast with the aqueous doubly charged ions of titanium, vanadium, and chromium, Mn^{++} has no reducing properties. In fact, the potential of the reaction

$$Mn^{+3} + e^- = Mn^{++}, \qquad \mathcal{E}^0 = 1.51 \text{ volts},$$

shows that it is very difficult to oxidize Mn^{++} to Mn^{+3} in aqueous solution.

When solutions of Mn^{++} are treated with alkali, a gelatinous precipitate of $Mn(OH)_2$ forms. This hydroxide and the oxide MnO are entirely basic.

It is not possible to obtain significant quantities of Mn^{+3} in aqueous solution, for as its reduction potential of 1.51 volts shows, it is an oxidizing agent powerful enough to evolve oxygen from water. Moreover, combination of the half-reactions

$$Mn^{+3} + e^- = Mn^{++}, \qquad \mathcal{E}^0 = 1.51 \text{ volts},$$
$$MnO_2 + 4H^+ + e^- = Mn^{+3} + H_2O, \qquad \mathcal{E}^0 = 0.95 \text{ volt},$$

to give

$$2Mn^{+3} + 2H_2O = Mn^{++} + MnO_2 + 4H^+, \qquad \Delta\mathcal{E}^0 = 0.56 \text{ volt},$$

shows that Mn^{+3} is unstable with respect to disproportionation to Mn^{++} and MnO_2. Consequently there is essentially no aqueous solution chemistry of manganese (III). The ion Mn^{+3} is stable in the solid state, however, and can be obtained by oxidation of $Mn(OH)_2$ under basic conditions:

$$2Mn(OH)_2 + \tfrac{1}{2}O_2 = Mn_2O_3 + 2H_2O.$$

The solid Mn_2O_3 is a completely basic oxide.

The chemistry of manganese (IV) is not extensive. Virtually the only stable compound of manganese in this state is MnO_2, a dark brown powder which is a nonstoichiometric compound that is always noticeably deficient in oxygen. In acidic media MnO_2 is a very powerful oxidizing agent:

$$MnO_2 + 4H^+ + 2e^- = Mn^{++} + 2H_2O, \qquad \mathcal{E}^0 = 1.23 \text{ volts}.$$

Quite a different situation obtains in basic solution:

$$MnO_2 + 2H_2O + 2e^- = Mn(OH)_2 + 2OH^-, \qquad \mathcal{E}^0 = -0.50 \text{ volt}.$$

These potentials illustrate the important generalization that oxyanions and oxides are most powerful oxidizing agents when in acid solution, and correspondingly, that it is easiest to *produce* these oxygenated species of high oxidation state in basic solution.

When MnO_2 is heated with KNO_3 and KOH, solid K_2MnO_4, or potassium manganate, is produced. This salt, in which manganese is the +6 oxidation state, is bright green, soluble in water, and is stable only in basic solutions. The half-reactions

$$MnO_4^- + 4H^+ + 2e^- = MnO_2 + 2H_2O, \qquad \mathcal{E}^0 = 2.26 \text{ volts},$$
$$MnO_4^- = MnO_4^- + e^-, \qquad \mathcal{E}^0 = 0.56 \text{ volt},$$

show that manganate ion is unstable with respect to disproportionation in acid solution:

$$3MnO_4^- + 4H^+ = 2MnO_4^- + MnO_2 + 2H_2O, \qquad \varepsilon^0 = 1.70 \text{ volts.}$$

As a consequence of this extreme instability in acidic solution, there is very little aqueous solution chemistry of manganese (VI).

Perhaps the best known compound of manganese is potassium permanganate, $KMnO_4$. This salt has a very intense purple color and is a very powerful oxidizing agent:

$$MnO_4^- + 8H^+ + 5e^- = Mn^{++} + 4H_2O, \qquad \varepsilon^0 = 1.51 \text{ volts.}$$

Even though permanganate is capable of oxidizing water to oxygen, the reaction is rather slow, and aqueous solutions of permanganate are important reagents in analytical chemistry. By addition of potassium permanganate to concentrated sulfuric acid, it is possible to produce the extremely unstable liquid Mn_2O_7. As we might have anticipated from the trend established among the other transition elements, compounds of manganese (VII) are entirely acidic.

Technetium and Rhenium

These two elements resemble each other and differ in a number of respects from manganese. The $+2$ oxidation state, important for manganese, is unknown in the chemistry of technetium and rhenium. The most important oxidation states of technetium are $+4$ and $+7$, while the $+3$, $+5$, and $+6$ states are difficult to prepare. For rhenium, the $+3$, $+4$, and $+7$ states are best known, while $+5$ and $+6$ states are difficult to prepare.

When technetium and rhenium are heated in air, the heptoxides Tc_2O_7 and Re_2O_7 are obtained. Both these compounds are low-melting solids (119 and 220°C, respectively). The oxides dissolve readily in water to give acidic solutions of TcO_4^- and ReO_4^-. These solutions possess only moderate oxidizing strength, in contrast to solutions of MnO_4^-. The lower oxides of technetium and rhenium can be made by heating the metals with the heptoxides, as for example in

$$Re + 3Re_2O_7 = 7ReO_3.$$

For both technetium and rhenium, the oxides MO_3 and MO_2 are known.

In contrast to manganese, the heavier elements of group VIIB form a number of relatively volatile halides. The known halides of technetium are $TcCl_4$, $TcCl_6$, and TcF_6. Rhenium forms a more extensive series of halides. The compounds ReX_4 are known; here X stands for F, Cl, Br, and I. Only fluorine, chlorine, and bromine form compounds of the formula ReX_5. The only known halides of rhenium (VI) are ReF_6 and $ReCl_6$, while ReF_7 is the only halide of rhenium (VII). Once again, the tendency for the heavy transition metals to form clusters in the halides of lower oxidation number is displayed. $ReCl_3$ actually consists of Re_3Cl_9 units, with the three Re atoms forming an equilateral triangle. Along each edge, and in the plane of the triangle, there is a Cl atom forming bridge bonds between Re atoms. In addition, two more Cl atoms are bonded to each Re, one above and one below the plane of the triangle.

16.7 IRON, COBALT, AND NICKEL

At this point we will temporarily abandon our practice of discussing the vertical groups of the transition metals, and instead treat the horizontal triad iron, cobalt, and nickel together. The reason for this is that the horizontal resemblances among these members of the first transition series are more noticeable than are their similarities to the heavier members of their vertical groups. These latter six elements, ruthenium, rhodium, and palladium in the second transition series, and osmium, iridium, and platinum in the third series, bear some strong resemblances to each other and will be treated as a group in the next section.

Table 16.7 Properties of the iron triad elements

	Fe	Co	Ni
Atomic number	26	27	28
Configuration	$3d^6 4s^2$	$3d^7 4s^2$	$3d^8 4s^2$
Ionization energy, kcal	182	181	176
Atomic radius, A	1.16	1.16	1.15
Melting point, °K	1812	1768	1728
Boiling point, °K	3150	3150	3160
ΔH_f^0 (atom), kcal	99.5	102	103
ΔH_f^0 (MO), kcal	−63.8	−57.2	−58.4

Iron, cobalt, and nickel are hard, high-melting, high-boiling metals of moderate reactivity. Table 16.7 summarizes some of their properties. They are all ferromagnetic to some degree. Their chemistry is principally confined to the lower oxidation states +2 and +3. Some higher oxidation states are known, but in accordance with the trend established by Ti, V, Cr, and Mn, these higher oxidation states are very unstable and have powerful oxidizing properties.

Iron

Iron constitutes 4.7% of the earth's crust and is second in abundance only to aluminum among the metals. This abundance, and its desirable mechanical properties in the impure condition, make iron an element of foremost technological importance. As is well known, metallic iron is produced by the reduction of iron oxide in a blast furnace. Some of the reactions that occur in the furnace reveal notable features of the chemistry of iron. One of the first changes that the oxide Fe_2O_3 undergoes occurs in the relatively cool regions (200°C) of the furnace:

$$3Fe_2O_3 + CO = 2Fe_3O_4 + CO_2.$$

The oxide of composition Fe_3O_4 also occurs naturally as magnetite, the magnetic iron oxide. It can be regarded as a mixed oxide of iron (II) and iron (III): $FeO \cdot Fe_2O_3$. At somewhat higher temperatures (350°C), further reduction of the iron oxide occurs:

$$Fe_3O_4 + CO = 3FeO + CO_2.$$

As the oxide drops further down into the blast furnace, it encounters higher temperatures and is finally reduced to the metal:

$$FeO + CO = Fe + CO_2.$$

The iron that is drawn from the blast furnace contains sulfur, phosphorus, and silicon impurities as well as 4% carbon, which is present as the carbide Fe_3C. To produce high-quality steels in which the carbon content is, in general, less than 1.5%, the molten iron is treated with air or oxygen until most of the carbon is burned out, and the other impurities are separated as oxides in a slag. The desired alloying metals are added, and the steel poured into molds to cool.

As pointed out in the summary of the steel manufacturing process, there are three important oxides of iron, FeO, Fe_2O_3, and Fe_3O_4. Each of these compounds has a marked tendency to have a nonstoichiometric composition, and each is rather readily oxidized or reduced into one of the other forms. We can rationalize these features in terms of the crystal structures of the solids. Imagine a cubic closest-packed lattice of oxide ions. If all the octahedral holes were filled with Fe^{++} ions, we would have a perfect rock-salt lattice, which is the structure of FeO. If a small number of the Fe^{++} ions are replaced by two-thirds as many Fe^{+3} ions, we would have an iron-deficient, but electrically neutral crystal. The actual composition of iron (II) oxide as usually prepared is $Fe_{0.95}O$. Conversion of two-thirds of the Fe^{++} ions to Fe^{+3} would give a composition $FeO \cdot Fe_2O_3$ or Fe_3O_4. In this mixed oxide, all Fe^{++} ions are in octahedral sites, but half of the Fe^{+3} ions are in octahedral sites and half are in tetrahedral sites. Finally, replacement of all the Fe^{++} ions by two-thirds as many Fe^{+3} ions gives the composition $Fe_{0.67}O$ or Fe_2O_3. Thus each of the oxides can change its composition in the direction of one or two of the others without the occurrence of a major structural change in the oxide lattice.

As the potential for the reaction

$$Fe^{++} + 2e^- = Fe, \qquad \varepsilon^0 = -0.44 \text{ volt},$$

shows, iron is a moderately good reducing agent, and it does in fact dissolve slowly in dilute acids. When treated with concentrated nitric acid, however, a protective oxide film forms and the metal becomes "passive" and does not dissolve. Although the anhydrous salts of iron (II) like $FeCl_2$ are colorless, the hydrated salts and aqueous solutions of Fe^{++} are pale green. In aqueous solution Fe^{++} can be oxidized by air to Fe^{+3}, as the following potentials show:

$$Fe^{+3} + e^- = Fe^{++}, \qquad \varepsilon^0 = 0.771 \text{ volt},$$
$$2Fe^{++} + \tfrac{1}{2}O_2 + 2H^+ = 2Fe^{+3} + H_2O, \qquad \Delta\varepsilon^0 = 0.46 \text{ volt}.$$

This oxidation of Fe^{++} is moderately rapid in neutral solution, but is somewhat slower in acidic solution. Because of air oxidation, solutions of Fe^{++} always contain some Fe^{+3} unless they have been freshly prepared and are acidic.

When solutions of Fe^{++} are treated with base, $Fe(OH)_2$ precipitates. Although this compound is white, it is darkened very rapidly by air oxidation. Although $Fe(OH)_2$ dissolves readily in acids and appears therefore to be basic, it does show some amphoteric behavior. Prolonged treatment with hot concentrated NaOH dissolves $Fe(OH)_2$, and when the solution is cooled, $Na_4[Fe(OH)_6]$ precipitates. Under most circumstances, however, $Fe(OH)_2$ behaves like a basic hydroxide.

Iron (III) exists in aqueous solution as the hydrated ion Fe^{+3}. Because of its large charge and small size, Fe^{+3} hydrolyzes, or acts as an acid, as the following equilibrium constants show:

$$[Fe(H_2O)_6]^{+3} = [Fe(H_2O)_5(OH)]^{++} + H^+, \qquad K = 9 \times 10^{-4},$$
$$[Fe(H_2O)_5(OH)]^{++} = [Fe(H_2O)_4(OH)_2]^+ + H^+, \qquad K = 5.5 \times 10^{-4}.$$

The reddish-brown color of solutions of Fe^{+3} is attributed to the hydrolysis products, as the solutions can be almost decolorized by the addition of nitric acid. On the other hand, as the pH of a solution of Fe^{+3} is raised, the color deepens until hydrated Fe_2O_3 precipitates from solution. As might be expected from the extreme acidity of Fe^{+3} in aqueous solution, hydrated Fe_2O_3 dissolves somewhat in base as well as in acids and is therefore slightly amphoteric.

Both Fe^{++} and Fe^{+3} form complexes with a large number of electron donors. The nature of some of these will be discussed in Section 16.11. At this point, however, a comparison of the reduction potentials

$$Fe^{+3} + e^- = Fe^{++}, \qquad \varepsilon^0 = 0.771 \text{ volt},$$
$$Fe(CN)_6^{-3} + e^- = Fe(CN)_6^{-4}, \qquad \varepsilon^0 = 0.36 \text{ volt},$$

provides a good example of how the relative stabilities of oxidation states can be affected by complex-ion formation. Aqueous ferric ion is a good oxidizing agent, but the ferricyanide ion, $Fe(CN)_6^{-3}$, is much less powerful. Apparently, formation of the complex ion stabilizes Fe(III) more than it does Fe(II).

While virtually all the solution chemistry of iron is confined to the +2 and +3 oxidation states, it is possible to prepare a compound of iron (VI). Treatment of Fe_2O_3 with strong base and chlorine produces a solution of the ferrate ion, $FeO_4^=$. Although stable in basic solution, in neutral or acidic media it decomposes according to

$$2FeO_4^= + 10H^+ = 2Fe^{+3} + \tfrac{3}{2}O_2 + 5H_2O.$$

The ferrate ion is a powerful oxidizing agent, even stronger than MnO_4^-.

Cobalt

This element occurs in nature as the sulfide Co_3S_4 and arsenide $CoAs_2$. Its ores also usually contain nickel and often iron and copper. The recovery of cobalt involves roasting the ores to convert them to the oxide CoO, which is then reduced with carbon, aluminum, or hydrogen. The metal itself is hard, has a bluish-white lustre, and is moderately reactive. The potential for the reaction

$$Co^{++} + 2e^- = Co, \qquad \varepsilon^0 = -0.28 \text{ volt},$$

shows that cobalt is a less active reducing agent than iron, but will nevertheless dissolve in dilute acids. Like iron, cobalt has two important oxidation states: $+2$ and $+3$. However, cobalt (III) is a much more powerful oxidant than iron (III). As the potential for the reaction

$$Co^{+3} + e^- = Co^{++}, \qquad \varepsilon^0 = 1.84 \text{ volts},$$

shows, Co^{+3} can oxidize water to oxygen, and does so quite rapidly. Consequently there is essentially no aqueous solution chemistry for the free Co^{+3} ion. Complexes of cobalt (III) are important, however, and are considerably weaker oxidants than the aqueous Co^{+3}, as is shown by

$$[Co(NH_3)_6]^{+3} + e^- = [Co(NH_3)_6]^{+2}, \qquad \varepsilon^0 = 0.1 \text{ volt}.$$

Many other complex ions of cobalt (III) are known, and some of these will be discussed in Section 16.11.

The hydroxide $Co(OH)_2$ is insoluble, and somewhat amphoteric. It dissolves readily in dilute acids, but very concentrated alkali is necessary to dissolve it as $[Co(OH)_4]^=$. Thus it behaves quite similarly to $Fe(OH)_2$. The oxide CoO, like FeO, has the rock-salt structure with the Co^{++} ions occupying the tetrahedral sites in the cubic closest-packed oxide lattice. Treatment with oxygen at high temperatures converts CoO to Co_3O_4, another cubic close-packed lattice of oxide ions with Co^{+2} at tetrahedral sites and Co^{+3} at octahedral sites. The simple oxide Co_2O_3, which would be analogous to Fe_2O_3 is not known, but the hydrate $Co_2O_3 \cdot H_2O$ is known. Thus while there are some similarities between the Co-O and Fe-O systems, they are not completely analogous.

Nickel

Nickel continues the trend of decreasing stability of higher oxidation states established by the other transition elements. The only important oxidation state that occurs in aqueous solution is +2, and the +3 and +4 states appear only in a few compounds. The electrode potential

$$Ni^{++} + 2e^- = Ni, \qquad \varepsilon^0 = -0.25 \text{ volt},$$

shows that from the thermodynamic standpoint, nickel is only slightly poorer as a reductant than cobalt is. In fact, nickel is quite corrosion resistant, for it is covered with a thin protective coating of oxide, and reacts only very slowly with oxidizing agents. It is particularly resistant to attack under alkaline conditions, and is often used to make crucibles or electrodes for use with basic media. Finally divided nickel can absorb large amounts of hydrogen gas, which enters the metallic lattice as atoms. As a result, porous nickel "sponge" is an excellent catalyst for the hydrogenation of organic compounds:

$$H_2 + C_2H_4 \xrightarrow{Ni} C_2H_6.$$

Monel, an alloy of nickel and copper, resists attack by fluorine, and is used to contain and handle this gas.

In Nature, nickel is frequently found as the sulfide NiS. Roasting this compound in air produces the oxide NiO, which can be reduced with carbon to give metallic nickel. Very pure nickel can be made by the carbonyl process. Crude metallic nickel reacts readily with carbon monoxide at 50°C to give the volatile nickel tetracarbonyl, $Ni(CO)_4$. The pure metal can be recovered merely by pyrolysing the carbonyl at approximately 200°C.

In aqueous solution the hydrated Ni^{+2} is green, and the salts of Ni^{+2} are green or blue. Like iron (II) and Co (III), nickel (II) forms many complex ions. Like $Fe(OH)_2$ and $Co(OH)_2$, $Ni(OH)_2$ is insoluble, but in contrast to the former substances, $Ni(OH)_2$ shows no amphoteric properties. The oxide NiO, like FeO and CoO, has the rock-salt structure.

When $Ni(OH)_2$ is treated with alkali and a moderately strong oxidizing agent like bromine, a black solid whose composition is close to $Ni_2O_3 \cdot H_2O$ is obtained. Strong oxidizing agents like Cl_2 acting on $Ni(OH)_2$ give a solid whose composition approaches NiO_2. Thus it is possible to prepare both the nickel (III) and (IV) oxides, although neither is pure. Both oxides are very strong oxidizing agents. In fact, Ni_2O_3 is used in the Edison cell, which employs the reaction

$$Fe + Ni_2O_3 \cdot H_2O + 2H_2O \underset{\text{charge}}{\overset{\text{discharge}}{\rightleftharpoons}} Fe(OH)_2 + 2Ni(OH)_2$$

and delivers about 1.3 volts.

16.8 THE PLATINUM METALS

Ruthenium, rhodium, palladium, osmium, iridium, and platinum are the elements of the $4d$- and $5d$-triads of group VIII, and are commonly known as the platinum metals. They are rather rare metals and have in common a general resistance to chemical attack. As Table 16.8 shows, the platinum metals have

Table 16.8 Properties of the platinum metals

	Ru	Rh	Pd	Os	Ir	Pt
Atomic number	44	45	46	76	77	78
Configuration	$4d^7 5s^1$	$4d^8 5s^1$	$4d^{10}$	$5d^6 6s^2$	$5d^9$	$5d^9 6s^1$
Ionization energy, kcal	170	172	192	200	200	210
Atomic radius, A	1.24	1.25	1.25	1.26	1.26	1.29
Melting point, °K	2770	2250	1823	2970	2720	2043
Boiling point, °K	4000	4000	3300	4500	4400	4100
ΔH_f^0 (atom), kcal	144	133	91	165	155	135

high melting and boiling temperatures and large enthalpies of vaporization. Apparently it is this great stability of the metallic lattice that is largely responsible for the inert nature of the metals.

The platinum metals differ somewhat in physical appearance and mechanical properties. Ruthenium and osmium are gray brittle metals that are very hard. Rhodium and iridium are white in appearance, and rhodium is rather soft and ductile while iridium is hard and brittle. Palladium and platinum are white metals of considerable lustre and are softer and more malleable than the other metals of the group. In this latter respect palladium and platinum resemble silver and gold, their neighbors in the periodic table.

Many of the generalizations that apply to the behavior of the other transition metals fail or are modified when applied to the platinum metals. The importance of the higher oxidation states, so obvious for the other heavy transition elements, is considerably diminished among the platinum metals. Ruthenium and osmium, members of the iron family, do form the oxides RuO_4 and OsO_4 in which the oxidation number of the metal is $+8$. However, Table 16.9 shows that none of the other platinum metals displays an oxidation number higher than $+4$ in its oxide. In general, oxidation states of $+4$ and lower are most important in the chemistry of the platinum metals.

Although the metals have in common a general inert nature, their response to the various oxidizing agents is rather different. Only osmium and ruthenium are readily attacked by oxygen and form volatile oxides. Only palladium will

Table 16.9 Principal oxides and halides of the platinum metals

	Ru	Rh	Pd	Os	Ir	Pt
F	RuF_5	RhF_3	PdF_3	OsF_5, OsF_6	IrF_3, IrF_6	PtF_4, PtF_6
Cl	$RuCl_3$	$RhCl_3$	$PdCl_2$	$OsCl_3$, $OsCl_4$	$IrCl_3$	$PtCl_2$, $PtCl_3$, $PtCl_4$
Br		$RhBr_3$				$PtBr_2$, $PtBr_3$, $PtBr_4$
I						PtI_2, PtI_3, PtI_4
M^{II}			PdO			
M^{III}		Rh_2O_3			Ir_2O_3	
M^{IV}	RuO_2	RhO_2		OsO_2	IrO_2	PtO_2
M^{VIII}	RuO_4			OsO_4		

dissolve in nitric acid, while platinum, osmium, and palladium are attacked by aqua regia, a mixture of nitric and hydrochloric acids. Osmium and ruthenium react with alkaline oxidizing agents to give Na_2OsO_4 and Na_2RuO_4, and similar treatment attacks platinum and palladium. Iridium combines rather readily with fluorine, but the other metals must be heated above 300°C before they react. Chlorine combines with all the metals at elevated temperatures, but ruthenium and palladium are particularly resistant and must be raised to a red heat before they react. In the use of platinum laboratory apparatus, it is important to remember that the platinum metals react with the semimetallic elements, particularly sulfur, phosphorus, arsenic, antimony, and lead.

The chemistry of the platinum metals is difficult to systematize, and in some respects is not well documented. Like the other heavy transition metals, these elements do not form simple monatomic cations. Moreover, only ruthenium and osmium form oxyanions. The oxides, halides, and sulfides are the only compounds that do not involve coordination or complexation of the metal atom. For a description of the complex compounds of the platinum metals, an advanced textbook of inorganic chemistry should be consulted.

16.9 COPPER, SILVER, AND GOLD

Copper, silver, and gold are moderately soft, very ductile and malleable metals that are excellent conductors of heat and electricity. These group IB elements show little resemblance to the alkali metals of group IA. Table 16.10 shows that copper, silver, and gold are moderately high-melting and high-boiling metals that have enthalpies of vaporization from two to four times greater than those of the alkali metals. The ionization energies of the atoms of the copper family are nearly twice those of the alkali atoms. Thus we might expect, and in fact find, that the metals of group IB are more inert than those of group IA. Copper, silver, and gold are sufficiently resistant to oxidation that they are sometimes found in the uncombined state in nature.

Table 16.10 Properties of the group IB elements

	Cu	Ag	Au
Atomic number	29	47	79
Electron configuration	$3d^{10}4s^1$	$4d^{10}5s^1$	$5d^{10}6s^1$
Ionization energy, kcal	178	174	213
Atomic radius, A	1.17	1.34	1.34
Melting point, °K	1356	1234	1336
Boiling point, °K	2855	2450	3080
ΔH_f^0 (atom), kcal	81.1	68.4	87.3
Oxidation states	+1, +2	+1, +2	+1, +3
ΔH_f^0 (MCl), kcal	−32.5	−30.4	−8.4
ΔH_f^0 (M$_2$O), kcal	−39.8	− 7.3	—
ΔH_f^0 (MO), kcal	−37.1	− 6.0	—

Table 16.10 gives the oxidation states displayed by the group IB elements. The $+1$ oxidation state is very important in the chemistry of silver, but less important for copper and gold. The $+2$ state is the most important oxidation state of copper, but occurs in only a few silver compounds, and does not appear at all in the chemistry of gold. The $+3$ state is not at all important for copper and silver, but is the most important state of gold. In the aqueous solution chemistry of these elements, the number of oxidation states is even further restricted. Only the $+2$ ion of copper and the $+1$ ion of silver exist in important concentrations in aqueous solutions.

Copper

The extraction and refinement of copper are relatively simple processes and reveal some of its more important chemical properties. The carbonate ores of copper can be reduced with carbon:

$$CuCO_3 \cdot Cu(OH)_2 + C = 2Cu + 2CO_2 + H_2O.$$

Sulfide ores are partially oxidized and then smelted to give a rather impure product:

$$Cu_2S \xrightarrow{O_2} Cu_2O + Cu_2S \xrightarrow{heat} Cu + SO_2.$$

The copper obtained from these reductions contains iron and silver impurities which can be removed by electrolysis. The impure copper is oxidized at the anode and the pure product recovered at the cathode. The half-reaction potentials

$$Ag^+ + e^- = Ag, \qquad \mathcal{E}^0 = 0.80 \text{ volt},$$
$$Cu^{++} + 2e^- = Cu, \qquad \mathcal{E}^0 = 0.34 \text{ volt},$$
$$Fe^{++} + 2e^- = Fe, \qquad \mathcal{E}^0 = -0.44 \text{ volt},$$

show that metallic silver is more difficult to oxidize than copper, and ferrous ion is more difficult to reduce than cupric ion. Thus by applying an appropriate potential to the electrolysis cell, it is possible to oxidize copper and iron, but not silver, and to reduce cupric ion, but not ferrous ion. The copper metal is recovered pure at the cathode, and the metallic silver is left as a sludge in the anode compartment.

In its most common compounds, copper is found in the $+1$ or the $+2$ oxidation states. The potentials for the half-reactions

$$Cu^+ + e^- = Cu, \qquad \mathcal{E}^0 = 0.52 \text{ volt},$$
$$Cu^{++} + 2e^- = Cu, \qquad \mathcal{E}^0 = 0.34 \text{ volt},$$

show that cuprous ion, Cu^+, is unstable in aqueous solution, for we have

$$2Cu^+ = Cu + Cu^{++}, \qquad \Delta\mathcal{E}^0 = 0.18 \text{ volt},$$

and thus

$$\frac{[Cu^{++}]}{[Cu^+]^2} = K = 1.2 \times 10^6.$$

Consequently there is no aqueous solution chemistry for uncomplexed copper (I) ions. On the other hand, complexes of copper (I) can be prepared without difficulty and many are stable in aqueous solution. For example, if a solution of Cu^{++} is boiled with excess chloride ion and metallic copper, the following reaction occurs:

$$Cu^{++} + Cu + 4Cl^- = 2CuCl_2^-.$$

If the resulting solution is diluted, the chloride-ion concentration decreases, and the insoluble CuCl precipitates:

$$CuCl_2^-(aq) + H_2O = CuCl(s) + Cl^-(aq), \qquad K = 6.5 \times 10^{-2}.$$

The cuprous halides CuCl, CuBr, and CuI are quite different from the alkali-metal halides. In the first place, the cuprous halides are only slightly soluble in water, as is shown by

$$CuCl = Cu^+ + Cl^-, \qquad K = 3.2 \times 10^{-7},$$
$$CuBr = Cu^+ + Br^-, \qquad K = 5.9 \times 10^{-9},$$
$$CuI = Cu^+ + I^-, \qquad K = 1.1 \times 10^{-12}.$$

Secondly, the cuprous halides have the zinc-blende crystal structure in which the coordination number is only four, while the alkali halides have either the rock-salt or cesium-chloride structure. Third, although the lattice energies of the alkali halides can be calculated quite accurately on the assumption that they are ionic crystals, the lattice energies of the cuprous halides are slightly greater than would be expected from the ionic model. This suggests that in contrast to the alkali halides, the cuprous halides exhibit a certain degree of covalent bonding.

Other well-known compounds of copper (I) are the oxide Cu_2O and the sulfide Cu_2S. Both these compounds can be made by direct combination of the elements at high temperature, both tend to be nonstoichiometric, and both are extremely insoluble in water. In these latter respects they differ noticeably from the corresponding compounds of the alkali metals.

The chemistry of copper in the $+2$ oxidation state is similar to the chemistry of the other $+2$ ions of the transition metals. The cupric ion is colored and reacts with a number of electron donors to form complex ions. Addition of base to a solution of Cu^{++} precipitates $Cu(OH)_2$. This hydroxide dissolves readily in acids and to a slight extent in excess base to form the anion $Cu(OH)_4^=$. Like the other sulfides of the transition metals, CuS is only very slightly soluble in water.

Cupric fluoride is an ionic compound which has the fluorite crystal lattice. In contrast, anhydrous cupric chloride and bromide consist of infinite chains of atoms with the arrangement shown in Fig. 16.1. This type of structure is a departure from those found in the dihalides of the alkaline-earth metals and the metals of the first transition series, and the low coordination number of copper in cupric chloride and bromide is interpreted by some chemists as an indication of some covalent bonding in these compounds. Cupric iodide is not known. Addition of I^- to Cu^{++} solutions results in the rapid, quantitative production of CuI and I_2.

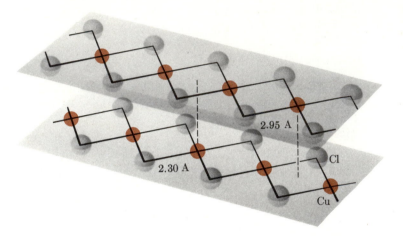

FIG. 16.1   The infinite chain structure of cupric chloride.

Silver

Both in solid compounds and in aqueous solutions, the normal oxidation state of silver is +1. The colorless salts $AgNO_3$ and $AgClO_4$ are readily soluble in water, but for the most part the simple binary compounds of silver are sparingly soluble. Thus addition of base to a solution of Ag^+ precipitates the brown oxide Ag_2O. This compound is predominately basic, but it does dissolve slightly in concentrated alkali to form $Ag(OH)_2^-$. Similarly, silver sulfide and the halides except the fluoride are quite insoluble:

$$Ag_2S = 2Ag^+ + S^=, \quad K = 10^{-50},$$
$$AgCl = Ag^+ + Cl^-, \quad K = 2.8 \times 10^{-10},$$
$$AgBr = Ag^+ + Br^-, \quad K = 5.0 \times 10^{-13},$$
$$AgI = Ag^+ + I^-, \quad K = 8.5 \times 10^{-17}.$$

Silver forms a number of complexes in aqueous solution. The following equilibrium constants indicate the range of stabilities:

$$AgCl_2^- = Ag^+ + 2Cl^-, \qquad K = 2.2 \times 10^{-6},$$
$$Ag(NH_3)_2^+ = Ag^+ + 2NH_3, \qquad K = 5.9 \times 10^{-8},$$
$$Ag(S_2O_3)_2^{-3} = Ag^+ + 2S_2O_3^=, \qquad K = 5.9 \times 10^{-14},$$
$$Ag(CN)_2^- = Ag^+ + 2CN^-, \qquad K = 1.8 \times 10^{-19}.$$

Combination of the reactions

$$AgCl(s) = Ag^+ + Cl^-, \qquad K = 2.8 \times 10^{-10},$$
$$AgCl_2^- = Ag^+ + 2Cl^-, \qquad K = 2.2 \times 10^{-6},$$

gives us

$$AgCl(s) + Cl^- = AgCl_2^-, \qquad K = 1.2 \times 10^{-4}.$$

This reaction and its equilibrium constant show that the slightly soluble AgCl can be dissolved in an excess of chloride ion.

Combination of the dissociation constant of the silver-ammonia complex ion with the solubility products of the silver halides yields the equilibrium constants for the following reactions:

$$AgCl + 2NH_3 = Ag(NH_3)_2^+ + Cl^-, \qquad K = 4.7 \times 10^{-3},$$
$$AgBr + 2NH_3 = Ag(NH_3)_2^+ + Br^-, \qquad K = 8.5 \times 10^{-6},$$
$$AgI + 2NH_3 = Ag(NH_3)_2^+ + I^-, \qquad K = 1.4 \times 10^{-9}.$$

Thus AgCl is moderately soluble in concentrated ammonia, but AgBr and AgI do not dissolve to an appreciable extent. Silver bromide and silver iodide are soluble in solutions of thiosulfate ion, $S_2O_3^=$, for the silver-thiosulfate complex ion has a very small dissociation constant. In the photographic process, the developed image is fixed by washing out unexposed grains of AgBr with a solution of sodium thiosulfate.

The electrode potential

$$Ag^+ + e^- = Ag, \qquad \varepsilon^0 = 0.799 \text{ volt},$$

shows that silver metal is rather difficult to oxidize to the aqueous ion. It is even more difficult to obtain higher oxidation states of silver. Treatment of aqueous silver nitrate with ozone yields a solution in which Ag^{++} exists as a short-lived species which is partially stabilized by complex formation with nitrate ion. The potential for the reaction

$$Ag^{++} + e^- = Ag^+, \qquad \varepsilon^0 = 2.0 \text{ volts},$$

is only approximate, but does show how strong an oxidizing agent Ag^{++} is.

One of the few known binary compounds of silver (II) is AgF_2, which is also a powerful oxidizing agent. Thus the $+2$ oxidation state, so important in the chemistry of copper, is not at all common for silver.

Gold

This element is the most ductile and malleable of metals and is an excellent conductor of heat and electricity. It is also quite inert to chemical attack, as the potentials for the reactions

$$Au^+ + e^- = Au, \qquad \varepsilon^0 \cong 1.7 \text{ volts},$$
$$Au^{+3} + 3e^- = Au, \qquad \varepsilon^0 \cong 1.5 \text{ volts},$$

suggest. In fact, these electrode potentials cannot be measured directly, since both the aurous ion Au^+ and the auric ion Au^{+3} are oxidizing agents powerful enough to oxidize water.

Aurous ion is also unstable with respect to disproportionation. Use of the estimated potentials gives

$$3Au^+ = Au^{+3} + 2Au, \qquad \Delta\varepsilon^0 \cong 0.2 \text{ volt},$$
$$\frac{[Au^{+3}]}{[Au^+]^3} = K \cong 10^{10}.$$

Thus for two reasons Au^+ does not exist as a simple cation in aqueous solution.

In both its oxidation states, gold forms stable complex compounds and ions. For example, gold can be oxidized readily in the presence of cyanide ion

$$Au + 2CN^- = Au(CN)_2^- + e^-, \qquad \varepsilon^0 = 0.60 \text{ volt}.$$

Combination of this half-reaction potential with that for reduction of Au^+ to the metal gives a standard potential and an equilibrium constant for

$$Au(CN)_2^- = Au^+ + 2CN^-, \qquad K \cong 5 \times 10^{-39}.$$

Gold also can be dissolved by nitric acid in the presence of chloride ion according to the reaction

$$Au + 3NO_3^- + 4Cl^- + 6H^+ = AuCl_4^- + 3NO_2 + 3H_2O.$$

It is the formation of the stable complex ion $AuCl_4^-$ that causes this reaction to proceed spontaneously.

16.10 ZINC, CADMIUM, AND MERCURY

These metals differ noticeably from the elements that precede them in the periodic table, and even among themselves display a considerable range of properties. As Table 16.11 shows, the metals of group IIB have rather low melting and boiling temperatures. Cadmium and particularly zinc are electropositive metals and in this respect resemble the elements of group IIA. Mercury, on the other hand, does not dissolve in hydrogen ion and is about as inert as silver and copper. Besides displaying the +2 oxidation state common to the group IIB elements, mercury forms the Hg_2^{++} ion, and compounds of this species have considerable stability. In contrast, Zn and Cd form the M_2^{++} ion only under very special conditions. Thus while cadmium and zinc resemble each other fairly closely, mercury displays rather different properties.

Table 16.11 Properties of the group IIB elements

	Zn	Cd	Hg
Atomic number	30	48	80
Configuration	$3d^{10}4s^2$	$4d^{10}5s^2$	$5d^{10}6s^2$
Ionization energy, kcal	217	207	240
Atomic radius, A	1.25	1.41	1.44
Melting point, °K	693	594	234
Boiling point, °K	1181	1040	630
ΔH_f^0 (atom), kcal	31.2	26.7	15.3
Ionic radius, M^{++}, A	0.74	0.97	1.10
ΔH_f^0 (MO), kcal	−83.17	−60.86	−21.68
ΔH_f^0 (MCl$_2$), kcal	−99.40	−93.00	−53.4
$\mathcal{E}^0$, M^{++} + 2e = M, volts	−0.763	−0.402	+0.854

Zinc occurs in a number of minerals, of which zinc blende, ZnS, calamine, $ZnCO_3$, and zincite, ZnO, are most important. Cadmium is found in these zinc ores and is recovered as a byproduct of the zinc reduction process. Zinc ores are usually roasted to produce the mixed oxides which are then reduced with carbon to give a mixture of zinc and cadmium metals. The cadmium and zinc are separated by distillation.

The principal source of mercury is the mineral cinnabar, HgS. To recover the metal, HgS is roasted to HgO, and when the oxide is heated to 500°C, it decomposes to give free mercury. Most of the world's supply of mercury comes from Spain, and the once copious ores in California and Nevada are all but exhausted. Unless substantial new deposits are found, all sources of mercury may be used up by the end of the 20th century.

As we have remarked, zinc and cadmium react with aqueous hydrogen ion, but mercury does not. All the metals react with oxygen and other nonmetals such as sulfur, phosphorus, and the halogens either at room temperature or upon moderate heating. The oxides ZnO, CdO, and HgO have very slight solubility in water, but all dissolve in solutions of strong acids. Zinc oxide also dissolves in strong bases according to the reaction

$$ZnO + 2OH^- + H_2O = Zn(OH)_4^=$$

and is therefore amphoteric. As might be expected from the larger size of the cadmium ion, CdO is not amphoteric, but is a basic oxide. Mercuric oxide is also basic, but very weakly so.

The halides of the group IIB metals show interesting trends in their properties. The halides of zinc have only moderately high melting temperatures, but when these compounds are fused, they conduct electricity. Consequently they are generally classified as ionic compounds. The structures of zinc chloride, bromide, and iodide can be pictured as close-packed lattices of halide ions with zinc ions occupying the tetrahedral holes. The cadmium halides are similar to the zinc halides, but have slightly different structures. The structures of the chloride, bromide, and iodide again can be thought of as a close-packed halide lattice, but the cadmium ions occupy the octahedral holes, rather than the tetrahedral holes.

While mercuric fluoride has a fluorite lattice that is associated with an ionic bonding, the other mercuric halides are covalently bonded. In the mercuric-chloride lattice, discrete, linear, symmetrical $HgCl_2$ molecules can be identified. This compound melts at 280°C and when fused does not conduct electricity. Moreover, in aqueous solution mercuric chloride remains largely as undissociated $HgCl_2$ molecules, for the equilibrium constants of the reactions

$$HgCl_2(aq) = HgCl^+ + Cl^-, \qquad K = 3.2 \times 10^{-7},$$
$$HgCl^+ = Hg^{++} + Cl^-, \qquad K = 1.8 \times 10^{-7},$$

are quite small.

Although zinc and cadmium display only the $+2$ oxidation state, both the $+1$ and $+2$ states are known for mercury. There is a wealth of evidence that the mercurous ion has the formula Hg_2^{++}; perhaps the most convincing is the direct observation of this structure in the x-ray analysis of mercurous salts, and the observation of a spectral feature of aqueous mercurous solution that can only be due to Hg_2^{++}. The conditions under which mercurous ion is formed can be understood with the aid of the following half-reaction potentials:

$$Hg_2^{++} + 2e^- = 2Hg(l), \qquad \varepsilon^0 = 0.789 \text{ volt},$$
$$2Hg^{++} + 2e^- = Hg_2^{++}, \qquad \varepsilon^0 = 0.921 \text{ volt},$$
$$Hg^{++} + 2e^- = Hg(l), \qquad \varepsilon^0 = 0.854 \text{ volt}.$$

Combination of the second and third half-reactions gives

$$Hg(l) + Hg^{++} = Hg_2^{++}, \qquad \Delta\varepsilon^0 = 0.066 \text{ volt},$$
$$K = 10^{[(2)(\Delta\varepsilon^0)]/0.059}$$
$$= 1.7 \times 10^2.$$

Thus mercurous salts have a slight tendency to disproportionate to Hg and Hg^{++}. If a solution of a mercurous salt is treated with a reagent that removes Hg^{++} from solution, the mercurous ion decomposes to form more Hg^{++} and mercury, and is eventually consumed. Thus we have reactions such as

$$Hg_2^{++} + H_2S = HgS + Hg + 2H^+,$$
$$Hg_2^{++} + 2OH^- = HgO + Hg + H_2O,$$

and consequently no mercurous oxide or sulfide exists. Among the mercurous salts that do exist are the halides, the nitrate, perchlorate, and sulfate.

16.11 TRANSITION-METAL COMPLEXES

Throughout our discussion of inorganic chemistry we have referred to the existence of complex ions such as BF_4^-, $Ag(NH_3)_2^+$, $Fe(CN)_6^{-3}$, and others. These complex ions, and neutral complex compounds as well, have distinctive properties that may be quite unlike those associated with their constituent molecules and ions. Because of their electronic structure, the transition metals form a large number of complex compounds, and a major part of current research in chemistry is devoted to the study of transition-metal complexes.

In general, a complex ion or compound consists of a central atom closely surrounded by a number of other atoms or molecules that have the property of donating electrons to the central atom. The central atom in a complex is sometimes called the *nuclear atom*, and the surrounding species are called coordinating groups or *ligands*. The nearest neighbor atoms to the nuclear atom constitute the *first or inner coordination sphere*, and the number of atoms in this first coordination sphere is the coordination number of the nuclear atom. A complex compound is distinguished from any other type of chemical compound by the fact that both the central nuclear atom and the ligands are capable of independent existence as stable chemical species.

As suggested above, the ligands in a complex compound, in general, donate electrons to the nuclear atom, which is usually an electron-deficient species. The word "donate" must not be overinterpreted, however, for in many instances it is not clear whether the ligand-nuclear atom interaction truly involves the sharing of electrons or is better described as Coulomb attraction between oppositely charged ions. In either case we can expect that the most stable complexes will be formed by small highly charged positive ions interacting with electron-donating atoms. This crude analysis accounts in some measure for the frequency with which transition-metal ions in particular form complexes with such species as NH_3, H_2O, Cl^-, and CN^-. There are many subtle features associated with the stability of complex ions, however, and we shall investigate some of them after discussing the geometric properties of complexes.

Stereochemistry

Complexes with coordination numbers from two to nine are known, but most exhibit two-, four-, or sixfold coordination and have structures with the geometry illustrated in Fig. 16.2. Twofold coordination occurs in the complexes of

Cu(I), Ag(I), Au(I) and some complexes of Hg(II); common examples are $Cu(CN)_2^-$, $Ag(NH_3)_2^+$, $Au(CN)_2^-$, and $Hg(NH_3)_2^{++}$. Fourfold coordination with tetrahedral geometry is common among complexes of the nontransition elements, but occurs with less frequency in transition-metal complexes. The ions $ZnCl_4^=$, $Zn(CN)_4^=$, $Cd(CN)_4^=$, and $Hg(CN)_4^=$ all have the tetrahedral configuration, but this geometry is otherwise fairly rare in the transition series.

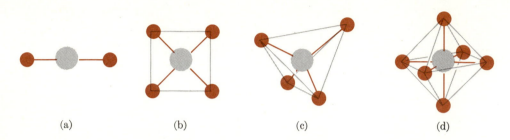

(a) (b) (c) (d)

FIG. 16.2   Common shapes for complex ions: (a) linear, (b) square planar, (c) tetrahedral, (d) octahedral. (After K. B. Harvey and G. B. Porter, *Physical Inorganic Chemistry*. Reading, Mass.: Addison-Wesley, 1963.)

Fourfold coordination with square planar geometry occurs in complexes of Pd(II), Pt(II), Ni(II), Cu(II), and Au(III). For most other ions this arrangement virtually never occurs. Sixfold coordination is the most common and occurs only in one geometric form, the octahedron.

A ligand that is capable of occupying one position in the inner coordination sphere and forming one coordinate bond to the nuclear atom is called a *unidentate* ligand. Examples are F^-, Cl^-, OH^-, H_2O, NH_3, and CN^-. When a ligand has two groups that are capable of bonding to the central atom, it is said to be *bidentate*. Common bidentate ligands are ethylene diamine, $NH_2CH_2CH_2NH_2$, where both nitrogen atoms can act as coordinating groups, and the oxalate ion, which has the structure

$$\left[\begin{array}{c} O \quad\quad\quad O \\ \diagdown\quad\quad\quad \diagup \\ C-C \\ \diagup\quad\quad\quad \diagdown \\ {}^-O \quad\quad\quad O^- \end{array} \right].$$

Because the two bonds from a bidentate ligand appear to enclose the metal atom in a pincerlike structure, the resulting compound is known as a *chelate* (Greek, *chele*-claw). Other ligands which have up to six coordinating groups are known, and the most common example is versene, or ethylene diamine

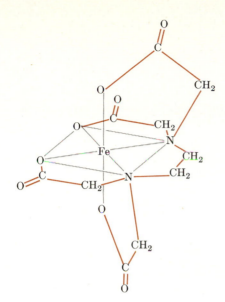

The ethylene diamine tetra-acetic complex of iron. The oxygen and nitrogen atoms occupy the corners of an octahedron with the iron atom at the center.

FIG. 16.3

tetra-acetic acid. The way in which versene occupies all six positions in the inner coordination sphere is shown in Fig. 16.3.

Several types of isomerism occur among complex ions. *Structural isomerism* is illustrated by the following example. There are three distinct compounds with the same formula: $Cr(H_2O)_6Cl_3$. One of these, violet in color, reacts immediately with $AgNO_3$ to precipitate all chlorine as AgCl. A second, light green in color, also reacts with $AgNO_3$, but only two-thirds of the chlorine is precipitated as AgCl. The third compound, dark green, releases only one-third of its chlorine to be precipitated as AgCl. On this basis the formulas might be written

$$[Cr(H_2O)_6]Cl_3 \qquad \text{(violet)},$$
$$[CrCl(H_2O)_5]Cl_2 \cdot H_2O \qquad \text{(light green)},$$
$$[CrCl_2(H_2O)_4]Cl \cdot 2H_2O \qquad \text{(dark green)},$$

where the species within the brackets are regarded as ligands bonded to the central chromium atom with some permanency. This assignment is substantiated by the fact that exposure of these compounds to drying agents results in the loss of zero, one, and two moles of water, respectively. Thus these structural isomers differ in the composition of their first coordination sphere and have noticeably different properties. Other similar examples are known, for instance

$$[Co(NH_3)_4Cl_2]NO_2 \qquad \text{and} \qquad [Co(NH_3)_4(Cl)(NO_2)]Cl.$$

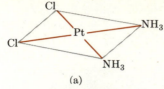

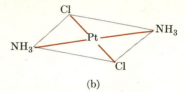

(a)   (b)

FIG. 16.4   Geometric isomers of dichlorodiammineplatinum(II): (a) *cis* isomer, (b) *trans* isomer.

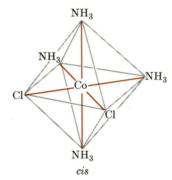

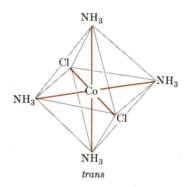

(a)   (b)

FIG. 16.5   Geometrical isomers of octahedral Ma_4b_2 complexes: (a) *cis* isomer, (b) *trans* isomer.

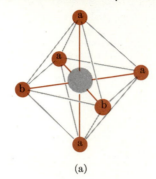

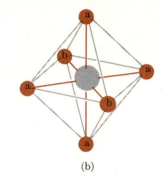

cis   *trans*

FIG. 16.6   The geometrical isomers of dichlorotetraaminecobalt(III) ion.

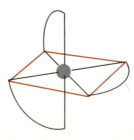

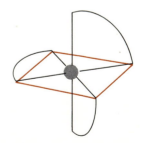

FIG. 16.7   Schematic drawing of the geometry of the optical isomers of an octahedral complex with bidentate ligands.

Besides the structural isomers, there are *geometrical isomers* which have coordination spheres of the same composition but different geometric arrangement. As a simple example of geometrical isomerism, consider the occurrence of *cis* and *trans* isomers of dichlorodiammineplatinum (II), shown in Fig. 16.4. *Cis* and *trans* isomers of square planar complexes of the type Ma_2b_2 can occur because although the ligands may be equidistant from the central atom M, they are not all equidistant from each other. Consequently it is possible to distinguish between ligands that lie next to each other on an edge of the square, and those that lie opposite each other on the square diagonal. In tetrahedral complexes, all four ligands are equidistant from each other, and *cis-trans* isomerism is not possible.

Geometrical isomerism is possible for octahedral complexes of the type Ma_4b_2, as Fig. 16.5 shows. Any two corners of an octahedron are *cis* to each other if they are linked by a single edge of the octahedron, while the *trans* positions lie on opposite sides of the metal atom. For example, two geometrical isomers of the complex ion $[Co(NH_3)_4Cl_2]^+$ exist: a *cis* isomer (violet) and a *trans* isomer (green), and they have the structures shown in Fig. 16.6.

Another important stereochemical feature of transition-metal complexes is *optical isomerism*. A molecule that lacks a plane or point of symmetry can exist in two nonequivalent forms that are mirror images of each other. These two forms are related as the right hand is to the left and cannot be superimposed on each other. Figure 16.7 illustrates the geometry of the optical isomers of a complex ion in which the ligands are bidentate. These optical isomers are identical in all respects except that one isomer rotates the plane of polarized light to the left, while the other rotates the plane of polarization to the right. Optical isomerism also occurs in organic molecules, and is discussed further in Section 17.8.

Nomenclature

Many complex ions, for example the ferrocyanide ion, $Fe(CN)_6^{-4}$, have acquired common names that are fairly descriptive of their composition. As more complicated complex ions have been synthesized, it has been necessary to adopt a systematic naming procedure. The following rules are sufficient to name many of the common complexes.

1. Ligands are assigned names as in Table 16.12. The names of anionic ligands end in *o*, while for neutral ligands the name of the molecule is used. Exceptions to the latter rule are water, ammonia, carbon monoxide, and nitric oxide, which are named as indicated in Table 16.12.

2. In naming a complex, the ligands are given first, with the Greek prefixes *di*, *tri*, *tetra*, etc., to indicate the number of identical ligands present.

3. The name of the central atom is given next, followed by its oxidation state designated by a Roman numeral enclosed in parentheses.

Table 16.12 Names of coordinating groups

Ligand	Name	Ligand	Name
H_2O	aquo	OH^-	hydroxo
NH_3	ammine	$C_2O_4^=$	oxalato
$O^=$	oxo	$SO_4^=$	sulfato
Cl^-	chloro	CO	carbonyl
CN^-	cyano	NO	nitrosyl

4. If the complex is a cation or neutral molecule, the name of the central atom is left unchanged. If the compound is a negative complex ion, the name of the central atom is made to end in *ate*. As examples we have

$[Ag(NH_3)_2]^+$,	diamminesilver (I) ion,
$[Zn(NH_3)_4]^{++}$,	tetraamminezinc (II) ion,
$[Co(NH_3)_3(NO_2)_3]$,	triamminetrinitrocobalt (III),
$[PtCl_6]^=$,	hexachloroplatinate (IV) ion,
$[Fe(CN)_6]^{-4}$,	hexacyanoferrate (II) ion,
$[Fe(CN)_6]^{-3}$,	hexacyanoferrate (III) ion.

For complexes that have common names, it is convenient to avoid using the sometimes cumbersome systematic names. For the more complicated and obscure complexes, however, use of the systematic name is necessary.

16.12 BONDING IN TRANSITION-METAL COMPLEXES

Like all other compounds, transition-metal complexes owe their stability to the lowering of energy that occurs when electrons move in the field of more than one nucleus. Therefore, the theories of bonding in the transition-metal complexes do not differ fundamentally from the theories used to discuss other chemical bonds. However, the bonding in transition-metal complexes does involve some new features that were not emphasized in our discussions of other systems. First, the *d*-orbitals of the transition-metal atom are involved in the bonding to the ligands. Second, it is important to take explicit account of the behavior of the nonbonding electrons. Third, it is interesting to examine not only the lowest electronic states, but also their excited electronic states, for it is the existence of these states that is responsible for the light absorption and color of the ions. Finally, the magnetic properties of transition-metal complexes are very important, and should be satisfactorily explained by the bonding theories. There are three important approaches to the problem of bonding in transition-metal complexes, and we shall now discuss each briefly.

In the crystal-field theory, the bonding between the central metal ion and its ligands is assumed to be purely electrostatic, due either to the attraction between oppositely charged ions or between the central positive ion and the negative end of dipolar molecules. This is an extreme picture that is probably never rigorously accurate, but it does have the virtue of simplicity. Because crystal-field theory *assumes* electrostatic bonding in complexes, it does not pretend to *explain* the nature of the metal-ligand bonds. The theory does, however, attempt to explain the effects of the ligands on the energies of the d-electrons of the metal ion and in this way helps us to understand the magnetic properties of complexes and their absorption spectra.

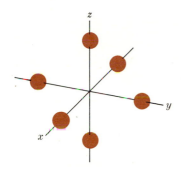

Six ligands of an octahedral complex defining a cartesian coordinate system.   **FIG. 16.8**

The conclusions of crystal-field arguments depend on the spatial arrangement of the ligands about the central transition-metal ion. Because sixfold coordination with regular octahedral geometry occurs so frequently in complex ions, let us begin our discussion to this case. Imagine a transition-metal ion in free space. In this condition, the energies of its five valence d-orbitals are the same, or as is often stated, the orbitals are degenerate. Now imagine six ligands placed symmetrically around the central ion so as to define a cartesian coordinate system, as shown in Fig. 16.8. As the ligands are brought close to the central ion, there is a general lowering of the energy of the entire system due to the electrostatic attraction between the metal ion and the ligands. Now the five d-orbitals of the metal ion are not spatially equivalent, as Fig. 16.9 shows. Two of them, $d_{x^2-y^2}$ and d_{z^2}, have their greatest electron density in directions that lie along the cartesian coordinate axes. The other three d-orbitals, d_{xy}, d_{xz}, and d_{yz}, have their greatest density in regions *between* the coordinate axes. The former pair of orbitals is often designated as d_γ- or e_g-orbitals, while the latter three are called the d_ϵ- or t_{2g}-orbitals. As the negative ligands are brought near the central ion, electrons in the e_g-orbitals feel a stronger electrostatic repulsion from the ligand than do electrons in the t_{2g}-orbitals, because the

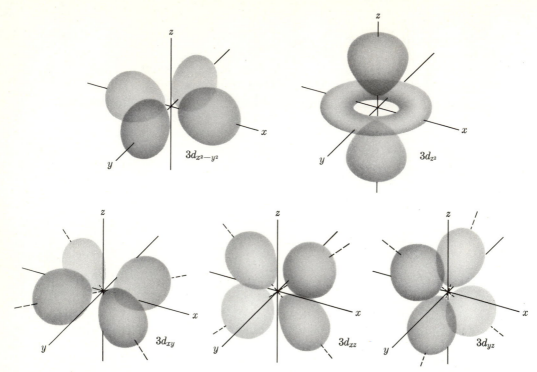

FIG. 16.9 The 3*d*-orbitals. (Adapted from K. B. Harvey and G. B. Porter, *Physical Inorganic Chemistry.* Reading, Mass.: Addison-Wesley, 1963.)

e_g-orbitals are more concentrated along the coordinate axes where the negative ligands are situated. Thus the presence of the ligands "splits" the *d*-orbitals into a higher energy pair of e_g-orbitals and a lower energy triplet of t_{2g}-orbitals, as Fig. 16.10 shows.

The magnitude of the crystal-field splitting of the *d*-orbital energies is usually designated by Δ_0. The magnitude of this quantity depends, according to crystal-field theory, only on the metal ion-ligand distance, the mean electron-nuclear separation for a *d*-electron, and the charge or dipole moment of the ligand. The smaller the metal-ligand distance, the larger the average *d*-electron–nucleus separation, and the larger the charge or dipole moment of the ligand, the larger is the splitting Δ_0. In general, the predicted values of Δ_0 based on crystal-field theory calculations are not highly accurate and give only fair estimates of splitting of the *d*-orbital energies.

Experimental values of Δ_0 can be derived from the absorption spectra of complex ions. In the simplest cases, the absorption of light by a complex ion is accompanied by the excitation of an electron in one of the lower t_{2g}-orbitals to an e_g-orbital. The energy that corresponds to the frequency of the light most

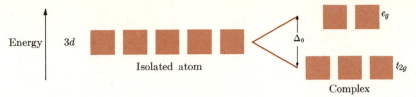

The splitting of the d-orbital energies by the octahedral ligand field.   **FIG. 16.10**

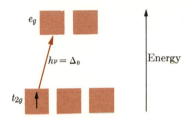

Schematic representation of the absorption of light by $[Ti(H_2O)_6]^{+3}$, showing the excitation of an electron from a t_{2g}- to an e_g-orbital.   **FIG. 16.11**

strongly absorbed is equal to Δ. For example, the hexaaquotitanium (III) ion, $[Ti(H_2O)_6]^{+3}$, has an absorption band in the visible region, and the absorption is strongest at wavelengths of approximately 5000 A. This absorption gives the ion its purple color and corresponds to the excitation of the single d-electron in Ti^{+3} from a t_{2g}- to an e_g-orbital, as shown in Fig. 16.11.

Table 16.13 Crystal-field splittings, Δ_0 (kcal/mole)

Metal ion	Ligand		
	H_2O	NH_3	CN^-
Ti (III) $3d^1$	58		
V (III) $3d^2$	51		
Cr (III) $3d^3$	50	62	75
Mn (III) $3d^4$	60		
Fe (III) $3d^5$	39		
Mn (II) $3d^5$	22		
Co (III) $3d^6$	53	65	97
Fe (II) $3d^6$	30		94
Co (II) $3d^7$	28	29	
Ni (II) $3d^8$	24	31	
Cu (II) $3d^9$	36	43	

Table 16.13 gives a few of the values of Δ_0 for the various transition-metal ions and ligands. While the value of Δ_0 is approximately constant for ions of a given charge with the same ligand, changing the ligand does change Δ_0, and thus

alters the absorption spectrum associated with the metal ion. It is this change in the crystal-field splitting of the d-orbital energies that is responsible for the color change that occurs when one ligand is replaced with another. From measurements of the absorption spectra, it is possible to arrange the common ligands in the order of the value of Δ_0 they induce in any metal ion. This *spectrochemical series* is as follows:

$$Br^- < Cl^- < F^- < OH^- < C_2O_4^= < H_2O < NH_3 < NO_2^- < CN^-,$$

where Δ_0 increases from left to right. The order of a closely related set of ions like the halides can be understood, for the smaller the ion, the smaller is the ligand–central ion separation, and the larger is the splitting, according to the crystal-field theory. It is important to note, however, that the crystal-field theory cannot explain the order of the entire spectrochemical series, and consequently the electrostatic picture of the bonding in transition-metal complexes must be an oversimplification.

The magnetic properties of the transition-metal complex ions can be understood with the aid of crystal-field theory. Compounds composed of molecules or ions that have unpaired electron spins tend to be drawn into a magnetic field. If, when the magnetic field is removed, the material retains a permanent magnetization, it is said to be *ferromagnetic*. If, however, the sample loses its magnetism when the field is removed, it is said to be *paramagnetic*. Materials that have no unpaired electrons tend to move out of magnetic fields and are designated as *diamagnetic*. Complex compounds of the transition-metal ions very often have unpaired electron spins and are therefore paramagnetic. For example, because the three t_{2g}-orbitals have the same energy in an octahedral complex, any central atom or ion that has three valence d-electrons will have one electron in each t_{2g}-orbital, and these electrons will have the same spin, just as the three p-electrons in the nitrogen atom have the same spin. The magnetic properties of such a complex, $[Cr(H_2O)_6]^{+3}$, for example, do in fact indicate the presence of three unpaired electron spins.

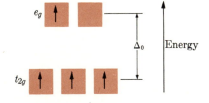

FIG. 16.12 Orbital occupancy scheme for a d^4 complex ion in which the crystal field splitting Δ_0 is small.

If the central atom has four d-electrons, the electron configuration to be expected is not so obvious. After the three t_{2g}-orbitals are half-filled, the fourth electron may be accommodated in one of the higher-energy e_g-orbitals, or it may enter one of the t_{2g}-orbitals at the cost of some energy associated with the coulomb repulsion of two electrons in the same orbital. Which of these two alternatives occurs depends on which is energetically the most favorable. If the crystal-field splitting Δ_0 is small, then the fourth electron will be accommodated in an e_g-orbital, as shown in Fig. 16.12. If the splitting of the orbitals is large, however, it is energetically more favorable for the fourth electron to enter an already half-filled t_{2g}-orbital than to go into a high-energy e_g-orbital.

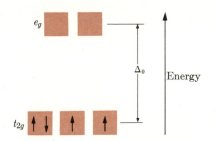

Energy

Orbital occupancy scheme for a d^4 complex ion in which the crystal field splitting Δ_0 is large.

FIG. 16.13

This situation is represented in Fig. 16.13. Thus if the splitting is small, there will be four unpaired electrons and if the orbital splitting is large, there will be only two electrons with unpaired spins. The former "high-spin" case has a larger paramagnetism associated with it than does the latter "low-spin" situation.

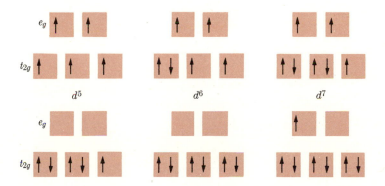

Orbital occupancy for the high- and low-spin states of the d^5-, d^6-, d^7-configurations.

FIG. 16.14

Our discussion can be extended to central ions with d^5, d^6, and d^7 valence-electron configurations. Figure 16.14 shows the orbital occupancy for the high- and low-spin states of these ions in octahedral crystal fields. The high-spin states are expected in complexes in which Δ_0 is small; that is, complexes which have ligands that lie in the first part of the spectrochemical series. Low-spin complexes are those whose ligands fall in the latter part of the spectrochemical series.

Some examples that illustrate the foregoing discussion are the following. The ion Fe^{+3} has a d^5-configuration, and as the aquo-complex, exists in the high-spin state $(t_{2g})^3(e_g)^2$. The crystal-field splitting induced by the cyanide ion is much larger than that produced by the water molecule, and it is not too surprising that in the ferricyanide ion, $[Fe(CN)_6]^{-3}$, iron is in the low-spin state $(t_{2g})^5$. Similarly, in $[CoF_6]^{-3}$, cobalt (d^6) is in the high-spin state $(t_{2g})^4(e_g)^2$, while in $[Co(NH_3)_6]^{+3}$, the low-spin configuration $(t_{2g})^6$ is found. Ions like $Ti^{+3}(d^1)$, $V^{+3}(d^2)$, and $Cr^{+3}(d^3)$, have unique d-electron configurations that are unaffected by the crystal-field splitting, and thus these ions are always found in a high-spin state.

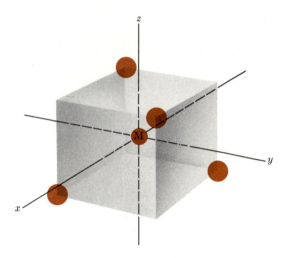

FIG. 16.15   The structure of a tetrahedral complex and its relation to a cube centered at the nuclear atom.

Recognition of the fact that the hexa-aquo complexes of Mn^{++} and Fe^{+3} exist in the high-spin $(t_{2g})^3(e)^2$-configuration leads to an explanation of why these ions are only very weakly colored. Manganous ion is a very pale pink, and ferric ion is a very pale violet in solutions acidic enough to prevent its hydrolysis and polymerization. In these ions, excitation of a t_{2g}-electron to an e_g-orbital requires a change in electron spin (see Fig. 16.14) during the process, and this is a highly improbable event. Thus these ions absorb only a very small fraction of the light incident on them and appear to be virtually colorless.

Tetrahedral geometry occurs in a few transition metal complexes such as $CoCl_4^=$, $MnBr_4^=$, and $FeCl_4^-$, so it is of interest to examine the d-orbital energy level pattern for this structure. To begin, consider Fig. 16.15, which shows that placing ligands at alternate corners of a cube centered on the nuclear atom produces a complex with tetrahedral geometry. If the coordinate axes are taken to be perpendicular to the faces of the cube, then it is easy to see that the d-orbitals divide into two groups. The d_{z^2}- and $d_{x^2-y^2}$-orbitals point directly at the faces of the cube and bisect the tetrahedral angle between the ligands. The d_{xy}-, d_{xz}-, and d_{yz}-orbitals have lobes that point directly at the edges of the cube, and thus are close to the ligands. Consequently, an electron in one of these latter three orbitals feels more repulsion from the electron clouds of the ligands than does an electron in the d_{z^2}- or $d_{x^2-y^2}$-orbital. As a result, splitting of the d-orbitals in a tetrahedral complex is as indicated in Fig. 16.16, with the magnitude of the splitting Δ_t generally less than the splitting in octahedral complexes with the same ligands. Because the splitting parameter Δ_t is small, there are no known low-spin tetrahedral complexes. The energy to be gained by moving an electron from the d_{xy}-, d_{xz}-, or d_{yz}-orbitals to either of the $d_{x_2-y_2}$- and d_{z_2}-orbitals is evidently always small compared to the energy lost through electron-electron repulsion.

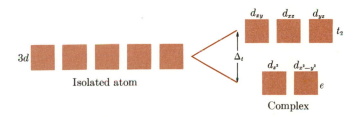

The splitting of the d-orbital energies by a tetrahedral ligand field.

FIG. 16.16

To find the orbital energy-splitting pattern characteristic of square-planar complexes it is easiest to begin by imagining an octahedral complex with the two e_g-orbitals lying above the three t_{2g}-orbitals. Then consider the effect of gradually withdrawing the two ligands which are along the z-axis while diminishing the metal-ligand distance along the x- and y-axes. The result of this so-called tetragonal distortion of the octahedron is shown in Fig. 16.17. Withdrawing the two ligands along the z-axis diminishes ligand-electron repulsion and thus lowers the energy of an electron in the d_{z^2}-orbital. Correspondingly, shortening of the metal-ligand electron-ligand repulsion, raises the energy of the $d_{x^2-y^2}$-

orbital, as is indicated in Fig. 16.17. The energy of the d_{xy}-orbital is also raised, since this orbital has its greatest density in the xy-plane and experiences greater repulsion from the ligands in the xy-plane as they are brought closer to the metal atom. In contrast, the d_{xz}- and d_{yz}-orbitals decrease in energy, since they point out of the xy-plane. The square-planar complex is the limiting case of a tetragonally distorted octahedral complex, with the two ligands along the z-axis completely removed.

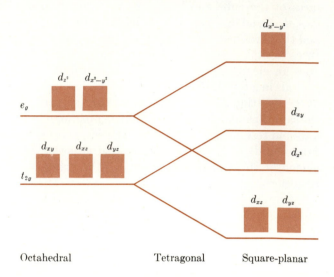

FIG. 16.17   Correlation between the energies of d-orbitals in octahedral, tetragonal, and square-planar complexes.

As Fig. 16.17 shows, a square-planar complex has two d-orbitals of low energy, two more of intermediate energy, and one ($d_{x^2-y^2}$) of quite high energy. Consequently, square-planar complexes are most likely to be formed with d^8- and d^9-ions, since in these situations the high-energy $d_{x^2-y^2}$-orbital is either empty or only half-filled, and the other four d-orbitals lie fairly low in energy. Complexes of the d^8-ions Pt(II), Pd(II), Au(III), Rh(I), and Ir(I) are usually square-planar, as are most complexes of Ni(II). Complexes of the d^9-ion Cu^{++} are either octahedral with substantial tetragonal distortion, or square-planar.

Valence-Bond Theory

In this approach, each ligand is assumed to donate a pair of electrons to the metal ion to form a *coordinate covalent* metal-ligand bond. The criterion for bond formation in an octahedral complex is that the metal ion have available six vacant equivalent valence orbitals, so that the valence bonds can be "made" from the overlap of orbitals on the metal with those on the ligands. Six equivalent orbitals on the metal ion can be formed as a d^2sp^3-hybrid set. Specifically, the orbitals that would be used for an ion of the first transition series are the two $3e_g$-, the $4s$-, and the three $4p$-orbitals. The three $3t_{2g}$-orbitals on the metal ion do not participate in the bonding. Thus formation of the complex $[Cr(NH_3)_6]^{+3}$ from Cr^{+3} and the ligands would be represented by

where the configuration on the left is that of the free metal ion, and on the right the twelve electrons shared with the ligands are indicated.

Application of the valence-bond theory to complexes of Co(III) and Ni(II) reveals some significant points. The configuration of the free Co^{+3} ion is

To form the vacant d^2sp^3 hybrid orbitals, two of the d-electrons must be relocated, and this is accomplished by placing them in two other half-filled d-orbitals to give the configuration

From this configuration, d^2sp^3 hybrid orbitals that accept twelve electrons from the ligands can be formed. The resulting complex has no unpaired electrons, and therefore should not be paramagnetic. Indeed, ions like $[Co(NH_3)_6]^{+3}$ and $[Co(CN)_6]^{-3}$ have no unpaired electrons, as the theory suggests.

The ion $[CoF_6]^{-3}$ is paramagnetic, however, despite being isoelectronic with the other complexes of Co(III). The rationalization of this observation is that the metal-ligand bonding in $[CoF_6]^{-3}$ is not covalent, but ionic, and consequently the cobalt ion in the complex retains the configuration it has in the free state with four electron spins unpaired. Thus just as crystal-field theory rationalizes the high- and low-spin complexes of Co(III) in terms of the magnitude of the splitting parameter Δ associated with the various ligands, valence-bond theory rationalizes the same phenomena by suggesting that some ligands are ionically bonded to the metal while others are bonded covalently.

The sharp distinction between ionic and covalent complexes that results from valence-bond theory is not particularly satisfactory. For example, with one exception, all the octahedral complexes of Ni(II) are paramagnetic, and since free Ni^{++} has the configuration

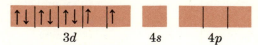

$$3d \qquad 4s \qquad 4p$$

this must mean that all octahedral complexes of Ni(II) are ionically bonded if the valence-bond theory is accepted. Various experiments make this conclusion seem very unlikely and force the conclusion that the simple valence-bond theory of metal-ligand bonding is not entirely satisfactory.

Ligand-Field Theory

In this approach, it is imagined that molecular orbitals are formed by the overlap of orbitals from the ligands with the atomic orbitals of the central atom. These molecular orbitals may be of a bonding, antibonding, or nonbonding character. Depending on the nature of the metal and the ligand, the bonding orbitals may be of a covalent type, where the electrons are shared approximately equally between the metal and ligands, or they may concentrate most electron density on the ligands, and thereby represent ionic bonding. The nonbonding orbitals in octahedral complexes are simply the d_{xy}-, d_{xz}-, and d_{yz}-orbitals of the central atom. The antibonding orbitals are similar to the bonding orbitals except that they lie higher in energy and have nodes or regions of low electron density between the central atom and the ligands.

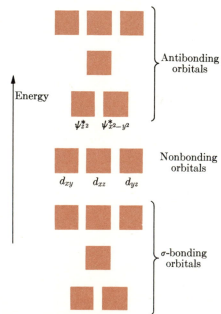

FIG. 16.18 Qualitative molecular-orbital energy-level diagram for octahedral complexes.

The energy-level pattern for the molecular orbitals of an octahedral complex is given in Fig. 16.18. The six orbitals of lowest energy are σ-bonding orbitals, and are filled by six-electron pairs from the ligands. The nonbonding t_{2g} atomic orbitals and the first two antibonding orbitals are filled according to the number of electrons available, the energy difference between the orbitals, and the magnitude of the repulsion between two electrons in the same orbital. In effect, the $\psi_{z^2}^*$ and $\psi_{x^2-y^2}^*$ antibonding orbitals in ligand-field theory take the place of the d_{z^2} and $d_{x^2-y^2}$ atomic orbitals used in crystal-field theory. Ligand-field theory differs from the valence-bond approach by recognizing the existence of the antibonding as well as the bonding and nonbonding orbitals. The orbital occupancy diagram in Fig. 16.19 shows how, by using the ligand-field energy-level scheme, we can explain the high-spin complexes of Ni(II) without having to invoke ionic bonding between the metal ion and the ligand. It is this ability to account for magnetic properties without forcing postulates of unreasonable forms for metal-ligand bonding that is one of the most satisfactory features of ligand-field theory.

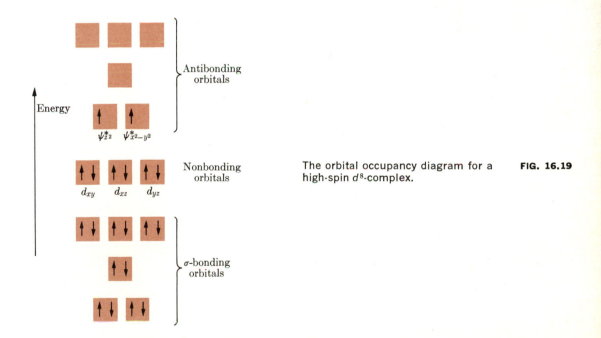

The orbital occupancy diagram for a high-spin d^8-complex.

FIG. 16.19

There are a number of consequences of the splitting of the energies of the d-orbitals of transition metal ions by the electric field of surrounding ligands. Consider first the apparent ionic radii as derived from the interionic distances in the metal (II) oxides MO. These compounds have the rock-salt structure, with each metal ion M^{++} surrounded by six oxide ions located at the corners of a regular octahedron. The variation in the apparent radii of the M^{++} ions is displayed in Fig. 16.20. If we look first at the ions $Ca^{++}(d^0), Mn^{++}(d^5)$, and $Zn^{++}(d^{10})$, all of which have spherical electron distributions, we see that there is a general decrease in ionic radius with increasing nuclear charge. However, the ions Ti^{++} and V^{++} are much smaller than would be expected by interpolating between Ca^{++} and Mn^{++}. The reason for this is quite straightforward. In V^{++}, the three d-electrons occupy the t_{2g}-orbitals which are principally directed between the coordinate axes where the surrounding oxide ions lie. The e_g-orbitals, which point directly toward the oxide ions, are empty in Ti^{++} and V^{++}, and therefore the oxide ions can reside closer to the metal ion in these compounds than in MnO or ZnO, where the e_g-orbitals are half- and completely filled, respectively.

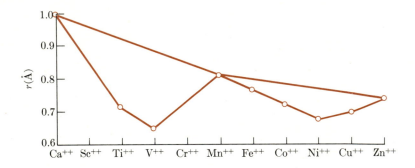

FIG. 16.20   Variation of the radius of the doubly charged ions of the first transition series.

A similar rationalization can be applied to behavior observed for the ions between Mn^{++} and Zn^{++}. In Fe^{++}, Co^{++}, and Ni^{++}, the e_g-orbitals are no more than half-filled, and the radii of these ions are less than the values interpolated from the radii of Zn^{++} and Mn^{++}, Thus the behavior of the d-electrons in the ligand field provides a very satisfactory explanation for the observed variation in ionic radii.

In Fig. 16.21 the variation of the enthalpy of hydration of the doubly charged ions of the first transition series is displayed. Considering only the spherical ions Ca^{++}, Mn^{++}, and Zn^{++}, we see the expected trend of increasing hydration enthalpy with diminishing size and increasing nuclear charge. However, the ions Ti^{++}, V^{++}, and Cr^{++} are more stable in aqueous solution than would be expected from interpolation between Ca^{++} and Mn^{++}. This extra stability is a result of the fact that in Ti^{++} and V^{++}, the d-electrons occupy t_{2g}-orbitals which are directed between, rather than at, the six nearest water molecules. In Cr^{++}, there is only one electron in a relatively unstable e_g-orbital. However, in Mn^{++}, both the e_g-orbitals are occupied, the electron distribution is spherical, and there is no extra ligand field stabilization energy derived from having the d-electrons avoid the ligand regions.

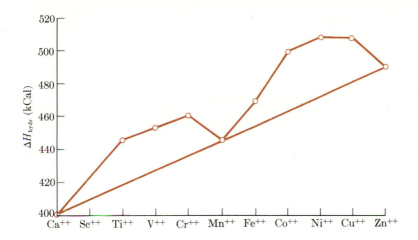

Variation of the hydration energies of the doubly charged transition metal ions.

FIG. 16.21

A similar argument applies to the variation in the hydration enthalpy observed between Mn^{++} and Zn^{++}. In Fe^{++}, Co^{++}, and Ni^{++}, the e_g-orbitals are never more than half-filled, and there is a ligand field stabilization derived from having most of the electrons in t_{2g}-orbitals. This effect is diminished in Cu^{++} which has a d^9 configuration, and is lost totally in Zn^{++}, where all d-orbitals are occupied. Ligand field stabilization effects resulting from favorable occupation of d-orbitals are observed in many other complexes of the transition metal ions.

Transition Metal Carbonyls

The carbonyls are compounds of the transition metal elements with carbon monoxide. They are rather remarkable substances in which the metal atom has a formal oxidation number of zero. It has already been mentioned that nickel carbonyl is formed readily by the direct reaction of carbon monoxide with metallic nickel at room temperature:

$$Ni(s) + 4CO(g) = Ni(CO)_4(g).$$

Iron carbonyl also can be made this way, but a temperature of 200°C and 100 atm of CO are required:

$$Fe(s) + 5CO(g) = Fe(CO)_5(g).$$

In some cases, one carbonyl can be used to make another, as in

$$WCl_6 + 3Fe(CO)_5 = W(CO)_6 + 3FeCl_2 + 9CO,$$

but more frequently, a metal halide and a reducing agent are used:

$$2VCl_3 + 2Mg\text{—}Zn + 12CO = 2V(CO)_6 + 2MgCl_2 + ZnCl_2.$$

The carbonyls which contain one metal atom with several carbon monoxide molecules are usually very volatile liquids at room temperature.

The structures of some metal carbonyls are shown in Fig. 16.22. The simplest description of the bonding and structure of the carbonyls of the first row of transition elements is that the metal atom receives a pair of electrons from each carbon monoxide molecule so that it attains completely filled $3d$-, $4s$-, and $4p$-orbitals. Thus in $Ni(CO)_4$ there are 10 electrons from Ni and 8 from CO to give 18 valence electrons, which corresponds to the complete valence shell in Kr. Similarly, $Fe(CO)_5$ and $Cr(CO)_6$ have 18 valence electrons. In $Mn_2(CO)_{10}$, each Mn is bonded to 5 CO molecules located at the corners of an octahedron, and the 2 $Mn(CO)_5$ units are joined by an electron pair bond between Mn atoms. Thus around each Mn there are 10 electrons from the CO molecules, 7 from the metal atom itself, and 1 from the other Mn atom, for a total of 18. In $Co_2(CO)_8$ there is a Co—Co electron pair bond, and in addition, 2 CO molecules form a pair of 3-center, 2-electron bonds between cobalt atoms. Thus around each Co there are 6 electrons from CO molecules directly bonded to it, 9 electrons of its own, 1 from the other Co, plus 1 electron from each of the 2 bridging CO molecules, or a total of 18. This simple approach does not explain the existence of the molecule $V(CO)_6$, which has 17 valence electrons.

In metal carbonyls, carbon monoxide molecules which are bonded only to one metal atom are oriented so that the M—C≡O structure is linear or nearly so. This is consistent with a metal-carbon σ-bond being formed by donation of a pair of (initially nonbonding) σ-electrons by CO to the metal atom. However, the stability of metal carbonyls and their spectral properties suggest that there

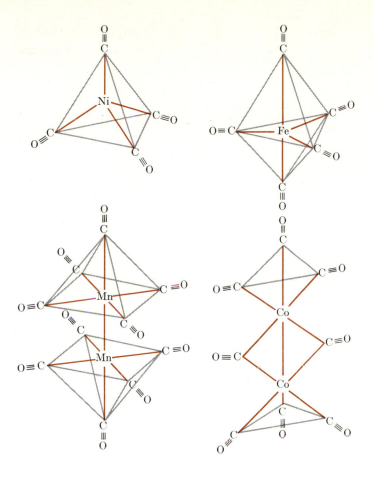

Structures of some transition metal carbonyls.

FIG. 16.22

is more to the bonding than just the simple M—C σ-bond. The additional bonding can occur as indicated in Fig. 16.23. One of the t_{2g}-orbitals of the metal atom, usually nonbonding, does in fact overlap one of the π^*-antibonding orbitals of CO in such a way that a π-bonding orbital can be formed between the carbon and metal atoms. If the metal t_{2g}-orbitals are occupied, there can be a strengthening of the metal-ligand interaction through this so-called back-bonding. The presence of electrons in what is still a C—O antibonding orbital should weaken the bonding in the carbon monoxide ligand, and there is clear spectroscopic evidence that this happens. Other ligands besides CO can engage in π-bonding with metal orbitals. For example, this is probably a major reason why CN^- complexes of transition metal ions are particularly stable.

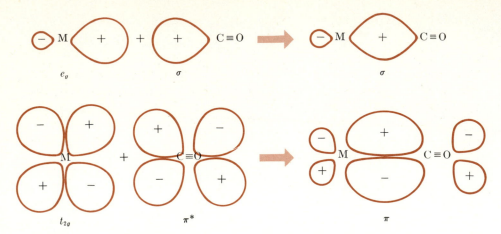

FIG. 16.23 Formation of σ- and π-bonds between a transition metal atom and carbon monoxide.

Transition Metal-Organic Compounds

Compounds of the transition metal elements with organic molecules have been known since 1830. However, only in the last two decades has it been possible to understand the nature of these substances. Many have been found to be valuable catalysts or intermediates in useful synthetic processes, and consequently the interest in organo-metallic compounds has become particularly intense.

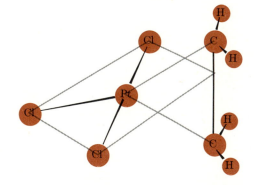

FIG. 16.24 The structure of the $PtCl_3C_2H_4^-$ anion.

One of the first organo-metallic substances made was a salt containing the anion $[PtCl_3C_2H_4]^-$. Its formation involves replacing one of the chloride ligands in the tetrachloroplatinate anion with ethylene, C_2H_4:

$$PtCl_4^= + C_2H_4 = [PtCl_3C_2H_4]^- + Cl^-.$$

The structure of this anion is shown in Fig. 16.24. The ethylene molecule has its axis perpendicular to the plane of the $PtCl_3$ group, and it occupies one of the ligand positions associated with square-planar coordination of the platinum.

The bonding in this and other olefin-transition metal complexes is closely related to the bonding in transition metal carbonyls. Figure 16.25 illustrates the basic mechanism of bond formation. The ethylene molecule has two electrons in a π-bonding orbital which can be donated to the metal atom. The constructive overlap between the positive lobe of the ethylene π-orbital and the positive lobe of a metal dsp^2 hybrid produces a molecular orbital which is bonding between the metal and ligand. In addition, there can be constructive overlap between the π^* antibonding orbital of ethylene and one of the d-orbitals such as d_{xz} which points out of the coordination plane. Any electrons which occupy this orbital in the atom can contribute to the metal-ligand bonding, and stabilize the complex.

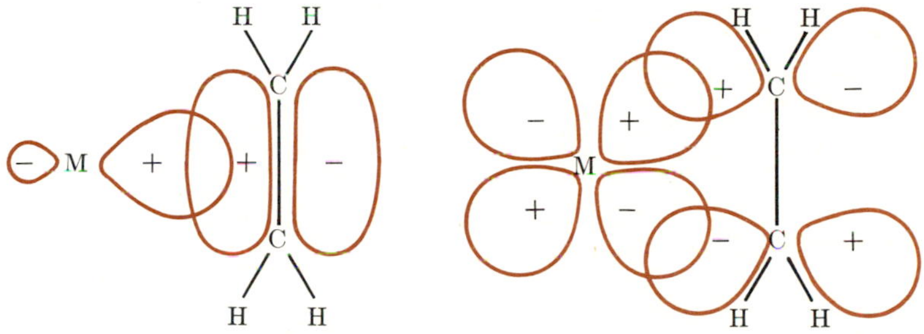

Formation of σ- and π-bonds between a metal atom and ethylene.

FIG. 16.25

The most stable transition metal organic compounds involve the cyclopentadiene anion, $C_5H_5^-$, which has the structure

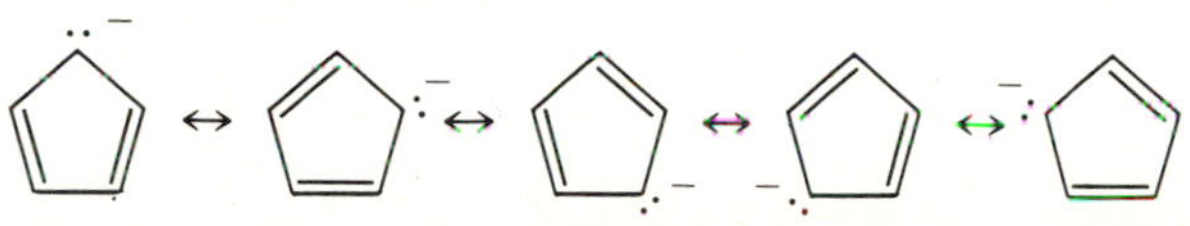

where there is a CH group at each apex of the regular pentagon. Thus the cyclopentadiene anion has six electrons in a system of delocalized π-orbitals, very much like benzene. These six electrons can be denoted to a transition metal ion to form stable metal-ligand bonds.

The first compound of this type to be recognized and its structure understood was ferrocene, bis- (π-cyclopentadienyl) iron (II). As Fig. 16.26 shows, ferrocene consists of a sandwich of iron between two cyclopentadiene rings. The rings are staggered with respect to each other, with the apices of one ring directly below the sides of the other. However, very little energy is required to rotate the rings into an eclipsed position, and in some sandwich compounds this arrangement is the more stable.

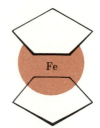

FIG. 16.26   Sandwich structure of ferrocene, $(C_5H_5)_2$ Fe.

It is of interest to note that in ferrocene, the iron atom is surrounded by 18 valence electrons, which corresponds to the electron configuration of krypton. That is, the valence shell of ferrocene contains 6 electrons from Fe^{++}, and 12 electrons donated by the two cyclopentadiene anions. The ion $(C_5H_5)_2Co^+$ is isoelectronic with ferrocene, and is a very stable sandwich of Co^{+3} between 2 cyclopentadiene anions. Dibenzene chromium, $(C_6H_6)_2Cr$ is a sandwich of neutral chromium atom between benzene rings. It also has 18 electrons around the central chromium atom, but is much less stable than ferrocene.

One of the most compelling reasons for the considerable interest of chemists in transition metal-organic compounds is the importance of these substances as catalysts. For example, a complex of rhodium (I) permits the addition of hydrogen to unsaturated organic molecules by a mechanism which involves metal-organic bond formation. This catalytic compound is $RhCl(PPh)_3$, where PPh_3 stands for triphenylphosphine, phosphorous with three phenyl (C_5H_6) groups bonded to it. The mechanism of the catalysis is as follows. Hydrogen adds to the complex, and occupies as separate atoms two of the coordination sites around the rhodium:

$$RhCl(PPh)_3 + H_2 = RhCl(PPh)_2(H)(H) + PPh_3.$$

A molecule of the olefin to be hydrogenated (for example, ethylene) then adds to the sixth coordination site:

$$RhCl(PPh)_2(H)(H) + C_2H_4 = RhCl(PPh)_2(H)(H)(C_2H_4).$$

The hydrogen atoms add to the ethylene molecule, which then separates as ethane, C_2H_6:

$$RhCl(PPh)_2(H)(H)(C_2H_4) = RhCl(PPh)_2 + C_2H_6.$$

The rhodium complex is then ready to add a new hydrogen molecule and begin another catalytic cycle.

Transition metal compounds are also important catalysts for polymerization reactions. Solid titanium trichloride is used to produce high quality polyethylene and polypropylene. The process is initiated by a reaction which produces an ethyl group (C_2H_5) bonded to a titanium atom at the surface of the catalyst. We can represent this surface atom by $Ti(C_2H_5)L_4$, where L represents the surrounding ligands. If, as the formula indicates, one of the coordination sites is vacant, an ethylene molecule can add to it:

$$Ti(C_2H_5)L_4 + C_2H_4 = Ti(C_2H_5)L_4(C_2H_4).$$

This facilitates addition of the C_2H_5 group to the C_2H_4, and the result is a longer carbon chain (C_4H_9) bonded to titanium, and a vacant coordination site:

$$Ti(C_2H_5)L_4(C_2H_4) = Ti(C_4H_9)L_4.$$

The metal atom is then ready to receive another ethylene molecule, and continue the carbon chain building process.

16.13 CONCLUSION

The transition metals constitute most of the known chemical elements and display an enormous range of chemical and physical properties. Some of the behavior of these elements can be understood through use of the general concepts that were applied to the representative metals. That is, along any of the transition series, there is a general decrease in electropositive character that is associated with the increase in nuclear charge and ionization energy. In their lower oxidation states, the transition metals form oxides and halides that are best described as ionic compounds. In aqueous solution, the $+2$ and $+3$ transition-metal ions tend to hydrolyze and produce acidic solutions. As was true for the representative metals, the higher oxidation states of the transition elements tend to be more acidic than the lower oxidation states. There are, however, some properties of transition metals that are not often encountered among the representative metals. The transition-metal ions are often colored, and we have seen that this phenomenon is a consequence of the splitting of the energies of the d-orbitals by the ligands surrounding the ions. The transition-metal ions are frequently paramagnetic, and this behavior also is a reflection of the energy separation of the d-orbitals. Indeed, there is a variety of chemical phenomena that have qualitative and quantitative explanations in terms of crystal-field theory and its refinements, as examination of the suggested reading will show.

SUGGESTIONS FOR FURTHER READING

Basolo, F., and R. C. Johnson, *Coordination Chemistry*. New York: W. A. Benjamin, 1964.

Cotton, F. A., and G. Wilkinson, *Advanced Inorganic Chemistry*, 3rd ed. New York: Interscience, 1972.

Day, M. C., and J. Selbin, *Theoretical Inorganic Chemistry*. New York: Reinhold, 1962.

Douglas, B. E., and D. H. McDaniel, *Concepts and Models of Inorganic Chemistry*. New York: Blaisdell, 1965.

Harvey, K. B., and G. B. Porter, *Introduction to Physical Inorganic Chemistry*. Reading, Mass.: Addison-Wesley, 1963.

Heslop, R. B., and P. L. Robinson, *Inorganic Chemistry*. New York: Elsevier, 1963.

Larsen, E. M., *Transitional Elements*. New York: Benjamin, 1965.

Phillips, C. S. G., and R. J. P. Williams, *Inorganic Chemistry*, Vols. 1 and 2. London: Oxford Univ. Press, 1965.

PROBLEMS

16.1 The metals Cu, Ag, and Au of group IB have properties that are very different from those of the alkali metals of group IA. Compare the enthalpies of vaporization and the ionization energies of elements in each of these two groups and suggest reasons why the elements of group IB are much less active reducing agents than the elements of group IA.

16.2 Write balanced equations to complete each of the following.

$$Au + CN^-(aq) + O_2 \rightarrow \qquad Cu^{++}(aq) + Cu + I^-(aq) \rightarrow$$
$$VO_2^+ + Zn \rightarrow \qquad Fe^{++}(aq) + O_2 \rightarrow$$
$$Co^{++}(aq) + NH_3(aq) + O_2 \rightarrow \qquad MnO_4^- + H^+ \rightarrow$$

16.3 Suggest a reason why copper is a much poorer reducing agent than its neighbors nickel and zinc.

16.4 Why is the complex ion $[CoF_6]^{-3}$ paramagnetic, while $[Co(CN)_6]^{-3}$ is diamagnetic?

16.5 Name the following complexes.

$$[Zn(NH_3)_4]^{++} \qquad Co(NH_3)_3Cl_3$$
$$[FeF_6]^{-3} \qquad [Fe(CN)_6]^{-3}$$
$$[Ag(CN)_2]^- \qquad [Cr(H_2O)_6]^{+3}$$

16.6 What series of steps could be used to synthesize potassium manganate and potassium permanganate from MnO_2?

16.7 Stainless steel is an alloy of iron with approximately 18% chromium and 10% nickel. Suggest a reason why this alloy is corrosion resistant.

16.8 Give some specific examples of the influence of the lanthanide contraction on the properties of the transition-metal elements.

16.9 Consider a solution that contains 0.10 mole of $Fe(NO_3)_3$ per liter. Assume that $[Fe(H_2O_6)]^{+3}$ is a monobasic acid and calculate the hydrogen-ion concentration in this solution. Now consider the second acid dissociation of the complex and calculate the hydrogen-ion concentration in the solution.

16.10 Discuss the reasons why the reduction potential for manganous ion is more negative than those of chromous and ferrous ions.

16.11 In the synthesis of compounds in which metallic elements are in high oxidation states, alkaline conditions are usually employed. Explain why.

16.12 For which of the following reactions is the increase in entropy the largest?

$$[Cu(NH_2CH_2CH_2NH_2)_2]^{++} + 4H_2O = [Cu(H_2O)_4]^{++} + 2NH_2CH_2CH_2NH_2$$
$$[Cu(NH_3)_4]^{++} + 4H_2O = [Cu(H_2O)_4]^{++} + 4NH_3$$

This question can be answered on the basis of the number of product molecules formed. Given that the bond energies are the same in both complexes, which will be the more stable with respect to dissociation?

CHAPTER 17

ORGANIC CHEMISTRY

The element carbon is unique: in combination with about a half-dozen other elements, it forms well over a million known compounds and thus it has a chemistry which is more extensive than any other element. Since the compounds of carbon are involved in all life processes, it is not surprising that there is enormous interest in the chemistry of this element. One of the most intriguing problems in science is that of uncovering the relations between molecular structure and physiological activity. But even apart from the biologically important compounds and their reactions, the richness and subtlety of the chemistry of carbon can account for the intensity with which it is studied.

In this chapter we can present only a brief introduction to organic chemistry. We shall be concerned principally with classes of organic compounds and the characteristic reactions of the compounds in each class. Our object is to demonstrate the basis for systematizing the subject of organic chemistry and to convey something of the nature of the problems that an organic chemist attempts to solve. First, to see how such a large number of distinct compounds can arise from the combination of so few elements, let us examine the molecular structures of the simplest carbon compounds, the *alkanes* or *paraffin hydrocarbons*.

17.1 THE ALKANES, OR PARAFFIN HYDROCARBONS

As their name implies, these are compounds of carbon and hydrogen only, and are comparatively inert (paraffin, from the Latin, means "small affinity"). The alkane of simplest structure is methane, CH_4. As was discussed in Section

11.5, there are four equivalent carbon-hydrogen bonds in this molecule, and the four hydrogen atoms lie at the apices of a regular tetrahedron with the carbon atom at the center. The equivalence of the hydrogen atoms is an important structural feature of methane, and should be noted carefully, for it is common practice to represent methane and other organic molecules by formulas that disguise this property. Figure 17.1 shows three representations of methane. The space-filling model and the ball-and-stick drawing show how the atoms are actually located in space and demonstrate the equivalence of the four hydrogen atoms. The conventional representation of methane is typographically convenient, but can be misleading. The five atoms in methane *do not* all lie in one plane, and the implication that there are two kinds of hydrogen atom, those that are opposite and those that are neighbors, is *incorrect* and must be ignored.

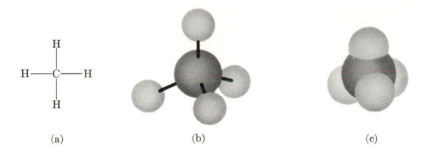

(a) (b) (c)

Three representations of the methane molecule: (a) conventional; (b) ball-and-stick model; (c) space-filling model.   **FIG. 17.1**

It is the ability of one carbon atom to form strong bonds to as many as four other carbon atoms that is largely responsible for the enormous number of organic molecules. This begins to be evident as we examine the other hydrocarbons. Next in order of complexity is ethane, C_2H_6. Three representations of its structure are given in Fig. 17.2. Once again the space-filling and ball-and-stick pictures show the geometric properties of the molecule, and once again the conventional picture must be interpreted with care. In ethane there are six geometrically equivalent hydrogen atoms, three bonded to each of the two carbon atoms that are themselves linked by an electron-pair bond. The bond angles in ethane are all very close to 109.5°, the tetrahedral angle. Consequently, the simplest description of the bonding in ethane is that each carbon atom forms four sp^3 hybrid bonds, three to hydrogen atoms and one to the other carbon atom.

It is sometimes profitable to regard ethane as a *derivative* of methane, formed conceptually by replacing one of the four hydrogens of methane with a CH_3

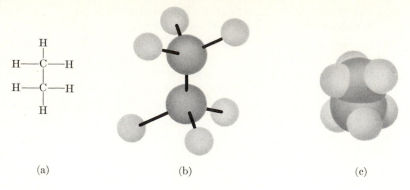

(a)   (b)   (c)

FIG. 17.2   Three representations of the ethane molecule: (a) conventional; (b) ball-and-stick model; (c) space-filling model.

fragment. We might represent this concept by

$$
\begin{array}{c}
\text{H} \\ | \\ \text{H—C—H} \\ | \\ \text{H}
\end{array}
\;+\;
\begin{array}{c}
\text{H} \\ | \\ \text{C—H} \\ | \\ \text{H}
\end{array}
\;\rightarrow\;
\begin{array}{c}
\text{H H} \\ |\;\; | \\ \text{H—C—C—H} \\ |\;\; | \\ \text{H H}
\end{array}
\;+\; \text{H.}
$$

A fragment of a molecule, such as CH_3, is called a *radical*, and since CH_3 is a fragment of methane, it is known as the *methyl radical*. We shall find that other radicals can be formed from other hydrocarbons; the general term applied to any fragment of an alkane is *alkyl radical*.

Propane, C_3H_6, can be derived conceptually by replacing one of the six equivalent hydrogen atoms of ethane by a methyl radical

$$
\begin{array}{c}
\text{H H} \\ |\;\; | \\ \text{H—C—C} \cdots \\ |\;\; | \\ \text{H H}
\end{array}
\;+\;
\begin{array}{c}
\text{H} \\ | \\ \text{C—H} \\ | \\ \text{H}
\end{array}
\;\rightarrow\;
\begin{array}{c}
\text{H H H} \\ |\;\; |\;\; | \\ \text{H—C—C—C—H.} \\ |\;\; |\;\; | \\ \text{H H H}
\end{array}
$$

The group CH_3CH_2, derived from ethane by removal of a hydrogen atom, is called the *ethyl radical*.

The structural representations of propane shown in Fig. 17.3 demonstrate that the eight hydrogen atoms are not equivalent, but fall into two groups: the six that are bonded to the exterior carbon atoms, and the two that are bonded to the interior carbon atom. Consequently, as we continue our conceptual process of generating paraffin hydrocarbons by replacing a hydrogen atom on propane by a CH_3 group, we find two ways of making the substitution. If any one of the six equivalent hydrogen atoms on the exterior carbon atoms is replaced by CH_3, the result is the molecule called normal butane, or *n*-butane,

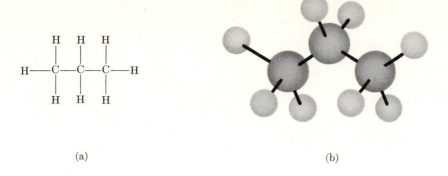

Two representations of the propane molecule: (a) conventional; (b) ball-and-stick model.   **FIG. 17.3**

whose structure is represented by

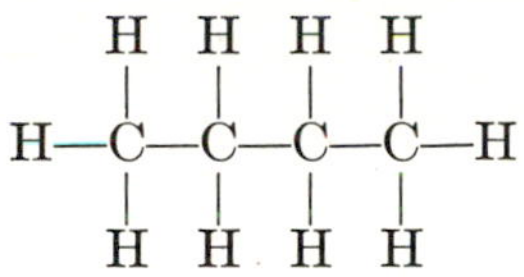

$$H-\underset{\displaystyle H}{\overset{\displaystyle H}{\underset{|}{\overset{|}{C}}}}-\underset{\displaystyle H}{\overset{\displaystyle H}{\underset{|}{\overset{|}{C}}}}-\underset{\displaystyle H}{\overset{\displaystyle H}{\underset{|}{\overset{|}{C}}}}-\underset{\displaystyle H}{\overset{\displaystyle H}{\underset{|}{\overset{|}{C}}}}-H \qquad \text{or} \qquad CH_3CH_2CH_2CH_3.$$

n-butane

On the other hand, replacement of either one of the two hydrogen atoms bonded to the internal carbon atom of propane gives the molecule known as isobutane.

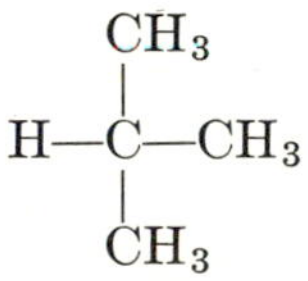

$$H-\underset{\displaystyle CH_3}{\overset{\displaystyle CH_3}{\underset{|}{\overset{|}{C}}}}-CH_3$$

isobutane

The molecules *n*-butane and isobutane both have the molecular formula C_4H_{10}, yet they are distinct compounds with different physical properties and slightly different chemical properties. They are examples of *positional isomers;* molecules that differ by the sequence in which their atoms are bonded to each other. Hydrocarbons like *n*-butane, in which no carbon atom is bonded to more than two other carbon atoms, are called *straight-chain hydrocarbons.* Isobutane, on the other hand, is an example of a *branched-chain hydrocarbon,* for one of its carbon atoms is bonded to *three* other carbon atoms.

Table 17.1 contains the names, formulas, and physical constants of the paraffin hydrocarbons containing five or fewer carbon atoms. The number of positional isomers increases rapidly as the number of carbon atoms increases. There are five isomers of C_6H_{14}, nine of C_7H_{16}, eighteen of C_8H_{18}, and seventy-

Table 17.1 Some saturated hydrocarbons

Name	Formula	MP(°C)	BP(°C)
Methane	CH_4	−183	−162
Ethane	C_2H_6	−172	− 89
Propane	C_3H_8	−187	− 42
n-butane	C_4H_{10}	−135	0
Isobutane	$(CH_3)_3CH$	−145	− 10
n-pentane	C_5H_{12}	−130	36
Isopentane	$CH_3CH_2CH(CH_3)_2$	−160	28
Neopentane	$(CH_3)_4C$	− 20	9.5

five of $C_{10}H_{22}$. Despite the multiplicity of compounds, all hydrocarbon molecules have two structural features in common: each carbon atom is bonded to four other atoms by four electron-pair bonds, and the angle between any two bonds is always close to the ideal tetrahedral angle, 109° 28′.

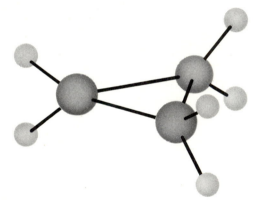

FIG. 17.4   The structure of cyclopropane.

There is another class of paraffin hydrocarbons that consists of molecules in which the carbon chain is formed into a ring. These are the cycloparaffins, and the first member of the series, cyclopropane, is pictured in Fig. 17.4. The carbon atoms in cyclopropane are at the apices of an equilateral triangle, and the hydrogen atoms lie above and below the plane of the three carbon atoms. In the four-carbon cycloalkane, cyclobutane, the carbon–carbon bond angles are nearly 90°, and the carbon skeleton forms a slightly puckered square. In cyclobutane and cyclopropane, then, there is considerable departure from the 109.5° bond angles found in other saturated hydrocarbons. In cyclopentane and other larger cyclic hydrocarbons, the atoms in the ring are arranged so that all the bond angles are near 109°.

Table 17.2 The simple alkyl radicals

$CH_3—$	$CH_3CH_2—$	$CH_3CH_2CH_2—$	CH_3CHCH_3
			|
Methyl	Ethyl	*n*-propyl	Isopropyl
$CH_3CH_2CH_2CH_2—$	$CH_3CHCH_2CH_3$	$(CH_3)_2CHCH_2—$	$(CH_3)_3C—$
	|		
n-butyl	*sec*-butyl	Isobutyl	*tert*-butyl

Nomenclature

The large number of organic molecules presents a formidable problem in nomenclature. Many familiar organic molecules have "common" names, based on their biological origin, the whim of their discoverer, or some other historical accident. To back up the common nomenclature, there are systematic methods for naming compounds which use the distinguishing structural features of a molecule for identification. Some of the formal rules of nomenclature are rather cumbersome, but fortunately we will need only the simplest of them to deal with the compounds we shall discuss.

First, the names of the straight-chain hydrocarbons are assigned as shown in Table 17.1. The suffix *ane* (from alk*ane*) is combined with a prefix which, after the C$_4$ hydrocarbons, is derived from the Greek expression for the number of carbon atoms in the chain. Branched hydrocarbons are regarded as alkyl derivatives of the straight-chain hydrocarbon; that is, as molecules obtained by replacing a hydrogen atom of a straight-chain hydrocarbon by an alkyl radial. For example, isobutane is a methyl derivative of propane, and its systematic name is methyl propane:

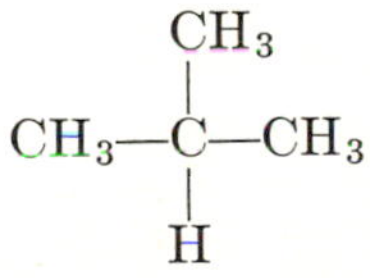

$$CH_3—\underset{\underset{H}{|}}{\overset{\overset{CH_3}{|}}{C}}—CH_3$$

isobutane or
methylpropane

To name other hydrocarbons in this way, the names of the alkyl radicals are needed; some of these are given in Table 17.2.

The name of a branched hydrocarbon is assigned as follows. The longest straight hydrocarbon chain in the compound is found and the molecule treated as a derivative of this hydrocarbon. As we noted above, isobutane contains a chain of three carbon atoms, and is considered to be a methyl-substituted propane. The next step is to number the chain starting at the end of the molecule which is closest to the branches. Thus we would have

$$\begin{array}{cccc} & CH_3 \\ & | \\ CH_3CHCH_2CH_3 \\ 1 & 2 & 3 & 4 \end{array} \qquad \begin{array}{cccccc} & & CH_3 \\ & & | \\ CH_3CH_2C &\!\!\!\!\!\!&\!\!\!\!\!\! CH_2CH_2CH_3 \\ & & | \\ & & CH_3 \\ 1 & 2 & & 4 & 5 & 6 \end{array}$$

The positions of the alkyl substituents are given by the number of the carbon atom to which they are attached. The names of these alkyl groups are prefixed to the name of the parent hydrocarbon. Thus our examples are named

$$\begin{array}{c} CH_3 \\ | \\ CH_3CHCH_2CH_3 \end{array} \qquad \begin{array}{c} CH_3 \\ | \\ CH_3CH_2C\!\!-\!\!CH_2CH_2CH_3 \\ | \\ CH_3 \end{array}$$

2-methylbutane   3, 3-dimethylhexane

The naming of compounds more complicated than hydrocarbons requires only minor extensions of these rules, as we shall see in the next sections.

17.2 FUNCTIONAL GROUPS

The multiplicity of hydrocarbon structures suggests that the number of molecules that can be constructed from the combination of carbon and hydrogen with nitrogen, oxygen, sulfur, etc., should be virtually limitless. The prospect of having to master the chemistry of a significant fraction of these compounds would be exceedingly depressing, were it not for the fact that organic compounds can be grouped into a few classes, members of which have very similar chemistry. The basis for this enormously helpful classification is the *functional group*, a group of atoms that occurs in many molecules and which confers on them a characteristic chemical reactivity, regardless of the form of the carbon skeleton. There are a number of different functional groups, but only about a half dozen that are of frequent occurrence. Thus instead of having to cope with the chemistry of a million compounds, we need only study the chemistry of a half-dozen functional groups.

Common Functional Groups

The functional groups of most frequent appearance and greatest importance occur in the following classes of molecules.

1. *Alcohols* contain the *hydroxyl group*, —OH, bonded to a hydrocarbon framework. The alcohols are named after the alkyl group to which the hydroxyl group is attached, or alternatively, the final *e* in the name of a hydrocarbon

is replaced by *ol*. Typical alcohols are

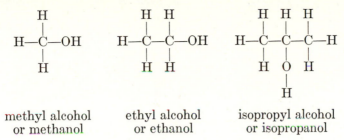

methyl alcohol   ethyl alcohol   isopropyl alcohol
or methanol   or ethanol   or isopropanol

2. *Acids* contain the *carboxyl group,*

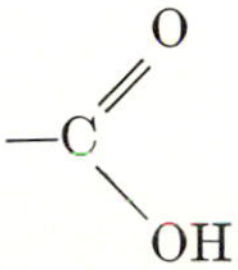

$$-C\!\!\begin{array}{c}\nearrow O \\ \searrow OH\end{array}$$

which is often represented by —COOH. As examples we have

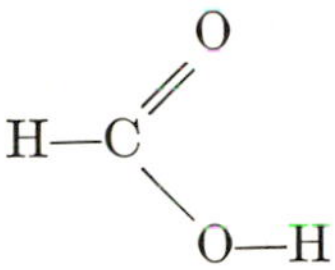

$$H\!-\!C\!\!\begin{array}{c}\nearrow O \\ \searrow O\!-\!H\end{array}$$

formic acid

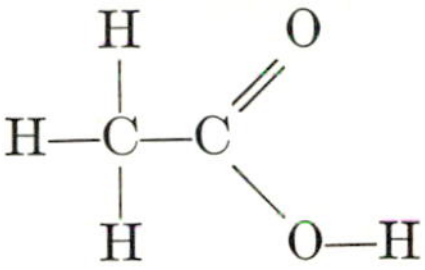

acetic acid

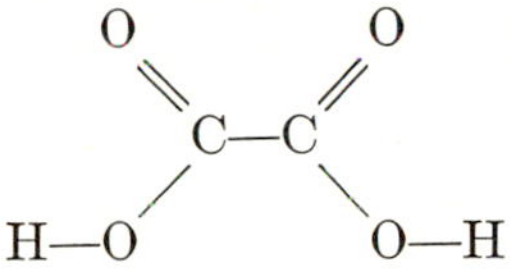

oxalic acid

The suffix *ic* is used to indicate the presence of the carboxyl group, or alternatively, the final *e* in the name of a hydrocarbon is replaced by *oic*. Thus formic and acetic acids can be called methanoic and ethanoic acids respectively.

3. *Aldehydes* and *ketones* contain the *carbonyl group,*

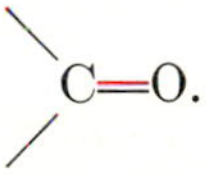

$$\begin{array}{c}\searrow \\ \nearrow\end{array}\!C\!=\!O.$$

In aldehydes, one of the two undesignated bonds is to a hydrogen atom; in ketones, both bonds link the carbonyl group to carbon atoms. Thus we have

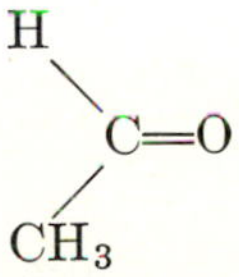

acetaldehyde

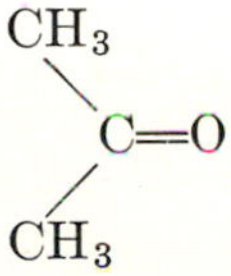

acetone
(a ketone)

4. *Alkenes* incorporate the *carbon–carbon* double bond,

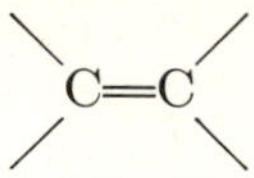

$$\text{C}=\text{C}$$

and are sometimes called *olefins*. A few examples are

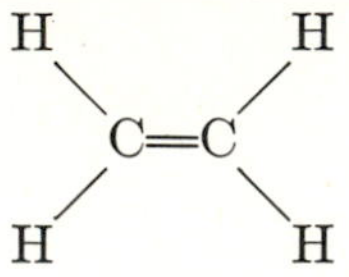

$$
\begin{array}{ccc}
\underset{\text{H}}{\overset{\text{H}}{}}\text{C}=\text{C}\underset{\text{H}}{\overset{\text{H}}{}} & \underset{\text{H}}{\overset{\text{CH}_3}{}}\text{C}=\text{C}\underset{\text{H}}{\overset{\text{H}}{}} &
\end{array}
$$

| ethylene or ethene | propylene or propene | cyclohexene |

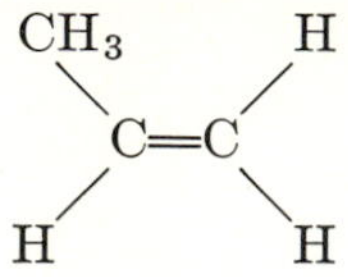

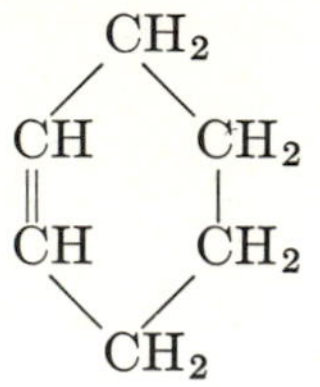

To name an alkene, the *ane* ending in the name of the corresponding alkane is replaced with *ene*. Sometimes the suffix *ene* is added to the name of the alkyl radical that contains the appropriate number of carbon atoms. Thus the two-carbon alkene is ethylene, since the two-carbon alkyl radical is the ethyl radical.

5. *Ethers* have an oxygen atom bonded to two carbon atoms, C—O—C, and in a sense are derivatives of alcohols. Two examples of ethers are

$$CH_3OCH_3 \qquad CH_3OCH_2CH_3.$$

dimethyl ether   methyl ethyl ether

6. *Esters* contain the group

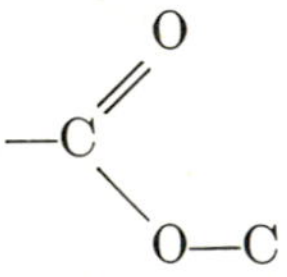

$$
\begin{array}{c}
\quad\quad \text{O}\\
\quad\quad \|\\
-\text{C}\\
\quad \diagdown\\
\quad\quad \text{O}-\text{C}
\end{array}
$$

and are the result of the reaction between an alcohol and an acid. Examples derived from acetic acid and methyl alcohol, and from formic acid and ethyl alcohol, respectively, are

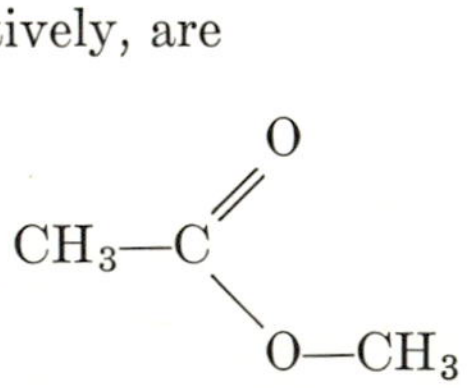

$$
\begin{array}{cc}
& \text{O} \\
& \| \\
CH_3-\text{C} & \\
& \diagdown \\
& \text{O}-CH_3
\end{array}
\qquad
\begin{array}{cc}
& \text{O} \\
& \| \\
H-\text{C} & \\
& \diagdown \\
& \text{O}-CH_2CH_3
\end{array}
$$

methyl acetate 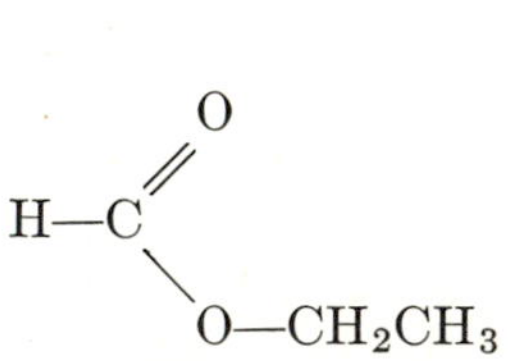   ethyl formate

The first part of the name of an ester is the name of the alkyl radical corresponding to the alcohol from which the ester is formed. The second part of the name uses the suffix *ate* with a prefix derived from the name of the acid constituent of the ester.

7. *Amines* contain the amino group, —NH_2. These molecules can be thought of as derivatives of ammonia:

$$CH_3NH_2 \qquad CH_3CH_2NH_2.$$

methylamine   ethylamine

There are other amines in which two or three alkyl groups are bonded to nitrogen:

$$(CH_3)_2NH \qquad (CH_3)_3N.$$

dimethylamine   trimethylamine

The naming of these compounds involves a straightforward combination of the names of the alkyl radicals with the suffix *amine*.

The great simplifying feature of organic chemistry is that the majority of the chemical reactions involve changes of the functional group only, with no change in the carbon framework of the molecule. This observation is sometimes called the principle of skeletal integrity, since the carbon skeleton of an organic molecule remains unchanged as the functional groups are altered.

There are a large number of reagents, organic and inorganic, which react with the various functional groups. Once again, however, there is a classification scheme that simplifies the chemistry considerably. Most reactions of the functional groups fall into one of the following classes.

1. *Displacement reactions*. These are processes in which one functional group is displaced (or replaced) by another.

2. *Addition-elimination reactions*. Often a functional group is modified by the direct *addition* of new atoms to it. The reverse process is also possible; a functional group is sometimes changed by losing or *eliminating* atoms.

3. *Oxidation-reduction reactions*. The name speaks for itself; some functional groups can be oxidized, others can be reduced, and still others can undergo both types of reactions.

There are other, less important, types of reactions of functional groups, and each of the three we have listed might be broken down into more subtle subdivisions. Nevertheless, we have now a scheme that helps considerably to organize the chemistry of organic molecules. Let us now examine the reactions of the functional groups and look for specific examples of these types of reaction.

17.3 REACTIONS OF ALCOHOLS

Table 17.3 gives the names, formulas, and physical properties of a few alcohols. Methanol, ethanol, and the propanols are colorless nonviscous liquids completely miscible with water, but as the carbon chain lengthens, alcohols approach the behavior of hydrocarbons and the water solubility decreases. The alcohols in

Table 17.3 Physical properties of alcohols

Name	Formula	MP(°C)	BP(°C)
Methyl alcohol	CH_3OH	− 97	65
Ethyl alcohol	CH_3CH_2OH	−114	78
n-propyl alcohol	$CH_3CH_2CH_2OH$	−126	97
Isopropyl alcohol	$(CH_3)_2CHOH$	− 89	82
n-butyl alcohol	$CH_3(CH_2)_3OH$	− 90	118
Isobutyl alcohol	$(CH_3)_2CHCH_2OH$	−108	108
sec-butyl alcohol	$CH_3CHOHCH_2CH_3$		100
tert-butyl alcohol	$(CH_3)_3COH$	25	83

Table 17.3 are of three different kinds, called *primary, secondary,* and *tertiary,* according to the number of alkyl radicals attached to the carbon atom bearing the —OH group. Thus we have

$$RCH_2OH \qquad R{-}\overset{|}{\underset{|}{C}HOH}_{R'} \qquad R{-}\overset{R''}{\underset{R'}{\overset{|}{C}}}{-}OH$$

a primary   a secondary   a tertiary
alcohol   alcohol   alcohol

where the symbol R is used to represent any alkyl group.

Alcohols are named systematically as derivatives of hydrocarbons, with the chain numbering chosen to give the atom bearing the functional group the lowest number, as illustrated by

$$\underset{\overset{|}{OH}}{CH_3CHCH_2CH_3}, \qquad \text{2-butanol; } not \text{ 3-butanol,}$$

$$\underset{\overset{|}{CH_3} \quad \overset{|}{OH}}{CH_3CHCH_2CHCH_3}, \qquad \text{4-methyl-2-pentanol; } not$$
$$\text{2-methyl-4-pentanol.}$$

The boiling points of the alcohols are higher than those of alkanes which have approximately the same molecular weight and number of electrons. This is a consequence of hydrogen bonding: the association between a hydrogen atom on one hydroxyl group with a pair of electrons on the hydroxyl group of another molecule. This hydrogen bonding suggests that alcohols can act as very weak acids and bases. In fact, alcohols do accept protons from the strongest acids, according to the reaction $ROH + H^+ = ROH_2^+$, but the equilibrium constants for such reactions are very small. The hydroxyl group is also very weakly acidic, as evidenced by

$$ROH + Na \rightarrow RONa + \tfrac{1}{2}H_2, \qquad ROH + NaOH \rightarrow RONa + \tfrac{1}{2}H_2O.$$

Compounds of the type RONa are called *alkoxides*, and consist of the ions RO^- and Na^+. The hydroxyl group bonded to an alkyl radical has only very limited acid-base properties: it acquires protons only from the strongest acids, and releases its proton to none but the strongest bases.

Displacement Reactions

The hydroxyl group can be displaced by a number of reagents. Typical of such reactions are

$$HBr + CH_3CH_2OH \rightarrow CH_3CH_2Br + H_2O,$$
$$\text{ethyl bromide}$$

$$HI + CH_3OH \rightarrow CH_3I + H_2O.$$
$$\text{methyl iodide}$$

Displacement reactions have been investigated very thoroughly, and their reaction mechanisms are known. For example, the rate of the reaction of HBr with a primary alcohol like ethanol is found to be proportional to the concentrations of H^+, Br^-, and the alcohol. That is,

$$\frac{d[CH_3CH_2Br]}{dt} = k_{exp}[H^+][Br^-][CH_3CH_2OH].$$

A mechanism consistent with this rate law is

$$CH_3CH_2OH + H^+ \overset{K}{=} CH_3CH_2OH_2^+ \quad \text{(fast equilibrium)};$$
$$CH_3CH_2OH_2^+ + Br^- \overset{k_2}{\longrightarrow} CH_3CH_2Br + H_2O \quad \text{(slow)}.$$

The second step is slow and rate-determining, and since it is an elementary process, its rate law is

$$\frac{d[CH_3CH_2Br]}{dt} = k_2[Br^-][CH_3CH_2OH_2^+]. \tag{17.1}$$

Since the first step in the mechanism is rapid, and the reactants and products are at equilibrium, $[CH_3CH_2OH_2^+] = K[H^+][CH_3CH_2OH]$, and substitution of this expression in Eq. (17.1) gives the experimentally observed rate law, with $k_{exp} = k_2K$.

Further substantiation of this mechanism comes from the observation that the rate of displacement of a hydroxyl group by the various halide ions depends on the identity of the ion. That is, for

$$ROH + H^+ + \begin{cases} F^- \\ Cl^- \\ Br^- \\ I^- \end{cases} \rightarrow RX + H_2O,$$

the reaction rate constants are in the order $F^- < Cl^- < Br^- < I^-$. The sensitivity of the reaction rate to the nature of the halogen shows that the halide ion is involved in the rate-determining step of the mechanism.

Other types of investigations of these displacement reactions have provided a convincing picture of the way in which the halide ion displaces the protonated hydroxyl group:

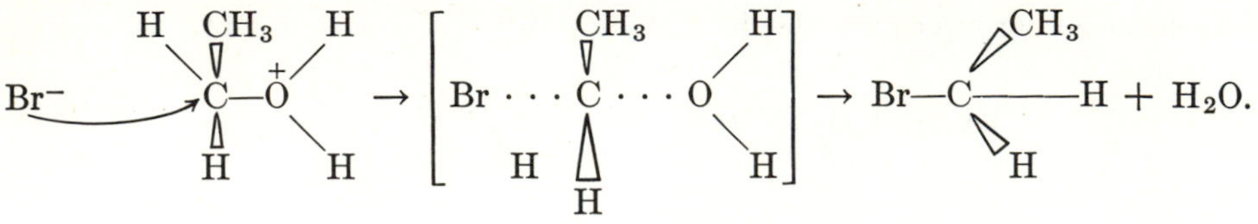

transition state or
activated complex

That is, the halide attacks the "back side" of the C—O bond, and causes the molecule to invert its geometric arrangement as the H_2O group leaves and the halide becomes attached to the carbon atom.

The mechanism we have discussed operates for the displacement reactions of primary and second alcohols. Tertiary alcohols behave somewhat differently. The rate of the reaction

$$\begin{array}{ccc} \begin{array}{c} CH_3 \\ | \\ CH_3\text{—}C\text{—}OH \\ | \\ CH_3 \end{array} + H^+ + Br^- \rightarrow \begin{array}{c} CH_3 \\ | \\ CH_3\text{—}C\text{—}Br \\ | \\ CH_3 \end{array} + H_2O \end{array}$$

tert-butyl alcohol *tert*-butyl bromide

is proportional to the concentration of H^+ and of alcohol, but *does not* depend on the concentration of the halide ion. That is

$$\frac{d[(CH_3)_3CBr]}{dt} = k_{\text{exp}}[H^+][(CH_3)_3COH].$$

Moreover, the rate of reaction does not depend on the nature of the halide ion: F^-, Cl^-, Br^-, and I^- all react with *tert*-butyl alcohol at the same rate. Consequently, the mechanism of the displacement reaction of a tertiary alcohol must have a rate-determining step that *does not* involve the halide ion. A mechanism consistent with these and other data is

$$\begin{array}{c} CH_3 \\ | \\ CH_3\text{—}C\text{—}OH \\ | \\ CH_3 \end{array} + H^+ \overset{K}{\rightleftharpoons} \begin{array}{c} CH_3 \\ | \\ CH_3\text{—}C\text{—}OH_2^+ \\ | \\ CH_3 \end{array} \quad \text{(rapid equilibrium),}$$

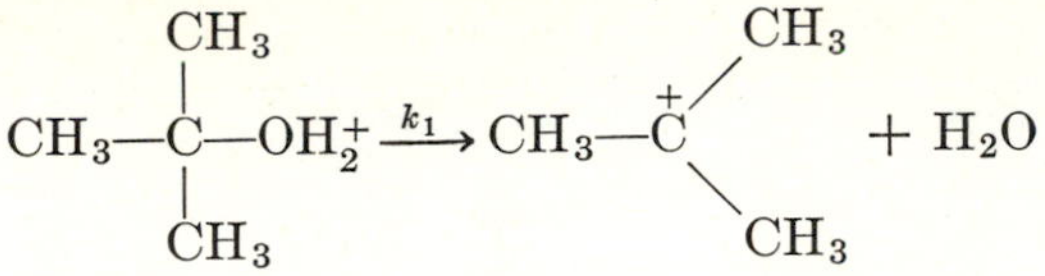

$$CH_3-\underset{\underset{\displaystyle CH_3}{|}}{\overset{\overset{\displaystyle CH_3}{|}}{C}}-OH_2^+ \xrightarrow{k_1} CH_3-\overset{+}{C}\overset{\displaystyle CH_3}{\underset{\displaystyle CH_3}{\Big\langle}} + H_2O \quad (slow),$$

$$CH_3-\overset{+}{C}\overset{\displaystyle CH_3}{\underset{\displaystyle CH_3}{\Big\langle}} + Br^- \rightarrow CH_3-\underset{\underset{\displaystyle CH_3}{|}}{\overset{\overset{\displaystyle CH_3}{|}}{C}}-Br \quad (fast).$$

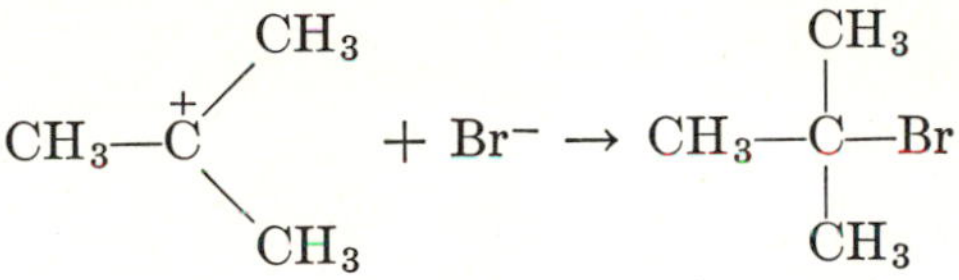

The species $(CH_3)_3C^+$ is called a *carbonium ion*. It is a stable but very reactive fragment and combines rapidly with the halide ion to give the final product. The rate of the reaction is equal to the rate of the slow step:

$$\frac{d[(CH_3)_3CBr]}{dt} = \frac{d[(CH_3)_3C^+]}{dt} = k_1[(CH_3)_3COH_2^+].$$

The reaction

$$[(CH_3)_3COH_2^+] = K[(CH_3)_3COH][H^+]$$

also holds, so the overall rate law is

$$\frac{d[(CH_3)_3CBr]}{dt} = k_1K[H^+][(CH_3)_3COH],$$

which is what is found by experiment, with

$$k_{exp} = k_1K.$$

The displacement reactions of alcohols are good examples of how the reactions of a functional group can be influenced by the nature of the carbon skeleton to which it is attached. Primary, secondary, and tertiary alcohols all undergo the displacement reaction with halides, but the reaction mechanism followed by tertiary alcohols is different from that followed by primary and secondary alcohols. The carbon skeleton thus may influence the rate and mechanism of a reaction of a functional group, but usually does not change its overall nature.

Before concluding the discussion of displacement reactions of alcohols, we might remark that alcohols can be *formed* from alkyl halides by a displacement reaction. Thus the process

$$CH_3CH_2Br + OH^- \rightarrow CH_3CH_2OH + Br^-$$

illustrates a convenient way to convert an alkyl halide to an alcohol.

The Elimination Reaction

The second major type of reaction that alcohols undergo is the elimination reaction. Two examples are

$$CH_3CH_2OH \xrightarrow{H_2SO_4} CH_2{=}CH_2 \; + \; H_2O,$$

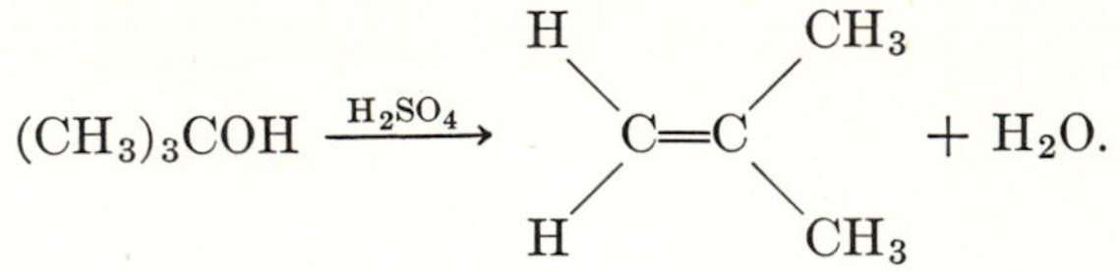

$$(CH_3)_3COH \xrightarrow{H_2SO_4} \quad \begin{array}{c} H \\ \diagdown \\ \end{array} C{=}C \begin{array}{c} CH_3 \\ \diagup \\ \end{array} \; + \; H_2O.$$

We see that the elimination reaction of alcohols is a "dehydration" reaction that forms an alkene. The dehydration of alcohols is a convenient method of synthesizing alkenes, and this reaction is important both in the laboratory and as an industrial process. In general, tertiary alcohols are easier to dehydrate than secondary alcohols, which in turn dehydrate more readily than primary alcohols. The facility with which tertiary alcohols dehydrate is a consequence of the ease with which these molecules form carbonium ions. Thus the mechanism of the dehydration of a tertiary alcohol is

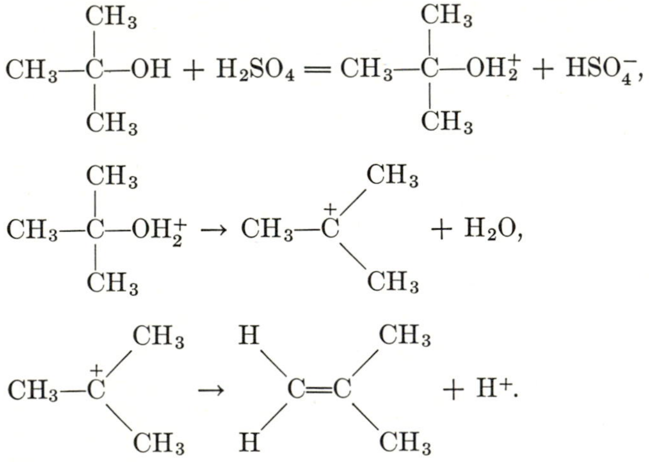

$$CH_3{-}\underset{\underset{CH_3}{|}}{\overset{\overset{CH_3}{|}}{C}}{-}OH + H_2SO_4 = CH_3{-}\underset{\underset{CH_3}{|}}{\overset{\overset{CH_3}{|}}{C}}{-}OH_2^+ + HSO_4^-,$$

$$CH_3{-}\underset{\underset{CH_3}{|}}{\overset{\overset{CH_3}{|}}{C}}{-}OH_2^+ \rightarrow CH_3{-}\overset{+}{C} \begin{array}{c} CH_3 \\ \diagup \\ \diagdown \\ CH_3 \end{array} + H_2O,$$

$$CH_3{-}\overset{+}{C} \begin{array}{c} CH_3 \\ \diagup \\ \diagdown \\ CH_3 \end{array} \rightarrow \begin{array}{c} H \\ \diagdown \\ \diagup \\ H \end{array} C{=}C \begin{array}{c} CH_3 \\ \diagup \\ \diagdown \\ CH_3 \end{array} + H^+.$$

That is, if the carbonium ion does not combine with a negative ion, it may lose a proton, and become an alkene. Secondary and primary alcohols do not form carbonium ions readily, and their dehydration follows a slower, more complicated reaction path.

Oxidation Reactions

The oxidation of alcohols can be accomplished by using a variety of oxidants, and is an important laboratory and industrial reaction. The oxidation of a secondary alcohol by dichromate ion in acidic aqueous solution is a moderately rapid reaction and produces a ketone as the final product. For example,

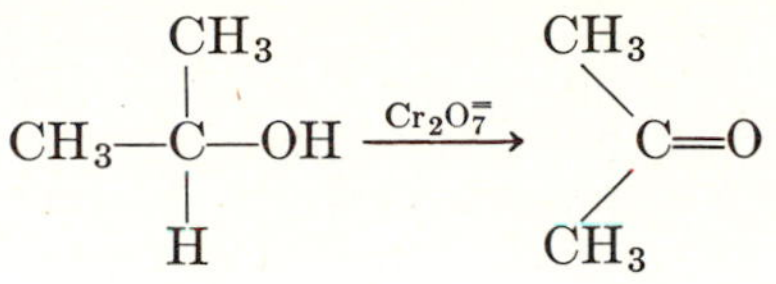

$$CH_3-\underset{\underset{H}{|}}{\overset{\overset{CH_3}{|}}{C}}-OH \xrightarrow{Cr_2O_7^=} \underset{CH_3}{\overset{CH_3}{>}}C=O$$

isopropanol　　　　　acetone

$$CH_3CH_2-\underset{\underset{H}{|}}{\overset{\overset{CH_3}{|}}{C}}-OH \xrightarrow{Cr_2O_7^=} \underset{CH_3CH_2}{\overset{CH_3}{>}}C=O$$

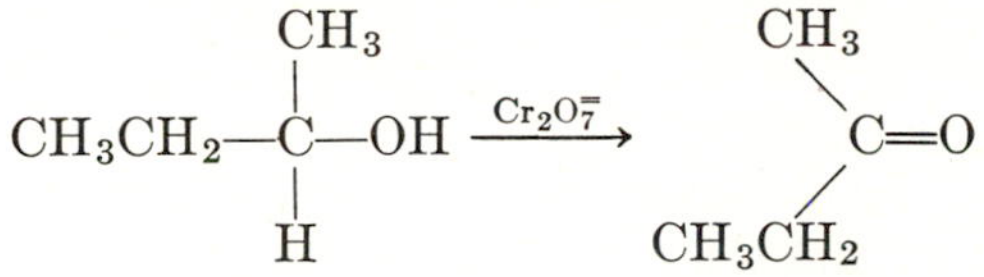

isobutanol　　　　　methyl ethyl ketone

When a primary alcohol is oxidized under the same conditions, the immediate reaction product is an aldehyde.

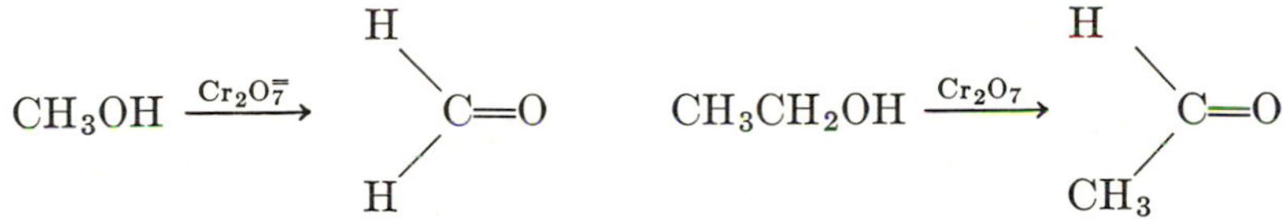

$$CH_3OH \xrightarrow{Cr_2O_7^=} \underset{H}{\overset{H}{>}}C=O \qquad CH_3CH_2OH \xrightarrow{Cr_2O_7} \underset{CH_3}{\overset{H}{>}}C=O$$

formaldehyde

The aldehydes are themselves susceptible to further oxidation, and must be distilled out of the reaction mixture as they are formed, to prevent their destruction.

We see from the examples that the oxidation of an alcohol produces a carbonyl compound: primary alcohols yield aldehydes, and secondary alcohols give ketones. Tertiary alcohols cannot be oxidized without destruction of their carbon skeleton, and these reactions will not be discussed.

17.4 THE REACTIONS OF ALKENES

Molecules that have included in their structure the carbon–carbon double bond are called alkenes. These compounds are said to be unsaturated because one of their principal reactions is the *addition* of reagents to the double bond. In contrast, alkanes are called *saturated hydrocarbons* because they do not undergo addition reactions. A few alkenes and their physical properties are listed in Table 17.4. The physical properties of the alkenes resemble those of the corresponding saturated hydrocarbons. Alkenes and alkanes with the same number of carbon atoms have similar boiling and melting points, and both types of hydrocarbon are insoluble in water.

Table 17.4 Physical properties of alkenes

Name	Formula	MP(°C)	BP(°C)
Ethylene	$CH_2{=}CH_2$	−169	−102
Propylene	$CH_3CH{=}CH_2$	−185	− 48
1-butene	$CH_2{=}CHCH_2CH_3$		− 6.5
Isobutene	$(CH_3)_2C{=}CH_2$	−141	− 7
cis-2-butene	$CH_3CH{=}CHCH_3$	−139	1
trans-2-butene	$CH_3CH{=}CHCH_3$		2.5

A discussion of the electronic nature of the carbon–carbon double bond was given in Section 11.7. As noted there, the double bond can be pictured as consisting of a σ-bond and a π-bond between two carbon atoms that are sp^2-hybridized. This qualitative description, while highly approximate, is consistent with the geometry of the olefinic group: the doubly bonded carbon atoms and the four other atoms attached to them all lie in a single plane, and all bond angles are near 120°, as Fig. 17.5 shows.

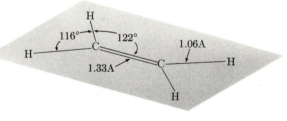

FIG. 17.5 The structure of ethylene.

As is also discussed in Section 11.7, the nature of the σ-π double bond leads to the possibility of *geometrical* isomers. Consequently, there are two distinct geometrical isomers of 2-butene:

cis-2-butene   *trans*-2-butene

The isomer *cis*-2-butene has both methyl groups on the same side of the double bond, while in *trans*-2-butene the methyl groups are on opposite sides. As Table 17.4 shows, the two isomers of 2-butene have nearly the same boiling and melting points. In addition, the two compounds have nearly the same molar free energies of formation: 71.9 kcal/mole for *cis*-2-butene, and 70.9 kcal/mole for *trans*-2-butene. Thus the *trans* isomer is the more stable by a small amount.

Addition Reactions of Olefins

The most important reactions of alkenes involve addition of reagents to the double bond. An olefin will rapidly consume bromine, as in the reactions:

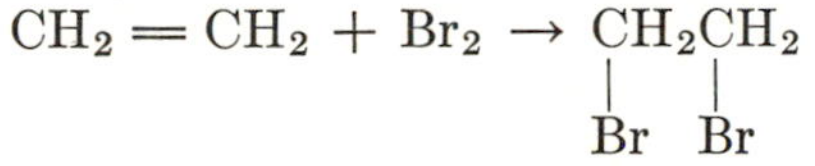

$$CH_2 = CH_2 + Br_2 \rightarrow CH_2CH_2$$
$$\qquad\qquad\qquad\quad \underset{Br}{|} \ \ \underset{Br}{|}$$

1, 2-dibromoethane

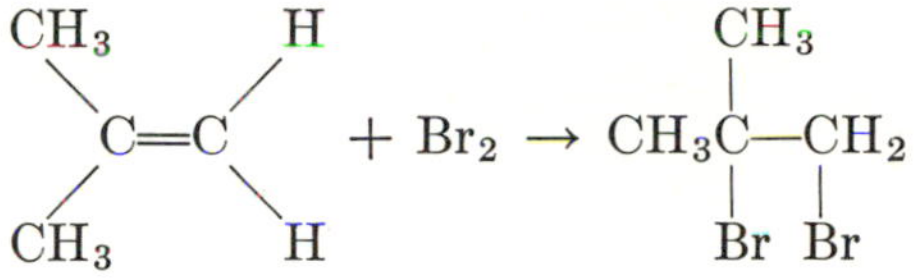

$$\underset{CH_3}{\overset{CH_3}{\diagdown}}C=C\underset{H}{\overset{H}{\diagup}} + Br_2 \rightarrow CH_3\overset{CH_3}{\underset{Br}{\overset{|}{C}}}-CH_2\underset{Br}{|}$$

isobutene   1, 2-dibromo-2-methylpropane

These reactions can be carried out merely by passing the hydrocarbon through a solution of bromine in water at room temperature. The addition of bromine to the double bond is the basis for a simple test that differentiates between an alkene and an alkane. The alkanes, saturated hydrocarbons, do not react with bromine except at elevated temperatures or under the influence of intense illumination by visible light. Consequently, if an unknown hydrocarbon is treated with bromine water, and the red color of bromine disappears, the presence of a double bond in the hydrocarbon is indicated.

An alkene can be converted to an alkane by addition of hydrogen to the double bond. These reactions are usually carried out by using a high pressure of hydrogen gas in the presence of a catalyst such as finely divided platinum, palladium, or nickel.

$$CH_3CH{=}CH_2 + H_2 \xrightarrow{Pd} CH_3CH_2CH_3$$

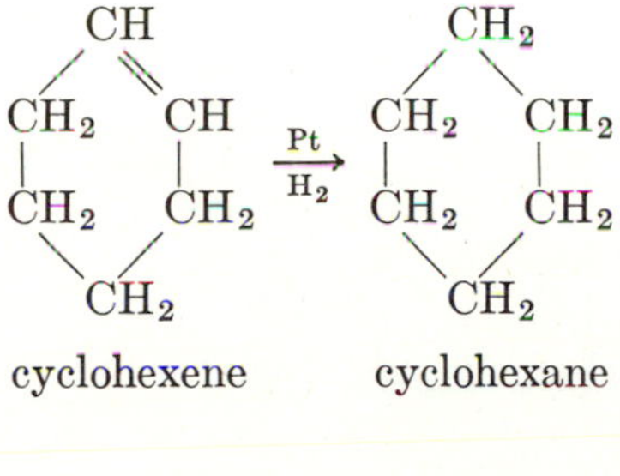

cyclohexene   cyclohexane

The halogen acids also add to the double bond. Two examples are

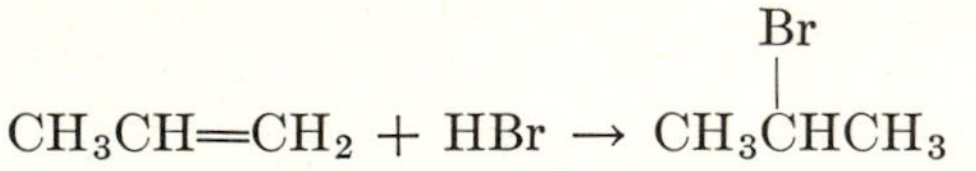

$$CH_3CH{=}CH_2 + HBr \rightarrow CH_3\overset{\overset{\displaystyle Br}{|}}{C}HCH_3$$

isopropyl bromide

$$(CH_3)_2C{=}CH_2 + HCl \rightarrow CH_3\overset{\overset{\displaystyle Cl}{|}}{\underset{\underset{\displaystyle CH_3}{|}}{C}}{-}CH_3$$

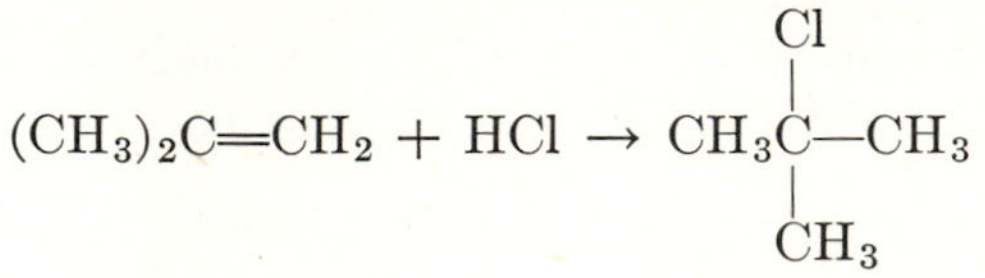

tert-butyl chloride

These reactions illustrate the addition of an acid to unsymmetrical molecules, where the two doubly bonded carbon atoms have different numbers of hydrogen atoms attached to them. Such reactions follow a predictable course: the hydrogen atom of the acid adds to that carbon atom which has attached to it the greater number of hydrogen atoms. The acid anion adds to that carbon atom which has the lesser number of hydrogens. This is known as Markovnikov's rule and is further illustrated by the hydration reaction

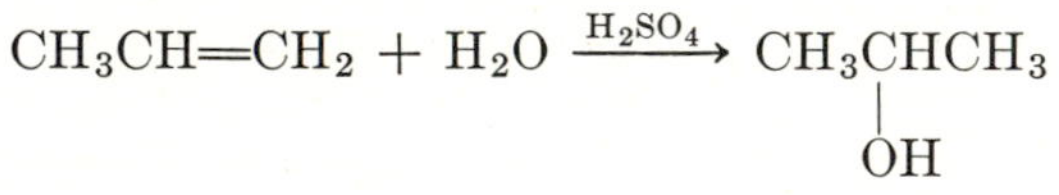

$$CH_3CH{=}CH_2 + H_2O \xrightarrow{\;H_2SO_4\;} CH_3\underset{\underset{\displaystyle OH}{|}}{C}HCH_3$$

There is considerable evidence that addition of acids to double bonds proceeds by a mechanism involving a carbonium ion formed by the attachment of a proton to the olefin,

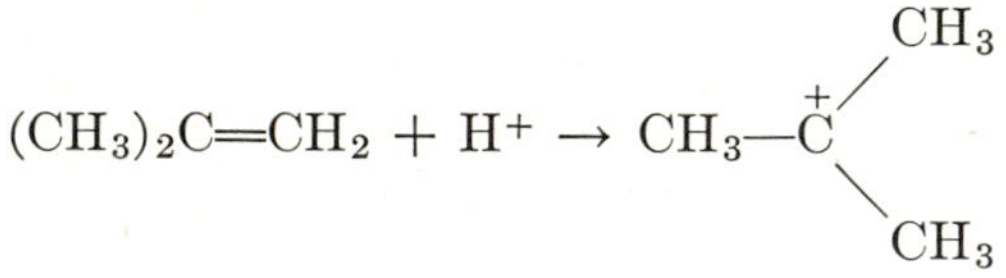

$$(CH_3)_2C{=}CH_2 + H^+ \rightarrow CH_3\overset{+}{C}\underset{\diagdown CH_3}{\overset{\diagup CH_3}{}}$$

The second step of the mechanism is

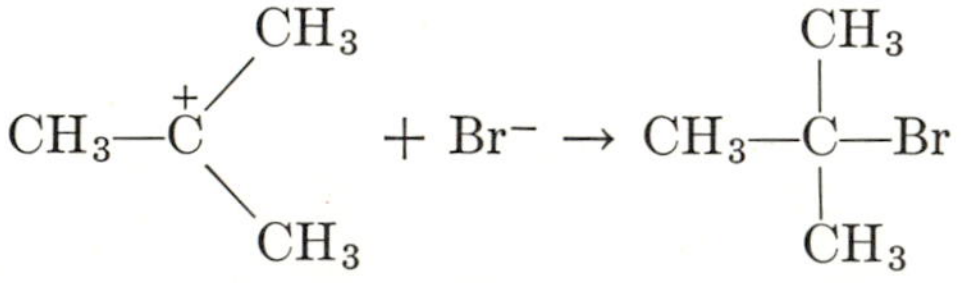

$$CH_3{-}\overset{+}{C}\underset{\diagdown CH_3}{\overset{\diagup CH_3}{}} + Br^- \rightarrow CH_3{-}\overset{\overset{\displaystyle CH_3}{|}}{\underset{\underset{\displaystyle CH_3}{|}}{C}}{-}Br$$

It is clear from this reaction that the carbon atom to which the *anion* becomes attached is the one that bears the positive charge in the carbonium ion. The position of the positive charge is determined by the position of attachment of the proton. For isobutene we have the following alternatives:

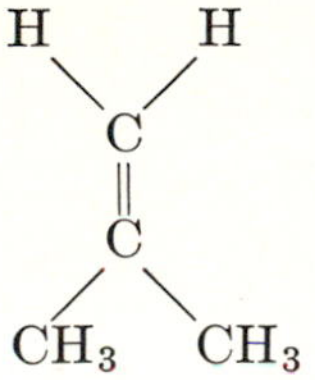

$$\underset{\underset{\displaystyle CH_3 \quad CH_3}{\diagdown \diagup}}{\overset{\overset{\displaystyle H \quad H}{\diagdown \diagup}}{\overset{C}{\underset{C}{\parallel}}}}$$

attachment of H$^+$ here gives $CH_3{-}\overset{\displaystyle CH_3}{\underset{+}{C}}{-}CH_3$,
a tertiary carbonium ion;

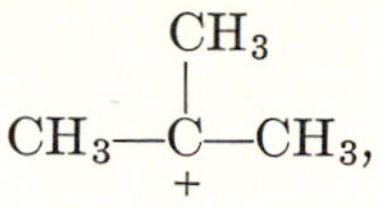

attachment of H$^+$ here gives $(CH_3)_2CHCH_2^+$,
a primary carbonium ion.

Now there is considerable evidence that the order of stability of carbonium ions is

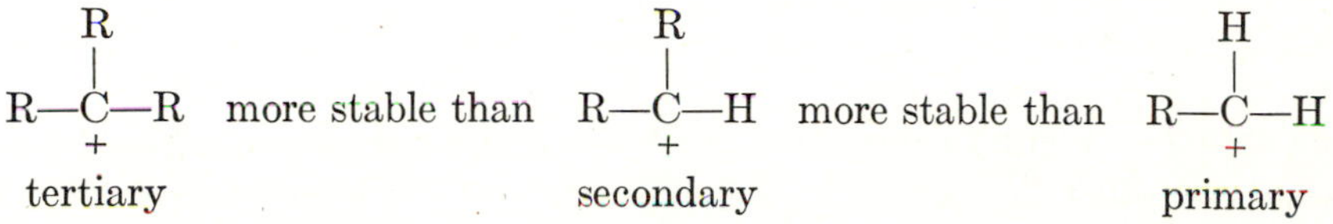

$$\underset{\text{tertiary}}{R{-}\overset{\displaystyle R}{\underset{+}{C}}{-}R} \quad \text{more stable than} \quad \underset{\text{secondary}}{R{-}\overset{\displaystyle R}{\underset{+}{C}}{-}H} \quad \text{more stable than} \quad \underset{\text{primary}}{R{-}\overset{\displaystyle H}{\underset{+}{C}}{-}H}$$

Recall, for example, that the carbonium ion plays a role in the substitution and dehydration of *tertiary* alcohols, which is one indication of the relative stability of tertiary carbonium ions. With the order of carbonium ion stabilities in mind, a little reflection leads us to conclude that the more stable carbonium ion is formed if the proton attacking a double bond adds to the carbon atom which has the greater number of hydrogen atoms attached to it. This, briefly, is the rationalization of the Markovnikov rule for addition reactions.

Oxidation Reactions

Alkenes react readily with a number of oxidizing agents. A simple test for the presence of the olefin group is the reaction with an aqueous acidic solution of permanganate ion. The purple color of permanganate ion disappears as the olefin is oxidized. The course of such a reaction is illustrated by the following examples:

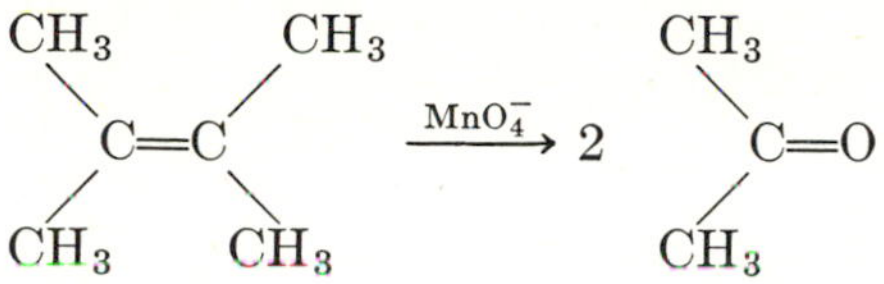

$$\underset{\underset{\displaystyle CH_3 \quad CH_3}{\diagdown \diagup}}{\overset{\overset{\displaystyle CH_3 \quad CH_3}{\diagdown \diagup}}{C{=}C}} \quad \xrightarrow{MnO_4^-} \quad 2 \underset{\displaystyle CH_3}{\overset{\displaystyle CH_3}{\diagdown}}C{=}O$$

acetone

$$\underset{\underset{\displaystyle CH_3 \quad CH_3}{\diagdown \diagup}}{\overset{\overset{\displaystyle H \quad H}{\diagdown \diagup}}{C{=}C}} \quad \xrightarrow{MnO_4^-} \quad 2CH_3C\overset{\displaystyle O}{\underset{\displaystyle OH}{\diagup}}$$

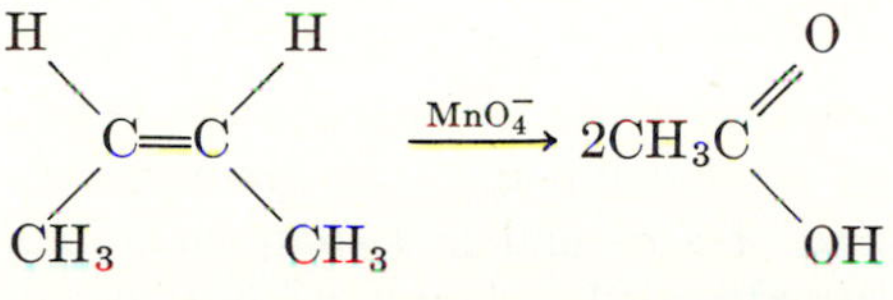

That is, the olefin is cleaved into two oxidized fragments. A carbon atom with *two* alkyl groups attached is converted to the carbonyl group of a ketone, while a carbon atom with one attached hydrogen becomes the carboxyl group of an acid. We can summarize the general features of the reaction by writing

$$\begin{matrix} H \\ \diagdown \\ C{=}C \\ \diagup \quad\quad \diagdown \\ R \quad\quad\quad R' \end{matrix} \begin{matrix} R'' \\ \diagup \\ \\ \\ \end{matrix} \xrightarrow{\;\text{MnO}_4^-\;} \begin{matrix} OH \\ \diagup \\ R{-}C \\ \diagdown\!\!\diagdown \\ O \end{matrix} + \begin{matrix} R'' \\ \diagup \\ O{=}C \\ \diagdown \\ R' \end{matrix}$$

17.5 CARBONYL COMPOUNDS

The carbonyl group occurs in aldehydes and ketones. Some of these compounds have considerable importance in the chemical industry; many tons of formaldehyde are used every year to make plastics, and large quantities of acetone and other ketones are consumed as paint and lacquer thinners. The large number of reactions that the carbonyl group can undergo also makes aldehydes and ketones valuable starting materials in laboratory syntheses.

FIG. 17.6   Schematic representation of the bonding and nonbonding valence orbitals of the carbonyl group.

The carbon-oxygen double bond that occurs in carbonyl compounds is intermediate in length and strength between the single bond in alcohols and the triple bond in carbon monoxide.

$$-\overset{\textstyle|}{\underset{\textstyle|}{C}}-OH \qquad\qquad \diagup\!\!\!\diagdown C{=}O \qquad\qquad C{\equiv}O$$

1.42 A	1.22 A	1.13 A
85 kcal	170 kcal	256 kcal

The detailed picture of the carbon-oxygen double bond is in many respects similar to that of the olefinic linkage. Conventionally the double bond is thought of as consisting of a σ- and a π-component linking the carbon and oxygen atoms which are regarded as being sp^2-hybridized. Thus we have the situation represented in Fig. 17.6. This description, while crude, is consistent with the geometry of aldehydes and ketones; the carbonyl group and the two atoms bonded to it lie in a single plane, and the bond angles about the carbonyl carbon atom are, in general, near 120°.

Because oxygen is more electronegative than carbon, the carbonyl group is polar with the oxygen atom negative. The extent of this polarity is suggested by comparison of the dipole moments of propylene and acetaldehyde:

$$CH_3CH\!=\!CH_2 \qquad CH_3CH\!=\!O$$
$$\mu = 0.35 \text{ D}, \qquad \mu = 2.65 \text{ D}.$$

These molecules are isoelectronic, but the aldehyde is considerably more polar than the alkene. The charge distribution in the carbonyl group can be represented by

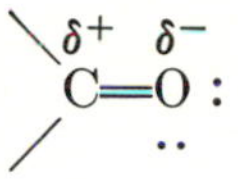

$$\underset{/}{\overset{\backslash}{C}}\!\!=\!\!\overset{\delta+}{\underset{}{}}\overset{\delta-}{O}:$$

The existence of this polarity and the presence of the pairs of nonbonded electrons on the oxygen atom suggest that aldehydes and ketones should be weak Lewis bases. As expected, these compounds can be protonated by strong acids, as in the reaction

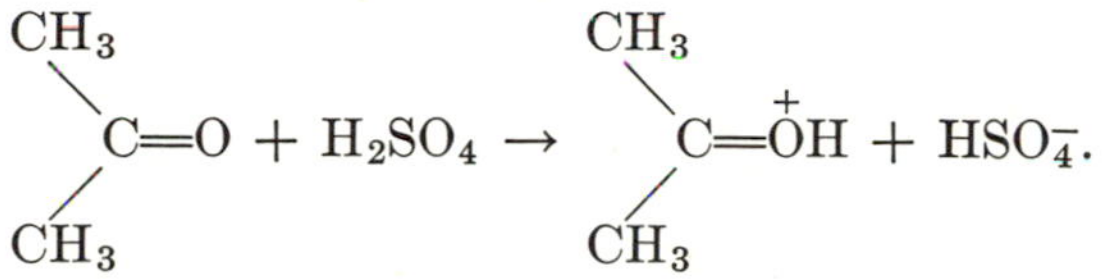

$$\underset{CH_3}{\overset{CH_3}{\diagdown}}C\!=\!O + H_2SO_4 \rightarrow \underset{CH_3}{\overset{CH_3}{\diagdown}}C\!=\!\overset{+}{O}H + HSO_4^-.$$

The name aldehyde is derived from the observation that these molecules can be prepared by *al*cohol *dehyd*rogenation at elevated temperatures,

$$CH_3CH_2OH \xrightarrow[250° \text{ C}]{Cu} CH_3CHO + H_2.$$

Individual aldehydes are named by combining the suffix *al* to the name of the longest straight-chain alkyl group. Thus we have

$$\begin{array}{cc} & \overset{\displaystyle CH_3}{\underset{\displaystyle |}{}} \\ CH_3CH_2CH_2CHO & CH_3CHCH_2CHO. \\ & 4 \quad 3 \quad 2 \quad 1 \\ \text{butanal} & \text{3-methyl butanal} \end{array}$$

The numbering of the carbon chain in such molecules always begins at the aldehyde group.

The designation ketone is derived from the name of the simplest such molecule, acetone. One systematic way to name individual ketones is to use the names of the two alkyl groups attached to the carbonyl carbon as in

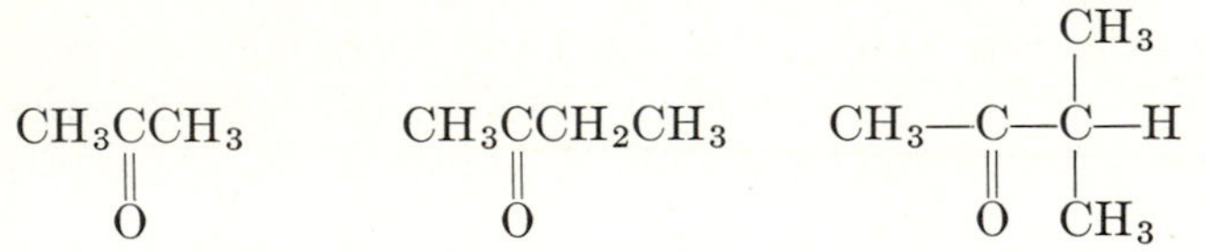

$$CH_3CCH_3 \qquad CH_3CCH_2CH_3 \qquad CH_3-\overset{\underset{|}{CH_3}}{\underset{\underset{CH_3}{|}}{C}}-\overset{}{C}-H$$
$$\underset{O}{\|} \qquad\qquad \underset{O}{\|} \qquad\qquad\quad \underset{O}{\|}$$

acetone or   methyl ethyl   methyl isopropyl
dimethyl ketone   ketone   ketone

For more complicated ketones the functional groups and substituents are located by number,

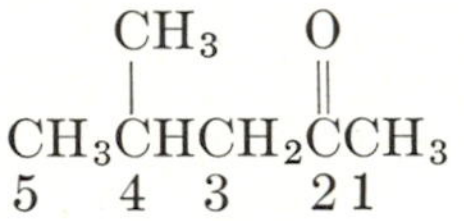

$$\underset{\underset{5\quad4\quad3\quad\ \ 2\,1}{CH_3CHCH_2CCH_3}}{\overset{\overset{CH_3\quad\ \ O}{|\qquad\ \ \|}}{}} \qquad \text{4-methyl-2-pentanone}$$

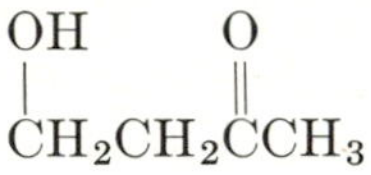

$$\underset{CH_2CH_2CCH_3}{\overset{\overset{OH\qquad O}{|\qquad\ \|}}{}} \qquad \text{4-hydroxy-2-butanone}$$

We see that the suffix -one indicates that the compound is a ketone, and the carbon chain is numbered from the end nearest the carbonyl group.

Addition Reactions

Additions formed the most important class of reactions of the olefinic double bond, and they are equally important for the carbonyl double bond. One reaction, characteristic only of the carbonyl group, is the bisulfite addition reaction,

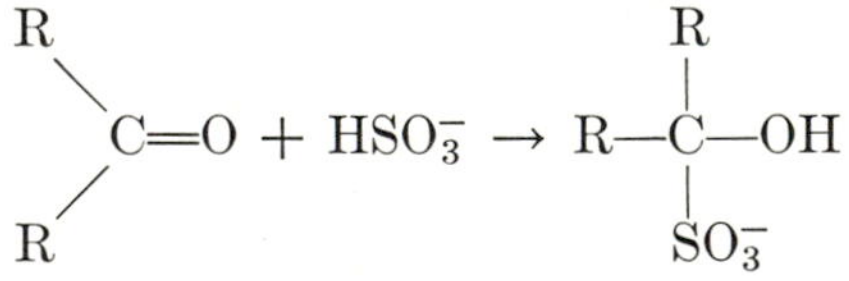

$$\underset{R}{\overset{R}{\diagdown}}C{=}O + HSO_3^- \rightarrow R-\overset{\overset{R}{|}}{\underset{\underset{SO_3^-}{|}}{C}}-OH$$

The bisulfite addition product is an ion that can be precipitated as a sodium salt; this reaction is used as a method for separating aldehydes and ketones from other organic substances in mixtures. After the bisulfite adducts are separated and crystallized, the aldehydes or ketones can be regenerated by treatment of the adducts with strong acid.

Two other addition reactions useful in demonstrating the presence of carbonyl groups in a compound are the following. Hydroxyl amine, NH_2OH, and hydrazone, NH_2NH_2, each add to the carbonyl bond, but the initial addition products lose water to give the final compounds.

$$\begin{matrix} R \\ \diagdown \\ C{=}O \\ \diagup \\ R \end{matrix} + NH_2OH \rightarrow R{-}\overset{\displaystyle R}{\underset{\displaystyle NHOH}{\overset{\displaystyle |}{\underset{\displaystyle |}{C}}}}{-}OH \xrightarrow{-H_2O} \begin{matrix} R \\ \diagdown \\ C{=}NOH \\ \diagup \\ R \end{matrix}$$

an oxime

$$\begin{matrix} R \\ \diagdown \\ C{=}O \\ \diagup \\ R \end{matrix} + NH_2NH_2 \rightarrow R{-}\overset{\displaystyle R}{\underset{\displaystyle NHNH_2}{\overset{\displaystyle |}{\underset{\displaystyle |}{C}}}}{-}OH \xrightarrow{-H_2O} \begin{matrix} R \\ \diagdown \\ C{=}NNH_2 \\ \diagup \\ R \end{matrix}$$

a hydrazone

These reactions are useful in identifying molecules because oximes and hydrazones are often crystalline compounds with characteristic melting points.

A very useful addition reaction of carbonyl molecules involves an organometallic compound known as a Grignard reagent. These substances are conventionally represented by the symbol RMgX, where R is an alkyl group, and X is a halogen atom. Grignard reagents are prepared by the reaction of an alkyl halide with metallic magnesium in an ether solvent,

$$CH_3CH_2Br + Mg \xrightarrow{ether} CH_3CH_2MgBr,$$

$$RX + Mg \xrightarrow{ether} RMgX.$$

This preparation must be carried out in the absence of water; one reason is that Grignard reagents react with water to give a hydrocarbon

$$RMgX + H_2O \rightarrow RH + \tfrac{1}{2}Mg(OH)_2 + \tfrac{1}{2}MgX_2.$$

This reaction can in fact be used to prepare a hydrocarbon from an alkyl halide.

Grignard reagents react with carbonyl compounds in the following general way:

$$R'MgX + \begin{matrix} R \\ \diagdown \\ C{=}O \\ \diagup \\ R \end{matrix} \rightarrow R{-}\overset{\displaystyle R}{\underset{\displaystyle OMgX}{\overset{\displaystyle |}{\underset{\displaystyle |}{C}}}}{-}R'$$

Water is then added to hydrolyze the addition product to an alcohol,

$$R{-}\overset{\displaystyle R}{\underset{\displaystyle OMgX}{\overset{\displaystyle |}{\underset{\displaystyle |}{C}}}}{-}R' + H_2O \rightarrow R{-}\overset{\displaystyle R}{\underset{\displaystyle OH}{\overset{\displaystyle |}{\underset{\displaystyle |}{C}}}}{-}R' + \tfrac{1}{2}Mg(OH)_2 + \tfrac{1}{2}MgX_2.$$

Thus the reaction of a Grignard reagent with a ketone leads to a tertiary alcohol. The reaction with aldehydes gives secondary alcohols,

$$\begin{array}{c}R\\ \diagdown\\ \diagup \;\; C{=}O\\ H\end{array} + R'MgX \;\rightarrow\; \begin{array}{c}R\\ |\\ H{-}C{-}R'\\ |\\ OMgX\end{array} \xrightarrow{H_2O} \begin{array}{c}R\\ |\\ H{-}C{-}R'\\ |\\ OH\end{array}$$

It is clear that the Grignard reaction provides a way of introducing any desired alkyl group into a molecule. Therefore the reaction is often useful in the synthesis of new molecules.

Oxidation-Reduction Reactions

Ketones are highly resistant to oxidation; they react only with the strongest oxidizing agents, and the result is the destruction of the carbon skeleton. Such reactions are seldom of value, and we will not consider them further. In contrast, aldehydes are very easily oxidized to carboxylic acids:

$$\begin{array}{c}R\\ \diagdown\\ \diagup \;\; C{=}O\\ H\end{array} \xrightarrow{Cr_2O_7^{=}} \begin{array}{c}O\\ \diagup\diagup\\ R{-}C\\ \diagdown\\ O{-}H\end{array}$$

This difference in response to oxidizing agents is the basis for qualitative tests that distinguish between aldehydes and ketones. Aldehydes, but not ketones, react with the complex ion $Ag(NH_3)_2^+$ to give a bright "mirror" of metallic silver, plated on the walls of the reaction vessel,

$$RCHO + 2Ag(NH_3)_2^+ + H_2O \rightarrow RCOO^- + 2Ag + 3NH_4^+ + NH_3.$$

Other functional groups like alcohols and olefins are not oxidized in this manner, so the test is quite specific to aldehydes.

Both aldehydes and ketones can be reduced to alcohols in a number of ways. For example,

$$\begin{array}{c}R\\ \diagdown\\ \diagup \;\; C{=}O\\ R\end{array} + H_2 \xrightarrow{Pt} \begin{array}{c}R\\ |\\ H{-}C{-}OH,\\ |\\ R\end{array} \qquad \begin{array}{c}R\\ \diagdown\\ \diagup \;\; C{=}O\\ R'\end{array} \xrightarrow[C_2H_5OH]{Na} \begin{array}{c}R\\ |\\ R'{-}C{-}OH,\\ |\\ H\end{array}$$

$$\begin{array}{c}CH_2{=}CH{-}CH_2\\ \diagdown\\ \diagup \;\; C{=}O\\ R'\end{array} + \begin{array}{c}CH_3\\ |\\ H{-}C{-}OH\\ |\\ CH_3\end{array} \xrightarrow{Al(OR)_3} \begin{array}{c}CH_2{=}CH{-}CH_2\\ \diagdown\\ R'{-}C{-}OH\;+\\ |\\ H\end{array} \begin{array}{c}CH_3\\ \diagdown\\ \diagup\;\;C{=}O\\ CH_3\end{array}$$

These examples show that reduction of an aldehyde yields a primary alcohol, while a secondary alcohol is the product of ketone reduction. The last reaction, catalyzed by aluminum isopropoxide, $Al(OR)_3$, is a very specific way of reducing the carbonyl group *without* simultaneously adding hydrogen to any carbon–carbon double bonds. Such highly specific reactions are very useful for effecting desired modifications of molecules that have more than one functional group.

17.6 SYNTHESES AND STRUCTURE DETERMINATIONS

Having discussed only a few reactions of organic molecules, we are in a position to see how the business of synthesizing new molecules is carried out. Suppose that we wished to make 3-hexanone, and 1-propanol was our only available organic compound. The simplest way to formulate a scheme for the synthesis of a compound is to work backward from the desired molecule. For the problem at hand, we know that 3-hexanone can be prepared by the oxidation of 3-hexanol,

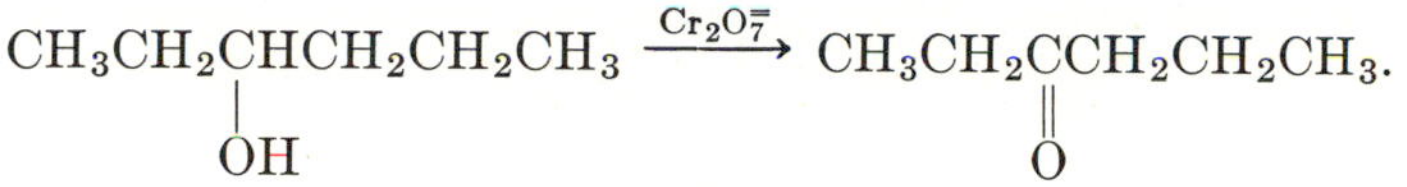

$$CH_3CH_2\underset{\underset{OH}{|}}{C}HCH_2CH_2CH_3 \xrightarrow{Cr_2O_7^=} CH_3CH_2\underset{\underset{O}{\|}}{C}CH_2CH_2CH_3.$$

Now 3-hexanol is a secondary alcohol, and thus can be prepared by the addition of a Grignard reagent to an aldehyde. The reaction we need is

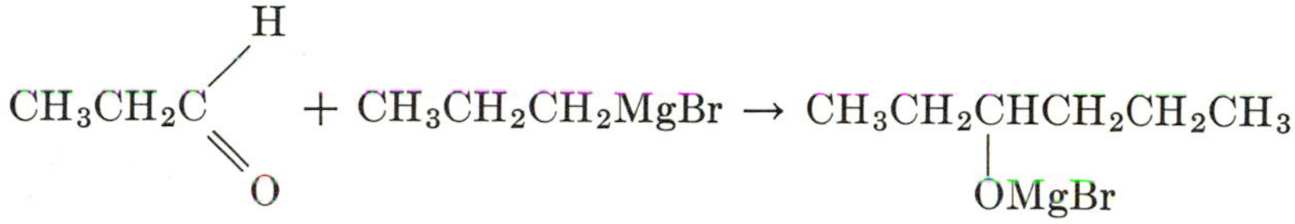

$$CH_3CH_2\underset{\underset{O}{\diagdown\!\!\|}}{C}\overset{\diagup H}{} + CH_3CH_2CH_2MgBr \rightarrow CH_3CH_2\underset{\underset{OMgBr}{|}}{C}HCH_2CH_2CH_3$$

propanal

$$CH_3CH_2\underset{\underset{OMgBr}{|}}{C}HCH_2CH_2CH_3 \xrightarrow{H_2O} CH_3CH_2\underset{\underset{OH}{|}}{C}HCH_2CH_2CH_3$$

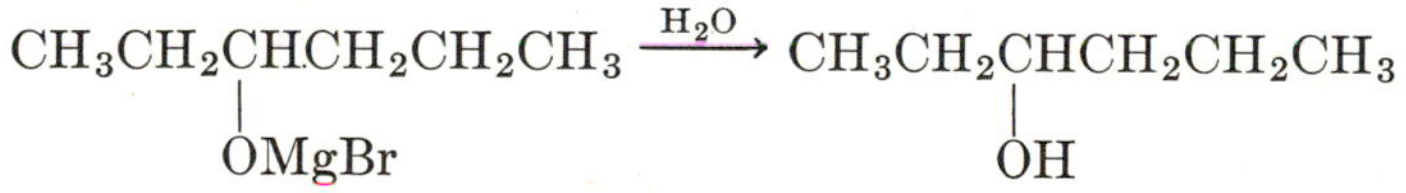

To carry out the preparation of 3-hexanol, then, we must have available propanal and *n*-propyl bromide. These can be obtained from 1-propanol in the following manner:

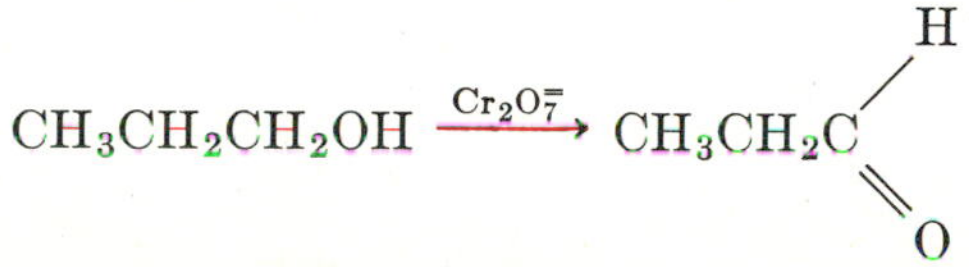

$$CH_3CH_2CH_2OH \xrightarrow{Cr_2O_7^=} CH_3CH_2\underset{\underset{O}{\diagdown\!\!\|}}{C}\overset{\diagup H}{}$$

$$CH_3CH_2CH_2OH \xrightarrow{HBr} CH_3CH_2CH_2Br$$

The propyl bromide is converted to the Grignard reagent by

$$CH_3CH_2CH_2Br + Mg \xrightarrow{ether} CH_3CH_2CH_2MgBr.$$

Thus we have found a set of reactions that can lead to the synthesis of 3-hexanone from 1-propanol.

As a second problem, let us attempt to prepare isobutene from 1-propanol and methyl bromide. The final product could be obtained by dehydration of *tert*-butyl alcohol,

$$
\underset{\underset{\text{OH}}{|}}{\overset{\overset{\text{CH}_3}{|}}{\text{CH}_3-\text{C}-\text{CH}_3}} \xrightarrow{\text{H}_2\text{SO}_4} \underset{\underset{\text{CH}_3}{}}{\overset{\overset{\text{CH}_3}{}}{\text{C}}}=\underset{\underset{\text{H}}{}}{\overset{\overset{\text{H}}{}}{\text{C}}} + \text{H}_2\text{O}.
$$

The necessary alcohol can be prepared from acetone and methyl Grignard reagent:

$$
\text{CH}_3\text{Br} + \text{Mg} \rightarrow \text{CH}_3\text{MgBr}
$$

$$
\text{CH}_3\text{MgBr} + \underset{\underset{\text{CH}_3}{}}{\overset{\overset{\text{CH}_3}{}}{\text{C}}}=\text{O} \rightarrow \underset{\underset{\text{OMgBr}}{|}}{\overset{\overset{\text{CH}_3}{|}}{\text{CH}_3-\text{C}-\text{CH}_3}} \xrightarrow{\text{H}_2\text{O}} \underset{\underset{\text{OH}}{|}}{\overset{\overset{\text{CH}_3}{|}}{\text{CH}_3-\text{C}-\text{CH}_3}}
$$

While acetone is the oxidation product of isopropanol, our starting material is 1-propanol. To obtain isopropanol, we take advantage of the Markovnikov rule:

$$
\text{CH}_3\text{CH}_2\text{CH}_2\text{OH} \xrightarrow{\text{H}_2\text{SO}_4} \text{CH}_3\text{CH}=\text{CH}_2 + \text{H}_2\text{O}
$$

$$
\text{CH}_3\text{CH}=\text{CH}_2 + \text{H}_2\text{O} \rightarrow \underset{\underset{\text{OH}}{|}}{\text{CH}_3\text{CHCH}_3}
$$

As a third example, let us plan the preparation of methyl cyclopentane, starting with cyclopentanone. The structures of the starting material and the products are similar enough so that we can proceed sequentially from reactants to product:

Because we are familiar with only a few reactions, these examples of synthesis have been very simple, and can only give some flavor of the problems solved by organic chemists. As the molecules to be synthesized become more complicated, the subtlety and interest of the problems increase.

Another type of problem that we can demonstrate using reactions familiar to us is the determination of the structure of a molecule from its chemical behavior. For example, there are two compounds, A and B, with the empirical formula C_3H_6. Compound A reacts readily with bromine to give a colorless product, but compound B does not. What are the structures of A and B? From the empirical formula C_3H_6 and the general rules of valence, we see there are only two possibilities,

$$CH_3CH{=\!\!=}CH_2 \qquad \overset{\displaystyle CH_2}{\underset{H_2C\text{------}CH_2}{\diagup\diagdown}}$$

propylene   cyclopropane

Since A reacts with bromine, but B does not, compound A must be an unsaturated molecule, and is therefore propylene.

A similar, but more challenging, problem is the following. A compound C_4H_8 reacts with bromine water and adds one mole of hydrogen per mole of compound upon catalytic hydrogenation. When C_4H_8 is treated with aqueous permanganate, acetone is found in the products.

To proceed with the structure determination, we note that the reactions with bromine and hydrogen show that the molecule is unsaturated, and thus must be one of the butenes:

$$CH_3CH_2CH{=\!\!=}CH_2, \qquad CH_3CH{=\!\!=}CHCH_3, \qquad (CH_3)_2C{=\!\!=}CH_2.$$

Of these molecules, the only one which would give acetone upon oxidation with permanganate is isobutene, or 2-methyl propylene, and consequently this must be the unknown C_4H_8 compound.

17.7 AROMATIC COMPOUNDS

In Section 11.8 we discussed the nature of the electronic structure of benzene, C_6H_6, and pointed out that this molecule offers a particularly good example of multicenter bonding. There is a vast number of organic molecules whose structures are based on that of benzene, and these are called *aromatic compounds*, as distinguished from *aliphatic* compounds which are related to alkanes. The multicenter bonding that exists in benzene and its derivatives confers characteristic chemical properties on aromatic compounds which are unique and very interesting. Before discussing the chemistry of aromatic compounds, let us review and extend our discussion of the bonding in benzene.

X-ray crystallography and spectroscopic studies show that benzene is a planar molecule with the form of a regular hexagon. All C—C bonds are identical, and all bond angles are 120°. There is no *single* valence formula made up

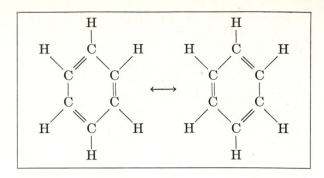

FIG. 17.7 Representation of benzene as a resonance hybrid of two extreme structures.

of electron-pair bonds localized between pairs of atoms that is consistent with these geometric data. Consequently it is conventional to represent benzene as a resonance hybrid of two extreme structures as shown in Fig. 17.7. As noted in Section 11.8, this representation is to be interpreted to mean that the six π-electrons are not localized between pairs of carbon atoms, but can each visit all six atoms in the ring. Thus benzene exhibits multicenter π-bonding. Another notation sometimes used to represent the multicenter π-bonding and complete equivalence of the C—C bonds is

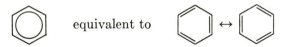

where we have followed the usual convention of omitting the hydrogen and the carbon atoms.

We have called attention to the fact that each type of chemical bond has a characteristic length which is nearly independent of the molecule in which the bond occurs. With this in mind we can compare the C—C bond length found in benzene with that in ethane and ethylene:

$$\text{H}_3\text{C—CH}_3 \qquad \bigcirc \qquad \text{H}_2\text{C}=\text{CH}_2$$

1.54 A   1.39 A   1.33 A

The C—C bond in benzene is shorter than a single bond but longer than a double bond. This is consistent with the idea that the six π-electrons of benzene are distributed among six C—C bond regions and suggests that we might think of the C—C bond in benzene as neither a single nor a double bond, but as a $1\frac{1}{2}$ bond.

We can pursue the relation between resonance and bond lengths a bit further by examining the resonance structures of naphthalene, C_{10}H_8. The important resonance structures are those in which all electrons are paired, and there are

Table 17.5 Bond lengths in naphthalene

Bond	1—2	2—3	1—9	9—10
Order	$1\frac{2}{3}$	$1\frac{1}{3}$	$1\frac{1}{3}$	$1\frac{1}{3}$
Length (A)	1.365	1.404	1.425	1.393

bonds only between neighboring atoms. For naphthalene there are only three such structures:

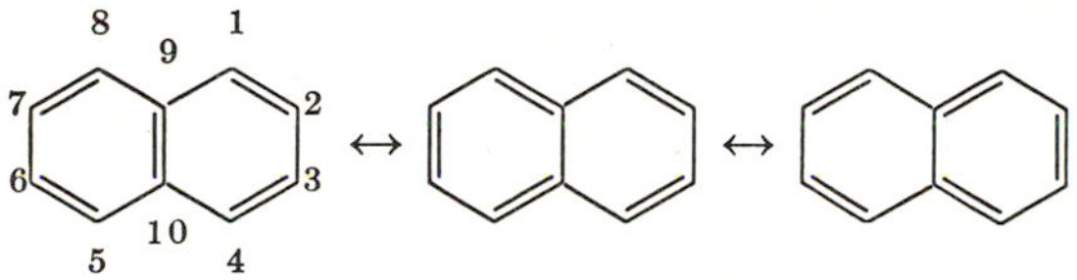

We can see that in two of these three structures the bond between carbon atoms 1 and 2 is double. Thus we can assign this bond a bond order of $1\frac{2}{3}$, and expect it to be longer than the double bond of ethylene (bond order 2), but shorter than the bond in benzene (order $1\frac{1}{2}$). Table 17.5 gives the bond order and bond length for the various bonds in naphthalene. It is clear that there is a definite correlation between the bond lengths and the bond orders assigned from the resonance structures. Thus resonance structures are of use in understanding the geometries of the aromatic compounds.

Let us now examine the thermochemical evidence that concerns the bonding in benzene. As Table 17.6 shows, the heat evolved when olefins of various structures are hydrogenated is nearly constant at 28.6 kcal/mole of double bonds. This fact is consistent with the approximate constancy of bond energies. In the hydrogenation process the H—H and C=C bonds are destroyed and replaced with two C—H bonds and one C—C bond. If these bond energies are nearly constant, the ΔH of hydrogenation per double bond should be nearly constant, as is observed. This reasoning suggests that if the bonds in benzene were in fact three ordinary double bonds, ΔH for the reaction shown at the right

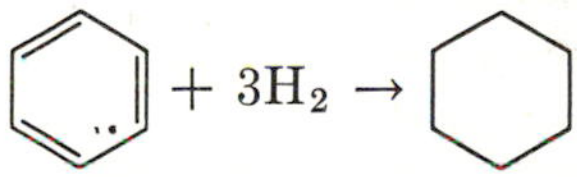

Table 17.6 Enthalpy of hydrogenation of olefins, ΔH_{hydrog} (kcal/mole)

Name	ΔH
Ethylene	−32.8
1-butene	−30.3
cis-2-butene	−28.6
trans-2-butene	−27.6
Isobutene	−28.4
2-methyl-1-butene	−28.5
Cyclohexene	−28.6

should be approximately $3 \times (-28.5) = -85.5$ kcal/mole of benzene. The experimental value for ΔH is in fact only -49.8 kcal, considerably less than expected. This means that benzene is more stable, or lower in energy, than it would be if the bonding consisted of three conventional double bonds.

There have been many attempts to use quantum mechanics qualitatively to explain the peculiar stability of the multicenter bonding in benzene, but the problem is complicated enough that no universally accepted explanation has been found. One quite logical and straightforward suggestion is that the multicenter bonding diminishes electron repulsion by permitting the π-electrons to be far from each other, but still be in regions between two nuclei. A structure that represents this situation is

where electrons of one spin are indicated by crosses, and those of opposite spin by circles. The fact that the electrons are not grouped as pairs, but still are in regions near and between nuclei leads to the extra stability of this situation. Whatever the source of the extra stability, however, the empirical observation that multicenter bonding leads to energy lowering is very useful. This observation is sometimes stated by saying that resonance between conventional structures leads to a decrease in the energy of a molecule. This language can be misleading, for it tends to suggest quite incorrectly that the resonance structures exist and the molecule oscillates between them. The real point is that the structure that actually exists has lower energy than any of the individual resonance structures we can draw using conventional localized electron-pair bonds.

Further evidence for the unique nature and considerable stability of benzene comes from its chemical behavior. Although benzene is unsaturated, it does not usually undergo addition reactions, which are characteristic of the alkenes we have studied. Instead, the characteristic reaction of benzene is substitution, or displacement of a hydrogen atom, as evidenced by

In these reactions, substitution, not addition or oxidation, occurs. There are many other examples that might be added. The point is that reagents that ordinarily add to or oxidize the double bond in alkenes leave the multicenter π-bond in benzene untouched and instead displace one of the hydrogen atoms attached to the benzene ring.

The Mechanism of Aromatic Substitution

The nitration of benzene and other substitution reactions of aromatic compounds have been carefully studied, and a rather detailed picture of the reaction mechanism has been generated. The nitration of benzene proceeds readily if a mixture of concentrated nitric and sulfuric acids is employed

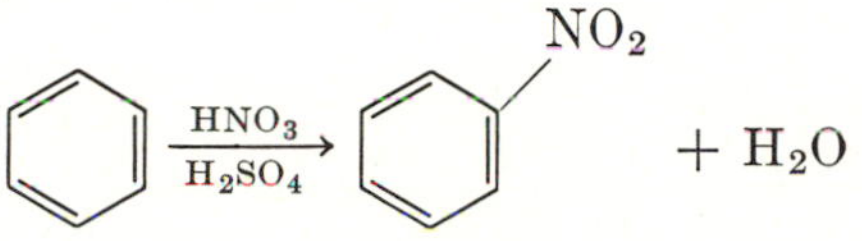

nitrobenzene

Freezing-point depression measurements show that for each mole of nitric acid dissolved in sulfuric acid, four moles of particles are formed. This observation can be explained by postulating that the reaction

$$2H_2SO_4 + HNO_3 \rightarrow H_3O^+ + 2HSO_4^- + NO_2^+$$

occurs when nitric acid dissolves in sulfuric acid. The existence of the nitronium ion, NO_2^+, in mixtures of nitric and sulfuric acids has been confirmed by spectroscopic studies, and the nitronium ion is believed to be the reagent responsible for the nitration of aromatic compounds. The first step of the nitration reaction is the attack of the nitronium ion on the π-electrons of the benzene ring.

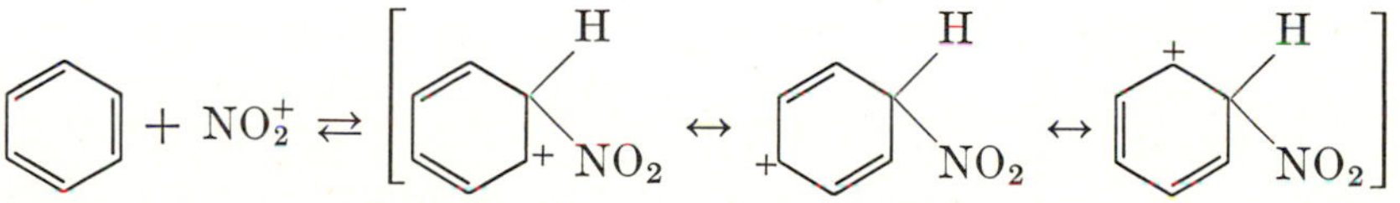

The fact that three resonance structures can be drawn for the intermediate addition product shows that multicenter π-bonding between five of the carbon atoms still exists in the intermediate molecule. The loss of a proton to some proton acceptor, for example HSO_4^-, leads to the restoration of the original π-bonding system.

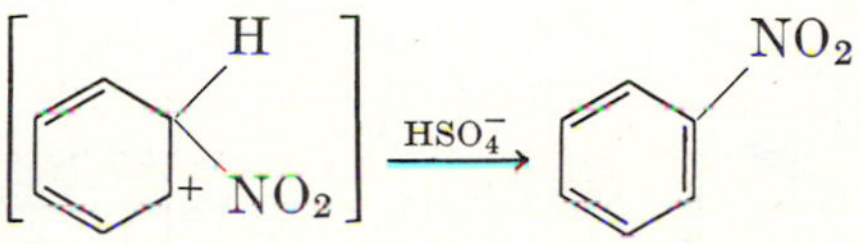

The important idea illustrated by the nitration mechanism is that attack on the benzene ring is generally accomplished by an electron-deficient species such as a positive ion, or the positive end of a polar molecule. Thus the mechanism of the catalyzed bromination of benzene is the following:

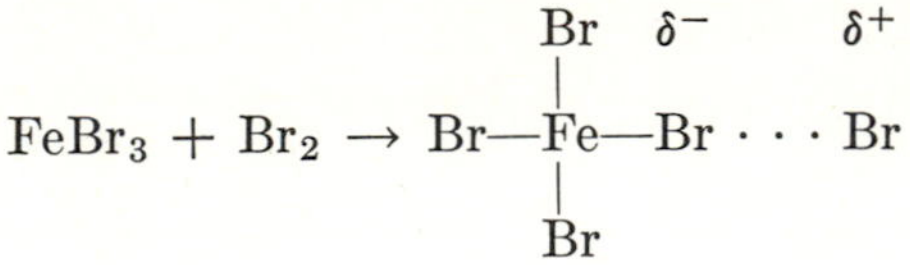

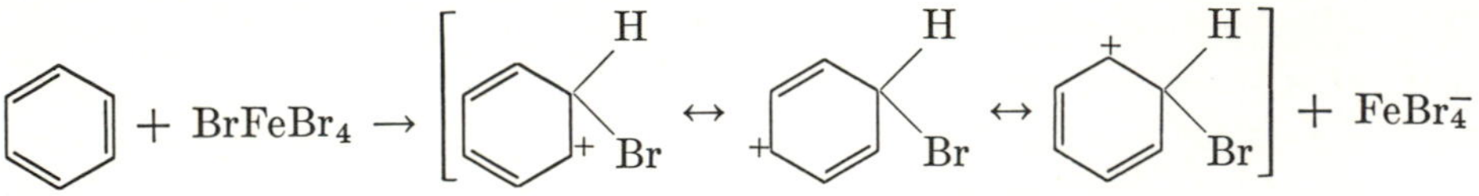

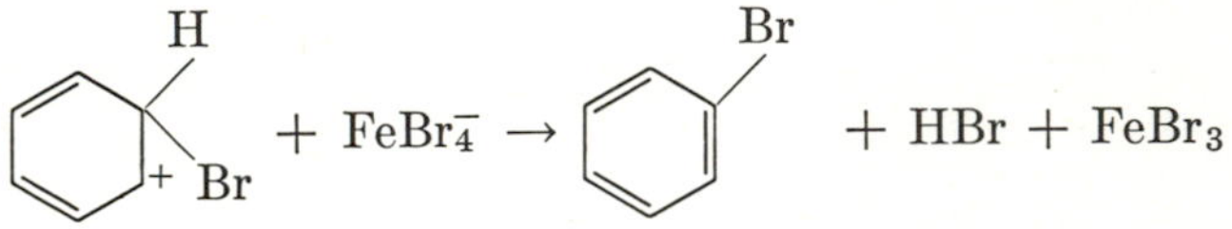

$$FeBr_3 + Br_2 \rightarrow \underset{\overset{|}{Br}}{\overset{\overset{\displaystyle Br}{|}}{Br-Fe-Br}} \cdots Br \quad {}^{\delta-} \quad {}^{\delta+}$$

When benzene derivatives undergo further substitution, a "directing effect" on the position of the new substituent is observed. Consider the following:

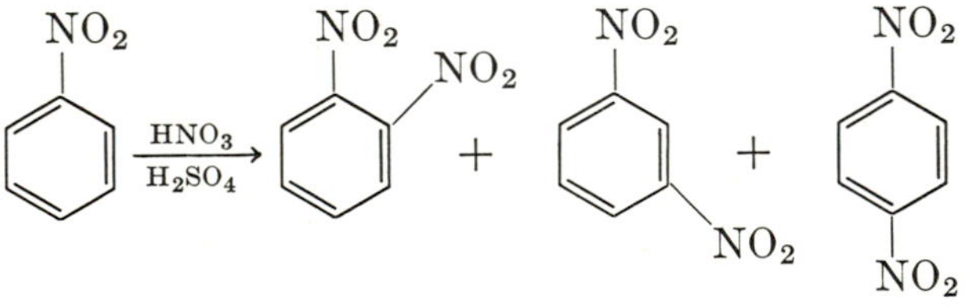

7%   88%   1%

The first isomer, ortho-dinitrobenzene, is formed in small amounts, while the most abundant product is meta-dinitrobenzene; para-dinitrobenzene is formed only in trace amounts. This pattern is quite general: whenever nitrobenzene undergoes further substitution by any reagent, the meta isomer is produced in greatest abundance.

To explain the directing effect of a nitro group on further substitution, we note first that the nitrogen atom in the nitro group has a formal charge of +1.

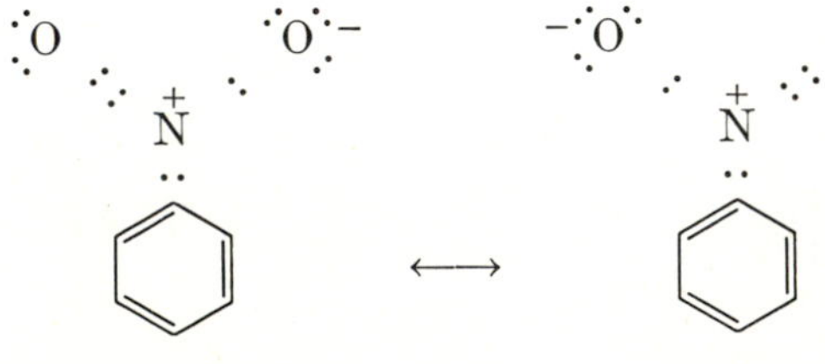

With this in mind, consider the electronic nature of the intermediate formed when NO_2^+ attacks nitrobenzene at the para position.

The resonance structures suggest that the electronic nature of the para intermediate is such that an electron deficiency exists on the carbon atom to which the already electron-deficient nitrogen atom is attached. Such extreme local electron deficiency is not energetically favorable; thus the energy of the para-substituted intermediate is relatively high. This means that the activation energy of the para-substitution reaction is relatively large, and therefore the rate of this reaction should be small, and the yield of para-dinitrobenzene relatively unimportant. A similar argument can be constructed for the intermediate formed by the ortho attack of NO_2^+ on nitrobenzene, as the reader can verify by writing the resonance structures of the intermediate.

Attack at the meta position leads to an intermediate with somewhat more favorable electronic properties.

In the resonance structures of this intermediate, the positive charge never occurs on the carbon atom to which the original nitro group is attached. Consequently, the extreme local electron deficiency encountered in ortho and para substitution is avoided, the meta intermediate is relatively more stable, and the rate of formation of the meta isomer is greater than that of the ortho and para isomers.

Quite the opposite pattern of substitution is observed when anisole is nitrated.

In this case the ortho and para isomers are favored, and the meta isomer appears only in very small amounts. The explanation makes use of the idea that the OCH_3 group of anisole can participate in the multicenter bonding of the benzene ring by donating electrons to it.

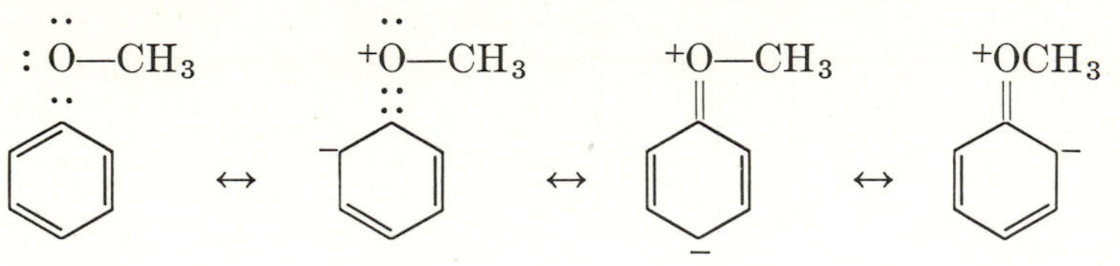

The resonance structures suggest that the electron density is increased, particularly at the positions ortho and para to the anisole group. Now consider the resonance structures of the intermediate formed by para attack of NO_2^+ on anisole.

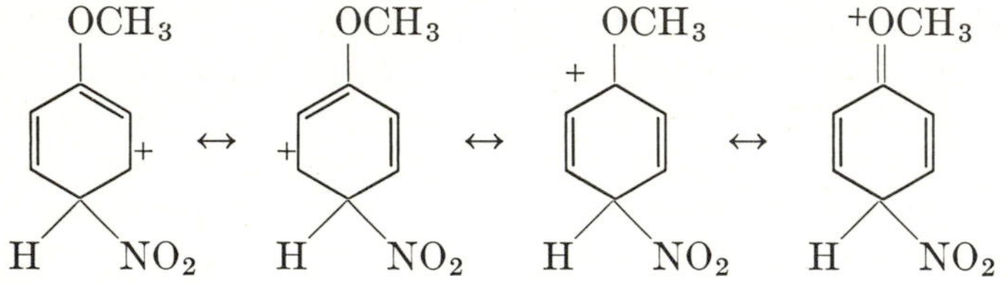

Note that the last resonance structure shows that in the para intermediate, the OCH_3 group can participate in multicenter bonding and donate electrons to the benzene ring. A similar conclusion can be reached concerning the ortho-substituted intermediate. In both cases the OCH_3 group relieves the electron deficiency in the ring brought about by the attack of the nitronium ion. The meta intermediate does not have this property.

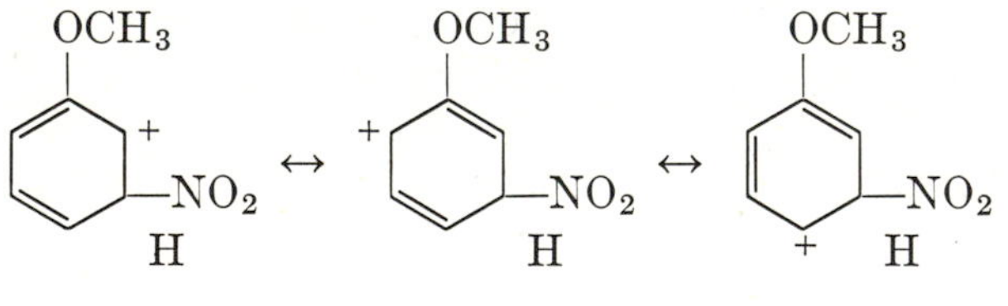

That is, the OCH_3 group cannot operate so as to remove positive charge from the ring. Because of this limitation, the energy of the meta intermediate is relatively high, and the rate of formation of the meta-substituted intermediate is relatively small.

The examples we have discussed illustrate a generalization that has been very useful in explaining the nature of the substitution reactions of aromatic molecules. Groups like —NO_2 which are electronegative or which tend to draw electrons from the benzene ring have a meta-directing effect. Groups like

—OCH_3 that can act as a source of electrons at the ortho and para positions exert an ortho-para directing effect. With this rule it is possible to predict the course of a variety of aromatic-substitution reactions.

In our discussion of organic chemistry we have encountered three types of isomeric molecules. First, there are *functional isomers*: molecules with the same atoms but with different arrangements of them so as to give different functional groups. As a simple example we have C_2H_6O:

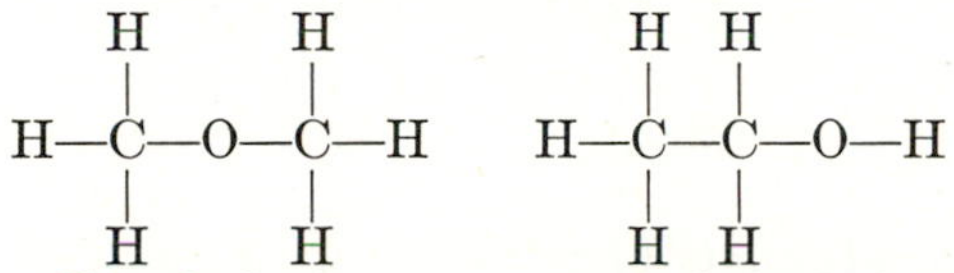

dimethyl ether   ethyl alcohol

A second type of isomerism occurs in *positional isomers*: molecules that have the same functional groups placed in different positions on the carbon skeleton. For example,

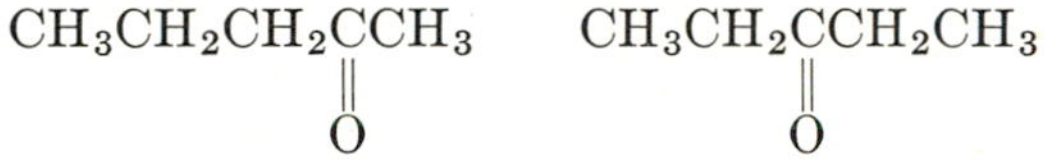

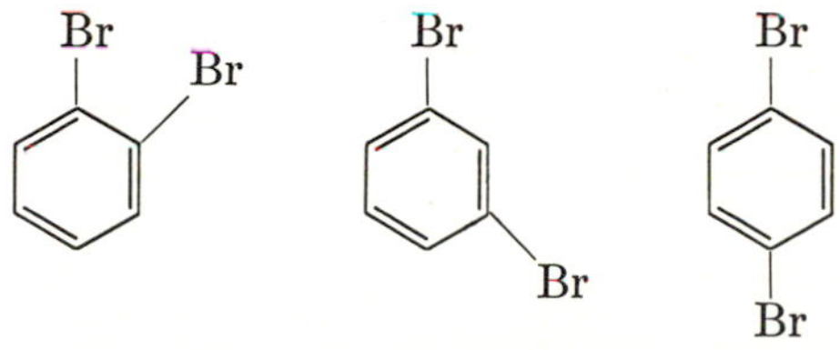

$$CH_3CH_2CH_2CCH_3 \qquad CH_3CH_2CCH_2CH_3$$

The third type of isomerism with which we are familiar is *geometrical isomerism*: geometrical isomers differ with respect to the location of groups attached to a pair of doubly bonded carbon atoms. The following are geometrical isomers.

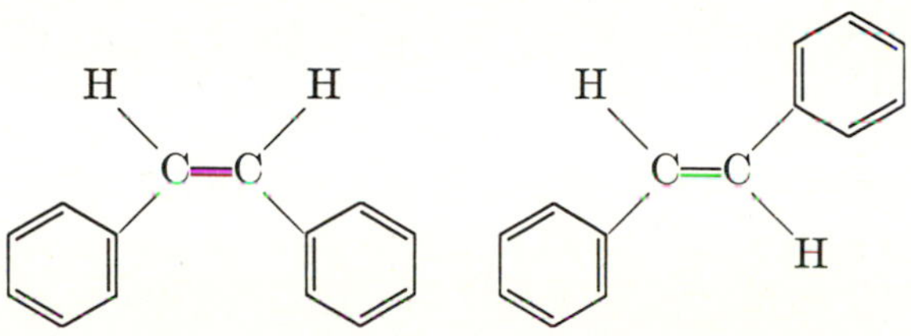

cis-stilbene
mp 1°C

trans-stilbene
mp 124°C

There is a fourth important type of isomerism which can arise if a molecule has no plane or point of symmetry. The isomeric molecules in this case are related as the right hand is to the left: they are mirror images of each other but cannot be superimposed. Lactic acid is an example:

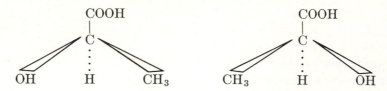

The carboxyl group and the central carbon atom lie in the plane of the paper, while the CH_3 and OH group project in front of the paper and the hydrogen atom behind. The two molecules shown are mirror images, but if we imagine an attempt to superimpose the two three-dimensional structures, we realize that it must fail. If the central carbon atoms, the COOH groups, and the H atoms are superimposed, the CH_3 group of one molecule is coincident with the OH group of the other, and no amount of twisting can change this. Two molecules related in this way are called *enantiomorphs* (Greek *enantios*, opposite; *morph*, form). Two enantiomorphs have identical physical properties except in one respect. Both isomers rotate the plane that contains the electric vector of polarized light, but one isomer rotates the plane in a clockwise direction, the other in a counterclockwise direction. Compounds that rotate the plane of polarized light are said to be *optically active*, and consequently enantiomorphs are also called *optical isomers*.

The requirement for optical isomerism is molecular asymmetry, and this can occur in several ways. Molecules such as lactic acid that have *four different groups* attached to a single carbon atom have neither a point nor a plane of symmetry, and can exhibit optical isomerism. Also, if a molecule has a coiled or helical structure, there are two ways in which the helix can exist—one having the sense of a right-handed screw thread, the other wound in the sense of a left-handed screw. These two forms of the same helix rotate the plane of polarized light in opposite directions and are optical isomers.

Is it possible to determine the absolute configuration of enantiomorphs; that is, to be able to tell which of the two structures belongs to the isomer that rotates the plane of polarization clockwise? This is a very difficult problem, but it was solved in 1951 by an x-ray crystal-structure determination. In this experiment, the optical isomer of glyceraldehyde that rotates the plane of polarization in a clockwise (positive) direction was found to have the structure

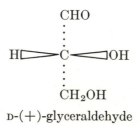

D-(+)-glyceraldehyde

In the name of this compound the (+) refers to the direction in which the plane of polarized light is rotated, and the D designates the structure in which the H and OH is in front, and the CHO and CH_2OH behind the plane of the paper. Once the absolute configuration of D-(+)-glyceraldehyde had been determined, it was possible to assign configurations to molecules that could be made from it. For example, D-(+)-glyceraldehyde can be converted *without disturbing the bonds at the asymmetric carbon atom* to the lactic-acid isomer that rotates polarized light in a negative sense.

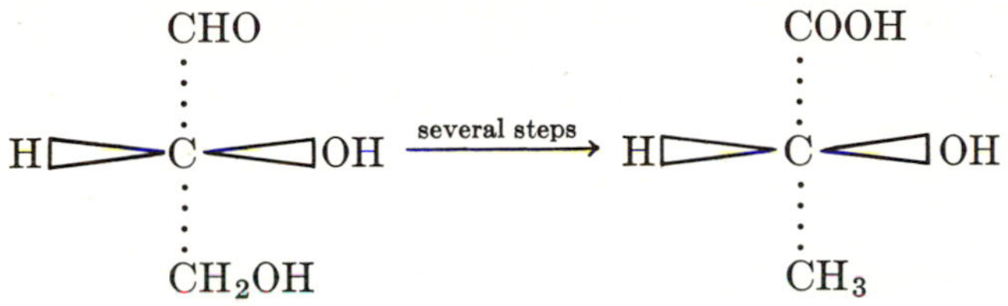

$$\text{CHO} \atop H{-}C{-}OH \quad \xrightarrow{\text{several steps}} \quad \text{COOH} \atop H{-}C{-}OH$$

$$CH_2OH \qquad\qquad CH_3$$

D-(−)-lactic acid

Thus the absolute configurations of the lactic-acid isomers are known.

The compounds in living organisms are for the most part optically active. As an example, we have the amino acids whose general formula is

$$\text{R} \atop NH_2CHCOOH$$

Proteins are giant molecules consisting of many amino acids linked together. The amino acids have an asymmetric carbon atom, and thus can exist in two enantiomorphic forms.

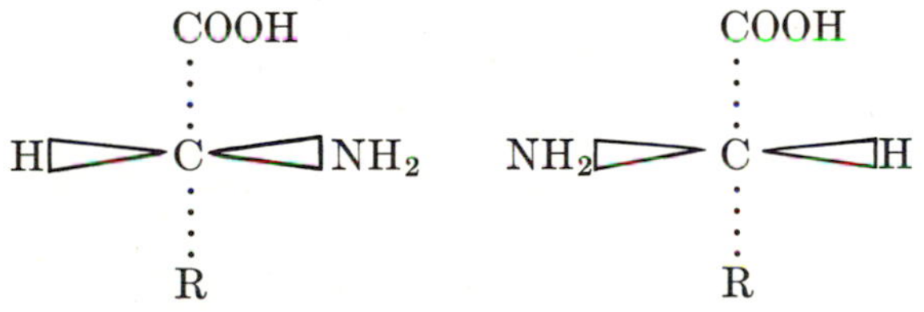

$$\text{COOH} \atop H{-}C{-}NH_2 \qquad\qquad \text{COOH} \atop NH_2{-}C{-}H$$

$$\text{R} \qquad\qquad\qquad \text{R}$$

D-amino acid   L-amino acid

The amino acids recovered from the hydrolysis of proteins are all L-amino acids. The reason for this specificity is not known, but it has important consequences. The various metabolic processes in the body are sensitive to the configurations of the molecules ingested as foods and drugs. Despite the similarity of optical isomers, the natural processes of the body consume one isomer and not the other. Apparently the reason for this is that the protein enzymes that catalyze the metabolic reactions are made from molecules with asymmetric carbon atoms, and as a result can only associate with, and catalyze the reactions of, molecules that have very specific geometric properties. There is much to be learned about this problem, and this accounts for a very active interest in the geometric properties of biologically important molecules.

17.9 INDUSTRIAL ORGANIC CHEMISTRY

One of the major technical developments of the 20th century is the use of organic chemicals to manufacture on a massive scale products which contribute to the comfort and variety of life. These important industrial organic processes range in chemical sophistication all the way from the controlled pyrolysis of crude petroleum to make low molecular weight hydrocarbons, to the subtle multistep syntheses of pharmaceutical agents. The products of industrial synthetic organic chemistry have had profound effects on the lifestyle in developed nations, and consequently it is appropriate to explore some of the processes used to manufacture these substances.

First, it is important to understand that a powerful and ubiquitous practical influence in all of industrual chemistry is the ultimate cost of the product. As a result, many industrial processes are carried out using reagents and reaction conditions which would not be used at all in a chemical laboratory. For example, expensive oxidizing agents such as permanganate and chromate ions, which are commonly used in laboratory syntheses, are never used in the manufacture of chemicals like acetic acid and acetone, whose annual production is measured in billions of pounds. Instead, cheap oxidants like oxygen and nitric acid are used. In general, processes are sought in which the desired chemical is produced continuously, rather than in batches, and can be readily separated from by-products with a minimum number of time and energy consuming purification steps. The possible utilization of by-products is also a very important consideration in the selection of an industrial synthetic process. It is clear that when a chemical is manufactured in amounts of 10^9 pounds yearly, a saving of only a fraction of a cent a pound can be economically significant.

Raw Materials—Coal and Oil

The important bulk organic chemicals are for the most part those which can be obtained from the inexpensive, readily available raw materials coal and crude oil. Let us examine the origin, nature, and uses of these important forms of fossil carbon.

Geological evidence indicates that coal had its origin in tree ferns and other vegetation growing in fresh water swamp areas approximately 300 million years ago. The action of the waters covered the dead plant material with sediment and allowed it to decompose in the absence of air. The pressure and increased temperature arising from the weight and insulation of the layers of sediment led to the conversion of the dead vegetation to coal. In this conversion, cellulose and lignin, the carbohydrate components of wood, lost oxygen and were converted to hydrocarbon substances. Thus, while wood has a composition by weight of 50 percent carbon, 6 percent hydrogen, and 43 percent oxygen, bituminous coal has 80 percent carbon, 6 percent hydrogen, and only 10 percent oxygen. All coals contain small but noticeable amounts of sulfur and nitrogen, and trace quantities of as many as 30 other elements.

The chemical structure of coal is exceedingly complex. However, the basic units are benzene rings that are fused together in the following manner:

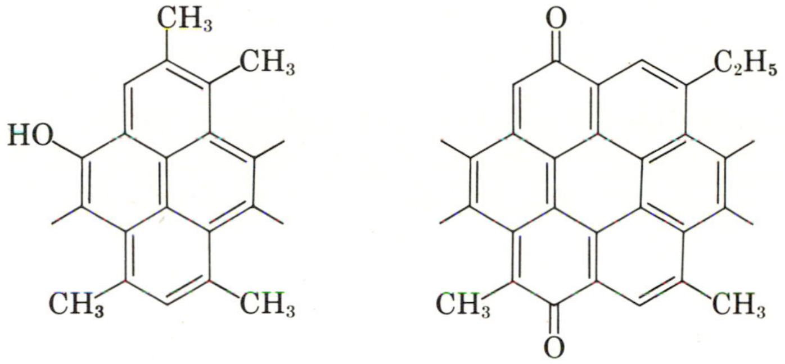

These fused aromatic systems are linked together by aliphatic hydrocarbon chains and also have hydrocarbon chain appendages which incorporate oxygen, nitrogen, and sulfur atoms. The number of rings in the aromatic clusters tends to be greater in anthracite than in the softer bituminous coals.

Given the complex molecular structure of coal, it is not surprising that it can be used to produce valuable industrial chemicals. Simple pyrolysis of bituminous coal at 1100°C yields a mixture of carbon monoxide and methane gases which can be used as fuel. In addition, there are produced quantities of methanol, ammonia, urea, and nitric acid which can be used in various industrial syntheses. Coal tar, the highest boiling fraction from the pyrolysis of coal, contains a variety of aromatic compounds, some of quite complicated structure. The solid carbon residue, or coke, is a valuable reducing agent in metallurgical processes. Pyrolysis of coal in the presence of hydrogen gas produces even greater amounts of simple and complex aromatic hydrocarbons, aliphatic hydrocarbons, and smaller amounts of coke. Even though important chemicals can be derived from coal, there has been little effort made to develop this type of use. Until 1974, the very low cost of petroleum made it the most attractive source of industrial organic chemicals, and the principal use for coal has been as a fuel.

Crude oil and natural gas are the other major sources of reduced fossil carbon. Natural gas is an almost pure hydrocarbon mixture, and it consists mostly of methane and ethane, with smaller amounts of propane and butane. The fact that its sulfur content is low makes natural gas a particularly attractive nonpolluting fuel. Crude oil also consists mostly of hydrocarbons, but it includes varying amounts of sulfur and oxygen containing compounds. Like coal, oil has a biological origin, and is usually found in areas which were marine sedimentary basins. Apparently, when marine plankton are precipitated and protected by sediment from contact with oxygen, they slowly decompose at moderate temperatures and pressures to yield a liquid hydrocarbon product.

Crude petroleum can be separated by distillation into several fractions. So-called petroleum ether (which contains little if any ether functional groups) boils from 20–70°C, and includes C_5 and C_6 hydrocarbons. The higher boiling fractions include ligroin, C_6–C_7 (70–100°C); straight gasoline, C_6–C_{12} (85–200°C); kerosene, C_{12}–C_{14} (200–275°C); heating oils, C_{15}–C_{18} (275°C); and the highest boiling fraction yields lubricating oils, paraffin wax, asphalt, and tar. Included in most of these fractions are alkanes, cycloalkanes, and aromatics in varying amounts depending on the source of the crude oil. Because there is in crude oil a variety of chemicals which have very similar physical properties, it is usually not economical to separate any one compound, particularly if it has more than four carbon atoms in it. Instead, in petroleum chemistry the emphasis is placed on finding reactions which convert natural mixtures of hydrocarbons to more desirable mixtures, or to the low molecular weight hydrocarbons (C_2–C_4) which can be separated readily. As we shall see, these latter compounds are the starting materials in a great many industrial syntheses.

Because of the availability of crude oil and the ease with which it can be refined and transported, petroleum has become the single most important energy source in the United States and certain other highly industrialized nations. However, the known world resources of petroleum are limited, and in view of the value of oil as a source of industrial organic chemicals, its indiscriminant use as a fuel is a great mistake.

Hydrocarbons

In order to convert abundant high molecular weight hydrocarbons to the more valuable light alkanes and olefins, the higher boiling petroleum fractions are "cracked," or subjected to temperatures of 700–900°C. In many cracking processes, catalytic agents, such as the transition metal oxides CrO_3, Mn_2O_3, and Fe_2O_3 supported on alumina, are used to promote the formation of particularly desirable hydrocarbon products. The result of cracking reactions is a mixture consisting mostly of C_2, C_3, and C_4 olefins along with methane and hydrogen. The individual components can be separated readily. Some of the major uses for these C_2–C_4 olefins are indicated below.

Acetylene is one of the most important light hydrocarbons in industry. It can be hydrated to give acetaldehyde which, in turn, can be oxidized to acetic acid, a reagent important in the manufacture of acetate polymers:

$$HC\equiv CH + H_2O \rightarrow CH_3CHO \xrightarrow{O_2} CH_3COOH.$$

Also, a large number of acids can be added to acetylene to produce the vinyl monomers used to make a variety of polymers:

$$
\begin{aligned}
HC\equiv CH + HCl &\rightarrow & H_2C\!=\!CHCl, & \quad \text{vinyl chloride;} \\
+ HCN &\rightarrow & H_2C\!=\!CHCN, & \quad \text{acrylonitrile;} \\
+ CH_3COOH &\rightarrow & H_2C\!=\!CHOOCH_3, & \quad \text{vinyl acetate;} \\
+ CO + CH_3OH &\xrightarrow{Ni(CO)_4} & H_2C\!=\!C(CH_3)COOCH_3, & \\
& & \text{methyl methacrylate.} &
\end{aligned}
$$

Addition of chloride to acetylene gives tetrachloroethane, a starting material for the production of cleaning solvents.

Ethylene is also a very important cracking product, and has a large number of uses. It can be polymerized directly to polyethylene or hydrated to give ethanol or ethylene glycol:

$$
\begin{aligned}
H_2C\!=\!CH_2 + H_2O &\rightarrow CH_3CH_2OH \\
HOCl &\rightarrow ClCH_2CH_2OH \xrightarrow{HCO_3^-} HOCH_2CH_2OH.
\end{aligned}
$$

Ethylene glycol is used as an antifreeze and cooling agent in engine radiators, and is one of the components in condensation polymers such as Dacron. Like ethylene, propylene can be polymerized directly or hydrated to give isopropanol:

$$CH_3CH\!=\!CH_2 + H_2O \rightarrow CH_3HOHCH_3.$$

The oxidation of isopropanol is the principal way in which acetone is manufactured

$$CH_3CHOHCH_3 \xrightarrow[300°C]{CuO} (CH_3)_2C\!=\!O + H_2.$$

The mixture of butenes collected from the cracking process can be isomerized with acid catalysts to isobutene:

$$CH_2\!=\!CHCH_2CH_3 + H^+ \rightarrow CH_3C^+HCH_2CH_3 \leftarrow H^+ + CH_3CH\!=\!CH\!-\!CH_3$$
$$\downarrow$$
$$(CH_3)_2CHCH_2^+ \rightarrow (CH_3)_2C\!=\!CH_2 + H^+.$$

Isobutene is the monomer used to make butyl rubber. In addition, isobutane and isobutene can be combined to produce isooctane, an important constituent of high-performance gasoline:

$$(CH_3)_2C\!=\!CH_2 + (CH_3)_3CH \xrightarrow{H_2SO_4} (CH_3)_2CHCH_2C(CH_3)_3.$$

Synthetic Polymers

By linking together low-molecular-weight organic compounds (monomers), it is possible to produce polymeric substances of very high molecular weight which have mechanical and chemical properties suitable for the manufacture of fibers, films, protective coatings, elastomers, tubing, containers, and insulating materials. Consequently, polymer manufacture is a very important component of the chemical industry. Through the understanding of the relationship between molecular structure and macroscopic properties, it has been possible to design synthetic polymers appropriate for many different types of application.

There are two major types of polymerization process in use in the chemical industry, addition reactions and condensation reactions. Addition polymerization almost always involves monomers with one or more carbon-carbon double bonds, such as ethylene, propylene, isobutene, and the various vinyl monomers. Addition polymerization frequently is initiated by free radicals formed by heat or irradiation, and it proceeds by the following mechanism:

Initiation:

$$I \rightarrow 2R\cdot, \quad R\cdot + CH_2=CHCl \rightarrow RCH_2\dot{C}HCl.$$

Propagation:

$$RCH_2\dot{C}HCl + CH_2=CHCl \rightarrow RCH_2CHClCH_2\dot{C}HCl.$$

Termination:

$$R(CH_2CHCl)_m CH_2\dot{C}HCl + R(CH_2CHCl)_n CH_2\dot{C}HCl \rightarrow R(CH_2CHCl)_{m+n+2}R$$

$$R(CH_2CHCl)_m CH_2\dot{C}HCl + R(CH_2CHCl)_n CH_2\dot{C}HCl \rightarrow$$
$$R(CH_2CHCl)_m CH=CHCl + R(CH_2CHCl)_n CH_2CH_2Cl$$

$$R(CH_2CHCl)_m CH_2\dot{C}HCl + RH \rightarrow R(CH_2CHCl)_m CH_2CH_2Cl + R\cdot.$$

The possible chain termination reactions are a combination of two chains, disproportionation to an alkane and olefin by hydrogen atom transfer, and abstraction of a hydrogen atom from a saturated molecule. Careful control of the polymerization conditions must be maintained in order to produce a polymer of desirable molecular weight and mechanical properties.

Addition polymerization can also be carried out using cationic initiators. This is the most common way to polymerize isobutene to butyl rubber:

$$AlCl_3 + H_2O \rightarrow AlCl_3OH^- + H^+,$$
$$H^+ + (CH_3)_2C=CH_2 \rightarrow (CH_3)_3C^+,$$
$$(CH_3)_3C^+ + CH_2=C(CH_3)_2 \rightarrow (CH_3)_3CCH_2C^+(CH_3)_2.$$

Addition of monomers to the chain continues until terminated by a reaction with an anion. Transition metal compounds, such as $TiCl_3$, which form complexes with olefins also are used to catalyze addition polymerization. The polymers produced in this manner have very regular geometric structures and tend to be more crystalline and higher melting than the same polymers produced by free radical reactions.

By proper selection of monomer, combination of two or more monomers (co-polymerization), and control of the molecular weight of the polymer, it is possible to create by addition polymerization substances which serve a great variety of purposes. Some addition polymers such as polyethylene, polypropylene, and polyvinyl chloride have common or commercial names which reveal their chemical composition. Other polymers for which the connection between the commercial name and molecular structure is less obvious or nonexistent include Teflon (C_2F_4), Lucite ($CH_2{=}C(CH_3)COOCH_3$), and Orlon or Acrilan ($CH_2{=}CHCN$).

In condensation polymerization reactions, parts of the monomer units are eliminated and the polymer has a composition which is different from that of the monomers used to make it. The simplest example is the combination of bifunctional acids and alcohols to make polyesters, with the elimination of water. In practice, it proves to be more advantageous to make polyesters by reacting a bifunctional acid which has already been esterified with methanol. The reaction then proceeds by displacement of the methanol. For example, consider the reaction of ethylene glycol and dimenthyl terephthalate:

$$CH_3O_2C{-}\langle\text{C}_6\text{H}_4\rangle{-}CO_2CH_3 + 2HOCH_2CH_2OH$$

$$HOCH_2CH_2O_2C{-}\langle\text{C}_6\text{H}_4\rangle{-}CO_2CH_2CH_2OH + 2CH_3OH.$$

Polymerization continues in a similar manner with the methanol being distilled from the reaction mixture as it is formed. The resulting polymer can be formed into a fiber known as Dacron, or a very strong film called Mylar.

Another important condensation polymerization is involved in the formation of nylon. The starting materials are adipic acid and hexamethylenediamine, and the reaction proceeds to form a polyamide:

$$n\,HOOC(CH_2)_4COOH + n\,H_2N(CH_2)_6NH_2 \rightarrow$$

$$\left[\begin{array}{c} H \quad O \qquad\quad O \quad H \\[-2pt] | \quad\ \| \qquad\quad\ \| \quad | \\[-2pt] {-}N{-}C(CH_2)_4C{-}N(CH_2)_6{-} \end{array} \right]_n + 2n\,H_2O.$$

Because of the formation of hydrogen bonds between chains, nylon tends to be crystalline, has a high melting point, and considerable tensile strength.

17.10 CONCLUSION

Organic chemistry is a vast but highly organized subject. Many reactions of organic compounds leave the carbon skeletons of the molecules largely unaltered and involve only the small number of atoms in a functional group. Each functional group has a set of characteristic reactions whose nature is only slightly dependent on the identity of the carbon skeleton to which the group is attached. Consequently, organic compounds can be classified and their reactions discussed in terms of functional groups. In this chapter we have discussed only a few of the more important reactions of the common functional groups. It should be clear from this introduction, however, that an extensive study of the detailed reactions of organic compounds is necessary if we are to understand the behavior of complex biological systems.

SUGGESTIONS FOR FURTHER READING

Allinger, N., and J. Allinger, *Structures of Organic Molecules*. Englewood Cliffs, N.J.: Prentice-Hall, 1965.

Cook, P. L., and J. W. Crump, *Organic Chemistry: A Contemporary View*. Lexington, Mass.: D. C. Heath and Co., 1969.

Morrison, R. T., and R. N. Boyd, *Organic Chemistry*, 3rd ed. Boston: Allyn and Bacon, 1973.

Stille, J. T., *Industrial Organic Chemistry*. Englewood Cliffs, N.J.: Prentice-Hall, 1969.

PROBLEMS

17.1 Write the structural formulas of the following compounds: (a) 2, 2-dimethylbutane; (b) 3-ethyl-2, 3, 4, 4-tetramethylheptane; (c) 2, 2, 4, 4-tetramethylpentane; (d) 4-methyl-3-hexanol; (e) 4, 4-dimethyl-2-pentanol; (f) 2, 3-dimethylbutanal; (g) methyl isopropyl ketone.

17.2 How would you accomplish the following conversions? (a) Acetaldehyde to 2-hexanol; (b) diisopropyl ketone to 2, 3, 4-trimethyl-3-pentanol; (c) *tert*-butyl bromide to 2, 2-dimethyl propanol.

17.3 Give a procedure by which 3-ethyl-2-pentene can be synthesized from ethanol and inorganic reagents.

17.4 Show how methyl ethyl ketone can be synthesized from ethanol and inorganic reagents.

17.5 Write the structural formulas of the hydration products of the following olefins: (a) propylene; (b) 2-methyl-2-pentene; (c) 1-methyl cyclohexene.

17.6 Give the name of the compound formed by the addition of HCl to (a) isobutene; (b) 3-methyl-2-pentene.

17.7 Show how to prepare 1, 2-dibromo-2-methyl propane from isopropyl alcohol and methyl bromide.

17.8 What are the products when the following hydrocarbons are oxidized with acidic permanganate solutions?   (a) 2-butene; (b) 3, 4-dimethyl-3-hexene; (c) 2-methyl-2-butene.

17.9 Draw the structure of the isomer of trinitrobenzene that would be easiest to synthesize from benzene, nitric acid and sulfuric acid.

CHAPTER 18

BIOCHEMISTRY

This topic includes all molecular phenomena associated with life processes, and is, therefore, a subject of enormous breadth, complexity, and challenge. Consequently, biochemistry is presently one of the most interesting and active areas of chemical research. In this chapter we shall find that the principles of stoichiometry, equilibrium, oxidation-reduction, chemical kinetics, and molecular structure discussed in earlier chapters are relevant to biochemical problems. In the last two decades in particular, application of chemical and physical techniques and principles to biological problems has produced an almost inestimable increase in our understanding of life processes.

18.1 THE CELL

Although we are principally concerned with the molecular phenomena associated with the life process, a brief discussion of the structure of the biological cell is in order. The cell is the smallest unit capable of effecting and regulating metabolism, energy conversion and storage, and molecular synthesis. In order to understand how these chemical processes are related, we must have some familiarity with cell structure and composition. Some of the chemical terms we shall use to describe the cell composition may not be familiar, but all are discussed in more detail subsequently in this chapter.

Cells occur in a variety of sizes and shapes, but the one shown in Fig. 18.1 displays the general features relevant to our discussion. This cell is approximately spherical, with a radius of about 2×10^{-3} cm. The gross volume of the cell is therefore approximately 3×10^{-8} cm^3. The actual cellular surface area is greater than that of a sphere of the same radius, because the cell membrane contains numerous folds and irregularities. Since delivery of nutrients and removal of waste products occur largely by diffusional transport through the cellular membrane, the surface-to-volume ratio of the cell is important. Should the radius of the cell become too large, and the surface-to-volume ratio too small, transport to and from the cell may not properly match the rate of chemical processes within the cell.

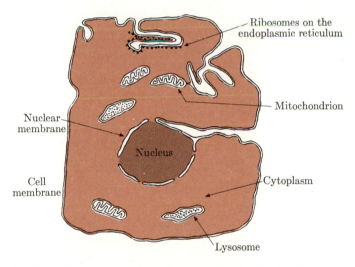

A schematic representation of a generalized animal cell.

FIG. 18.1

The cell is approximately 80 percent water, by weight. The nonaqueous substances generally average 14 percent protein, 2 percent lipids or fatty materials, 1 percent materials related to starch, 2 percent ribonucleic acid, and 1 percent deoxyribonucleic acid. These percentages vary according to cell function, and in addition to the major substances mentioned, there are minor amounts of physiologically important constituents like sodium ion ($\sim 0.02 \ M$) and potassium ion ($\sim 0.1 \ M$).

A membrane approximately 100 A thick surrounds the cell. In animal cells, this membrane has a "sandwich" structure consisting of a lipid layer between two layers of protein which are each approximately 25 A thick. The membrane is flexible and readily permeable to small molecules.

There are a number of substructures within the cell. The *nucleus* is separated from the main aqueous body or *cytoplasm* by a double-layered, slightly porous protein membrane. The nucleus serves as the site for storage and transmission of hereditary characteristics. This genetic control is accomplished by deoxyribonucleic acid (DNA) molecules, whose general structure we will study in Section 18.6.

Animal cells contain *mitochondria*, small rod-shaped particles concerned with chemical degradation of fats, part of the carbohydrate metabolism, and production of the energy-rich compound adenosine triphosphate (ATP).

The *lysosomes* are intracellular particles somewhat smaller than the mitochondria. They contain enzymes that catalyze the degradation of various complex molecules in the cell. The controlled use of these enzymes permits the cell to digest large molecules and membrane material.

The cell contains a highly folded, maze-like membrane called the *endoplasmic reticulum*. In certain areas of this membrane there are attached granular particles of about 300 A diameter called *ribosomes*. The ribosomes are made up of lipid, protein, and ribonucleic acid (RNA), and are of extreme interest, since they are the sites at which protein synthesis occurs.

We see that within the cell there are regions which are associated with quite different types of metabolic, synthetic, and transport processes. The operation of the cell involves concerted, coupled action by these components, each of which is itself a highly complex chemical system. The aim of biochemistry is to achieve an understanding of the operation of these systems by studying the structure and chemistry of lipids, polysaccharides, proteins, nucleic acids, and other cell constituents.

18.2 BIOCHEMICAL ENERGETICS

From our sketch of cell structure and function we can see that the energy-releasing metabolic processes occur at sites which are removed from the places where the often energy-consuming synthetic reactions go on. Moreover, much of the food metabolism occurs at a slower rate and at a different time from the energy-consuming muscle action. If the energy released by metabolism is not to be lost as heat, it must be stored as internal energy of molecules which can easily be transported to the appropriate places and used at the proper time and rate. The molecule of central importance in this energy storage and transport is adenosine triphosphate, or ATP. Its structure, shown in Fig. 18.2, should be examined carefully. It consists of a polyphosphate chain attached by an ester linkage to the sugar ribose. The nitrogen-containing base adenine is attached to the ribose fragment at another position. This base–sugar–phosphate structure is important not only because it occurs in ATP, but because it is the fundamental unit in nucleic acids, and in a number of other molecules involved in metabolism and synthesis.

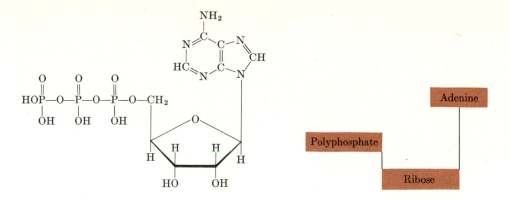

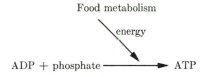

The structure of adenosine triphosphate (ATP).

FIG. 18.2

In the course of oxidative metabolism, ATP is formed from adenosine *di*phosphate (ADP) and inorganic phosphate ions:

$$\text{Food metabolism}$$
$$\searrow \text{energy}$$
$$\text{ADP + phosphate} \longrightarrow \text{ATP}$$

Thus the energy for the ADP-to-ATP conversion comes from the oxidative metabolism of foods. This "stored" energy can be recovered when needed by the hydrolysis of ATP:

$$\text{ATP} + \text{H}_2\text{O} \rightarrow \text{ADP} + \text{phosphate}, \qquad \Delta G^0 \cong -8 \text{ kcal},$$

or by other reactions in which a phosphate group is transferred from ATP to another molecule. Because free energy decreases upon conversion of ATP to ADP or to adenosine monophosphate (AMP), the phosphorus–oxygen bonds in these molecules are often referred to as "energy-rich" or "high-energy" bonds. This terminology, in common use in the biochemical literature, may be somewhat confusing, since a biochemist's "high-energy bond" is, in fact, a weak bond, and *not* one with a high bond-dissociation energy.

There is one other property of ATP which is important to its biochemical function. Although ATP is thermodynamically unstable with respect to hydrolysis, this reaction is very slow. Consequently, ATP is kinetically stable, and its energy-releasing reactions occur only when the appropriate enzyme catalyst is provided. This is the means by which the energy-release processes are controlled.

As a simple example of how the energy stored in ATP can be used, consider the following. The ΔG^0 of the esterification reaction

$$RCOOH + R'OH \rightarrow \overset{\overset{\displaystyle O}{\|}}{RC}-OR' + H_2O, \qquad \Delta G^0 = +2 \text{ kcal}$$

is positive, and consequently the ester synthesis does not proceed to completion. A more complete conversion can be obtained by coupling the esterification to the hydrolysis of ATP. The coupling can occur by the following two-step process:

$$RCOOH + ATP \rightarrow \overset{\overset{\displaystyle O}{\|}}{RC}-O-\overset{\overset{\displaystyle O}{\|}}{\underset{\underset{\displaystyle OH}{|}}{P}}OH + ADP, \qquad \Delta G^0 \cong -3 \text{ kcal},$$

$$\overset{\overset{\displaystyle O}{\|}}{RC}-O-\overset{\overset{\displaystyle O}{\|}}{\underset{\underset{\displaystyle OH}{|}}{P}}-OH + R'OH \rightarrow \overset{\overset{\displaystyle O}{\|}}{RC}-OR' + H_3PO_4, \qquad \Delta G^0 \cong -3 \text{ kcal}.$$

The overall reaction

$$RCOOH + R'OH + ATP \rightarrow RCOOR' + H_3PO_4 + ADP, \qquad \Delta G^0 \cong -6 \text{ kcal}$$

has a negative free-energy change, and proceeds largely to completion. Thus, ATP formed by an exothermic metabolic reaction can be used subsequently to carry out a necessary synthetic or degradative reaction whose energetics may be quite unfavorable. The general phenomenon of the coupling of endo- and exoenergetic reactions is extremely important in biochemical systems, as the following discussion will demonstrate.

Oxidation-Reduction Reactions

In the cell, oxidation of foodstuffs releases energy which is used subsequently for macromolecular synthesis, transport of matter, and muscle action. The complete oxidation of a carbohydrate food like glucose releases an enormous amount of energy:

$$C_6H_{12}O_6 + 6O_2 \rightarrow 6H_2O + 6CO_2, \qquad \Delta H = -686 \text{ kcal}.$$

In biochemical systems, oxidations of complex molecules do not take place in a single step. In contrast, they occur by a series of a dozen or more reactions, which successively break down the original molecule into smaller, more highly oxidized species, and eventually produce carbon dioxide and water. The energy released in at least some of these individual metabolic steps must be used directly or indirectly to convert ADP to ATP.

Even the individual steps in the metabolic sequence do not directly involve molecular oxygen. Oxygen is such a powerful oxidizing agent that the energy released in even the partial oxidation of an organic substrate molecule is so large that it could not be used or stored efficiently. For example, the direct oxidation of malic acid to oxaloacetic acid would release 46 kcal of free energy:

$$
\begin{array}{c}
\text{COOH} \\
| \\
\text{HCOH} \\
| \\
\text{CH}_2 \\
| \\
\text{COOH} \\
\text{Malic} \\
\text{acid}
\end{array}
\quad + \tfrac{1}{2}\text{O}_2 \rightarrow
\begin{array}{c}
\text{COOH} \\
| \\
\text{C}{=}\text{O} \\
| \\
\text{CH}_2 \\
| \\
\text{COOH} \\
\text{Oxaloacetic} \\
\text{acid}
\end{array}
\quad + \text{H}_2\text{O}, \quad \Delta G^0 = -46 \text{ kcal.}
$$

The large driving force of this reaction would be largely wasted if it were to occur directly. Instead, in the mitochondria of the cell, a much weaker oxidant, nicotinamide adenine dinucleotide (NAD^+), carries out the oxidation of the substrate:

$$
\begin{array}{c}
\text{COOH} \\
| \\
\text{HCOH} \\
| \\
\text{CH}_2 \\
| \\
\text{COOH}
\end{array}
\quad + \text{NAD}^+ \rightarrow
\begin{array}{c}
\text{COOH} \\
| \\
\text{C}{=}\text{O} \\
| \\
\text{CH}_2 \\
| \\
\text{COOH}
\end{array}
\quad + \text{NADH} + \text{H}^+.
$$

Because NAD^+ is not a particularly strong oxidant, its reduced form NADH can be oxidized by a slightly stronger oxidant found in the cell, flavin adenine dinucleotide (FAD). Concurrently, ADP and inorganic phosphate (designated by P_i) is converted to ATP:

$$
\text{NADH} + \text{ADP} + \text{P}_i + \text{H}^+ + \text{FAD} \rightarrow \text{NAD}^+ + \text{ATP} + \text{FADH}_2.
$$

In effect, the driving force derived from the NADH oxidation and FAD reduction is used to form energy-rich ATP.

The reduced form FADH_2 is itself oxidized by an oxidant stronger than FAD. The sequence of oxidation-reduction reactions continues for five more steps, and in each successive step, a stronger oxidant is used. Molecular oxygen appears as the oxidant only in the last of these steps.

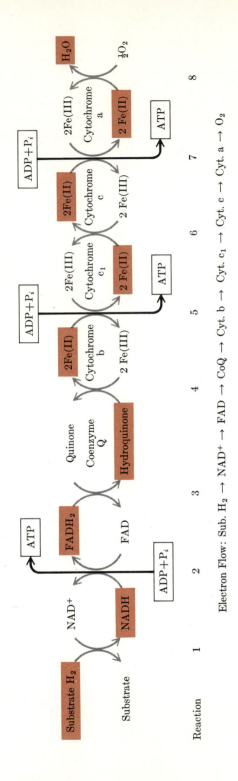

Electron Flow: Sub. $H_2 \rightarrow NAD^+ \rightarrow FAD \rightarrow CoQ \rightarrow Cyt. b \rightarrow Cyt. c_1 \rightarrow Cyt. c \rightarrow Cyt. a \rightarrow O_2$

FIG. 18.3   The chain by which substrates derived from foodstuffs are oxidized and their electrons transferred to oxygen with the production of ATP. The reduced substances are indicated by color.

The total sequence of oxidation-reduction reactions which couple the oxidation of a foodstuff substrate to the reduction of molecular oxygen is shown in Fig. 18.3. Four of the oxidant-reductant pairs are cytochromes, in which ferric or ferrous ion is complexed in a porphyrin ring and is bound to a protein, as shown in Fig. 18.4. Differences in the structure of the porphyrin ring and the protein account for the differing oxidative power of the cytochromes.

Figure 18.3 shows that three of the sequential oxidation-reduction reactions involve the conversion of ADP and inorganic phosphate to ATP. Thus, by this sequence of *coupled* oxidation-reduction reactions, the energy released by the oxidation of the carbohydrate substrate can be stored in a usable form as ATP.

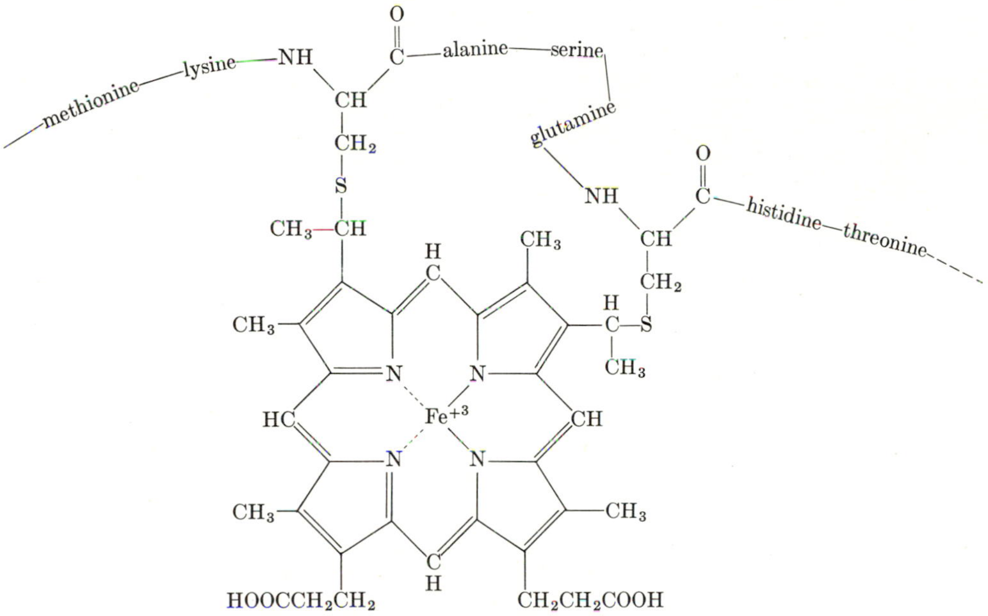

Part of the structure of cytochrome c, showing the iron atom complex in the porphyrin ring system which is in turn attached to a polypeptide chain.

FIG. 18.4

The intermediates NAD^+, FAD, etc., are left unchanged, and the net reaction is schematically

$$\text{Substrate } H_2 + 3ADP + 3P_i + \tfrac{1}{2}O_2 \rightarrow \text{Substrate} + 3ATP + H_2O.$$

In subsequent discussions of metabolic processes, we shall frequently encounter oxidations by NAD^+. It is important to realize that these can lead, by the steps of Fig. 18.3, to reduction of oxygen and production of ATP.

Lipid is a name applied to the cellular components which are water-insoluble, but which can be extracted with organic solvents like ether, benzene, and chloroform; consequently, this classification includes a large number of molecules whose structures and functions are at best distantly related. A major part of a lipid extract, however, consists of substances which yield, upon hydrolysis, long-chain aliphatic acids called fatty acids. We shall restrict our discussion of lipids to these substances. This group is further classified as follows:

1. *Simple lipids.* This group includes *fats*, which are esters of fatty acids and glycerol, $CH_2OHCHOHCH_2OH$, and *waxes*, in which fatty acids are esterified with alcohols of high molecular weight.

2. *Compound lipids.* This includes molecules in which glycerol is *esterified* with fatty acids and phosphoric acid, and in addition, fatty-acid *esters* of sugar molecules.

Fatty acids are carboxylic acids, RCOOH, of high molecular weight, in which the alkyl group R may be saturated, unsaturated, cyclic, or branched-chain. Acids in which R is an unbranched open chain are by far the most common. In virtually all the naturally occurring acids, there is an even number of carbon atoms. The formulas, common names, and occurrence of some of the fatty acids are given in Table 18.1. The most abundant saturated acid in animal fats is palmitic (C_{16}), with stearic (C_{18}) second in importance. Oleic (C_{18}) and palmitoleic (C_{16}) are the two most frequently occurring unsaturated acids.

The most abundant unsaturated fatty acids have the formula

$$RCH{=}CH(CH_2)_7COOH,$$

where seven carbon atoms separate the carboxylic acid and ethylenic functional groups. The R group itself may be unsaturated. The presence of double bonds introduces the possibility of *cis-trans* isomerism. The *cis* configuration is the one found in virtually all the naturally occurring acids.

Table 18.1 Some common fatty acids

Common Name	Systematic Name	Formula	Source
Butyric	Butanoic	C_3H_7COOH	Butter
Caprylic	Octanoic	$C_7H_{15}COOH$	Coconut oil
Palmitic	Hexadecanoic	$C_{15}H_{31}COOH$	Palm oil
Stearic	Octadecanoic	$C_{17}H_{35}COOH$	Mutton fat
Palmitoleic	9-Hexadecenoic	$C_{15}H_{29}COOH$	Butter
Oleic	9-Octadecenoic	$C_{17}H_{33}COOH$	Olive oil
Linoleic	9, 12-Octadecadienoic	$C_{17}H_{31}COOH$	Soybean oil
Linolenic	9, 12, 15-Octadecatrienoic	$C_{17}H_{29}COOH$	Linseed oil

Nearly 10 percent of the body weight of a mammal may be in the form of fats, or triglyceryl esters of the fatty acids. The general formula for these compounds is

$$
\begin{array}{ccc}
CH_2 & CH & CH_2 \\
| & | & | \\
O & O & O \\
| & | & | \\
C{=}O & C{=}O & C{=}O \\
| & | & | \\
R & R' & R''
\end{array}
$$

The three $\overset{\displaystyle O}{\underset{\displaystyle \|}{R{-}C{-}}}$ groups, which represent the fatty-acid residues, may be the same or different. Triglycerides of saturated acids tend to have higher melting temperatures than those of unsaturated acids. Animal fats are relatively rich in saturated triglycerides, and this is the reason they are solids at room temperature. Vegetable oils like corn and safflower oil have a greater percentage of unsaturated triglycerides, and consequently are liquids at room temperature. Frequently these liquid vegetable oils are hydrogenated in order to produce a more saturated solid fat for table use.

Waxes are also esters of fatty acids, but the alcohol is a long-chain ($\sim C_{30}$) aliphatic primary or secondary monoalcohol. As an example, we have beeswax, which is largely an ester of palmitic acid with myricyl alcohol, $CH_3(CH_2)_{29}OH$:

$$
CH_3(CH_2)_{14}\underset{\displaystyle \underset{\displaystyle O}{\|}}{C}{-}OH(CH_2)_{29}CH_3.
$$

Lipid Function

As was mentioned in Section 18.1, lipids occur in the cell membrane, sandwiched between two layers of protein. The lipid layer exerts some selectivity and control over the transport of substances to and from the cell. Molecules which dissolve readily in organic solvents readily pass through the lipid layer of the membrane. Molecules which are only water-soluble cannot easily diffuse through the lipid layer, and must enter and leave the cell either in close association with lipid-soluble substances, or through pores in the membrane which themselves exert some selectivity on the size and charge of the molecules that can pass.

Lipids are the principal constituents of the protective tissue which insulates warm-blooded animals against a low-temperature environment. In plants, waxes serve to protect surfaces of leaves and stems against water, and attack by insects and bacteria.

The main function of fats is to serve as the major and most efficient repository of energy. Complete combustion of 1 gram of fat produces approximately

9 kcal, which is considerably more than 5.6 kcal/gm obtained from protein, or 4.2 kcal/gm produced by carbohydrates. The high heat of combustion of fats is a consequence of their being nearly entirely hydrocarbon, whereas in proteins and particularly in carbohydrates, the hydrocarbon skeleton is already partially oxidized.

The first step in the metabolism of fats is the hydrolysis (in the intestine) of triglycerides to glycerol and fatty acids. The reaction

$$
\begin{array}{c}
\quad\quad\quad O \\
\quad\quad\quad \| \\
CH_2-O-C-R \\
| \\
\quad\quad\quad O \quad\quad\quad\quad\quad\quad\quad CH_2OH \\
\quad\quad\quad \| \quad\quad\quad\quad\quad\quad\quad\quad | \\
CH-O-C-R' \; + 3H_2O \rightarrow \; CHOH \; + \text{Fatty acids} \\
| \quad\quad\quad\quad\quad\quad\quad\quad\quad\quad\quad | \\
\quad\quad\quad O \quad\quad\quad\quad\quad\quad\quad CH_2OH \\
\quad\quad\quad \| \\
CH_2-O-C-R''
\end{array}
$$

Glycerine

is catalyzed by water-soluble enzymes called lypases or esterases. The water-insoluble lipids are emulsified by bile acids, and the hydrolysis catalyzed by the water-soluble enzyme occurs at the interface of the lipid drop and the aqueous digestive fluid. The hydrolysis products then are carried to the cells, where they undergo oxidative metabolism. The glycerol enters the carbohydrate-metabolism scheme which we shall study in Section 18.4. The acids, which are the major energy source, are oxidized by a stepwise process, the details of which we shall now discuss.

Fatty Acid Oxidation

In the first stage of utilization of fatty acids, these molecules are systematically and repeatedly shortened by two carbon atoms at a time by the series of reactions indicated schematically in Fig. 18.5. A key substance in this cyclic degradation scheme is called Coenzyme A (CoA or SCoA), a molecule whose structure is given in Fig. 18.6. Note that part of this molecule is made up of the adenine, ribose sugar, and polyphosphate groups that occur in ADP, NAD, and FAD. For the present purposes, however, the important functional group on CoA is the sulfhydryl or thio-alcohol group SH. In the first step of fatty-acid degradation, a thioester, the sulfur analog of an ordinary ester, is formed between CoA and the fatty acid. This reaction is coupled to and driven by an ATP-to-AMP conversion.

In the second step of the degradation, the acid residue in the thioester is oxidized or dehydrogenated by FAD at the positions immediately adjacent to the carbonyl group. The $FADH_2$ produced by this dehydrogenation can be

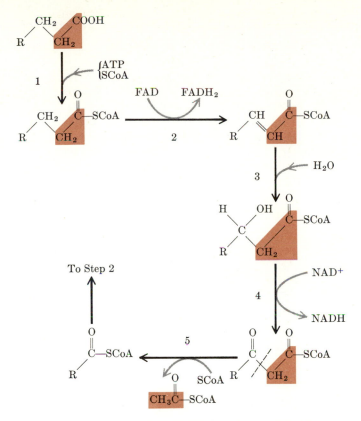

The sequence of steps by which fatty acids are shortened by two carbon units. The carboxyl and α-carbon atoms are removed as an acetyl-SCoA molecule which undergoes subsequent oxidation. The fatty-acid residue reenters the degradative cycle.

FIG. 18.5

oxidized by oxygen through the cytochrome series with concomitant production of ATP. In step three, the ethylenic bond is hydrated catalytically to an alcohol which is then oxidized in the fourth step to a ketone by NAD^+. In the fifth step, this ketone is enzymatically cleaved at the indicated bond, with the introduction of another molecule of SCoA. We have at this point two thioesters. One, acetyl CoA, or

$$CH_3\underset{\underset{O}{\|}}{C}-SCoA,$$

contains the acetyl group,

$$CH_3\underset{\underset{O}{\|}}{C}-,$$

which came from the first two carbon atoms of the original fatty acid. The other thioester,

$$\underset{\displaystyle R'C—SCoA,}{\overset{\displaystyle O}{\overset{\displaystyle \|}{}}}$$

contains the original fatty-acid chain shortened by two carbon atoms. This thioester is ready to begin the degradation cycle again, and does so repeatedly. In each passage through the cycle, the acid chain is reduced in length by two carbon atoms, and each time one molecule of acetyl CoA is formed.

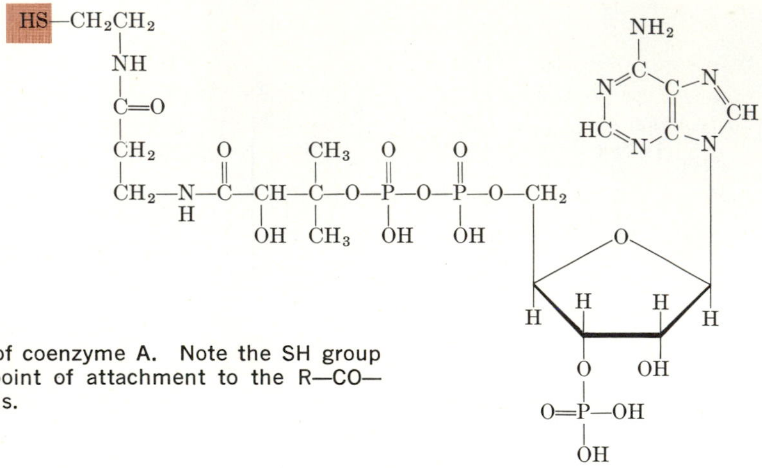

FIG. 18.6   The structure of coenzyme A. Note the SH group which is the point of attachment to the R—CO— function of acids.

The acetyl CoA units from acid degradation are subsequently oxidized to CO_2 in the Krebs citric-acid cycle, and it is in these latter steps that most of the energy associated with fat metabolism is released, or stored as ATP. The Krebs cycle is also responsible for the oxidation of acetyl CoA produced from the degradation of carbohydrates, and will be discussed in detail in Section 18.4. It has been calculated that through degradation and Krebs oxidation, one mole of a C_{16} fatty acid leads to the production of 130 moles of ATP. This stored energy corresponds to a stored energy of about 45 percent of the entire energy of combustion of a C_{16} acid; the rest is dissipated as heat.

18.4 CARBOHYDRATES

Carbohydrates occupy a most important position in the chemistry of life processes. They are formed in plants by photosynthesis, and thus are the major product of processes by which inorganic molecules and energy from the sun are incorporated into living things. The carbohydrate cellulose, which is a very high molecular-weight polymer of glucose sugar units, is a major structural component of plants. In animals, carbohydrate metabolism is a very important source of energy. Nucleic acids, which control the replication processes within the cells, are polymers in which the repeating unit contains a sugar molecule, and are consequently closely related to the carbohydrates.

Carbohydrates are polyhydroxyaldehydes or ketones of the empirical formula $C_nH_{2n}O_n$. The simplest such molecules are called monosaccharides, and if n is 5 to 8, these substances have a sweet taste. Molecules in which two to ten monosaccharide units are linked are called oligosaccharides (Greek *oligas*, few), and the term polysaccharide is applied to polymeric molecules which may contain several thousand monosaccharide units.

Monosaccharides

The most important monosaccharides are the five- and six-carbon sugar molecules called pentoses and hexoses, respectively. There are a number of ways in which the structures of these molecules can be displayed. The five-carbon sugars D-ribose and D-2-deoxyribose, which are found in nucleic acids, are shown below as they occur in the form of five-membered rings.

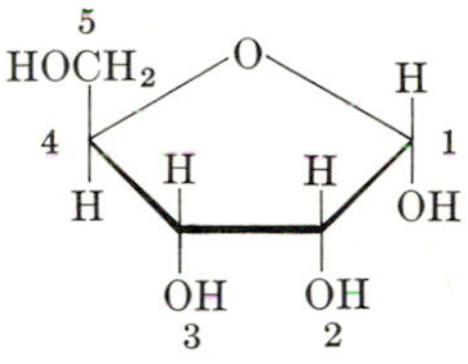

D-ribose 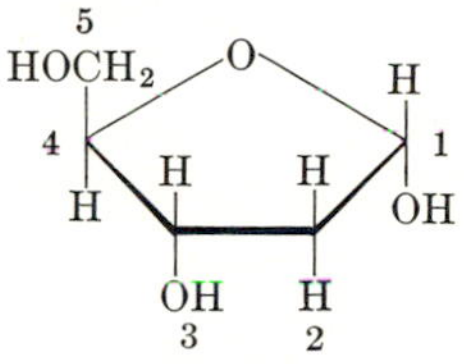   D-2-deoxyribose

In this formula, a carbon atom is understood to be at each apex of the ring, except where an oxygen atom is indicated. It is important to note the spacial relationships of the OH groups that are revealed by these formulas. The five-membered ring is planar, and in both molecules, all OH groups bonded to ring carbon atoms lie below the plane of the ring, while the CH_2OH group lies above this plane.

A free sugar exists as an equilibrium mixture of ring and open-chain forms. Thus D-glucose, the repeating unit in starch, can exist in either of the following two forms:

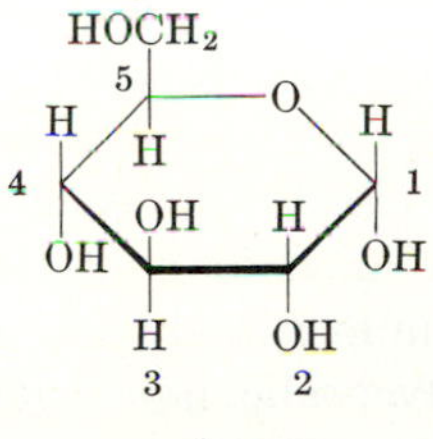

D-glucose

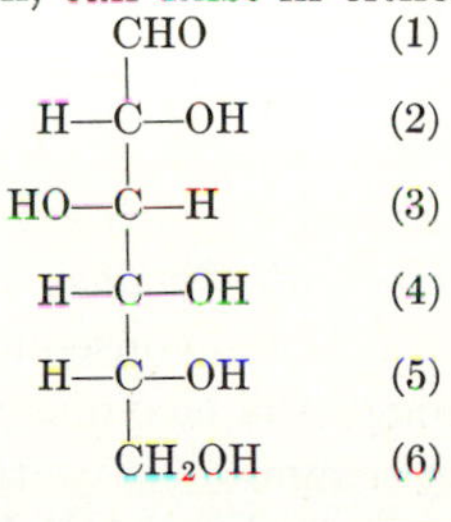

In the open-chain form, the functional group on the first carbon is an aldehyde. When glucose assumes the ring form, the hydroxyl of the 5-carbon adds to the aldehyde carbon, completes the ring, and converts the aldehyde oxygen to an OH group. The orientation of the OH group at the 1-carbon in the ring form is very important. The configuration shown above, where the OH at the 1-carbon and at the 4-carbon are on the same side of the ring, is called α-glucose. The molecule which has these two OH groups on opposite sides of the ring is called β-glucose. The two forms can interconvert through ring opening to the aldehyde structure, followed by ring closure:

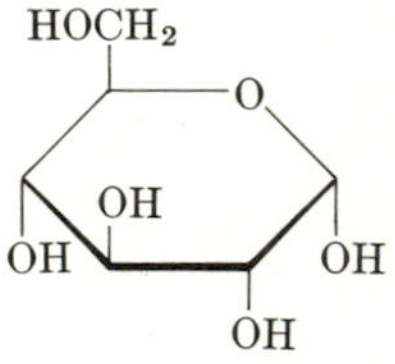

α-glucose   β-glucose

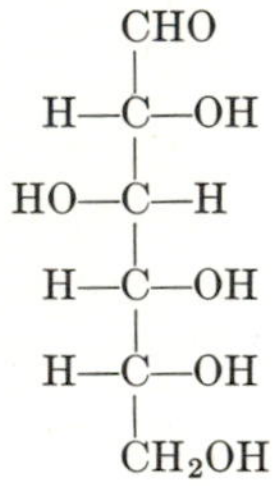

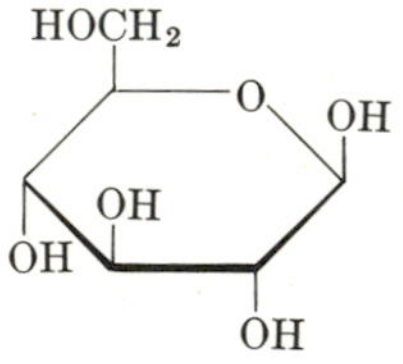

Starch, a polysaccharide which is readily digestible by humans, is a polymer of α-glucose, whereas cellulose, which cannot be digested by humans, is a polymer of β-glucose.

Polysaccharides

Three of the most important polysaccharides are starch, cellulose, and glycogen. Starch is a foodstuff produced in plants, cellulose is the structural material of plants, and glycogen is the form in which glucose is stored in animal cells. All of these substances are polymers of glucose, and differ from each other in molecular weight, the nature of the linkage between glucose molecules, and the degree of polymer chain branching.

Cellulose is a long-chain polymer of about 3000–4000 glucose units. Cotton is approximately 90 percent cellulose. The strong fibrous nature of this and other such plant material is a consequence of the long-chain structure of the cellulose molecule.

Figure 18.7 shows part of the cellulose structure. We see that adjacent glucose units are linked by an oxygen bridge between carbons 1 and 4. This bridge is called a *glycosidic linkage*. We see also that in each case, the 1-carbon has the β-configuration, and consequently cellulose is said to have a β-glycosidic linkage. Even though the enzymes in the human body can cleave an α-glycosidic linkage, they cannot cleave the β-glycosidic bonds in cellulose; consequently, humans cannot digest cellulose. Some bacteria found in the intestines of herbivorous animals do possess the enzymes necessary to break down cellulose to glucose, and consequently these animals can use cellulose as a food.

As Fig. 18.8 shows, in the starch molecule the glucose units are joined by an α-glycosidic linkage between the 1- and 4-carbon atoms of successive rings. The hydrolysis of the α-glycosidic linkage is catalyzed by enzymes secreted by the human salivary and pancreatic glands, and consequently starch can be used as a food.

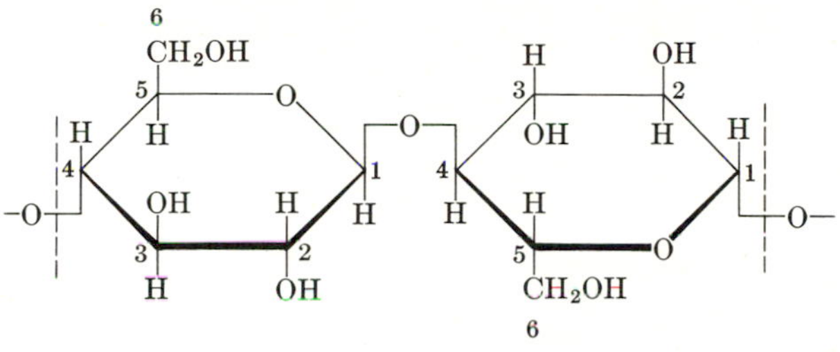

Part of the cellulose chain. Note the β-glycosidic link between carbons 1 and 4 of adjacent rings.   FIG. 18.7

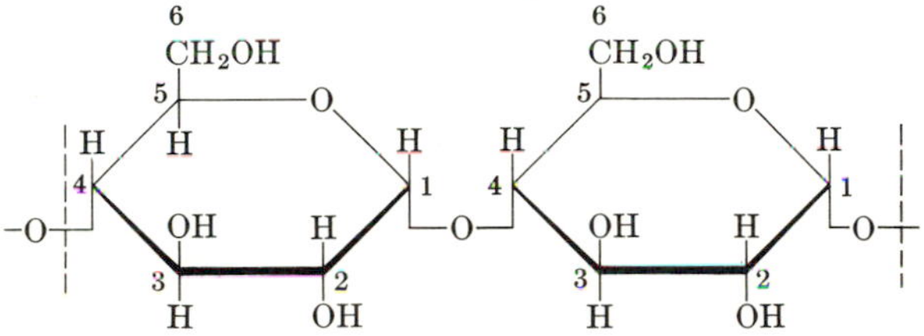

Part of the chain structure of starch. Note the α-glycosidic link between carbons 1 and 4. At other places in the chain these are 1–6 glycosidic bonds.   FIG. 18.8

Two kinds of starch molecules occur in nature. About 10 to 20 percent consists of long unbranched-chain molecules, and is called amylose. The other component, named amylopectin, is a highly branched polymer in which most monomers are joined by 1,4 linkages, with chain branching occurring through 1,6 linkages, as Fig. 18.9 shows. The linear polymer amylose is soluble in hot water, whereas amylopectin is not. Both forms represent plant-energy storage which can be used directly by humans.

Glycogen is the counterpart in animal tissue to the amylopectin of plants. Like amylopectin, glycogen is a polymer of glucose with 1,4-α-glycosidic linkages, and considerable chain branching through 1,6 linkages. The molecular weight of glycogen ranges from 2×10^5 to 10^8. While all animal tissue contains glycogen, the liver is the principal site in which it is stored.

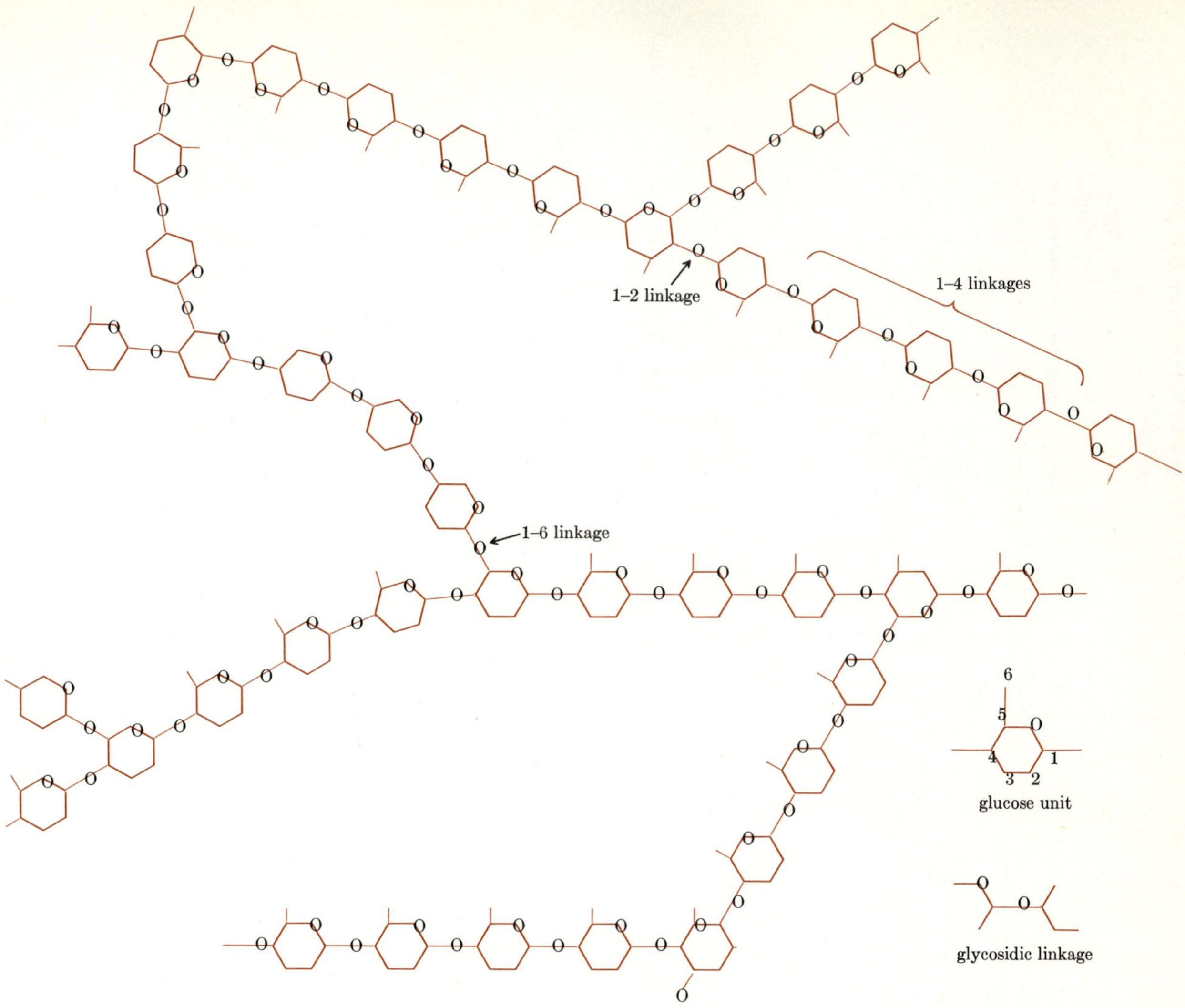

FIG. 18.9   Representation of a segment of the amylose structure.

1-4 linkages

1-2 linkage

1-6 linkage

6
5
4 3
1
2

glucose unit

glycosidic linkage

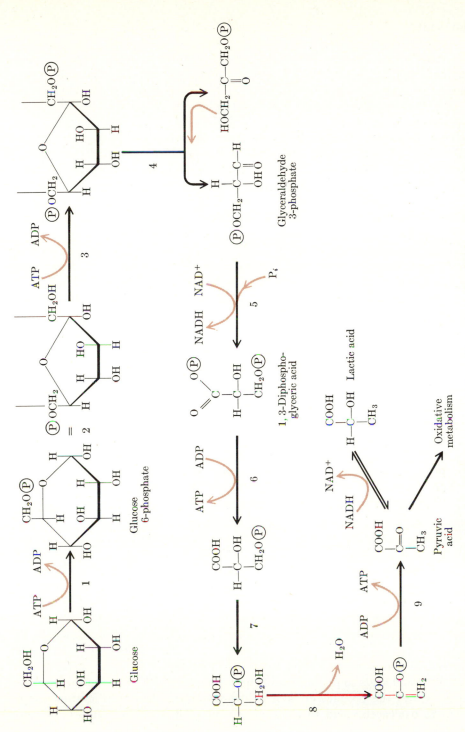

FIG. 18.10 The Embden-Meyerhof pathway of glycolysis. The symbol P stands for the phosphate group —PO₃H.

The necessary first step in the utilization of polysaccharides by animals is the enzymatic hydrolysis of starch or glycogen to free glucose sugar molecules. The glucose molecules are then broken down and oxidized in two major stages. In the Embden-Meyerhof glycolytic pathway, glucose is converted by a series of steps to two molecules of pyruvic acid, $CH_3COCOOH$. This decomposition is accomplished anaerobically, that is, without oxygen. The pyruvic acid then enters the Krebs citric-acid cycle, which has as its ultimate products CO_2, H_2O, and ATP. The oxidants which participate directly in the Krebs cycle are themselves oxidized by a chain of redox couples which has molecular oxygen as its ultimate oxidant.

The major features of the Embden-Meyerhof glycolytic pathway are given in Fig. 18.10. Glucose is converted to a phosphate ester by ATP, isomerized enzymatically to a five-membered ring, and then another phosphate group is added by ATP. The resulting diphosphate is broken into two three-carbon fragments which are both converted to 1,3-diphosphoglyceric acid. This molecule then loses one phosphate group, and restores one mole of ADP to ATP. After an isomerization and dehydration, the last phosphate group is eliminated, and another molecule of ATP formed; the product pyruvic acid is then ready to proceed to the Krebs oxidative cycle. If no oxygen is available, as might be the case following brief but violent muscle action, pyruvic acid is reduced to lactic acid by NADH. When oxygen becomes available, this lactic acid is reoxidized to pyruvic acid, and then enters the Krebs oxidative cycle.

There are, in the Embden-Meyerhof pathway, two oxidation-reduction reactions. In step 5, NAD^+ is the oxidant, and NADH is produced. If lactic acid is formed from pyruvic acid in the last step, however, an equivalent amount of NADH is consumed. Hence, if lactic acid is formed, there is no net oxidation or reduction associated with glycolysis. There is, nevertheless, a net production of ATP. A total of two moles of ATP per mole of glucose are consumed by steps 1 and 3, but in steps 6 and 9, two moles of ATP per mole of three-carbon fragment, or four moles of ATP per mole of glucose are produced. Thus the decomposition of glucose to pyruvic acid is accompanied by formation of energy-rich ATP.

The Krebs citric-acid cycle is sketched in Fig. 18.11. This cycle converts the products of glycolysis to carbon dioxide and water, and is also the pathway by which fragments from fatty acids are oxidized. It is, therefore, of central importance in the metabolic scheme.

Pyruvic acid enters the Krebs cycle by losing CO_2 and being converted to an acetyl group CH_3CO attached to the sulfur of Coenzyme A. The oxidation required in this step is effected by NAD^+, which, as we have discussed, is reoxidized and leads to production of three molecules of ATP. The acetyl CoA, which is also the product of fatty-acid breakdown, then transfers its acetyl group to oxaloacetic acid to form citric acid. Then follow dehydration and

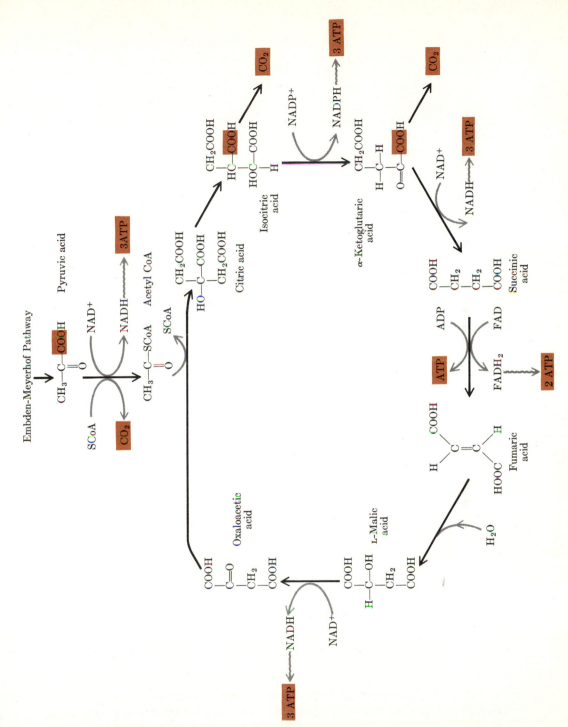

FIG. 18.11 The Krebs citric acid cycle. The enzymes which catalyze these reactions are in the cell mitochondria.

hydration reactions which convert citric acid to isocitric acid. An oxidation by $NADP^+$, a phosphate ester of NAD^+, liberates CO_2, leads to the eventual production of 3 ATP, and forms α-ketoglutaric acid. Another oxidation, this time by NAD^+ aided by CoA and other agents, liberates another CO_2 molecule, and produces succinic acid. At this stage, the two-carbon fragment which entered the cycle as an acetyl group has been oxidized to CO_2, and the remaining steps of the cycle serve to restore the oxaloacetic acid with which the cycle began. The succinic acid is dehydrogenated to fumaric acid. Addition of water gives malic acid, and oxidation of this molecule by NAD^+ finally produces oxaloacetic acid, which is then available to begin the cycle again.

For every acetyl group that enters the Krebs cycle, 12 molecules of ATP are produced. In addition, three molecules of ATP and one molecule of CO_2 are formed in converting the pyruvic acid from glycolysis to the acetyl CoA that enters the cycle. Consequently the overall carbohydrate oxidation reaction can be written

$$C_3H_4O_3 + \tfrac{5}{2}O_2 + 15ADP + 15H_3PO_4 \rightarrow 3CO_2 + 2H_2O + 15ATP \cdot H_2O.$$

The oxygen enters the reaction indirectly through the oxidants NAD^+, $NADP^+$, and FAD.

18.5 PROTEINS

We have already pointed out that proteins constitute most of the nonaqueous component of the cell. Even considering this abundance, the variety of functions performed by proteins is extremely impressive. Enzyme molecules which are such specific catalysts for so many synthetic and degradative reactions of the life cycle are proteins, as are many of the regulatory hormones. Proteins are components of the peri- and intra-cellular membranes, serve as antibodies to foreign antigens, perform the oxygen-carrying function in the blood, and constitute some of the chromosomal material. Thus the form, regulation, and reproduction of living things are dominated by the proteins.

Amino Acids

Proteins are polymers of α-amino acids. As we found in Chapter 17, the α-amino acids have the general structure

$$\begin{array}{c} H \\ | \\ R-C-COOH, \\ | \\ NH_2 \end{array}$$

in which the amino group and the radical R are attached to the first (α) carbon atom removed from the carboxylic acid group. There are 20 amino acids which occur in protein molecules, and the individual properties of these acids are dictated by the nature of the R group. The unique features of the different proteins are a consequence of the total number, variety, and sequence of the amino acids which occur in the polymer chain, and as well, of the spacial configuration of the chain itself.

The structural formulas, common names, and three-letter abbreviations for the 20 amino acids are given in Fig. 18.12. We see that the acids can be considered to be derivatives of glycine, NH_2CH_2COOH, in which various R groups have been substituted for one of the α-hydrogen atoms. In a number of the acids, the R group is entirely an aliphatic or, in one case, aromatic hydrocarbon. In the other acids, the R radical contains a potentially reactive functional group. Serine, threonine and tyrosine have an OH group which can esterify with organic acids or with phosphoric acid. Glutamic and aspartic acids have a second acid functional group, while lysine and arginine have a second amino group. The highly reactive SH or sulfhydryl function in cysteine is very significant, since two of these can form a disulfide bond —S—S—, and thereby link together two protein chains. As was mentioned in Chapter 17, all of the amino acids except glycine have four different substituents on the α-carbon atom, and consequently are optically active. Of the two possible arrangements of atoms at the asymmetric center, only the L-structure has been found in natural proteins.

Amino acids are linked together to form proteins by the *peptide bond*. This linkage can be pictured as the result of the condensation of the carboxyl group of one acid with the amino group of another, accompanied by the elimination of water.

$$H-\underset{\underset{H}{|}}{\overset{\overset{H}{|}}{N}}-\underset{\underset{R}{|}}{C}-\overset{\overset{O}{\|}}{C}-\boxed{OH + H}-\underset{\underset{N}{|}}{N}-\overset{\overset{O}{\|}}{C}-\overset{\overset{O}{\|}}{C}-OH \rightarrow H_2N-\underset{\underset{R}{|}}{C}-\overset{\overset{O}{\|}}{C}-\underset{\underset{H}{|}}{\overset{\overset{H}{|}}{N}}-\overset{\overset{O}{\|}}{C}-OH + H_2O$$

Amino acid	Amino acid	Dipeptide

It is the link between the carbonyl carbon and the amino nitrogen that is called the peptide bond.

Continuation of the condensation process to link together many amino acids produces a *polypeptide*. The repeating unit in the polypeptide chain

$$-\underset{\underset{H}{|}}{\overset{\overset{H}{|}}{N}}-\underset{\underset{R}{|}}{C}-\overset{\overset{O}{\|}}{C}-$$

is referred to as an amino acid residue, since it contains what is left of the amino acid after the elements of water are eliminated. Usually molecular chains of 70 or fewer amino acids are referred to as polypeptides, while larger naturally-occurring molecules are called proteins.

Side-Chain (R-Group) Characteristic	Chemical structure	Amino acid	Symbol
Aliphatic, nonpolar	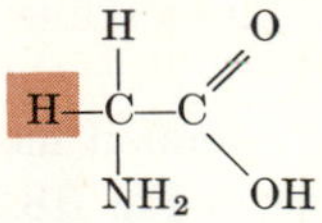	Glycine	Gly
	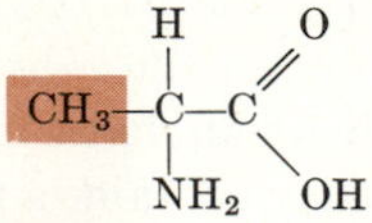	Alanine	Ala
	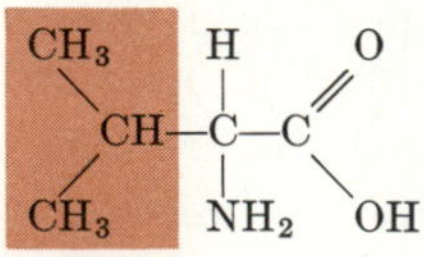	Valine	Val
	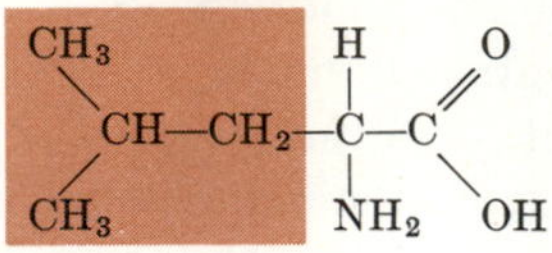	Leucine	Leu
	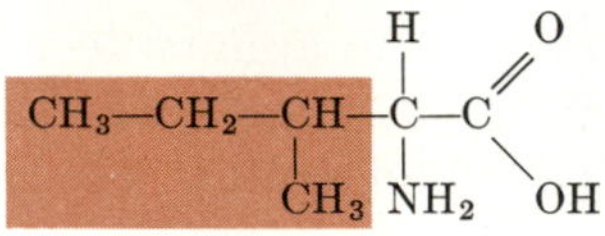	Isoleucine	Ileu
Alcoholic, aliphatic and aromatic	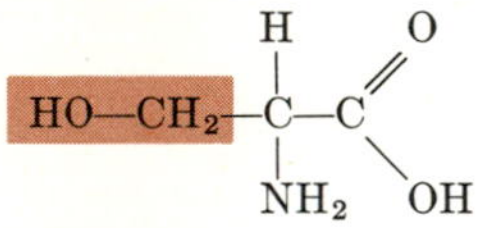	Serine	Ser
	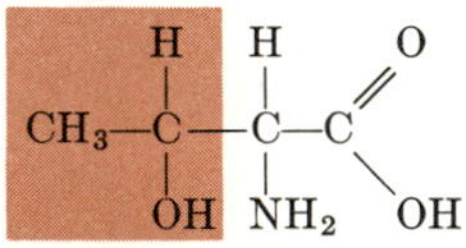	Threonine	Thr
Aromatic	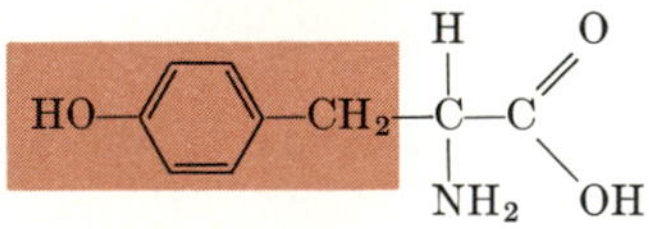	Tyrosine	Tyr
	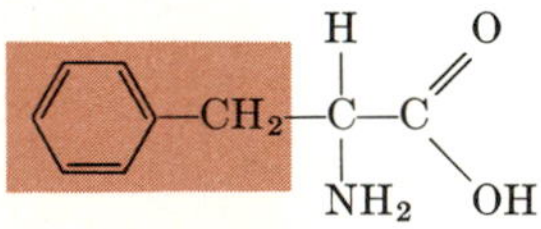	Phenylalanine	Phe
	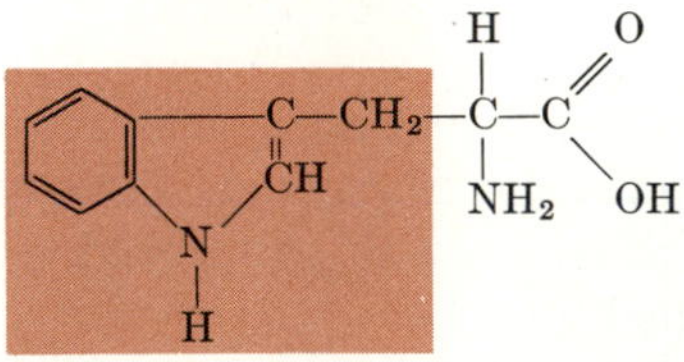	Tryptophan	Try

FIG. 18.12   The common amino acids, their structures, and their symbols.

Side-Chain (R-Group) Characteristic	Chemical structure	Amino acid	Symbol
Carboxylic (acidic)	$\underset{HO}{\overset{O}{\parallel}}C-CH_2-\underset{NH_2}{\overset{H}{C}}-\underset{OH}{\overset{O}{C}}$	Aspartic	Asp
	$\underset{HO}{\overset{O}{\parallel}}C-CH_2-CH_2-\underset{NH_2}{\overset{H}{C}}-\underset{OH}{\overset{O}{C}}$	Glutamic	Glu
Amine bases (basic)	$NH_2-CH_2-CH_2-CH_2-CH_2-\underset{NH_2}{\overset{H}{C}}-\underset{OH}{\overset{O}{C}}$	Lysine	Lys
	$NH_2-\underset{NH}{\overset{}{C}}-NH-CH_2-CH_2-CH_2-\underset{NH_2}{\overset{H}{C}}-\underset{OH}{\overset{O}{C}}$	Arginine	Arg
	$HC=C-CH_2-\underset{NH_2}{\overset{H}{C}}-\underset{OH}{\overset{O}{C}}$; N, NH, $\underset{H}{C}$	Histidine	His
Sulfur-containing	$HS-CH_2-\underset{NH_2}{\overset{H}{C}}-\underset{OH}{\overset{O}{C}}$	Cysteine	Cys
	$CH_3-S-CH_2-CH_2-\underset{NH_2}{\overset{H}{C}}-\underset{OH}{\overset{O}{C}}$	Methionine	Met
Amides	$\underset{NH_2}{\overset{O}{\parallel}}C-CH_2-\underset{NH_2}{\overset{H}{C}}-\underset{OH}{\overset{O}{C}}$	Asparagine	Asp
	$\underset{NH_2}{\overset{O}{\parallel}}C-CH_2-CH_2-\underset{NH_2}{\overset{H}{C}}-\underset{OH}{\overset{O}{C}}$	Glutamine	Gln
Imino	CH_2-CH_2 ; CH_2 $CH-\underset{OH}{\overset{O}{C}}$; $\underset{H}{N}$	Proline	Pro

Protein Structure

There are four levels at which the structure of proteins can be described. The *primary* structural feature is the amino-acid sequence. The *secondary* structure refers to the spacial configuration of the amino-acid chain; frequently this is a helical structure. The *tertiary* structure is a description of how the helix is folded and bent. Finally, the *quarternary* structure arises from the association of individual proteins to form distinct complex super molecules. Each of these features can be very important to the biological function of the protein. To establish the complete structure of a protein at all four levels is an enormously difficult problem, and the techniques used to learn about each structural level differ very greatly.

The first step toward establishing the primary structure of a protein or polypeptide is to obtain an analysis of the amino acids present in a pure sample. The preparation of the purified protein may in itself be a difficult task. An approximate molecular weight for the protein can be obtained by several physicochemical methods which include measurement of the rate of sedimentation in a centrifuge, osmotic-pressure determinations, and light-scattering studies. Then weak-acid hydrolysis may be used to cleave the peptide bonds and produce free amino acids, which can then be identified and analyzed quantitatively. With the quantitative amino-acid composition known, one has an empirical formula for the protein.

At this point, however, the primary-structure determination has just begun. The number of different sequences of acids that is possible for even a very small molecule is very great. If we were dealing with a molecule in which there were only ten amino acids, all different, there would be 10! or 3,628,800 possible sequences. While there are some polypeptides of biological importance with only nine different amino acid residues, the relatively small insulin molecule of molecular weight 5733 has 51 acid residues, and the muscle protein myoglobin contains 153 residues. Other proteins have molecular weights up to the order of 7×10^6, and the number of possible acid sequences in such molecules is difficult to imagine.

A number of chemical techniques have been used to determine the amino-acid sequence in several polypeptides and proteins. One of the important reagents used is dinitrofluorobenzene. This molecule attaches itself to the free amino group at the end of a polypeptide:

N-terminal acid   C-terminal acid

If the resulting adduct is hydrolyzed, the peptide bonds break, but the dinitrobenzene remains attached to the amino group of the N-terminal acid. This acid can be separated and identified by chemical analysis.

It is also possible to establish the identity of the acid at the carboxyl end of the polypeptide chain quite simply. The enzyme carboxypeptidase removes the C-terminal amino acid from a protein along with only very much smaller amounts of the other acids. This separated acid is then easily identified. If the enzyme is left in contact with the peptide, and the identities of the liberated amino acids are studied as a function of time, the sequence of several of the residues near the end of the chain may be determined.

The determination of the acid sequence of the protein hormone insulin exemplifies some of the other features of this general problem. Treatment of insulin with dinitrofluorobenzene followed by hydrolysis showed that there were *two different* dinitrobenzene derivatives formed. This indicated that there are two N-terminal acids, and therefore, two parallel polypeptide chains in the molecule. These so-called A and B polypeptide chains were separated by oxidizing the disulfide (—S—S—) links between them to —SO₃H groups. The two chains were isolated, and then partially hydrolyzed to intermediate-sized peptides containing two to five amino-acid residues. These peptides were separated, and their amino-acid sequences determined by the difluorobenzene method.

The sequence determination was completed in the following manner. The acid hydrolysis cleaves the protein chain randomly into small groups of acid residues. However, if one finds that there are present the three-acid sequences

Gly-Ser-His,   Ser-His-Leu,   and   His-Leu-Val,

the overlapping suggests that there is a five-acid sequence

Gly-Ser-His-Leu-Val.

Evidence which corroborates and extends this order is found in the sequences of the four- and five-acid fragments. By making use of such overlapping short sequences, the sequence of the entire chain can be deduced.

The complete primary structure of beef insulin, determined in this way by F. Sanger in 1953, is given in Fig. 18.13. We see that the A and B chains are held together in two places by disulfide links between cysteine residues in the separate chains. In addition, there is a disulfide link between two acids in the A chain. Although these acids may seem to be separated, the coiling of the peptide chain in fact allows them to be close enough to form the disulfide bond.

The acid sequence has been determined for insulin taken from swine, sheep, and whales. The molecules from the four sources are identical except for the three amino acids in positions 8, 9, and 10 of the A chain. Evidently this variation is the chemical basis for some of the antigenic differences between insulin from different animal sources, which had been observed well before the molecular structures were determined.

The complete amino-acid sequences are known for a small but growing number of proteins, including ribonuclease, myoglobin, and hemoglobin. Ribonuclease is an enzyme consisting of one chain of 124 acid residues. Myoglobin is a protein found in muscle tissue, and has 153 acid residues. Human hemoglobin contains two identical α-chains (141 residues) and β-chains (146 residues). The determination of each of these primary structures was a long and difficult job and is only the first step in finding an explanation of their biological function in terms of their molecular structure.

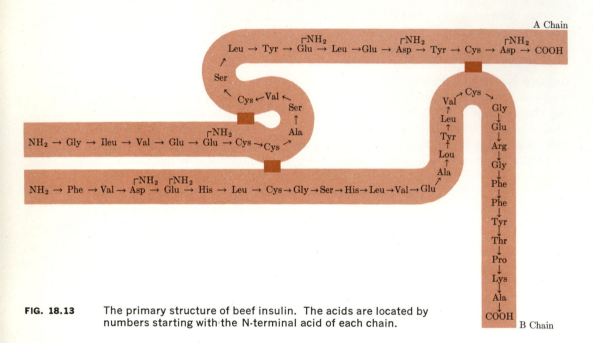

FIG. 18.13 The primary structure of beef insulin. The acids are located by numbers starting with the N-terminal acid of each chain.

Secondary Protein Structure

Figure 18.14 shows part of a polypeptide chain in a fully extended conformation. This conformation, known as the β-form, is only one of many which are consistent with the C—C and C—N bond distances and angles, since it is possible to fold the chain back on itself by rotating two segments about a C—C or C—N bond. The fully extended β-form of the polypeptide chain is thought to occur in the insoluble, fibrous proteins like β-keratin, a major component of hair and nail.

It is thought that in most proteins the polypeptide chain is largely in the form of the α-helix structure proposed by Pauling and Corey in 1951. These workers examined theoretically the properties of several conformations of the protein chain, and selected the α-helix as the one which would have the lowest energy, and still be consistent with the known bond angles and lengths in the amino-acid residues. Figure 18.15 shows two representations of the α-helix. The stability of this structure is derived from the fact that it allows the maximum possible number of hydrogen bonds to be formed between the amino hydrogen on one acid and the carbonyl oxygen of a residue in the subsequent turn of the helix.

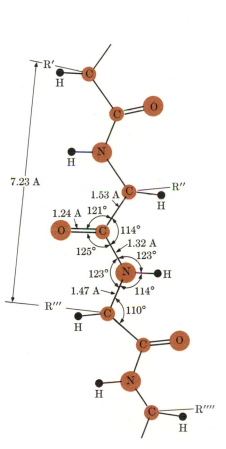

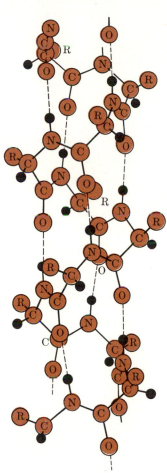

18.14

18.15

Part of a polypeptide chain in the extended, or β-conformation. [Reprinted by permission from the Royal Society and from Dr. Linus Pauling, *Proc. Roy. Soc.* **B141**, 10 (1953).]

The right-handed α-helix structure of a polypeptide chain. The dashed lines represent hydrogen bonds. [Reprinted by permission from The Royal Society and from Dr. Linus Pauling, *Proc. Roy. Soc.* **B141**, 10 (1953).]

FIG. 18.14

FIG. 18.15

The deduction of the α-helix structure is a good example of how detailed knowledge of the structure of small molecules can facilitate prediction of the structures of complicated molecules. The fundamental dimensions of the peptide group as measured by x-ray investigations of small peptides is shown in Fig. 18.16. The C—C bond and the C—N bond which involves the α-carbon have lengths of 1.53 and 1.47 A respectively, which are normal distances for single bonds between these atoms. However, the length between the nitrogen and the carbon atom of the carbonyl group is 1.32 A, quite a bit smaller than the 1.47 A expected for a C—N single bond. This shortening suggests that the peptide link has partial double-bond character, which corresponds to the following resonance description:

As a result, the peptide group has a planar conformation. That is, all bonds involving the amino nitrogen and the carbonyl carbon lie in the same plane. In postulating possible structures for proteins, Pauling and Corey ruled out any conformation which seriously violated this constraint.

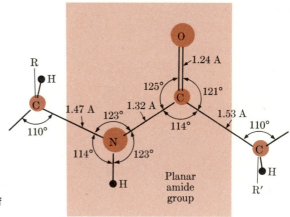

FIG. 18.16 The fundamental dimensions of the peptide group.

Close inspection of the α-helix shows that each peptide group is in a plane essentially tangent to a cylinder coaxial to the helix. Each peptide group is connected by a hydrogen bond to the third peptide group along the chain in either direction. Except near the ends of the helix, every carbonyl oxygen and amino nitrogen is involved in hydrogen bonding. The helix itself is coiled very tightly, with no room in the center for any occluded molecules. The sense of the helices so far found in proteins is that of a right-handed screw.

If the entire polypeptide chain were in the form of the α-helix, the molecules would have the shape of long, relatively narrow rigid rods. A number of physico-chemical techniques show, however, that many proteins are globular and nearly round, and that others are much shorter and thicker than their chain length and the pure helical structure would imply. Thus there must be folding of the α-helix in most proteins, and perhaps little or no α-helix in others. X-ray crystal studies of a number of proteins are beginning to make these tertiary structural features clear.

Figure 18.17 shows a schematic representation of the myoglobin molecule which was derived from x-ray studies. The tube represents the space occupied

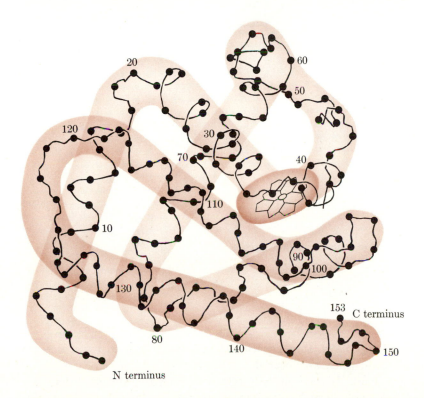

The secondary and tertiary structure of the myoglobin molecule. Note that the helical conformation of the peptide chain is lost at the bends in the molecule. Note also the porphyrin-ring structure attached to the chain near residue 40.

FIG. 18.17

by the chain which we see is folded into a complicated form. There are regions in which the tube is relatively straight for 30 to 40 A, and in these regions the chain has the α-helix conformation. At the regions in and around the bends in the tube, which constitute about 30 percent of the peptide, the chain is in some nonhelical form. Hemoglobin also has chains that are folded into a similarly complicated shape. The relation between the tertiary structural features and biological function of the protein is a completely unsolved problem.

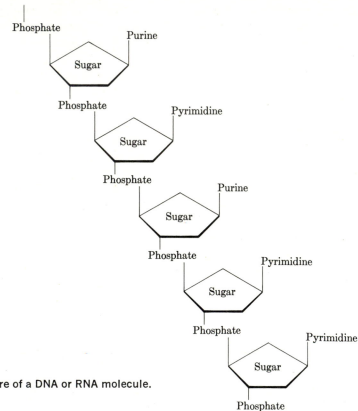

FIG. 18.18   The general structure of a DNA or RNA molecule.

18.6 THE NUCLEIC ACIDS

Biologists have long known that genetic information is carried by structures called chromosomes which are located in the nucleus of the cell, and whose subunits are the genes. It has been only relatively recently, however, that biochemists have been able to make substantial progress in elucidating the molecular structure of the chromosomal material. It is known that genes are made up of the macromolecule deoxyribonucleic acid (DNA), and that this molecule carries the information needed to direct protein synthesis, and preserves and transmits this information during cell division. Another related type of molecule, ribonucleic acid, is present throughout the cell, and is even more directly involved in protein synthesis. Before considering these molecules in detail, let us examine some general features of their structure.

Nucleic acids are polymers in which the repeating units are sugar molecules linked by phosphate bridges. This general structure is indicated in Fig. 18.18. In ribonucleic acid the sugar is ribose, and in deoxyribonucleic acid the sugar is deoxyribose. The structures of these two sugar molecules are

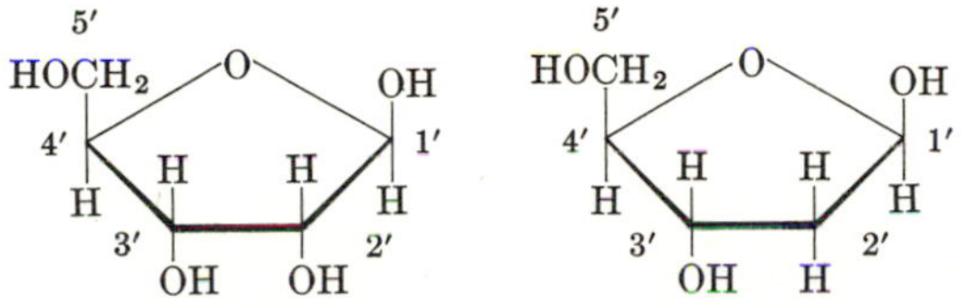

Ribose   Deoxyribose

We can see that these structures differ only in that deoxyribose has two hydrogens attached to the 2'-carbon, whereas ribose has both H and OH at this position. The numbers designating the carbon atoms in these sugar molecules are usually primed to distinguish them from the numbers used to locate atoms in the organic bases which also occur in DNA and RNA.

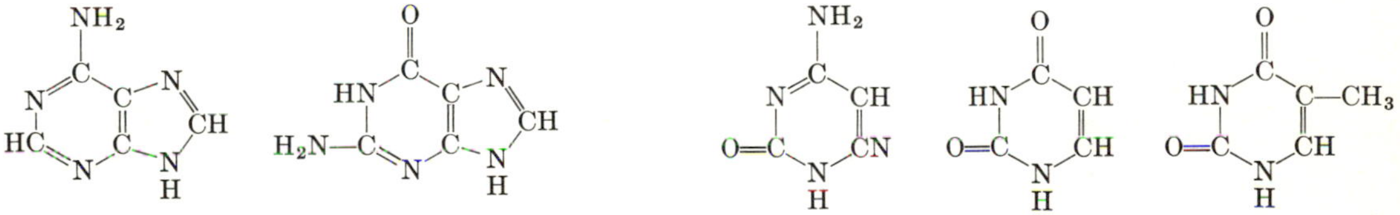

Adenine   Guanine   Cytosine   Uracil   Thymine

(a)   (b)

The structures of the (a) purine and (b) pyrimidine bases that occur in DNA and RNA.   **FIG. 18.19**

Attached to each sugar unit in DNA and RNA is an organic base of the type designated purine or pyrimidine. The structures of these bases are given in Fig. 18.19. The base uracil occurs predominantly in RNA, whereas the closely related thymine is found in DNA. Cytosine is present in both nucleic acids. The two purine bases, adenine and guanine, occur in both DNA and RNA.

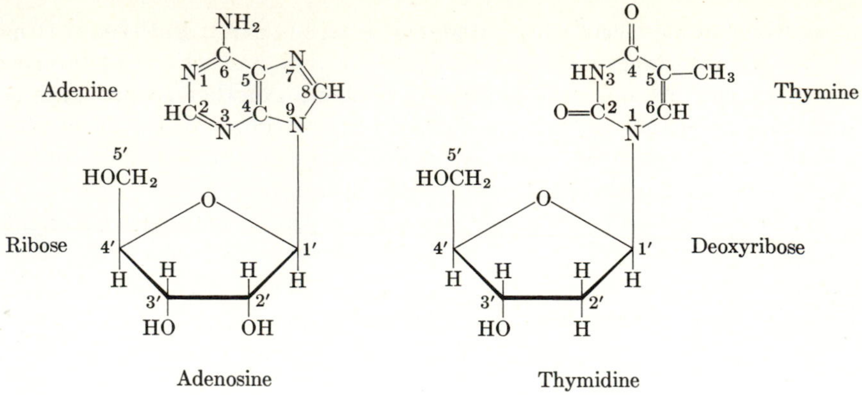

FIG. 18.20   The structures of the nucleosides adenosine and thymidine.

The way in which the purine and pyrimidine bases are attached to ribose and deoxyribose is shown in Fig. 18.20. Such a combination of a sugar and a purine or pyrimidine base is called a *nucleoside*. In the pyrimidine nucleosides the sugar and base are joined by a β-glycosidic link from the 1'-carbon of the pentose to the 1-nitrogen of the pyrimidine base, as is shown for thymidine in Fig. 18.20. In the purine nucleosides, the 1'-carbon of the sugar is connected through the β-glycosidic link to the 9-nitrogen of the purine base, as for example, in the molecule adenosine.

The combination

Base–Sugar–Phosphate

is called a *nucleotide*, and is just a phosphate ester of a nucleoside. The structural formulas of two nucleotides are given in Fig. 18.21. The phosphate groups in

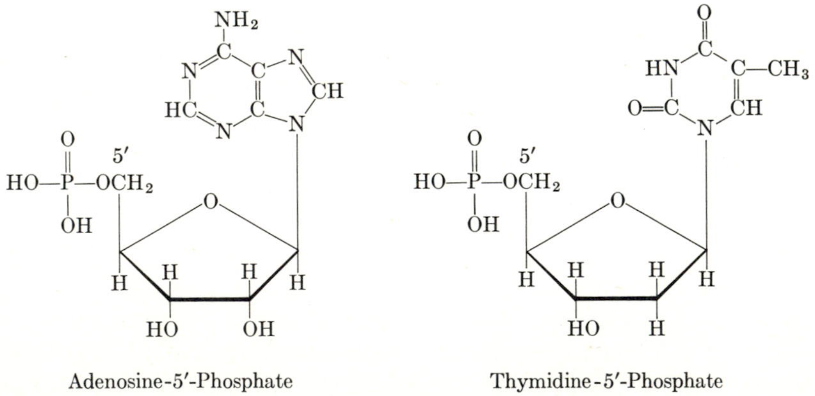

Adenosine-5'-Phosphate   Thymidine-5'-Phosphate

FIG. 18.21   The nucleotides, or phosphate esters of adenosine and thymidine.

these molecules are, as shown, attached to the 5'-carbon of the sugar ring, but the 3'-carbon of both sugars or the 2'-carbon of ribose are also possible points of attachment. Nucleic acids are polynucleotides, in which the phosphate groups link the 5'-carbon of one sugar with the 3'-carbon of the next. A partial structure of the polynucleotide chain in DNA is shown in Fig. 18.22.

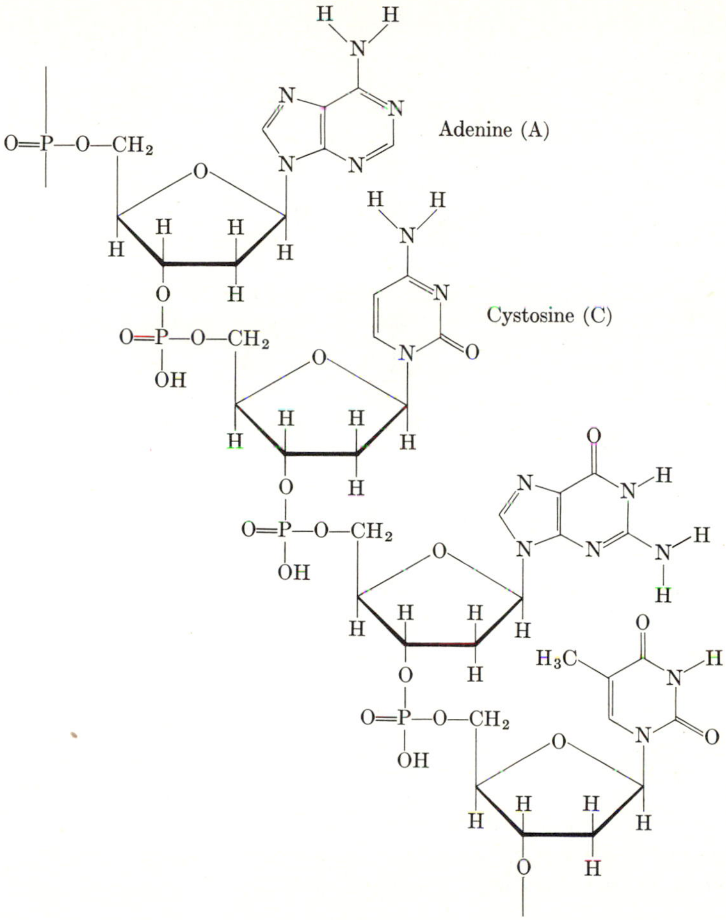

Part of the polynucleotide chain of a DNA molecule.

FIG. 18.22

The Structure of DNA

In 1953, J. D. Watson and F. H. C. Crick proposed that DNA had a structure in which two parallel polynucleotide strands were wound into a double helix, a conformation which would be stabilized by numerous hydrogen bonds between bases attached to the two strands. In part, this proposal was based on x-ray studies of DNA by M. H. F. Wilkins and R. Franklin which were consistent with a helical conformation, but in large measure it was suggested by observations of the frequency of occurrence of the purine and pyrimidine bases.

Prior to 1953, studies of the base composition of DNA showed that whatever the frequency of the individual bases, the molar ratio of adenine to thymine was unity, and the same was true for the molar ratio of guanine to cytosine. This observation made it appear that the DNA structure was one in which adenine was specifically *paired* with thymine, and guanine was specifically paired with cytosine.

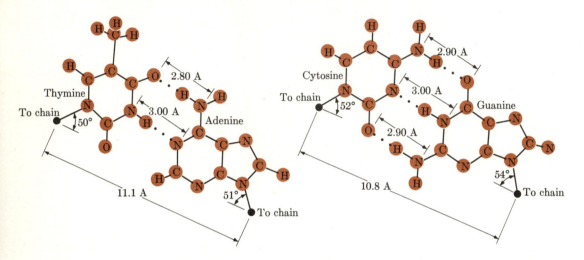

FIG. 18.23   Structure and critical dimensions of the base pairs thymine–adenine and cytosine–guanine.

A study of the molecular models for these base pairs finally suggested why this pairing occurs, and what its consequences might be. Figure 18.23 shows the basis of the explanation. The structures of thymine and adenine are complementary in that they can be fitted together in the same plane so that two hydrogen bonds can be formed between them. At the same time, the atoms by which the bases are attached to their sugar molecules, the 1-nitrogen of thymine and the 9-nitrogen of adenine, are at opposite ends of the molecular complex. The same situation holds for cytosine and guanine, except that the base association in this case is accompanied by formation of three hydrogen bonds. A very significant point is that the "end-to-end" distances given in Fig. 18.23 are nearly the same for the A–T and G–C pairs.

These considerations led to the proposal that DNA consists of two parallel helical polynucleotide chains that are held together by hydrogen bonds between purine bases of one chain and pyrimidine bases of the other, and vice versa. The schematic structure of the DNA double helix is shown in Fig. 18.24. Each

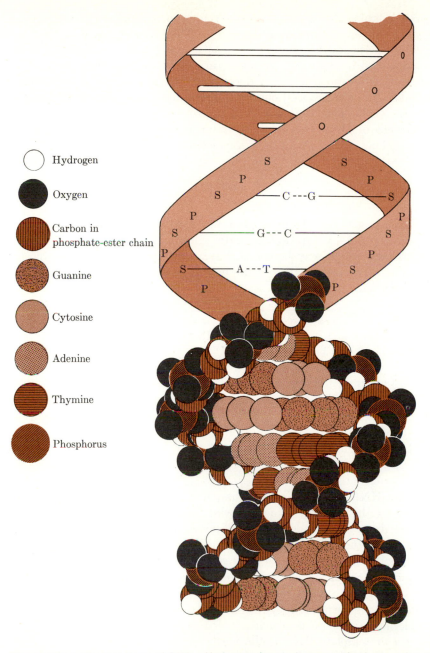

Hydrogen

Oxygen

Carbon in
phosphate-ester chain

Guanine

Cytosine

Adenine

Thymine

Phosphorus

The double-helix structure of DNA. Hydrogen bonds between the base pairs adenine–thymine and guanine–cytosine hold the sugar–phosphate strands together. [Reprinted by permission from American Cancer Society and Dr. L. D. Hamilton, Brookhaven National Laboratory; Ca, *A Bulletin of Cancer Progress*, **5**, 163 (1955). Upper portion of drawing simplified for clarity.]

FIG. 18.24

strand of the double helix consists of a sugar–phosphate "backbone" with bases extending inward toward the axis of the helix. The bases lie in planes that are approximately perpendicular to the helix axis, and planes of successive base pairs are separated by 3.4 A. A complete turn of the helix occurs every 34 A.

The sizes of DNA molecules vary depending on the type of cell from which they are taken. One fairly well characterized sample had a molecular weight of about 1.4×10^8, and thus involved approximately 400,000 nucleotides. Autoradiographs of DNA from the bacterium *escherichia coli* show molecules about 0.4 mm long, which would have a molecular weight of about 10^9.

The Structure of RNA

While ribonucleic acid is a polynucleotide like DNA, the size and structure of naturally occurring RNA show much greater variation. RNA occurs frequently as a single strand which may be coiled, but not in any simple, easily characterized manner. The molar ratios of the base pairs do not show the regularity displayed by DNA.

There are at least three distinct forms of RNA which have different roles in protein synthesis. These three molecular species are known as *transfer* or *soluble* RNA, *messenger* RNA, and *ribosomal* RNA, and they differ considerably in molecular weight and base composition.

Transfer RNA is the smallest known type of ribonucleic acid, and usually consists of 70 to 80 nucleotides with a molecular weight of about 25,000. The biological role of transfer RNA is to pick up individual amino acids and carry them to the sites of protein synthesis. Since each amino acid is recognized and carried by only one type of transfer RNA, there are at least 20 distinct transfer RNA molecules. The sequence of the 77 nucleotides in the RNA molecule responsible for the transfer of alanine was first determined in 1965 by R. W. Holley.

All transfer RNA molecules contain as terminal segments the three nucleotides containing the bases cytosine, cytosine, and adenine in that order. On the terminal ribose the 2' and 3' hydroxyl groups are free. The amino acid to be transferred by the RNA becomes attached by forming an ester linkage at either of these two hydroxyl groups in a reaction driven by ATP, and catalyzed by an enzyme. It is the specificity of the enzyme that assures that the proper amino acid becomes attached to the appropriate transfer RNA molecule. In the subsequent steps of protein synthesis, the amino acid is recognized by the *t*-RNA molecule to which it is attached.

Messenger or template RNA is the most recently discovered form of ribonucleic acid. Its biological function is to carry the genetic information contained in one portion of one strand of a DNA molecule in the nucleus to the ribosomes, which are the site of protein synthesis. The transcription of the information on the DNA molecule to messenger RNA evidently involves a partial uncoiling of the DNA helix, and then specific base pairing between the DNA and the nucleotides of the messenger RNA as the RNA is synthesized.

Ribosomal RNA has the highest molecular weight (about 1.2×10^6) of the cellular ribonucleic acids. It is concentrated in the ribosomes of the cell, where it participates in protein synthesis in some manner which is obscure at the present time.

18.7 BIOLOGICAL FUNCTIONS OF THE NUCLEIC ACIDS

The DNA molecule has two major functions: it contains the information necessary to replicate and synthesize new DNA for the chromosomes of daughter cells, and it stores and supplies the information necessary for protein synthesis. This information is contained in DNA as a genetic code expressed by the *sequence* of the bases adenine, quanine, cytosine, and thymine. Since DNA directs the synthesis of the enzymes which in turn catalyze the cell reactions, it is of central importance in physiological chemistry.

Replication of DNA

The double helical structure of DNA suggests the way in which this molecule may replicate. Because of the specific base pairing of adenine with thymine and guanine with cytosine, the two strands of the helix are complementary, and a particular base sequence in one strand implies a specific sequence in the other. In replication, the two polynucleotide chains may unwrap, either partially or fully, and act as templates upon which free deoxyribonucleotides can be deposited and linked in a complementary pattern. The result is two DNA molecules identical to the first. This process is indicated schematically in Fig. 18.25 on page 782.

Experiments in which bacteria have been allowed to replicate in a medium containing nitrogenous compounds entirely labeled with the isotope N^{15} have shown that the DNA in the first-generation daughter cells is 50 percent labelled with N^{15}. This suggests that in the daughter cells, the DNA had one helical strand from the parent cell which contained only N^{14}, and one newly synthesized strand which had only N^{15}-labelled bases. When these daughter cells were transferred to an all-N^{14} medium and allowed to replicate, their first generation had some DNA which contained entirely N^{14}, and some which had equal amounts of N^{14} and N^{15}. The pure-N^{14} DNA evidently came from the N^{14} strand in the parent serving as the template for synthesis of a new strand containing only N^{14}-labelled bases, while the N^{14}-N^{15} DNA came from combination of the N^{15} strand with new N^{14} nucleotides. These results are consistent with the DNA replication mechanism outlined above.

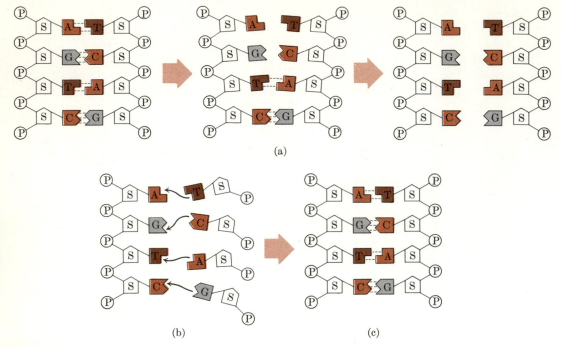

FIG. 18.25 A schematic representation of the DNA replication process. In step (a), the double-stranded helix separates. In step (b), free nucleotides complementary to those in the DNA strand are delivered and selected by base pairing, and then linked in (c) to give a completed DNA molecule. A, T, G, and C stand for adenine, thymine, guanine, and cytosine respectively, the sugar–phosphate chain by S — P — S.

Nucleic Acids and Protein Synthesis

As we have already remarked, the DNA in the cell nucleus contains in its base sequence the information necessary to direct specific protein synthesis. The sites of protein synthesis are the ribosomes, which are located on the endoplasmic reticulum. The genetic information is carried from the nuclear DNA to the ribosomes by messenger RNA. In addition, transfer RNA delivers the amino acids for protein synthesis to the ribosomes, and serves as the label by which each acid is recognized.

The relation between the molecules and processes involved in protein synthesis is indicated schematically in Fig. 18.26. In the cell nucleus the DNA serves as a template upon which ribonucleotides are deposited and messenger RNA synthesized. The base sequence in the *m*-RNA is complementary to the sequence in the part of the DNA template used.

As the transcription of the DNA code to *m*-RNA goes on in the nucleus, in the body of the cell specific enzymes attach amino acids to transfer RNA. The *m*-RNA and *t*-RNA move to the ribosomes where they interact. The *m*-RNA

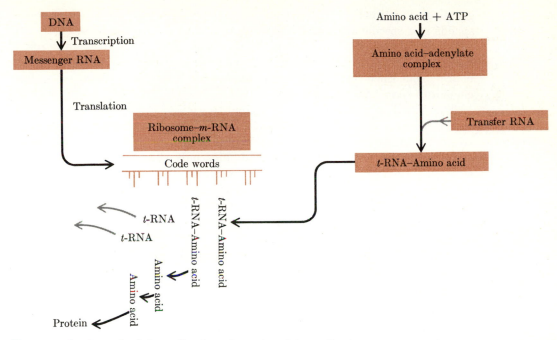

The general scheme for information transfer and protein synthesis.

selects the certain t-RNA and its amino acid by a base-pairing mechanism between the two molecules. Another selected t-RNA and its acid complexes on the m-RNA adjacent to the first acid, and a peptide bond is formed. The first t-RNA then leaves, freed of its acid. The process continues as the specific t-RNA and its acid appropriate for the third position in the peptide chain arrives, is recognized by base-pairing, and complexes with m-RNA. Acids continue to be added in this manner until a base sequence in the m-RNA is encountered that causes the synthesis to stop. It is known that polypeptide growth starts from the amino or nitrogen end of the peptide chain, with successive acids being added to the carboxyl end. In at least some cells, the growth of a chain is always triggered by a molecule of the amino acid methionine which has had a formyl group HCO— attached to the amino nitrogen. This formyl group blocks growth of the peptide chain at the nitrogen end, but allows subsequent acids to be added to the carboxyl end of the molecule.

The Genetic Code

We have seen that there are only four different kinds of bases in DNA molecules, and that the sequence of these bases must be able to determine uniquely the sequence of 20 amino acids in a protein chain. Since there are fewer types of bases than amino acids, it must be that various groupings of bases constitute genetic code words for the different acids. The code word in DNA which corresponds to a specific acid must consist of more than a base pair, since there are from four bases only 4^2 or 16 distinct types of pairs. This is not enough words to specify the 20 amino acids separately. The words of the code could consist of base triplets, since there are 4^3 or 64 distinct combinations possible, more than enough to specify 20 amino acids.

The actual code words that correspond to each of the 20 amino acids have been discovered by feeding synthetic RNA of known base composition to cells, and detecting the new polypeptides formed as a result. For example, in an early experiment a synthetic RNA containing only the base uracil ("poly U") was used, and a small amount of a polypeptide which contained only the amino acid phenylalanine was found. This suggested that the RNA base code word for phenylalanine is the sequence of three uracils, UUU. Similar experiments with other synthetic RNA's led to the discovery of the triplets for the other acids.

In other, separate work, it was discovered that trinucleotides of only three uracil bases (UUU) or three adenine bases (AAA) or three cytosines (CCC) were able to cause the binding of the transfer RNA of respectively phenylalanine, lysine, and proline to the ribosomal sites where polypeptide synthesis occurs. In contrast, the dinucleotide UU showed very little such activity. This observation constitutes a quite direct demonstration that the genetic code is indeed made up of triplets of bases.

18.8 CONCLUSION

From this brief introduction to biochemistry we can see that the beginnings of an understanding of the life process in terms of molecular properties and structure are well established. As this understanding deepens and becomes more complete, it may become possible to treat disease and aging as chemical problems, and to perform specific chemical alterations and construction of genes. The construction of specifically designed organisms is an inevitable result of this work. The area of biochemistry introduces, therefore, not only problems of immense scientific challenge, but opens up possibilities which have moral and humanistic implications that exceed any yet encountered by mankind.

SUGGESTIONS FOR FURTHER READING

Bennett, T. P., and E. Frieden, *Modern Topics in Biochemistry*. New York: Macmillan, 1966.

Kay, E. R. M., *Biochemistry*. New York: Macmillan, 1966.

Kopple, K. D., *Peptides and Amino Acids*. New York: Benjamin, 1966.

Light, R. J., *A Brief Introduction to Biochemistry*. New York: Benjamin, 1968.

Snell, F. M., S. Shulman, R. P. Spencer, and C. Moos, *Biophysical Principles of Structure and Function*. Reading, Mass.: Addison-Wesley, 1965.

Watson, J. D., *Molecular Biology of the Gene*. New York: W. A. Benjamin, 1965.

18.1 For the following classes of molecules give a brief description of the chemical composition, molecular structure and conformation, and the principal biochemical role:

a) lipids;
b) proteins;
c) carbohydrates;
d) nucleotides;
e) enzymes;
f) cytochromes.

18.2 Briefly describe the principal function of each of the following animal cell components:

a) lysosomes;
b) mitrochondria;
c) nucleus;
d) ribosomes;
e) membrane.

18.3 Explain the significance of the following:

a) the Krebs or citric acid cycle;
b) the Embden-Meyerhof glycolytic pathway;
c) coupled reactions in biochemistry;
d) purine-pyrimidine base pairing in DNA and RNA;
e) the base sequence in nucleic acids;
f) polyphosphate groups;
g) ATP.

18.4 List the four principal types of nucleic acids, and describe briefly their structure and function.

18.5 For each of the following molecules, describe briefly a principal biological function:

a) bile acids;
b) Coenzyme A;
c) NAD;
d) glycogen;
e) glucose;
f) triglycerides;
g) the porphyrin ring structure.

CHAPTER 19

THE NUCLEUS

Even though the nuclei retain their identities in chemical processes, and even though nuclear properties other than charge influence chemical behavior only in indirect and subtle ways, the nature of the nucleus is an important subject to chemists. The abundance of the elements and their origin is a problem in nuclear structure and reactivity. The synthesis of new elements not found in nature has been carried out primarily by chemists. The use of both radioactive and stable isotopes has aided in the determination of the mechanisms of chemical reactions and complex biochemical processes. Many of the problems that are associated with the use of nuclear reactions as sources of energy are chemical in nature. Thus there are ample reasons for all chemists to be familiar with nuclear properties and phenomena. In this chapter we shall examine the aspects of the nucleus that are of most importance in chemistry.

19.1 THE NATURE OF THE NUCLEUS

To begin, let us review some definitions and notation. Nuclei are composed of protons and neutrons, and thus these particles are often referred to as *nucleons*. The description of a particular nucleus is given in terms of its charge Z and its *mass number* A, which is the sum of its neutrons and protons. To represent a nucleus, the chemical symbol is written with a subscript equal to Z and a superscript equal to A. Thus

$$_8O^{16}, \quad _8O^{17}, \quad _8O^{18},$$

represent three *isotopes* of oxygen: nuclei with the same charge but different mass numbers.

Now we can discuss the general properties of the nucleus—its size, mass, shape, and the type of forces that hold it together.

Nuclear Size

The first indication of the size of the nucleus was the Rutherford α-particle scattering experiment, which we discussed in Section 10.2. The qualitative result of Rutherford's experiment is that α-particles can approach to within 10^{-12} cm of the center of an atom and still be scattered away by a force given by Coulomb's law. If, however, the energy of the bombarding α-particles is increased sufficiently, the intensity pattern of the scattered α-particles changes in a way which indicates that Coulomb's law of repulsion fails when α-particles come very close to the atomic center. The scattering pattern and other data indicate that the potential energy of an α-particle as a function of the distance from the atomic center can be represented as in Fig. 19.1. As the α-particle approaches the nucleus, there is an initial repulsive Coulomb force that causes the potential energy to rise until the α-particle is close enough to feel the very strong attractive nuclear forces. At this distance, which we might take as the nuclear radius, the potential energy drops abruptly. The increase in potential energy that an α-particle experiences as it enters or leaves the nucleus is often called the *Coulomb barrier*.

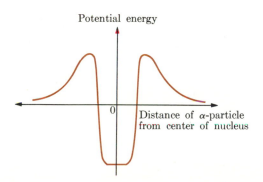

Potential energy

Distance of α-particle from center of nucleus

The potential energy of an α-particle as a function of its distance from the center of a nucleus.

FIG. 19.1

Because a neutron is uncharged, it experiences no Coulomb repulsion when it approaches a nucleus. Instead, the potential energy of a neutron remains essentially constant until it falls abruptly at a distance somewhat less than

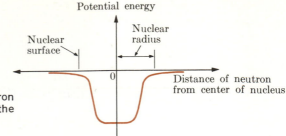

FIG. 19.2 The potential energy of a neutron as a function of its distance from the center of a nucleus.

10^{-12} cm from the atomic center. This behavior is represented in Fig. 19.2. As far as a neutron is concerned, the nucleus is a potential energy "well" with rather steep sides. Because the neutron experiences an abrupt change in potential energy at the nuclear surface, the pattern of neutrons scattered by nuclei can be used to determine the size of the nucleus. A large number of nuclear radii have been determined by neutron scattering, and the results can be summarized by the following equation:

$$R = R_0 A^{1/3},$$
$$R_0 = 1.33 \times 10^{-13} \text{ cm.} \tag{19.1}$$

In Eq. (19.1), R is the nuclear radius, A is the mass number, and R_0 is a constant common to all nuclei.

We can draw an interesting conclusion from the dependence of the nuclear radius on the mass number. The nuclear volume V should be proportional to R^3 or, according to Eq. (19.1), to A:

$$V \propto R^3 \propto A.$$

Thus the nuclear volume is directly proportional to the total number of neutrons and protons in the nucleus. This fact suggests that protons and neutrons pack together somewhat like hard spheres and make the total nuclear volume equal to the sum of the volumes of individual protons and neutrons. We shall encounter other evidence that is consistent with this simple picture, but it should not be overinterpreted. The nucleons in the nucleus are not stationary, stacked like oranges, but they contribute to the nuclear volume as though they were.

Nuclear Shape

A perfectly spherical nucleus exerts an electrical force on the atomic electrons that is given exactly by the Coulomb law expression. However, if the protons in the nucleus are not grouped in a spherical shape, the nucleus is said to have an electric quadrupole moment, and the surrounding electrons feel, in addition to the Coulomb attraction, a small electric quadrupole force. While the effects of the nuclear quadrupole are small, they can be detected, and a measure of the quadrupole moment and the shape of the nucleus obtained. Several nuclei have zero quadrupole moment and thus are spherical: $_1$H, $_8$O, $_{20}$Ca, $_{28}$Ni, $_{50}$Sn, and $_{82}$Pb are a few examples. Nuclei which have atomic numbers close to one of those in the preceding list are either spherical or very nearly so, while the majority of other nuclei are slightly prolate spheroids (football shaped). The departure from spherical shape is never very extreme: the ratio of the semi-major axis to the semiminor axis is generally less than 1.2.

Nuclear Masses

The unit of nuclear mass is the atomic mass unit, or amu, which is defined to be exactly $\frac{1}{12}$ of the mass of a $_6$C^{12} atom. On this scale, a neutron has a mass of 1.00866544, while the mass of a hydrogen atom (proton plus electron) is 1.00782522. In a discussion of nuclei we might expect to be concerned primarily with nuclear masses, but it is the masses of *atoms* (nucleus and electrons) that are determined experimentally and tabulated. This does not introduce any serious complications, as we shall see subsequently.

Because both the neutron and the hydrogen atom have a mass of nearly 1 amu, the masses of the various isotopes of atoms are all near integral values. In fact, the observation of integral masses was the basis for the original suggestion that nuclei are made up of neutrons and protons. A careful comparison of the mass of any atom with the sum of the masses of its constituent hydrogen atoms and neutrons reveals an interesting mass deficiency. Consider, for example, the atom $_8$O^{16}, which has a mass of 15.994915 amu. In contrast, the mass of eight neutrons and eight hydrogen atoms together is 16.131925 amu. Therefore, $_8$O^{16} is lighter than we might expect by (16.131925 − 15.994915) amu or 0.137010 amu. By the mass-energy equivalence expressed by the Einstein relation

$$E = mc^2, \tag{19.2}$$

the mass deficiency of the oxygen atom can be attributed to the energy evolved or lost by the system when eight neutrons, protons, and electrons are formed into an $_8$O^{16} atom. In Eq. (19.2), c is the velocity of light; if it is expressed in centimeters/second, and the mass m in grams, the units of energy E are ergs. Thus 1 gram-mass is equivalent to about 9×10^{20} ergs, or 2.1×10^{10} kcal. Another mass-energy equivalence that is used more commonly is the following: 1 amu is equivalent to 931.4 million electron-volts energy (931.4 Mev). The energy released when one oxygen atom is formed from its neutrons, protons, and electrons is, therefore, 0.137010×931.4 or 127.6 Mev *per atom*. To appreciate how large an amount of energy this is, compare it to the 5.2-ev energy released when two oxygen atoms form a strong chemical bond with each other. In general, the energies associated with nuclear processes are roughly a million or more times as great as the energies involved in chemical phenomena.

By comparing the measured mass of an atom with the sums of the masses of its constituent neutrons, protons and electrons, we can calculate the total energy E_b which binds the nucleus together. More instructive than the total binding energy, however, is the binding energy per nucleon, E_b/A, which is plotted in Fig. 19.3 as a function of mass number. After an abrupt rise among the lightest nuclei, the binding energy per nuclear particle changes only slightly and has a value of approximately 8 Mev per nucleon. The nuclei of maximum stability have mass numbers of about 60 or charges of about 25. Because of the maximum in the binding energy per particle that occurs near mass 60, the *fission* of a very heavy nucleus to a pair of nuclei of approximate mass 60 is a process that releases energy. Similarly, the *fusion* of two of the lightest nuclei is also accompanied by release of energy.

Because there are only slight variations in the binding energies per nucleon for elements of mass number greater than 20, we can say that, to a first approximation,

$$\frac{E_b}{A} \cong \text{constant}$$

or

$$E_b \cong \text{constant} \times A.$$

That is, the total binding energy of a nucleus is approximately proportional to the number of nucleons. This observation suggests that the forces that bind the nucleons together are of short range; that is, one nucleon exerts attractive forces only on its nearest neighbors. If the nuclear forces were of long range, each of A nucleons would be attracted to $A - 1$ others, and the total nuclear binding energy would be proportional to $A(A - 1)$, instead of to A.

Besides being of short range, the *attractive* forces between nucleons are independent of charge. There is, however, a Coulomb repulsion between protons, so the *net* binding energy of two protons is less than that of two neutrons. When a correction is made for the Coulomb repulsions between protons, the *attractive* nuclear binding energy is found to be 14.1 Mev per particle; it is the Coulomb repulsion between protons that reduces this to the *net* value of 8 Mev per particle we discussed earlier.

Now let us suppose that in the nucleus, the neutrons and protons interact as though they were close-packed spheres, each having 12 nearest neighbors. This would mean that there exists $\frac{12}{2}$, or 6, nucleon bonds or attractions per particle, since two particles are required to make one bond. Thus we might interpret the binding energy of 14.1 Mev per particle to mean that the energy of attraction between a single pair of nucleons is 14.1/6 or 2.3 Mev. There is one nucleus, $_1\text{H}^2$, in which there is only one nucleon-nucleon attraction, and the binding energy is in fact 2.2 Mev. Thus the crude idea that nucleons interact only with their 12 nearest neighbors seems to be substantiated, at least approximately.

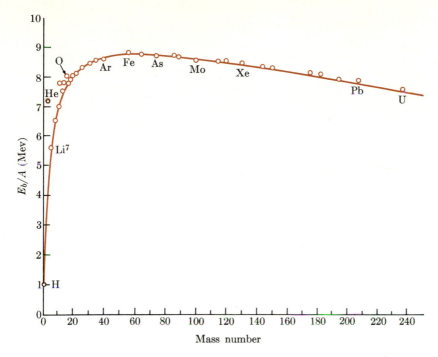

Binding energy per nucleon as a function of mass number for some stable nuclei.

FIG. 19.3

While they tell us nothing about the origin of nuclear forces, the foregoing observations do suggest the qualitative nature of the nucleon-nucleon interaction, and how it affects the stability of the nucleus. In 1935 Von Weizsäcker devised a semiempirical expression for the nuclear binding energy which is based largely on the qualitative picture we have outlined. The total nuclear binding energy is given, to a good approximation, by

$$E_b = 14.1A - 13A^{2/3} - \frac{0.6Z^2}{A^{1/3}}. \tag{19.3}$$

The first term on the right-hand side expresses the fact that the attractive binding energy is 14.1 Mev per particle. Nucleons at the nuclear surface do not have their full complement of 12 nearest neighbors, however, and therefore do not contribute a full 14.1 Mev to the nuclear binding energy. The number of surface nucleons is proportional to the surface area, which in turn varies as the nuclear radius squared, or as $A^{2/3}$. Consequently, the term $-13A^{2/3}$ appears in the binding energy equation; it is negative to represent the loss of binding energy due to surface effects. Finally, there is the Coulomb repulsion between protons, which also constitutes an energy loss, and its effect is represented by the term $-0.6Z^2/A^{1/3}$. Coulomb repulsion increases as the square of the number of protons, and the energy loss is inversely proportional to the nuclear radius, or to $A^{1/3}$.

While Eq. (19.3) shows how some of the gross nuclear properties affect the binding energy, it does not, by any means, provide a complete picture of nuclear binding. Other more subtle effects are important. For example, nuclei with even numbers of neutrons and even numbers of protons seem to be particularly stable. Table 19.1 shows the distribution of stable nuclei.

Table 19.1 Frequency of occurrence of stable nuclear types

A	N	Z	Number of nuclei
Even	Even	Even	166
	Odd	Odd	8
Odd	Even	Odd	57
	Odd	Even	53

There are only eight stable nuclei which have an odd number of both neutrons and protons, and most stable nuclei are of the even-even type. These observations suggest that there is a separate pairing of neutrons and of protons that affects nuclear stability. There is an even more specific effect having to do with numbers of protons and neutrons. Nuclei which have the "magic" neutron or proton numbers 2, 8, 20, 28, 50, 82, 126 are particularly stable and abundant in nature. The existence of these magic numbers suggested a "shell model" of the nucleus: an energy-level scheme somewhat analogous to the orbital energy-level scheme used for atomic electrons, and this idea has led to successful predictions of a number of nuclear properties.

19.2 RADIOACTIVITY

We have already mentioned one form of natural radioactivity, the spontaneous fission of a very heavy nucleus into two more stable fragments of mass number near 60. Spontaneous fission is rather uncommon, and most spontaneously radioactive nuclei decay by emitting either an α-particle, a positive or negative β-particle, a γ-ray, or by capturing an orbital electron. Some simple rules for predicting the occurrence and nature of radioactivity can be generated by reference to Fig. 19.4. By plotting the charge Z as a function of the number of neutrons N for all the nonradioactive nuclei, we find that the stable nuclei fall in a well-defined belt. For nuclei lighter than $_{20}Ca^{40}$, most stable nuclei have equal numbers of neutrons and protons. Among the heavier elements, the most stable nuclei contain more neutrons than protons. It appears that the cause of this behavior is the excessive Coulomb repulsion that occurs in nuclei of high charge, and this can be diminished somewhat by increasing the neutron number and increasing nuclear size.

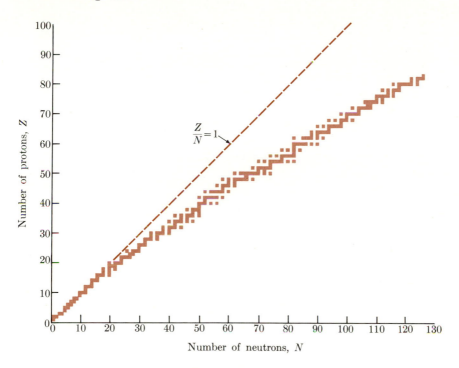

The number of protons as a function of the number of neutrons for the stable nuclei.

FIG. 19.4

Beta-Decay Processes

Nuclei that lie outside of the belt of stability are radioactive and decay in a manner that forms a nucleus which lies in the stable region. Nuclei that lie to the right of the stability belt in Fig. 19.4 are neutron rich and achieve stability by emitting negative β-particles, or electrons. This process can be pictured as a transformation of a neutron in the nucleus to a proton and an electron which is emitted. The nucleus that results has one more proton and one less neutron than its parent and lies closer to the belt of stability. In contrast, nuclei which lie to the left of the stable region in Fig. 19.4 must diminish their positive charge to achieve stability. Two processes are possible: the first is the capture of an orbital electron (K-capture) followed by conversion of a proton to a neutron;

the second possibility is the emission from the nucleus of a positron, or positive electron, which results in the conversion of a nuclear proton to a neutron. As examples of the foregoing processes we have:

$_6C^{10}$ $_6C^{11}$	$\boxed{_6C^{12}\ \ _6C^{13}}$	$_6C^{14}$ $_6C^{15}$
positron emission	stable	negative β-emission

$_8O^{14}$ $_8O^{15}$	$\boxed{_8O^{16}\ \ _8O^{17}\ \ _8O^{18}}$	$_8O^{19}$
electron capture	stable	negative β-emission

A spontaneous β-decay process releases energy, and although the nucleus does not undergo a change in mass number, there is a decrease in its mass. As an illustration, consider

$$_6C^{14} \rightarrow\ _7N^{14} +\ _{-1}\beta^0.$$

To calculate the energy released in this process, we have only to compare the mass of a $_6C^{14}$ *atom* with the mass of a $_8N^{14}$ *atom*, because in the decay, a carbon atom with six orbital electrons is converted to a nitrogen *ion* with six orbital electrons *and* a β-particle. The total mass of these products is, therefore, equal to the mass of a $_7N^{14}$ atom. The mass of $_7N^{14}$ is 14.003074 amu and that of $_6C^{14}$ is 14.003242. The mass difference, 1.68×10^{-4} amu, corresponds to 0.155 Mev, the total energy released in the β-decay process.

Calculation of the energetics of a positron-decay process requires some care. We can write the decay of $_6C^{11}$ as

$$\underset{\substack{\text{nucleus} +\\ \text{6 electrons}}}{_6C^{11}} \quad \rightarrow \quad \underset{\substack{\text{nucleus} + \text{6 electrons}\\ + \text{positron}}}{_5B^{11} +\ _{+1}\beta^0.}$$

Thus the total mass of the products is equal to the mass of a $_5B^{11}$ *atom*, plus the mass of the extra orbital electron not used by boron, plus the mass of the positron. The energy equivalent to the electron or positron mass is 0.511 Mev, so the energy of the positron-emission process is

$$(\text{mass}\ _6C^{11} - \text{mass}\ _5B^{11}) \times 931 - 2 \times 0.511\ \text{Mev},$$

or

$$(11.011443 - 11.009305)(931) - 1.022 = 0.156\ \text{Mev}.$$

Because a positron decay produces two "extra" particles equivalent to 1.022-Mev energy, the energy released by positron emission is less than the atomic-mass differences. Unless the atomic-mass difference exceeds 1.022/931 or 1.098×10^{-3} amu, spontaneous positron emission is not possible.

After an orbital electron-capture process such as

$$_8O^{15} \xrightarrow[\text{capture}]{\text{electron}} {_7N^{15}},$$

the correct number of electrons is associated with the newly formed nucleus, so the energy evolved can be calculated directly from the atomic masses. For the example at hand, we have 15.003072 for the mass of $_8O^{15}$, and 15.000108 for the mass of $_7N^{15}$. Consequently, the energy released is 2.76 Mev.

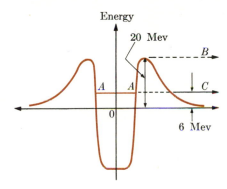

FIG. 19.5

Energy diagram for α-particle emission. Level *A-A* represents the energy of the α-particle in the nucleus. Level *B* represents the kinetic energy the α-particle would have if it passed over the Coulomb barrier. Level *C* represents the kinetic energy of an α-particle that has tunneled through the Coulomb barrier.

Alpha-Decay Processes

With few exceptions decay by α-particle emission occurs only among elements with mass numbers greater than 200. A typical example of α-decay is

$$_{92}U^{238} \rightarrow {_{90}Th^{234}} + {_2He^4}.$$

We see that the nuclear mass number decreases by four, and the nuclear charge by two units in this process. A particularly intriguing feature of α-decay is the observation that the energies of the emitted α-particles all lie between 3 and 9 Mev. The reason this is interesting can be understood with the help of Fig. 19.5, which shows the potential energy of interaction between an α-particle and a nucleus. Apparently, in order to be emitted, an α-particle in the nucleus should have enough energy to surmount the Coulomb potential-energy barrier, and after the α-particle has left the nucleus, the Coulomb repulsion should accelerate it to a kinetic energy equal to the barrier height, which is 20 Mev or more. This amount of energy is much greater than the largest value observed experimentally. This discrepancy can be removed if the behavior of the α-particle is described in terms of quantum mechanics. From this point of view, there is a finite probability that the α-particle will escape from the nucleus, even though it has insufficient energy to "pass over" the potential-energy barrier. In effect, the α-particle behaves as though it can "tunnel" through the barrier at an energy level below its maximum and thus acquire an energy which is less than 20 Mev as it departs from the nucleus. The mathematical analysis of this "tunneling" phenomenon leads to the prediction that the narrower the potential-energy barrier, the more probable and frequent is the emission of the α-particle. Because the width of the nuclear barrier decreases as energy increases, we can expect that those nuclei which undergo α-decay most frequently also emit the most energetic α-particles. This correlation between emission frequency and energy is observed experimentally.

Gamma-Decay Processes

Frequently the daughter nucleus formed by α- or β-decay of its parent is produced in an excited state. The newly formed nucleus releases this excitation energy by emitting a γ-ray; that is, electromagnetic radiation of extremely short wavelength. A nucleus can have only discrete energies which are determined by its structure. Therefore, it can only emit γ-rays that have energies equal to the difference in the energy of two nuclear levels. Consequently, an excited nucleus has a discrete emission spectrum of γ-rays, just as an atom has a characteristic emission spectrum of visible and ultraviolet radiation. By determining the energies of emitted γ-rays, the energy-level pattern of a nucleus can be deduced, at least in part. For example, consider the data summarized in Fig. 19.6. The nucleus $_{92}U^{238}$ emits α-particles which have energies of either 4.18 Mev or 4.13 Mev. When a 4.13-Mev α-particle is emitted, the daughter nucleus, $_{90}Th^{234}$, is left in an excited state which has an energy $(4.18 - 4.13)$ or 0.05 Mev greater than the state reached when a 4.18-Mev α-particle is emitted. Consequently, it might be expected that the excited nucleus should emit a 0.05-Mev γ-ray, and this is indeed found experimentally. In other decay pro-

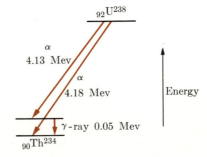

FIG. 19.6   The decay scheme of $_{92}U^{238}$ to $_{90}Th^{234}$; α-particles of 4.18 and 4.13 Mev and a γ-ray of 0.05 Mev are observed.

cesses, several excited states of the daughter nucleus are formed, and a number of γ-rays of different energies are emitted. In such cases, complete analysis of the energies of the emitted particles and γ-rays allows the nuclear chemist to construct a detailed energy-level pattern for the daughter nucleus.

Interaction of Radiation with Matter

The α-, β-, and γ-particles emitted in nuclear decay processes are highly energetic species which can cause substantial chemical alterations in the matter with which they interact. All three types of radiation cause the electronic excitation and ionization of atoms and molecules. The electrons produced by these primary ionization processes, in general, have high kinetic energy and can themselves cause further ionization and excitation.

The electronic excitation of a molecule may lead to its dissociation into atoms or free radicals, or may cause it to react directly with other molecules. The atomic and molecular ions produced by radiation are also usually very reactive, and this may have important chemical consequences. For example, irradiation of a mixture of H_2 and D_2, a very simple chemical system, induces a number of chemical reactions which lead to the formation of HD. Some of the more important reactions are

$$H_2 + \alpha \longrightarrow H_2^+ + \alpha + e^- \text{ (fast)},$$
$$e^- + H^2 \rightarrow H_2^+ + 2e^-,$$
$$H_2^+ + D_2 \rightarrow H_2D^+ + D,$$
$$H_2D^+ + H_2 \rightarrow H_3^+ + HD,$$
$$H_2D^+ + e^- \text{(slow)} \rightarrow HD + H.$$

In more complex chemical systems, the variety of ions, excited molecules, atoms, and free radicals which are produced by the primary nuclear radiation and secondary electrons can be much greater, and the resulting chemical changes more complicated. Biochemical systems, with their very complex and often delicate molecules, are particularly susceptible to deleterious alteration by radiation.

The human body is fairly well protected from the effects of certain types of radiation, so long as the radiation sources are not ingested and incorporated in vital organs. Even the most energetic α-particles can travel only a very short distance in condensed matter before they lose their initial kinetic energy and become converted to harmless helium atoms. Thus α-particles are stopped without effect by the outermost layer of the skin. However, if an α-particle emitter like Pu^{239} is ingested, it tends to concentrate in the bones where its α-emission can interfere with the red blood cell production. Consequently, plutonium is one of the most deadly poisons known, and has a maximum tolerated dose of only 0.7 microgram.

In a similar manner, the skin can protect the human body from the more serious effects of β-particles of moderate energy. However, severe burns can result from exposure to intense external β-radiation, and ingestion of β-emitters such as Sr^{90} and H^3 can be very serious. The skin does not provide protection from x-rays, γ-rays, and neutrons, which do penetrate the body, and can induce changes in internal organs.

It is clear that some measure of the amount of radioactivity in a given chemical system is needed. Since the emission of an α-, β-, or γ-particle corresponds to the alteration or "decay" of a nucleus, the activity of a source can be expressed in terms of the number of decays which occur per unit time. The standard unit of radioactivity is the *curie*, which corresponds to 3.70×10^{10} disintegrations per second. The natural radiation in 1 cm^3 of air is about 10^{-10} microcurie (mostly radon), while milk contains approximately 5×10^{-8} microcuries of K^{40} in 1 cm^3. The human body has a total activity of about 0.1 microcurie, due to ingestion of naturally occurring K^{40} and C^{14}. In contrast, the total activity in the projected large breeder nuclear reactors would be approximately 10^{10} curies.

While the curie is the unit of nuclear radioactivity, other units are used to measure the amount of radiation necessary to produce a given effect in matter. The first of these units is the *roentgen*, which is defined as the quantity of x-rays or γ-radiation which produces in 1 cm^3 of air positive ions of total charge equal to one electrostatic unit. This corresponds to the creation of 2.1×10^9 singly charged ions, since the fundamental electronic charge is 4.8×10^{-10} esu. A luminous dial watch produces approximately 30 milliroentgens (mr) per year, while a dental or chest x-ray involves 5000 mr of radiation. The roentgen is applied only to x- and γ-radiation. To measure the effect of all types of radiation, the *rad* unit is used. One rad is defined as the amount of radiation which will deposit 100 ergs of energy in each gram of material. The energy deposited in 1 gm of water or tissue by 1 roentgen is approximately 90 ergs. Thus 1 roentegen is equivalent to a radiation dose of 0.9 rad in body tissue.

A dose of several hundred rad over the entire human body results in death within a few weeks. For doses of 100 rads, the immediate death rate is effectively zero, but such heavy exposures may have delayed ($\sim$20 years) consequences, such as leukemia, cancer, and a general acceleration of the aging process. It is very difficult to define the maximum dose of radiation which can be tolerated without a substantial chance of long term damage. It has never been established that doses of less than 50 rads actually do lead to cancer in humans, but this may only be a consequence of the limited number of cases of this type that have been studied. By extrapolating the effects of high dosages down to lower dose rates, the International Commission on Radiological Protection has established maximum permissible doses (MPD) of radiation for the limited group of people who work with radioactive material. Currently, the MPD is 5 rads per year. However, for the population as a whole, it is recommended that exposure be limited to less than 0.2 rad per year. Nuclear reactors used for energy sources must be designed to be consistent with this exposure limitation.

In 1919, Rutherford achieved the first artificial transmutation of an element by bombarding a sample of nitrogen with α-particles from a radium source. The reaction was

$$_7N^{14} + {_2}He^4 \rightarrow {_8}O^{17} + {_1}H^1,$$

and Rutherford was able to detect the emitted protons. Because α-particles available from naturally radioactive sources have a limited range of energies, they can induce relatively few nuclear reactions. The development of particle accelerators like the cyclotron and its various modifications has made it possible to produce relatively intense beams of energetic particles, and a very large number of nuclear reactions have been studied. One of the best known achievements in this area has been the synthesis of the trans-uranium elements. Here are four such nuclear reactions which illustrate the use of four different bombarding particles:

$$_{92}U^{238} + {_1}H^2 \rightarrow {_{93}}Np^{238} + 2{_0}n^1,$$
$$_{92}U^{238} + {_2}He^4 \rightarrow {_{94}}Pu^{239} + 3{_0}n^1,$$
$$_{92}U^{238} + {_6}C^{12} \rightarrow {_{98}}Cf^{246} + 4{_0}n^1,$$
$$_{92}U^{238} + {_7}N^{14} \rightarrow {_{99}}Es^{247} + 5{_0}n^1.$$

The compound nucleus formed by the combination of the target and bombarding nuclei is energy-rich, and this results in the ejection of one or more neutrons. Heavy nuclei formed by bombardment with $_2He^4$, $_6C^{12}$, or $_7N^{14}$ followed by neutron emission have a deficiency of neutrons, and therefore subsequently undergo electron-capture or positron-emission processes.

The quantities of material that can be transmuted by charged-particle bombardment is always severely limited by the intensity of the particle beam and sometimes by the amount of target material. The first synthesis of $_{101}Md^{256}$ was achieved by bombarding 10^9 atoms of $_{99}Es^{253}$ with α-particles, and only 13 atoms of $_{101}Md^{256}$ were detected in the products. To synthesize radioactive isotopes in quantity, neutron-capture reactions are useful, for high fluxes of neutrons exist in nuclear reactors. For example, tritium, $_1H^3$, can be produced by two neutron reactions:

$$_5B^{10} + {_0}n^1 \rightarrow {_1}H^3 + 2{_2}He^4,$$
$$_3Li^6 + {_0}n^1 \rightarrow {_1}H^3 + {_2}He^4.$$

The nucleus $_{27}Co^{60}$, whose decay product emits γ-rays useful in cancer therapy, is also a product of a neutron-capture reaction,

$$_{27}Co^{59} + {_0}n^1 \rightarrow {_{27}}Co^{60}.$$

The most famous of nuclear reactions is the fission of $_{92}U^{235}$, induced by neutron capture. There are no unique products of this reaction; the fission produces fragments whose mass numbers range from approximately 70 to 160.

One fission process for $_{92}U^{235}$ is

$$_{92}U^{235} + _0n^1 \rightarrow _{38}Sr^{90} + _{54}Xe^{143} + 3_0n^1.$$

Approximately 50 other modes of fission occur, and varying amounts of energy and different numbers of emitted neutrons are associated with each process. On the average, however, approximately 200 Mev of energy and 2.5 neutrons are released in the fission of $_{92}U^{235}$. The intense radioactivity of the fission products makes their chemical identification difficult. Nevertheless, these data have been obtained and are displayed in Fig. 19.7 by plotting the logarithm of the percentage yield as a function of the mass number. The two maxima of the yield curve show that the majority of the fissions occur asymmetrically to yield two fragments of rather different mass numbers.

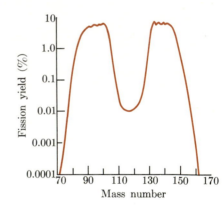

FIG. 19.7   Yield curve for the fission of $_{92}U^{235}$ induced by slow neutrons.

Energy from Nuclear Reactions

Because more than one neutron is emitted per fission process on the average, it is possible for the fission of U^{235} to be self-maintaining. If neutrons are not lost in other ways, the fission of one nucleus can induce the fission of two or three others, and so on. When carried out in a controlled manner in a nuclear reactor, this chain fission process is a valuable energy source, for the 200 Mev per fission process amounts to 5×10^9 kcal/mole of U^{235}. As of 1972, there were 30 nuclear power plants in the United States, and these generated 4 percent of the nation's electrical power. However, another 130 nuclear power plants were either under construction or licensed by the Atomic Energy Commission.

In a conventional nuclear reactor used for power generation, the fuel is uranium oxides which have been enriched to contain approximately 3 percent U^{235}. The enrichment is necessary, since the natural abundance of the fissionable U^{325} is only 0.7 percent of the heavier U^{238}, which is not fissionable. The uranium oxides are hermetically sealed in tubes of a zirconium alloy, and assemblies of these tubes are mounted within a heavy-walled steel vessel, as shown in Fig. 19.8. Water at a pressure of 150 atm is used to remove the heat generated in the fuel rods, and transfer it to a steam generator. The water also serves to moderate the energy of the fast neutrons emitted by the fission process, and convert them to slow neutrons which are more effective in inducing fission of U^{235}.

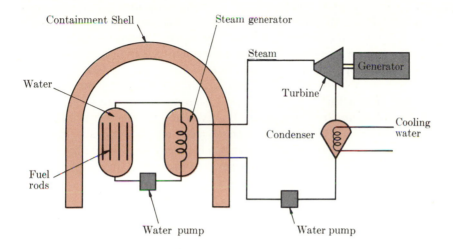

Schematic diagram of a pressurized water nuclear reactor.

FIG. 19.8

In the steam generator, heat is transferred from the pressurized water to a secondary water system which operates at 50 atm. Because of the lower pressure in the secondary system, the water in it is converted to steam at approximately 260°C, and this steam is used to drive a turbine-electric generator combination. The effluent from the turbine is condensed and pumped back to the steam generator.

Application of the second law of thermodynamics shows that the maximum efficiency η with which the heat delivered to the steam generator can be converted to useful work is given by

$$\eta = \frac{T_h - T_c}{T_h},$$

where T_h is the absolute temperature of the steam generator, and T_c is the temperature of the condenser. Taking T_h and T_c as 530°K and 330°K, respectively, we find that η is 0.38. Because of inevitable friction and heat losses, the actual operating efficiency is only about 32 percent. Power plants which burn oil or coal generate steam at approximately 800°K, and therefore have significantly higher efficiencies than conventional pressurized water nuclear power plants. Significant increases in the thermal efficiency of pressurized water reactors are not foreseeable, since the necessary increase in temperature would require the primary water system to be pressurized at impractical levels to prevent boiling.

Besides its low thermal efficiency, the pressurized water reactor has the disadvantage that it consumes U^{235}, an isotope which has a rather small terrestrial abundance. The estimated reserves of U^{235} are small enough so that this resource might be exhausted in 30 years if as much as 50 percent of the nation's electricity were generated by pressurized water reactors. Consequently, so-called fast breeder reactors, which generate more fissionable material than they consume, are under development as power sources.

The fast breeder reactor is fueled with a mixture of plutonium and the abundant U^{238}. When Pu^{239} absorbs a fast neutron, it undergoes fission, and produces a pair of lighter nuclei and, on the average, 2.5 fast neutrons. The fission chain reaction is maintained if one of these neutrons is absorbed by another Pu^{239} nucleus. If another of these fast neutrons is absorbed by a U^{238} nucleus, it induces a sequence of transformations which result in the *production* of a Pu^{239} nucleus. That is, the sequence

$$_{92}U^{238} + _{0}n^{1} \rightarrow _{92}U^{239},$$
$$_{92}U^{239} \rightarrow _{93}Np^{239} + _{-1}\beta^{0},$$
$$_{93}Np^{239} \rightarrow _{94}Pu^{239} + _{-1}\beta^{0}$$

exactly replaces the plutonium consumed in the first fission reaction. Since two or more neutrons are produced in each fission, the reactor can produce more Pu^{239} than it consumes. Reactors currently under development are expected to operate so as to double the amount of their fissionable material in periods of 5 to 20 years.

Since it is not necessary to moderate the energy of the neutrons in a breeder reactor, water is not used as a coolant and heat transfer agent. Instead, either helium gas or liquid sodium is used for this prupose. Since liquid sodium boils at 880°C, the reactor can be operated at a high temperature, and steam generated at approximately 800°K. Therefore, the fast breeder reactor cooled by liquid sodium has a thermal efficiency which is comparable to that of the best conventionally fueled steam power plants. Based on considerations of fuel availability and efficiency alone, breeder reactors could, if necessary, supply electric power at present consumption rates for tens of thousands of years.

There are very substantial problems associated with the use of breeder reactors. The amount of plutonium in the reactor at any time may be approximately 10^{6} gm, and since the maximum dose of plutonium tolerated by the human body is less than 10^{-6} gm, it is extremely important to construct a reactor in such a way that the fuel and other radioactive materials will be contained even in the event of a serious accident. In addition, the recovery of the plutonium formed in the breeding process necessitates periodic removal of the fuel rods and transport to a processing plant where the plutonium can be separated from highly radioactive fission products. Finally, there is the necessity of safely disposing or storing these radioactive waste products until their activity has diminished to a safe level. Since the storage periods required may be of the order of one thousand years, this problem is extremely formidable.

Because of the problem of the safe disposal of radioactive waste products associated with nuclear fission reactors, there is considerable interest in developing methods of using nuclear *fusion* as an energy source. As Fig. 19.3 shows, the binding energy per nucleon in light nuclei such as $_1H^2$, $_1H^3$, $_3Li^6$, and $_3Li^7$ is rather small. However, there is a general increase in the binding energy per nucleon as the mass number of the nucleus increases. Thus we can expect that the fusion of two of these light nuclei to form a heavier nucleus will be accompanied by release of energy. As important examples, we have the fusion reactions of the "heavy" hydrogen isotopes, deuterium ($_1H^2$) and tritium ($_1H^3$):

$$_1H^2 + {}_1H^2 \rightarrow {}_2He^3 + {}_0n^1 + 3.27 \text{ Mev}$$
$$\rightarrow {}_1H^3 + {}_1H^1 + 4.03 \text{ Mev},$$
$$_1H^2 + {}_1H^3 \rightarrow {}_2He^4 + {}_0n^1 + 17.6 \text{ Mev}.$$

While the energy released in each of these fusion reactions is much smaller than the average of 200 Mev released by the fission of U^{235} and Pu^{239}, the abundance of deuterium is so great that controlled fusion could supply the energy needs of the Earth for many millions of years.

In order for two deuterium nuclei to fuse into a heavier nucleus, they must collide with sufficient kinetic energy to pass over or tunnel through the potential energy barrier imposed by their mutual Coulomb repulsion. The practical consequence of this requirement is that if fusion is to occur in a homogeneous hot gas of deuterons, the effective temperature of the gas must be approximately $10^8 °K$. This corresponds to an average energy of about 10^{-2} Mev for the deuterium nuclei. Such very high temperatures are reached in stars where fusion reactions of light nuclei are the source of the radiated energy, and in nuclear fission explosions used to trigger nuclear fusion bombs. However, attaining these temperatures in a controlled manner which would permit conversion of the energy from fusion reactions into useful work has proved to be an exceedingly difficult problem.

Since a gas at $10^8 °K$ is totally ionized to electrons and nuclei, it can be confined by strong magnetic fields which prevent charged particles from moving perpendicular to the magnetic field direction. In addition, the actual generation of these confining magnetic fields can be used to heat a partially ionized gas up to the ignition temperature for fusion. In this manner, ignition of the reaction

$$_1H^2 + {}_1H^3 \rightarrow {}_2He^4 + {}_0n^1 + 17.6 \text{ Mev}$$

has been achieved in a number of laboratories since 1963. However, interactions between the ionized gas (the plasma) and the confining magnetic fields have produced plasma instabilities which have limited the length of time and the density of the ionized gas that can be contained. It is generally believed that in order to be able to extract useful energy from a thermonuclear plasma, the product of the ion density (in ions/cm^3) and confinement time (in sec) must exceed 10^{14} ion sec/cm^3. As of 1973, values of 6×10^{11} ion sec/cm^3 (3×10^{13} ions/cm^3, 2×10^{-2} sec confinement) had been achieved, and devices intended to reach the critical value of 10^{14} ion sec/cm^3 were under construction.

Even when useful ion temperatures, densities, and confinement times are achieved, there will remain the problem of converting the fusion energy into useful work. Since the energy from the most easily ignited fusion reaction

$$_1H^2 + {}_1H^3 \rightarrow {}_2He^4 + {}_0n^1 + 17.6 \text{ Mev}$$

appears principally as kinetic energy of the neutron, one proposal is to surround the fusion reactor with molten lithium. In this lithium blanket the reactions

$$_3Li^7 + {}_0n^1 \text{(fast)} \rightarrow {}_3Li^6 + 2{}_0n^1,$$
$$_3Li^6 + {}_0n^1 \rightarrow {}_2He^4 + {}_1H^3 + 4.8 \text{ Mev}$$

would occur, with the energy evolved producing a temperature increase in the molten lithium. The heated lithium would be used to generate steam, and the tritium would be extracted and used as fusion fuel. Such a mode of operation would consume lithium, but the supply of this element is sufficient to meet the energy needs for as much as a million years. It has also been proposed to convert the kinetic energy of the proton released in the reaction

$$_1H^2 + {}_1H^2 \rightarrow {}_1H^3 + {}_1H^1 + 4.3 \text{ Mev}$$

directly to electrical energy. If this could be achieved, greater efficiency of energy conversion would be realized. In addition, it would not be necessary to generate tritium fuel from lithium, and thus the energy supply would be determined by the virtually infinite amount of deuterium in sea water.

In addition to the enormous energy resources that could be made available by controlled nuclear fusion, there are other attractive features of this process. No long-lived radioactive materials are produced in quantity, so the problem of waste disposal is virtually nonexistent. The tritium which would be produced in the lithium blanket would eventually be consumed, and the amount present in the reactor at any time would be very much smaller than the amount of radioactive materials in a fission reactor. A fusion reactor would be intrinsically much safer than fission reactors, since there would be no danger of nuclear explosions, and in the event of failure of any component, the fusion reaction could be terminated essentially instantaneously.

The proton is the most abundant nucleus in the Universe, and is the principal constituent of the visible stars. Stars are formed by the gravitational collapse of enormous clouds of gaseous hydrogen atoms and other matter. If enough mass is involved, the energy released by the collapse is large enough to initiate nuclear reactions. After nuclear reactions start to occur, they produce enough energy to raise the internal temperature and pressure sufficiently to counteract the gravitational forces and stop the star from contracting. Then there follows a long period in which nuclear reactions convert the hydrogen in the star to heavier elements. The analysis of this element-building process in stars is of considerable interest, since it may reveal clues to the origin and early evolution of the Universe.

The first step in the stellar conversion of hydrogen to heavier elements is postulated to be

$$_1H^1 + {}_1H^1 \rightarrow {}_1H^2 + {}_{+1}\beta^0.$$

Although this reaction has never been detected in the laboratory, there are indirect experiments which indicate that it can occur. The second step is the reaction of a deuteron with a proton to form $_2He^3$:

$$_1H^2 + {}_1H^1 \rightarrow {}_2He^3 + \gamma.$$

Two $_2He^3$ nuclei can make $_2He^4$ by the process

$$_2He^3 + {}_2He^3 \rightarrow {}_2He^4 + 2{}_1H^1.$$

If we combine these three reactions in a way which eliminates intermediate nuclei (multiply the first and second by two, then add), we get

$$4_1H^1 \rightarrow {}_2He^4 + 2{}_{+1}\beta^0 + 26.7 \text{ Mev.}$$

Thus the net result is the conversion of four protons into an α-particle, two positrons, and a considerable amount of energy. This set of reactions is the principal source of energy in stars of approximately the same mass as the sun.

In older and more massive stars, considerable amounts of helium nuclei accumulate, and in the hot, dense interior regions, the reaction

$$3_2He^4 \rightarrow {}_6C^{12} + \gamma$$

is the most probable way in which carbon nuclei are formed. Once present, the carbon nuclei can serve as catalysts for the conversion of protons to α-particles.

The sequence of reactions which accomplishes this is

$$_6C^{12} + {_1H^1} \rightarrow {_7N^{13}} + \gamma$$
$$_7N^{13} \rightarrow {_6C^{13}} + {_{+1}\beta^0}$$
$$_6C^{13} + {_1H^1} \rightarrow {_7N^{14}} + \gamma$$
$$_7N^{14} + {_1H^1} \rightarrow {_8O^{15}} + \gamma$$
$$_8O^{15} \rightarrow {_7N^{15}} + {_{+1}\beta^0}$$
$$_7N^{15} + {_1H^1} \rightarrow {_6C^{12}} + {_2He^4}$$
$$\overline{4_1H^1 \qquad \rightarrow {_2He^4} + 2_{+1}\beta^0.}$$

The net result is the same as the three-step conversion of protons to α-particles discussed previously. However, the individual steps of the carbon-catalyzed process have high reaction rates, and the overall sequence is responsible for most of the transformation of hydrogen in stars which are much more massive than the sun.

The relative abundances of most of the elements in the solar system can be determined from observations of the light emitted by the sun and examination of meteorites. These elemental abundance data are an important clue to the reactions which occur in stars and which may have occurred as the Universe originated. In Fig. 19.9, the relative numbers of atoms of the elements in the solar system are displayed as a function of atomic number. There are several striking facts revealed by this graph. Hydrogen is more abundant than all other elements put together, with helium second. The next three elements, lithium, beryllium, and boron have very small abundances, while some of the other light elements like carbon, nitrogen, oxygen, and neon are very prominent. There is a general decrease in the abundances as the atomic number increases, but the element iron interrupts this trend with a particularly high abundance. Finally, the elements of even atomic number are present in greater amounts than those of odd atomic number.

Some of these observations can be explained by using the simplest concepts of nuclear stability and reactions. As we have discussed, protons are converted to helium nuclei in the hot dense regions of a star. Since $_2He^4$ is a very stable nucleus with both the neutron and proton number at the "magic" value of two, it does not react with either protons or neutrons to form stable heavier nuclei. The reaction

$$_2He^3 + {_2He^4} \rightarrow {_4Be^7} + \gamma$$

does occur, but in a star the Be^7 can be rapidly destroyed by

$$_4Be^7 + {_1H^1} \rightarrow {_5B^8} + \gamma,$$
$$_5B^8 \rightarrow 2_2He^4 + {_{+1}\beta^0},$$

since the B^8 nucleus is unstable. The only way to synthesize elements heavier

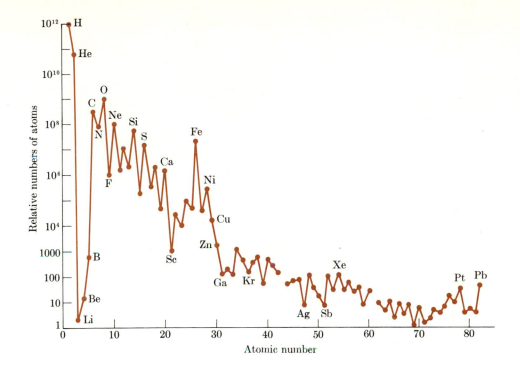

Relative numbers of atoms in the solar system as a function of atomic number. Note the **FIG. 19.9**
gaps that occur at $_{43}$Tc and $_{61}$Pm which are not found in Nature.

than $_2$He^4 appears to be by the previously mentioned reaction,

$$3_2\text{He}^4 \rightarrow {}_6\text{C}^{12} + \gamma,$$

followed by

$$_6\text{C}^{12} + {}_2\text{He}^4 \rightarrow {}_8\text{O}^{16}.$$

Thus the reason for the large amounts of C^{12} and O^{18} in the sun is clear: these
elements are formed directly from the abundant $_2$He^4. The elements lithium,
beryllium, and boron are of low abundance because they cannot be formed by
direct reactions from lighter nuclei. The origin of these elements is evidently
the fragmentation of heavier nuclei as a result of violent collisions with cosmic
ray particles.

The temperature and density of the sun are not great enough to promote
the production of elements much heavier than carbon. The fact that these
elements are present suggests that they may have been produced in the violent
nuclear processes which occurred at the origin of the Universe, or that they are
the result of the processing of hydrogen by much more massive stars which have
since disappeared. In these massive stars, the temperatures are high enough
so that carbon "burning" reactions can occur and produce elements of high

atomic number. The important processes are

$$_{6}C^{12} + {_6}C^{12} \rightarrow {_{10}}Ne^{20} + {_2}He^4 + 4.6 \text{ Mev}$$
$$\rightarrow {_{11}}Na^{23} + {_1}H^1 + 2.24 \text{ Mev}.$$

As massive stars evolve, their temperatures rise so high that oxygen burning occurs and a variety of nuclei appear:

$$_{8}O^{16} + {_8}O^{16} \rightarrow {_{14}}Si^{28} + {_2}He^4 + 9.6 \text{ Mev}$$
$$\rightarrow {_{15}}P^{31} + {_1}H^1 + 7.7 \text{ Mev}$$
$$\rightarrow {_{16}}S^{31} + {_0}n^1 + 1.5 \text{ Mev}.$$

Still heavier nuclei are formed by fusion of $_2He^4$ with abundant species like Si^{28}.

By such fusion reactions, elements up to Fe^{56} can be built from lighter nuclei. The binding energy per nucleon reaches a maximum at Fe^{56}, and fusion reactions with this nucleus will absorb, rather than release, energy. Moreover, the Coulomb repulsion which inhibits nuclear fusion increases as the atomic number increases, and synthesis of elements of high atomic number by this process would require temperatures in excess of those attained in stars. Some other mechanism must be responsible for formation of elements heavier than Fe^{56}.

The primary reaction leading to synthesis of the heavy elements is the absorption of one or more neutrons, followed by emission of a β-particle. Signifying a nucleus of mass A and charge Z by (A, Z), the process may be written as

$$(A, Z) + 3{_0}n^1 \rightarrow (A + 3, Z) \rightarrow (A + 3, Z + 1) + {_{-1}}\beta^0.$$

In the dense, neutron-rich stellar medium, the heavier elements are built up by a series of such steps. The nuclei with neutron numbers near the magic values of 50, 82, or 126 are particularly abundant because their greater stability tends to lower the rate at which they are destroyed. Similarly, nuclei with even numbers of protons are more stable to β-emission processes than those with odd atomic numbers and are found in greater abundance.

The final stage in the evolution of many massive stars is a violent explosion during which some of the less-stable nuclei may be produced. These explosions also disperse the stellar matter, which eventually may condense with large amounts of hydrogen to form a new star. Thus the heavier elements formed in massive stars can appear in smaller stars where the temperatures and density are insufficient to synthesize them from hydrogen.

19.4 RADIOACTIVE DECAY RATES

The spontaneous decay of radioactive nuclei is a first-order process; the number of disintegrations per second is proportional to the number of nuclei present. Thus we can write for the decay rate, $-dN/dt$,

$$-\frac{dN}{dt} = \lambda N,$$

where λ is called the *decay constant* of the nucleus, and N is the number of nuclei in the sample. We can write this expression in the form

$$-\frac{dN}{N} = \lambda \, dt,$$

which shows that the *fraction* of nuclei, dN/N, that decay in a length of time dt is a constant. Integration of this last expression gives

$$\ln \frac{N}{N_0} = -\lambda t, \qquad (19.4)$$

where N is the number of nuclei left at time t, and N_0 is the number of nuclei present at time zero. Rather than report the rate of a nuclear disintegration in terms of its decay constant, it is more convenient to give the half-life of the process; that is, the time it takes for half of the sample present at any given time to decay. To see the relation between the half-life and the decay constant, substitute $N = \frac{1}{2}N_0$, and $t = t_{1/2}$ in Eq. (19.4) to give

$$\ln \tfrac{1}{2} = -\lambda t_{1/2},$$
$$2.3 \log 2 = \lambda t_{1/2},$$
$$t_{1/2} = 0.693/\lambda.$$

Thus if the half-life of a nucleus is known, its decay constant can be calculated and vice versa.

Radiometric Dating

The measured decay rates of certain naturally cocurring radioactive nuclei can be used to date minerals—that is, to determine the time at which the mineral sample was solidified. To illustrate the ideas involved in ratiometric dating, let us discuss the so-called rubidium-strontium method for determining the ages of mica and feldspar minerals.

Naturally occurring rubidium contains 28 percent Rb^{87}, which decays to Sr^{87} by emission of a β-particle:

$$_{37}Rb^{87} \rightarrow {_{38}}Sr^{87} + {_{-1}}\beta^0$$

The half-life of Rb^{87} is 4.7×10^9 years, which is comparable to the estimated age of the universe. Let us assume that at the time a sample of mica solidified, it contained no Sr^{87}, and that in the time since solidification, all Sr^{87} formed by decay of Rb^{87} has been trapped in the rigid crystal lattice of the mica. If we also assume that no Sr^{87} from any other source entered the crystal, then it is clear that by measuring the ratio of Sr^{87} to Rb^{87} and knowing the half-life or decay constant of Rb^{87}, we can calculate the length of time that the mica has been solidified.

The mathematical details are quite straightforward. Let P stand for the number of parent Rb^{87} nuclei in the sample, and D stand for the number of daughter Sr^{87} nuclei. Then at any time

$$D + P = P_0$$

where P_0 is the number of parent nuclei at the time when the mineral solidified. According to Eq. (19.4), we have

$$\ln \frac{P}{P_0} = -\lambda t.$$

Substituting for P_0, and rearranging, we get

$$t = \frac{1}{\lambda} \ln \left(1 + \frac{D}{P} \right). \tag{19.5}$$

The decay constant λ for Rb^{87} is given by

$$\lambda = 0.693/t_{1/2} = 1.47 \times 10^{-11} \text{ yr}^{-1}.$$

Thus by measuring D/P, the age of the sample can be calculated.

There are other parent-daughter pairs whose relative abundance can be used to date minerals that do not contain rubidium. The properties of the most important of these are summarized in Table 19.2. The potassium-argon method is valuable because potassium is abundant and widespread in the Earth's crust. The isotope K^{40} decays in two different ways: 89 percent of the nuclei form Ca^{40} by β-decay, and 11 percent form Ar^{40} by electron capture. The formation of Ca^{40} cannot be used to date minerals, since this isotope is very abundant, and extraneous sources of it mask the Ca^{40} formed from K^{40}. However, certain minerals have the property of trapping the radiogenic Ar^{40}, and preventing argon from the atmosphere from entering. Thus reliable age estimates can be from measurements of the K^{40} to Ar^{40} ratio. The uranium-lead methods provide valuable consistency tests on dating procedures. The U^{238} radioactivity series involves eight α-decay steps and six β-decay steps, and ends with the stable daughter nucleus Pb^{206}. The U^{235} series involves seven α-decay steps and four β-decay steps, and terminates with the stable nucleus Pb^{207}. If the ages cal-

Table 19.2 Methods of radiometric age determination

Parent nucleus	Half life (years)	Daughter nucleus
U^{238}	4.51×10^9	Pb^{206}
U^{235}	0.713×10^9	Pb^{207}
K^{40}	1.30×10^9	Ar^{40}
Rb^{87}	47.0×10^9	Sr^{87}

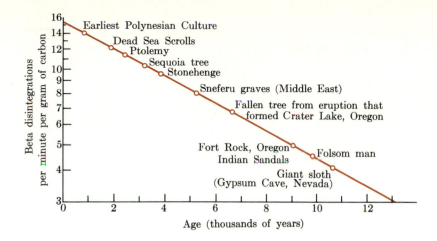

Beta-disintegration rate as a function of age for some objects dated by C^{14} activity.

FIG. 19.10

culated from U^{238}/Pb^{206} and U^{235}/Pb^{207} agree, it is highly likely that no contamination or loss of isotopes has occurred, and that the radiometric age is the true age of the mineral.

As an example of the dating of more recently formed objects, let us consider the technique of C^{14} dating of archeological materials. In the atmosphere, nitrogen is constantly bombarded by cosmic neutrons and converted to $_6C^{14}$,

$$_7N^{14} + _0n^1 \rightarrow _6C^{14} + _1H^1.$$

This carbon is oxidized to carbon dioxide and eventually ingested by plants which are in turn consumed by animals. The $_6C^{14}$ nucleus is radioactive, and emits a low-energy β-particle in a process that has a half-life of 5730 years. Through natural balance of $_6C^{14}$ intake and radioactive decay, living organisms reach a stationary level of $_6C^{14}$ radioactivity which amounts to 15.3 ± 0.1 disintegrations per minute per gram of carbon. When life ceases, intake of $_6C^{14}$ stops, and the radioactivity decays with a half-life of 5730 years. By carefully measuring the decay rate of a sample of wood, for instance, it is possible to tell when the tree died. In this way, an absolute time scale for dating archeological objects of ages between 1000 and 10,000 years has been developed, as shown in Fig. 19.10.

19.5 APPLICATIONS OF ISOTOPES

Although the decay of a single radioactive nucleus may seem to be an insignificant event, the amount of energy released in such a process is often large enough to be detected easily. Consequently, measurement of nuclear radioactivity is the most sensitive technique available for the detection of atoms. This sensitivity can be used to advantage in a number of ways. For example, consider the method of radioactivation analysis. The absorption of a neutron by any nucleus produces an "activated" or energy-rich species that decays by a process characteristic of the nucleus involved. The various isotopes of the elements differ considerably in their ability to absorb a neutron. Consequently, by irradiating a mixture of nuclei with neutrons, it is possible to selectively activate certain elements, detect their presence, and measure their concentration by ascertaining the intensity of the induced radioactivity. The sensitivity of activation analysis depends on the neutron flux available for irradiation, the ability of a nucleus to absorb a neutron, and on the energy of the decay process. It is possible to detect as little as 10^{-10} gm of copper, sodium, or tungsten by activation analysis, and the method can be applied to a number of other elements with somewhat reduced sensitivity.

Another application of radioactive isotopes, in which the emphasis is more on specificity than on sensitivity, occurs in the study of reaction rates and mechanisms. With the aid of a radioactive isotope, it was possible to determine the rate at which iron ions change their oxidation states in an aqueous solution of ferric and ferrous perchlorate. In a mixture of ordinary ferric and ferrous ions, the exchange reaction

$$Fe^{++} + Fe^{+3} \rightarrow Fe^{+3} + Fe^{++}$$

goes on all the time, but it is impossible to observe, since the products are chemically the same as the reactants. It was possible to observe the reaction through use of $_{26}Fe^{55}$, a positron emitter with a half-life of four years. In the mixture, the radioactive isotope was present initially as ferrous ion, and as time passed, extraction of samples of the solution, followed by separation of the Fe^{++} from the Fe^{+3}, and determination of their radioactivities showed that the Fe^{+3} became increasingly radioactive. By studying the rate of exchange of radioactivity as a function of concentration, one term in the rate law for the exchange reaction was shown to be

$$\text{rate of exchange} = k[Fe^{++}][Fe^{+3}].$$

Thus the use of radioactive or "tagged" atoms makes the exchange of oxidation states observable, and the reaction rate measurable.

Use of radioactive isotopes in tracer experiments such as the one just described is advantageous, since the detection of the nature and intensity of the radioactivity is a simple method of qualitative and quantitative analysis. Some elements do not have a radioactive isotope of convenient half-life, and in these instances stable isotopes must be used in tracer experiments. As an example, let us consider the problem of determining the course of an esterification reaction such as

$$C_6H_5C\overset{\displaystyle O}{\underset{\displaystyle OH}{\big\backslash}} + CH_3OH \rightarrow C_6H_5C\overset{\displaystyle O}{\underset{\displaystyle O^*CH_3}{\big\backslash}} + H_2O.$$

Does the starred oxygen atom come from the alcohol or from the acid? The problem was solved by synthesizing methyl alcohol in which the oxygen was abnormally enriched with the O^{18} isotope. This "labeled" methyl alcohol was then used in the esterification reaction, and the isotopic composition of the product ester examined with a mass spectrometer. The mass spectrum showed that the ester was enriched with the O^{18} isotope, and consequently, the oxygen atom in the ester linkage must come from the alcohol, not the acid.

These simple examples indicate the types of applications of radioactive and stable isotopes in chemical research. There have been countless elaborations and variations on these ideas to deal with more complicated chemical systems. In particular, much of what we know of the chemistry of biological systems has been deduced from experiments with radioactive and stable isotopes. For more detailed discussions of these applications, the reader is referred to the reading list at the end of the chapter.

SUGGESTIONS FOR FURTHER READING

Choppin, G. R., *Nuclei and Radioactivity*. New York: W. A. Benjamin, 1964.

Harvey, B. G., *Introduction to Nuclear Physics and Chemistry*. Englewood Cliffs, New Jersey: Prentice-Hall, 1962.

Kaplan, I., *Nuclear Physics*, 2nd ed. Reading, Mass.: Addison-Wesley, 1963.

Overman, R. T., *Basic Concepts of Nuclear Chemistry*. New York: Reinhold, 1963.

PROBLEMS

19.1 Write equations that represent each of the following processes: (a) positron emission by $_{51}Sb^{120}$; (b) negative beta emission by $_{16}S^{35}$; (c) alpha emission by $_{88}Ra^{226}$; (d) electron capture by $_4Be^7$.

19.2 The masses of $_{11}Na^{22}$ and $_{10}Ne^{22}$ atoms are 21.994435 and 21.991385 amu, respectively. Is it energetically possible for Na^{22} to decay to Ne^{22} by positron emission?

19.3 What is the alpha activity in disintegrations per min for a 0.001-gm sample of Ra^{226} ($t_{1/2} = 1620$ years)?

19.4 A radioisotope decays at such a rate that after 68 min, only $\frac{1}{4}$ of the original amount remains. Calculate the decay constant and half-life of the radioisotope.

19.5 Articles found in the Lascaux Caves in France have a C^{14} disintegration rate of 2.25 disintegrations per minute per gram of carbon. How old are these articles?

19.6 The only stable isotope of fluorine is F^{19}. What type of radioactivity would you expect from each of the isotopes F^{17}, F^{18}, F^{20}, and F^{21}?

19.7 When an electron and positron encounter each other, they are annihilated and two photons of equal energy formed. Calculate the wavelength of these photons.

APPENDIXES.
ANSWERS

APPENDIXES

Avogadro's Number

The Avogadro number is of fundamental importance in chemistry, and a variety of methods have been used to determine its value. Here we cite only a few.

Radioactive decay and gaseous viscosity are two phenomena that can be used to give Avogadro's number, accurate to within a few percent. The α-decay of a mole of Ra^{226},

$$_{88}Ra^{226} \rightarrow \; _{86}Rn^{222} + \; _{2}He^{4}$$

produces 1.35×10^{-11} mole of helium/sec, while the number of disintegrations is 8.15×10^{12}/sec. The ratio of these two rates is Avogadro's number, but the accuracy of the result is limited by the difficulty of collecting and measuring the extremely small amounts of helium produced by the decay.

The rigorous kinetic theory of a gas of hard spheres of diameter σ shows that the viscosity is

$$\eta = \frac{5}{16} \frac{\sqrt{\pi M R T}}{\pi N \sigma^2},$$

while the volume of a mole of liquid is

$$V = N \frac{\pi}{3} \sigma^3,$$

where M is the molecular weight, R is the gas constant, T is the absolute temperature, and N is Avogadro's number. Simultaneous solution of these equations yields N and σ in terms of measured values of V and η. However, the accuracy of the result is limited by the underlying assumption that molecules behave like hard spheres.

One of the most reliable methods of determining N involves combining the density of a crystal, its formula weight, and its interatomic spacing, as determined by x-rays. This method has been discussed in Section 3.3. The diffraction pattern of x-rays of grazing incidence on a plate that has been ruled with accurately spaced lines gives the wavelength λ of the x-rays. The x-ray diffraction pattern of a crystal then gives the interatomic spacing d through the Bragg equation $n\lambda = 2d \sin \theta$, where θ is the diffraction angle. Since one mole of a crystal like NaCl contains $2N$ atoms, we have for the volume V of 1 mole of NaCl

$$V = \frac{M}{\rho} = 2Nd^3,$$

where M is the molecular weight, and ρ is the density. Knowledge of d allows calculation of N with an uncertainty of less than 0.01%.

Another reliable method for determining N is through measurement of the Faraday constant $\mathfrak{F}$ and the electronic charge e, since

$$\mathfrak{F} = Ne.$$

The most direct way to determine the Faraday constant is to measure the number of coulombs (ampere-seconds) necessary to deposit one equivalent by electrolysis. The determination of the electronic charge e can be carried out by Millikan's method, as discussed in Section 10.1. However, this technique does not give a particularly accurate result for e, and other procedures described below are preferred.

The electronic charge, in combination with other physical constants, determines a number of quantities that can be measured precisely. The problem is to combine these measurements so as to yield separate values of the constants. For example, the natural orbital frequency of an electron in a magnetic field, the so-called cyclotron resonance frequency ν, is given by

$$\nu = \frac{1}{2\pi c}\left(\frac{e}{m}\right)\mathcal{3C},$$

where $\mathcal{3C}$ is the magnetic field and c is the velocity of light. The value of e/m derived from measurements of ν and $\mathcal{3C}$ can be combined with the measured value of the Rydberg constant

$$\mathcal{R} = \frac{2\pi^2 m e^4}{h^3 c}$$

to give

$$\left(\frac{e}{m}\right)\mathcal{R} = \frac{2\pi^2 e^5}{ch^3}.$$

Only one more measurement of a quantity that involves e and Planck's constant h is necessary to give values of both constants. The most straightforward, although not the most precise, method of separating the constants is to use the photoelectric effect:

$$h(\nu - \nu_0) = (\tfrac{1}{2}mv^2)_{\text{max}},$$

where ν_0 is the photoelectric threshold frequency. The kinetic energy of the electrons is frequently measured by a device that involves the interaction of their charge e with an electric field E. The electric field necessary to repel the most energetic electrons and prevent them from leaving the photoelectric surface is given by

$$(\tfrac{1}{2}mv^2)_{\text{max}} = eE.$$

Thus the photoelectric experiment determines h/e, and this, combined with the Rydberg constant and the cyclotron frequency, gives values for e, h, and m. Combination of e and $\mathcal{F}$ gives Avogadro's number N.

Table B.1 Physical constants

Avogadro's number	6.02217×10^{23} particles/mole
Electronic charge	4.80325×10^{-10} esu
	1.60219×10^{-19} coulomb
Electron mass	9.10956×10^{-28} gm
Atomic mass unit	1.66053×10^{-24} gm
Gas constant	8.31434×10^7 erg/mole deg
	1.9872 cal/mole deg
	0.08206 liter atm/mole deg
Faraday constant	96486.7 coulomb/mole
	23061 cal/volt mole
Boltzmann constant	1.38062×10^{-16} erg/deg
Planck constant	6.6262×10^{-27} erg sec

Table B.2 Energy conversion factors

	ergs/molecule	kJ/mole	kcal/mole	electron volts/molecule
ergs/molecule	1	6.0222×10^{13}	1.4393×10^{13}	6.2415×10^{11}
kJ/mole	1.6605×10^{-14}	1	0.23901	1.0364×10^{-2}
kcal/mole	6.9478×10^{-14}	4.184	1	4.336×10^{-2}
electron volts/molecule	1.6022×10^{-12}	9.6487×10^1	23.061	1

SI UNITS AND CONVERSION FACTORS

To make them meaningful, measurements of physical quantities must be expressed as multiples of appropriate standards or units. Thus the same length can be expressed as 10 inches, 0.83333 feet, or 0.2540 meters, where the words inches, feet, and meters refer to different standard lengths. As the various physical quantities we know came to be recognized and measured, units were defined for them usually to suit the immediate convenience of the experimenter. As a result, we have such units as light-years and nautical miles for length, electron volts, joules, and calories for energy. Consequently, it is necessary to have conversion factors such as 2.54 cm/inch to allow results expressed in one unit to be compared with findings expressed in other units.

As the understanding of physical science increased, it became clear that all physical quantities such as volume, density, and energy could be expressed in terms of a few fundamental entities such as length, mass, and time. This led to the idea of a *system* of units where, for example, the unit of energy would be equal to the energy associated with a unit mass moving a unit distance in unit time. An important such system of units is the centimeter, gram, second, or CGS system, where the units for derived quantities such as force and energy are defined in terms of the base units for length, mass, and time. Thus the (unnamed) unit velocity in the CGS system is the cm/sec, and the unit of acceleration is the cm/sec^2.

Since force is defined as the product of mass and acceleration, the unit of force is the gm cm/sec^2, which for convenience is called the dyne. The product of force and distance is work, and consequently the unit of work or energy is the dyne-cm, or one gm cm^2/sec^2. Again for convenience, the unit of work or energy is given a special name, the erg.

To include electrical quantities in the CGS system, it is only necessary to use Coulomb's law. The force f between two equal charges q separated by a distance r is given by $f = q^2/r^2$. Thus the electrostatic unit (esu) is defined as the amount of charge that produces a force of one dyne on an identical charge at one cm distance in vacuum. One of the most attractive aspects of the CGS-esu system is the simplicity with which electrostatic quantities can be included in mechanical problems. For example, since the esu as defined has units of (dyne)$^{1/2}$ cm, use of esu and cm in the expression $\phi = q^2/r$ gives the electrostatic potential energy ϕ directly in units of dyne-cm or ergs.

Unfortunately, there are several different systems of units. For example, the MKS (meter-kilogram-second) system of units offers certain advantages over CGS in dealing with macroscopic systems. The FPS (foot-pound-second) has been popular among engineers in English speaking countries. There are also units such as the atmosphere, the calorie, and the electron-volt which have no obvious connection to a system of units. The existence of data expressed in these various units has necessitated the use of conversion factors which allow the results measured in one set of units to be compared or combined with data expressed in other units. To eliminate the necessity for the excessive use of unit conversion factors, the International System of Units (SI) was defined by the General Conference on Weights and Measures in 1960. This system has been widely adopted in Europe, and its use in the United States is being encouraged by the National Bureau of Standards.

The SI is constructed from seven fundamental units, given in Table C.1. These base units are, as far as possible, defined in terms of naturally occurring physical phenomena which can be reproduced and referred to with high precision. Thus the second is defined as 9,192,631,700 cycles of the radiation associated with a certain electronic transition of the cesium atom, and the meter is defined as 1,650,763.73 wavelengths in vacuum of a certain orange-red spectral line of Kr^{86}. The Kelvin degree is defined as 1/273.16 of the interval between absolute zero and the triple point of water. The fundamental unit of mass is the kilogram, the mass of a platinum-iridium alloy cylinder kept in Paris. Thus in contrast to the foregoing base units, the kilogram is defined in terms of an artifact, not a naturally occurring phenomenon. The base unit of electrical current, the ampere, is defined in terms of the force between parallel wires produced by current they carry. We see that in contrast to the CGS system, the base electrical unit in the SI is a current, not a charge.

Table C.1 SI base units

Physical quantity	Name of unit	Symbol
length	meter	m
mass	kilogram	kg
time	second	s
electric current	ampere	A
thermodynamic temperature	kelvin	K
luminous intensity	candela	cd
amount of substance	mole	mol

Table C.2 gives the special names and symbols for certain derived units in the SI. The unit of energy is the joule or km m^2/sec^2. The unit of pressure is the pascal, or newton (force) per square meter. The utility of expressing all measurements in one such system of units can be illustrated by the following example.

Suppose a gas undergoes an expansion and does 1 liter-atm of work. It is desired to compare this work with the energy change of some chemical reaction expressed in the conventional unit of kilocalories. To convert 1 l-atm to calories, it is necessary to have the appropriate conversion factor tabulated, or to re-

Table C.2 Certain SI derived units

Physical quantity	Name	Symbol	Definition
force	newton	N	$\mathrm{kg\ m\ s^{-2}}$
pressure	pascal	Pa	$\mathrm{kg\ m^{-1}\ s^{-1}}$ ($= \mathrm{N\ m^{-2}}$)
energy	joule	J	$\mathrm{kg\ m^2\ s^{-2}}$
power	watt	W	$\mathrm{kg\ m^2\ s^{-3}}$ ($= \mathrm{J\ s^{-1}}$)
electric charge	coulomb	C	$\mathrm{A\ s}$
potential difference	volt	V	$\mathrm{kg\ m^2\ s^{-3}\ A^{-1}}$ ($= \mathrm{J\ C^{-1}}$)
electric resistance	ohm	Ω	$\mathrm{kg\ m^2\ s^{-3}\ A^{-2}}$ ($= \mathrm{V\ A^{-1}}$)
frequency	hertz	Hz	$\mathrm{s^{-1}}$

member that 1 atm $= 1.013 \times 10^6$ dynes/cm^2, and 1 cal $= 4.184$ J. Then

$$1\ \text{l-atm} = 10^3\ \text{cm}^3 \times 1.013 \times 10^6\ \text{dyne/cm}^2 = 1.013 \times 10^9\ \text{dyne-cm}$$
$$= 1.013 \times 10^2\ \text{J} = 1.013 \times 10^2\ \text{J} \times 1/4.184\ \text{cal/J}$$
$$= 24.21\ \text{cal}.$$

Now if the measurements had been made in SI units of pressure (pascals) and volume (cubic meters), we would have for the same quantities

$$1.013 \times 10^5\ \text{Pa} \times 10^{-3}\ \text{m}^3 = 101.3\ \text{N m} = 101.3\ \text{J}.$$

Since in an exclusive SI tabulation of data, all energies, including heats of reaction, would be expressed in terms of joules, there would be no need to carry the calculation further to calories.

In using one base unit of, for example, length to describe dimensions ranging from atomic to astronomic size, it is useful to have prefixes which designate a multiple of the base unit. Thus distances between various cities are specified in terms of thousands of meters or kilometers (km), and atomic dimensions are given as a multiple of 10^{-9} m or the nanometer (nm). Table C.3 contains a list of these prefixes and their symbols.

As the SI units are adopted, they will replace a number of conventional units whose use will gradually diminish. However, because the older literature con-

Table C.3 Prefixes for fractions and multiples

Fraction	Prefix	Symbol	Multiple	Prefix	Symbol
10^{-1}	deci	d	10	deka	da
10^{-2}	centi	c	10^2	hecto	h
10^{-3}	milli	m	10^3	kilo	k
10^{-6}	micro	μ	10^6	mega	M
10^{-9}	nano	n	10^9	giga	G
10^{-12}	pico	p	10^{12}	tera	T
10^{-15}	femto	f			
10^{-18}	atto	a			

Table C.4 Units to be abandoned

Physical quantity	Name	Symbol	Definition
length	inch	in	2.54×10^{-2} m
mass	pound	lb	0.45359237 kg
pressure	atmosphere	atm	101,325 N m^{-2}
energy	calorie	cal	4.184 J

tains data expressed in these units, it will be necessary to be able to re-express these data in SI units. Table C.4 gives a list of these units which will eventually be abandoned, and their definitions in terms of SI units.

There is a special aspect of SI (and MKS) units that should be mentioned. Because static charge is not defined directly in terms of force and distance in these systems, Coulomb's law involves a proportionality constant between q^2/r^2 in terms of C^2/m^2 and f in terms of newtons. The required expression is

$$f(N) = \frac{1}{4\pi\epsilon_0} \frac{q^2}{r^2} = 8.9874 \times 10^9 \frac{q^2}{r^2} \left(\frac{C^2}{m^2}\right),$$

where the proportionality constant which is conventionally written as $1/4\pi\epsilon_0$ has the value 8.9874×10^9 N m^2/C^2. Thus the force between two electrons (1.602×10^{-19} C) one Angstrom (10^{-10} m) apart is

$$f = 8.9874 \times 10^9 \frac{(1.602 \times 10^{-19})^2}{(10^{-10})^2} = 2.31 \times 10^{-8} \text{ N.}$$

The same calculation done in CGS units is

$$f(\text{dynes}) = \frac{e^2}{r^2} = \frac{(4.803 \times 10^{-10} \text{ esu})^2}{(10^{-8} \text{ cm})^2} = 2.31 \times 10^{-3} \text{ dynes.}$$

This is the same result, since 1 N $= 10^5$ dynes. The calculation done in SI or MKS units is only very slightly more complicated by the inclusion of the proportionality constant.

APPENDIX D

Some Fundamental Operations of the Calculus

Often it is possible to put the equation that relates two variables x and y into the form

$$y = f(x),$$

which is read: y is some function (f) of x. The quantity x is called the independent variable, and y is the *dependent variable*, inasmuch as the equation expresses the fact that y depends in some fashion on x.

It is frequently of interest to learn what change Δy, in y, is produced by a given change Δx, in x. The equation

$$y + \Delta y = f(x + \Delta x)$$

states that a change Δx in x does produce some change Δy in y. The question to be answered is, what is the relation between Δy and Δx?

When y is a linear function of x, as in

$$y = f(x) = a + bx,$$

where a and b are constants, the answer is simple. We write

$$y + \Delta y = a + b(x + \Delta x),$$
$$y = a + bx,$$

and subtract the second equation from the first to get

$$\Delta y = b(x + \Delta x) - bx,$$
$$\Delta y = b \, \Delta x.$$

Thus the quantity

$$\Delta y / \Delta x = b,$$

which is the change in y per unit change in x, is for this function a constant, and is equal to the slope of the straight line represented by $y = a + bx$. Because the equation represents a straight line, the slope $b = \Delta y / \Delta x$ is a constant, independent of x and y.

Suppose we apply the same operation to the function

$$y = f(x) = x^2,$$

which represents a parabola. Then we have

$$y + \Delta y = (x + \Delta x)^2$$

The method by which we found dy/dx for the function $y = x^2$ is called the delta process. Application of the delta process shows that for

$$y = x^3, \qquad \frac{dy}{dx} = 3x^2,$$

and, in general, for

$$y = x^n, \qquad \frac{dy}{dx} = nx^{n-1}.$$

The derivatives of other important functions are listed in Table D.1.

Table D.1 Some important derivatives

$$\frac{d}{dx}(x) = 1 \qquad\qquad \frac{d}{dx}(e^u) = e^u \frac{du}{dx}$$

$$\frac{d}{dx}(au) = a\frac{du}{dx} \qquad\qquad \frac{d}{dx}(\sin u) = \cos u \frac{du}{dx}$$

$$\frac{d}{dx}(u^n) = nu^{n-1}\frac{du}{dx} \qquad\qquad \frac{d}{dx}(\cos x) = -\sin u \frac{du}{dx}$$

$$\frac{d}{dx}(\ln u) = \frac{1}{u}\frac{du}{dx} \qquad\qquad \frac{d}{dx}(uv) = u\frac{dv}{dx} + v\frac{du}{dx}$$

u and v denote functions of x; a is a constant.

The following are noteworthy properties of derivatives. If c is any constant, and

$$y = cx^n, \text{ then } \frac{dy}{dx} = (c)nx^{n-1}.$$

If we have y not a function of x, as in

$$y = c, \text{ then } \frac{dy}{dx} = 0.$$

Finally, to find dy/dx when y is the *product* of two functions $f(x)$ and $g(x)$, as in

$$y = f(x) \cdot g(x),$$

we write

$$\frac{dy}{dx} = f(x)\frac{dg(x)}{dx} + g(x)\frac{df(x)}{dx}.$$

To demonstrate this relation, let us differentiate

$$y = x^n$$

by writing it as

$$y = x^\ell x^m, \qquad \ell + m = n.$$

We get

$$\frac{dy}{dx} = x^\ell m x^{m-1} + x^m \ell x^{\ell-1}$$

$$= mx^{n-1} + \ell x^{n-1}$$

$$= nx^{n-1},$$

which is the answer found by more direct differentiation of x^n.

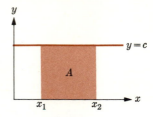

FIG. D.1

It is often necessary to find the area under a curve $y = f(x)$ between two values of the independent variable x. When $f(x)$ has one of two simple forms, this presents no difficulty. As shown by Fig. D.1, if $y = c$, a constant, then the area under $y = c$ between x_1 and x_2 is simply $A = c(x_2 - x_1)$. Also, if we have the expression $y = cx$, which represents a straight line through the origin, then the area A between $x = x_1 = 0$ and x_2 is $A = \frac{1}{2}cx_2$, as Fig. D.2 suggests. For situations in which $y = f(x)$ represents a curve, a different approach is required in order to find the required area.

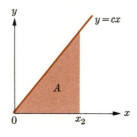

FIG. D.2

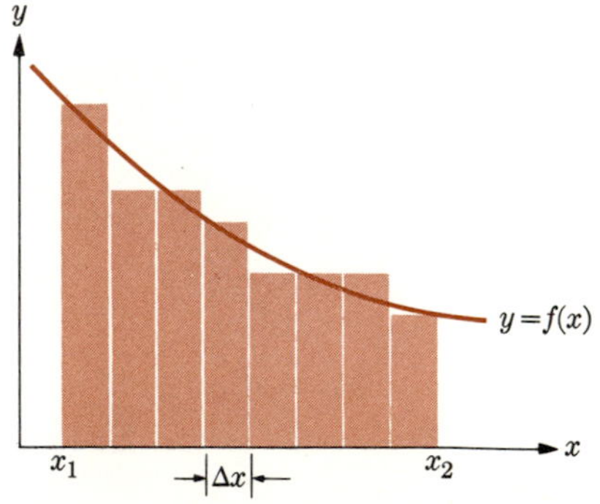

FIG. D.3

This type of problem and its possible solution is sketched in Fig. D.3. The area under the curve is approximated by a series of rectangles of equal width Δx and height y_i, for the ith rectangle. Thus if ΔA_i is the area of the ith rectangle,

$$A \cong \sum_{i=1}^{n} \Delta A_i = \sum_{i=1}^{n} y_i \, \Delta x.$$

It seems intuitively obvious that the approximation should improve if we were to increase the number of rectangles and make each one narrower. The area of a rectangle of height y is

$$\Delta A = y \, \Delta x = f(x) \, \Delta x;$$

and since

$$\lim_{\Delta x \to 0} \frac{\Delta A}{\Delta x} \equiv \frac{dA}{dx} = f(x),$$

we can write

$$dA = f(x) \, dx.$$

The interpretation of dA is that it is the area of a rectangle of height $f(x)$ and infinitesimal width dx. Consequently, we can express the total area as the sum (or integral) of the infinitesimal contributions

$$A = \lim_{\Delta x \to 0} \sum \Delta A_i = \lim_{\Delta x \to 0} \sum y_i \, \Delta x,$$

$$A = \int dA = \int_{x_1}^{x_2} f(x) \, dx.$$

That is, the integral is the sum of small areas between x_1 and x_2 in the *limit* where Δx approaches zero.

To evaluate the integral, we note from above the following relation:

$$A = \int dA.$$

Thus, since dA is the differential of A, and the integral sign restores dA to A, the integral sign must stand for the operation of reverse differentiation or antidifferentiation. With this in mind, we can evaluate

$$\int f(x) \, dx.$$

If the function $f(x)$ were itself the derivative with respect to x of another function $F(x)$, that is, if

$$f(x) = \frac{d}{dx} F(x),$$

then we could make the substitution

$$\int f(x)\, dx = \int \frac{dF}{dx}\, dx = \int dF.$$

The last form shows that the integral sought is just F, so we might write

$$A = \int f(x)\, dx = F,$$

where

$$f(x) = \frac{dF}{dx}.$$

We conclude that in order to evaluate the integral of $f(x)$, we must find the function F whose derivative dF/dx is equal to $f(x)$. To check this we start with our answer

$$A = F(x), \qquad f(x) = dF/dx,$$

and differentiate

$$\frac{dA}{dx} = \frac{dF}{dx},$$
$$f(x) = f(x).$$

There is one very important problem remaining, however. Besides the answer already found, the result

$$A(x) = F(x) + c,$$

where c is any constant, is apparently equally valid, for differentiation gives us

$$\frac{dA}{dx} = \frac{dF}{dx} + 0,$$
$$f(x) = f(x).$$

Thus the integral we have found is *indefinite* in that any constant may be added to it.

This problem has arisen because we have ignored the fact that what we are seeking is the area under the curve $y = f(x)$ *between* two limits x_1 and x_2. If we regard $A(x_2)$, that is, $A(x)$ evaluated at $x = x_2$, as the area under the curve up to the limit x_2 plus the constant c, and $A(x_1)$ as the area up to x_1 plus the same constant, then we can write

$$A(x_2) - A(x_1) = A = F(x_2) + c - F(x_1) - c,$$
$$A = F(x_2) - F(x_1).$$

That is, the area A under $y = f(x)$ between x_2 and x_1 is just the difference between the integral evaluated at the "upper limit" x_2 and the "lower limit" x_1. This is compactly denoted by

$$\int_0^A dA = \int_{x_1}^{x_2} f(x)\, dx = \int_{x_1}^{x_2} dF,$$

$$A\Big|_0^A = A - 0 = F(x)\Big|_{x_1}^{x_2} = F(x_2) - F(x_1).$$

Thus, when the limits of integration are specified, the numerical result is not arbitrary, and we have a *definite* integral.

From the table of derivatives, we can deduce a few specific formulas for the evaluation of integrals. Since

$$\frac{d}{dx} x^n = n x^{n-1},$$

we must have

$$\int_{x_1}^{x_2} x^m\, dx = \frac{x^{m+1}}{m+1}\bigg|_{x_1}^{x_2}.$$

Also from

$$\frac{d(\sin x)}{dx} = \cos x,$$

$$\frac{d(\cos x)}{dx} = -\sin x$$

we deduce that

$$\int_{x_1}^{x_2} \sin x\, dx = -\cos x\bigg|_{x_1}^{x_2},$$

$$\int_{x_1}^{x_2} \cos x\, dx = \sin x\bigg|_{x_1}^{x_2}.$$

Because

$$\frac{d \ln x}{dx} = \frac{1}{x},$$

$$\int_{x_1}^{x_2} \frac{dx}{x} = \int_{x_1}^{x_2} d \ln x = \ln x\bigg|_{x_1}^{x_2}.$$

Finally, we have the very important relation

$$\frac{d}{dx} e^{ax} = a e^{ax},$$

so

$$\int_{x_1}^{x_2} e^{ax}\, dx = \frac{1}{a} \int_{x_1}^{x_2} d(e^{ax}) = \frac{1}{a}\, e^{ax}\Big|_{x_1}^{x_2}.$$

Other important integrals are given in Table C.2.

Table D.2 Some important integrals

$\int df(x) = f(x) + c$	$\int \dfrac{du}{u} = \ln u + c$
$\int af(x)\, dx = a \int f(x)\, dx$	$\int e^u du = e^u + c$
$\int (u \pm v)\, dx = \int u\, dx \pm \int v\, dx$	$\int \sin u\, du = -\cos u + c$
$\int u^n\, du = \dfrac{u^{n+1}}{n+1} + c,\ \ n \neq -1$	$\int \cos u\, du = \sin u + c$
$\int_a^b f(x)\, dx = -\int_b^a f(x)\, dx$	$\int_a^c f(x)\, dx = \int_a^b f(x)\, dx + \int_b^c f(x)\, dx$

u and $f(x)$ are functions of x; a, b, and c are constants.

APPENDIX E

Some Useful Mathematical Approximations

The form $1/(1 + x)$ occurs frequently in equations that describe physical phenomena. It can be replaced by an infinite series of powers of x with alternating sign

$$\frac{1}{1+x} = 1 - x + x^2 - x^3 + \cdots$$

As we can verify by algebraic division,

$$
\begin{array}{r}
1 - x + x^2 \\
\hline
1 + x \,\overline{)\,1} \\
1 + x \\
\hline
-x \\
-x - x^2 \\
\hline
x^2 \\
x^2 + x^3 \\
\end{array}
$$

If x is less than unity, each higher power will be a smaller magnitude than the one preceding, and for x very small, it may be sufficiently accurate to write

$$\frac{1}{1+x} \cong 1 - x, \qquad x \ll 1,$$

which is a useful approximation. Similar analysis of $1/(1-x)$ shows that

$$\frac{1}{1-x} \cong 1 + x, \qquad x \ll 1.$$

A useful approximation to square roots can be derived in the following manner. Consider that

$$\left(1 \pm \frac{x}{2}\right)^2 = 1 \pm x + \frac{x^2}{4}.$$

If $x \ll 1$, we can write

$$\left(1 \pm \frac{x}{2}\right)^2 \cong 1 \pm x, \qquad x \ll 1.$$

Taking the square root of both sides and transposing gives

$$(1 \pm x)^{1/2} \cong 1 \pm \frac{x}{2}, \qquad x \ll 1,$$

which is the desired approximation. For example, the square root of ten is

$$10^{1/2} = (9 + 1)^{1/2} = 3(1 + \tfrac{1}{9})^{1/2}$$
$$\cong 3(1 + \tfrac{1}{18}) = 3.167$$

approximately, while the more accurate value is 3.1623.

Other frequently used approximations are derived from infinite series. The series

$$e^{\pm x} = 1 \pm x + \frac{x^2}{2!} \pm \frac{x^3}{3!} + \frac{x^4}{4!} + \cdots$$

is valid for any value of x, but if x is much less than unity, the higher powers of x can be neglected to give simply

$$e^{\pm x} \cong 1 \pm x, \qquad x \ll 1.$$

If we take the natural logarithm of both sides we get

$$\ln e^{\pm x} \cong \ln (1 \pm x)$$

or

$$\ln (1 \pm x) \cong \pm x, \qquad x \ll 1,$$

which is another useful approximation.

ANSWERS TO SELECTED PROBLEMS

Chapter 1

1.1 Sb_2O_5 **1.2** CH_2, C_3H_6 **1.4** Fe_2O_3 **1.6** SnF_4
1.7 (d) Tl   **1.9** 199, MO   **1.11** 0.74   **1.13** 95.3%
1.15 BH_3, B_2H_6, 27.6, 1.51 gm   **1.17** 0.377

Chapter 2

2.1 23.1 cc   **2.3** 274, $HgCl_2$ **2.5** 9 vol O_2, CSH_4, molecular
2.7 0.48   **2.9** He
2.11 N_2; 5.4×10^9, 5.4×10^6, 54; He, 7.1×10^9, 7.1×10^6, 71
2.13 1.16×10^7 cm/sec, 1.45×10^4 cm/sec
2.15 $\lambda \gg$ distance between walls, $P < 10^{-4}$ mm Hg   **2.17** H_2
2.19 $C_P/C_V = 1.67$ for monatomic gases only

Chapter 3

3.3 $V = 16\sqrt{2}\,R^3$, 0.74
3.5 Octahedral face planes contain all Na^+ or all Cl^-.
3.6 $\sqrt{6}\,R$, $(\sqrt{3} - \sqrt{2})R/\sqrt{2}$ **3.9** (a) 0; (b) 0.002 gm/cc
3.11 0.53A

Chapter 4

4.1 $2.92\ m$, $2.57\ M$ **4.3** 40 ml   **4.5** 3.85

4.7 333 gm   **4.9** 470, Hg_2Cl_2

4.11 (a) ethanol = 0.51

b) P(ethanol) = 22.7, P(methanol) = 43.5, ethanol = 0.34

4.13 absorbed, $\Delta H > 0$

Chapter 5

5.2 reaction (b)

5.4 (a) increased   (b) decreased   (c) increased   **5.6** 0.137

5.8 2.31×10^{-4}, greater, decrease   **5.10** 202 gm

5.12 $SO_2Cl_2 = 0.67$ atm, $SO_2 = 0.86$ atm, $Cl_2 = 1.86$ atm

5.14 $CO = H_2O = 1.36$ atm, $CO_2 = H_2 = 4.3$ atm

Chapter 6

6.1 1.1×10^{-10} **6.3** $0.69 \times 10^{-3}\ M$, $K_{sp} = 6.1 \times 10^{-12}$

6.5 (a) $1.34 \times 10^{-4}\ M$ (b) $1.8 \times 10^{-7}\ M$ (c) $1.8 \times 10^{-5}\ M$

6.7 $[SO_4^=] = 1.1 \times 10^{-9}\ M$, $BaSO_4$, $[Ba^{++}] = 4.6 \times 10^{-7}\ M$

6.9 $[Pb^{++}] = 2.9 \times 10^{-10}$, $[IO_3^-] = 0.03\ M$

6.11 (a) $[OH^-] = 4.2 \times 10^{-4}\ M$ (b) $[NH_3] = 6.5 \times 10^{-5}\ M$, $[OH^-] = 3.5 \times 10^{-5}\ M$

6.13 $[H_3O^+] = 0.11\ M$, $[SO_4^=] = 0.94 \times 10^{-2}\ M$, $[HSO_4^-] = 9.1 \times 10^{-2}\ M$

6.14 $[NH_3] = 0.15\ M$, $[OH^-] = 0.05\ M$, $[NH_4^+] = 5.4 \times 10^{-5}\ M$

6.17 $[OH^-] = [HCN] = 2.72 \times 10^{-3}\ M$ **6.19** $4.1 \times 10^{-4}\ M$

6.21 $[OH^-] = 4.4 \times 10^{-11}\ M$ **6.23** 1.2×10^{-5}

6.25 $[CO_3^=] = 0.05\ M$, $[HCO_3^-] = 3.23 \times 10^{-3}$ **6.27** pH = 10.3

6.29 $1.23 \times 10^{-5}\ M$, $1.36 \times 10^{-5}\ M$ **6.31** pH = 2.12, no

6.32 (a) $2.1 \times 10^{-5}\ M$ (b) $2.5 \times 10^{-10}\ M$ (c) $10^{-13}\ M$

Chapter 7

7.4 stronger: $Cr_2O_7^=$, MnO_4^-; unchanged: Cl_2, Fe^{+3}

7.6 $\Delta\mathcal{E}^0 = 0.13$ volt, $K = 1.6 \times 10^2$

7.8 (a) negative (b) Ni^{++}, Zn (c) no, yes (d) less positive (e) zinc, 0.52 volt

7.10 (a) $Co + Cl_2 = 2Cl^- + Co^{++}$ (b) -0.27 volt (c) increase (d) 1.67 volts

7.12 $H^+ < 0.1\ M$, $H^+ = 1.9 \times 10^{-4}\ M$

7.15 872 sec, 0.430 gm, 0.443 gm

Chapter 8

8.1 $\Delta H_f[Ca(OH)_2] = -235.4$ kcal   **8.3** $\Delta E = 1238$ kcal, $\Delta H_f = 18.3$ kcal

8.7 (a) $\Delta S_{sys} = 4.57$ eu $= -\Delta S_{surr}$, (b) $\Delta S_{sys} = 4.57$ eu, $\Delta S_{surr} = 0$

8.9 $\Delta S(H_2O) = -4.94$ eu, $\Delta S_{surr} = 5.13$ eu, $\Delta S_{tot} = 0.2$ eu, yes

8.11 (a) Ag (b) B(398°K) (c) 2Br(g) (d) Ar(0.1 atm)

8.13 $\Delta G^0 = -8.33$, K $= 1.3 \times 10^6$, enthalpy

8.15 $K_{298} = 1.86 \times 10^{12}$, $K_{600} = 3.96 \times 10^3$

8.17 3.38 kcal, 0   **8.19** $\Delta H = 13.1$ kcal

8.21 $\Delta G^0 = -2.12 \times 10^5$ joules $= -50.7$ kcal, $K = 1.94 \times 10^{37}$

Chapter 9

9.1 (a) $\sec^{-1}$ (b) liter/mole-sec (c) $\text{liter}^2/\text{mole}^2$-sec   **9.4** -1

9.6 $-[d[NH_4^+]/dt] = k_3 K_1 K_2 [HNO_2][NH_4^+]$

9.7 (a) $k_2[E] \ll k_{-1}[D]$ (b) $k_2[E] \gg k_{-1}[D]$

9.9 first order, $k = 5 \times 10^{-4} \sec^{-1}$ **9.11** 12.4 kcal   **9.12** ΔE

Chapter 10

10.1 1.24×10^{15} cps, 2412 A

10.3 5263 A, 2.35 eV

10.9 $\psi^2(2p_z) \propto \cos^2 \theta$, $\cos \theta = 0$ in xy-plane

10.11 Na, Na$^+$, Zn, N, O$^=$

10.13 1.63×10^{-11} ergs $= 10.1$ ev, 1.29×10^{-8} ergs $= 8.05 \times 10^3$ ev

Chapter 11

11.1 LiF has smallest separation and is most stable.

11.3 MgO, 1110 kcal; CaO, 970 kcal; SrO, 900 kcal; BaO, 840 kcal

11.5 tetrahedral, very distorted tetrahedral, linear, bent, octahedral, square planar, linear

11.7 Build-up of electron density between and near nuclei; electron sharing is not sufficient.

11.11 sp^2, sp^3, $sp^3 d^2$, $sp^3 d$

11.13 Linear molecule, sp hybrid at carbon atoms, carbon atoms linked by one σ-bond and two π-bonds

11.15 12.8 D

Chapter 12

12.1 (a) F_2^+ (b) NO (c) BO (d) NO (e) Be_2^+ **12.4** nonlinear

12.5 linear: (a), (b), (f), (g)   **12.7** linear: (a), (b), (d)

Chapter 13

13.3 Increasing nuclear charge increases bonding energy of extra electron. In nitrogen an extra electron must be added to a p-orbital that is already half-filled.

13.5 Ti^{++}, Mn^{++}, O$^=$, Se$^=$, Tl$^+$ **13.6** HCl, PH$_3$, H$_2$O

13.7 Higher oxidation state (a) increases (b) decreases in stability

13.8 (a) acidic (b) small molecules (c) covalently

13.9 Increasing metallic bond strength, ionization energy, and ionic size makes ΔH_f^0 more positive.

Chapter 14

14.1 Increasing [OH$^-$] should make Mg a better reductant, and the same effect should occur for Ba, but less marked.

14.3 Compare Li$^+$, Be^{++}, Mg^{++}, Al^{+3} **14.5** 2.14 A   **14.7** 10^{-54}

14.9 Least stable BeSO$_4$, most stable BaSO$_4$

Chapter 15

15.1 As atomic number increases, so do van der Waals attractions, and volatility goes down.

15.3 Properties indicate incipient metallic behavior, which is appropriate for this region of the periodic table.

15.6 N_3^-, OCN^- are isoelectronic with CO_2, and are linear.　　**15.8** pH = 10.1

15.9 X-ray determination of O—O bond length to distinguish between O_2 (1.21 A) and O_2^+ (1.12 A)

15.11 3.2×10^{26}, 5×10^{45}, 3.2×10^{27}　　　　　　　**15.12** A kinetic effect

15.15 $SOCl_2$ is pyramidal, SO_2Cl_2 tetrahedral.

Chapter 16

16.1 The values of ΔH_{sub}, I_1 for IB metals are uniformly greater than those of IA metals, and this accounts for difference in activity.

16.3 Cu is a poorer reductant than Ni because of difference in I_2, Cu is poorer than Zn because of difference in ΔH_{sub}.

16.5 tetraamminezinc(II) ion, hexafluoroferrate(III) ion, dicyanoargentate(I) ion, trichlorotriamminecobalt(III), hexacyanoferrate(III) ion, hexaaquochromium(III) ion

16.9 $[H^+] = 9 \times 10^{-3}$ M, $[H^+] = 9.5 \times 10^{-3}$ M

16.12 Second reaction has greater ΔS, first complex is the more stable.

Chapter 17

17.6 (a) *tert*-butyl chloride　(b) 3-chloro-3-methyl pentane

17.8 (a) acetic acid　(b) methyl ethyl ketone　(c) acetic acid and acetone

17.9 1,3,5-trinitrobenzene

Chapter 19

19.2 yes

19.4 0.0204 min^{-1}

19.7 0.024 A

19.3 2.16×10^9 events/min

19.5 1.5×10^4 years

INDEX

INDEX

Carbonic acid, 243–250, 624
Carbonium ion, 758, 762
Carbonyl compounds, inorganic, 735–737
  organic, 764–769
Carboxyl group, 751
Catalyst, 400–406
Cathode rays, 412
Cell, biological, 790–792
  nucleus of, 792
Cellulose, 804
Cesium, 594–602
Cesium-chloride lattice, 495
Chain reactions, 383–385
Charles' law, 38–40
Chelate, 718
Chlorate ion, 676
Chloric acid, 676
Chlorine, 668–680
  dioxide, 674
  heptoxide, 674
  hexoxide, 674
  monoxide, 674
Chlorous acid, 675
Chromate ion, 698
Chromic ion, 679–680
Chromium, 686, 696–699
Chromium subgroup, 696–699
Chromous ion, 699
Citric-acid cycle, 802, 808–810
Closest-packed structures, 108–112
  octahedral site in, 112, 115
  structures related to, 112
  tetrahedral sites in, 112, 115
Coal, 782–784
Cobalt, 703, 706
Coenzyme A, 800–802
Colligative properties, 345–349
Collisions, rate of, 387–389
Complex ions, bonding in, 722–735
  equilibria among, 250
  nomenclature, 721–722
  stereochemistry, 717–721
Compressibility factor, 67
Concentration, units of, 148–150
Concentration cell, 279
Coordinate-covalent bond, 717
Coordination, number, 110, 717
  sphere, 717

Copper, 686, 709–712
Copper subgroup, 709–714
Corrosion, 289–291
Coulomb barrier, 833
Covalent bonds, 496–528
  coordinate, 717
  multicenter, 525–528
  multiple, 522–525
Crystal-field theory, 723–730
Crystal-lattice geometry, 492–496
Crystals, covalent network, 94–95
  ionic, 91–93, 485–496
  metallic, 95–96
  molecular, 93–94
  sizes and shapes, 88–90
Cubic closest packing, 109
Cupric ion, 710–711
Cuprous ion, 710–711
Cyanide ion, 523
Cyclobutane, 748
Cyclopropane, 748
Cytochromes, 797
Cytoplasm, 792
Cytosine, 821

Dalton, J., atomic theory, 2–10
  Law of Partial Pressures, 44
  rule of greatest simplicity, 8
DeBroglie, L., and wave-particle duality, 430
Defects in solids, 119–122
Definite Proportions, Law of, 3
Deoxyribose, 721
Derivatives, table of, 871
Detailed balancing, 387
Diagonal relationship in periodic table, 572, 633
Diamagnetism, 726
Diamond, 621–622
Diamond-crystal lattice, 94
Dichromate ion, 698
Diffusion of gases, 57–59, 74–77
Diffusion-controlled reactions, 398–399
Dinitrogen pentoxide, 643
  tetroxide, 642
  trioxide, 642
Dipole moments, 518–520
  table of, 519
Displacement reactions, 755–757

Rate of reaction, 366–368
  in condensed phases, 397–400
  dependence on temperature, 394–397
  and equilibrium, 385–394
Reduced mass, 480
Reducing agent, 257
Reduction-potential diagram, 646
Representative elements, 571
Resonance, 526–528
Reversibility, 319
Rhenium, 700, 702–703
Rhodium, 707–709
Ribonuclease, 816
Ribose, 821, 823
Ribosomes, 792
RNA, *see* Nucleic acids
Rock-salt lattice, 115
Rubidium, 594–602
Ruthenium, 707–709
Rutherford's scattering experiment, 418–421

Saccharides, 803–810
Salt bridge, 259
Scandium, 686, 689
Schottky defect, 119
Schrödinger equation, 433
Second law of thermodynamics, 323
Selenium, 656, 667
Silanes, 626
Silicates, 627–633
Silicon, 621, 625–630
Silicon dioxide, 626
Silicones, 627–630
Silver, 709, 712–714
Sodium, 594–602
Sodium-chloride crystal structure, 93, 115
Solids, amorphous, 86, 88
  crystalline, 86, 88
Solubility, of ideal solutes, 164–167
  product, 205
  of slightly soluble salts, 204–212
Solute, definition of, 148
Solutions, 147–167
  ideal, 151–162
    boiling points of, 152

    freezing points of, 152
    vapor pressures of, 152
  nonideal, 162–164
    solid, 148
Solvay process, 610
Solvent, 148
Specific heat, 12
Spectra, atomic, 425–430
  molecular, 479–484
  x-ray, 428
Spectrochemical series, 726
Spectrograph, 425
Speed distribution, Maxwell-Boltzmann, 59–64
Spontaneous processes, 318–322
Standard atmosphere, 34
Standard cell potential, 271
  and equilibrium constant, 279
  sign convention for, 272
Standard half-cell potentials, 269–276
  conventions for use of, 274
    definition and measurement, 269–276
    and standard free-energy change, table of, 275
Standard temperature and pressure (STP), 23, 42
Starch, 804–810
State function, 301
State of a system, macroscopic, 327
  microscopic, 328
Statistical mechanics, 332, 352–358
Steady-state approximation, 379–383
Stereochemistry, of complex ions, 717
  of organic molecules, 779
Steric factor, 392
Strontium, 602–611
Sublimation temperature, 147
Sulfanes, 662
Sulfite ion, 664
Sulfur, 656, 660–667
  dioxide, 662
    electronic structure, 527
  hexafluoride, 666
  trioxide, 663
    electronic structure, 527, 560
Sulfuric acid, 663

LIST OF THE ATOMIC WEIGHTS OF THE ELEMENTS

Element	Symbol	Atomic Number	Atomic Weight
Actinium	Ac	89	(227)
Aluminum	Al	13	26.98
Americium	Am	95	(243)
Antimony	Sb	51	121.75
Argon	Ar	18	39.948
Arsenic	As	33	74.92
Astatine	At	85	(210)
Barium	Ba	56	137.34
Berkelium	Bk	97	(249)
Beryllium	Be	4	9.012
Bismuth	Bi	83	208.98
Boron	B	5	10.81
Bromine	Br	35	79.909
Cadmium	Cd	48	112.40
Calcium	Ca	20	40.08
Californium	Cf	98	(251)
Carbon	C	6	12.011
Cerium	Ce	58	140.12
Cesium	Cs	55	132.91
Chlorine	Cl	17	35.453
Chromium	Cr	24	52.00
Cobalt	Co	27	58.93
Copper	Cu	29	63.54
Curium	Cm	96	(247)
Dysprosium	Dy	66	162.50
Einsteinium	Es	99	(254)

Element	Symbol	Atomic Number	Atomic Weight
Mercury	Hg	80	200.59
Molybdenum	Mo	42	95.94
Neodymium	Nd	60	144.24
Neon	Ne	10	20.183
Neptunium	Np	93	(237)
Nickel	Ni	28	58.71
Niobium	Nb	41	92.91
Nitrogen	N	7	14.007
Nobelium	No	102	(253)
Osmium	Os	76	190.2
Oxygen	O	8	15.9994
Palladium	Pd	46	106.4
Phosphorus	P	15	30.974
Platinum	Pt	78	195.09
Plutonium	Pu	94	(242)
Polonium	Po	84	(210)
Potassium	K	19	39.102
Praseodymium	Pr	59	140.91
Promethium	Pm	61	(147)
Protactinium	Pa	91	(231)
Radium	Ra	88	(226)
Radon	Rn	86	(222)
Rhenium	Re	75	186.23
Rhodium	Rh	45	102.91
Rubidium	Rb	37	85.47
Ruthenium	Ru	44	101.1